COMPUTERIZED ENGINE CONTROLS

Tenth Edition

Steve V. Hatch

Australia • Brazil • Mexico • Singapore • United Kingdom • United States

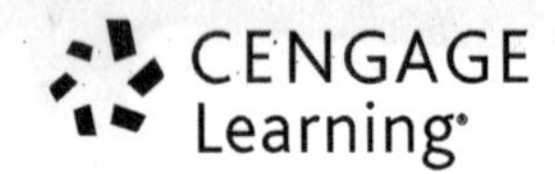

Computerized Engine Controls, Tenth Edition
Steve V. Hatch

SVP, GM Skills & Global Product Management: Dawn Gerrain

Product Director: Matthew Seeley

Product Team Manager: Erin Brennan

Senior Director, Development: Marah Bellegarde

Senior Product Development Manager: Larry Main

Senior Content Developer: Meaghan Tomaso

Product Assistant: Maria Garguilo

Vice President, Marketing Services: Jennifer Ann Baker

Marketing Manager: Jonathon Sheehan

Senior Production Director: Wendy Troeger

Production Director: Andrew Crouth

Senior Content Project Manager: Cheri Plasse

Senior Art Director: Benjamin Gleeksman

Cover image(s): Steve V. Hatch

Library of Congress Control Number: 2015957627

ISBN: 978-1-3054-9754-2

Cengage Learning
20 Channel Center Street
Boston, MA 02210
USA

Printed in the United States of America
2 3 4 5 6 22 21 20 19 18

Contents

Preface

The application of electronics has made automotive technology exciting, fast paced, and, certainly, more complicated. Technological advancements continue to add complexity to the modern automobile at record-setting rates. Almost all systems on the automobile are now controlled by electronic control modules. Autonomous (self-driving) cars and trucks already exist and are being experimented with on public roads. They will likely soon be available in dealership showrooms for consumers to buy. Today's automobiles already have collision avoidance systems, lane departure/lane sway warning systems, and parking assist systems that can control the electric steering system in order to parallel park the vehicle precisely in a tight parking space. Ultimately, the advancements in technology require automotive service technicians to be trained in the electronic principles used in automotive technology and to continue to actively pursue upgrade training throughout their careers. Those who do this will find the task challenging, but achievable and rewarding.

This text was written in response to a widely recognized need within the industry: to help both entry-level students and experienced professional technicians to acquire a strong grasp of how computerized automotive systems operate and how to diagnose problems with them. While this text focuses mainly on electronic *engine* control systems, it will also help readers to understand the principles that underlie any vehicle system that is under the control of a computer.

Computerized Engine Controls is written with the assumption that readers are familiar with the basic operating principles of the internal combustion engine.

CHAPTER AIDS

Chapters in this edition contain the following features:

- *Objectives.* Objectives are provided at the beginning of each chapter to help the reader identify the major concepts that are presented.
- *Key Terms.* Terms that are unique to computerized engine control systems are listed at the beginning of each chapter as key terms, and they appear in **boldface** type at their first use in the chapter. These key terms are also defined in the glossary.
- *Diagnostic & Service Tips.* These tips offer advice that can be helpful to the technician when diagnosing and servicing vehicles, as well as when addressing customer concerns.
- *Chapter Articles.* These short segments provide additional nice-to-know information about the technical topics covered in the chapter.
- *Summary.* Each chapter contains a summary that reviews the major concepts presented.
- *Diagnostic Exercise.* Following the chapter summary, a diagnostic exercise is provided that presents an important chapter concept in the context of a real-life scenario.

- *Review Questions.* Review questions are provided at the end of each chapter to help readers to test their recall and comprehension of the material, as well as to reinforce the concepts covered. All of these review questions have been written in a multiple-choice format—the type of question that would be found on an ASE test.

WARNINGS AND CAUTIONS

Personal safety concerns about specific computerized engine control systems are highlighted where applicable. This text follows industry standards for using the following terms:

- **Warnings** indicate that failure to observe correct diagnostic or repair procedures could result in personal injury or death.
- **Cautions** indicate that failure to observe correct diagnostic or repair procedures could result in damage to tools, equipment, or the vehicle being serviced.

Students should understand that, while working with computerized controls is not inherently dangerous, failure to observe recognized safety practices is. There are, unfortunately, many more injuries and accidents in the automotive repair business than there should be. Good safety practices, if learned early in a student's career, can literally be lifesaving later on.

NEW TO THIS EDITION

This textbook is now divided into two sections. Section One contains all of the generic chapters, Chapter 1 through Chapter 11. The principles in these chapters apply to all modern vehicles, regardless of manufacturer. Section Two contains all of the manufacturer-specific chapters, Chapter 12 through Chapter 16, with one chapter dedicated to each of the following:

- General Motors Computerized Engine Controls
- Ford Motor Company Computerized Engine Controls
- Chrysler Corporation Computerized Engine Controls
- European (Bosch) Computerized Engine Controls
- Asian Computerized Engine Controls

Other changes for the 10th edition of *Computerized Engine Controls* include the following:

- The material on wideband air-fuel ratio sensors in Chapter 3 has been greatly expanded.
- Chapter 5 and Chapter 6 from the 9th edition ("Diagnostic Concepts" and "Diagnostic Equipment") have now been combined into one chapter, Chapter 5, "Introduction to Diagnostic Concepts and Diagnostic Equipment" due to the fact that there was some overlap between the two original chapters. It is difficult to discuss one of these topics without getting into a discussion of the other. Chapter 5 is an introductory diagnostic chapter. Chapter 11 discusses diagnostics at a much deeper level.
- A new chapter, Chapter 10, "Modern Systems that Interact with the Engine Control System," has been added to this edition. This chapter is designed to be a somewhat light-hearted discussion that introduces the reader to many of the modern electronic systems found on newer automobiles.
- Chapter 11 now includes a discussion of how to diagnose whether a loss of a reference voltage to the sensors is a fault of the computer or a fault of a sensor/wire shorting it to ground. Also, a discussion of how to narrow down a parasitic draw problem (when the battery is discharging with the ignition turned off) without having to remove fuses (which

resets computer timers) is now included in Chapter 11.

- Appendix A (which was Appendix B in the previous edition) is a list of many automotive-related websites. A new Appendix B has been added to this edition—this is a list of many automotive-related apps for smartphones and tablets.

SUGGESTIONS ON HOW TO USE THIS TEXT

Different manufacturers often follow the same basic principles. You can learn a great deal by placing your initial focus on the information that is applicable to most, if not all, vehicles. Therefore, you should first study the generic chapters in Section One. Once you have mastered the information in Section One, read the chapters in Section Two according to your specific interests.

SUPPLEMENTS INSTRUCTOR RESOURCES

The Instructor Resources, now available online, contain an instructor's manual including answers to all end-of-chapter questions, lab sheets, chapter tests powered by Cognero with hundreds of test questions, and an image gallery with all photos and illustrations from the text.

To access these Instructor Resources, go to login.cengagebrain.com, and create an account or log into your existing account.

Acknowledgments

I am, once again, very honored to be able to present the current revision of this textbook. There are many people upon whom I depend greatly and without whom this textbook would not be the success it is today.

I would like to thank my family, my friends, and my colleagues who have encouraged me. I would also like to thank my students, both at the entry level and at the professional level, who, while in my classes over the years, have affirmed the need for a textbook such as this and have provided me with the positive feedback that has continued to encourage me in this endeavor. I also enjoy the feedback of real-world diagnostic situations provided to me by the professional technicians in my classes that reinforce the theories of this textbook. I would specifically like to thank a few of my co-instructors/colleagues for their assistance in reviewing and/or gathering information about specific portions of this textbook:

- Randy Cowan
- Tim Freeman
- Frank Johnson
- Tom Smith
- Larry Wilkenson

I would like to thank William K. Bencini of Colorado State University in Pueblo, Colorado, who allowed me to use several of his photos of Honda i-VTEC components in the book. And I would like to thank Chris Chesney of CARQUEST Technical Institute for the information he provided to me and for the fuse voltage drop charts in Chapter 11.

I would also like to thank my wife, Geralyn, for her continued patience with the many hours that this effort has required of me.

Finally, many thanks to the following for their critical reviews of this edition of the text:

George McNitt
SCCC/ATS
Liberal, KS

Tim Mulready
University of Northwestern Ohio
Lima, OH

Randell Peters, PhD
Indiana State University
Terre Haute, IN

Michael Huneke
Texas State Technical College
Waco, Texas

John F. Kennedy
University of Northwestern Ohio
Lima, OH

The publisher would also like to thank the author, Steve V. Hatch, for providing the cover photo and the photographs used throughout this edition of the textbook.

About the Cover Photo

A tablet is communicating with the PCM for diagnostic purposes using an OBDLink wireless adapter connected to the DLC at the lower left.

SECTION ONE

GENERIC CHAPTERS

Chapter 1

A Review of Electricity and Electronics

OBJECTIVES

Upon completion and review of this chapter, you should be able to:

- ❑ Understand the conceptual differences between the terms *electrical* and *electronic.*
- ❑ Understand how a compound is different from an element.
- ❑ Define the difference between an element and a compound.
- ❑ Describe the importance of an atom's valence ring as it pertains to electrical theory.
- ❑ Understand the relationship between voltage, resistance, and amperage.
- ❑ Define circuit types in terms of series circuits and parallel circuits.
- ❑ Understand the construction and operation of semiconductors such as diodes and transistors.
- ❑ Define the difference between an analog voltage signal and a digital voltage signal.
- ❑ Describe the relationship between a variable frequency, variable duty cycle, and variable pulse width.

KEY TERMS

Amp or Ampere
Amperage
Analog
Armature
Capacitor
Clamping Diode
Compound
Digital
Diode
Dual In-Line Package (DIP)
Electrical
Electromotive Force
Electronic
Element
Free Electrons
H-Gate
Integrated Circuit (IC)
Molecule
Negative Ion
Ohm
Ohm's Law
Permeability
Positive Ion
Reluctance
Resistance
Semiconductors
Solenoid
Transistor
Valence Ring
Volt
Voltage or Voltage Potential
Voltage Drop

The earliest automobiles had little in the way of **electrical** systems, but as the automobile has become more complicated and as more accessories have been added, electrical and **electronic** systems have replaced mechanical methods of control on today's vehicles. Additional electronic control systems have made and will continue to make the automobile comply with government standards and consumer demands. Today, most major automotive systems are controlled by computers.

This increased use of electrical and electronic systems means two things for the automotive service technician: First, to be effective, all service technicians need skills in electrical diagnosis and repair, almost regardless of the technician's service specialty; second, technicians with such skills will command significantly greater financial rewards and will deserve them.

There are several principles by which electrical systems operate, but they are all fairly simple; learning them is not difficult. As each principle is introduced to you through your reading or in class, ask questions and/or read until you understand it. Review the principles often and practice the exercises that your instructor assigns.

ELECTRICAL CIRCUITS VERSUS ELECTRONIC CIRCUITS

The differences between electrical circuits and electronic circuits are not always clear-cut. This has led to some confusion about the use of terms and how an electronic system differs from an electrical system. Perhaps the comparisons in the following table will help.

Think of electrical circuits as the *muscle* and electronic circuits as the *brain.* Electrical circuits have been used in the automobile since the first one came off the assembly line, but electronic circuits have only been added to the automobile in more recent years. For example, interior lighting circuits began on the automobile as simple electrical circuits without any electronic control. But, more often than not, interior lighting systems on today's vehicles are controlled electronically by a computer.

Even though the use of solid-state components may often be used as a criterion to identify an electronic circuit, solid-state components, such as *power transistors,* may also be used in an electrical circuit. A power transistor is a type of transistor designed to carry larger amounts of **amperage** than are normally found in an electronic circuit. A power transistor is essentially a highly reliable relay.

Ultimately, an electrical circuit is a circuit that performs work through a load device. An electronic circuit is used to intelligently control an electrical circuit. Therefore, an electrical circuit may or may not be under the control of an electronic circuit.

It should also be noted that a component identified as an *electronic* device always needs a proper power (positive) and ground (negative) just to power up properly, whether it is a small **integrated circuit** (IC) chip or a complex, sophisticated computer. If either one is lacking it cannot do its assigned job properly.

Electrical Circuits	Electronic Circuits
Do physical work: heat, light, and electromagnetism used to create movement.	Communicate information: voltages or on/off signals.
Use electromechanical devices: motors, solenoids, relays.	Use solid-state devices (semiconductors) with no moving parts, such as transistors and diodes.
Operate at relatively high current or amperage.	Operate at relatively low current or amperage.
Have relatively low resistance (ohms).	Have relatively high resistance (ohms).
May or may not be controlled by an electronic circuit.	Are used to control electrical circuits.

ELECTRON THEORY

Molecules and Atoms

A study of electricity begins with the smallest pieces of matter. All substances—air, water, wood, steel, stone, and even the various substances that our bodies are made of—are made of the same bits of matter. Every substance is made of units called **molecules.** A molecule is a unit formed by combining two or more atoms; it is the smallest unit that a given substance can be broken down to and still exhibit all of the characteristics of that substance. For example, a molecule of water, or H_2O, is made up of two atoms of hydrogen and one atom of oxygen (H is the chemical symbol for hydrogen and O is the chemical symbol for oxygen). If a molecule of water is broken down into its component atoms, it is no longer water.

As molecules are made up of atoms, atoms are in turn made up of:

- Electrons, or negatively charged particles
- Protons, or positively charged particles
- Neutrons, or particles with no charge; at the level of atomic activity concerning us here, neutrons just add mass to the atom

The smallest and lightest atom is the hydrogen atom. It contains one proton and one electron (Figure 1–1); it is the only atom that does not have a neutron. The next smallest and lightest atom is the helium atom. It has two protons, two neutrons, and two electrons (Figure 1–2). Since the hydrogen atom is the smallest and lightest, and since it has one electron and one proton, it is given an atomic number of 1. Since helium is the next lightest, it has an atomic number of 2. Every atom has been given an atomic number that indicates its relative size and weight (or its mass) and the number of electrons, protons, and neutrons it contains. An atom usually has the same number of electrons, protons, and neutrons.

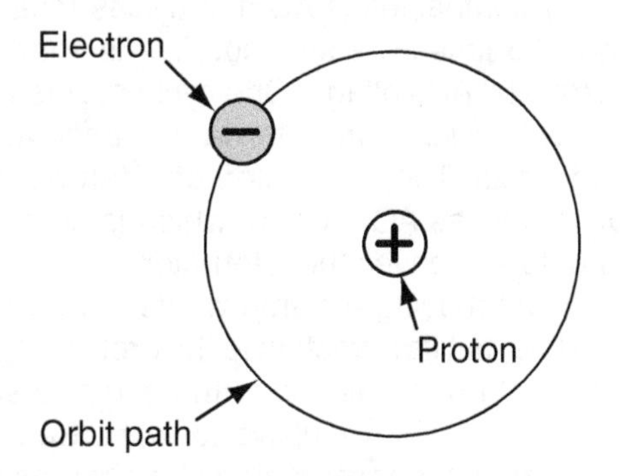

Figure 1–1 Hydrogen atom.

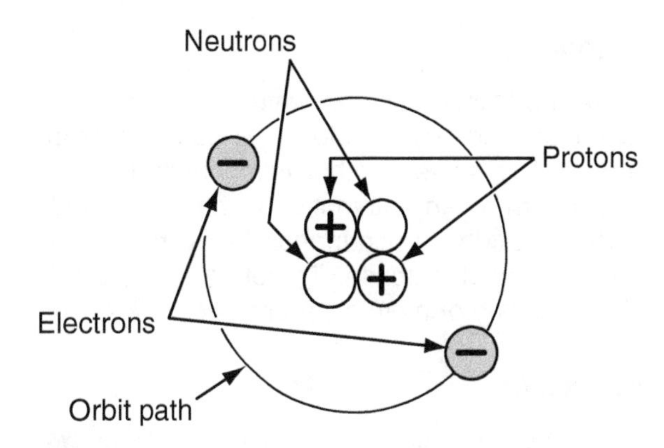

Figure 1–2 Helium atom.

Elements

Once the three different bits of matter are united to form an atom, two or more atoms combine to form a molecule. If all of the atoms in the molecule are the same, the molecule is called an **element.** Which element it is depends on how many protons, neutrons, and electrons the atoms contain. There are more than a hundred different elements. Some examples of elements are gold, lead, iron, and sodium. Examples of other elements that are of concern to an automotive technician include hydrogen, carbon, nitrogen, oxygen, and silicon. An element, then, is a pure substance whose molecules contain only one kind of atom.

Compounds

A substance such as water, which contains hydrogen and oxygen atoms, is called a **compound.** Examples of other compounds that are of concern to an automotive technician include carbon dioxide, carbon monoxide, hydrocarbons, and oxides of nitrogen. Therefore, compounds consist of two or more elements.

Molecules

A molecule consists of a minimum of two atoms that are chemically bonded together; it is electrically stable, with a neutral charge. A molecule may contain two or more identical atoms and therefore be an element, such as an oxygen molecule (O_2), or it may have atoms of two or more elements and therefore be a compound, such as water (H_2O).

Atomic Structure and Electricity

Notice in Figure 1–1 and Figure 1–2 that the protons and neutrons are grouped together in the center of each atom, which is called the *nucleus* of the atom. The electrons travel around the nucleus of the atom in an orbit, similar to the way that the Earth travels around the sun. But because an atom usually has several electrons orbiting around its nucleus, the electrons form in layers, rather than all of them traveling in the same orbit (Figure 1–3). Some, however, share the same orbit, as seen in Figure 1–3. For the purposes of this text, only the electrons in the last layer are of any real importance. This layer is often called the *outer shell* or **valence ring.** The student should realize that we are speaking very loosely here when we describe electrons in shells having orbits. For our purposes, this simple explanation (a model once called the *Rutherford atom)* satisfactorily conveys the nature of the electron.

As mentioned, electrons are negatively charged and protons are positively charged. You have probably heard or know that like charges repel and unlike charges attract. Electrons are always moving; in fact, they are sometimes said to move at nearly the speed of light. These characteristics work together to explain many of the behaviors of an atom that make current flow. *Current* is defined as a mass of **free electrons** moving in the same direction.

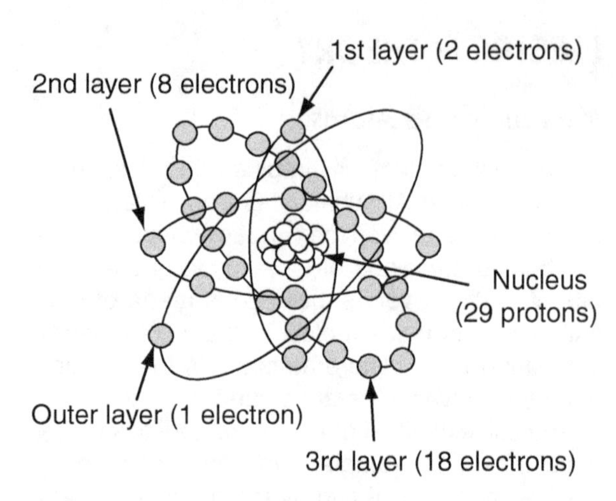

Figure 1–3 Layers of electrons around a copper atom nucleus.

There are two types of current: *direct current (DC)* and *alternating current (AC).* Direct current always flows in one direction. Current from a battery is the best example. Most of the devices in an automobile use DC. Circuits with AC repeatedly switch the polarity of the circuit so that current flow (electron movement) reverses direction repeatedly. The power available from commercial utility companies is AC and cycles (changes *polarity)* 60 times per second. This is known as 60 Hertz (Hz) AC voltage. One cycle occurs when the current switches from forward to backward to forward again. The car's alternator (an AC generator) produces AC current, which is converted to DC before it leaves the alternator.

The fast-moving electron wants to move in a straight line, but its attraction to the proton nucleus makes it act like a ball tied to the end of a string twirled around. The repulsive force between the electrons keeps them spread as far apart as their attraction to the nucleus will allow.

The fewer electrons there are in the outer shell of the atom and the more layers of electrons there are under the outer shell, the weaker is the bond between the outer electrons and the nucleus. If one of these outer electrons can somehow be broken free from its orbit, it will travel to a neighboring atom and fall into the outer shell there, resulting in two unbalanced atoms. The first atom is missing an electron. It is now positively charged and is called a **positive ion.** The second atom has an extra electron. It is negatively charged and is called a **negative ion.** Ions are unstable. They want either to gain an electron or to get rid of one so that they are balanced.

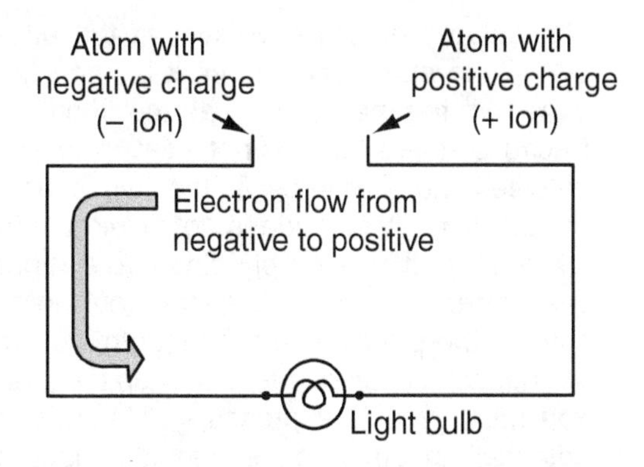

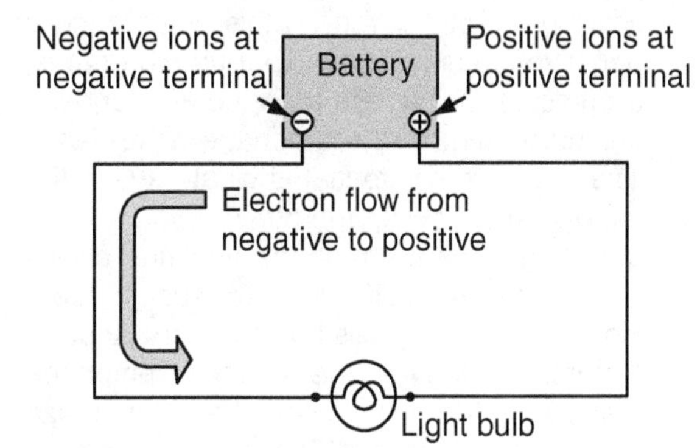

Figure 1–4 Negative versus positive potential.

ELECTRICAL THEORY

Voltage Potential

An atom that is a positive ion has positive *potential.* It has more positive charge than negative charge because it has more protons than electrons. Suppose that this atom is at one end of a circuit (Figure 1–4). Further suppose that there is a negative ion at the other end of the circuit in that this atom has an extra electron, thus giving it a negative potential. Because of the *difference* in potential at the two ends of the circuit, an electron at the negatively charged end will start moving toward the positively charged end. The greater the difference in potential (the greater the number of opposite-charged ions) at each end of the circuit, the greater the number of electrons that will start to flow. This potential difference between the two charges is commonly known as **voltage potential.**

An example can be created by attaching something between the two ends of a circuit that will produce positive and negative ions. This is what a battery or generator does in a circuit (Figure 1–4). If you connect both ends of a copper wire to a battery, the voltage potential will cause electron flow through the wire. However, because the wire will not be able to handle the electron flow that the battery can provide, it will burn open very quickly. Therefore some kind of **resistance,** or opposition to a steady electric current, is needed in the wire. Actually, this resistance has two functions. It limits current flow so as to keep the wire from burning open and it also turns the current flow into some type of useful work—heat, light, or electromagnetism.

It is the voltage potential that makes current flow. Actually, three factors must be present for an electrical circuit to work properly. These three factors are voltage potential, resistance to flow, and current flow, as demonstrated in the following example:

> Suppose that you have a glass of your favorite lemonade sitting on your patio table on a nice summer day. Suppose that there is a straw sitting in the glass of lemonade. There is atmospheric pressure acting on the lemonade in the glass and therefore at the lower end of the straw. There is also

atmospheric pressure present at the upper end of the straw (remember, it is just sitting there on the patio table at the moment). Because there is no potential difference between the pressures at the two ends of the straw, the lemonade is not flowing in the straw. Now, if you simply close your mouth over the upper end of the straw but do not change the pressure in your mouth, you will still not get the lemonade to flow up the straw. You must provide a pressure difference (or potential) in order to get the lemonade to flow up the straw. You do this by creating a negative pressure within your mouth (that is, a pressure that is less than the atmospheric pressure acting on the lemonade in the glass). The pressure differential is what causes the lemonade to flow up the straw. If you want to get a larger flow of lemonade, you must create a larger pressure difference (or suck harder). However, this only tells half the story. Another factor that influences the volume of lemonade that you get to flow up the straw is the size of the straw. Suppose that you replace your normal-size straw with a slender coffee stir stick/straw. This small straw would limit the volume of lemonade that you could get to flow up the straw. The same is true in an electrical circuit. In an electrical circuit, the amount of current that flows is dictated by two factors—how much voltage potential is applied to the circuit and how much resistance to flow is present in the circuit. This is, in essence, what is commonly known as **Ohm's law,** described later in this chapter. You can also apply this same principle to a fuel injector in that the flow rate of a fuel injector is dictated by the amount of pressure difference between the two ends of the injector and the orifice size (restriction) of the injector.

Magnetism

Magnetism is closely tied to the generation and use of electricity. In fact, one of the prevailing theories is that magnetism is caused by the movement and group orientation of electrons. Some materials strongly demonstrate the characteristics associated with magnetism and some do not. Those that strongly demonstrate the characteristics of magnetism, such as iron, are said to have high **permeability.** Those that do not, such as glass, wood, and aluminum, are said to have high **reluctance.**

Lines of Force

It is not known whether there actually is such a thing as a *magnetic line of force.* What is known, however, is that magnetism exerts a force that we can understand and manipulate if we assume there are magnetic lines of force. Magnetic force is linear in nature, and it can be managed to do many kinds of work. By assigning certain characteristics to these lines of force, we can explain the behavior of magnetism. Magnetic lines of force:

1. Have a directional force (north to south outside the magnet)
2. Want to take the shortest distance between two poles (just like a stretched rubber band between the two points from which it is held)
3. Form a complete loop
4. Are more permeable to iron than air
5. Resist being close together (especially in air)
6. Resist being cut
7. Will not cross each other (they will bend first)

Magnetic lines of force extending from a magnet make up what is commonly called a magnetic field and more correctly called magnetic *flux* (Figure 1–5). If a magnet is not near an object made of permeable material, the lines of force will extend from the north pole through the air to the south pole (characteristic 1). The lines of force will continue through the body of the magnet to the north pole to form a complete loop (characteristic 3). Every magnet has a north pole and

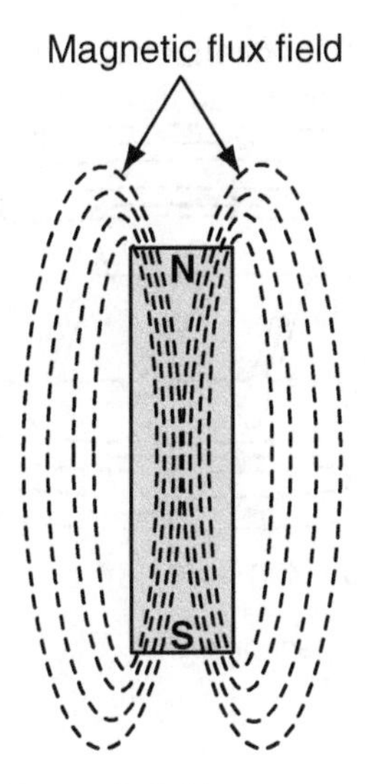

Figure 1–5 Magnetic field.

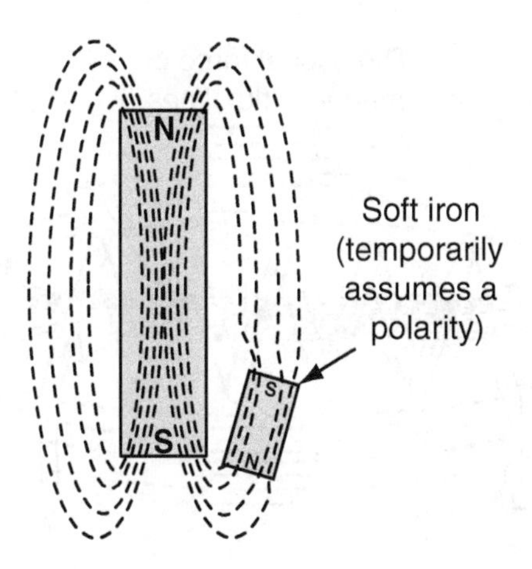

Figure 1–6 Magnetic field distortion.

a south pole. The poles are the two points of a magnet where the magnetic strength is greatest. As the lines of force extend out of the north pole, they begin to spread out. Here you see opposition between characteristics 2 and 5. The lines of force want to take the shortest distance between the poles, but they spread out because of their tendency to repel each other (characteristic 5). The result is a magnetic field that occupies a relatively large area but has greater density near the body of the magnet.

Because the body of the magnet has high permeability, the lines of force are very concentrated in the body of the magnet (characteristic 4). This accounts for the poles of the magnet having the highest magnetic strength.

If there is an object with high permeability near the magnet, the magnetic lines of force will distort from their normal pattern and go out of their way to pass through the object (Figure 1–6). The tendency for the lines of force to pass through the permeable object is stronger than their tendency to take the shortest route. The lines of force will, however, try to move the object toward the nearest pole of the magnet.

Electromagnets

Early researchers discovered that when current passes through a conductor, a magnetic field forms around the conductor (Figure 1–7). This principle makes possible the use of electromagnets, electric motors, generators, and most of the other components used in electrical circuits.

If a wire is coiled with the coils close together, most of the lines of force wrap around the entire coil rather than going between the coils of wire

Figure 1–7 Lines of force forming around a conductor. If you place your left hand on the wire with your thumb pointing in the direction of electron flow, your fingers will be pointing in the direction of the directional force of the magnetic lines of force. When thinking of conventional current flow, the same would apply for the right hand.

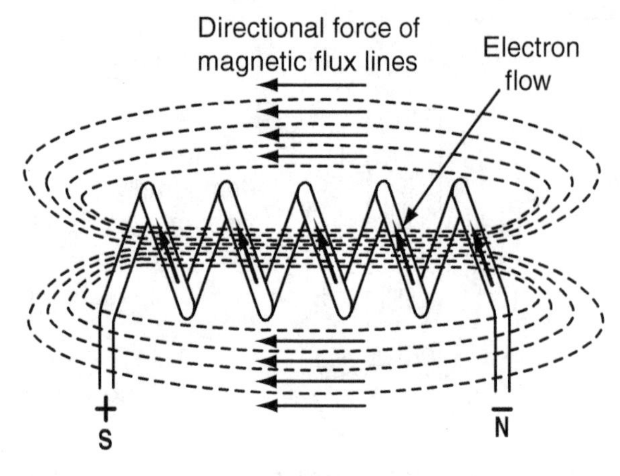

Figure 1–8 Magnetic field around a coil. Place your left hand, with thumb extended, around the coil with your fingers pointing in the direction of the electron flow through the coil. Your thumb will point to the north pole of the magnetic field.

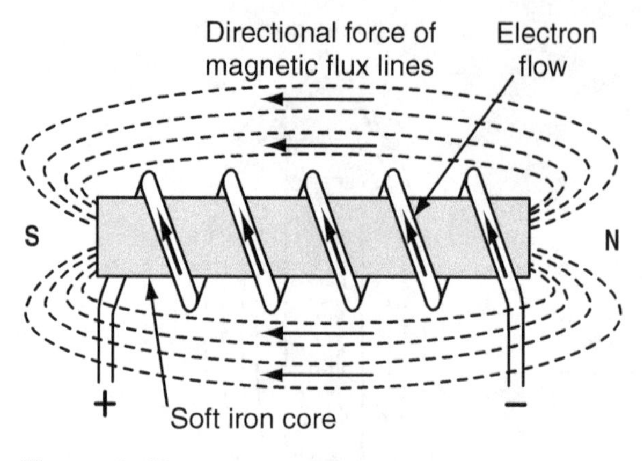

Figure 1–10 Electromagnet.

(Figure 1–8). This is because if they do try to wrap around each loop in the coil, they must cross each other, which they will not do (characteristic 7) (Figure 1–9).

If a highly permeable core is placed in the center of the coil, the magnetic field becomes much stronger because the high permeability of the core replaces the low permeability of the air in the center (Figure 1–10). If the core is placed toward one end of the coil (Figure 1–11), the lines of force exert a strong force on it to move it toward the center so that they can follow a shorter path. If the core is movable, it will move to the center of the coil. A coil around an off-center, movable, permeable core is a **solenoid**. A spring is usually used to hold the core off center. When the current flow is switched on to create the electromagnet, the magnetic field is stronger than the spring tension and will overcome the spring tension in order to move the iron core. This solenoid can be used to do physical work such as to engage a starter drive to a flywheel or to lock or unlock a door. But for most subjects to which this textbook pertains, there will be a valve attached to the movable iron core. This solenoid-operated valve may be spring loaded normally closed (N/C) and open when electrically energized, or it may be spring loaded normally open (N/O) and close when electrically energized.

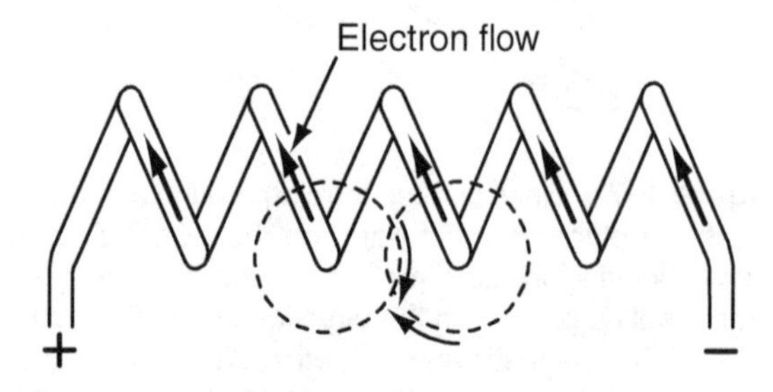

Figure 1–9 Magnetic lines of force cannot cross.

Motors

In an electric motor, current is passed through a conductor that is looped around the **armature** core (Figure 1–12). The conductor loops are placed in grooves along the length of the core. The core is made of laminated discs of permeable, soft iron that are pressed onto the armature shaft. The soft iron core causes the magnetic field

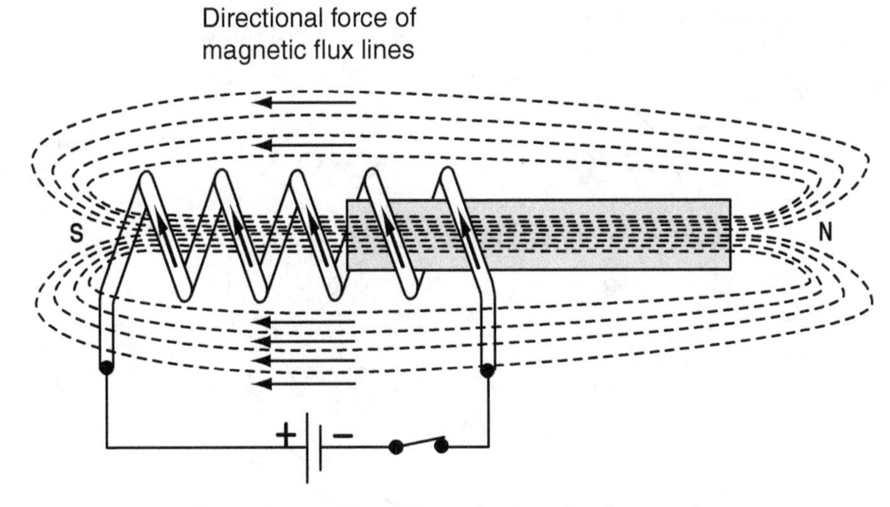

Figure 1–11 Solenoid.

that forms around the conductor to be stronger due to its permeability. There are several conductor loops on the armature, but only the loop that is nearest the center of the field poles has current passing through it. The loops are positioned so that when one side of a loop is centered on one field pole, its other side is centered on the other field pole.

The field poles are either permanent magnets or pieces of soft iron that serve as the core of an electromagnet. If electromagnets are used, an additional conductor (not shown in Figure 1–12) is wound around each field pole, and current is passed through these field coils to produce a magnetic field between the field poles. The motor frame that the poles are mounted on acts as the magnet body.

Looking at the armature conductor near the north field pole in Figure 1–12, you see that its magnetic field extends out of the armature core and that it has a clockwise force. The magnetic field between the field poles has a directional force from north pole to south pole. At the top of the armature conductor, the field it has produced has a directional force in the same direction as that of the lines of force between the field poles. The lines of force in this area are compatible, but combining these two fields in the same area produces a high-density field. Remember that magnetic lines of force resist being close together.

At the bottom of the armature conductor, the lines of force formed around it have a directional force opposite to those from the north field pole. The lines of force will not cross each other, so some lines from the field pole distort and go up and over the conductor into the already dense portion of the field above the conductor, and some just cease to exist. This produces a high-density field above the conductor and a low-density field below it. The difference in density is similar to a difference in pressure. This produces a downward force on the conductor.

The other side of the armature loop, on the other side of the armature, is the same except that the current is now traveling the opposite way.

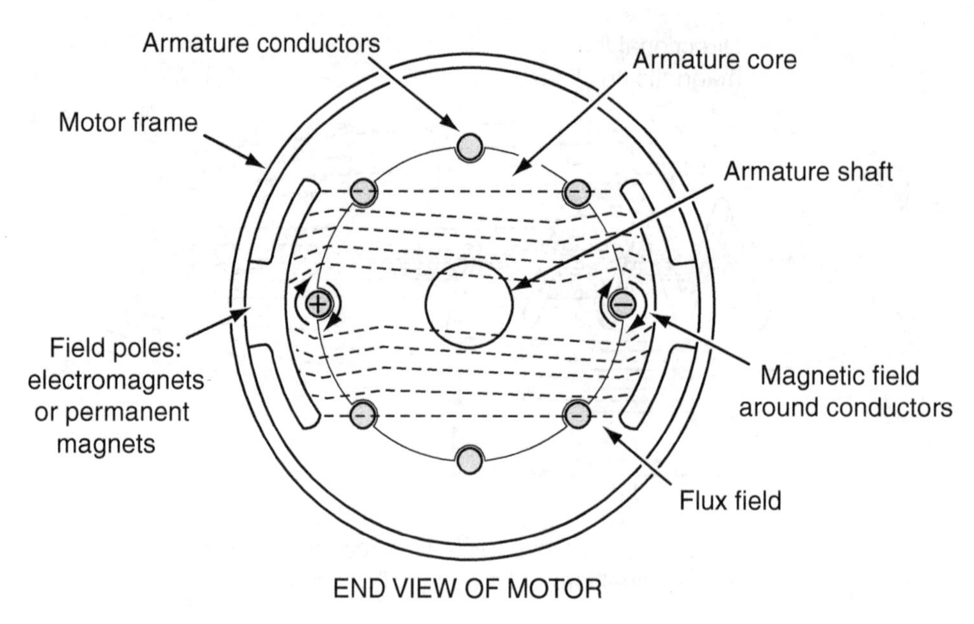

Figure 1–12 Electric motor.

The loop makes a U-shaped bend at the end of the armature. The magnetic field around this part of the conductor has a counterclockwise force. Here, the lines of force around the conductor are compatible with those between the field poles under the conductor, but they try to cross at the top. This produces an upward force on this side of the armature loop. The armature rotates counterclockwise. To change the direction in which the armature turns, either change the direction that current flows through the armature conductors or change the polarity of the field poles.

Magnetic Induction

Passing voltage through a wire causes a magnetic field to form around the wire. However, if lines of force can be formed around a conductor, a voltage is produced in the wire and current starts to flow. This assumes, of course, that the wire is part of a complete circuit. Lines of force can be made to wrap around a conductor by passing a conductor through a magnetic field (Figure 1–13). This phenomenon occurs because of characteristic 6. As the conductor passes through the magnetic field, it cuts each line of force. Because the lines of force resist being cut, they first wrap around the conductor, much like a blade of grass would if struck by a stick (Figure 1–14). This principle is used in generators

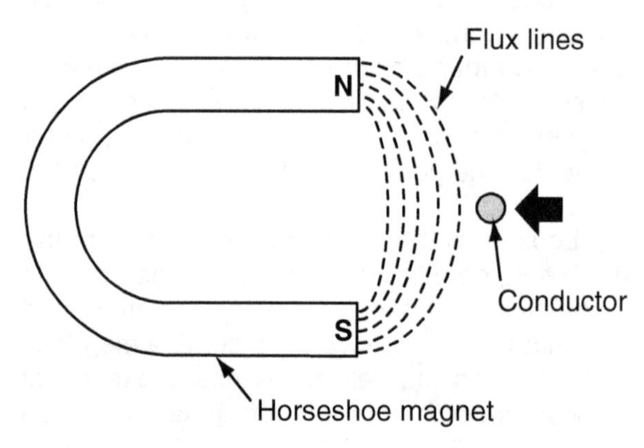

Figure 1–13 Magnetic induction.

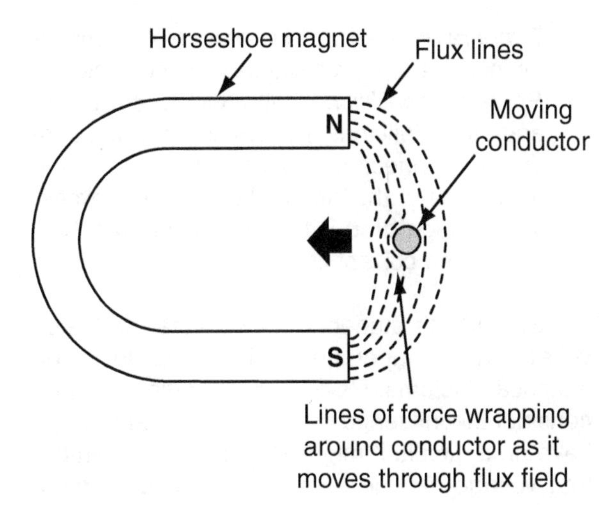

Figure 1–14 Cutting lines of force.

to produce voltage and current flow. The principle will work regardless of whether:

- The conductor is moved through a stationary magnetic field, as in a DC generator
- A magnetic field is moved past stationary conductors, as in an AC generator
- The lines of force in an electromagnetic field are moved by having the circuit producing the magnetic field turned on and off, as in an ignition coil

Note that in each case, movement of either the lines of force or the conductor is needed. A magnetic field around a conductor where both are steady state will not produce voltage. The amount of voltage and current produced by magnetic induction depends on four factors:

1. The strength of the magnetic field (how many lines of force there are to cut). A tiny amount of voltage is induced in the wire by each line of force that is cut.
2. The number of conductors cutting the line of force. Winding the conductor into a coil and passing one side of the coil through the magnetic field cuts each line of force as many times as there are loops in the coil.
3. How fast and how many times the conductors pass through the magnetic field.
4. The angle between the lines of force and the conductor's approach to them.

Amperage

Amperage is a measure of the amount of current flowing in a circuit. One **ampere** (amp) equals 6,250,000,000,000,000,000 (6.25 billion billion) electrons moving past a given point in a circuit per second. This is often expressed as one *coulomb.*

Voltage

A **volt** is a measure of the force or pressure that causes current to flow; it is often referred to as **voltage.** The difference in potential is voltage. The most common ways of producing voltage are chemically, as in a battery, or by magnetic induction, as in a generator. A more accurate but less-used name is **electromotive force.** Note that volts are what drive the electrons through the circuit; voltage is the measurement of that force. Similarly, **amps** are the number of electrons moving; amperage is the measurement of that number.

Resistance

The fact that voltage is required to push current through a circuit suggests that the circuit offers resistance. In other words, you do not have to push something unless it resists moving. Resistance limits the amount of amperage that flows through a circuit (Figure 1–15). The unit of measurement of resistance to flow is an **ohm.** If a circuit without enough resistance is connected across a reliable voltage source, wires or some other component in the circuit will be damaged by heat because too much current flows.

As mentioned, a bond exists between an electron and the protons in the nucleus of an atom. That bond must be broken for the electron to be freed so that it can move to another atom.

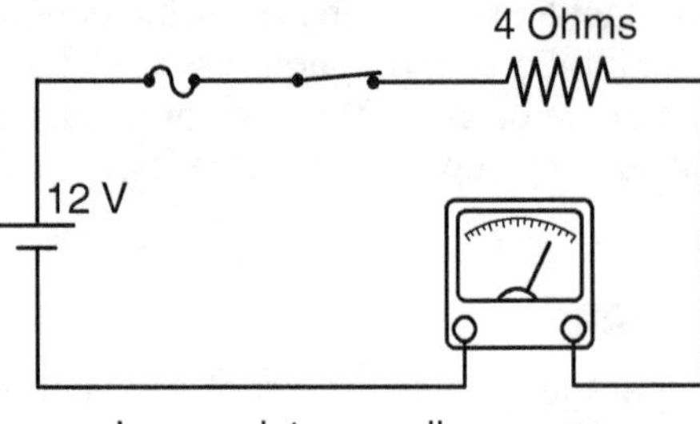

Low resistance allows more current to flow.

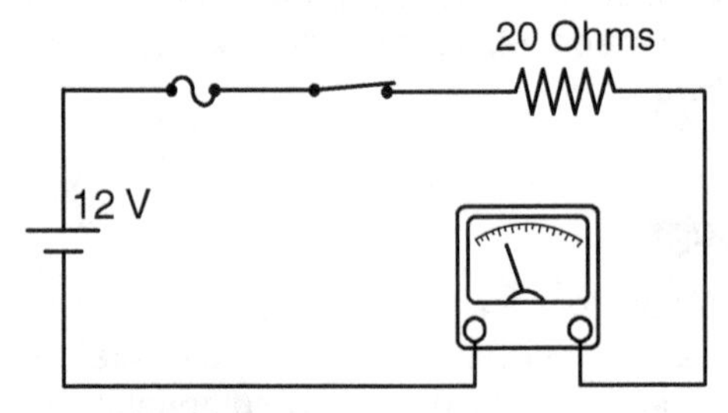

Higher resistance allows less current to flow.

Figure 1–15 Resistance versus current flow.

Breaking that bond and moving the electron amount to doing work. Doing that work represents a form of resistance to current flow. This resistance varies from one conductive material to another, depending on the atomic structure of the material. For example, lead has more resistance than iron, and iron has more resistance than copper. It also varies with the temperature of the conductor. Loose or dirty connections in a circuit also offer resistance to current flow. Using current flow to do work (to create heat, light, or a magnetic field to move something) also amounts to resistance to current flow. There are five things that will influence resistance within a wire and/or circuit:

- *Atomic structure* of the material—for example, copper wiring versus aluminum wiring.
- *Wire gauge*—a smaller gauge of wire (higher number) increases circuit resistance.
- *Wire length*—as the wire's length is increased, its resistance also increases.
- *Temperature*—as the temperature of a conductor increases, the resistance also increases.
- *Physical condition*—cuts, nicks, corrosion, and other deficient physical aspects of a conductor, connector, or load will increase the resistance of the circuit, thus creating excessive voltage drop within the circuit, resulting in reduced current flow.

Sometimes students get confused between voltage and amperage while doing tests on electrical systems. Review these definitions and consider the influence that voltage and amperage have on a circuit. It might also help to remember that voltage can be present in a circuit without current flowing. However, current cannot flow unless voltage is present.

Voltage Drop

When current is flowing through a circuit, voltage is lost, or used up, by being converted to some other energy form (heat, light, or magnetism). This loss of voltage is referred to as **voltage drop.** Every part of a circuit offers some resistance, even the wires, although the resistance in the wires should be very low (Figure 1–16). The voltage drop in each part of the circuit is proportional to the resistance in that part of the circuit. *The total voltage dropped in a circuit must equal the source voltage.* In other words, all of the voltage applied to a circuit must be converted to another energy form within the circuit. If excessive voltage drop occurs somewhere in a circuit due to unwanted resistance, the amperage flowing in the circuit will be reduced. This also reduces the voltage drop across the intended load component because excessive voltage drop is occurring elsewhere in the circuit. This reduces the load component's ability to function properly. Many good computers have been replaced simply because of problems involving excessive voltage drop in the power and ground circuits that are used to power up the computer.

It should be stressed that, in order for voltage to be dropped or used up, current must be flowing

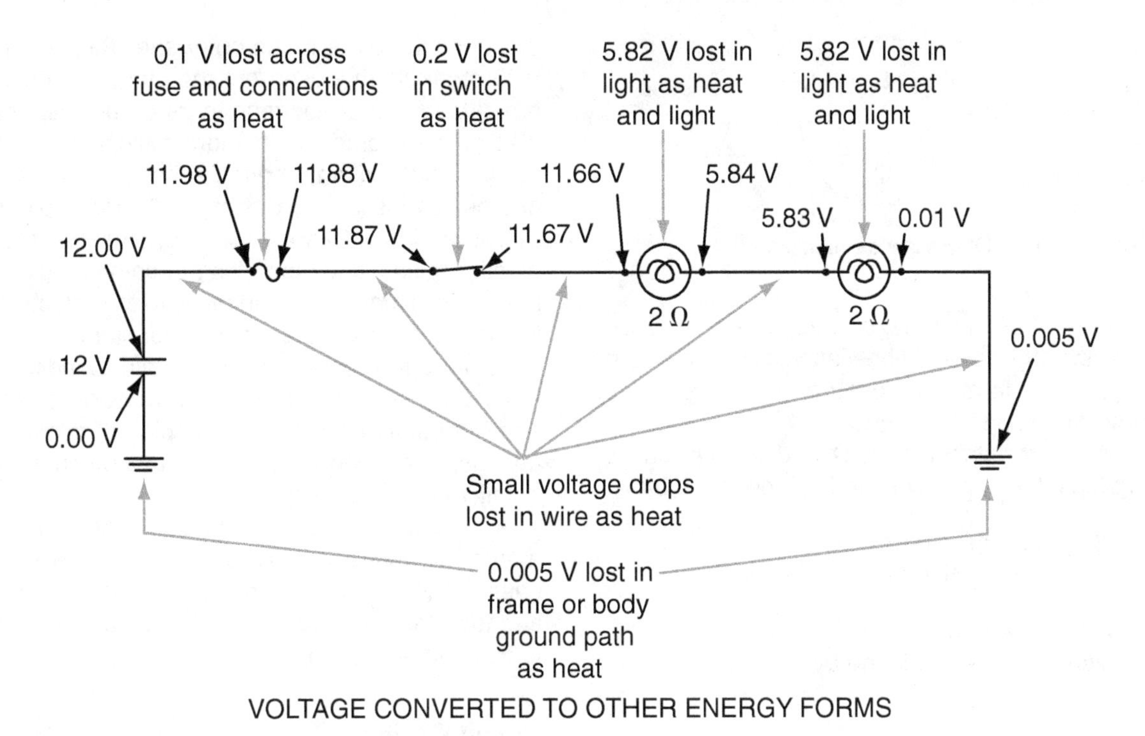

Figure 1–16 Voltage drop.

in the circuit. If current is not flowing in the circuit (as in the case of a burned fuse or other open), voltage will not be dropped, but rather source voltage will be present regardless of resistance all the way from the battery to the positive side of the open, and a ground measurement of zero volts will be present on the ground side of the open as well.

Ohm's Law

Ohm's law defines a relationship between amperage, voltage, and resistance. Ohm's law says that it takes 1 volt (V) to push 1 amp (A) through 1 ohm (Ω) of resistance. Ohm's law can be expressed in one of three simple mathematical equations:

$$E = I \times R$$
$$I = E/R$$
$$R = E/I$$

where: E = electromotive force or voltage
I = intensity or amperage
R = resistance or ohms

The simplest application of Ohm's law enables you to find the value of any one of the three factors—amperage, voltage, or resistance—if the other two are known. For example, if the voltage is 12 V and the resistance is 2 Ω (Figure 1–17), the current flow can be determined as follows:

$$I = E/R \text{ or}$$
$$I = 12\text{ V}/2\ \Omega = 6\text{ amps}$$

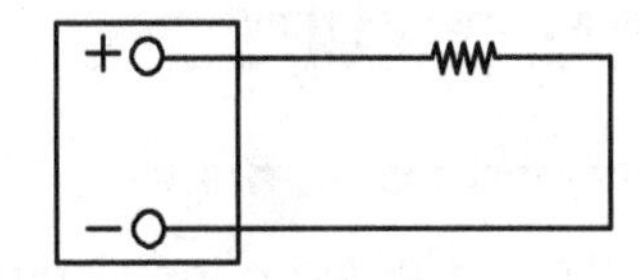

Figure 1–17 A simple series circuit.

Figure 1–18 Ohm's law calculation.

(The Greek letter Ω, or omega, is often used as a symbol or an abbreviation for ohms, and amps is the term often used as an abbreviation for amperes.)

If the resistance is 4 Ω and the current is 1.5 amps, the voltage applied can be found by:

$$E = I \times R \text{ or}$$
$$E = 1.5 \text{ amp} \times 4\ \Omega = 6\ V$$

If the voltage is 12 V and the current is 3 amps, the resistance can be found by:

$$R = E/I \text{ or}$$
$$R = 12\ V/3 \text{ amps} = 4\ \Omega$$

Perhaps the easiest way to remember how to use these equations is to use the diagram in Figure 1–18. To find the value of the unknown factor, cover the unknown factor with your thumb, and multiply or divide the other factors as their positions indicate.

There are many other applications of Ohm's law, some of which are quite complex. (A more complicated application is covered later in this chapter.) An automotive technician is rarely required to apply Ohm's law directly to find or repair an electrical problem on a vehicle. But knowing and understanding the relationship of the three factors is a must for the technician who wants to be able to diagnose and repair electrical systems effectively.

ELECTRICAL CIRCUITS

Conductors and Insulators

Previously, in discussing the electrons in the outer shell of an atom, it was said that the fewer electrons there are in the outer shell the easier it is to break them loose from the atom. If an atom has five or more electrons in its outer shell, the electrons become much more difficult to break away from the atom, and the substance made up of those atoms is a very poor conductor—so poor that it is classed as an insulator. Rubber, most plastics, glass, and ceramics are common examples of insulators. Substances with four electrons in their outer shell are poor conductors but can become good ones under certain conditions. Thus, they are called **semiconductors.** Silicon and germanium are good examples of semiconductors. (Semiconductors are covered in more detail later in this chapter.)

Conductors are substances made up of atoms with three or fewer electrons in their outer shell. These electrons are called free electrons because they are loosely held and can be freed to travel to another atom.

Circuit Design

The following three elements are essential to the operation of all electrical circuits:

1. *Voltage source,* such as a battery or generator, provides voltage to the power circuit.
2. *Load,* such as a motor or light, performs the function the circuit was designed to perform and also provides resistance to limit current.
3. *Circuitry or current conductors* complete the circuit between the voltage source and the load.

Also, most electrical circuits will have some type of component to provide each of the following functions:

- *Circuit protection,* such as a fuse or circuit breaker, serves to open the circuit in the event of excessive current flow.
- *Circuit control,* such as a switch or relay (electrically operated switch), provides the ability to control when current flows in the circuit.

- *Ground*—meaning the metal of the vehicle's body, chassis, and engine—is commonly used in place of running a copper wire on the negative side of the circuit (referred to as a negative ground).

These components are often represented by symbols, as shown in Figure 1–19. You should note that any circuit protection devices, circuit control devices, and grounds constitute part of the circuitry or current conductors when used.

Circuit Types

There are two distinct types of electrical circuits, plus combinations of the two.

Series Circuits. In a series circuit, there is only one path for current flow, and all of the current flows through every part of the circuit. Parts A and B of Figure 1–19 show simple series circuits. Even though there is only one load in each of these circuits, they qualify as series circuits because there is only one path for current flow. Figure 1–20 shows a better example of a series circuit. Not only is there just one path for current flow, there are also two loads in series with each other. When there are two or more loads in series, the current must pass through one before it can pass through the next. The characteristics of a series circuit include the following:

- Current flow is the same at all parts of the circuit.
- Resistance units are added together for total resistance.
- Current flow decreases as resistance units are added.
- All of the voltage will be used up by all of the resistance in the circuit if current is flowing. There will be no voltage left after the last resistance. Furthermore, each ohm of resistance will share equally in the voltage drop.
- An open in any part of the circuit disrupts the entire circuit.

The following problems apply Ohm's law to a series circuit. Refer to Figure 1–21, which shows a compound series circuit.

Series Circuit—Problem 1. Assume that the resistance of R1 is 2 Ω and that R2 is 4 Ω; find the total current flow. In a series circuit, the resistance value of each unit of resistance can be added together because all of the current passes through each resistor.

$$I = E/R \text{ or}$$
$$I = E/(R1 + R2) \text{ or}$$
$$I = 12\text{ V}/(2\ \Omega + 4\ \Omega) \text{ or } 12\text{ V}/6\ \Omega = 2\text{ A}$$

Series Circuit—Problem 2. Assume the resistance values are unknown in Figure 1–21, but that the total current flow is 3 amps. To find the total resistance:

$$RT = E/I \text{ or}$$
$$RT = 12\text{ V}/3\text{ amps} = 4\ \Omega$$

Series Circuit—Problem 3. Find the voltage drop across R1, applying the same resistance values as in problem 1. Each ohm of resistance value shares equally in the total voltage drop in a series circuit. Therefore, how much voltage is dropped (used up) by each ohm of resistance in a circuit is also numerically equal to how many amps are flowing in the circuit. As an example, if a 12 V circuit has a resistance total of 6 V, we know that there the current flow is 2 amps (12/6 = 2). But the same math is also used to calculate how much voltage is dropped by each measured ohm of resistance to flow. If each ohm of resistance shares equally in the voltage drop, then 12 V divided by 6 Ω indicates that each ohm of resistance will drop 2 V. Thus, because you are concerned with the voltage drop across R1 in Figure 1–21, multiply the resistance of R1 by the circuit's amperage.

$$\text{Voltage drop across R1} = R1 \times I \text{ or}$$
$$2\ \Omega \times 2\text{ A} = 4\text{ V (voltage drop across R1)}$$

Parallel Circuits. In a parallel circuit, the conductors split into branches with a load in each

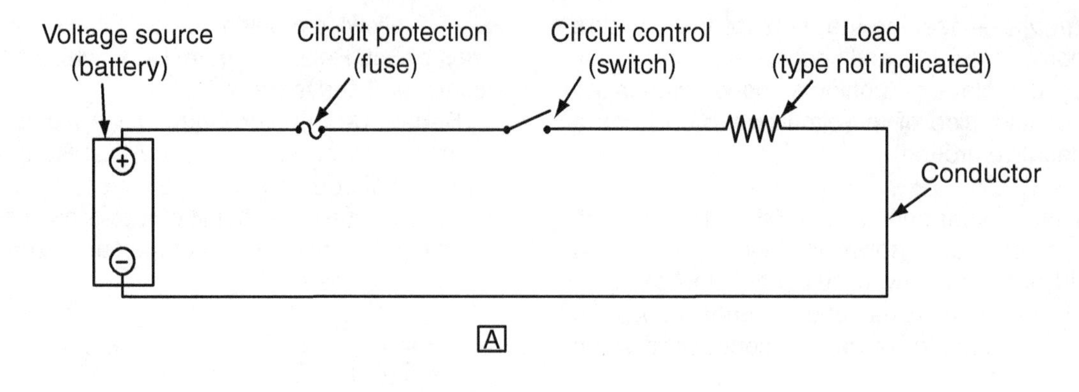

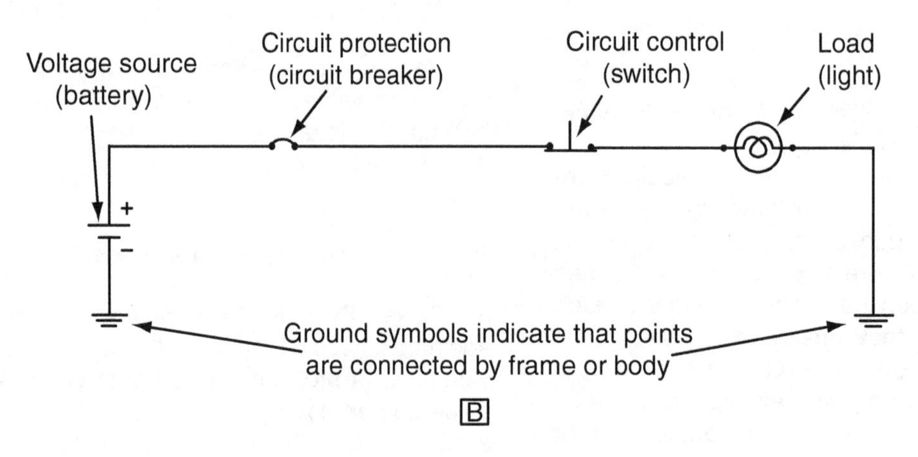

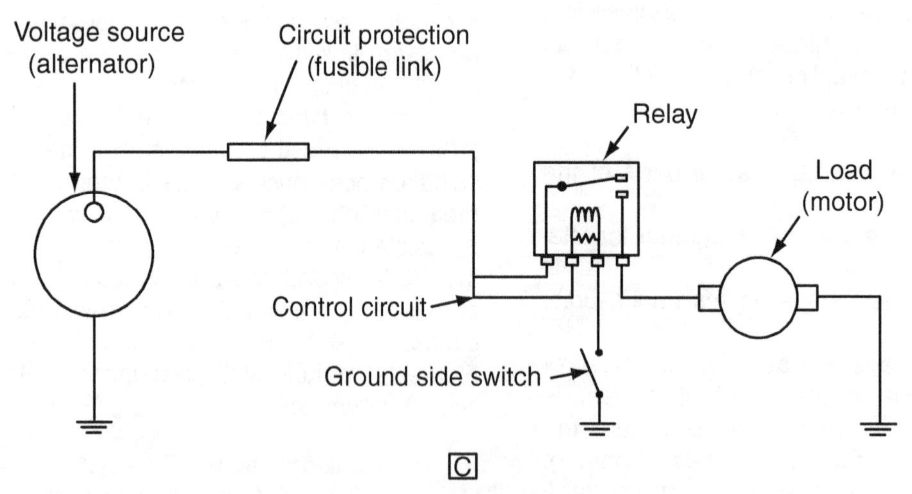

Figure 1–19 Circuit components.

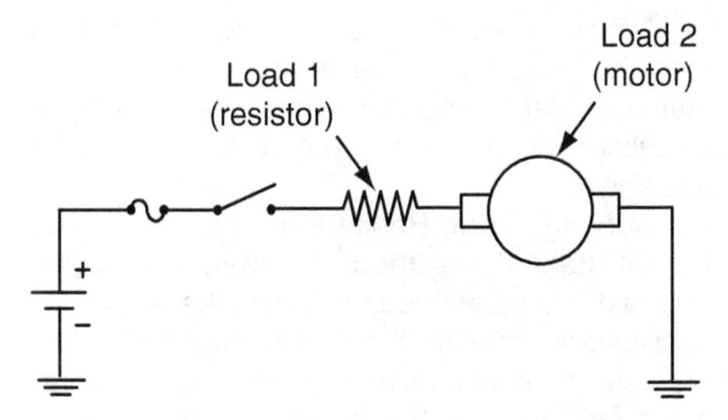

Figure 1–20 Series circuit.

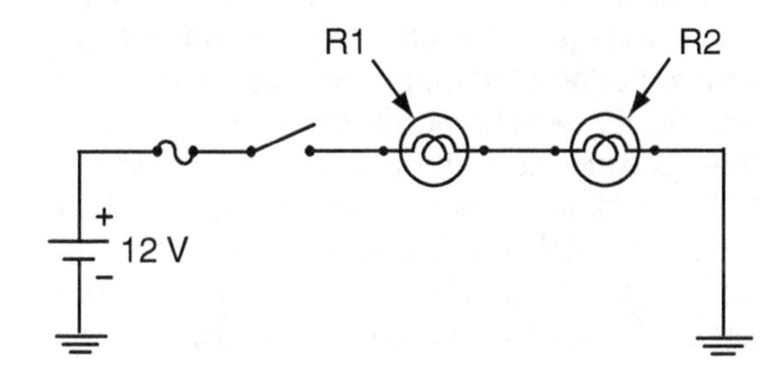

Figure 1–21 Series circuit.

branch (Figure 1–19, Part C, and Figure 1–22). Some current will flow through each branch, with the most current flowing through the branch with the least resistance. Characteristics of a parallel circuit are:

- Current varies in each branch (unless resistance in each branch is equal).
- Total circuit current flow increases as more branches are added.
- Total circuit resistance goes down as more branches are added and will always be less than the lowest single resistance unit in the circuit.
- Source voltage is dropped across each branch.
- An open in one branch does not affect other branches.

There are three possible mathematical formulas for calculating the total resistance in a parallel circuit depending on the number of

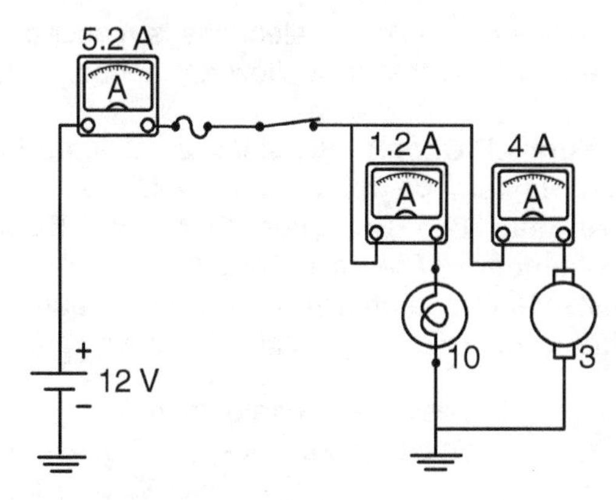

Figure 1–22 Parallel circuit.

branches and how the resistances relate to each other. There is also a shortcut method that can be used in place of any of the three formulas, as demonstrated. The following problems apply Ohm's law to a parallel circuit.

Parallel Circuit—Problem 1: Calculating Total Resistance for a Parallel Circuit with Two Branches. The product of R1 multiplied by R2 divided by the sum of R1 plus R2 equals the resistance total (RT). This is mathematically stated as:

$$(R1 \times R2)/(R1 + R2) = RT$$

In Figure 1–23, assume that R1 is 3 Ω and R2 is 4 Ω. Then calculate the resistance total as follows:

$$(3\ \Omega \times 4\ \Omega)/(3\ \Omega + 4\ \Omega) = 12\ \Omega\ /7\ \Omega = 1.71\ \Omega$$

Notice that the resistance total is less than the least resistor value. This is because the total

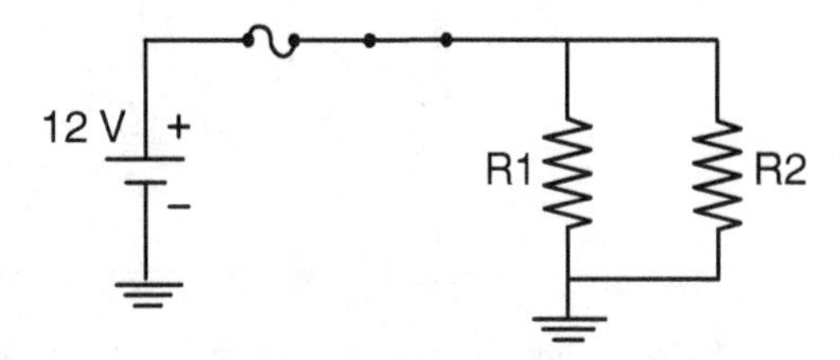

Figure 1–23 Parallel circuit with two branches.

resistance to the flow of electrons is reduced as more branches that allow flow are added to the circuit.

Parallel Circuit—Problem 2: Calculating Total Resistance for a Parallel Circuit with More than Two Branches that Have Equal Resistances. The resistance total is equal to the resistance of one branch divided by the number of branches. This is mathematically stated as:

(Resistance of one branch)/
(Number of branches) = RT

In Figure 1–24, if each of the resistances in each of the four branches is 3 Ω of resistance (all are equal), the total resistance can be calculated by dividing 3 (the resistance value of one branch) by 4 (the number of branches). Thus, the resistance total would equal 0.75 Ω. Notice, again, that the resistance total is less than the least resistor value.

Parallel Circuit—Problem 3: Calculating Total Resistance for a Parallel Circuit with More than Two Branches that Have Dissimilar Resistances. The resistance total is equal to the reciprocal of the sum of the reciprocals of each of the branch's resistance values. This is mathematically stated as:

1/((1/R1) + (1/R2) + (1/R3) + (1/R4)
+ (and so on . . .)) = RT

In Figure 1–24, if R1 = 2 Ω, R2 = 3 Ω, R3 = 4 Ω, and R4 = 6 Ω, then:

RT = 1/((1/2) + (1/3) + (1/4) + (1/6)) or
RT = 1/((0.5) + (0.33) + (0.25) + (0.16)) or
RT = 1/1.24 = 0.806 Ω

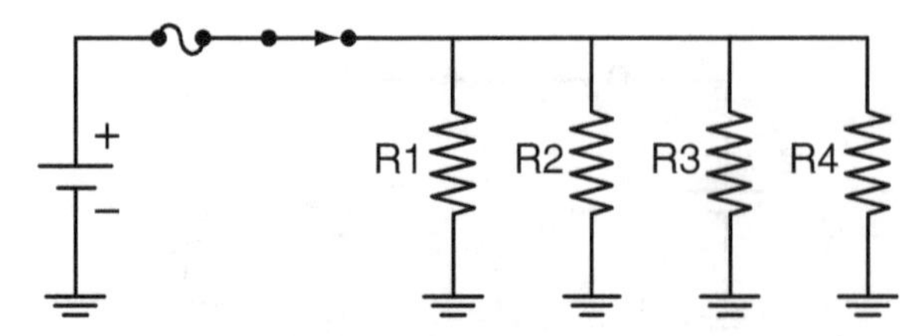

Figure 1–24 Parallel circuit with more than two branches.

Notice, again, that the resistance total is less than the least resistor value. Also, this complex mathematical formula is best done using a calculator with a memory feature. But there is an easier way.

Solving Total Resistance for a Parallel Circuit Using the Method of Assumed Voltages. This is a "cheat" method for finding the resistance total of any parallel circuit, whether it has two branches or more than two branches and whether the resistances of the branches are equal or dissimilar.

Assume any convenient voltage, be it your source voltage or any other voltage that is convenient to perform the math with. Calculate the current flow for each branch at the assumed voltage. Then add up all of the current flows to find the total current flow. Then divide the assumed voltage by the total current flow to find the resistance total. Let us try applying it to Figure 1–24 with the same resistance values as used earlier.

Let us assume a source voltage of 12 V. Using Ohm's law, we can calculate the current flow easily for each branch as follows:

12 V divided by 2 Ω (R1) = 6 amps
12 V divided by 3 Ω (R2) = 4 amps
12 V divided by 4 Ω (R3) = 3 amps
12 V divided by 6 Ω (R4) = 2 amps

Then add up the current of each branch.

6 amps + 4 amps + 3 amps + 2 amps
= 15 amps

Now divide the assumed voltage by the total current.

12 V divided by 15 amps = 0.8 Ω (RT)

Now let us assume another convenient voltage of 60 V. Again, using Ohm's law, we can easily calculate the current flow for each branch as follows:

60 V divided by 2 Ω (R1) = 30 amps
60 V divided by 3 Ω (R2) = 20 amps

60 V divided by 4 Ω (R3) = 15 amps
60 V divided by 6 Ω (R4) = 10 amps

Then add up the current of each branch.

30 amps + 20 amps + 15 amps + 10 amps
= 75 amps

Now divide the assumed voltage by the total current.

60 Ω divided by 75 amps = 0.8 Ω (RT)

Therefore, any voltage that is convenient to use with the particular resistance values will work for this shortcut method. This makes it easy enough to calculate the resistance total for any parallel circuit, thereby eliminating the need for a calculator.

Series-Parallel Circuits. Some circuits have characteristics of a series as well as those of a parallel circuit. There are two basic types of series-parallel circuits. The most common is a parallel circuit with at least one resistance unit in series with all branches (Figure 1–25). All of the current flowing through the circuit in Figure 1–25 must pass through the indicator light before it divides to go through the two heating elements, which are in parallel with each other.

To solve for current, resistance, or voltage drop values in a series-parallel circuit, you must identify how the resistance units relate to each other, then use whichever set of formulas (series circuit or parallel circuit) applies. For example:

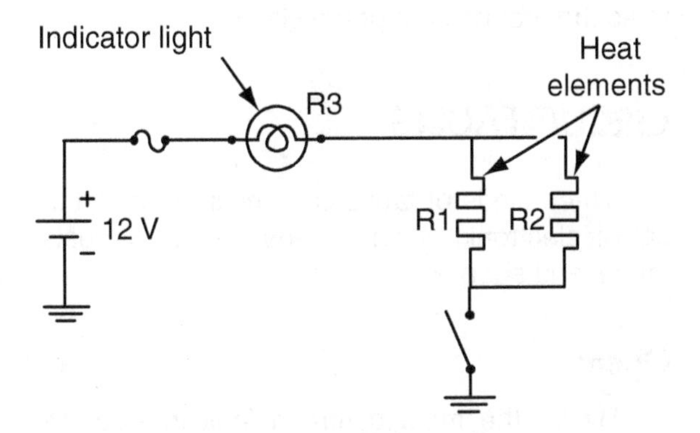

Figure 1–25 Series-parallel circuit.

Series-Parallel Circuit—Problem 1. In Figure 1–25, assume a resistance of 4 Ω for the indicator light (R3), 20 Ω for R1, and 30 Ω for R2. Find the total resistance for the circuit.

Branches R1 and R2 are in parallel with each other and in series with R3. First, use the product over the sum method (or the method of assumed voltages may also be used) to solve for the combined resistance of R1 and R2.

R = R1 × R2/R1 + R2 or
R = 20 Ω × 30 Ω /20 Ω + 30 Ω
R = 600 Ω/50 Ω = 12 Ω

Now add the total resistance of R1 and R2 (12 Ω) to the resistance of R3 (4 Ω).

RT = R(R1 and R2) + R3 or 12 Ω + 4 Ω = 16 Ω

Series-Parallel Circuit—Problem 2. Find the total current flow for the circuit.

IT = E/RT or
IT = 12 V/16 Ω= 0.75 amp

Before you can find the current flow through R1 or R2, you must find the voltage applied to R1 and R2. Remember that R3 is in series with both R1 and R2. Also remember that in a series circuit, a portion of the source voltage is dropped across each resistance unit. Therefore, because there is a voltage drop across R3, there is not a full 12 V applied to R1 and R2.

Series-Parallel Circuit—Problem 3. Find the voltage drop across R3.

V3 = IT × R3 or
V3 = 0.75 amp × 4 Ω = 3 V

If 3 V are dropped across R3 from a source voltage of 12 V, 9 V are applied to R1 and R2.

Series-Parallel Circuit—Problem 4. Find the current flow through each of the following resistors.

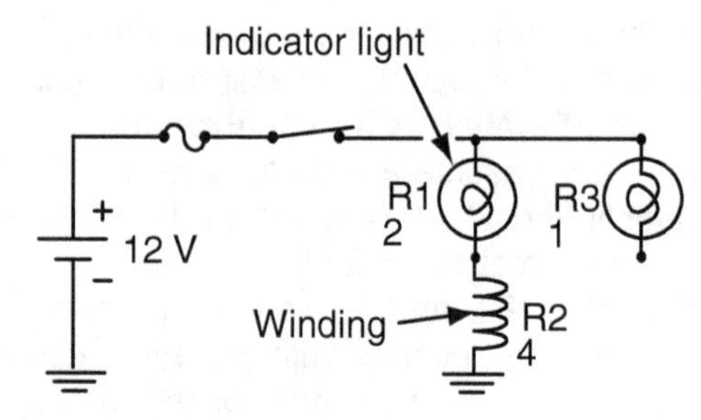

Figure 1–26 Series string in parallel.

The current flow through R1 is calculated as follows:

$$I1 = E/R1 \text{ or}$$
$$I1 = 9\ V/20\ \Omega = 0.45 \text{ amp}$$

The current flow through R2 is calculated as follows:

$$I2 = E/R2 \text{ or}$$
$$I2 = 9\ V/30\ \Omega = 0.30 \text{ amp}$$

The second type of series-parallel circuit is a parallel circuit in which at least one of the branch circuits contains two or more loads in series (Figure 1–26). This type of circuit may be referred to as a *series string in parallel.* Finding the total current, resistance, or voltage drop for this circuit is easy. Identify the branch, or branches, that have multiple loads in series and add up the resistances within that branch. Then treat the circuit like any other parallel circuit. When troubleshooting this kind of circuit in a vehicle, failing to recognize the branch of a circuit you are testing as having loads in series could cause confusion.

CAUTION: If you are not absolutely sure about the circuit you are testing, consult the correct wiring schematic for the vehicle.

POLARITY

Just as a magnet has polarity—a north pole and a south pole that determine the directional force of the lines of force—an electrical circuit has polarity. Instead of using north and south to identify the polarity, an electrical circuit's polarity is identified by positive and negative. An electrical circuit's polarity is determined by its power source. The best example is a battery (Figure 1–27). The side of the circuit that connects to the positive side of the battery is positive and on most vehicles is the insulated side. The voltage on this side is usually near the source voltage. Most of the voltage is dropped in the load where most of the work is done. The negative side of the circuit carries the current from the load to the negative side of the battery. There is usually very little resistance in the negative side of the circuit, and the voltage is near zero. Because this side of the circuit has the same potential as the vehicle's frame and sheet metal, this side of the circuit is not usually insulated, but rather is connected to the metal of the engine, chassis, or body and is therefore referred to as the ground side of the circuit.

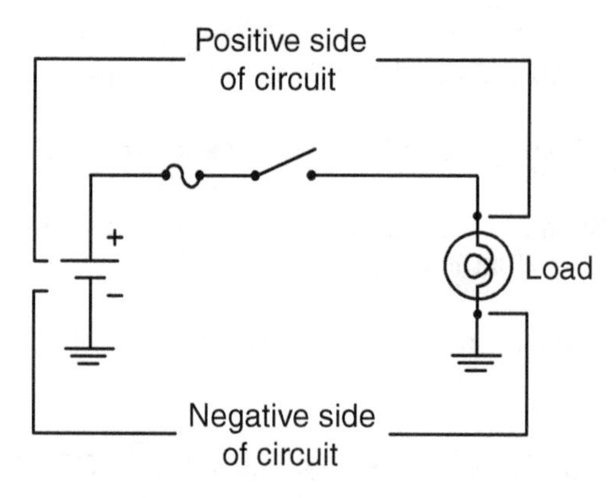

Figure 1–27 Circuit polarity.

CIRCUIT FAULTS

Three kinds of faults can occur in an electrical or electronic circuit: opens, excessive resistance, and shorts.

Opens

By far the most common fault in electrical and electronic circuits is an open circuit, or *open.*

An open circuit means that there is a point in the circuit where resistance is so high that current cannot flow. An open can be the result of a broken wire, a loose or dirty connection, a blown fuse, or a faulty component in the load device. Experienced technicians know that wires rarely break except in applications in which the wire experiences a lot of flexing. Most opens occur in connections, switches, and components. A switch in a circuit provides a way to conveniently open the circuit. Of course, when a circuit is opened deliberately, it is not a fault.

Excessive Resistance

A loose or dirty connection or a partially cut wire can cause excessive resistance in a circuit. Under these conditions, the circuit can still work, but not as well as it could, because the additional resistance reduces current flow through the circuit (Figure 1–28). Excessive resistance can also result from a faulty repair or modification of a circuit in which a wire that is too long or too small in diameter has been installed. The location of excessive resistance in a circuit can be easily found with a series of voltage drop tests, which are discussed in Chapter 5.

Shorts

A short is a fault in a circuit that causes current to bypass a portion of a circuit. The term *short* as used in electrical terminology means that the current is taking a shortcut rather than following the path it is supposed to take.

If, in an electromagnetic load, the windings overheat, resulting in melting of the material that insulates one winding from the next, the current may bypass just a few of the windings, thus reducing the circuit's intended resistance (Figure 1–29A). If only a relatively small number of the coil's loops are shorted, the increase in current flow might not be enough to further damage the circuit, although it might blow a fuse if the circuit is fuse protected. The device might even continue to work, but probably at reduced efficiency. But if the shorted load is controlled by a computer, the resulting increase in current flow may be substantial enough to destroy the computer's ability to control this circuit (depending

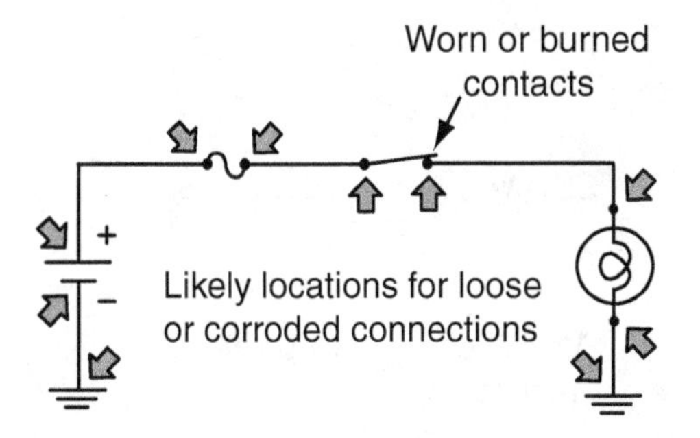

Figure 1–28 Possible locations in a circuit for an open or high resistance.

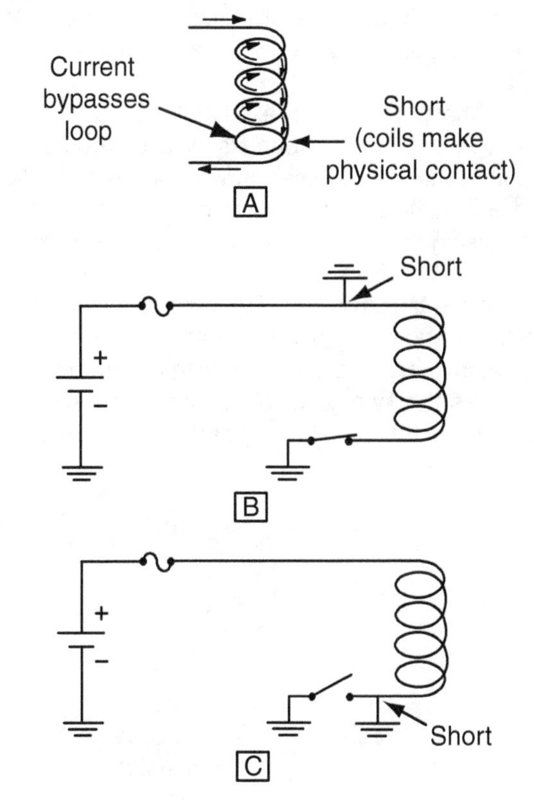

Figure 1–29 Shorts.

upon the current protection that may or may not be designed into the computer), resulting in the need to replace the computer as part of the repair. Or the current may bypass the entirety of the load, thus reducing the circuit's resistance to near zero ohms (Figure 1–29B). Also, a short may bypass a switch or relay-operated switch and thus create an "always on" condition in which the circuit cannot be properly controlled (Figure 1–29C).

Short to Power. Figure 1–29B demonstrates a short to power (positive) and a short that also allows current to fully bypass the load. This would likely quickly blow the fuse due to the increased current. In this example, the circuit will not operate at all until the problem is corrected and the fuse is replaced. If an improper fuse has been installed with an amperage rating that is higher than the manufacturer-recommended amperage rating, a fire could result. ***Do not ever*** replace a fuse with one other than the fuse recommended by the manufacturer unless the manufacturer directs you to do so through a technical service bulletin (TSB).

A short finder is an excellent tool for finding this type of short; it consists of a self-resetting circuit breaker that temporarily replaces the fuse and a sensitive inductive meter that can sense the pulsing of the magnetic field from the battery to the point of the short.

Short to Ground. Figure 1–29C demonstrates a short to ground (negative) and a short that also bypasses the switch that controls the load. This short would not blow the fuse but instead would keep the circuit energized continuously.

ELECTRICAL COMPONENTS

Fixed Resistors

A fixed resister is a component that has a fixed electrical resistance designed into it (Figure 1–30). Carbon is a substance commonly used to form a fixed resistor. A fixed resistor can be designed with many different levels of fixed resistance, different tolerances, and different wattage ratings.

Variable Resistors

A variable resistor has three terminals (Figure 1–30). The outermost two terminals have a specific fixed resistance designed between them. The center terminal connects to an electrically conductive wiper that wipes across the length of the resistive material, therefore changing the resistance between the center terminal and each of the outermost terminals.

A variable resistor may be used to control current flow in a load circuit, in which case it is known as a *rheostat*. It may also be used as a *sensor* to sense physical position and mechanical action as described in Chapter 3. It may operate in a linear fashion (straight-line movement) or in a rotary fashion. Generally, rheostats have a high-wattage design and a resistance material that consists of a wound resistance wire. Sensors have a low-wattage design and a resistance material that consists of carbon. Either type has a physical wear factor due to the wiping action of

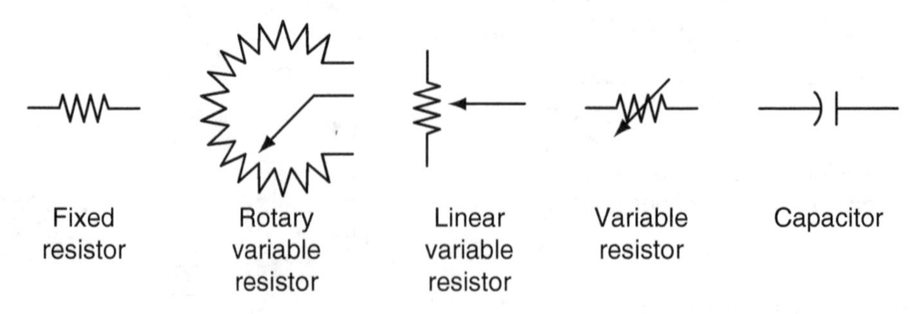

Figure 1–30 Symbols for common electrical components.

the wiper across the resistance material and will commonly fail over time.

Capacitors

A **capacitor** is an electrical component that will store and give up electrons according to the voltages it is connected between (Figure 1–30). It does not have electrical continuity through it and, therefore, does not complete the circuit it is connected into. The number of electrons that can be stored in it is known as its capacitance, usually rated in micro-farads. A farad is 6.28 billion billion electrons or one amp's worth of electrons.

A capacitor consists of two electrical elements separated by a dielectric insulator. A simple capacitor can be made by layering a sheet of wax paper (nonconductive) in between two sheets of aluminum foil (conductive), and then rolling them up into a tight roll. One piece of foil will not touch the other piece of foil. Therefore, it will not complete a circuit that the two pieces of foil are connected into. However, because of the close proximity of the two pieces of foil to each other, it will take on and give off electrons according to the voltages it is connected between and according to its physical size. In essence, a capacitor is a storage battery for electrons.

SEMICONDUCTORS

Semiconductors are the basis of today's *solid-state electronics,* electronic devices such as computers and amplifiers that can control the most complex systems without having any moving parts. As mentioned previously in this chapter, a semiconductor is an element with four valence electrons. The two most-used semiconductor materials for solid-state components are silicon and germanium. Of these two, silicon is used much more than germanium. Therefore, most of this discussion will apply to silicon.

As previously stated, an atom with three or fewer electrons in its outer shell easily gives them up. If an atom has more than four but fewer than eight electrons in its outer shell, it exhibits a tendency to acquire more until it has eight. If there are seven electrons in the outer shell, the tendency to acquire another one is stronger than if there are only six. Once there are eight, it becomes very stable; in other words, it is hard to get the atom to gain or lose an electron. In a semiconductor material—for example, a silicon crystal—the atoms share valence electrons in what are called covalent bonds (Figure 1–31). Each atom positions itself so that it can share the valence electrons of neighboring atoms, giving each atom, in effect, eight valence electrons. This lattice structure is characteristic of a crystal solid and provides two useful characteristics:

1. Impurities can be added to the semiconductor material to increase its conductivity; this is called *doping.*
2. It becomes *negative temperature coefficient,* meaning that its resistance goes down as its temperature goes up. (This principle is put to use in temperature sensors, as discussed in Chapter 3.)

Doping

All of the valence electrons in a pure semiconductor material are in valence rings containing eight electrons (Figure 1–31). With this atomic structure, no electrons can be easily freed, and there are no holes to attract an electron even if some were available. The result is that this material has a high resistance to current flow. Adding very small amounts of certain other elements can greatly reduce the semiconductor's resistance. Adding trace amounts (about 1 atom of the doping element for every 100 million semiconductor atoms) of an element with either five or three valence electrons can create a flaw in some of the covalent bonds.

Adding atoms with five valence electrons (referred to as *pentavalent atoms)* such as arsenic, antimony, or phosphorus achieves a crystal structure as shown in Figure 1–32. In Figure 1–32, phosphorus is the doping element and silicon is the base semiconductor element. Four of the

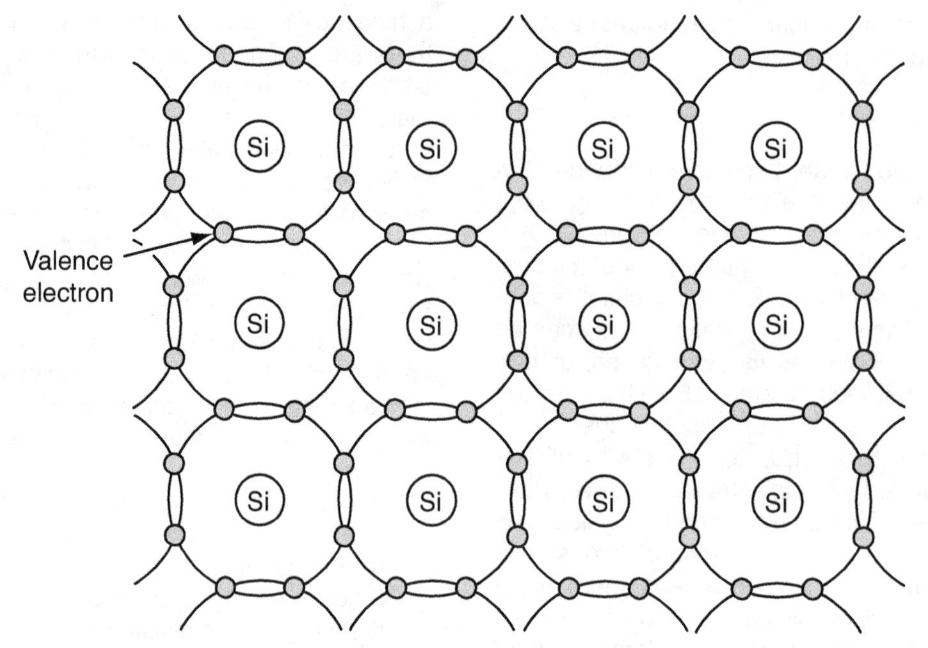

Figure 1–31 Covalent bonds.

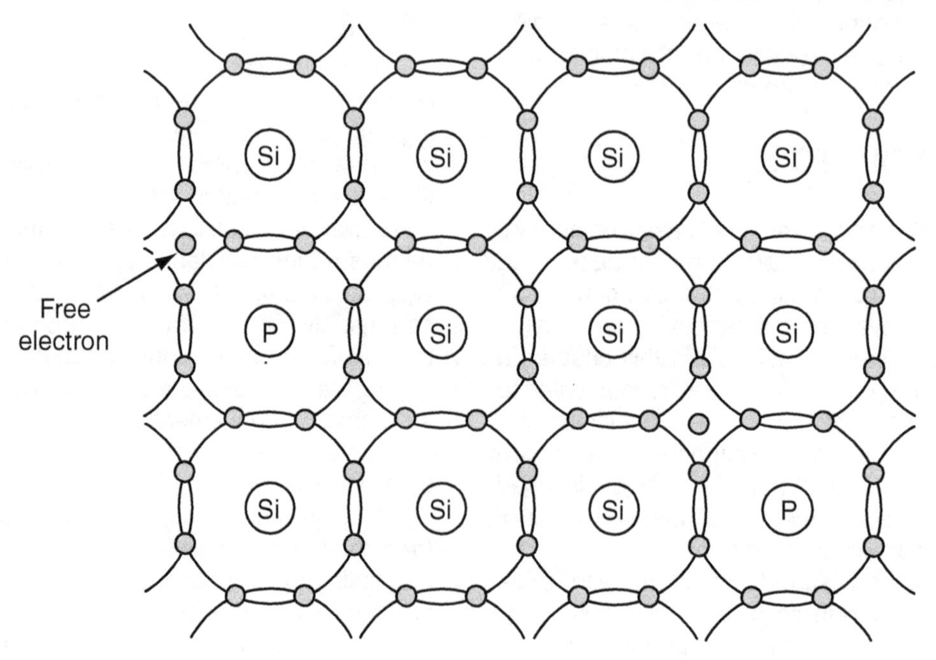

Figure 1–32 Silicon doped with phosphorus.

phosphorus atom's five valence electrons are shared in the valence rings of neighboring silicon atoms, but the fifth is not included in the covalent bonding with neighboring atoms; it is held in place only by its attraction to its parent phosphorus atom—a bond that can be broken easily. This doped semiconductor material is classified as an N-type material. Note that it does not have a negative charge because the material contains the same number of protons as electrons. It does have electrons that can be easily attracted to some other positive potential.

Adding atoms with three valence electrons (referred to as *trivalent atoms)* such as aluminum, gallium, indium, or boron achieves an atomic structure as seen in Figure 1–33. In Figure 1–33, boron is used as the doping material in a silicon crystal. The boron atom's three valence electrons are shared in the valence ring of three of the neighboring silicon atoms, but the valence ring of the fourth neighboring silicon atom is left with a hole (electron deficiency) instead of a shared electron. Remember that a valence ring of seven electrons aggressively seeks an eighth electron. In fact, the attraction to any nearby free electron is stronger than the free electron's attraction to its companion proton in the nucleus of its parent atom. Thus this material is classified as a P-type.

Doping Semiconductor Crystals

In actuality, a PN junction is not produced by placing a P- and an N-type semiconductor back to back. Rather, a single semiconductor crystal is doped on one side with the pentavalent atom's opposite sides and on the other with trivalent atoms. The center of the crystal then becomes the junction. The doping is done by first bringing the semiconductor crystal to a molten temperature. In a liquid state, the covalent bonds are broken. The desired amount of doping material is then added. As the semiconductor crystal cools, the covalent bonds redevelop with the doping atoms included.

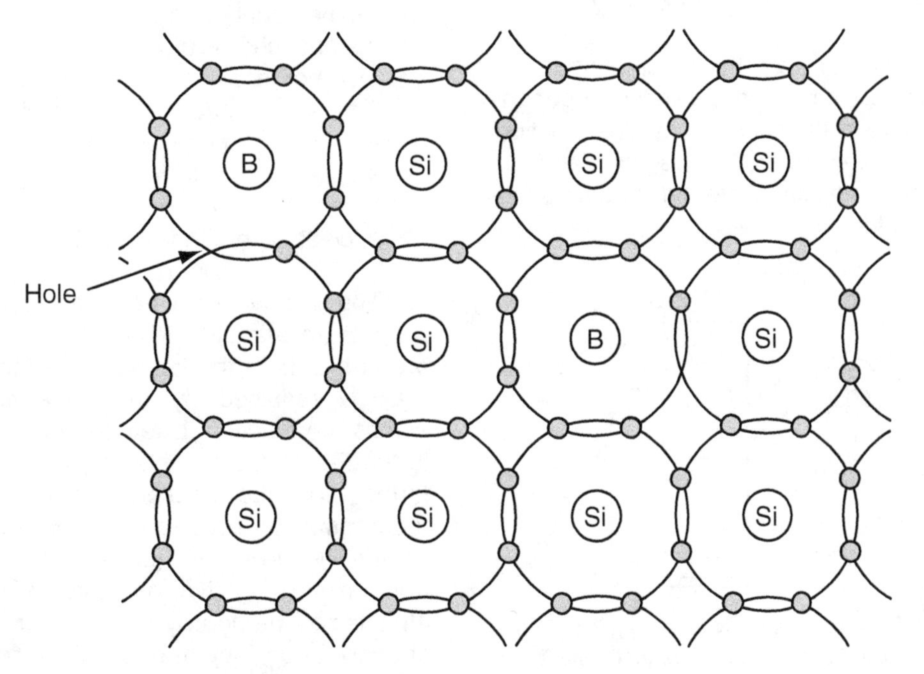

Figure 1–33 Silicon doped with boron.

PN Junction

Figure 1–34 shows a P-type crystal and an N-type crystal separated from each other. This figure shows only the free electrons in the N-type material and the holes (electron deficiencies) in the P-type material that result from the way in which the doping atoms bond with the base semiconductor material. If the P-type and the N-type are put in physical contact with each other, the free electrons near the *junction* in the N-type cross the junction and fill the first holes they come to in the seven-electron valence rings near the junction in the P-type (Figure 1–35). The junction is the area that joins the P-type and N-type. This action quickly creates a zone around the junction in which:

- There are no more free electrons in that portion of the N-type.
- There are no more free holes (electron deficiencies) in that portion of the P-type.
- The valence rings near the junction in both the N- and P-type have eight shared electrons, so they are reluctant to gain or lose any more.

Another way to state the information in the preceding list is to say that there are no longer any current carriers in the zone around the junction. This zone is often referred to as the *depletion zone.* It is also sometimes referred to as a boundary layer. In addition:

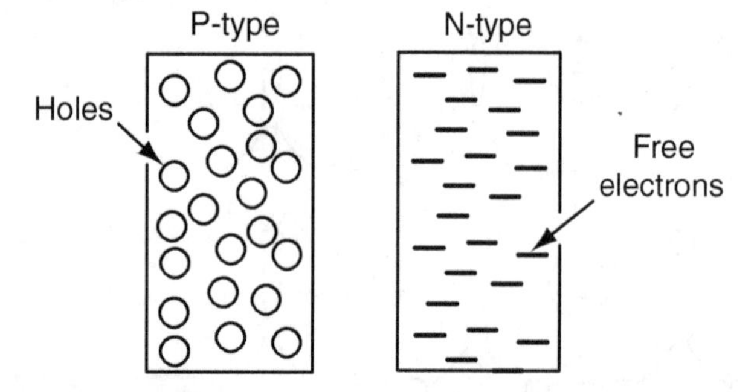

Figure 1–34 P-type and N-type crystals. Note: For simplicity, only the holes and free electrons in the valence ring are shown.

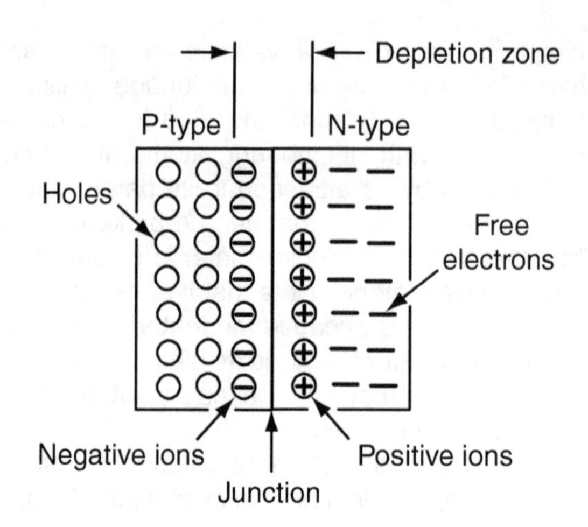

Figure 1–35 PN junction. Note: For simplicity, only the holes and free electrons in the valence ring are shown.

- The phosphorus atoms near the junction in the N-type have each lost an electron, which makes them positive ions (the free electron that crossed the junction to fill a hole in the P-type abandoned a proton in the nucleus of the phosphorus atom).
- The boron atoms near the junction in the P-type have each gained an electron, which makes them negative ions (the electron that dropped into the valence ring around the boron atom is not matched by a proton in the nucleus of the boron atom).

The positive ions near the junction in the N-type are attracted to the free electrons that are farther from the junction, but they are more strongly repulsed by the negative ions just across the junction. Likewise, the negative ions in the P-type are attracted to the holes that are farther from the junction, but those holes are more strongly repulsed by the positive ions on the N-side of the junction. The extra electrons in the negative ion atoms are somewhat attracted to the nearby holes, but they are more strongly bound by the covalent bonding into which they have just dropped.

Keep in mind that a layer of negative ions along the junction in the P-type and a layer of positive ions along the junction of the N-type have been created. The opposing ionic charges on the two sides of the depletion zone create an electrical potential of about 0.6 V (0.3 V for germanium). This potential, often referred to as the *barrier potential,* cannot be measured directly, but its polarity prevents current from flowing across the junction unless it is overcome by a greater potential.

Diodes

If two of these semiconductor materials, one N-type and one P-type, are placed back to back, the simplest semiconductor device, known as a **diode,** is formed. A diode operates as an electrical one-way check valve with no moving parts; it will allow current to pass in one direction only. If a positive electrical potential is applied to the positive end and a negative electrical potential is applied to the negative end and the applied electrical pressure (voltage) is greater than 0.6 V, the applied polarity will cause the diode to gain continuity across the PN junction. In this condition the diode is said to be *forward biased* (Figure 1–36). The higher negative potential introduced by the forward bias voltage at the negative side of the crystal repels the free electrons in the N-type material. They move toward the junction, canceling the charge of the positive ions. At the same time, the higher positive potential introduced by the forward bias voltage on the positive side of the crystal repels the free holes in the P-type. They move toward the junction, canceling the charge of the negative ions. With the barrier potential overcome, current easily flows across the junction, with electrons moving toward the external positive potential and holes moving toward the external negative potential. When the forward bias voltage is removed, barrier potential redevelops and the diode again presents high resistance to current flow.

Conventional Current Flow versus Electron Current Flow

Before electrons were known, Benjamin Franklin surmised that current was a flow of positive charges moving from positive to negative in a circuit. Franklin's belief became so accepted that even after electrons were discovered and scientists learned that current flow consists of electrons moving from negative to positive, the old idea was hard to give up. As a result, the idea of positive charges moving from positive to negative is still often used. It is referred to as *conventional current flow* or less formally as *hole flow.*

The conventional current flow theory has gotten a boost in recent years because it helps explain how semiconductors work: Positively charged holes move from positive charges to negative charges as electrons move from negative charges to positive charges.

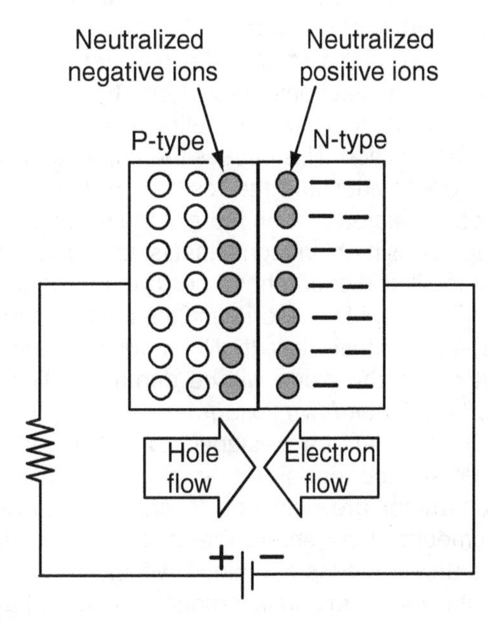

Figure 1–36 Forward bias voltage applied to a diode.

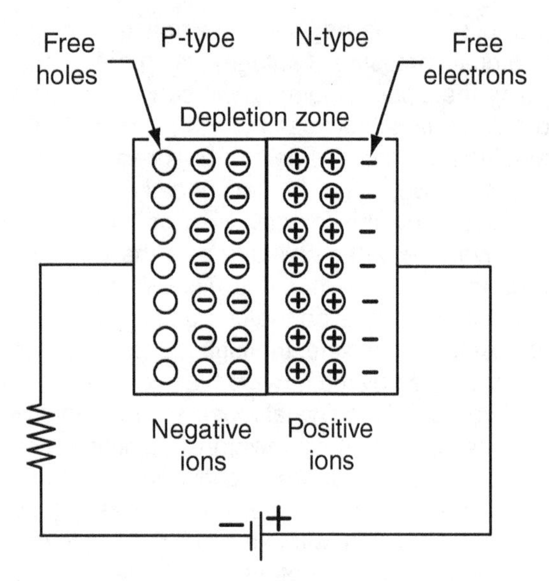

Figure 1–37 Reverse bias voltage applied to a diode.

If this polarity is reversed, the diode acts as an insulator. In this condition the diode is said to be *reverse biased* (Figure 1–37). If a reverse bias voltage is applied, with a negative potential applied to the P-side and a positive potential applied to the N-side, the positive potential attracts free electrons away from the junction, and the negative potential attracts holes away from the junction. This causes the depletion zone to be even wider and the resistance across the junction to increase even more. If the reverse bias voltage goes high enough—that is, above 50 V for most rectifier diodes (those designed to conduct enough current to do work, and the most common type)—current will flow. It will rise quickly, and in most cases the diode will be damaged. This is called the *breakdown voltage.*

Diodes can be designed to carry various amounts of current. For example, the diodes in an alternator are designed to carry a substantial amount of current; at the other extreme, the miniaturized diodes contained within an IC chip can only carry very small amounts of current and yet will still function electrically as diodes. The amount of current that a diode (or any other type of semiconductor with one or more PN junctions) can safely handle is determined by such things as its physical size, the type of semiconductor and doping material used, its heat dissipating ability, and the surrounding temperature. If the circuit through which forward bias voltage is applied does not have enough resistance to limit current flow to what the semiconductor can tolerate, it will overheat and the junction will be permanently damaged (open or shorted).

Diode Symbols

The symbol most commonly used to represent a diode is an arrow with a bar at the point (Figure 1–38A). The point always indicates the direction of current flow using conventional theory. With electron theory the flow is opposite the direction that the arrow is pointing. The arrow side of the symbol also indicates the P-side of the diode, often referred to as the *anode.* The bar at the end of the arrow's point represents the N-side and is often called the *cathode.* Thus, a better way to look at a diode symbol within an electrical schematic and know how it affects the circuit is to do as follows: If the polarity applied to the diode is more positive at the anode end (or more negative at the cathode end), it will operate as a conductor; likewise, if the polarity applied to the diode is more negative at the anode end (or more positive at the cathode end), it will operate as an insulator. Figure 1–38B shows a modified diode symbol that represents a zener diode.

On actual diodes, the diode symbol can be printed on one side to indicate the anode and cathode ends (Figure 1–38C), or a colored band can be used instead of the symbol (Figure 1–38D). In this case the colored band will be closer to the cathode end. Figure 1–38E shows a power diode. A power diode is one large enough to conduct larger amounts of current to power a working device. It will be housed in a metal case, which can serve as either the anode or the cathode connector and will also dissipate heat away from the semiconductor crystals inside. The polarity of a power diode can be indicated by markings, or in

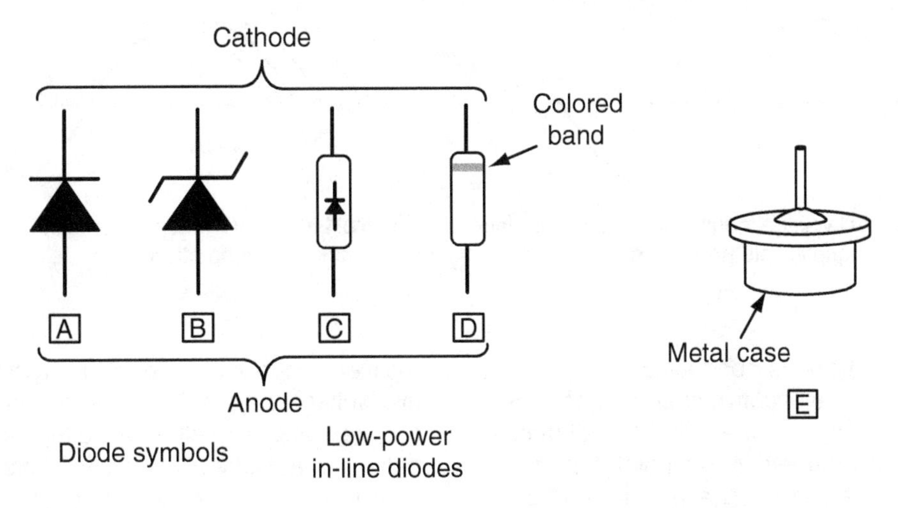

Figure 1–38 Diodes.

Zener Diode

A zener diode is one in which the crystals are more heavily doped. Because of this the depletion zone is much narrower, and its barrier voltage becomes very intense when a reverse bias voltage is applied. At a given level of intensity, the barrier voltage pulls electrons out of normally stable valence rings, creating free electrons. When this occurs, current flows across the diode in a reverse direction without damaging the diode. The breakdown voltage at which this occurs can be controlled by the amount of doping material added in the manufacturing process. Zener diodes are often used in voltage-regulating circuits.

the case of specific part number applications such as for an alternator, it may be sized or shaped so that it can be installed only one way.

Diode Applications

The best-known application of diodes in the automobile is their use to *rectify AC voltage to DC voltage* within the alternator. The winding in Figure 1–39 represents one of the stator windings within the alternator. The alternator's rotating electromagnetic field windings induce an AC voltage pulse, within each stator winding with each passing of a north and south pole of the electromagnetic field windings (Figure 1–40). In Figure 1–39, when point A is positive and point B is negative, the positive charge at point A is blocked by the

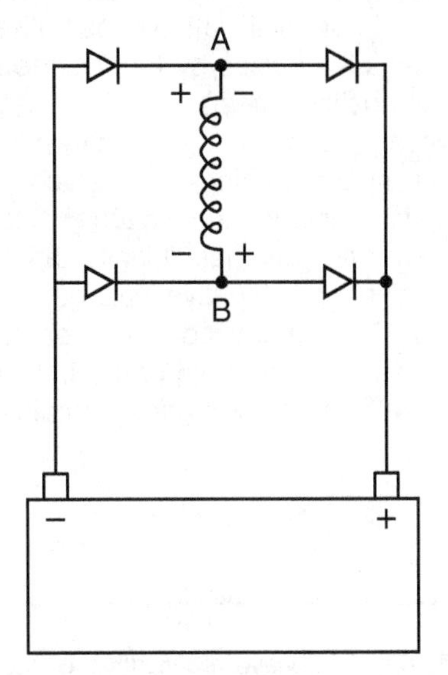

Figure 1–39 Full wave AC-to-DC rectification.

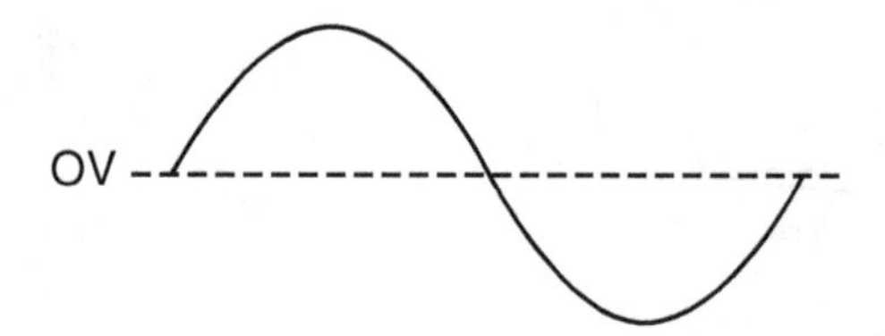

Figure 1–40 AC voltage induced in a stator winding with one north magnetic field and one south magnetic field passing by.

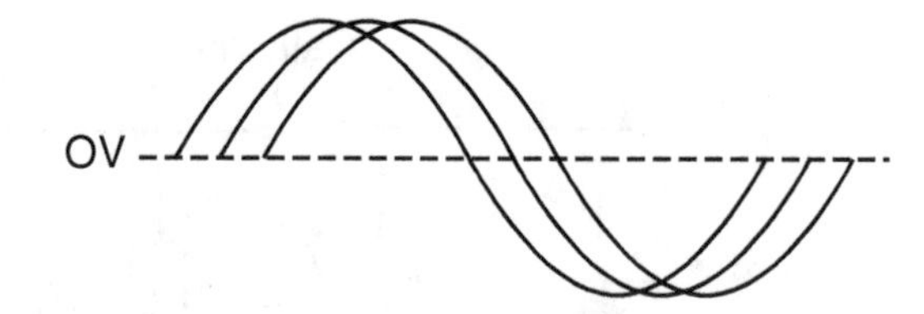

Figure 1–42 AC voltage pulses induced in the three stator windings of an alternator.

diode on the left and is conducted by the diode on the right, and the negative charge at point B is blocked by the diode on the right and conducted by the diode on the left. When point A is negative and point B is positive, the negative charge at point A is blocked by the diode on the right and is conducted by the diode on the left, and the positive charge at point B is blocked by the diode on the left and conducted by the diode on the right. Thus, the AC voltage pulses are applied to the vehicle's battery as positive DC voltage pulses, because the negative AC voltage pulse has now been flipped upright as a positive voltage pulse (Figure 1–41). This is known as full wave AC-to-DC rectification. (Half wave rectification would simply use one diode to block the negative voltage pulses, turning them into wasted heat energy, whereas full wave rectification uses both the positive and negative pulses by turning the negative pulses into positive ones.)

However, because the alternator contains three stator windings into which the rotating field windings induce voltage pulses, one immediately after another, initially three sets of these positive and negative voltage pulses are induced into the stator windings for each passing of a north and a south pole of the field windings (Figure 1–42). Then six diodes contained in the rectifier bridge are used to fully rectify the negative voltage pulses into positive voltage pulses (Figure 1–43). The effective voltage applied to the battery is actually a relatively stable DC voltage that only varies slightly around 14.2 V, as shown by the solid waveform depicted in Figure 1–44. The slight variation in this waveform is known as alternator AC ripple and is best measured at the alternator output terminal with an AC voltmeter or a lab scope. If a diode is open or shorted, or if a stator winding is open, the amplitude (difference between the high and the low voltage) of this signal will be excessive. Excessive AC ripple causes problems with a computer's internal and external communication and can be a reason why good computers are unnecessarily replaced. AC ripple should never exceed 200 mV.

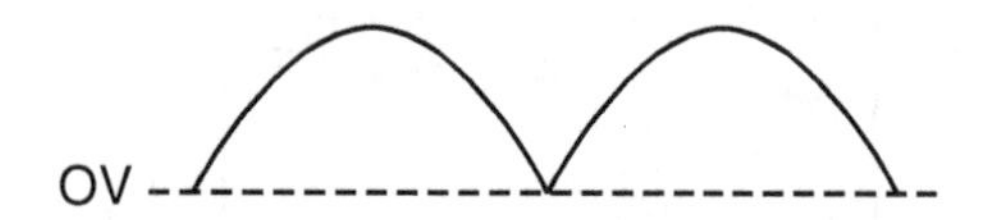

Figure 1–41 Full wave rectification of the AC voltage to a DC voltage.

Another automotive application of diodes is to *control current flow paths in circuits that*

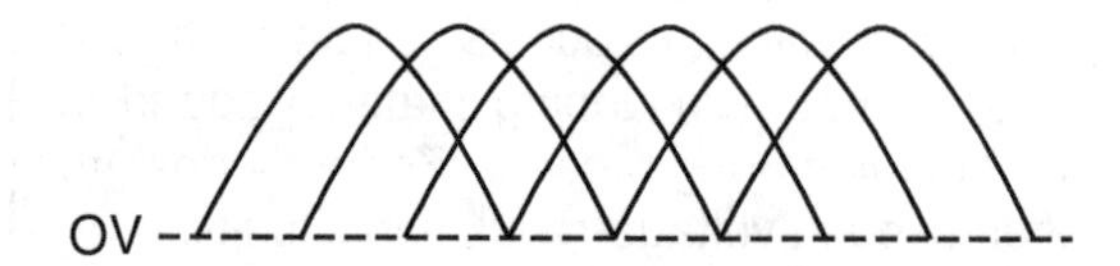

Figure 1–43 Voltage pulses of three stator windings, all rectified to positive voltage pulses.

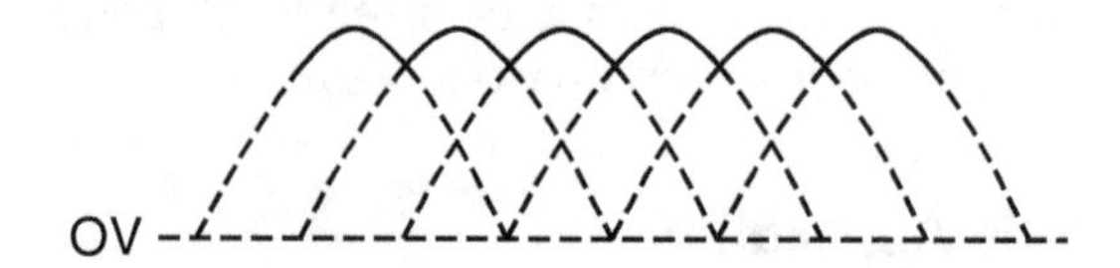

Figure 1–44 The resulting DC voltage after AC-to-DC rectification as it is applied from the alternator's output terminal to the battery.

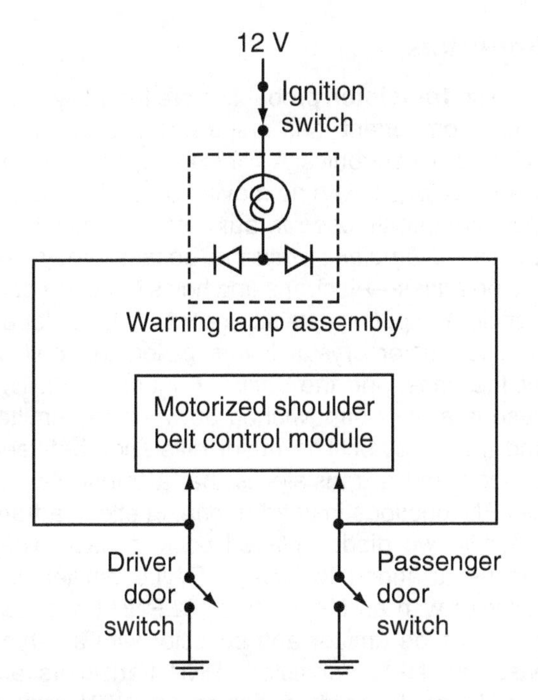

Figure 1–45 Diodes used to control current flow paths in circuits that share a common load.

share a common load component. An example of this is shown in Figure 1–45. The driver's door switch closes when the door is opened to alert the module that it should remove the motorized shoulder belt from the driver. Similarly, the passenger's door switch closes when the passenger door is opened to alert the module that it should remove the motorized shoulder belt from the passenger. However, either switch will ground the "door ajar" lamp in the instrument cluster to illuminate it. Without diodes, opening one door would cause a ground signal that would back up through the other door's circuit and would cause the module to believe falsely that both doors had been opened. By using diodes within the "door ajar" lamp assembly, the circuits can be effectively isolated from each other even though either circuit can illuminate the warning lamp.

A third automotive application that uses diodes is *voltage spike suppression,* though the diode is only one of three methods used.

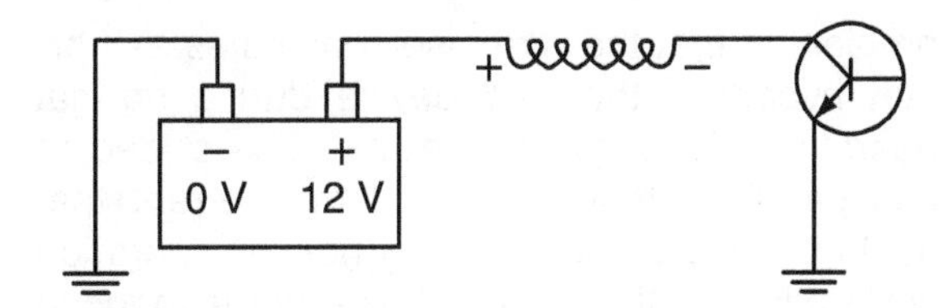

Figure 1–46 Electromagnetic load controlled by a computer's transistor.

Figure 1–46 shows a winding that represents an electromagnetic load that is switched on and off. When current flow is switched on, current ramps up relatively slowly to the fully saturated level as dictated by Ohm's law, just as turning on a water valve will only *allow* the water pressure that is present to begin pushing water through the pipe. Therefore, the resulting electromagnetic field does not build quickly enough to induce a voltage spike within the electromagnetic winding as it expands across it. However, when the driver (transistor within the computer that controls the load) is turned off, current collapses very quickly, just as when you shut off a valve in a water pipe, the valve forcibly and immediately stops the flow of water. In an electrical circuit, when the driver is turned off, the current stops rapidly. As a result, the electromagnetic field collapses so rapidly across its own winding that it induces a voltage potential within the winding of the opposite polarity of the originally applied voltage. In many circuits, this voltage spike is suppressed at the actuator's winding in order to keep it from reaching and damaging the on-board computers.

In order to suppress the voltage spike created each time an electromagnetic load is de-energized, a circuit must be placed in parallel to the electromagnetic load that will allow the voltage potential to dissipate. The voltage potential, of course, simply consists of an excess of electrons at the negative end of the winding and a deficiency of electrons at the positive end. If another wire were simply connected between the two ends of the winding, the voltage spike would be allowed to dissipate each time it was induced in the winding. However, this also creates another problem in that the wire also constitutes a short that bypasses the load device during normal operation. So a component must be added to this parallel circuit to restrict the current that could bypass the load during normal operation while allowing the voltage spike that is induced in the winding each time it is de-energized to dissipate.

Three different methods are commonly used. If a diode is placed in this parallel circuit with the cathode end placed toward the positive side of the circuit, the diode blocks the current flow during normal operation (Figure 1–47). This prevents any current from bypassing the load, but it will allow the voltage spike to dissipate through the diode each time the circuit is de-energized due to the reversed polarity of the voltage spike. This is commonly called a **clamping diode.** Another method is to place a calculated resistance in the parallel circuit that allows a minimal amount of current flow during normal operation but has a resistance value that is low enough to allow the bulk of the voltage spike to dissipate. The diode or resistor is commonly seen in electrical schematics in spike suppression relays. A third method is to place a capacitor in the parallel circuit, which temporarily absorbs the voltage spike potential, such as in many Ford idle air control solenoids.

There are many other applications of diodes in today's vehicles, including solid-state voltage regulators and electronic modules. As the simplest semiconductor, the diode is a basic but effective building block of today's electronic systems.

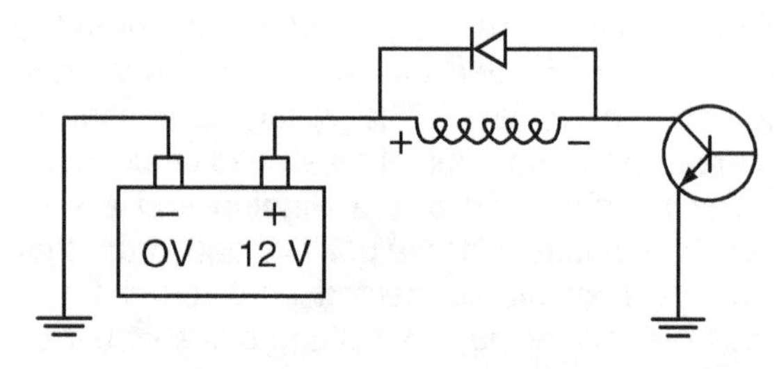

Figure 1–47 A diode used to suppress the voltage spike, known as a spike diode or a clamping diode.

Transistors

The **transistor,** probably more than any other single component, has made possible the world of modern electronics. A transistor operates as an electrical switch with no moving parts. Transistors most commonly used in automotive applications are called *bipolar transistors* because they use two polarities—electrons and holes (electron deficiencies). Bipolar transistors contain three doped semiconductor crystal layers called the *collector,* the *base,* and the *emitter* (Figure 1–48). The base is always sandwiched between the emitter and the collector. The major difference between a diode and a transistor is that a transistor has two PN junctions instead of one. In effect, a transistor is two diodes placed back to back. They can be arranged to have a P-type emitter and collector with an N-type base (a PNP transistor) or an N-type emitter and collector with a P-type base (an NPN transistor). PNP transistors are used for positive side switching and NPN transistors are used for negative (ground) side switching. Because automotive computers control most automotive load components by switching the ground side, the NPN is the more common transistor; it is discussed in the following paragraphs.

In an NPN transistor, the emitter is heavily doped with pentavalent atoms; its function is to emit free electrons into the base. The base is lightly doped with trivalent atoms and is physically much thinner than the other sections. The collector is slightly less doped than the emitter but is more doped than the base.

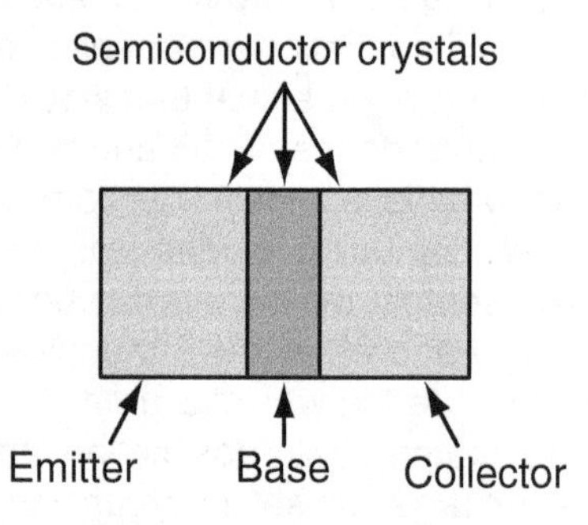

Figure 1–48 Components of a bipolar transistor.

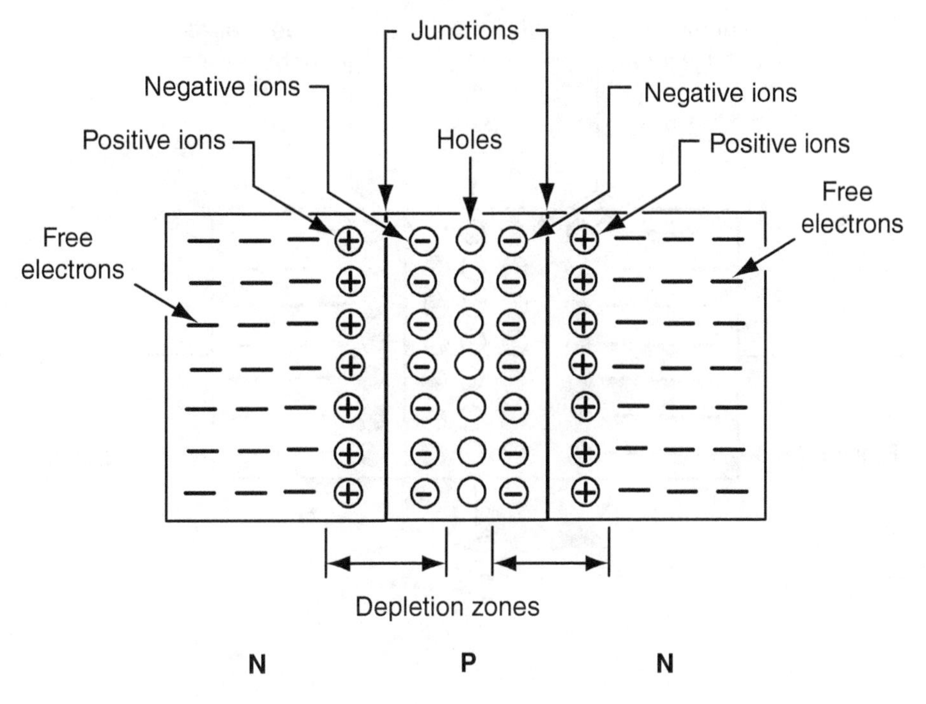

Figure 1–49 Unbiased transistor.

Recalling the discussion of the PN junction, you know that a barrier potential forms at each junction (Figure 1–49). An NPN transistor can be forward biased by applying an external voltage of at least 0.6 V to both the base and the emitter, with the positive potential to the base (the P portion) and the negative potential to the emitter. With a positive potential also applied to the collector, the diode will gain continuity through it from the collector to the emitter, thereby completing the circuit for whatever load is connected to the collector (Figure 1–50).

Because the base is lightly doped, it does not have many holes into which the free electrons can drop. The holes that do exist are quickly filled. Because the base crystal is thin and has so few holes for free electrons to drop into, the majority of the free electrons coming from the emitter cross the base into the collector. The collector readily accepts them because the free electrons in the collector are attracted to the positive potential at the collector electrode and leave behind a lot of positive ions. Because there are so few holes that allow free electrons to drop into valence rings in the base, the current in the base–emitter circuit is quite small compared to the current in the collector–emitter circuit.

If the biasing voltage is removed from the base, the barrier potential is restored at the junctions, and both base and collector currents stop flowing. The base current flow caused by the forward-biasing voltage between the base and the emitter allows current to flow between the collector and the emitter.

A transistor is also much like a relay in that a relatively small current through the base circuit controls a larger current through the collector. The amount of current through the base is determined by the amount of voltage and resistance in the base circuit. In fact, by carefully controlling the voltage applied to the base, the amount of current flow through the collector can also be controlled.

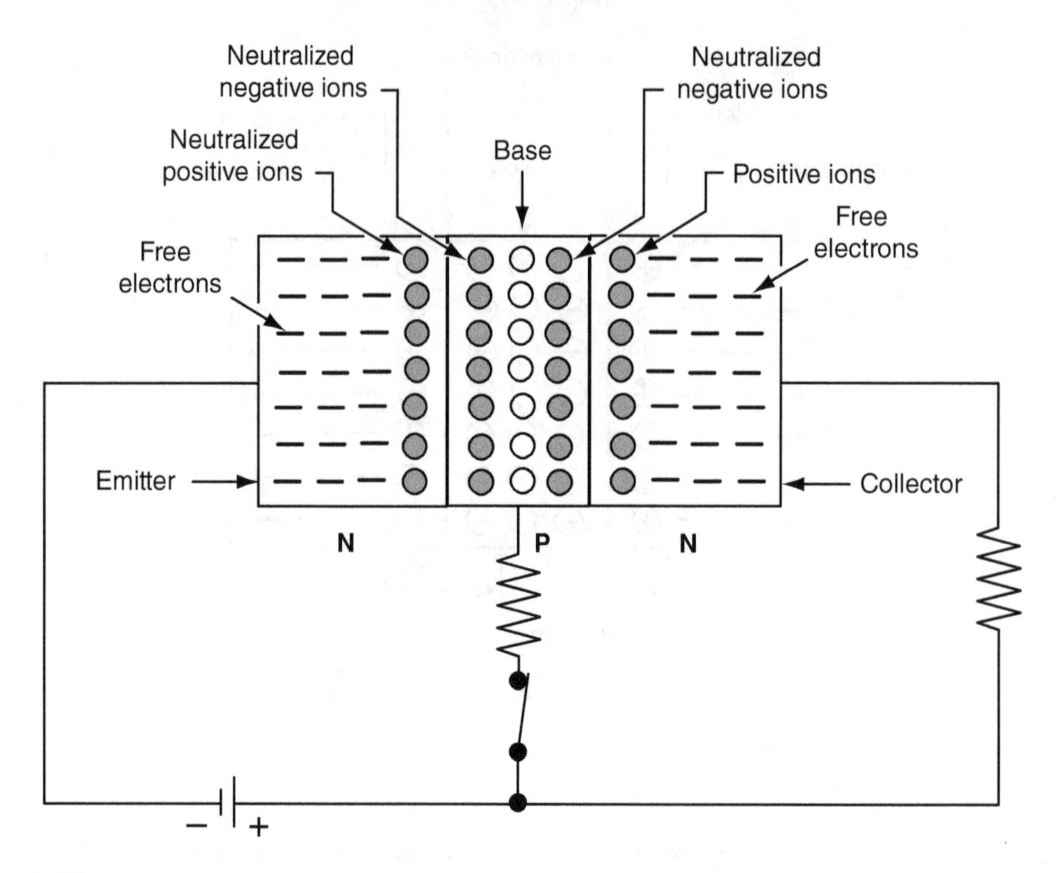

Figure 1–50 Forward-biased NPN transistor.

If enough voltage is applied to the base to just start reducing barrier potential, a relatively small amount of current will begin to flow from the emitter to the collector. This is called *partial saturation.* As base voltage is increased, collector current increases. When enough voltage is applied to the base, full saturation is achieved. There is usually a small voltage spread between minimum and full saturation, and once full saturation is achieved, increasing voltage at the base will not increase collector current. If too much voltage is applied, the transistor will break down.

Transistors, therefore, act like switches that can be turned on by applying power to the base circuit. There are many types of transistors, but the most common fall into two categories: power and switching.

Power transistors are larger because the junction areas must be larger to pass more current across them. Passing current across the junctions produces heat; therefore, a power transistor must be mounted on something, ordinarily called a *heat sink,* which can draw heat from the transistor and dissipate it into the air. Otherwise, it will likely overheat and fail.

Switching transistors are much smaller. They are often used in information processing or control circuits, and they conduct currents ranging from a few milliamps (thousandths of an ampere) down to a few micro-amps (millionths of an ampere). They are most often designed for extremely fast on and off cycling rates, and the base circuits that control them are designed to turn them on at full saturation or to turn them off.

Effect of Temperature on a PN Junction

As the temperature of a semiconductor device goes up, the electrons in the valence ring move at a greater speed. This causes some to break out of the valence ring, creating more free electrons and holes. The increase in the number of free electrons and holes causes the depletion layer to become thinner, reducing the barrier potential. Barrier potential voltages of 0.7 for silicon and 0.3 for germanium semiconductors are true at room temperature only. At elevated temperatures, the lower barrier potential lowers the external bias voltage required to cause current to flow across the junction and raises the current flow.

This means that the expected operating temperature range of a semiconductor device must be considered when the circuit is being designed, and that when in operation, the operating temperature must be kept within that temperature limit. If the operating temperature goes higher, unless some kind of compensating resistance is used, current values will go up, and the semiconductor might be damaged.

Transistor Symbols and Transistor Operation

In electrical schematics, transistors are represented by symbols. Figure 1–51 shows the symbols that represent the components of a transistor: the collector, the base, and the emitter. When these components are arranged as in Figure 1–52, an NPN transistor is represented. (Most electrical schematics will put a circle around these three components, but exceptions will be seen in some schematics.) An NPN transistor can easily be identified because the emitter always points outward (simply pretend that the NPN acronym stands for *NEVER POINTING IN)*. In Figure 1–52, the IC chip controls the signal to the base leg of the transistor in order to turn it on

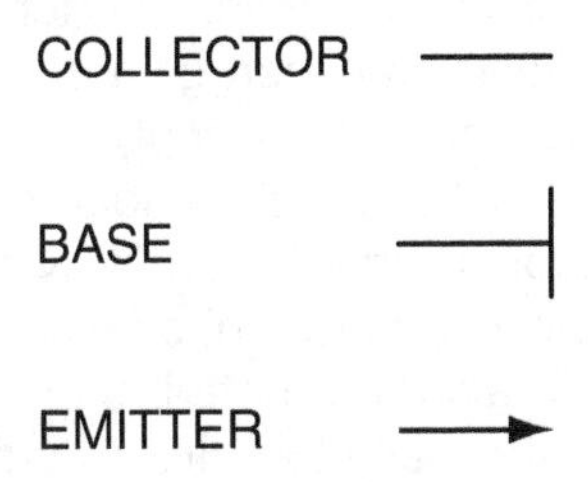

Figure 1–51 The components that make up a transistor.

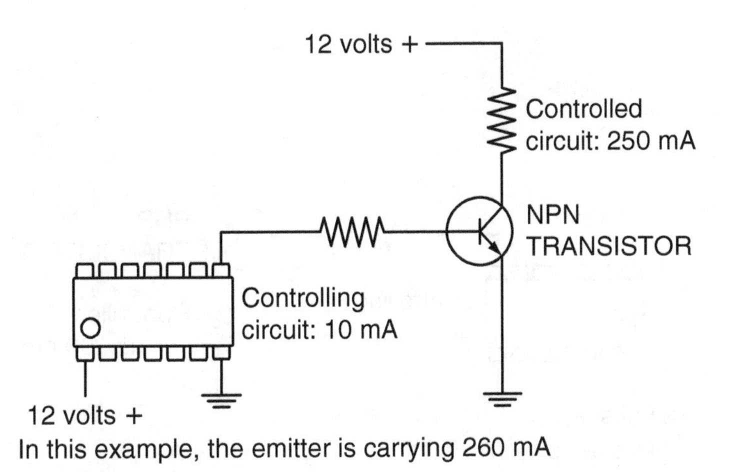

Figure 1–52 An NPN transistor circuit.

and off. This controlling signal must be of a positive polarity since the base is also positive (simply match the polarity of the applied voltage potential to the polarity of the base—or the middle letter of the transistor's acronym). But remember that the transistor's controlling circuit must also be completed through the emitter. Therefore, if the base takes a positive potential to forward bias the transistor, then the emitter must be connected to a negative potential. The negative potential applied to the emitter is always connected, thereby allowing the transistor to be controlled solely by the switching on and off of the positive potential applied to the base. Of course, if the emitter of an NPN transistor is always connected to a negative potential by reason of the controlling circuit, then it must also be connected to the same negative potential by reason of the controlled circuit. That is, when the positive potential is switched on at the base, the collector will then gain electrical continuity with the emitter, thus applying the emitter's negative potential to whatever load is connected to the collector. Therefore, an NPN transistor is used for negative (or ground) side switching.

It should also be noted that the emitter carries both the base-to-emitter current flow of the controlling circuit and the collector-to-emitter current flow of the controlled circuit. Therefore, in Figure 1–52, if 10 milliamps is flowing through the base and emitter in order to forward bias the transistor and 250 milliamps is flowing from the collector through to the emitter once the transistor gains continuity, then the emitter is carrying a total of 260 milliamps.

Likewise, when these components are arranged as in Figure 1–53, a PNP transistor is represented. A PNP transistor can be identified by the fact that the emitter always points inward. With a PNP transistor, a negative potential is applied to the base to forward bias the transistor, and the emitter is always connected to a positive potential. Therefore, because the emitter has a positive potential applied by reason of the controlling circuit, it will also apply the same positive potential to and through the collector to the load when the transistor gains electrical continuity through it. Therefore, a PNP transistor is used for positive side switching of a load component. Although this is done less often than ground side switching, it is sometimes done for safety reasons, as explained in Chapter 3. Also, sometimes a computer must be able to reverse polarity in a circuit, as with a reversible DC motor, and will therefore use both NPN and PNP transistors to control the motor. An **H-gate** uses both types of transistors for this express purpose.

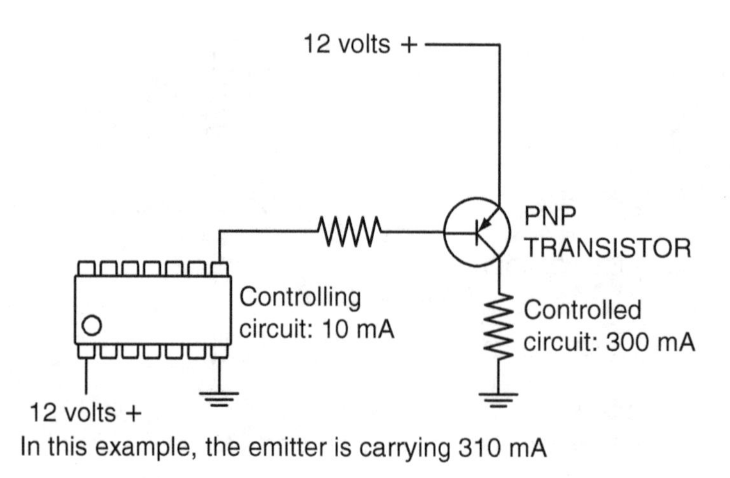

Figure 1–53 A PNP transistor circuit.

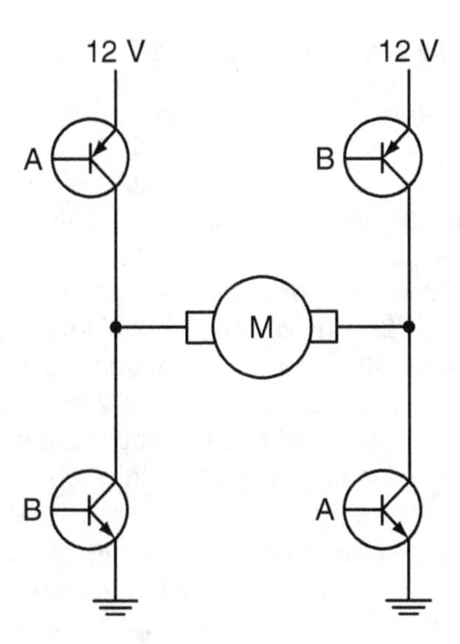

Figure 1–54 An H-gate.

H-Gate Operation

An H-gate is defined as two pairs of transistors designed to be able to reverse the polarity applied to a load component, thereby reversing the direction of current flow through the load. In Figure 1–54, the load component is shown on the center leg of a circuit that looks like the letter H. If transistors A are turned on, current flows through the motor in one direction. If transistors B are turned on, the direction of current flow through the motor is reversed. The computer turns on the transistors in pairs only, either transistors A or transistors B. This is common in DC motors and stepper motors (discussed in Chapter 3) that are used to control the engine's idle speed. H-gates are also used in controlling motorized shoulder safety belts and temperature/mode doors in heating, ventilation, and air conditioning systems. In an electrical schematic, an H-gate is more likely to be depicted as in Figure 1–55 or Figure 1–56.

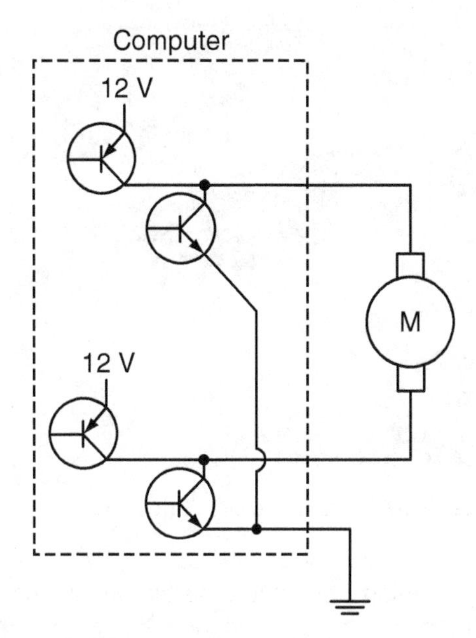

Figure 1–55 Electrical schematic depicting an H-gate.

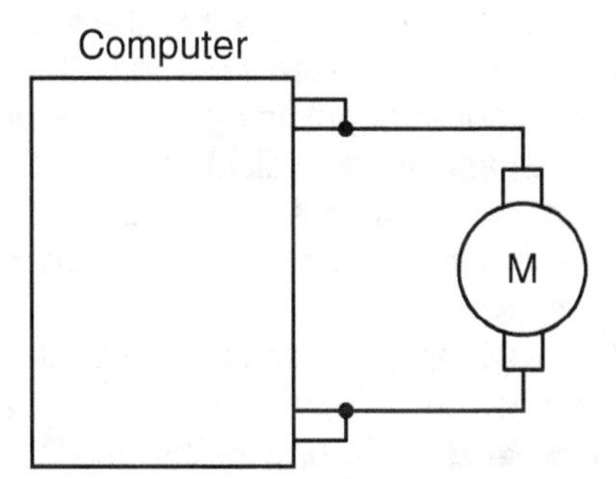

Figure 1–56 Electrical schematic depicting an H-gate.

INTEGRATED CIRCUITS

Thanks to scientific research, manufacturers of electronic systems produce microscopic transistors, diodes, and resistors. As a result, complete circuits are produced containing thousands of semiconductor devices and connecting conductor paths on a chip as small as two or three millimeters across. An IC chip, as shown in Figure 1–57, can operate with current values as low as a few milliamps or less, and can process information, make logic decisions, and issue commands to larger transistors. The larger transistors control circuits that operate on larger current values. Personal computers and the computers in

Figure 1–57 Integrated circuit (IC) chip.

today's vehicles became possible because of the development of ICs. Because the components in an IC are so small, they cannot tolerate high voltages. Care must be taken to avoid creating high-voltage spikes such as those produced by disconnecting the battery while the ignition is on. Care must also be taken when handling an electronic component containing ICs to avoid exposing it to an electrostatic discharge (ESD) such as those you sometimes experience when touching something with ground potential after walking across a carpet.

Many ICs are mounted in a chip with two rows of legs (one row of legs on each side of the chip), known as a **dual in-line package (DIP).** These legs provide terminals that are large enough to connect to a circuit board.

Logic Gates

If two semiconductor elements placed back to back create the simplest semiconductor, the diode, and if three of these elements placed back to back form a transistor, when more than three of these elements are placed back to back, combinations of transistors known as logic gates are formed. Logic gates form the decision-making circuits within an IC chip; they are discussed in greater depth in Chapter 2.

Stepping Up the Amperage

In Figure 1–52, the logic gates within the IC chip can only carry a small amount of current, about 10 milliamps, which is enough to forward bias the transistor that controls a load that is external to the computer. (This transistor is known as a *driver.*) The transistor, in turn, carries the 250 milliamps (or 1/4 amp) that the electromagnetic load draws (assuming the load's resistance is about 48 Ω in a 12 V circuit). If this load is the winding of a fuel pump relay, when the transistor turns on current flow through the relay's winding, the relay's electrical contacts will close and supply about 5 amps to the fuel pump motor. Thus, the IC chip is actually controlling a 5-amp load even though it is only capable of handling a few milliamps within itself. It is all about stepping up the amperage. The circuit in Figure 1–53 would operate similarly.

THE DIGITAL AGE

Voltage Characteristics

There are two ways of categorizing voltage: AC versus DC, as described in the early portion of this chapter, and analog versus digital.

AC versus DC Voltage. On an oscilloscope, a DC voltage will appear as a voltage that can operate in a range from zero volts to a positive voltage, but never goes below the ground level (zero volts), whereas an AC voltage goes both above and below ground.

Analog versus Digital Voltage. An **analog** voltage will appear on an oscilloscope as a voltage that is in a constant state of change, changing between a low value and a high value. While it will at times be at the extreme low or the extreme high value, at other times it will be at a level somewhere in between the two extremes. You could equate this to the dimmer switch that might be used to control the lighting in your dining room—it can be adjusted fully off, fully on and bright, or to a dimmed lighting level somewhere in between. An analog voltage may be either AC or DC.

A **digital** voltage, on the other hand, will be one of two voltage values (the two extremes) but will never be at a value in between. It is usually in either an "off" condition or an "on" condition, as with a standard light switch that allows the lights to be turned fully off or fully on and bright, but will not allow adjustment to some lighting level in between. It could also be one voltage value versus another voltage value. A digital voltage may also be either AC or DC.

Early computers made analog adjustments on the output side in response to analog inputs on the input side. By the late 1970s, analog computers were giving way to digital computers, and the latter began appearing on the automobile. The IC chips, transistorized circuits, and logic gates discussed earlier (and to be discussed further, throughout this textbook) are indicative of those computers that are known as digital computers.

The Four Digital Pulse Trains

You need to be familiar with the four rapidly switched voltage signals that are switched in a digital fashion. These are known as digital pulse trains. Each of these signals has two characteristics that can be measured: the signal's on-time and the frequency of the signal (the definition of frequency as it is used here is cycles per second, which is measured in hertz). Of these four digital signals, the last three retain a certain relationship that will enhance your understanding of them as you complete this portion. The four digital pulse trains are:

Binary Code or Serial Data. This digital pulse train is less closely related than the other three and will be discussed in depth in Chapter 2.

Variable Digital Frequency. Used by certain sensors on the input side of a computer, this allows for greater precision in reporting information to a computer. The frequency of this signal varies, but the on-time of the signal is stable. The frequency of this signal should be measured with a frequency counter known as a hertz meter. (Frequency is a count of how many full on *and* off cycles occur in the time frame of one second.) While you could measure the on-time of this signal, it is generally not beneficial to do so.

Variable Duty Cycle. This is used by a computer as a method of controlling some load components. The on-time of this signal varies, but the frequency of the signal is stable. The on-time of this signal should be measured with a duty cycle meter in order to measure the signal's on-time as a percentage of the total cycle. (In the absence of such a meter, a DC voltmeter or a dwell meter may be used, although the resulting measurement will have to be converted to percent of duty cycle on-time.) While you could measure the frequency of this signal, it is generally not beneficial to do so.

Variable Pulse Width. This is used by a computer as a method of controlling some load components. Both the on-time and the frequency of this signal will vary, but, like a variable duty cycle, it is of most importance to measure the on-time of the signal. However, the on-time is generally measured as a real time measurement (as milliseconds), unlike the method used to measure the on-time of a variable duty cycle.

The relationship of a variable frequency, a variable duty cycle, and a variable pulse width can be described as follows: If the frequency of a consistent digital pulse train was measured at 10 Hz (10 cycles per second), it could be derived from that measurement that a full on/off cycle would be 100 ms long (one second or 1,000 ms divided by the frequency). If a duty cycle meter then showed that the duty cycle on time was 6 percent, the pulse width of the signal would be 6 ms (6 percent of 100 ms).

SUMMARY

In this chapter we have looked at the basics of electricity and electrical relationships. We have also discussed the various types of electrical circuits and looked at the various methods that may be used to calculate their resistance. Then we took an in-depth look at the basic elements of all electronic systems, including diode construction and operation and transistor construction and operation. We also briefly discussed ICs and their use in today's vehicles. We then finished the chapter with a look at the relationship of variable frequency, variable duty cycle, and variable pulse width (three of the four digital pulse trains).

While much of this information, especially in the early part of this chapter, may seem quite basic considering the scope of this textbook, it is critically important that the reader be familiar and comfortable with *all* of the material discussed in this chapter before continuing further into this textbook. Because today's vehicles have so many electronic systems on-board, because today's vehicles no longer have any systems that are not controlled by a computer, and because this trend will continue to escalate in tomorrow's vehicles, it is critically important that students have a solid foundation in their understanding of the electrical concepts discussed in this chapter before continuing to study the use of on-board computers in controlling engine performance or any other electronic system.

▲ DIAGNOSTIC EXERCISE

A technician has used his ohmmeter to measure the resistance of a light bulb's filament and finds that it measures 6 Ω. He then proceeds to power the light bulb using a 12 V battery as the source. With the bulb now illuminated, he uses his ammeter to measure the current flowing through the bulb. He is surprised to find that the current is significantly lower than he had calculated using the Ohm's law formula. Why would this be so?

Review Questions

1. Two technicians are discussing electrical and electronic systems as used on the automobile. *Technician A* says that an electrical system may or may not be under the electronic control of a computer depending upon the system and how old the vehicle is. *Technician B* says that modern vehicles use electronic control systems to control virtually every system in the vehicle. Who is correct?
   A. *Technician A* only
   B. *Technician B* only
   C. Both technicians
   D. Neither technician
2. Which of the following has a negative charge?
   A. Neutron
   B. Electron
   C. Proton
   D. All of the above
3. Which of the following best describes the electron layer of an atom that is referred to as the atom's valence ring?
   A. The inner shell
   B. The second shell
   C. The third shell
   D. The outer shell
4. Which of the following best describes the *electrical pressure* that pushes electrons through a circuit?
   A. Resistance
   B. Amperage
   C. Voltage
   D. Capacitance
5. Which of the following best describes the *current flow of electrons* through a circuit?
   A. Resistance
   B. Amperage
   C. Voltage
   D. Capacitance

6. *Technician A* says that voltage drop always occurs in a circuit, regardless of whether or not current is flowing. *Technician B* says that if the circuit is complete and current is flowing, most of the voltage drop should occur across the load. Who is correct?
   A. *Technician A* only
   B. *Technician B* only
   C. Both technicians
   D. Neither technician
7. Which of the following would increase the resistance of a wire?
   A. A wire of a shorter length
   B. An increase in the temperature of the wire
   C. A larger diameter (lower gauge number) of wire
   D. All of the above
8. Which of the following, while used in most circuits, is not essential to have an operating circuit?
   A. Voltage source
   B. Switch
   C. Load component
   D. Circuitry or current conductors
9. A series circuit is connected to a 12 V battery and is flowing 1/2 amp of current. How much resistance is present in the circuit?
   A. 6 Ω
   B. 12 Ω
   C. 18 Ω
   D. 24 Ω
10. A parallel circuit has two branches. One branch has a load with 10 Ω of resistance. The other branch has a load with 30 Ω of resistance. What is the total resistance of the parallel circuit?
    A. 7.5 Ω
    B. 15 Ω
    C. 20 Ω
    D. 40 Ω
11. A parallel circuit has three branches. One branch has a load with 18 Ω of resistance. Another branch has a load with 22.5 Ω of resistance. The third branch has a load with 30 Ω of resistance. What is the total resistance of the parallel circuit?
    A. 7.5 Ω
    B. 18 Ω
    C. 22.5 Ω
    D. 70.5 Ω
12. A series circuit has two loads. The first load has 10 Ω of resistance and the second load has 30 Ω of resistance. What is the total resistance of this circuit?
    A. 7.5 Ω
    B. 15 Ω
    C. 20 Ω
    D. 40 Ω
13. Which of the following could affect the ability of an electrical circuit to operate properly?
    A. Open wire or connector
    B. Short to power or ground
    C. Unintended high resistance within the circuit
    D. All of the above
14. If a circuit's current flow is too high, which of the following could be at fault?
    A. An open exists in the circuit.
    B. A short exists in the circuit.
    C. Unintended high resistance exists in the circuit.
    D. The circuit has a loose connection, causing an intermittent open.
15. A diode may be used for all except which of the following?
    A. Switching current flow on and off
    B. AC-to-DC rectification
    C. Voltage spike suppression
    D. Controlling current flow paths in circuits that share a common load

16. What type of component is used by a computer to perform ground side switching of a load component?
    A. Diode
    B. NPN transistor
    C. PNP transistor
    D. Capacitor
17. What is contained in an H-gate to allow it to reverse current flow within a load device?
    A. Two pairs of resistors
    B. Two pairs of diodes
    C. Two pairs of transistors
    D. One pair of transistors
18. *Technician A* says that the terms *AC voltage* and *analog voltage* have the same meaning and are therefore interchangeable. *Technician B* says that the terms *DC voltage* and *digital voltage* have the same meaning and are therefore interchangeable. Who is correct?
    A. *Technician A* only
    B. *Technician B* only
    C. Both technicians
    D. Neither technician
19. Which of the digital pulse trains varies both the on-time and the frequency of the signal?
    A. Binary code (serial data)
    B. Variable digital frequency
    C. Variable duty cycle
    D. Variable pulse width
20. Why should you, as a student, be familiar and comfortable with *all* of the material discussed in this chapter?
    A. Because today's vehicles have many electronic systems on board
    B. Because virtually all systems on modern vehicles are controlled electronically
    C. Because the trend of adding electronic systems on vehicles will continue to escalate in the future
    D. All of the above

Chapter 2

Computers in Cars

OBJECTIVES

Upon completion and review of this chapter, you should be able to:

- ❑ Recognize why all modern vehicles are equipped with a computerized engine control system as well as many other computerized control systems.
- ❑ Describe the major components of a computer.
- ❑ Be able to list the major exhaust emissions.
- ❑ Describe what is meant by a stoichiometric air-fuel ratio.
- ❑ Understand why a PCM will try to maintain a stoichiometric air-fuel ratio when in closed loop.
- ❑ Describe the difference between open-loop and closed-loop operation.

KEY TERMS

Baud Rate
Binary Code
Bit
Byte
Clock Oscillator
Closed Loop
Corporate Average Fuel Economy (CAFE)
Driveability
Engine Calibration
Environmental Protection Agency (EPA)
Feedback
Footprint
Greenhouse Gas (GHG)
Hexadecimal
Interface
KAM
Logic Gates
Memory
Microprocessor
National Highway Traffic Safety Administration (NHTSA)
Open Loop
Powertrain Control Module (PCM)
Quad Drivers/Output Drivers
PROM
RAM
Reference Voltage Regulator
ROM
Signal
Stoichiometric
Voltage Signal

WHY COMPUTERS?

In 1963, positive crankcase ventilation systems were universally installed on domestic cars as original equipment. People in the service industry felt that dumping all of those crankcase vapors into the induction system would plug up the carburetor and be harmful to the engine. Instead, engine life doubled.

In 1968, exhaust emission devices were universally applied to domestic cars. Compression ratios began to go down; spark control devices denied vacuum advance during certain driving conditions; thermostat temperatures went up;

air pumps were installed; heated air intake systems were used during engine warmup; and air-fuel ratios began to become leaner. Through the 1970s, evaporation control systems, exhaust gas recirculation (EGR) systems, and catalytic converters were added. Although some of the emissions systems actually tended to improve **driveability** and even improve fuel mileage, for the most part, in the early days of emission controls, driveability and fuel mileage suffered in order to achieve a dramatic reduction in emissions.

In 1973 and 1974, when domestic vehicle fuel economy was at its worst, the United States also experienced an oil embargo and an energy shortage. The federal government responded by establishing fuel mileage standards in addition to the already established emission standards. By the late 1970s, car manufacturers were hard-pressed to meet the ever-more-stringent emissions and mileage standards, and the standards set for the 1980s looked impossible (Figure 2–1 and Figure 2–2). To make matters worse, not only did the consumer's car get poor fuel mileage, it had poor driveability (idled rough, often hesitated or stumbled during acceleration if the engine was not fully warmed up, and had little power). What made the situation so difficult was that the three demands—lower emissions, better mileage, and better driveability—were largely in opposition to each other using the technology available at that time (Figure 2–1).

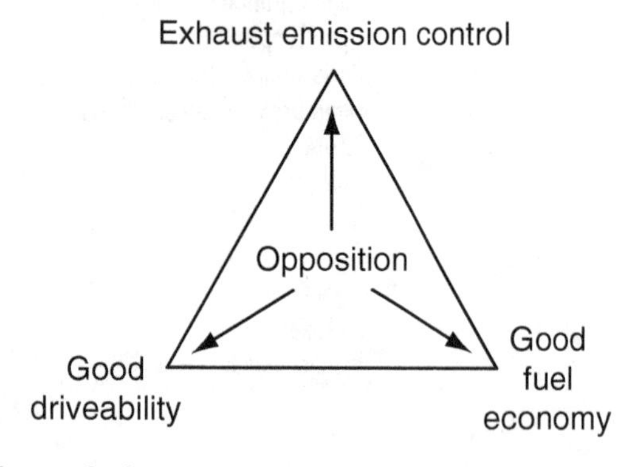

Figure 2–1 Opposition of exhaust emissions, driveability, and fuel economy.

What was needed was a much more precise way to control engine functions, or **engine calibration.** The automotive industry had already looked to **microprocessors**—processors contained on integrated circuits. In 1968, Volkswagen introduced the first large-scale-production, computer-controlled electronic fuel injection (EFI) system, an early version of the Bosch D-Jetronic system. In 1975, Cadillac introduced a computer-controlled EFI system. In 1976, Chrysler introduced a computer-controlled electronic spark control system, the Lean-Burn system. And in 1977, Oldsmobile introduced a computer-controlled electronic spark control system known as MISAR (micro-processed-sensing automatic regulation). All of these systems had three things in common: They all controlled only one engine function, they all used an analog computer (a mechanism that continuously varies within a given range with time needed for the change to occur), and none of them started any landslide movement toward computer controls.

By the late 1970s, the electronics industry had made great strides with digital microprocessors. These small computers, with their comparatively low cost, compact size and weight, great speed, and application flexibility, proved to be ideal for the industry. Probably the most amazing feature of the digital computer is its speed. To put it into perspective, one of these computers controlling functions on an eight-cylinder engine running at 3,000 RPM can send the spark timing command to fire a cylinder; reevaluate input information about engine speed, coolant temperature, engine load, barometric pressure, throttle position, air-fuel mixture, spark knock, and vehicle speed; recalculate air-fuel mixture, spark timing, and whether to turn on the EGR valve, canister purge valve, and the torque converter clutch; and then take a short nap before sending the commands, all before the next cylinder fires. The computer is so much faster than even the fastest engine that it spends most of its time doing nothing but counting time on its internal clock.

At last the automotive manufacturers had the technology to precisely monitor and control, to

EMISSION REQUIREMENTS FOR PASSENGER CARS (PC)
Measured in Grams per Mile (GPM)

	Hydrocarbons (HC)		Carbon Monoxide (CO)		Oxides of Nitrogen (NO_x)	
	Federal	California	Federal	California	Federal	California
1978	1.50	0.41	15.0	9.0	2.0	1.5
1979	0.41	0.41	15.0	9.0	2.0	1.5
1980	0.41	0.39	7.0	7.0	2.0	1.0
1981	0.41	0.39	3.4	7.0	1.0	0.7
1982–1995 Tier 0	0.41	0.39	3.4	7.0	1.0	0.4
1993–2003 Tier 1	0.41	0.25	3.4	3.4	0.4 [1]	0.4

Tier 0 passenger cars phased out as follows:						
	1992	1993	1994	1995	1996	1997
Federal	100%	100%	60% max	20% max	0%	
California	100%	60% max	20% max	0%		

Tier 1 passenger cars phased in as follows:					
	1993	1994	1995	1996	1997-2003
Federal		40% min	80% min	100%	100%
California	40% min	80% min	100%	100%	100%

1994 and newer vehicles also include the following vehicles (as defined by their maximum emissions):			
	HC	CO	NO_X
TLEV (Transitional Low-Emission Vehicle)	0.41	3.4	0.4
LEV (Low-Emission Vehicle)	0.41	3.4	0.2
ULEV (Ultra-Low-Emission Vehicle)	0.41	1.7	0.2
ZEV (Zero-Emission Vehicle)	0.00	0.0	0.0

Tier 2 vehicles are being phased in (beginning in 2004) and will for the first time put all Light Duty Vehicles (LDVs) under the same standard, including passenger cars, SUVs, minivans, and light pickup trucks. Tier 2 requires reduced sulfur in gasoline and also reduces the NO_xstandard as follows:

	2004	2005	2006	2007	2008	2009
Max NO_X in GPM	0.60	0.20 (phased-in reduction)			0.07 50% of all LDVs	0.07 100% of all LDVs

[1] Diesel-fueled vehicles allowed 1.0 GPM through the 2003 model year.

Figure 2–2 Federal and California emission standards.

instantly and automatically adjust, while driving, enough of the engine's calibrations to make their vehicles comply with the government's demands for emissions and fuel economy, while still satisfying the consumer's demands for better driveability. This brought about what is probably the first real revolution in the automotive industry's recent history; other changes have been evolutionary by comparison.

After some experimental applications in 1978 and 1979 and some limited production in 1980, the whole domestic industry was using comprehensive computerized engine controls in 1981. At the same time, General Motors alone produced more computers than anyone else in the world and used half of the world's supply of computer parts.

As the years have progressed, the U.S. has continually tightened up both the emission standards and the fuel economy standards. The **Environmental Protection Agency (EPA)** is responsible for setting the emission standards in place. See Figure 2–2 and Figure 2–3 for past and present U.S. emission standards. Tier 2 emission standards were phased in between 2004 and 2009. It should be noted that Tier 3 emission standards are scheduled to be phased in beginning in 2017 and should be fully phased in by 2025. Both Tier 2 and Tier 3 emission standards also seek to reduce the sulfur content in gasoline.

The **National Highway Traffic Safety Administration (NHTSA)** is responsible for determining fuel economy goals, known as the **Corporate Average Fuel Economy (CAFE)** standards. See Figure 2–4, Figure 2–5, and Figure 2–6 for past, present, and future CAFE standards. The CAFE standards for 2012 through 2025 define a vehicle's standards according to its **footprint**. The vehicle's footprint is defined as the vehicle's wheelbase times its average track width and is indicated in square feet. Using this formula, a small-footprint passenger car is equal to or less than 41 sq ft. An example would be a Honda Fit at 39.85 sq ft or a Smart car at only 27.8 sq ft. These small cars are required to have a CAFE rating of 41 MPG in 2016 and 60 MPG in 2025. (The window sticker may indicate a lesser MPG rating than the CAFE standard due to the fact that the EPA changed its formulas for calculating window sticker MPG ratings to be more real-world beginning in the 2008 model year. This change did not affect how the CAFE standards are calculated.) Under the CAFE standards, a large passenger car, equal to or greater than 55 sq ft, such as a Mercedes S550 with a footprint of 55.27 sq ft, is required to achieve 31 MPG in 2016 and 46 MPG in 2025. As shown in Figure 2–5 and Figure 2–6, the combination of small, medium, and large passenger cars must achieve an average of 37.8 MPG in 2016 and 55.3 to 56.2 MPG in 2025. (Note that under current law, the NHTSA is not permitted to set CAFE standards for more than five years at a time. Thus, the 2017 through 2025 CAFE standards are projected, but have not been finalized yet.)

Likewise, small, medium, and large footprint light trucks must achieve a combined average of 28.8 MPG in 2016 and 39.3 to 40.3 MPG in 2025. Across a manufacturer's entire fleet, the combination of all passenger cars and light trucks sold must average 34.1 MPG in 2016 and 48.7 to 49.7 MPG in 2025.

Also, the EPA sets guidelines for **greenhouse gases (GHGs)**. These guidelines include the carbon dioxide (CO_2) emitted from a vehicle's tailpipe. The CO_2 specification for model year 2025 is 163 grams per mile (GPM) maximum (Figure 2–7). If this were to be achieved solely through increased fuel economy, which, in turn, reduces CO_2 emissions, the required average fuel economy across all passenger cars and light trucks would be 54.5 MPG in the 2025 model year. However, some of this reduction in CO_2 emissions is expected to occur in other areas such as improvements in the refrigerants used in air-conditioning systems and improvements in system design to reduce refrigerant leakage.

If a manufacturer misses the CAFE standard across its fleet, the manufacturer pays a fine or civil penalty. Likewise, if a manufacturer exceeds the CAFE standard across its fleet, the manufacturer can generate credits in a given model year and has several options for using those credits, including credit carry-back, credit carry-forward,

FEDERAL TIER 2 EXHAUST EMISSION STANDARDS[1][2]

Standard	Emission Limits at 50,000 miles			Emission Limits at Full Useful Life (120,000 miles)		
	NO_X[4] (g/mi)	NMOG[5] (g/mi)	CO (g/mi)	NO_X[4] (g/mi)	NMOG[5] (g/mi)	CO (g/mi)
Bin 1	-	-	-	0	0	0
Bin 2	-	-	-	0.02	0.01	2.1
Bin 3	-	-	-	0.03	0.055	2.1
Bin 4	-	-	-	0.04	0.07	2.1
Bin 5	0.05	0.075	3.4	0.07	0.09	4.2
Bin 6	0.08	0.075	3.4	0.1	0.09	4.2
Bin 7	0.11	0.075	3.4	0.15	0.09	4.2
Bin 8	0.14	0.100/0.125[6]	3.4	0.2	0.125/0.156	4.2
Bin 9[3]	0.2	0.075/0.140	3.4	0.3	0.090/0.180	4.2
Bin 10[3]	0.4	0.125/0.160	3.4/4.4	0.6	0.156/0.230	4.2/6.4
Bin 11[3]	0.6	0.195	5	0.9	0.28	7.3

(1) The information in this chart was derived from the Environmental Protection Agency at EPA.gov.
(2) Tier 2 emission standards were phased in between 2004 and 2009.
(3) The Tier 2 emission standards were formed into 8 permanent and 3 temporary "certification bins" of which Bins 9 through 11 were temporary and expired in 2006 for light-duty vehicles and in 2008 for heavy-duty vehicles.
(4) Vehicle manufacturers may certify a particular vehicle to any bin from Bin 1 through Bin 8, but the average NOX emissions of a manufacturer's entire light-duty vehicle fleet have to meet the average NOX standard of Bin 5.
(5) NMOG (Non-methane organic gases) now replaces NMHC (non-methane hydrocarbons) due to the popular use of ethanol in oxygenated gasoline blends.
(6) Emission specifications with two numbers have a separate initial certification standard (1st number) and another in-use standard (2nd number).

Figure 2–3 Federal Tier 2 exhaust emission standards.

credit transfers, and credit trading. In years past, several European manufacturers have chosen to pay millions of dollars in civil fines rather than meet the CAFE standards. On the other hand, the EPA does not allow manufacturers to pay civil fines in lieu of meeting the emission standards.

Ultimately, there are several reasons why computers are being used to control so many systems on the automobile, including both engine management systems and non-engine systems. When compared to mechanical methods of control, computers provide:

- Precise control
- Fast response
- Dependability
- A potential to add features and functions
- Enhanced diagnostic capabilities

1978–2011 CORPORATE AVERAGE FUEL ECONOMY (CAFE) STANDARDS[1]

Model Year	Passenger Cars (MPG)	Light Trucks[2][3]		
		Two-Wheel Drive (MPG)	Four-Wheel Drive (MPG)	Two- & Four-Wheel Drives Combined (MPG)
1978	18.0			
1979	19.0	17.2	15.8	
1980	20.0	16.0	14.0	
1981	22.0	16.7	15.0	
1982	24.0	18.0	16.0	17.5
1983	26.0	19.5	17.5	19.0
1984	27.0	20.3	18.5	20.0
1985	27.5	19.7	18.9	19.5
1986	26.0	20.5	19.5	20.0
1987	26.0	21.0	19.5	20.5
1988	26.0	21.0	19.5	20.5
1989	26.5	21.5	19.0	20.5
1990	27.5	20.5	19.0	20.0
1991	27.5	20.7	19.1	20.2
1992	27.5			20.2
1993	27.5			20.4
1994	27.5			20.5
1995	27.5			20.6
1996–2004	27.5			20.7
2005	27.5			21.0
2006	27.5			21.6
2007	27.5			22.2
2008	27.5			22.5
2009	27.5			23.1
2010	27.5			23.5
2011	30.2			24.1

(1) CAFE standards have been occasionally revised downward by the federal government.

(2) Light trucks are defined as those having a gross vehicle weight rating (GVWR) of 6,000 lbs or less for model years 1979 and older vehicles and 8500 lbs or less for 1980 and newer vehicles.

(3) For model years 1982–1991, manufacturers could comply with the two-wheel and four-wheel drive standards or could choose to combine all of their light trucks and comply with the combined standard.

Figure 2–4 1978–2011 CAFE standards.

**2012–2016
CORPORATE AVERAGE FUEL ECONOMY
(CAFE) STANDARDS[1]**

MY	Combined Small-to-Large Footprint[2] Passenger Cars (MPG)	Combined Small-to-Large Footprint[2] Light Trucks (MPG)	Passenger Cars and Light Trucks Combined (MPG)
2012	33.3	25.4	29.7
2013	34.2	26	30.5
2014	34.9	26.6	31.3
2015	36.2	27.5	32.6
2016	37.8	28.8	34.1

**GALLONS/100 MILES EQUIVALENT
UNDER THE CAFE STANDARDS[1]**

MY	Combined Small-to-Large Footprint[2] Passenger Cars	Combined Small-to-Large Footprint[2] Light Trucks	Passenger Cars and Light Trucks Combined
2012	2.9988	3.9370	3.3634
2013	2.9277	3.8472	3.2783
2014	2.8624	3.7622	3.1931
2015	2.7628	3.6298	3.0699
2016	2.6483	3.4766	2.9329

[1] The information in these charts is derived from the National Highway and Safety Adminitration's website—NHTSA.gov.

[2] The vehicle's "footprint" as defined by the EPA/NHTSA equates to the vehicle's wheelbase times its average track width. Specific CAFE standards are also set for specific size passenger car and light truck footprints.

Figure 2–5 2012–2016 CAFE standards.

It is through these advantages that manufacturers have been able to meet the stringent requirements of the past. Through many additional technologies, many of which are discussed later in this textbook, manufacturers will likely continue to meet the ever-more-stringent requirements that have been set forth by the NHTSA and the EPA.

2017–2025 CORPORATE AVERAGE FUEL ECONOMY (CAFE) STANDARDS(1)(3)

MY	Combined Small-to-Large Footprint(2) Passenger Cars (MPG)	Combined Small-to-Large Footprint(2) Light Trucks (MPG)	Passenger Cars and Light Trucks Combined (MPG)
2017	39.6–40.1	29.1–29.4	35.1–35.4
2018	41.1–41.6	29.6–30.0	36.1–36.5
2019	42.5–43.1	30.0–30.6	37.1–37.7
2020	44.2–44.8	30.6–31.2	38.3–38.9
2021	46.1–46.8	32.6–33.3	40.3–41.0
2022	48.2–49.0	34.2–34.9	42.3–43.0
2023	50.5–51.2	35.8–36.6	44.3–45.1
2024	52.9–53.6	37.5–38.5	46.5–47.4
2025	55.3–56.2	39.3–40.3	48.7–49.7

GALLONS/100 MILES EQUIVALENT UNDER THE CAFE STANDARDS(1)

MY	Combined Small-to-Large Footprint(2) Passenger Cars	Combined Small-to-Large Footprint(2) Light Trucks	Passenger Cars and Light Trucks Combined
2017	2.49–2.53	3.40–3.44	2.82–2.85
2018	2.40–2.43	3.33–3.38	2.74–2.77
2019	2.32–2.35	3.27–3.33	2.65–2.70
2020	2.26–2.23	3.20–3.27	2.57–2.61
2021	2.14–2.17	3.00–3.07	2.44–2.48
2022	2.04–2.07	2.87–2.92	2.33–2.36
2023	1.95–1.98	2.73–2.79	2.22–2.26
2024	1.87–1.89	2.60–2.67	2.11–2.15
2025	1.78–1.81	2.48–2.54	2.01–2.05

(1) The information in these charts is derived from the National Highway and Safety Administration's website—NHTSA.gov.

(2) The vehicle's "footprint" as defined by the EPA/NHTSA equates to the vehicle's wheelbase times its average track width. Specific CAFE standards are also set for specific size passenger car and light truck footprints.

(3) These standards are not final, due to the statutory requirement that the NHTSA set average fuel economy standards not more than 5 model years at a time.

Figure 2–6 Projected CAFE standards for 2017–2025.

2016–2025
CO_2 STANDARDS GRAMS-PER-MILE (GPM)[1][3]

MY	Combined Small-to-Large Footprint[2] Passenger Cars	Combined Small-to-Large Footprint[2] Light Trucks	Passenger Cars and Light Trucks Combined
2016	225	298	250
2017	213	295	243
2018	203	287	234
2019	193	278	223
2020	183	270	214
2021	173	250	200
2022	164	238	190
2023	157	226	181
2024	150	214	172
2025	143	204	163[4]

[1] The information in this chart is derived from the National Highway and Safety Administration's website—NHTSA.gov.

[2] The vehicle's "footprint" as defined by the EPA/NHTSA equates to the vehicle's wheelbase times its average track width. Specific CAFE standards are also set for specific size passenger car and light truck footprints.

[3] The EPA and the NHTSA are responsible for separate sets of standards for passenger cars and light trucks, under their respective legal authority. The EPA has made itself responsible for setting CO_2 emissions standards while the NHTSA is responsible for setting CAFE standards.

[4] At 163 GPM of CO_2, the equivalent CAFE standard would be 54.5 MPG if this level were achieved solely through improvements in fuel efficiency and tailpipe CO_2 emissions. However, a portion of these improvements will be made through improvements in air conditioning leakage and through the use of alternative refrigerants, which will not contribute to fuel economy.

Figure 2–7 Projected CO_2 standards for 2016–2025.

HOW COMPUTERS WORK

Contrary to some "informed" opinions and many suggestions from popular movies, computers cannot think for themselves. When properly programmed, however, they can carry out explicit instructions with blinding speed and almost flawless consistency.

Communication Signals

A computer uses voltage values or voltage pulses as communication signals; thus, voltage is often referred to as a **signal** or a **voltage signal.** There are two types of voltage signals: analog and digital (Figure 2–8). An analog signal's voltage is continuously variable within a given range, and time is needed for the voltage to change. An analog signal is generally used to convey information about a condition that changes gradually and continuously within an established range. Temperature-sensing devices usually return analog signals. Digital signals also vary but not continuously, and time is not needed for the change to occur. Turning a switch on and off creates a digital signal; voltage is either there or it is not. Digital signals are often referred to as *square wave signals.*

Binary Code. A computer's internal components communicate with each other using **binary code**, or a series of digital signals that equate to a binary numbering system made up of ones and zeros; voltage above a given threshold value converts to 1, and voltage below that converts to 0. Each one or zero represents a **bit** (or binary digit) of information. Eight bits equal a **byte** (or binary term, sometimes referred to as a *word)* to an 8-bit computer. All communication between the microprocessor, the memories, and the input and output interfaces is in binary code, with each exchange of information being in the form of a byte of binary code. (Computers also communicate externally with other computers using the same concept—see Chapter 8.) This form of communication using binary code is also referred to as serial data.

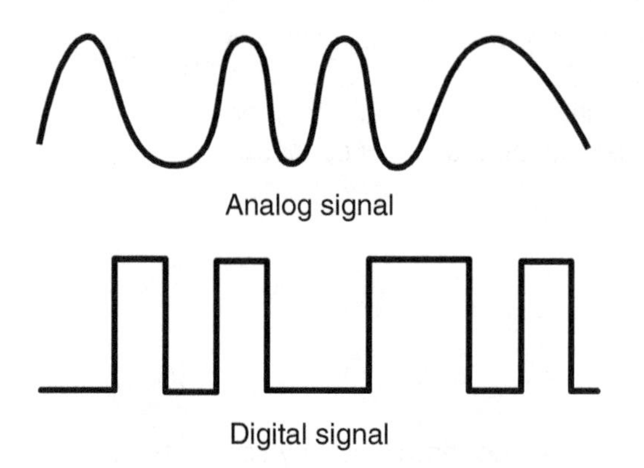

Figure 2–8 Analog signals can be constantly variable. Digital signals are either on/off or high/low.

In order to understand the binary (base 2) numbering system, it is first important to understand the decimal (base 10) system that you are already accustomed to. Base 10 has ten single digits (0 through 9). A single digit representing the value of 10 does not exist, but instead another column must be added. In all bases, the value of the column most to the right is ones. In a base 10 numbering system, each column's value (as it progresses from right to left) is ten times the previous column's value. Therefore, the number 8,763 is actually eight groups of thousands (8 × 1,000), seven groups of hundreds (7 × 100), six groups of tens (6 × 10), and three ones (3 × 1), thus adding up to what we know to be eight thousand, seven hundred, sixty-three.

Any base that you have reason to create can be created using these same principles. For example, if you had reason to create a base 7 numbering system, you would have seven single digits (0 through 6); you would not have a single digit representing the value of 7 but would, rather, have to create another column, and beginning with the right-hand column that represents ones, each column's value would be seven times the previous column's value (1s, 7s, 49s, and so on). However, these *column values* are base 10 values, and therefore they give you a way to convert the base 7 number to a decimal (base 10) number.

With a base 2 (binary) numbering system, you have only two single digits, 0 and 1. There is no single digit representing the value of 2, so another column must be created. And in a base 2 numbering system, each column's value will be two times the previous column's value. Since the first column is always ones, the column values are (right to left) 1s, 2s, 4s, 8s, 16s, and so on. These column values are base 10 values, though, so you can convert a base 2 number to decimal (base 10) by multiplying the 1 or the 0 by the column's value above it and then adding up the sum of it all. In Figure 2–9A, $0 \times 128 = 0$; $1 \times 64 = 64$; $1 \times 32 = 32$; $1 \times 16 = 16$; $1 \times 8 = 8$; $0 \times 4 = 0$; $1 \times 2 = 2$; and $0 \times 1 = 0$. Then add it up: $0 + 64 + 32 + 16 + 8 + 0 + 2 + 0 = 122$. In reality, you simply add up the column value of any column with a 1 in it and ignore the columns that have a 0.

In Figure 2–9B, a 1 is shown in every column, adding up to the maximum value that can be had when a binary number is only carried out to eight places: 255. This means that if you began with seven zeros and a one (representing a decimal value of 1) and increased the value by one at a time, you would have 255 different combinations of zeros and ones. But this ignores the decimal value of zero (eight zeros in binary). With zero also added in, there are actually 256 different combinations of zeros and ones when carried to eight places. Restated, an 8-bit computer has 256 different words in its language. All you really need to do to find how many different combinations of zeros and ones are available to a computer (how many words are in its language) is to take the last column's value that the binary is carried out to and double it.

What this means is that an 8-bit computer can divide an input voltage into 256 different binary communications for its internal decision making. For example, a throttle position sensor that operates in a range of zero to five volts would have a new binary number assigned to its voltage with every 0.02 V change in the analog input signal. Or an 8-bit computer may be able to assign 256 different steps to an idle air control stepper motor (from step 0 through step 255).

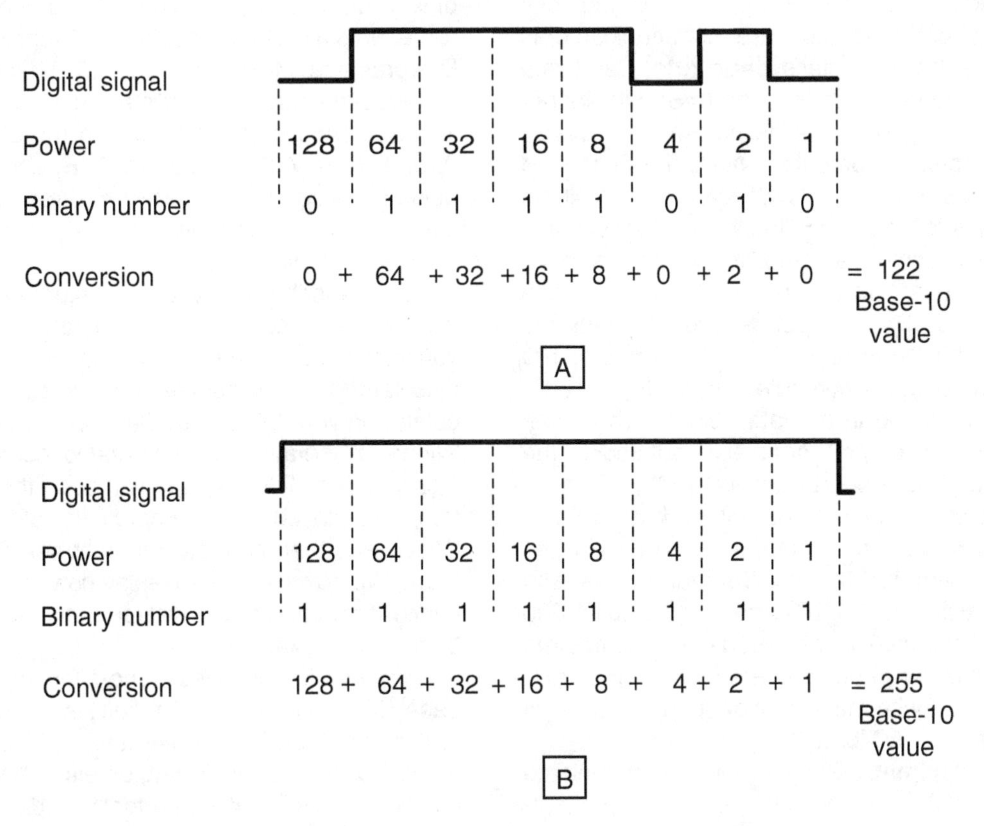

Figure 2–9 Binary numbers.

When carried to 16 places (a 16-bit computer), the last column's value is 32,768, thus allowing for 65,536 different words in a 16-bit computer's vocabulary. When carried to 24 places (a 24-bit computer), the last column's value is 8,388,608, thus allowing for 16,777,216 different words in a 24-bit computer's vocabulary. And when carried to 32 places (a 32-bit computer), the last column's value is 4,294,967,295, thus allowing for 8,589,934,590 different words in a 32-bit computer's vocabulary. And a 64-bit computer...well, you get the idea.

Automotive computers used 8-bit microprocessors for many years, but as these computers have taken on more responsibility and have had to communicate more rapidly, 16-bit and 32-bit computers have become commonplace in the automobile. And 64-bit computers are now commonplace in home computers.

Baud Rate. The frequency of the bits of binary code as they are communicated is known as the computer's **baud rate.** Baud rate is expressed in bits per second (b/S), kilobits per second (Kb/S), or megabits per second (Mb/S). For example, a computer with a baud rate of 10,400 b/S (or 10.4 Kb/S) can communicate 10,400 bits of binary code information per second.

Baud rates in automotive computers have gone up considerably since those used in the early 1980s. The computers used by General Motors in the early 1980s had a baud rate of 160 b/S (some of these computers were still found on applications even in the late 1980s), while their P-4 electronic control module introduced in the late 1980s has a baud rate of 8,192. Ford introduced a computer in 1988 with a baud rate of 12,500 b/S. Modern computers have baud rates that are even faster, with General Motors and Chrysler using 10,400 b/S and Ford using 41,600 b/S as of the mid- to late 1990s. And computers with even faster baud rates are found in the automobile today, with the fastest baud rates between 1 Mb/S and 22.5 Mb/S.

Hexadecimal. Scan tools communicate with the PCM and the other on-board computers using binary code. In reality, this binary code is converted to **hexadecimal** values for ease in recording information. As a result, scan tools may display some of their available information in the hexadecimal numbering system.

Think of hexadecimal as a form of *shorthand* for binary code. Hexadecimal numbering is the process of assigning a single digit to represent the base 10 value for each four bits of binary code. All binary code, whether 8 bits, 16 bits, 24 bits, 32 bits, or 64 bits, is subdivided into groups of four bits. Since the decimal (base 10) values for the four columns of binary code are (from right to left) 1, 2, 4, and 8, as shown in Figure 2–10, having a 1 in each of the four columns would result in a maximum decimal value of 15. Values from 0 through 9 are assigned the single digit 0 through 9 accordingly. For values above 9, the single digit is an assigned letter that represents the decimal value represented by the four bits of binary information, as follows: A represents 10; B represents 11; C represents 12; D represents 13; E represents 14; F represents 15. Therefore, the hexadecimal numbering system is actually a base 16 numbering system containing the digits 0, 1, 2, 3, 4, 5, 6, 7, 8, 9, A, B, C, D, E, and F. Hexadecimal values are generally referred to as "hex codes" and may be identified by a dollar sign ($) in front of the value.

If, as a technician, you encounter hexadecimal values on your scan tool that you need to interpret, you may do so manually or you may use a scientific calculator. The easiest method is to open the calculator on your PC. It is available with all Microsoft Windows operating systems and is located under "Accessories." The first time you open this calculator it is displayed as a standard calculator. Click on "View," and then click on "Programmer" (Windows 7 and Windows 10). The display now shows a calculator that can convert values between decimal, binary, and hexadecimal.

If you click on "Dec" (short for "decimal" or base 10), you will have a 10-digit keypad displayed (0 through 9) and may enter a value as a decimal value. Follow this by clicking on either "Bin" (short for "binary") or "Hex" (short for "hexadecimal") for conversion to either of these formats.

Base-10 column values	8	4	2	1	8	4	2	1	8	4	2	1	8	4	2	1
Binary numbers	0	1	1	0	1	1	1	1	0	0	1	1	1	1	0	1
Hex Codes	**6**				**F**				**3**				**D**			

Binary-to-Hex Conversion

Binary		*Hexadecimal Equivalent*	*Binary*		*Hexadecimal Equivalent*
0000	=	0	1000	=	8
0001	=	1	1001	=	9
0010	=	2	1010	=	A
0011	=	3	1011	=	B
0100	=	4	1100	=	C
0101	=	5	1101	=	D
0110	=	6	1110	=	E
0111	=	7	1111	=	F

Figure 2–10 A byte consisting of 16 bits of binary code, subdivided into groups of four bits, and then converted to hexadecimal using the conversion chart.

If you click on "Bin," you will have a 2-digit keypad displayed (only 0 and 1) and you may enter a value as a binary value. Follow this by clicking on either "Dec" or "Hex" for conversion to either of these formats.

If you click on "Hex," you will have a 16-digit keypad displayed (0 through 9 and A through F), and you may enter a value as a hexadecimal value. Follow this by clicking on either "Dec" or "Bin" for conversion to either of these formats.

Therefore, if you see hexadecimal values on your scan tool that you wish to interpret, simply open the Microsoft Windows calculator and enter the value as a hex value. Then click on "Dec" and the decimal value will be displayed. Also, there are programmer calculator apps available for smartphones and tablets that can be used to decipher hexadecimal and binary values.

Internal Circuits of the Computer

Microprocessor or Central Processing Unit (CPU). The microprocessor (an IC chip), also referred to as the central processing unit (CPU), is the heart of a computer and has the task of making all decisions to be made by the computer. The CPU records input information. It then compares this information with the internal programs. This results in a specific response on the output side of the computer. It then sends the proper commands to control the output circuits appropriately. In essence, the CPU performs the data processing of the computer.

Interface. The microprocessor makes all of the decisions, but it relies on several support functions, one of which is the **interface.** A computer has an input interface and an output

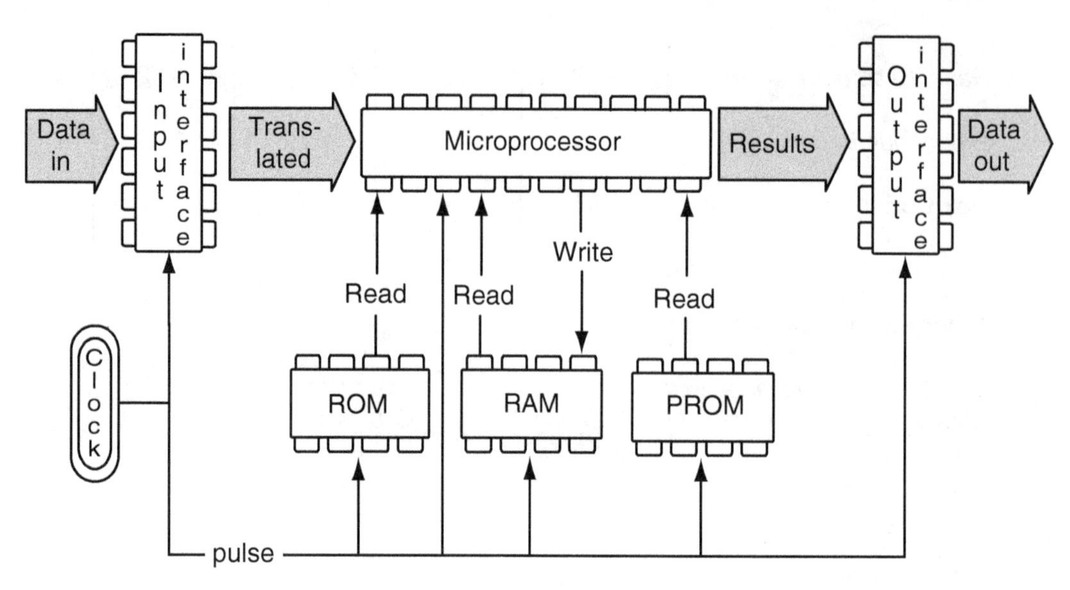

Figure 2–11 Interaction of the microprocessor and its support system.

interface (Figure 2–11). The interface has two functions: It protects the delicate electronics of the microprocessor from the higher voltages of the circuits attached to the computer, and it translates input and output signals. The input interface translates all sensor input data (including DC analog signals, AC analog signals, and variable digital frequency signals) to digital binary code. It is sometimes referred to as the analog-to-digital, or A/D, converter. The output interface, or D/A converter, translates digital binary signals to analog to control the actuators and other output signals. The transistors that control the outputs are located in the output interface and are called *drivers.*

Memories. The microprocessor of a computer performs the calculations and makes all of the decisions, but it cannot store information. The computer is therefore equipped with information storage capability called **memory.**

There are two ways that computer memories are classified: An IC memory chip may be in the form of either read-only memory **(ROM)** or random access memory **(RAM),** and this memory chip may be classified as either volatile or nonvolatile.

If the memory is classified as any form of ROM, the CPU cannot change or write new information into it, but rather can only read from it. But if the memory is classified as RAM, the CPU can write new information into it and can read it back. If the memory is classified as being volatile, the memory will be erased when power is removed from the computer (although many volatile memories are kept alive for some period of time following a battery disconnect by capacitors within the computer designed for this purpose). And if the memory is classified as being nonvolatile, it is retained indefinitely when power is removed.

In basic computer theory, the automotive computer has three types of memory: a ROM, a programmable read-only memory **(PROM),** which is simply another form of ROM, and a RAM (Figure 2–12). (Some manufacturers change these concepts slightly by adding additional memories and combining others.)

ROM. The ROM IC chip contains permanently stored information that instructs (programs) the microprocessor on what to do in a very basic sense and in response to input data.

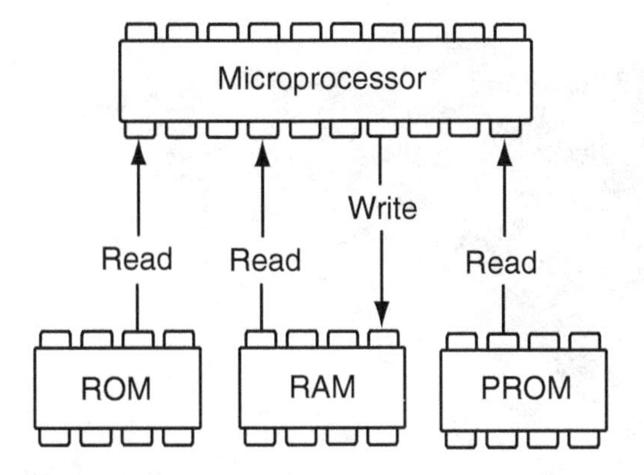

Figure 2–12 The three types of memory within a computer.

For example, the ROM instructs the CPU on the order in which it should look at the sensor values and how to react to these values when controlling actuators on the output side of the computer. In an engine computer, the ROM is nonvolatile. The ROM IC chip is soldered into the computer and is not easily removed.

PROM. The PROM IC chip differs from the ROM in that it is programmed with regard to a specific vehicle. Information stored in the PROM is stored in the form of electronic lookup tables or charts (similar in concept to the tables or charts that one might find in the back of a chemistry or math textbook, except that they are recorded electronically). These tables are also sometimes referred to as *maps.* These charts enable the computer to respond to a specific combination of input conditions when making a decision concerning an actuator. For example, Figure 2–13 shows an ignition timing map that shows how much spark advance the computer has been programmed to add in response to various combinations of engine speed and load.

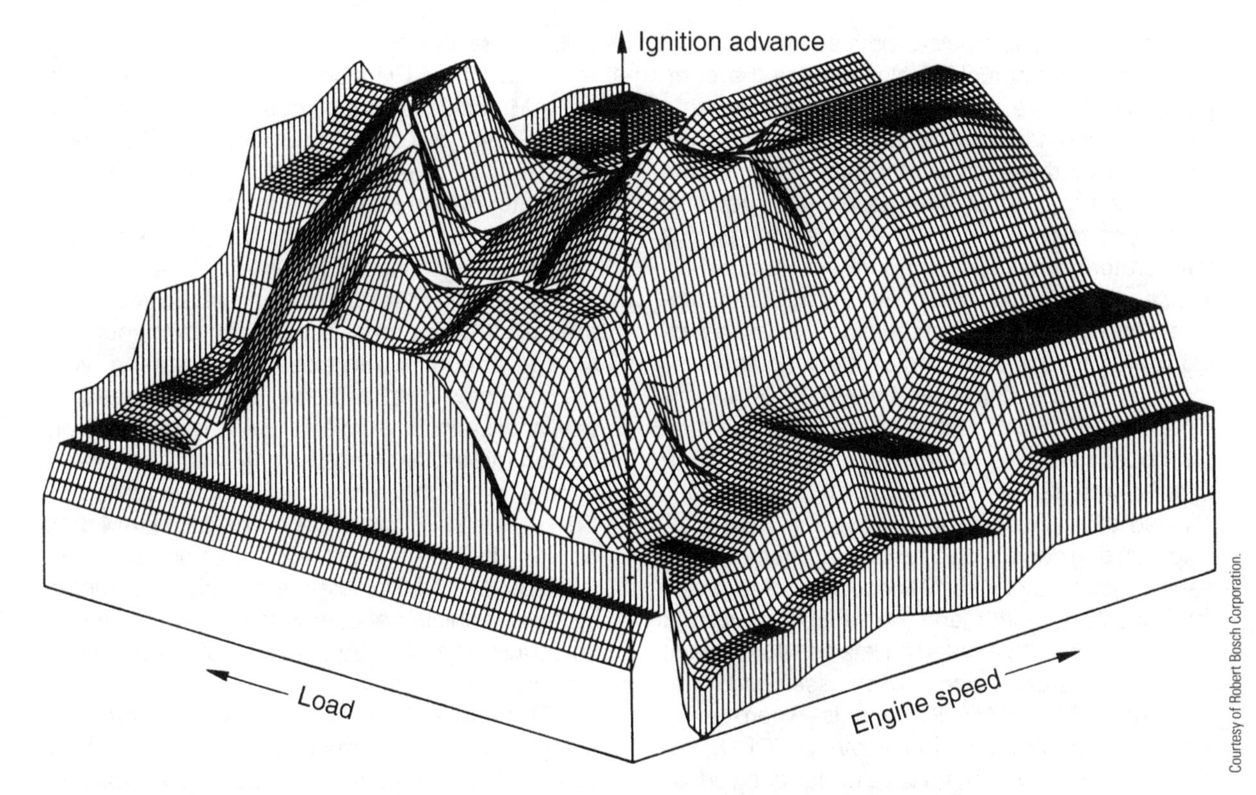

Courtesy of Robert Bosch Corporation.

Figure 2–13 Ignition timing map. *Reprinted with permission from Robert Bosch Corporation.*

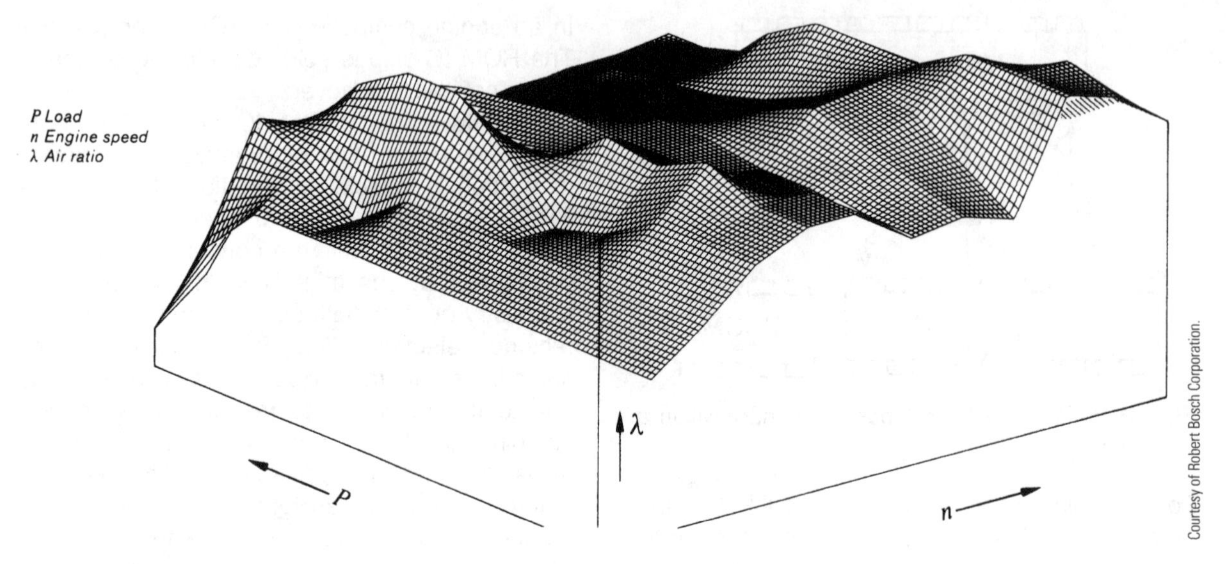

Figure 2–14 Pulse width control map.

Figure 2–14 shows a pulse width control map that shows how much fuel injector on-time the computer should give the fuel injectors according to various combinations of engine speed and load. And both of these control systems also add in other pieces of input information not shown in these charts, including throttle position, engine temperature, intake air temperature, and so on. These electronic charts allow the CPU to precisely control a system on the output side according to various combinations of many inputs.

In an engine computer, the PROM is nonvolatile. The PROM IC chip may be reprogrammed on modern vehicles in order to update the computer's strategies as needed. Some early models allowed the technician to replace the PROM IC chip with one containing a revised program—these early systems included General Motors vehicles throughout the 1980s and Ford EEC I, II, and III systems from 1978 through as late as 1984 (known by Ford as the *engine calibration assembly).*

RAM. The RAM IC chip is where information is stored temporarily by the CPU. For example, as each sensor's value is received by the CPU, the CPU records it in RAM. The CPU will update these values as new information is received. The CPU also records strategy results in RAM, including adaptive strategy values such as fuel trim (discussed in Chapter 4). Diagnostic trouble codes are also recorded in RAM when the computer determines that a fault exists.

In an engine computer, the RAM is volatile. (There are two kinds of RAM: volatile and nonvolatile. A volatile RAM, sometimes called keep-alive memory **(KAM)**, must have a constant source of voltage to continue to exist and is erased when disconnected from its power source. In automotive applications, a volatile RAM is usually connected directly to the battery by way of a fuse or fusible link so that when the ignition is turned off, the RAM is still powered. A nonvolatile RAM does not lose its stored information if its power source is disconnected. Vehicles with digital odometers store the vehicle's accumulated mileage in a nonvolatile RAM.) The RAM is also soldered in place and not easily removed.

The terms *ROM, PROM,* and *RAM* are fairly standard throughout the computer industry; however, vehicle manufacturers do not all use the same terms in reference to computer memory.

For example, as mentioned earlier, instead of using the term *PROM,* Ford calls its equivalent unit an *engine calibration assembly.* Another variation of a *PROM* is an *E-PROM.* An E-PROM has a housing with a transparent top that will let light through. If ultraviolet light strikes the E-PROM, it reverts back to its pre-programmed state. To prevent memory loss, a cover must protect it from light. The cover is often a piece of tape. An E-PROM is sometimes used where ease of changing stored information is important. General Motors uses one as a PROM on some applications because it has more memory capacity.

Another variation of RAM: Ford divided the RAM into two volatile memories with their EEC-IV system. One portion, still called RAM, is powered through the ignition switch. The other portion, called KAM, is powered by a fuse straight from the battery. Therefore, when the ignition is turned off, the KAM remains powered, and its stored information is retained, but the RAM loses its memory. Information stored in this RAM is not critical to retain when the ignition is turned off, including information such as the last value obtained from the throttle position sensor—as soon as the ignition is switched on again, the CPU will regather this information. Other types of memory unique to specific manufacturers are introduced in appropriate chapters.

Clock Oscillator. To maintain an orderly flow of information into, out of, and within the computer, a quartz crystal, called a **clock oscillator,** is used to produce a stream of continuous, consistent voltage pulses (Figure 2–15). The clock oscillator is a small silver, oval can, as shown in Figure 2–15 between the IC chip and the transistor. The frequency of these clock pulses determines the computer's clock speed. Once the clock speed is established, the computer's baud rate can be established, thereby establishing an orderly flow of information within, or from, the computer.

Reference Voltage Regulator. The **reference voltage regulator** looks like a transistor, is approximately 1/4 inch square, and has three legs (electrical terminals) attached to it. The three terminals are voltage in, ground, and voltage out. The ground terminal is common to both the input and output legs. If a voltage above 9.6 V is provided to the input terminal, then the regulator, using a rapid pulse width modulation, outputs a steady 5 V. (On some late 1970s and early 1980s vehicles, this output was 9 V. But this value became standardized worldwide at 5 V by 1984.) This 5 V reference signal, sometimes referred to as *VREF,* is provided to the internal circuits of the computer, including the **logic gates** contained within the CPU (discussed later in this chapter), and is for communication between the CPU, the interfaces, and the memories. The computer also provides this 5 V signal to external sensor circuits—those sensors that require a 5 V signal in order to operate. (One exception: On modern Chrysler vehicles, the computer supplies an 8 V signal to the vehicle speed sensor, the crankshaft position sensor, and the camshaft position sensor.)

Figure 2–15 Clock oscillator on control module board.

Quad Drivers/Output Drivers. Internally, the PCM has many devices, including analog-to-digital converters, signal buffers, counters, timers, and special drivers. It controls most components, as car computers have for many years, by completing the circuit to ground (the components' power source usually comes from the ignition switch or some other source). These power transistors are solid-state on/off switches, often called **quad drivers/output drivers.** If the switches are surface-mounted in a group of four, they are quad driver modules; if the switches are surface-mounted in a group of up to seven, they are output driver modules. Certain switches may go unused on a particular vehicle, depending on its model year, accessory list, and so on.

External Circuits of the Computer

Inputs. Inputs are defined as circuits that operate as informants. That is, they *inform* the computer of the various conditions that exist within the vehicle. As a result, these circuits are generally high-resistance circuits that operate at ultralow current flows. Their primary purpose is to communicate values to the computer through voltage signals. Input circuits include:

- Sensor circuits: A sensor commonly generates a voltage or changes resistance in some manner so as to modify a voltage that is then sent to the computer. Common sensors include the oxygen sensor in the exhaust stream, the throttle position sensor, the engine coolant temperature sensor, and many other sensors. Sensor circuits also include those circuits that monitor the operation of a simple switch, such as those switches inside an automatic transmission that, in combination, can inform the engine computer of the gear that the transmission is operating in. Other switches may monitor power steering pressure, whether the brake pedal is depressed, or whether air conditioning has been selected.
- Diagnostic monitor circuit: Beginning in the mid- to late 1980s, manufacturers began monitoring output circuits with input circuits known as **feedback** circuits. For example, both GM and Ford ran a fuel pump monitor circuit off the circuit between the fuel pump relay and the fuel pump motor so that when the computer's driver (transistor) energized the fuel pump relay, the feedback circuit would inform the computer about whether the relay had actually been energized successfully. Similarly, Ford ran a feedback circuit off the negative terminal of the ignition coil known as the ignition diagnostic monitor (IDM) so that the computer could monitor whether the ignition module successfully fired the ignition coil. The use of such feedback circuits enhances the computer's ability to aid the technician in diagnosis of these systems.
- Informational input circuits from other computers: Computers share information with each other. While these input circuits do not come directly from a sensor, they are used to communicate sensor values or other important information, thus operating as output circuits of the computer sending the information and as input circuits of the computer receiving the information. On older vehicles, these circuits were usually dedicated to one message, such as a cooling fan control module informing the engine computer that the cooling fan relay had been energized so that the idle speed could be adjusted to compensate for the additional load. On newer vehicles, these circuits are commonly known as serial data buses, and they can carry a variety of information over one or two wires (see Chapter 8). This signal technically becomes an input to the computer that is receiving the information.
- Informational input signals from a driver or passenger: These circuits may monitor a switch that the driver can operate on the instrument panel, such as a trip or trip reset switch or a switch that requests air conditioner operation or rear defogger operation. On modern vehicles, even horn buttons no longer energize the horn relay directly, but instead will simply make a request of a computer to energize the horn relay when the driver operates the switch.
- Informational input signals from a technician: Technicians may talk to a computer in several different ways. On older vehicles, for example, they may ground a computer-monitored circuit to request diagnostic trouble codes or to request that some other diagnostic func-

tion be carried out. On modern vehicles the diagnostics are carried out most often by the technician connecting a scan tool that can talk directly to the on-board computers via binary code.

Outputs. Outputs are defined as circuits that allow the computer to make a change in the performance of a circuit or system. For example, the computer may energize a fuel pump relay with the ultimate goal of energizing the fuel pump itself, or it may energize a fuel injector or an A/C compressor clutch. Because most output components are load components, their sole purpose is to turn electrical current flow into some type of work—heat, light, or electromagnetism. Therefore, these circuits tend to flow much higher current than an input circuit does. Output circuits include:

- Actuator circuits: These circuits allow the computer to control a light bulb (such as a warning light on the instrument panel or the interior courtesy lights), a heating grid (such as the rear window defogger or that found behind the glass of a heated mirror), or an electromagnetic device such as a relay, a motor (may be controlled via a relay or may be directly controlled, depending on how much current the motor draws), or a solenoid (such as a door lock/unlock solenoid or a solenoid-operated valve such as a fuel injector).
- Informational output circuits to other computers: As explained earlier under "Informational input circuits from other computers," computers share information with each other, either through dedicated signal wires or through serial data bus circuits. This signal technically becomes an output of the computer that is sending the information.
- Informational output signals to a driver or passenger: The computer at times will inform the driver of certain conditions, such as an indicator that reminds the driver that the rear defogger is turned on or a bulb on the instrument panel (malfunction indicator lamp) that will illuminate to tell the driver that the computer is aware that a problem exists within the system.
- Informational output signals to a technician: When a technician makes a request of the computer for diagnostic trouble codes, the computer will respond by talking back to the technician in some way. On older vehicles, the computer may flash the malfunction indicator lamp (check engine light), may flash an LED that is mounted on the computer itself, or may pulse voltage that can be monitored with an analog voltmeter so as to count code pulses. On modern vehicles, the computer communicates the same information and more via a scan tool through binary code.

Data Buses. When computers communicate with other control modules or control panels in the form of binary code or serial data, they communicate through circuits called *serial data buses* or *data links.* Some data buses transmit data in only one direction, although most transmit bidirectionally. What makes a data bus different from an ordinary circuit is that it is not dedicated to just one voltage signal, but can allow for multiple communications between multiple components.

Logic Gates

Definition. Logic gates are defined as the programming circuitry within a digital computer that determines what output reaction to make in response to a given combination of input values. The CPU, therefore, does not actually "think," but rather makes decisions based upon a combination of the input values and the information stored in the memory IC chips. The logic gates are sometimes referred to as the *logic circuits.*

Logic Gate Basic Construction. If two semiconductor materials placed back to back make up a diode and three semiconductor materials placed back to back make up a transistor (see Chapter 1), then more than three semiconductor materials placed back to back create *combinations of transistors* known as logic gates. Also, as described in Chapter 1, an NPN transistor requires a *positive* polarity applied to its base in order to become forward biased and complete the circuit it controls; we will now refer to this *positive* polarity as a binary *one.* And likewise, a PNP transistor requires a *negative* polarity (or ground) applied to its base in order to become forward biased and complete the circuit it controls; we will now refer to this *negative* polarity as a binary *zero.*

It is also important to understand that all logic gates receive 5 V reference voltage from the computer's internal reference voltage regulator; this is the voltage that allows the logic gates to operate. This voltage signal is fed through a resistor and then to the logic gates. If this reference voltage is grounded out by an external switch, the voltage is pulled low (to zero volts) by the closing of the switch (Figure 2–16). This represents a binary zero to the logic gates. But if the external switch is open and is not grounding out the reference voltage, then the logic gates receive a 5 V signal (Figure 2–17). This represents a binary one.

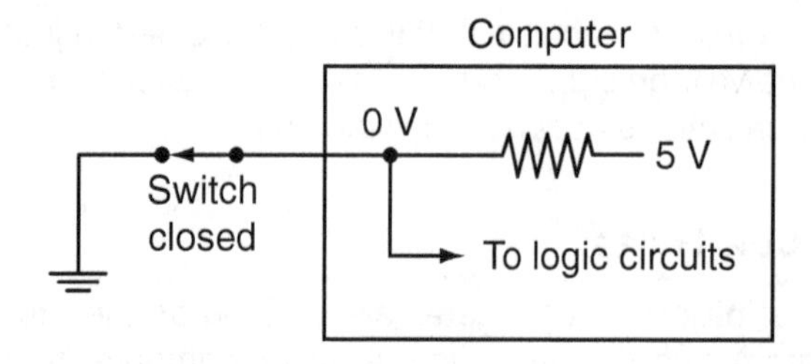

Figure 2–16 Grounding-type switch grounding out the reference voltage to the logic circuits—results in a binary zero.

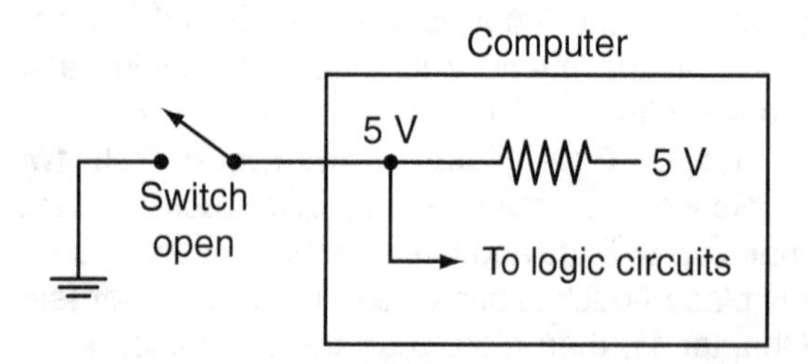

Figure 2–17 Grounding-type switch open, releasing the reference voltage from ground—results in a binary one.

Finally, if the voltage applied to the computer is less than 9.6 V, the reference voltage regulator quits producing its 5 V output and the logic circuits no longer function properly. We will discuss this in more depth later.

The Need to Understand Logic Gates. Logic gates are the programming circuits in digital computers that began replacing analog computers in the automobile in the late 1970s. Today, digital computers are everywhere—from the automobile and heavy-duty diesel truck to the modern PC, laptop, or smartphone that you use to get your e-mail. The following explanation of logic gates is not given with the intent of training you to be able to design and program computers. It is simply to take some of the magic out of the "magic box," as a computer was once called. Students/technicians who can visualize that an automotive computer makes decisions based on real electrical circuitry (okay, so they are microscopic circuits, but they are still real) will take a more efficient approach to diagnostics than many of their counterparts within the industry.

It is also critical to understand the dependence that logic gates have on the reference voltage regulator, and in turn, the dependence that the reference voltage regulator has on proper voltage and ground being supplied to the computer in order to reduce the chance of replacing a good computer at an unnecessary cost to you or your customer. (About 85 percent of computers that are replaced by technicians do not need to be replaced.)

Also, bear in mind that more and more computers are being added to the vehicle, with one manufacturer now having as many as 124 computers on board (2011 Audi R8), depending on how many features the vehicle is ordered from the factory with. Of course, not all of these are major computers; many are small computers having few inputs and controlling few outputs, but they still operate on the principles discussed in this book. Therefore, it is becoming increasingly important for you to understand how computers are designed to operate internally in order to better facilitate your diagnostic approach on today's electronic vehicles.

Basic Logic Gates. There exist three basic gates out of which all of the more complex gates are built—the *AND gate,* the *OR gate,* and the *NOT gate.* Let us discuss these three gates and then we will take a look at a simple computer schematic that uses them. Each gate has a symbol, a definition, and a truth table. Bear in mind that you will not find a physical component in a computer that looks like the symbol—it is simply a symbol that represents more complex circuitry. With the discussion of each gate, there is an electrical schematic that depicts what the gate accomplishes electrically.

AND Gate. An AND gate is depicted with the symbol shown in Figure 2–18. It has a flat input side with two inputs, A and B, and a curved output side with one output, C. An AND gate is defined as a gate in which the output, C, will be a binary 1 only when both inputs, A and B, are a binary 1. The truth table in Figure 2–18 demonstrates the four possible combinations of the two inputs and the results. Figure 2–19 shows the more complex circuitry that an AND gate represents. It is like two switches in series— the only time that light bulb C will light is when both transistors are forward biased (or when both transistors have a binary 1 applied to their bases).

A simple example of the use of an AND gate would be a motorized shoulder belt—only one input is typically required to move the belt off the occupant (to the A pillar). However, two inputs are typically required to move the belt back across the occupant (to the B pillar)—the door must be closed and the ignition switch must be turned to "run." If either one of these input conditions is not met, the belt will not move to the B pillar. An AND gate requires two conditions to be met before the computer will perform the assigned task.

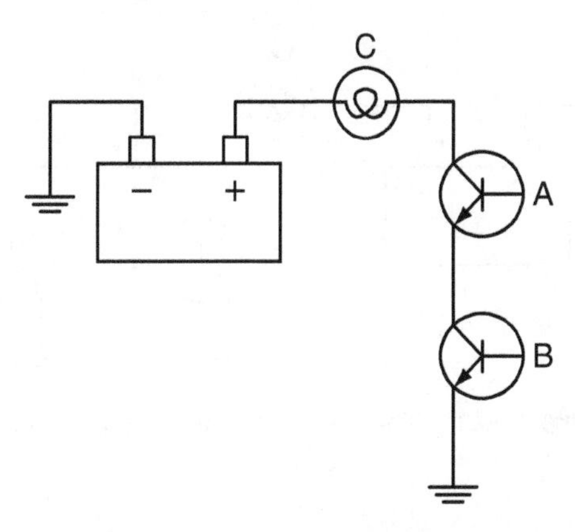

Figure 2–19 The circuitry represented by an AND gate.

OR Gate. An OR gate is depicted with the symbol shown in Figure 2–20. It has a concave input side with two inputs, A and B, and an output side that is curved and comes to a slight point with one output, C. An OR gate is defined as a gate in which the output, C, will be a binary 1 when either input, A or B, is a binary 1. The truth table in Figure 2–20 demonstrates the four possible combinations of the two inputs and the results. Figure 2–21 shows the more complex circuitry that an OR gate represents. It is like two switches in parallel—if either transistor is forward biased (or both), light bulb C will light (or when either transistor has a binary 1 applied to its base).

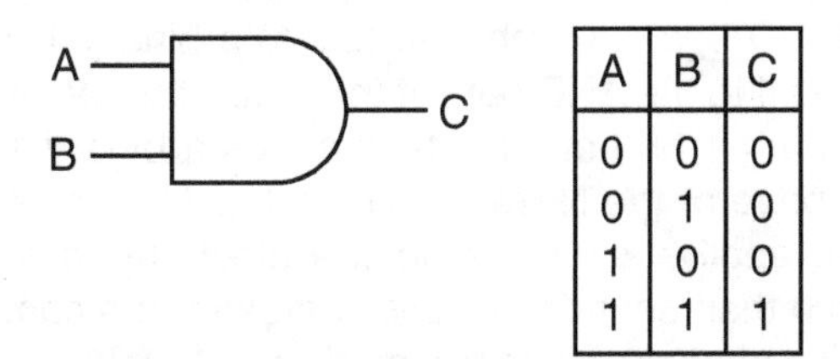

A	B	C
0	0	0
0	1	0
1	0	0
1	1	1

Figure 2–18 The symbol and truth table for an AND gate.

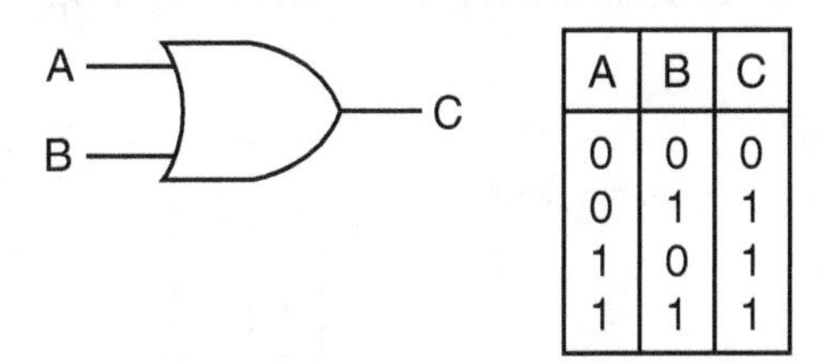

A	B	C
0	0	0
0	1	1
1	0	1
1	1	1

Figure 2–20 The symbol and truth table for an OR gate.

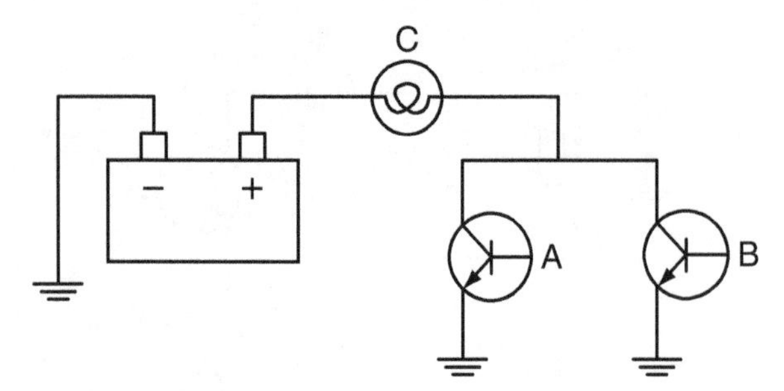

Figure 2–21 The circuitry represented by an OR gate.

A simple example of the use of an OR gate would be a cruise control system that is currently engaged. The cruise control computer will disengage the system if the driver either steps on the brake pedal (thus activating the stop lamp switch) or presses the CANCEL switch. If either one of these input conditions is met, the computer will respond. An OR gate requires only one of two conditions to be met before the computer will perform the assigned task.

NOT Gate. A NOT gate is depicted with the symbol shown in Figure 2–22. It is shown as a triangle with a circle after it. It has a single input, A, and a single output, B. The triangle represents the input side and the circle represents the output side. It is actually an inverter in that it inverts whatever the input value is to the opposite value. As shown in the truth table in Figure 2–22, if the input is a binary 0, then the output is a binary 1. And if the input is a binary 1, then the output is a binary 0. Figure 2–23 shows the more complex circuitry that a NOT gate represents. When transistor A is turned off, 5 V is available to B. But when transistor A is forward biased (or has a binary 1 applied

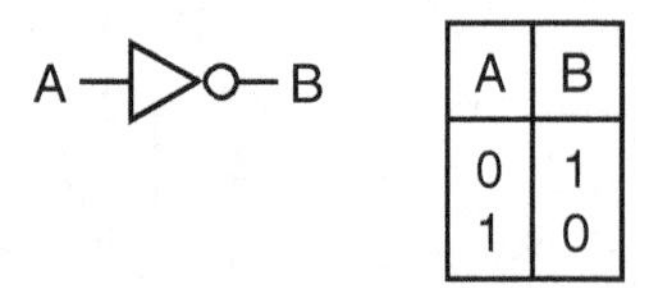

A	B
0	1
1	0

Figure 2–22 The symbol and truth table for a NOT gate.

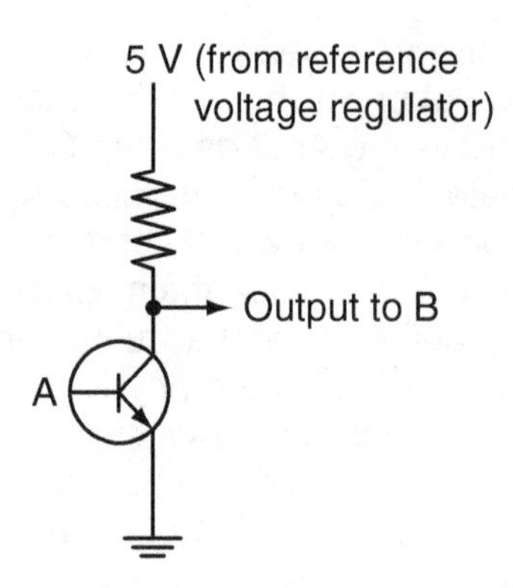

Figure 2–23 The circuitry represented by a NOT gate.

to its base), the voltage to B is grounded out and therefore reduced to zero volts (or a binary 0).

Basic Gates Example. Figure 2–24 shows a simple computer, using all three types of gates, which might be used to control an audible electronic chime underneath the instrument panel. The headlight switch, when turned on, sends a voltage of a positive polarity (a binary 1) to an OR gate. When the ignition key is in the lock cylinder, it completes a path to ground, thus sending a signal of a negative polarity (a binary 0) to a NOT gate, which inverts it to a binary 1 and relays it to the OR gate. (When the key is *not* placed in the lock cylinder, no ground is applied to this circuit; thus, reference voltage is not pulled low, resulting in a binary 1 to the NOT gate, which inverts it to a binary 0 before relaying it to the OR gate.) If the OR gate receives a binary 1 from either source, it will send a binary 1 to the AND gate. If the driver's door is opened, the door open switch grounds the 5 V reference and sends a binary 0 to a NOT gate, which inverts it to a binary 1 and relays it to the AND gate. If the AND gate receives a binary 1 on both inputs, it sends a binary 1 to the base of the NPN transistor. The transistor, in turn, applies a ground to the electronic chime, which then sounds audibly. Therefore, the combination of gates within this module determines that if either the headlights are turned on or the key is in the ignition lock cylinder *and* the driver's door

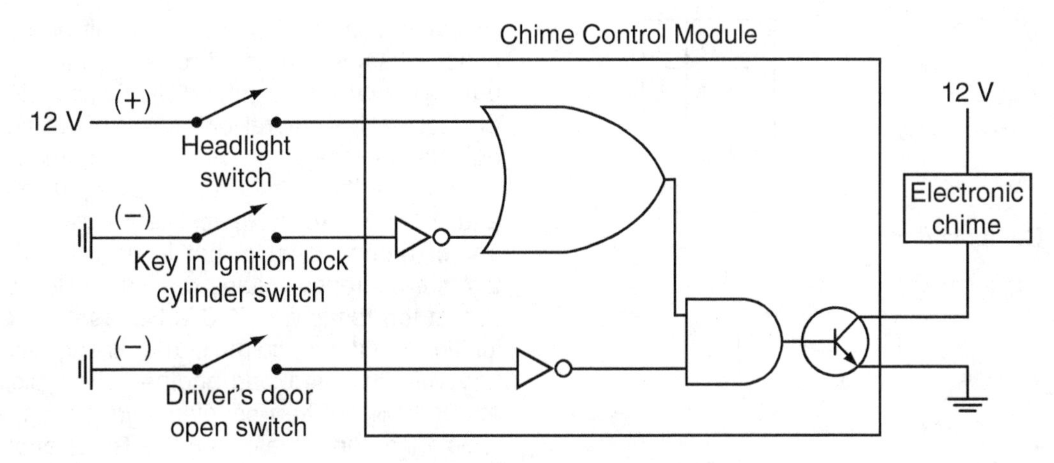

Figure 2–24 A schematic of a typical control module showing the internal logic gates.

is also opened, the chime will sound audibly to alert the driver. This example involves a simplistic computer, but it demonstrates how logic gates are used to determine the existence of the proper combination of input conditions before performing an output function. Most major computers in an automobile have tens of thousands of logic gates that provide for the computer's programming.

Other Logic Gates. There are other gates that are built out of the three basic gates already discussed. Some of these are shown as follows:

NAND Gate. A NAND gate is simply an AND gate with the inverter function of a NOT gate placed immediately after it. The symbol and truth table for a NAND gate are shown in Figure 2–25. The symbol is that of an AND gate with the circle from a NOT gate placed after it. Figure 2–25 also shows the basic gates that a NAND gate is comprised of.

NOR Gate. A NOR gate is simply an OR gate with the inverter function of a NOT gate placed immediately after it. The symbol and truth table for a NOR gate are shown in Figure 2–26. The symbol is that of an OR gate with the circle from a NOT gate placed after it. Figure 2–26 also shows the basic gates that a NOR gate is comprised of.

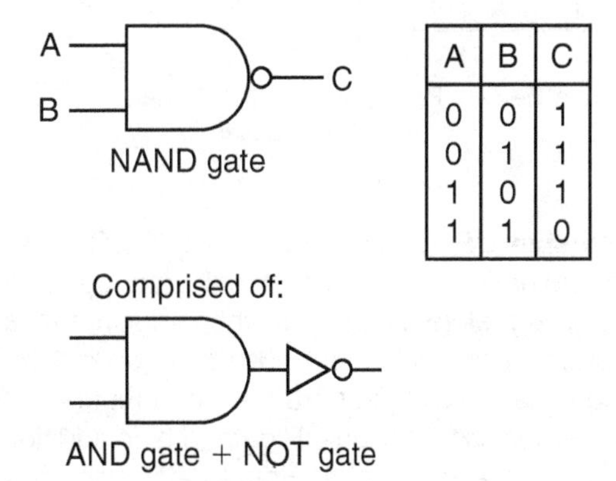

A	B	C
0	0	1
0	1	1
1	0	1
1	1	0

Figure 2–25 NAND gate.

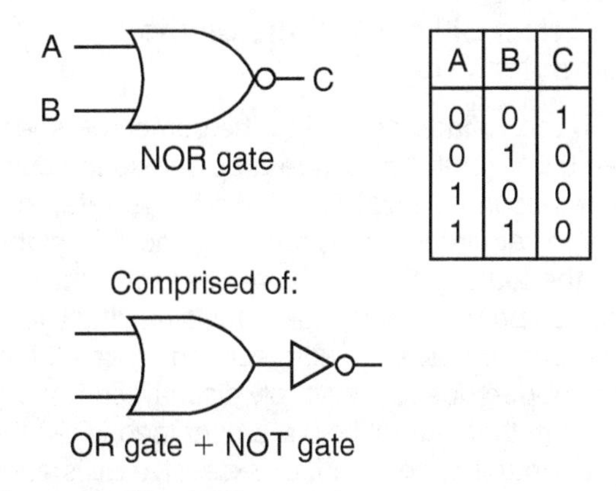

A	B	C
0	0	1
0	1	0
1	0	0
1	1	0

Figure 2–26 NOR gate.

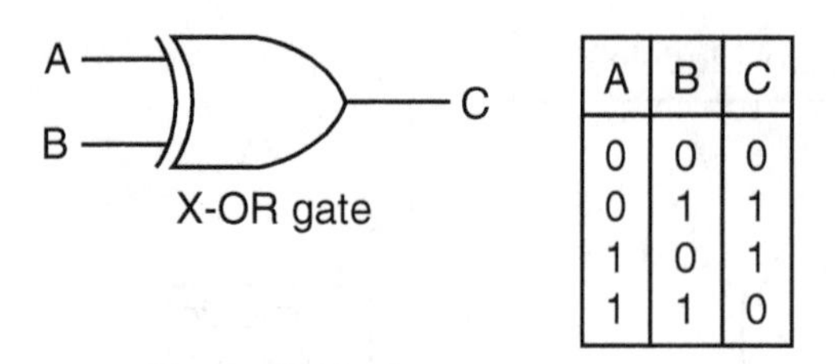

A	B	C
0	0	0
0	1	1
1	0	1
1	1	0

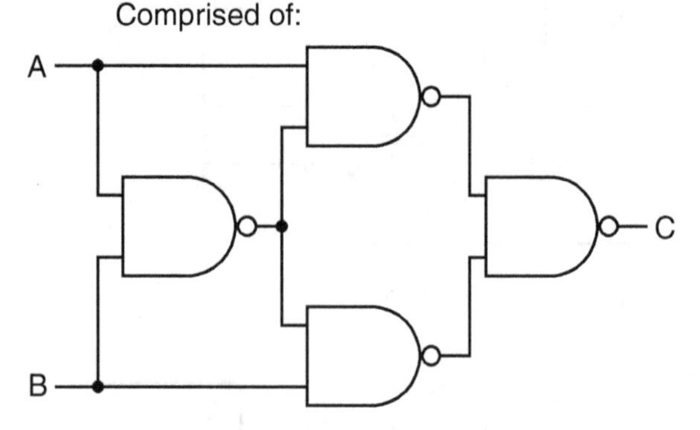

Figure 2–27 Exclusive-OR (X-OR) gate.

Exclusive-OR (X-OR) Gate. An X-OR gate is a combination of gates that will produce a high output signal (a binary 1) when the inputs are different from each other. The symbol and truth table for an X-OR gate are shown in Figure 2–27. The symbol is that of an OR gate with an additional concave line ahead of it. Figure 2–27 also shows that an X-OR gate is comprised of a combination of NAND gates.

The Effect of Low Voltage on the Computer's Logic Gates

If the voltage applied to the computer is less than 9.6 V (or if the voltage drop across the computer becomes less than 9.6 V), the reference voltage regulator quits producing the 5 V signal that the logic gates (and some sensors) rely on, at which point the computer begins making wrong decisions or may even quit responding at all. This can happen due to a severely discharged battery, a battery that cannot hold a proper load when the starter motor is engaged, or excessive resistance in the computer's power and/or ground circuits. However, this can also be the result of a combination of some unwanted resistance in the computer's power and/or ground circuits, coupled with the fact that it is natural for the load of the starter to pull battery voltage low when starting the engine. Restated, if the battery is fully charged at 12.6 V and there is excessive resistance in the power and/or ground circuits that is dropping another 2 V, the reference voltage regulator will still function at the remaining 10.6 V. But as soon as the ignition key turns and the starter is engaged, battery voltage is reduced another 2 V, typically to about 10.6 V. When coupled with the unwanted resistance, the remaining 8.6 V is not enough to allow the regulator to output its 5 V, and the logic gates quit functioning. If you suspect this as a possibility when the computer does not seem to operate properly, test the battery's open circuit voltage and perform a load test. Then simply measure the voltage at a sensor that also receives this 5 V reference voltage (the throttle position sensor, for instance) with the ignition key turned to "run" and also while energizing the starter circuit. If the reference voltage is not correct, perform the appropriate voltage drop tests on the computer's power and ground circuits (discussed in Chapter 5). For this type of fault, replacing the computer will not help. The power and/or ground circuits must be properly diagnosed and repaired.

FUNCTIONS OF THE ENGINE COMPUTER

The engine computer, known today as the **powertrain control module (PCM),** controls (or may control) several different systems as discussed in the following pages.

Fuel Management

The PCM controls the amount of fuel being delivered to the engine in proportion to the volume of air that the engine is drawing in. This ultimately controls the air-fuel ratio. This was true of the early systems with electronically controlled feedback carburetors, and it remains true in modern fuel injection systems. The PCM's goal is to control the air-fuel ratio to produce the fewest exhaust emissions and the best fuel economy, and still provide the desired driveability.

During cold engine operation and heavy load conditions, as at wide-open throttle, the PCM adds fuel to the combustion chamber to create a rich air-fuel ratio for best driveability. During warm engine operation at light to moderate load, the PCM controls the fuel to maintain an air-fuel ratio of approximately 14.7:1 (14.7 parts air to 1 part fuel as measured by weight). This air-fuel ratio is commonly known as a **stoichiometric** air-fuel ratio. A stoichiometric air-fuel ratio is the air-fuel ratio whereby all of the fuel and all of the oxygen entering the cylinder are both fully consumed during the combustion process. This air-fuel ratio provides for low emissions, good fuel economy, and still enough power for light or moderate load conditions. Many modern PCMs will run the air-fuel ratio even leaner than this under certain cruise conditions to improve fuel economy even further.

Idle Speed Control

In controlling the engine's idle speed electronically, a PCM can precisely control the idle speed to maintain the best compromise between a speed at which the engine is prone to stalling and one that wastes fuel and produces an annoying lurch as an automatic transmission is pulled into gear. It can also control idle speed to compensate for the load of the air-conditioning clutch as it cycles on and off, as well as to compensate for the load of the power steering pump in a parking situation. It can also adjust idle speed in an effort to compensate for low charging system voltage, low engine temperature, and engine overheating.

Electronic Ignition Timing

Ignition timing is well known to be a major factor in both fuel economy and driveability. It is also a critical factor in the control of exhaust

Stoichiometry

A stoichiometric air-fuel ratio is one where all of the fuel and all of the oxygen in the cylinder is consumed during combustion. Following combustion there is neither any fuel nor any oxygen left over. A leaner mixture results in unused oxygen being left over after the burn and a richer mixture results in unburned fuel being left over after the burn.

When straight gasoline is in the fuel tank, the stoichiometric air-fuel ratio is close to 14.7 parts air (as measured by weight, not volume) to each one part of fuel. With gasoline that contains ethanol, this ratio changes due to the fact that ethanol molecules contain oxygen atoms. Therefore, ethanol is referred to as an *oxygenated* fuel.

A stoichiometric air-fuel ratio is the one that results in the least amount of toxic emissions leaving the combustion chamber. It is also the ratio that allows the catalytic converter to work at its peak efficiency in further reducing exhaust emissions. When these systems were first invented, 14.7:1 was the ideal air-fuel ratio for standard gasoline. Since that time, many changes have been made to fuels, including different blends for different seasons of the year, different altitudes, and even different areas, depending on their compliance with federal air quality standards. And, while *oxygenated* and other special fuels have changed the numbers of the ratio somewhat, the objective remains the same: to maintain a stoichiometric air-fuel ratio so that the resulting exhaust emissions are kept at their optimal state.

A stoichiometric air-fuel ratio provides just enough air and fuel for the most complete combustion, resulting in the smallest possible total amount of leftover combustible materials: oxygen, carbon, and hydrogen. This maximizes the production of CO_2 and H_2O and minimizes the potential for producing CO, HC, and NO_x. It also provides good driveability and fuel economy.

The most power is obtained from a slightly rich air-fuel ratio of about 12.5:1, although the best fuel economy is obtained from a slightly lean air-fuel ratio of about 16:1.

A leaner mixture also increases engine temperature as a result of slower burn rate due to the fact that it takes longer for the flame to propagate from one fuel molecule to the next. Although a lean mixture does not burn as hot as a slightly rich mixture, a lean mixture can do more heat damage to the engine than a rich mixture due to the longer "residence time" of the combustion taking place in the cylinder.

emissions. With the PCM in control of the ignition timing, timing does not have to be dependent on engine speed and load alone. It can be adjusted to meet the greatest need during any driving condition. For example, following a cold start, it can be advanced for driveability. During light load operation with a partially warmed-up engine, it can be retarded slightly to hasten engine warmup and reduce exhaust emissions. During acceleration or wide-open throttle (WOT) operation, it can be adjusted for maximum torque.

Emission Systems

The PCM influences and/or controls several emission systems. For example, the ability of the PCM to maintain the air-fuel ratio at a stoichiometric value allows the catalytic converter to perform its job efficiently. The PCM also plays a direct role in controlling other emission systems, including the secondary air injection system, the EGR system, the evaporative/canister purge system, and others. See Chapter 4 for a further discussion of emission systems.

Torque Converter Clutch Control

In an automatic transmission, most PCMs control a clutch in the torque converter, whether or not the transmission is fully electronic. This function eliminates converter slippage and heat production and thus contributes to fuel economy when the vehicle is under cruise conditions.

Electronic Automatic Transmission Control

In most modern applications, the PCM controls the upshifts and downshifts of the automatic transmission. The valve body is now under the PCM's control. The PCM uses throttle position and vehicle speed in order to precisely control the shift points.

Transmission Upshift Light

In some manual transmission applications, the computer activates an upshift light, which alerts the driver when to upshift to obtain maximum fuel economy.

Air-Conditioning Control

Some computerized engine control systems feature an air-conditioning (A/C) clutch cut-out function that is controlled by the PCM. The PCM disengages the air-conditioning clutch during wide-open throttle operation. This function contributes to driveability under heavy load conditions. In some systems, the PCM may also disengage the A/C clutch when power steering pressure is high, as when turning into a parking space, to avoid engine stall.

Turbocharger Boost Control

To increase performance, many manufacturers offer a turbocharged engine option. In most cases, the computer controls the amount of boost that the turbocharger can develop. Many manufacturers use an air-to-air intercooler on selected turbo applications. This lowers manifold air temperature by 120°F to 150°F and allows for even more aggressive turbo boost and spark advance due to the denser air charge.

Speed Control

Many PCMs also have the added responsibility of controlling the vehicle speed control (cruise control) function. While many early electronic speed control systems were controlled by dedicated modules, today they are more often controlled by one of the more major computers, most often the PCM.

CONTROLLING EXHAUST GASES

The computer of a computerized engine control system has a mission: to reduce emissions, improve fuel mileage, and maintain good driveability. The priority varies, however, under certain driving conditions. For example, during warmup, driveability has a higher priority than mileage, and at full throttle the computer will give the performance aspect of driveability a higher priority than emissions or mileage.

The Seven Exhaust Gases

There are seven major exhaust gases that exit from an automobile's tailpipe: nitrogen (N_2), oxygen (O_2), hydrocarbons (HC), water vapor (H_2O), carbon dioxide (CO_2), carbon monoxide (CO), and oxides of nitrogen (NO_x).

1. *N_2:* Nitrogen makes up about 78 percent of atmospheric air. Most of the nitrogen in the air simply goes through the engine unchanged and exits the tailpipe as nitrogen.
2. *O_2:* Oxygen forms the basis for all combustion. It constitutes about 21 percent of our air and is the element that supports all flame. The burning of oxygen and fuel releases the chemical energy that does all the work of the engine and the vehicle. A properly operating feedback control system results in a slightly fluctuating amount of residual oxygen in the exhaust. A very rich mixture would have so much fuel that all the oxygen would be consumed; a very lean mixture would leave too much oxygen unused in the exhaust.
3. *HC:* Gasoline is a hydrocarbon compound, as is engine oil. When hydrocarbons burn properly, the hydrogen and carbon atoms separate; each combines with oxygen to form either water or carbon dioxide. If for any reason the gasoline in the cylinder fails to burn, it is pumped into the exhaust system as a raw HC molecule, a most undesirable emissions result. Hydrocarbons may also be emitted to the atmosphere by reason of simple evaporation, as would happen if the gas cap were not properly reinstalled after refueling.
4. *H_2O:* Water vapor forms the bulk of the exhaust product of most engines, even those running very poorly. The major source of energy in the fuel is the hydrogen, which readily combines chemically with the atmospheric oxygen drawn in with the intake air, releasing heat energy and forming water vapor. This is the most benign product of the engine's combustion, and it is of concern only if the engine is run so little that condensed water is left on the interior surfaces or in the exhaust system, providing a favorable environment for rust.
5. *CO_2:* Carbon dioxide forms in the burning of the carbon portion of the fuel. Carbon dioxide is a harmless portion of the exhaust. It is the same gas used by plants in the respiration cycle, and it is the same gas used to carbonate beverages. While a little carbon dioxide is harmless, environmental concerns have focused recently on the sheer amount of the gas produced by internal combustion engines. Some people have expressed fears that a significant increase in the level of atmospheric carbon dioxide could have a greenhouse effect on the Earth's atmosphere, eventually changing the climate.
6. *CO:* Carbon monoxide is a very deadly and poisonous gas, odorless and colorless, and lethal in almost undetectably small concentrations. During normal combustion, each carbon atom tries to take on two oxygen atoms (CO_2). If, however, there is a deficiency

of oxygen in the combustion chamber, some carbon atoms are only able to combine with one oxygen atom and thus produce CO. The carbon monoxide actively tries to obtain another oxygen atom wherever it can find one, and this oxygen-scavenging property of CO is what makes it so lethal. If breathed into the lungs, it not only provides no oxygen, it removes what oxygen it finds there. An overly rich air-fuel ratio is the only cause of CO production.

7. *NO_x:* Air is made up of about 78 percent nitrogen and 21 percent oxygen. Under normal conditions, nitrogen and oxygen do not chemically unite. When raised to a high enough temperature, however, they unite to form a nitrogen oxide compound, NO or NO_2. These compounds are grouped into a family of compounds referred to as oxides of nitrogen: NO_x. While oxides of nitrogen can start to form at lower temperatures, NO_x production begins to become critical at about 2,500°F, with production increasing as temperatures go higher.

Catalytic Converter

The catalytic converter is the single most effective emissions control device on the modern automobile. The catalytic converter's role in controlling exhaust emissions is directly affected by the PCM's ability to properly control the air-fuel ratio. Because the catalytic converter needs a stoichiometric air-fuel ratio or richer to reduce NO_x emissions back to nitrogen and oxygen, as shown in Figure 2–28, and because it also needs a stoichiometric air-fuel ratio or leaner in order to oxidize hydrocarbon and carbon monoxide emissions, the ability of the PCM to hold the air-fuel ratio close to stoichiometric improves the efficiency of the catalytic converter. Figure 2–29 also shows how the efficiency level of the catalytic converters is affected by changes in the air-fuel ratio.

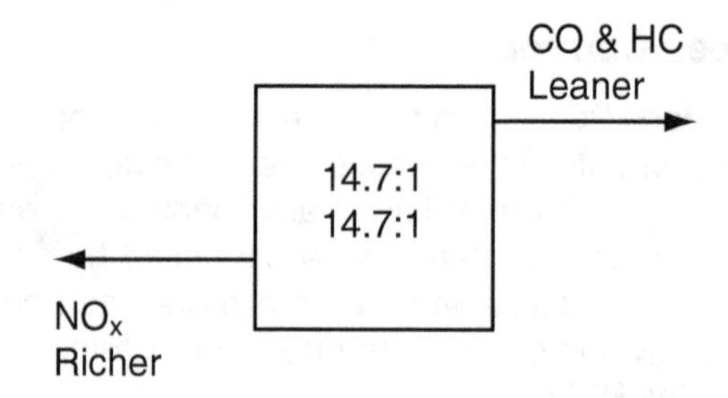

Figure 2–28 Three-way catalytic converter air-fuel ratio requirements.

Other Exhaust Emissions from Gasoline Engines

- Aldehydes
- Ammonia
- Carboxylic acids
- Inorganic solids
  - lead (from leaded fuel)
  - soot
- Sulfur oxides (from fuel impurities)

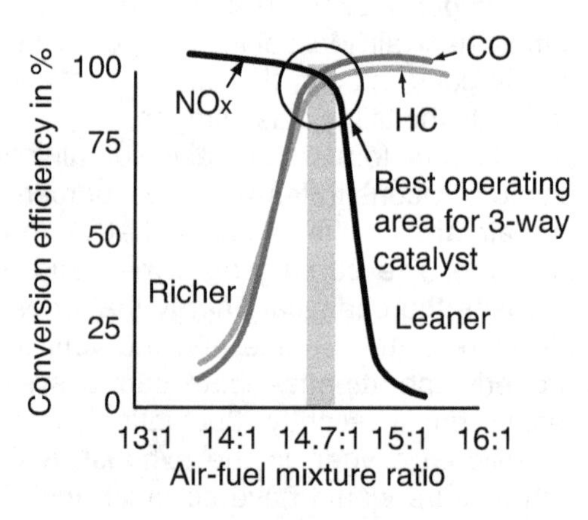

Figure 2–29 Three-way catalytic converter efficiency.

CLOSED-LOOP AND OPEN-LOOP OPERATION

Closed-Loop Operation

Made possible by the development and use of the oxygen sensor, a feature of all comprehensive computerized engine control systems is the ability to operate the engine in a closed-loop mode. The term **closed loop** refers to the fact that the engine computer *monitors the results of its own control* through a feedback circuit. With the use of a feedback circuit, a computer can monitor what actually happened as a result of the last command it issued to an actuator. Without a feedback circuit, the computer must *assume* that it is controlling the actuator properly and that the actuator is responding properly. As computer technology has moved forward, increasing numbers of output circuits have been monitored with feedback circuits and sensors.

In engine control systems, "closed loop" refers to the use of the oxygen sensor to monitor the results of the last commands issued to the fuel injector(s) (or mixture control solenoid in a feedback carburetor). Closed loop is a mode that allows the engine computer to maintain an air-fuel ratio that stays very close to stoichiometric. When operating in closed loop, the engine computer responds to the other critical sensors in bringing the air-fuel ratio close to the ideal. These sensors include throttle position, engine load, barometric pressure, engine RPM, engine coolant temperature, and intake air temperature. Then the engine computer uses the oxygen sensor to fine-tune the air-fuel ratio. The oxygen sensor reports to the computer a voltage representing the amount of oxygen that was left over after the burn during the most recent combustion. The computer sends a command to the fuel-metering control device to adjust the air-fuel ratio toward a stoichiometric ratio. This adjustment usually causes the air-fuel ratio to cross stoichiometric, and it starts to move away from stoichiometric in the opposite direction. The oxygen sensor sees this and reports it to the computer. The computer again issues a command to adjust back toward stoichiometric. With the speed at which this cycle occurs, the air-fuel ratio never gets very far from stoichiometric in either direction. This cycle repeats continuously.

Closed loop is the most efficient operating mode, and the computer is programmed to keep the system in closed loop as much as possible. However, the following three criteria must be met before the system will go into closed-loop control:

1. The oxygen sensor must reach operating temperature, about 315°C/600°F for a zirconia narrow-band sensor.
   - Newer vehicles use a heater to bring the oxygen sensor up to its operating temperature more quickly at engine startup. This allows the engine to get into closed-loop control more quickly following a cold engine start.
2. The engine coolant temperature must reach a criterion temperature.
   - This varies somewhat but tends to be around 65°C/150°F to 82°C/180°F on older vehicles.
   - On many vehicles built since about the 2000 model year, this may be as low as 0°C/32°F, which allows the engine to get into closed-loop control more quickly following a cold engine start. This also effectively removes the engine coolant sensor as a requirement for closed-loop operation except on cold days during the winter months.
3. A predetermined amount of time must elapse from the time the engine was started.
   - This varies from 1 or 2 minutes on older vehicles to 10 to 20 seconds on most modern vehicles.

Ultimately, closed-loop operation, using feedback information from the oxygen sensor to more precisely control the air-fuel ratio, is no longer reserved for warm engine operation as it once was. Manufacturers are now programming the computers to put the system into closed-loop operation more quickly following a cold engine start to reduce cold engine exhaust emissions substantially. (Cold engine operation has always been considered a "dirty" operating mode that creates substantial excessive exhaust emissions.)

Because the programmed goal of closed-loop fuel control is to maintain a stoichiometric air-fuel ratio, an operating condition such as hard acceleration forces the system out of closed loop due to the fact that it needs a richer air-fuel ratio than closed loop can offer.

Open-Loop Operation

Open-loop operation is used during periods when a stoichiometric air-fuel ratio is not appropriate, such as cold engine warm-up (see "Closed-Loop Operation") or at WOT. Open-loop operation is also implemented during limp-in mode operation when a critical sensor has been lost. During this operational mode, the computer uses input information from several sensors to determine what the air-fuel ratio should be. These sensors may include throttle position, engine load, barometric pressure, engine RPM, engine coolant temperature, and intake air temperature.

Once the necessary information is processed, the computer sends the appropriate command to the mixture control device. The command does not change until one of the inputs changes. In this mode, the computer does not use oxygen sensor input and therefore does not know if the command it sent actually achieved the most appropriate air-fuel ratio for the prevailing operating conditions.

THE PCM AND EXHAUST EMISSIONS

The PCM uses closed-loop control to achieve a stoichiometric air-fuel ratio for two primary reasons:

1. At this air-fuel ratio, the engine produces the smallest amount of toxic exhaust emissions (CO, HC, and NO_x) and maximizes the production of nontoxic emissions (H_2O and CO_2) within the combustion chamber.
2. At this air-fuel ratio, the catalytic converter, which is the single most effective emissions control device, operates at its peak efficiency, thereby further reducing exhaust emissions.

ATTITUDE OF THE TECHNICIAN

Historically, many automotive technicians have had a negative attitude about manufacturers' continuous changes to cars. Changes mean always having to learn something new, having to cope with a new procedure, and so on. There is no doubt that these changes will continue to be a challenge to technicians. But there is another way to look at it: Every time new knowledge or a new skill is required, it represents a new opportunity to get ahead of the pack. Most of us are in this business to make money; the more you know that other people have not yet bothered to learn, the more money you can make. Successful, highly paid automotive technicians *earn* the money they get because of what they *know.* Their knowledge enables them to produce more with equal effort. More importantly, it enables them to do things that other people cannot do. Therefore, it is becoming increasingly important to pursue this knowledge, both through textbooks such as this one and through technical classes as they are offered in your area.

Computerized automotive control systems represent the newest and probably the most significant and most complicated development ever to occur in the history of automotive service. (It is complicated not because it is more difficult, but because it involves more processes.) We have seen only the beginning. In recent years, every major component and function of the automobile has come under the control of an electronic control module, and engineers will undoubtedly continue to find new ways to use electronic control as we move forward. Those who appreciate the capability of these systems and learn everything they can about them will be able to service them and do very well while other people scratch their heads and complain.

✔ SYSTEM DIAGNOSIS AND SERVICE

Approaching Diagnosis

Diagnosing problems on complicated electronic systems takes a different approach than many automotive technicians are used to. The flat-rate pay system has encouraged many of us to take shortcuts whenever possible; however, shortcuts on these systems get you into trouble. Use the service manual and follow the procedures carefully. It is very useful to spend some time familiarizing yourself with the diagnostic guides and charts that manufacturers present in the service manuals you will be using.

Pre-Diagnostic Inspection

Although modern computerized engine control systems offer a high level of self-diagnostic capabilities, it is always a good idea to begin diagnostics with a purposeful inspection of the basics. Inspect the fuses in both the under-hood junction box and the instrument panel box both visually and electrically. A visual test of each fuse, with it removed, can quickly identify whether a fuse contact is corroded, resulting in a poor connection. A quick electrical test can be done using either a grounded test light or high-impedance test light (logic probe). Start the engine and conscientiously listen for vacuum leaks. Perform a visual inspection of components and wiring, including the wiring connectors. Keep in mind that while the vast majority of system faults will be identified by a computer, it is still possible to have a fault occur that the computer cannot identify.

SUMMARY

In this chapter, we have begun building a conceptual foundation for understanding why vehicle manufacturers use computers in their vehicles and what the computers are supposed to do. The principal objectives of an engine computer are to control exhaust emissions to a legally acceptable level, to optimize the vehicle's driveability, and to get the best fuel economy consistent with the driving conditions.

We have learned about the major components of a computer, including the microprocessor as the central element of any control module. We also discussed the various types of memories and how they are used in a computer. We covered the digital voltage signals, known as binary code or serial data, which a computer uses to communicate both internally and externally. And we took an in-depth look at logic gates, or combinations of transistors, that allow the computer to respond appropriately to particular combinations of input values.

We have learned about the various elements in a car's exhaust that are toxic: hydrocarbons (or unburned fuel), carbon monoxide, and oxides of nitrogen. We have seen how residual oxygen is used as a measure of the computer's mixture control success and how even water vapor and carbon dioxide are produced as a result of the hydrogen and carbon in the fuel.

The concept of stoichiometry has been introduced, a concept that will play a role in each successive chapter. A stoichiometric air-fuel ratio not only allows the engine to produce the fewest toxic emissions, but also allows the catalytic converter to keep emissions to a minimum. The only way to accurately maintain this ratio is through the precision control of a computer.

We have learned the concepts of open loop and closed loop. Open loop describes the conditions under which the computer determines air-fuel mixture and ignition timing based on information stored in its memory about engine temperature, engine load, engine speed, and other system parameters. Closed loop, in contrast, is the feedback mode in which the computer controls air-fuel mixture by adding in the output signal from the oxygen sensor.

Finally, we have learned the importance of following precise diagnostic steps if this type of computer system is to be diagnosed. Without such a sequence, the technician can be misled by other problems or by countermeasures the

Excessive HC Production

Over-advancing ignition timing by 6 degrees can cause HC production to go up by as much as 25 percent and NO_x production to increase by as much as 20 to 30 percent. A thermostat that opens at 160°F can cause HC production to go up as much as 100 to 200 parts per million.

computer employs to solve some other problem that has affected the system.

▲ DIAGNOSTIC EXERCISE

In an uninformed and socially irresponsible attempt to improve his car's performance, a do-it-yourselfer disconnected, disabled, or removed most of the emissions controls on his car. He disabled the EGR system and removed the catalytic converter and rear oxygen sensor.

What will the consequences be for the car? Will its driveability, power, or fuel economy improve? What should the repair/diagnostic technician's approach be?

Review Questions

1. *Technician A* says that computerized engine control systems are intended to reduce exhaust emissions. *Technician B* says that computerized engine control systems are designed to improve fuel mileage and driveability. Who is correct?
   A. *Technician A* only
   B. *Technician B* only
   C. Both technicians
   D. Neither technician
2. Two technicians are discussing why computers have replaced mechanical control methods for controlling both engine management systems and non-engine systems. Technician A says that computers can provide more precise control, faster response, and increased dependability than mechanical methods of control. *Technician B* says that computers provide the potential to add additional features and functions and also offer increased diagnostic capabilities. Who is correct?
   A. *Technician A* only
   B. *Technician B* only
   C. Both technicians
   D. Neither technician
3. An analog voltage is a voltage signal that does which of the following?
   A. It is continuously variable within a given range and requires time to change.
   B. It varies, but not continuously, and requires no time to change.
   C. It is often referred to as a square wave.
   D. Both B and C.
4. A digital voltage is a voltage signal that does which of the following?
   A. It is continuously variable within a given range and requires time to change.
   B. It varies, but not continuously, and requires no time to change.
   C. It is often referred to as a square wave.
   D. Both B and C.
5. How many combinations of zeros and ones are available to an 8-bit computer?
   A. 64
   B. 128
   C. 256
   D. 65,536
6. The binary code 01101110 equals what decimal (base-10) numerical value?
   A. 5
   B. 110
   C. 124
   D. 421

7. How many combinations of zeros and ones are available to a 16-bit computer?
   A. 64
   B. 128
   C. 256
   D. 65,536
8. Which of the following components in an engine computer is a nonvolatile memory that contains information specific to a particular vehicle or model?
   A. A/D converter
   B. ROM
   C. RAM
   D. PROM
9. Which of the following components in an engine computer translates analog input signals to binary code?
   A. A/D converter
   B. ROM
   C. RAM
   D. PROM
10. Which of the following components in an engine computer is a volatile memory used by the microprocessor to store temporary information?
    A. A/D converter
    B. ROM
    C. RAM
    D. PROM
11. Concerning an AND gate, which of the following would cause the output (C) to be a binary 1?
    A. When input A is a binary 0 and input B is a binary 1
    B. When input A is a binary 1 and input B is a binary 0
    C. When both inputs, A and B, are a binary 1
    D. All of the above
12. Concerning an OR gate, which of the following would cause the output (C) to be a binary 1?
    A. When input A is a binary 0 and input B is a binary 1
    B. When input A is a binary 1 and input B is a binary 0
    C. When both inputs, A and B, are a binary 1
    D. All of the above
13. Which of the following could cause the reference voltage regulator and the logic gates within a computer to function improperly or not at all?
    A. Low battery voltage
    B. Battery does not hold a proper load when the starter motor is engaged
    C. Excessive resistance within the computer's power and/or ground circuits
    D. All of the above
14. What air-fuel ratio is required to allow a three-way catalytic converter to work effectively?
    A. 12:1
    B. 13.7:1
    C. 14.7:1
    D. 17:1
15. Why does the PCM try to maintain a stoichiometric air-fuel ratio?
    A. To minimize the potential for producing toxic emissions (CO, HC, and NO_x) in the combustion chamber
    B. To maximize the production of H_2O and CO_2 in the combustion chamber
    C. To help the three-way catalytic converter work most effectively
    D. All of the above
16. *Technician A* says that any time a computer begins making incorrect decisions and starts controlling a system improperly, it should automatically be replaced. *Technician B* says that if a computer begins making incorrect decisions and starts controlling a system improperly, replacing the computer may not be the appropriate repair. Who is correct?
    A. *Technician A* only
    B. *Technician B* only
    C. Both technicians
    D. Neither technician

17. *Technician A* says that closed-loop operation is designed to allow the engine computer to monitor the results of its own fuel control through an oxygen sensor mounted in the exhaust stream. *Technician B* says that closed loop is the best engine-operating mode for wide-open throttle performance because of the increased precision that closed loop provides. Who is correct?
    A. *Technician A* only
    B. *Technician B* only
    C. Both technicians
    D. Neither technician
18. Closed-loop operation requires all except which of the following?
    A. The oxygen sensor must reach operating temperature.
    B. The engine coolant temperature must be at or above a criterion temperature.
    C. The vehicle speed must be at least 20 MPH.
    D. A predetermined amount of time must have elapsed since the engine was started.
19. *Technician A* says that the continued changes in automotive design are a challenge to every technician. *Technician B* says that if you pursue learning through the reading of textbooks and through taking technical classes, new technology can become your advantage. Who is correct?
    A. *Technician A* only
    B. *Technician B* only
    C. Both technicians
    D. Neither technician
20. *Technician A* says that a fault code indicating a coolant temperature sensor problem could be the fault of a defective coolant-temperature sensor or its circuit. *Technician B* says that a fault code indicating a coolant temperature sensor problem could be the result of a defective thermostat in the engine cooling system. Who is correct?
    A. *Technician A* only
    B. *Technician B* only
    C. Both technicians
    D. Neither technician

Chapter 3

Common Components for Computerized Engine Control Systems

OBJECTIVES

Upon completion and review of this chapter, you should be able to:

- ❑ Define the features common to most computerized engine control systems.
- ❑ Explain the functional concepts of each of the most common sensing devices.
- ❑ Understand the differences between a speed density system and a mass air flow system.
- ❑ Explain the operation of the most common types of actuators: solenoids, relays, motors, and stepper motors.
- ❑ Define how a computer's control of an actuator's on-time through pulse-width modulation differs from duty cycle control.

KEY TERMS

Actuator
Barometric Pressure
Duty Cycle
Gallium Arsenate Crystal
Hall Effect Sensor
Hertz (Hz)
Homogeneous Air-Fuel Charge
Hot-Wire Mass Air Flow Sensor
Light-Emitting Diode (LED)
Magnetic Resistance Element (MRE) Sensor
Piezoelectric
Piezoresistive
Potentiometer
Potentiometer Sweep Test
Schmitt Trigger
Sensor
Speed Density Formula
Stepper Motor
Stratified Air-Fuel Charge
Tach Reference Signal
Thermistor
Variable Reluctance Sensor (VRS)
Volumetric Efficiency (VE)
Wide-Open Throttle (WOT)
Zirconium Dioxide (ZrO_2)

Among the various car manufacturers, vehicle models, and model years, there are many different computerized engine control systems. Although there are significant differences among the various systems, when compared, they are actually more alike than they are different. Some of the more common components and circuits are discussed in this chapter to avoid needless duplication when discussing the specific systems. As you read the following chapters, you may find it useful to refer to this chapter to clarify how a particular component or circuit works.

COMMON FEATURES

Computers

The comprehensive automotive engine computer (known as the powertrain control module, or PCM) is a special-purpose, small, highly reliable, solid-state digital computer protected inside a metal box. It receives information in the form of voltage signals from several **sensors** and other input sources. With this information, which the PCM rereads several thousand times per second, it can make "decisions" about engine calibration functions such as air-fuel mixture, spark timing, exhaust gas recirculation (EGR) application, and so forth.

These decisions appear as commands sent to the **actuators**—the solenoids, relays, and motors that carry out the output commands of the PCM. Commands usually amount simply to turning an actuator on or off. In most cases, the ignition switch provides voltage to the actuators either directly or through a relay. The PCM controls the actuator by using one of its internal solid-state switches (power transistors and quad drivers are two types) to ground the actuator's circuit. Whereas in most actuator circuits the PCM completes the actuator's ground side, occasionally the PCM may be circuited to complete the power side. If the wire that completes the circuit between the actuator and the PCM were to short to the engine block or to chassis ground and the PCM was providing power to the actuator, this short could destroy the PCM's driver (Figure 3–1). But if the PCM uses this circuit to complete the actuator's ground path, a short to ground would not overload the PCM's circuits. However, it would keep the actuator energized (Figure 3–2). In certain instances where safety would be compromised by a short that could energize the actuator, the PCM may be circuited to provide power to the actuator.

Also, the PCM may be circuited to complete both the power and ground circuits when it needs to be able to reverse the circuit's polarity. Examples include reversible DC motors and the windings of many **stepper motors.** (See "H-Gate" in Chapter 1.)

It should also be noted that beginning in the late 1990s, automotive manufacturers began building many of their automotive computers with the ability to sense current levels in their output circuits. Often, if the computer senses that current is rising high enough to endanger the computer, it will shut itself down in a self-protective mode. Thus, a shorted winding or the installation of an

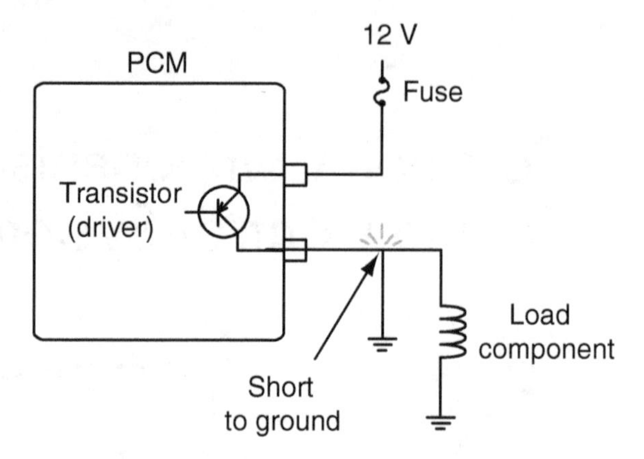

Figure 3–1 Computer providing power to an actuator. A short to ground can destroy the computer's driver.

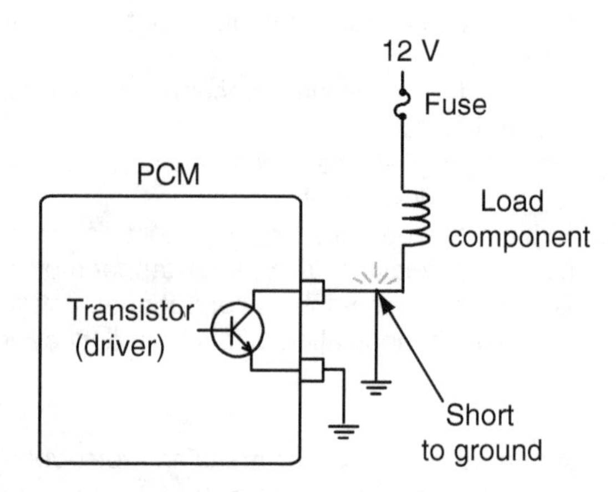

Figure 3–2 Computer completing an actuator's ground circuit. A short to ground keeps the actuator energized but does not overload the computer.

incorrect part number in an output circuit does not have the same potential to destroy the computer's driver as with older computers. However, never assume that a computer has this self-protective strategy programmed into it. It should always be presumed that a current overload could result in damage to the computer.

CAUTION: To continue a precaution that you should already be familiar with, when you work on computer-controlled vehicles, you must be very careful to prevent damage to components by static electricity through the fingers. While it takes about 4,000 V of static discharge for a person to feel even a slight zap, less than 100 V of static discharge—1/40th of what can be felt—can be enough to damage a computer chip or memory unit.

CAUTION: No attempt should be made to open a computer's housing. The housing protects the computer from static electricity. Opening it or removing any circuit boards from it outside carefully controlled laboratory conditions will likely result in damage to some of its components. Certain late-model systems use computers with a special kind of memory that can be reprogrammed by exposing that component to light, usually by removing a piece of opaque tape. If you do not intend to do such reprogramming, or you do not know what the effect will be, do not open a computer just to see what the internal parts look like. Over time, a technician can expect to find a completely failed computer that can be opened up without doing any additional harm.

PCM Location. Over the years, most manufacturers have located the engine computer within the passenger compartment rather than in the engine compartment. It is sometimes mounted under the instrument panel, behind either kick panel, or under a front seat. This placement helps to protect it from the harsh environment of the engine compartment with its extreme temperature changes. Other systems such as Chrysler's systems, Ford's early Microprocessor Control Unit (MCU) system, and modern General Motors systems, located the PCM in the engine compartment. These PCMs have cases that are designed to protect the computer from the engine compartment's environment. In fact, late-model Chrysler engine computers have cases that also shield the computer's internal electronics from radio frequency interference (RFI) and electromagnetic frequency interference (EFI). Many engine computers are located on the passenger compartment side of the firewall under the instrument panel and attach to a harness connector on the engine compartment side of the firewall. Many heavy-duty applications mount the engine computer on the engine block because the computer is matched to the engine rather than to the vehicle.

Engine Calibration. Because of the wide range of vehicle sizes and weights, engine and transmission options, axle ratios, and so forth, and because many of the decisions the computer makes must be adjusted for those variables, the manufacturers use some type of engine calibration unit (Figure 3–3). The calibration unit is a chip that contains information that has to be specific to each vehicle. For example, the vehicle's weight affects the load on the engine, so in order to optimize ignition timing, ignition timing calibration must be programmed for that vehicle's weight. In a given model year, a manufacturer might have more than a hundred different vehicle models, counting different engine and transmission options, but might use fewer than a dozen different computers. This is made possible by using an engine calibration unit specific to each vehicle.

The engine calibration unit may be referred to as the PROM, or some other term may be used. On early computer-controlled engines, some manufacturers made the calibration unit removable and others made it a permanent part of the computer. On modern vehicles this calibration

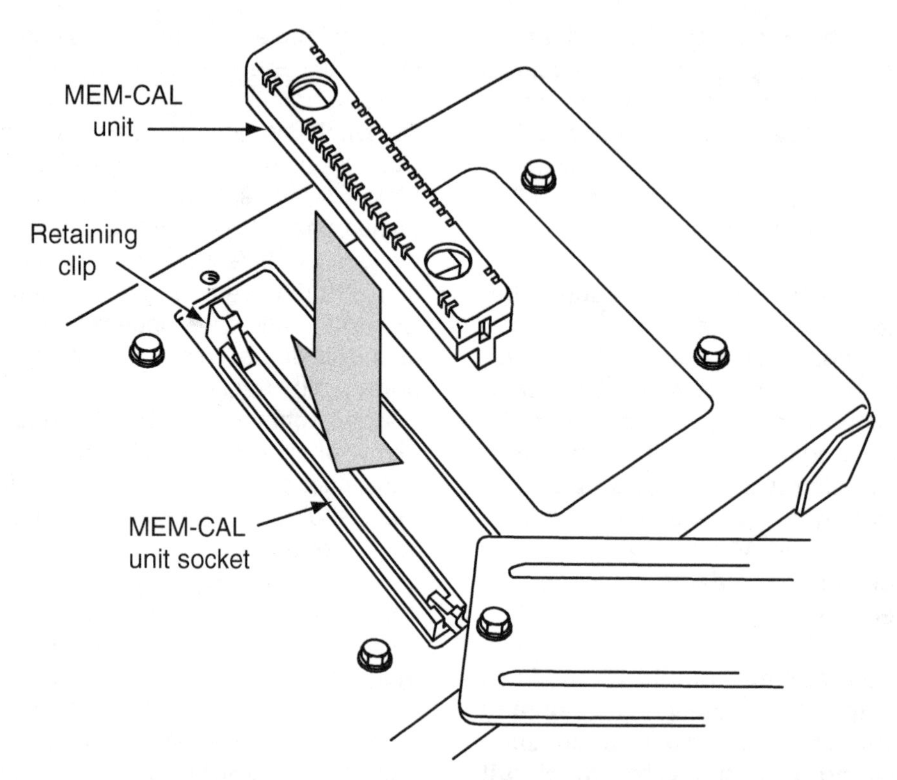

Figure 3–3 General Motors electronic control module and PROM.

unit is likely to be reprogrammable (a process sometimes referred to as *reflashing*).

5 V Reference

With few exceptions, most manufacturers use a 5 V reference voltage from the PCM, which is sent to the sensors that require a reference voltage to be able to operate. (Some late 1970s and early 1980s systems used a 9 V reference.) Within the electronics industry, 5 V has been almost universally adopted as a standard for information-transmitting circuits. This voltage value is high enough to provide reliable transmission and low enough not to damage the tiny circuits on the chips in the computer. Of course, the use of a computer-industry standard voltage makes parts specification more economical for the car makers. It should be noted that other voltages can be used in specific situations. For example, some Chrysler engine computers will send an 8 V reference to the system's Hall effect sensors.

SENSING DEVICES

Exhaust Gas Oxygen Sensors

Purpose of an Oxygen Sensor. The purpose of an oxygen sensor, known also as either an O_2 sensor or O_2S, is to sense the amount of free oxygen in the exhaust after combustion has occurred (Figure 3–4). This free oxygen is a direct result of the air-fuel ratio. A rich mixture yields little free oxygen, with most of it being consumed during combustion. A lean mixture yields more free oxygen because not all of the oxygen is consumed during combustion. With an oxygen sensor, mounted either in the exhaust manifold or in the exhaust pipe ahead of the catalytic converter, measuring the amount of

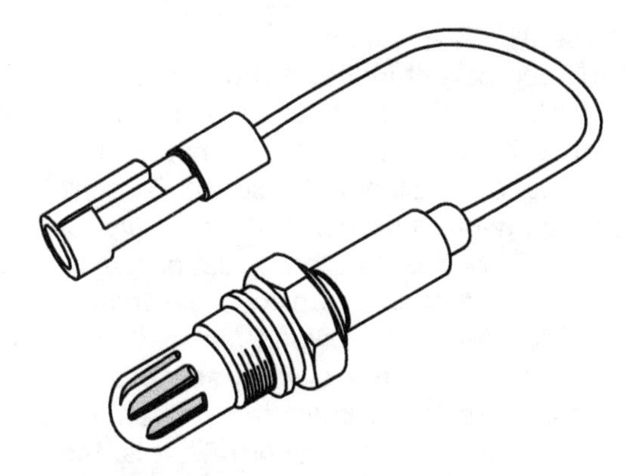

Figure 3–4 Oxygen sensor.

oxygen in the exhaust indirectly reveals the air-fuel ratio of the combustion in the cylinder. This holds true provided that any oxygen in the exhaust is truly from the combustion chamber and that there is no atmospheric air entering the exhaust pipe ahead of the oxygen sensor due to an exhaust leak or a malfunctioning secondary air injection system.

The O_2S is the only sensor that allows the PCM to monitor the results of its own control as it relates to control of the air-fuel ratio. Also commonly known as an exhaust gas oxygen (EGO) sensor, this sensor allows the PCM to operate in the mode known as *closed loop*, whereby the PCM reacts to exhaust gas oxygen levels by adjusting the fuel being delivered to the cylinders in order to maintain a stoichiometric air-fuel ratio.

Oxygen Sensor Types. There are three types of oxygen sensors: the zirconia O_2 sensor, the titania O_2 sensor, and the wide-band air-fuel ratio sensor. These sensors are discussed in this order in the paragraphs that follow because this is the order in which they were introduced to the automotive industry. Any of these oxygen sensors may also be used with an electric heating element to heat them up to operating temperature more quickly and to sustain the proper operating temperature during normal operation.

Zirconia Oxygen Sensors. Zirconia O_2 sensors are generally considered to be the standard O_2 sensor in the automotive industry since they were the type used originally in automotive applications and they continue to be used in many applications today. A zirconia O_2 sensor is a voltage-generating device that consists of a ceramic element shaped like a thimble that has atmospheric air available to the interior side while the exterior side is exposed to exhaust gases (Figure 3–5). The ceramic coating has **zirconium dioxide (ZrO_2)** embedded in it.

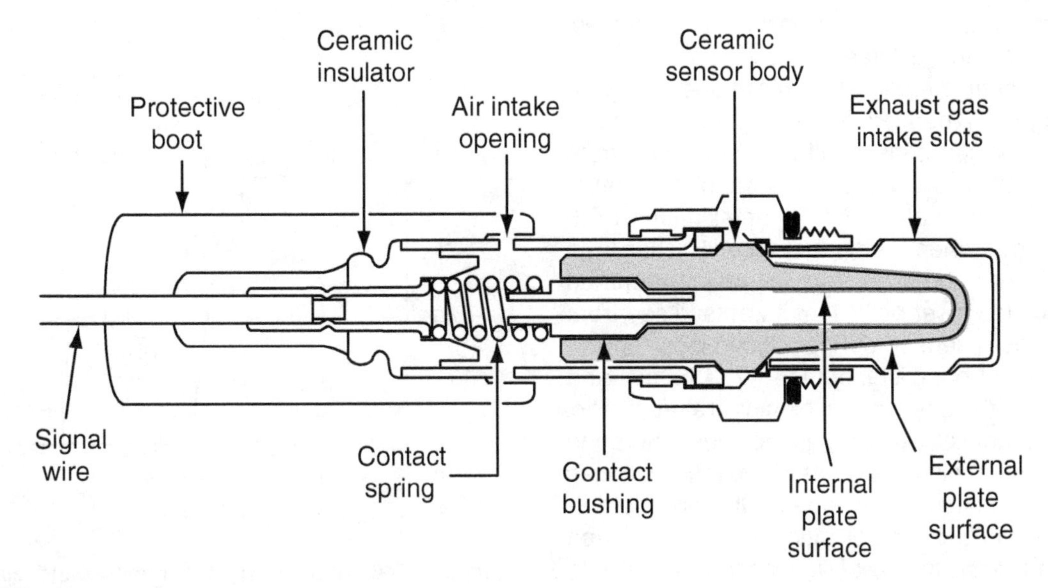

Figure 3–5 Exhaust gas oxygen sensor.

Zirconium dioxide is a white crystalline compound that becomes oxygen-ion conductive at approximately 600°F. The inner and outer surfaces of the ceramic body are coated with separate super-thin gas-permeable films of platinum. The platinum coatings serve as two electrodes. The outer electrode surface is covered with a thin, porous ceramic layer to protect against contamination from combustion residue. This body is placed in a metal shell similar to a spark plug shell except that the shell has a louvered nose that encloses the exhaust-side end of the ceramic body. When the shell is screwed into the exhaust pipe or manifold, the louvered end extends into the exhaust passage. The outer electrode surface contacts the shell; the inner electrode connects to a wire that goes to the computer. Ambient oxygen is allowed to flow into the hollow ceramic body and to contact the inner electrode.

The oxygen sensor has an operating temperature range of 572°F to 1,562°F (300°C to 850°C). When the zirconium dioxide reaches 572°F, it becomes oxygen-ion conductive and collects negatively charged oxygen ions on both the ambient air side and the exhaust side from the available oxygen atoms. Since the level of oxygen in the ambient air is greater than that in the exhaust stream, a positive voltage is generated. The greater the difference (as with the exhaust from a richer air-fuel ratio), the higher the voltage signal generated. Voltage generated by a zirconia oxygen sensor will normally range from a minimum of 0.1 V (100 mV) when the air-fuel ratio is leaner than 14.7:1 to a maximum of 0.9 V (900 mV) when the air-fuel ratio is richer than 14.7:1. The computer has a preprogrammed value, called a *set-point,* that it wants to see from the oxygen sensor during closed-loop operation. The set-point is about 0.45 V (450 mV) and equates to the desired stoichiometric air-fuel ratio.

A zirconia O_2 sensor is sometimes referred to as a switch. This is because its voltage switches rapidly to the rich or lean sides of the stoichiometric value as the air-fuel ratio switches between rich and lean. In reality, it responds quickly to small changes in the air-fuel ratio, with 900 mV representing approximately 14.3 or 14.4 parts air to each part of fuel and 100 mV representing about 15.0 or 15.1 parts air to each part of fuel. Because it can measure only a narrow range of air-fuel ratios, a zirconia O_2 sensor is referred to as a *narrow-band* oxygen sensor (Figure 3–6).

The PCM has two terminals for the oxygen sensor circuit, each of which is electrically connected to one of the oxygen sensor's two electrodes. One is for the oxygen sensor's positive signal and the other is for the oxygen sensor's negative signal (signal ground). That is, the voltage-sensing circuits within the computer must use two wires to measure the oxygen sensor's voltage, similar to how you would use a voltmeter to do the same thing. Although some oxygen sensors complete both circuits using insulated wiring, resulting in a two-wire oxygen sensor, many early applications used the metal of the engine block (vehicle ground) to complete the signal ground circuit, resulting in a single-wire oxygen sensor.

By the late 1980s, most manufacturers were using heated oxygen sensors on many applications. The operation of the oxygen sensor itself is no different, but a heating element is added to

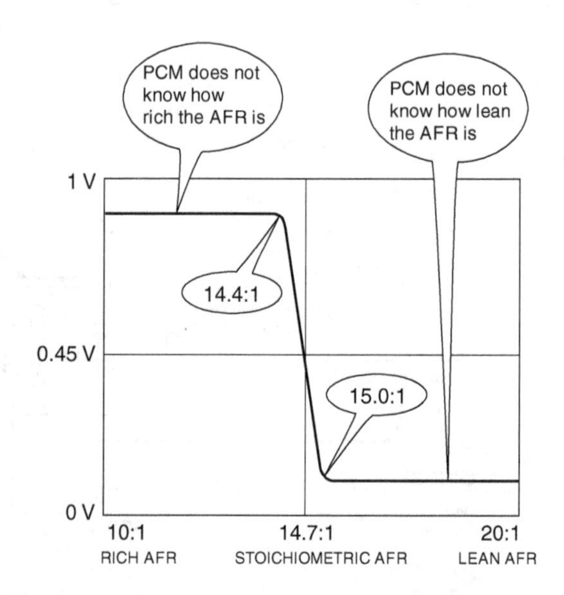

Figure 3–6 Operation of a narrow-band zirconia oxygen sensor.

(and placed within) the oxygen sensor to heat it to operating temperature more quickly and to help it maintain proper operating temperature when contacted by cooler exhaust gases, such as at idle. The addition of a heating element also allows it to be placed further downstream in the exhaust system so that it can sample a more accurate average of the exhaust gas produced by all cylinders. The addition of a heating element results in the addition of two current-carrying wires to the oxygen sensor, one that provides power for the heater and one that provides a ground. As a result, what was previously a single-wire oxygen sensor now has three wires and what was previously a two-wire oxygen sensor now has four wires.

With early heated O_2 sensors, the heater circuit was not part of the computer control system, but on many modern applications the PCM controls the heater circuit with a pulse-width modulated signal, thus allowing it to control the temperature of the O_2 sensor.

Titania Oxygen Sensors. Titania oxygen sensors respond faster than zirconia oxygen sensors to changes in the air-fuel ratio, but they are also more expensive. As a result, titania oxygen sensors were used on a very limited number of vehicles as follows:

- 1987–1990 Jeep Cherokee, Jeep Wrangler, and Eagle Summit
- 1986–1993 Nissan trucks with 3.0L engines
- 1991–1994 Nissan Maxima with 3.0L engines
- 1991–1994 Nissan Sentra with 2.0L engines

On a Nissan, a titania oxygen sensor can be identified by the presence of a red wire in the sensor's wiring harness.

A titania oxygen sensor works more like a coolant temperature sensor in that it modifies an applied reference voltage by changing its resistance internally in response to changes in air-fuel ratios. Instead of a gradual change in resistance as with other resistance changing sensors, it changes its resistance rapidly from about 950 ohms when the air-fuel ratio is rich to around 21,000 ohms when the air-fuel ratio is lean. Therefore, as with a zirconia O_2 sensor, a titania O_2 sensor is also sometimes referred to as a switch. It also is considered a narrow-band oxygen sensor in that it responds quickly to a narrow range of air-fuel ratios, as in Figure 3–6. Titania sensors were used from the late 1980s up until 1994, primarily on Nissan vehicles, but a few late 1980s Jeeps also used them. Titania sensors can switch faster than zirconia sensors, but they are more expensive. A titania O_2 sensor does not use an ambient air vent as a zirconia O_2 sensor does and, therefore, it is better for dirty environments such as four-wheeling in mud.

The PCM supplies a titania O_2 sensor with a reference voltage. On Nissan applications, this reference voltage was 1 V, thereby allowing this sensor to operate between 0 and 1 V and in a fashion similar to a zirconia oxygen sensor. Jeep and Eagle applications used a 5 V reference. The resistance of the O_2 sensor is in series with a fixed resistor that is internal to the PCM (Figure 3–7 and Figure 3–8). Some of the reference voltage is dropped across the internal fixed resistor and some of the voltage is dropped across the O_2 sensor. As the resistance of the O_2 sensor changes, total circuit resistance is changed, which reallocates how many volts are dropped by each ohm of resistance and changes the voltage drop across the internal fixed resistor.

If the fixed resistor within the PCM is on the positive side, as in Figure 3–7, when the air-fuel mixture is rich and the sensor's resistance is low (around 950 ohms), the voltage signal the PCM sees is also low. And when the air-fuel mixture is lean and the sensor's resistance is high (around 21,000 ohms), the voltage signal the PCM sees is also high. Therefore, with the sensor circuited in this manner, the voltage signal operates backwards from that of a zirconia O_2 sensor. In effect, the voltage measured by the PCM on the signal wire equates to the voltage drop across the O_2 sensor.

If the fixed resistor within the PCM is on the ground side, as in Figure 3–8, when the air-fuel mixture is rich and the sensor's resistance is low (around 950 ohms), the voltage signal the PCM

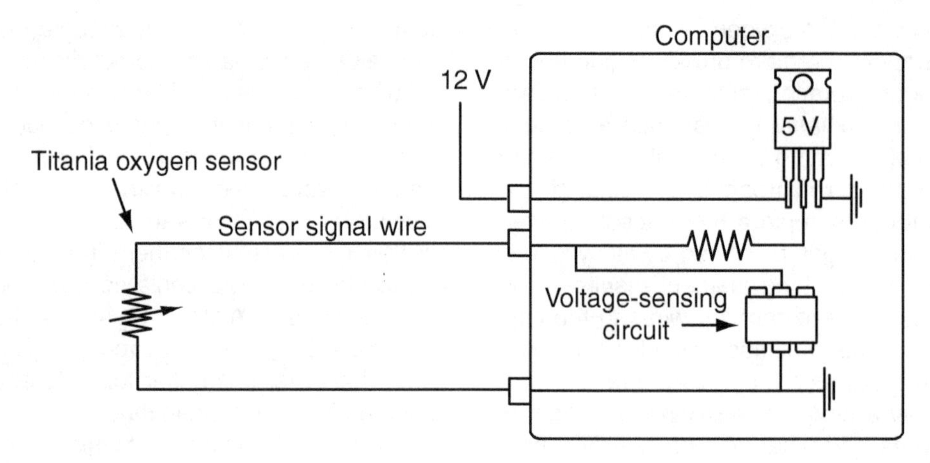

Figure 3–7 A titania oxygen sensor circuit with the fixed resistor on the positive side.

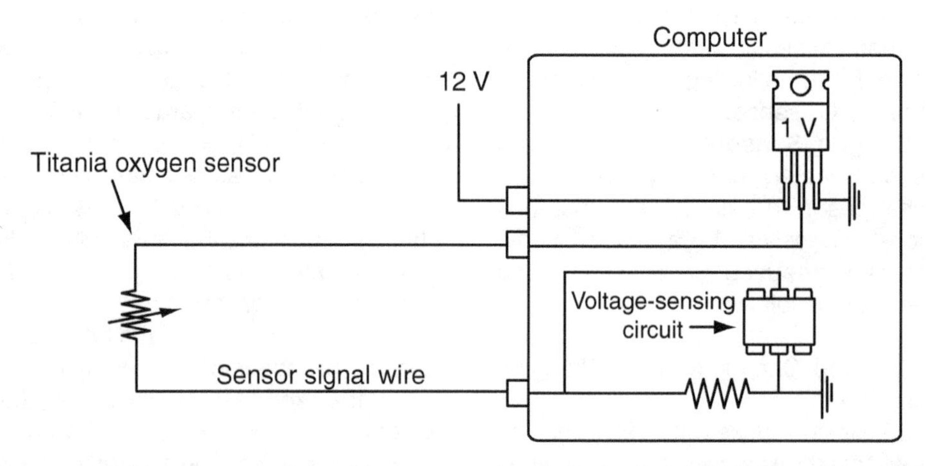

Figure 3–8 A titania oxygen sensor circuit with the fixed resistor on the ground side.

sees is high. And when the air-fuel mixture is lean and the sensor's resistance is high (around 21,000 ohms), the voltage signal the PCM sees is low. Therefore, with the sensor circuited in this manner, the voltage signal operates similarly to that of a zirconia O_2 sensor. In effect, the voltage measured by the PCM on the signal wire equates to the voltage drop across the fixed resistor within the PCM.

Wideband Oxygen Sensors. A wideband oxygen (WBO_2) sensor is designed to be able to measure an air-fuel ratio across a much broader spectrum than standard zirconia and titania O_2 sensors. A WBO_2 sensor can measure air-fuel ratios accurately throughout the entire range, from about 10:1 (rich air-fuel condition) all the way to straight ambient air (extremely lean). This type of sensor does not tend to switch abruptly in the manner of traditional zirconia sensors, but rather can adjust its signal slightly in response to very small changes in the air-fuel ratio, thus allowing the PCM to determine *exact* air-fuel ratios. Other names for these sensors include *air-fuel ratio sensor* or *AFR sensor*, *air-fuel sensor* or *A/F sensor*, *linear air-fuel sensor* or *LAF sensor*, *universal exhaust gas oxygen sensor* or *UEGO sensor*, and *wide range air fuel sensor* or *WRAF sensor*.

Most engines over previous decades provided a **homogeneous air-fuel charge** to their cylinders, meaning that the air and fuel were both evenly mixed and evenly distributed throughout the entirety of the combustion chamber. These engines generally used a zirconia or titania oxygen sensor that would cross-count while operating in closed loop, thereby allowing the PCM to maintain a stoichiometric air-fuel charge as an average. Many modern engine designs use a **stratified air-fuel charge** which is, overall, much leaner than a homogeneous air-fuel charge. This stratified charge is homogeneous and close to stoichiometric in the immediate area near the spark plug. But there is additional air in the cylinder outside of this area, which allows the engine to run much leaner without producing excessive NO_X emissions and without the lean misfire generally associated with lean mixtures. The air pocket outside of the combustion area also helps to insulate the combustion from the relatively cool cylinder walls, thus reducing premature quenching of the flame front. As a result, many modern engines are designed to run an excessively lean air-fuel ratio, particularly during cruise conditions, and need an oxygen sensor that can measure the results of this overall lean condition in the exhaust. Standard zirconia and titania oxygen sensors were not capable of providing information to the PCM that could indicate when the air-fuel ratio became leaner than about 15.1:1. Wideband O_2 sensors solve this problem because of their unique ability to measure lean conditions all the way lean to straight air. It is important to consider that a stratified charge engine does not then necessarily operate at stoichiometric. Instead, the PCM may use the input from a wideband O_2 sensor to slightly oscillate the air-fuel ratio while averaging a very lean overall condition.

Patented by Robert Bosch Corporation in 1994 as the LSU4 O_2 sensor, after a dozen years wideband sensors were only found on a few vehicles. Today they are commonly found on many vehicles as pre-cat sensors, with the rear oxygen sensor still being a zirconia sensor. However, some manufacturers are now also beginning to use wideband O_2 sensors as post-cat sensors. If an engine is equipped with gasoline direct injection (GDI), wideband O_2 sensors are generally used.

A wideband O_2 sensor will typically have five wires entering the sensor body. (Some early air-fuel ratio sensors had only four wires.) The sensor may have an additional wire (a sixth wire) going to its connector if a *calibration resistor* is used (Figure 3–9).

A wideband O_2 sensor has a planar construction (instead of a thimble construction). Think of the term *planar* referring to the layering of components, as in building a sandwich. While there are several variations of wideband O_2 sensors, the following description depicts the typical components, wiring, and function of such a sensor.

Figure 3–9 shows that closest to the exhaust gases lies a zirconium oxygen pump cell with a small gap (or diffuser) that allows some exhaust gas into a diffusion chamber. On the other side of the diffusion chamber lies a Nernst cell, essentially a zirconium dioxide diaphragm (named after Walther Nernst, 1864–1941, winner of the Nobel Prize in chemistry in 1920 "in recognition of his work in thermochemistry"). On the other side of the Nernst cell is an ambient air pocket.

As shown in Figure 3–9, one wire is dedicated to the pump cell current, another wire is dedicated to measure Nernst cell voltage, and a third wire is common for the other side of both the pump cell and the Nernst cell and is part of the circuitry used for both functions.

Two other wires serve as the heater positive and heater ground wires. These sensors require a heater because their operating temperature is typically between 1,292°F and 1,472°F, much hotter than a zirconia O_2 sensor needs to operate. The heater circuit is pulse width modulated

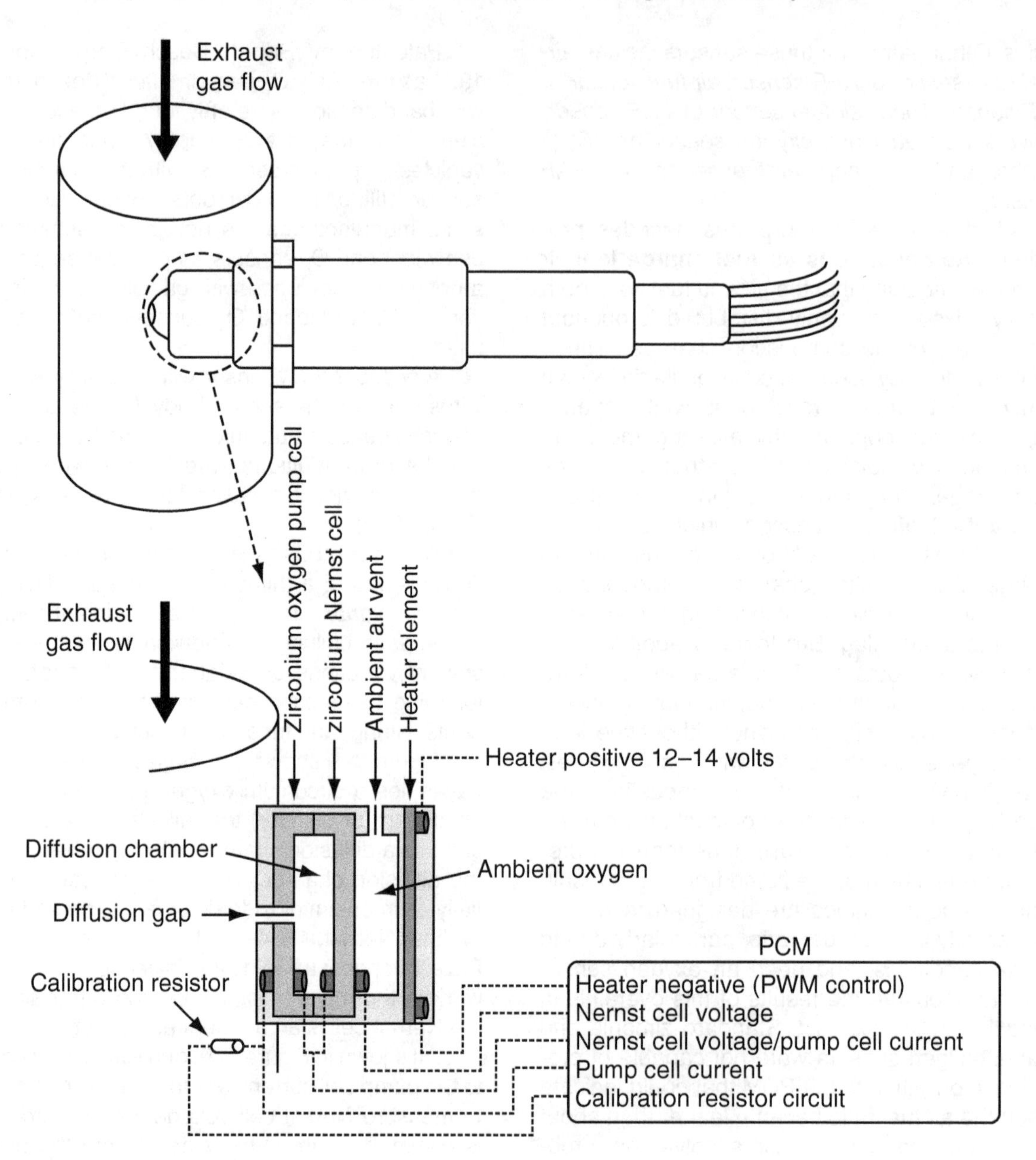

Figure 3–9 Wideband oxygen sensor.

by the PCM and may carry up to 8 amps. This PWM signal allows the PCM to precisely control the sensor's temperature. The PCM monitors the resistance of the Nernst cell in order to know the cell's temperature. Thus, through this feedback, a closed-loop heater control function is established.

A sixth wire may be used with a calibration resistor to allow the PCM to calibrate the sensor to the vehicle's electronics. Before this feature was added, these sensors had to be precisely calibrated at the factory. The use of this calibration circuit has allowed the manufacturers to reduce the cost of the wideband O_2 sensors.

In its functional concept, the Nernst cell can produce a small voltage ranging from 100 mV to 900 mV according to the difference in oxygen levels on the two sides of the membrane, similar to a standard zirconia O_2 sensor. However, in this instance, the PCM tries to maintain the voltage difference across the Nernst cell at a steady 450 mV. It does this through the use of the oxygen pump cell. The PCM controls the current through the pump cell through the use of an H-gate (see Chapter 1), which it uses to reverse the current's polarity as needed. The current is controlled through a high-frequency pulse width modulation (PWM) that allows the PCM to precisely control current flow levels. It is by controlling current flow to the pump cell that the PCM can vary the amount of oxygen atoms in the diffusion chamber.

The Nernst cell and the oxygen pump cell are wired together in a manner that allows current in the oxygen pump cell to control and maintain a balanced level of oxygen in the diffusion chamber. If the exhaust gas entering the diffusion chamber is stoichiometric, the Nernst cell produces 450 mV and the PCM sends no current through the pump cell. However, if the exhaust gas is lean, the Nernst cell produces a voltage less than 450 mV. In this case, the PCM sends a positive current through the pump cell which, in turn, effectively decreases the number of oxygen atoms in the diffusion chamber until the Nernst cell voltage is back at 450 mV. And if the exhaust gas is rich, the Nernst cell produces a voltage above 450 mV. In this case, the PCM sends a negative current through the pump cell which, in turn, effectively increases the number of oxygen atoms in the diffusion chamber until the Nernst cell voltage is back at 450 mV. By monitoring the level of positive or negative current required to maintain the Nernst cell's voltage at 450 mV, the PCM can precisely monitor air-fuel ratios through the entire range from quite rich to extremely lean.

In reality, the preceding explanation is overly simplistic. The pump cell can use the diffusion chamber to create chemical reactions that are quite complex. For example, how can the pump cell increase the number of oxygen atoms in the diffusion chamber when the exhaust is rich and lacks free oxygen? The pump cell and diffusion chamber can chemically extract the needed oxygen atoms from the carbon dioxide (CO_2) and the water (H_2O) in the exhaust, both of which are byproducts of engine combustion.

The underlying principle for a wideband O_2 sensor to be able to operate with such precision is that during the combustion process no atoms are created or destroyed. Rather, they are all conserved and only change chemical composition during the combustion process. Therefore, such a sensor can be used to allow the PCM to evaluate the total air-fuel mixture going into the combustion chamber by measuring what is leaving the combustion chamber. Then a reverse calculation can be performed in order to know what went in.

Also, within the PCM, the positive or negative current is converted to a voltage that is used to indicate the precise air-fuel ratio when read with a scan tool. For example, Toyota uses a reference voltage of 3.3 V DC to indicate that the air-fuel ratio is stoichiometric. Based on this reference voltage, a rich air-fuel ratio of 12:1 is indicated on the Toyota scan tool (Techstream) as a voltage of 2.40 V DC, whereas a lean air-fuel ratio of 19:1 is indicated on the Toyota scan tool as a voltage of 4.00 V DC. When using a generic scan tool, these voltage values are divided by 5 in order to meet the OBD II requirement that an O_2 sensor can be read as a voltage between zero and one volt (this requirement has since been retracted). Thus, a 12:1 air-fuel ratio that would read on the Toyota scan tool as 2.40 V DC would read on a generic scan tool as 0.48 V DC. And a 19:1 air-fuel ratio that would read on the Toyota scan tool as 4.00 V DC would read on a generic scan tool as 0.80 V DC. A stoichiometric air-fuel ratio that would read on a Toyota scan tool as 3.30 V DC would read on a generic scan tool as 0.66 V DC.

(Honda uses 2.7 V DC, and both Bosch and GM use 2.6 V DC as their reference voltages.)

Wideband O_2 sensors have some special considerations and precautions:

- If exhaust pressure is modified from the original calibration, wideband O_2 sensor operation will be adversely affected. For example, if exhaust backpressure is increased, more exhaust gases are forced through the diffuser or diffuser gap into, and then out of, the diffusion chamber, which, in turn, affects the accuracy of the sensor to measure air-fuel ratios.
- Never disconnect a wideband O_2 sensor while the heater circuit is live—the sensor will likely be damaged.
- Never apply battery voltage directly to the heater—without a precise PWM control signal the heater will overheat quickly and damage the sensor.

An advantage of wideband oxygen sensors is how quickly they reach operating temperature. In recent years the EPA has tightened up cold start emissions limits, thus making it critical for the engine to enter into closed-loop control as quickly as possible. A heated zirconia thimble-type sensor took up to 1 minute to get hot enough for the sensor to function. The planar-type wideband sensors can reach operating temperature in as little as 10 to 12 seconds. And through the use of these sensors, manufacturers can now run the engines leaner than when zirconia and titania sensors were used. Ultimately, manufacturers are using wideband oxygen sensors to meet the more stringent emission requirements placed on the industry by the EPA.

Oxygen Sensor Testing. Oxygen sensors are subject to contamination under certain conditions. Contaminating either electrode surface will shield it from oxygen and adversely affect its performance. (The outside electrode is more vulnerable because it is exposed to exhaust gases.) Slight contamination may result in a lazy oxygen sensor. Prolonged exposure to a rich fuel mixture in the exhaust can cause it to become carbon fouled. This fouling can sometimes be burned off by operating in a lean condition for 2 or 3 minutes. The oxygen sensor will also be fouled by exposure to:

- lead from leaded gasoline
- vapors from a silicone-based gasket-sealing compound
- antifreeze residue from coolant leaking into the combustion chamber
- engine oil in the exhaust

Zirconia Oxygen Sensor Testing. While a voltmeter may be used to test a zirconia O_2 sensor, the most effective method is to use a lab scope. (While a voltmeter can display the minimum and maximum values, it cannot display a sensor's response time as a scope can.) When the air-fuel mixture is forced rich it should produce more than 800 mV and when forced lean its output should drop to less than 175 mV. To identify a lazy sensor, the response time should also be tested—it should switch between 300 mV and 600 mV in either direction in less than 100 ms. If a zirconia O_2 sensor's voltage goes negative, replace it—there is more oxygen on the exhaust side of the membrane than on the ambient air side. Do not attempt to test a zirconia O_2 sensor with an ohmmeter—you will likely damage it. Also perform a visual check to make sure the sensor's housing is not cracked.

Titania Oxygen Sensor Testing. To test a titania O_2 sensor, you may use an ohmmeter, but the most effective method is to use a lab scope, applying the same specifications as with a zirconia sensor (although possibly in the reverse direction concerning how the voltage relates to rich and lean air-fuel mixtures, depending upon the system). As with a zirconia sensor, perform a visual check to make sure the sensor's housing is not cracked.

Wideband Oxygen Sensor Testing. When testing a wideband oxygen sensor, a scan tool should be used. There really aren't any valid tests that can be performed with a voltmeter, ohmmeter, or lab scope that are useful in validating/disputing a computer's diagnostic code. With the scan tool connected and the PCM's data stream displayed, you may be able to observe a current parameter of pump cell current or a voltage parameter based on the pump cell's current. Snap accelerate and decelerate the engine and observe the sensor's response. Because wideband oxygen sensors were not used prior to the implementation of OBD II, in all cases where they are used the PCM will determine a failure of the sensor, the sensor's heater element, or the sensor's circuitry. But, as always, visually inspect the sensor for any physical damage as well.

Diagnostic & Service Tip

Oxygen Sensors and Mixture. Technicians should remember that while the oxygen sensor's signal is central to the control of air-fuel mixture, it does not directly sense that mixture. All the oxygen sensor can respond to is the difference in oxygen in the two different gases (ambient air and exhaust gas). Anything other than combustion that allows a change in the oxygen content of the exhaust will thus make the sensor's signal inaccurate. If there is an air leak into the exhaust, perhaps around the exhaust manifold, the sensor will report a lean mixture and the computer will try to correct for it by enriching the mixture—exactly the wrong tactic. If a spark plug fails to deliver the ignition spark, the air in that cylinder's charge will not be consumed but will go into the exhaust. The sensor will again perceive an overly lean condition, and the computer will try to enrich the air-fuel mixture. For the computer's sensors, you must remember exactly what each one is measuring before you can know what the possibilities are for corrective measures.

Thermistors

A **thermistor** is a resistor made from a semiconductor material. Its electrical resistance changes greatly and predictably as its temperature changes. At −40°F (−40°C), a typical thermistor can have a resistance of 100,000 ohms. At 212°F (100°C), the same thermistor can likely have a resistance between 100 ohms and 200 ohms. Even small changes in temperature can be observed by monitoring a thermistor's resistance. This characteristic makes it an excellent means of measuring the temperature of such things as water (engine coolant) or air.

There are negative temperature coefficient (NTC) and positive temperature coefficient (PTC) thermistors. The resistance of the NTC thermistor goes down as its temperature goes up, while the resistance of the PTC type goes up as its temperature goes up. Most thermistors that are used as temperature sensors in automotive computer systems are the NTC type. A good example is an engine coolant temperature sensor, common to most computerized engine control systems. The coolant temperature sensor consists of an NTC thermistor in the nose of a metal housing (Figure 3–10). The sensor's housing is screwed into the engine (usually in the head or thermostat housing) with its nose extending into the water jacket so that the thermistor element will be the same temperature as that of the coolant. The PCM sends a 5 V reference voltage through a fixed resistance and then to the

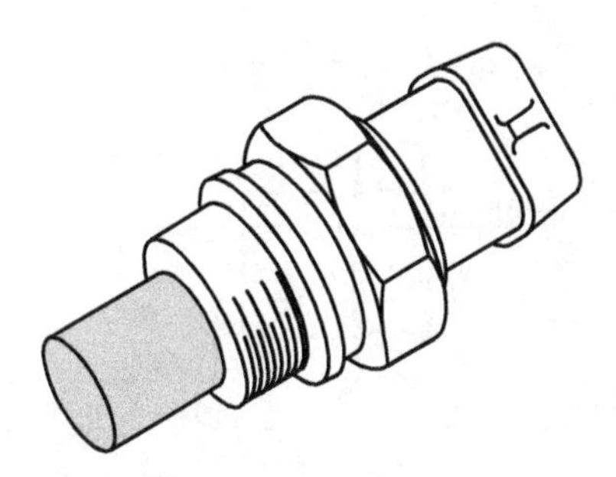

Figure 3–10 Coolant temperature sensor.

sensor (Figure 3–11). A small amount of current flows through the thermistor (usually shown as a resistor symbol with an arrow diagonally across it) and returns to ground through the PCM. This is a voltage-divider circuit (current flows through the first resistance and then through the second resistance) and is commonly used as a temperature-sensing circuit. Because the resistance of the thermistor changes with temperature, the voltage drop across it changes also. The computer monitors the voltage drop across the thermistor and, using preprogrammed values, converts the voltage drop to a corresponding temperature value.

Many modern systems use a more complex engine coolant temperature sensor called a *range-switching* or *dual-range sensor.* As the engine warms up, the thermistor's resistance decreases, causing the voltage drop across the sensor to decrease. At about 110°F to 125°F (Chrysler), the computer lowers the effective resistance within the computer by adding another resistive circuit in parallel to it. This instantly increases the voltage drop across the sensor by effectively transferring it from the internal resistance. As the engine continues to warm up, the voltage drop across the sensor decreases a second time. This allows the computer to see more voltage change as the engine warms up than the supplied 5 V reference signal would normally allow, thereby increasing the accuracy of the sensor input. This is also used to extend the temperature-sensing range of the engine coolant temperature sensor on those engines that utilize a coolant over-temp protection strategy.

If an open circuit develops in either of these temperature-sensing circuits (either by reason of the sensor or the circuitry itself), the computer sees full reference voltage and interprets this as a signal representing about −40°F. This is due to the fact that without any current flow in the circuit, the fixed resistor within the computer will not drop any voltage.

Potentiometers

A **potentiometer** is another application of a voltage divider circuit. It is formed from a carbon resistance material (or, in higher-wattage potentiometers, a wound resistance wire) that has reference voltage supplied to one end and is grounded at the other end. It has a movable center contact (or wiper) that senses the voltage at a physical point along the resistance material (Figure 3–12). At any position of the wiper, there is always some resistance both before and after the point of wiper contact. That is, some of the reference voltage is dropped before the point of wiper contact and the remainder of the reference voltage is dropped after

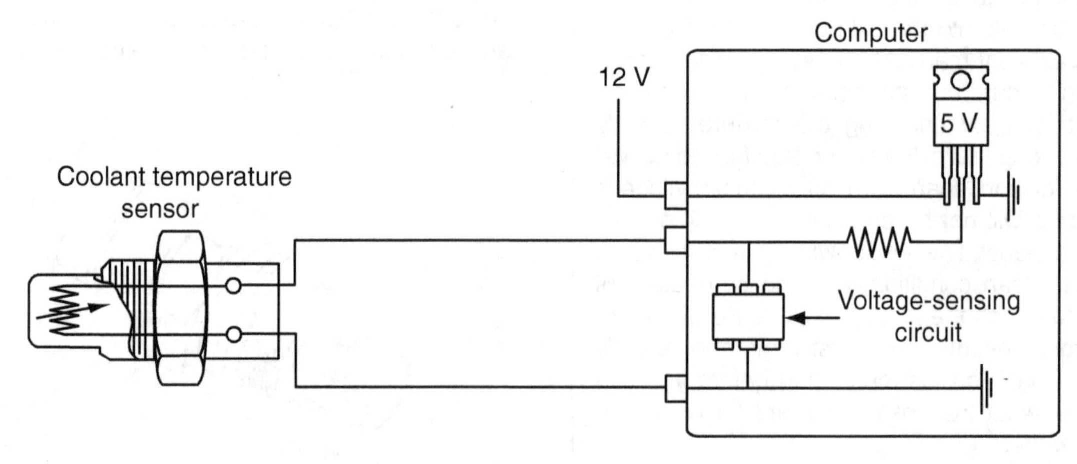

Figure 3–11 Temperature-sensing circuit.

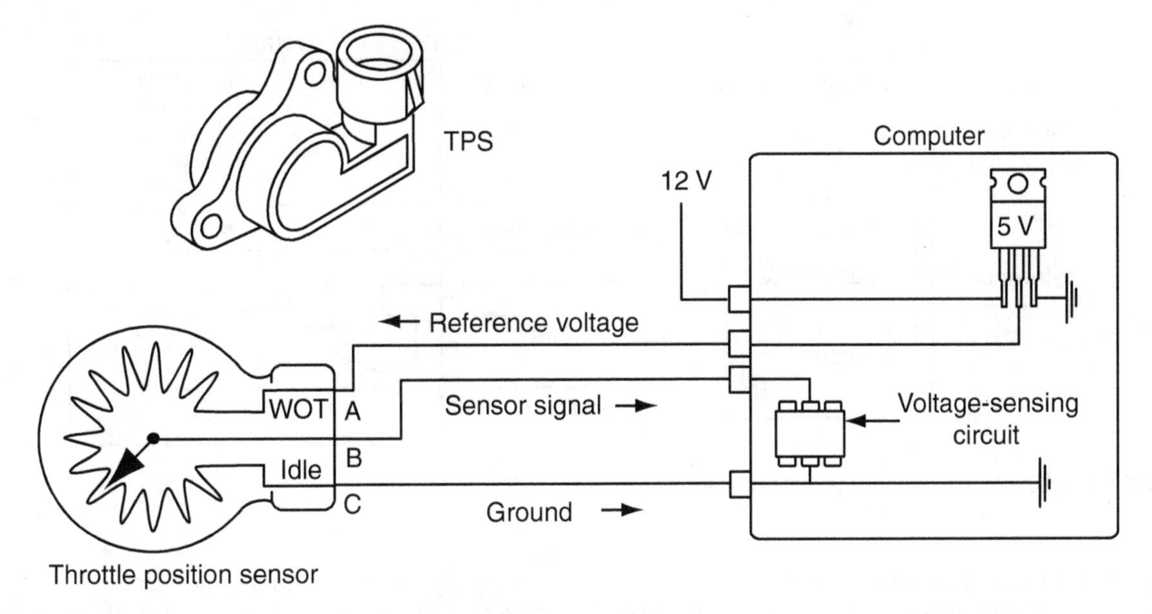

Figure 3–12 Potentiometer used as a throttle position sensor.

the point of contact. As the wiper slides along the resistance material, the voltage signal changes. This voltage signal equates to the voltage drop that occurs after the point of wiper contact and is fed to the PCM as the sensor's voltage signal. A potentiometer measures physical position and is used to sense either linear or rotary motion.

For example, when a potentiometer is used as a throttle position sensor, as depicted in Figure 3–12, the potentiometer would typically be mounted on one end of the throttle shaft. As the throttle is opened, the throttle shaft rotates and moves the wiper along the resistance material. The computer sends a 5 V reference voltage to point A. If the wiper is positioned near point A (**wide-open throttle [WOT]** on most applications), there will be a low voltage drop between points A and B (low resistance), and a high voltage drop will exist between points B and C. When the wiper is positioned near point C (idle position on most applications), there is a high voltage drop between points A and B and a low drop between points B and C. The computer monitors the voltage drop between points B and C and interprets a low voltage, 0.5 for example, as idle position. A high voltage, around 4.5, will be interpreted as WOT. Voltages between these values will be interpreted as a proportionate throttle position.

A potentiometer has a higher wear factor than any other type of sensor due to the frequent physical movement of the wiper across the resistance material. As a result, it is quite common for a potentiometer to develop points along the wiper's travel that result in a loss of electrical contact. The term **potentiometer sweep test** refers to the testing of a potentiometer throughout its range of operation to determine if any electrical opens exist. Additionally, if a potentiometer loses its ground, the computer will sense full reference voltage in all positions of the wiper because the voltage-sensing circuits of the computer do not load the potentiometer circuit.

Pressure Sensors

Pressure sensors are commonly used to monitor intake manifold pressure and/or **barometric pressure (BP).** Intake manifold pressure is a direct response to throttle position and engine speed, with throttle position being the most

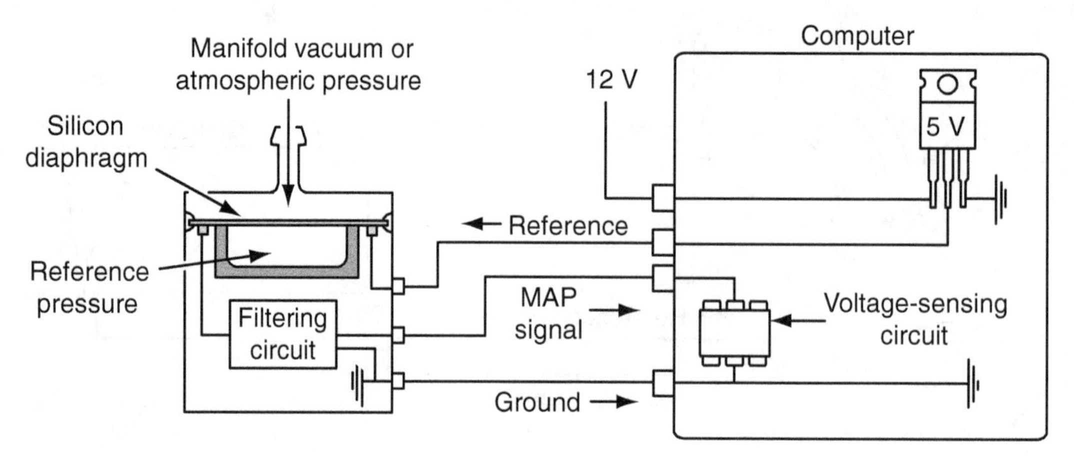

Figure 3–13 Silicon diaphragm pressure sensor.

significant factor. The greater the throttle opening, the greater the manifold pressure becomes (lower vacuum). At WOT, manifold pressure is nearly 100 percent of atmospheric pressure, reduced only by the slight energy consumed by friction with the intake channel walls. Intake manifold pressure can therefore be translated as close to engine load; this is a critical factor in determining many engine calibrations. Barometric pressure is the actual ambient pressure of the air at the engine. It also affects manifold pressure and in most systems is considered when calculating engine calibrations such as air-fuel mixture and ignition timing.

Intake manifold pressure (usually a negative pressure or *vacuum* unless the vehicle is equipped with a turbocharger or supercharger) and/or BP are, on most systems, measured with a silicon diaphragm that acts as a resistor (Figure 3–13). The resistor/diaphragm (about 3 mm wide) separates two chambers. As the pressure on the two sides of the resistor/diaphragm varies, it flexes. The flexing causes the resistance of this semiconductor material to change. The computer applies a reference voltage to one side of the diaphragm. As the current crosses the resistor/diaphragm, the amount of voltage drop that occurs depends on how much the diaphragm is flexed. The signal is passed through a filtering circuit before it is sent to the computer as a DC analog signal. The computer monitors the voltage drop and, by using a look-up chart of stored pressure values, determines from the returned voltage the exact pressure to which the diaphragm is responding.

Two different pressure sensor designs use the **piezoresistive** silicon diaphragm: the absolute pressure and the pressure differential (delta) sensors (Figure 3–14). In one design, the chamber under the diaphragm is sealed and contains a fixed reference pressure. The upper chamber is exposed to either intake manifold pressure or to atmospheric pressure. If the upper chamber is connected to the intake manifold, the sensor functions as a manifold absolute pressure (MAP) sensor. Absolute, as used in the term MAP, refers to a sensor that compares a varying pressure to a fixed pressure. The output signal (return voltage) from this sensor increases as pressure on the variable side of the diaphragm increases (wider throttle opening for engine speed). If the upper chamber is exposed to atmospheric pressure, the sensor functions as a BP sensor. Modern PCMs take an initial reading from the MAP sensor during the initial key-on, engine start sequence that is then used as a BP reference for the next driving trip. The barometric value is also updated using the MAP sensor each time the throttle is opened to the WOT position.

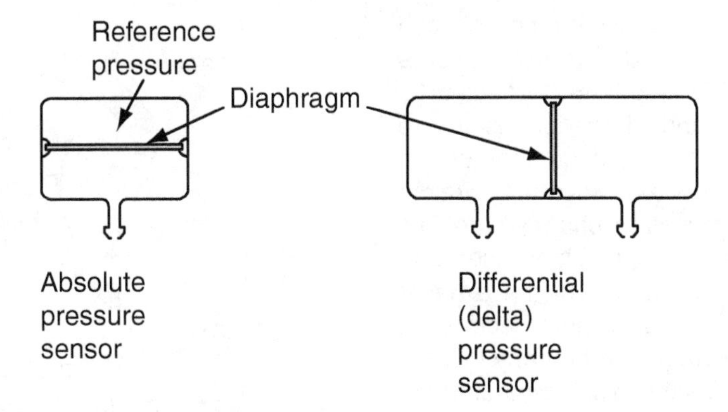

Figure 3–14 Two types of pressure sensors.

The differential pressure sensor combines the functions of both the MAP and the BP sensors. Instead of using a fixed pressure, one side of the resistor/diaphragm is exposed to BP and the other side is connected to intake manifold pressure. The output signal is the result of subtracting manifold pressure from BP. The output signal is opposite to that of the MAP sensor, however, because as manifold pressure decreases (higher vacuum), output voltage increases. The pressure differential sensor had limited use on some early General Motors Computer Command Control systems.

Other types of pressure sensors are discussed in the appropriate chapters.

Speed Density Formula. To know how much fuel to meter into the cylinder, the microprocessor must know how much air, as measured by weight (more accurately, by mass), is in the cylinder. Since the microprocessor has no way to actually weigh the air going into a cylinder, some other method must be used to determine this value. Many systems use a mathematical *calculation* called the **speed density formula.**

$$EP \times EGR \times VE \times MAP/AT = \text{Air density in cylinder}$$

where: EP = engine parameters
EGR = EGR flow
VE = volumetric efficiency
MAP = manifold absolute pressure
AT = air temperature

To make this calculation, all the factors that influence how much air gets into the cylinders must be considered. Throttle position and engine speed affect air intake, and engine temperature (measured as coolant temperature) affects how much heat the air gains as it passes through the induction system and thus affects the air's density in the cylinder. These factors as a group are called *engine parameters.* The microprocessor gets these pieces of information from sensors that are common to all systems.

The amount of air getting into the cylinders is reduced by the amount of exhaust gas that the EGR valve meters into the induction system. With one exception (Ford), early 1980s automotive computers made an assumption about the rate of EGR flow. This was accomplished by allowing the PCM to take a very passive approach to controlling the EGR system. The PCM then used estimates (based on engine parameters) of the EGR system's flow rate that were stored in the computer's memory. Early 1980s Ford products allowed the PCM to monitor the EGR valve's operation according to the aggressive control strategies programmed into the PCM.

Because the EGR flow rate affects the air-fuel ratio, modern systems use some type of sensor and/or sensor/actuator–based strategy to allow the PCM to monitor the actual flow rate of the EGR system.

Volumetric Efficiency. How much air can be drawn into and exhausted from the cylinders (or how well the engine breathes) is affected by the diameter, length, and shape of the intake manifold runners; the size and number of the intake and exhaust valves; valve lift, timing, and duration; the combustion chamber design; the cylinder size and compression ratio; and the diameter, length, and shape of the exhaust manifold runners. These groups of factors all affect the engine's **volumetric efficiency (VE).** A VE value for any given set of engine parameters is stored in the computer's memory. In principle, VE is the amount of air that gets into the cylinder by the time the intake valve closes, divided by the theoretical amount displaced by the moving piston. Ordinarily, this is less than 100 percent, but VE also varies with engine speed and load. With careful use of valve overlap, VE can sometimes go just above 100 percent for certain engine speeds. Maximum engine torque, of course, occurs just at the highest VE for a given engine. Volumetric efficiency can also be increased through the use of supercharging or turbocharging.

Air Temperature. The air's temperature affects its density, and therefore how much air enters the cylinders. Most systems use an air temperature sensor to measure the temperature of the intake air. Some early systems used estimates of air temperature that were based on a fairly predictable relationship between coolant temperature and air temperature, the parameters of which were stored in memory. In modern systems, intake air temperature is considered a critical input on those systems that use the speed density formula.

You also know that BP pushes air past the throttle plate(s) and manifold pressure pushes that air past the intake valve into the cylinder. The last known values that must be determined in order to know how much air the engine will inhale during any given set of operating conditions are BP and manifold pressure. The MAP sensor, as you will see in later chapters, almost always provides both pressure values. A MAP sensor identifies those systems that use the speed density formula.

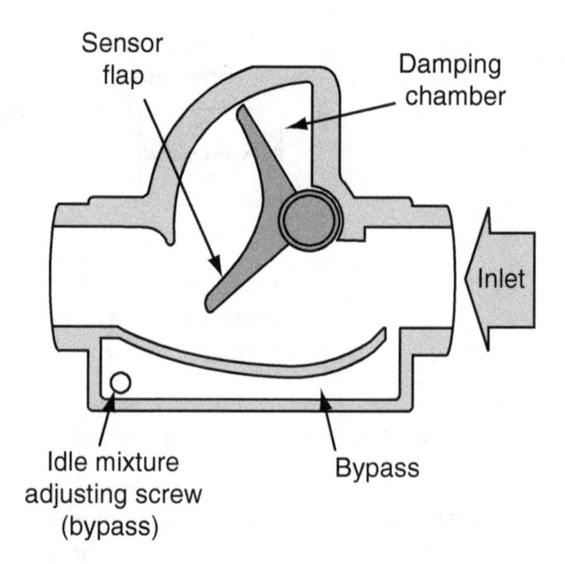

Figure 3–15 Vane airflow (VAF) sensor.

It should also be noted that some older European, Asian, and U.S. vehicles that used the speed density formula used a vane air flow (VAF) meter instead of a MAP sensor. However, these systems did use a full-time BP sensor as well. The VAF meter was mounted in the intake air stream and used a potentiometer to monitor a spring-loaded door that opened according to the volume of intake air that the engine was drawing through it (Figure 3–15). It also contained an NTC thermistor located just ahead of the door to monitor intake air temperature, sometimes known as a *vane air temperature (VAT)* sensor (Figure 3–16). A VAF meter, although no longer used, also identifies the use of the speed density formula.

Measuring Air Mass

Port fuel injection, or multipoint fuel injection, has been the predominant fuel-metering system prior to the introduction of gasoline direct injection (GDI). Although a multipoint fuel injection system offers several distinct advantages over other types of fuel-metering systems, it has one shortcoming. A multipoint injection system provides less opportunity (in time duration) for the fuel to vaporize

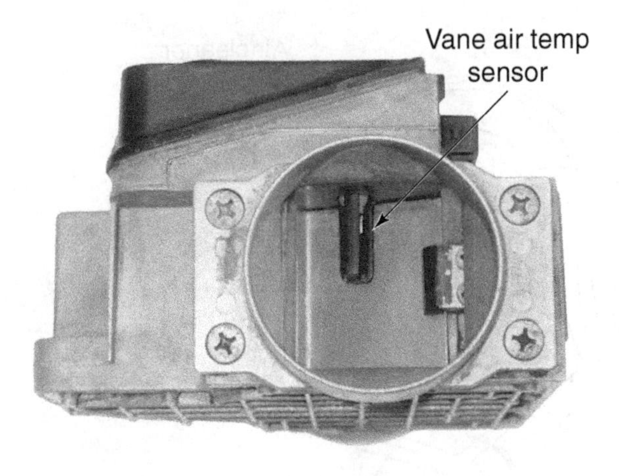

Figure 3–16 Vane airflow (VAF) meter with vane air temperature (VAT) sensor.

than does a system where the fuel is introduced earlier in the intake channels, into the center of the intake manifold, and only the fuel that vaporizes before combustion occurs is burned. In the past, this problem was probably most apparent as a lean stumble when accelerating from idle and was the result of the high manifold pressure that comes from the sudden increase in throttle opening. With the throttle open, atmospheric pressure forces air into the manifold faster than the engine can use it. The turbulence produced during the intake and compression strokes (and the increased oxygen density on turbocharged applications) help to atomize the fuel and hasten its vaporization. This condition is only completely overcome, however, by spraying in additional fuel to compensate for the failure of the heavier hydrocarbon molecules to vaporize. The capabilities of a digital computer allow for the calculation of just the right amount of fuel to avoid a lean stumble without sacrificing economy or increasing emissions.

For an engine at a given operating temperature and an identified set of atmospheric conditions (air temperature, pressure, and humidity), there is one precise amount of fuel that should be injected into the intake port for each manifold pressure value within the operating range. This amount will provide an evaporated 14.7 to 1 air-fuel ratio in the combustion chamber with the lowest possible leftover unvaporized, unburned hydrocarbons.

Hot-Wire Mass Air Flow (MAF) Sensor. The MAP sensor used with the speed density formula has been reasonably effective in enabling the microprocessor to accurately determine the mass of the air that goes into the cylinder. Keep in mind that the microprocessor quantifies both the air and the fuel by weight when calculating air-fuel ratios. The VAF meter, as used on older vehicles, was also fairly effective in providing the microprocessor with the needed information to calculate the air's mass. Each of these two methods has, however, been unable to measure one factor that affects the air's mass—humidity.

The **hot-wire mass air flow (MAF) sensor**, introduced in the mid-1980s, provides information to the microprocessor that accounts for the flow rate and density, including those factors that affect the air's density: air temperature, BP, and the air's humidity.

A MAF sensor can determine the intake air's mass by *measuring* the quantity of air molecules that enter the engine (as opposed to a speed density application, in which the computer *calculates* how much air enters the engine). It does this by measuring the air's ability to cool. Air that flows over an object that is at a higher temperature than the air carries heat away from the object. The amount of heat carried away depends on several factors: the difference in temperature between the air and the object, the object's ability to conduct heat, the object's surface area, the air's mass, and the air's flow rate.

The MAF sensor, sometimes referred to as a *hot-wire sensor,* is placed in the air duct that connects the air cleaner to the throttle body so that all of the air entering the induction system must pass through it (Figure 3–17). Within the MAF assembly, intake air flows across both a thermistor and a *hot wire* that is kept heated to a predetermined temperature. A MAF sensor essentially measures how much current it takes to keep the hot wire at its assigned temperature. This assigned temperature is not an absolute temperature, but

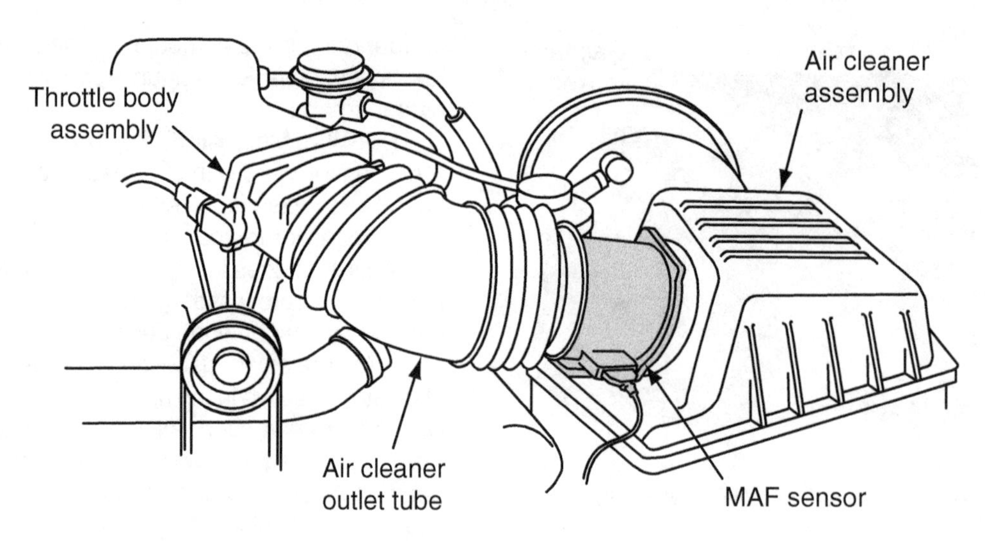

Figure 3–17 Mass airflow (MAF) sensor location.

rather a temperature that is a specific number of degrees hotter than the intake air as measured by the thermistor. Restated, the temperature of the hot wire is indexed to the temperature of the intake air. Because of this, a cold air molecule on a December day will not fool the sensor into believing that a greater volume of air molecules flowed through the sensor as opposed to a warm air molecule on a hot day in July or August.

In one design, the thermistor and heated wire are located in a smaller air sample tube (Figure 3–18). The thermistor measures the incoming air temperature. The heated wire is maintained at a predetermined temperature above the intake air's temperature by a small electronic module mounted on the outside of the sensor's body. The volume of air flowing through the air sample tube is proportional to the total volume of air flowing through the entire sensor.

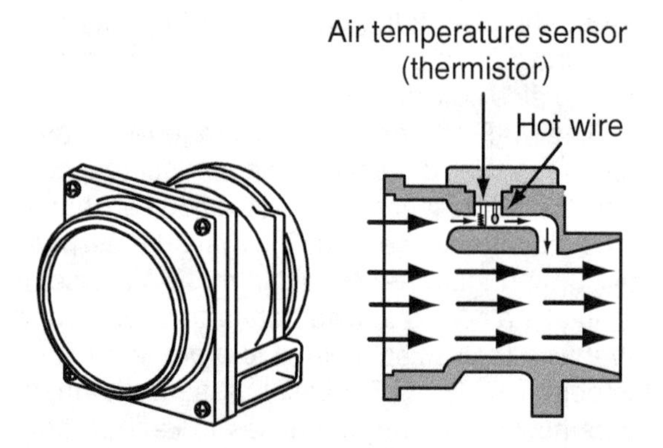

Figure 3–18 Components of a hot-wire–type mass airflow (MAF) sensor.

In this design, the heated wire is actually one of several resistors in a circuit. Figure 3–19 represents a simplified MAF balanced bridge circuit. The module supplies battery voltage to the balanced bridge. Resistors R1 and R3 form a series circuit in parallel to resistors R2 and R4. The voltage at the junction between R1 and R3 is equal to the voltage drop across R3. Because the values of R1 and R3 are fixed, the voltage at this junction is constant; it will vary, but only as battery voltage varies. Resistors R1 and R2 have equal values. With no air flowing across R4, its resistance is equal to R3, and the voltage at the junction between R2 and R4 is equal to the voltage at the R1/R3 junction.

As the volume of air flowing across R4 increases, its temperature drops, and because it is a PTC resistor, its resistance goes down with its temperature. As the air's mass increases, R4's

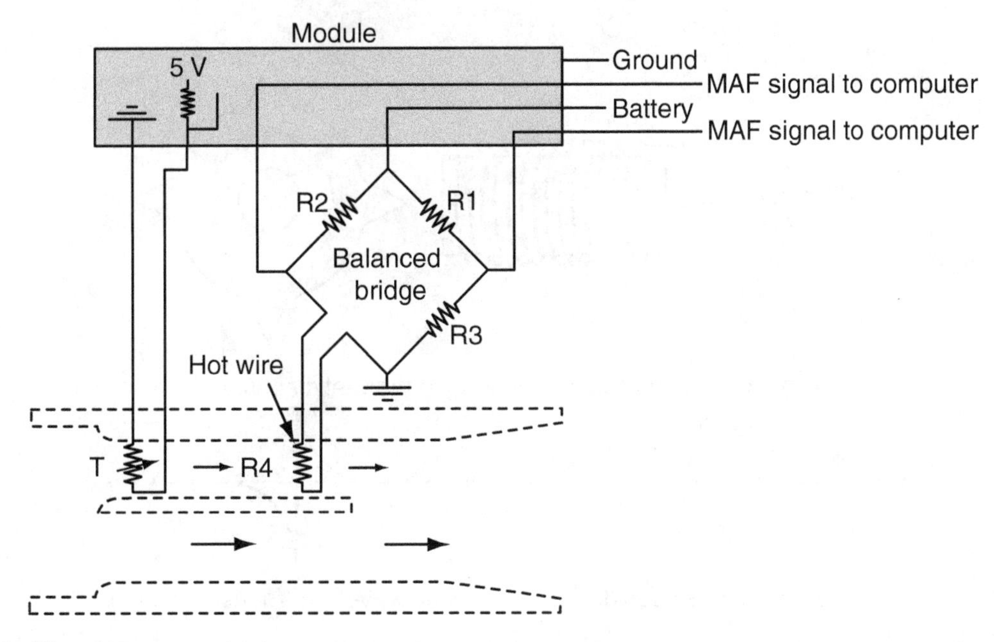

Figure 3–19 MAF sensor circuit.

resistance goes down even more. As the resistance of R4 goes down, a smaller portion of the voltage applied to the circuit formed by R2/R4 is dropped across R4. Therefore, a larger portion of the voltage is dropped across R2. So as the resistance of R4 goes down, the voltage at the R2/R4 junction goes down. The computer reads the voltage difference between the R1/R3 junction and the R2/R4 junction as the MAF sensor value. By looking at a table in the computer's memory, the microprocessor converts the voltage value to a MAF rate.

A MAF sensor does not actually measure BP or humidity. But because it is an air molecule counter, it will be accurate for the total mass of the intake air regardless of what might affect the air's density, including BP and humidity.

Another advantage of a MAF sensor is that a MAF sensor tends to maintain a higher level of accuracy as the engine wears than the MAP sensor in a speed density system. This is because an engine tends to develop less manifold vacuum with wear and age. When compared to BP as a reference, a MAP sensor on a speed density application tends to interpret this lower level of manifold vacuum as increased engine load, thereby causing the PCM to mistakenly add too much fuel to the incoming air. But a MAF sensor retains its accuracy as the engine wears, continuing to measure accurately the mass of air that is being drawn into the engine, thus allowing the PCM to deliver the correct amount of fuel in spite of the engine wear.

Variable Reluctance Sensors

A **variable reluctance sensor (VRS)**, also sometimes known as a *permanent magnet sensor* or *magnetic sensor*, consists of a coil of wire wrapped around a permanent magnet (Figure 3–20). Magnetic sensors are voltage-generating devices. A permanent magnet sensor operates on the principle that magnetic fields can cause electron flow as they expand or collapse across a coil of wire. A series of iron teeth and notches are passed near the magnet. As an iron tooth approaches the magnet, the magnet's magnetic field strength is increased and a voltage

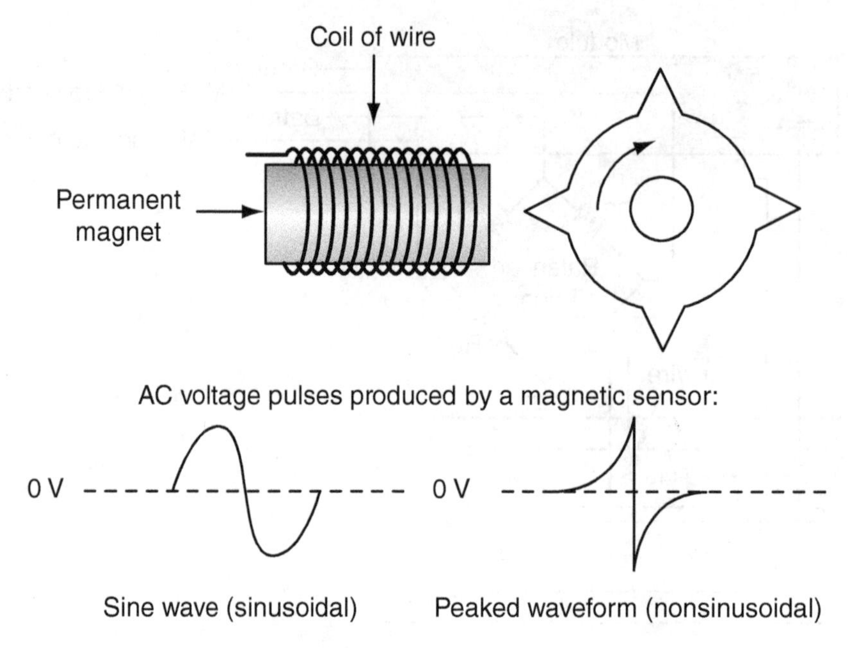

Figure 3–20 Variable reluctance sensor, also known as a permanent magnet sensor.

potential is generated within the coil of wire that is wrapped around the magnet. Then, as the iron tooth pulls away from the magnet, the magnet's magnetic field strength is decreased and a voltage potential of the opposite polarity is generated within the coil of wire. As a result, a magnetic sensor produces an AC voltage. The computer is not so concerned with how much voltage is produced (as long as enough voltage is produced to make the signal recognizable to the computer); instead, it is more concerned with the timing of the AC voltage pulse and/or the frequency of the voltage pulses. A magnetic sensor is a two-wire sensor, although some applications will also add a wire for a grounded shield, visible as a third wire at the first electrical connector near the sensor.

Testing of a magnetic sensor consists of using an ohmmeter to measure the resistance of the sensor's coil of wire and comparing the result to the sensor's specification. Then a voltmeter should be used (on an AC voltage scale) to test the sensor's ability to produce a voltage signal while the sensor is operating. Also, if the sensor has a metallic body, an ohmmeter should be used to verify that the sensor's coil of wire is not grounded (shorted) to the sensor's frame. (This step is not important with plastic-bodied magnetic sensors.)

Permanent magnet sensors are commonly used as ignition pickups and vehicle speed sensors, and they are used in many other applications as well.

Hall Effect Sensors

A **Hall effect sensor** is a semiconductor sensor that is sensitive to magnetism. The "Hall effect" was discovered by Edwin Herbert Hall (1855–1938) in 1879. A Hall effect sensor may be configured in two forms: A linear Hall effect sensor that outputs an analog voltage and a digital Hall effect switch that switches its output signal on and off in a digital fashion.

Linear Hall Effect Sensor. A linear Hall effect sensor consists of a permanent magnet and a **gallium arsenate crystal.** This sensor can be used to sense physical position and mechanical action, as a potentiometer does. But a linear Hall effect sensor does not have a physical wear factor as a potentiometer does, and therefore it is sometimes referred to as a *noncontact sensor*. As a magnet is moved near the Hall effect sensor, an analog voltage is output. The analog output voltage of a linear Hall effect sensor may be either AC, shown in Figure 3–21, or DC, shown in Figure 3–22, depending upon the mechanical action and magnetic influence.

Digital Hall Effect Sensor. A digital Hall effect sensor, sometimes referred to as a *Hall effect switch*, is often used to sense engine speed and crankshaft position and, therefore, provide a **tach reference signal** to the PCM. It provides a signal each time a piston reaches top dead center; this signal serves as the primary input on which ignition timing is calculated. Because of the frequency with which this signal occurs and the critical need for accuracy, the digital Hall effect sensor is a popular choice due to its ability to digitize its output signal. On most applications, a separate digital Hall effect sensor is used to monitor camshaft position as well.

A digital Hall effect sensor consists of a permanent magnet, a Hall effect crystal with its related circuitry, and a shutter wheel (Figure 3–23). The permanent magnet is mounted so that there is a small space between it and the gallium arsenate crystal. The shutter wheel, rotated by a shaft, alternately passes its vanes through the narrow space between the magnet and the crystal (Figure 3–24).

When a vane is between the magnet and the crystal, the vane intercepts the magnetic field and thus shields the crystal from it. When the vane moves out, the gallium arsenate crystal is

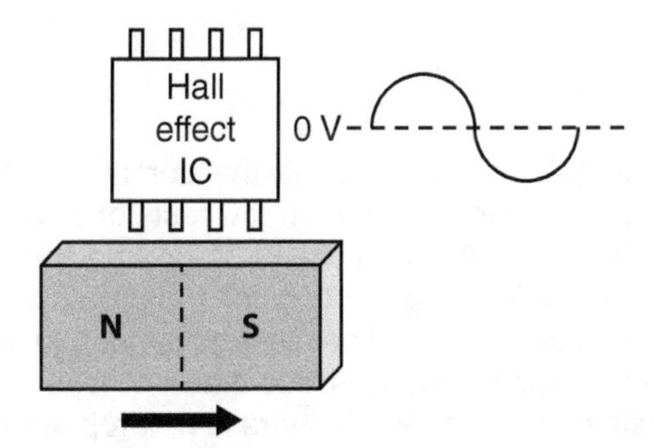

Figure 3–21 Linear Hall effect sensor producing an AC analog voltage.

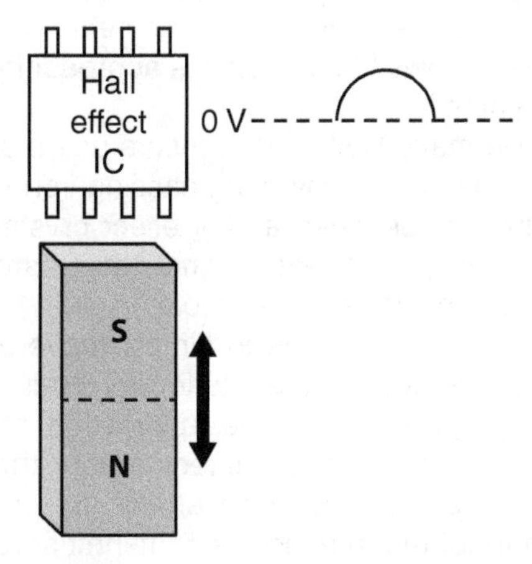

Figure 3–22 Linear Hall effect sensor producing a DC analog voltage.

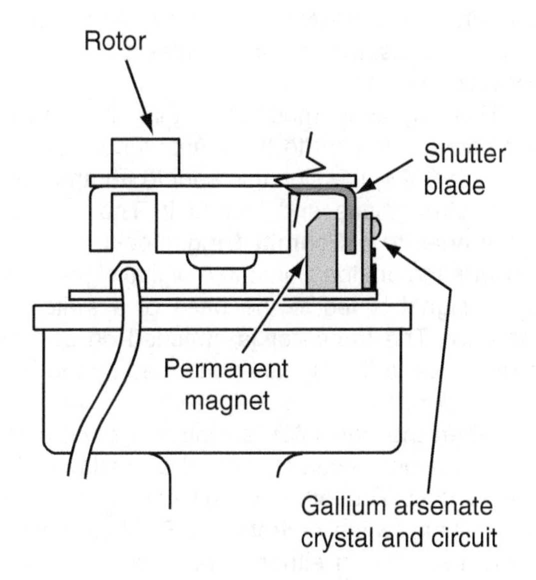

Figure 3–23 Digital Hall effect sensor.

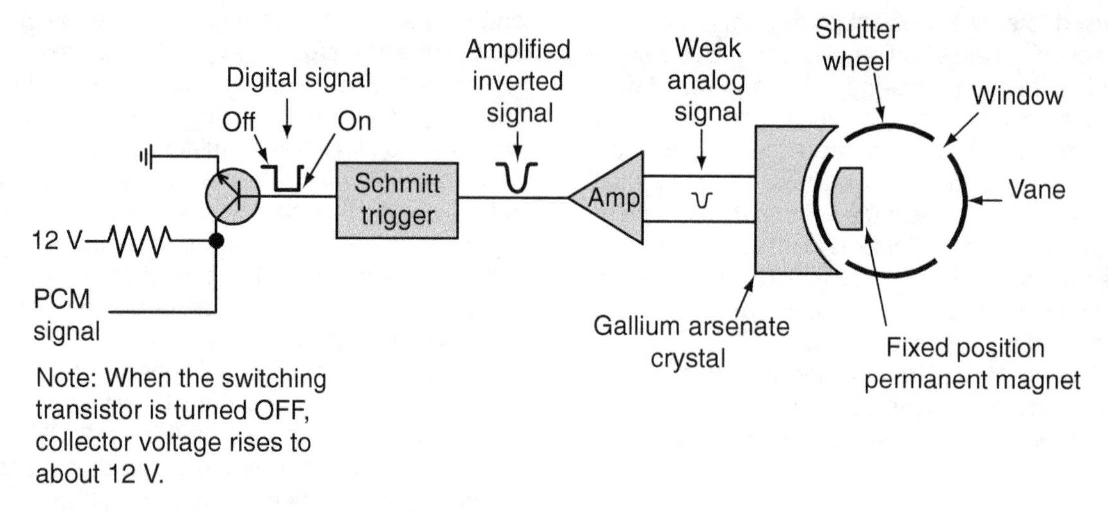

Figure 3–24 Digital Hall effect sensor circuit.

invaded by the magnetic field. A steady current is passed through the crystal from end to end. When the magnetic lines of force from the permanent magnet invade the crystal, the current flow across the crystal is distorted. This will result in a weak voltage potential being produced at the crystal's top and bottom surfaces—negative at one surface and positive at the other. As the shutter wheel turns, the crystal provides a weak high/low-voltage signal.

This signal is modified within the sensor and before it is sent to the computer. As shown in Figure 3–24, the signal is sent to an amplifier, which strengthens and inverts it. The inverted signal goes to a **Schmitt trigger** device, which converts the analog signal to a digital signal. The digital signal is fed to the base of a switching transistor. The transistor is switched on and off in response to the signals generated by the Hall effect sensor assembly.

When the transistor is switched on, it completes another circuit to ground and allows current to flow. The other circuit can come from the ignition switch or from the PCM as shown in Figure 3–24. In either case it has a resistor between its voltage source and the transistor. A voltage-sensing circuit in the computer connects to the circuit between the resistor and the transistor. When the transistor is on and the circuit is complete to ground, a voltage drop occurs across the resistor. The voltage signal to the voltage-sensing circuit is less than 1 V. When the transistor is switched off, there is no drop across the resistor, and the voltage signal to the voltage-sensing circuit is near source voltage. The PCM will monitor the voltage level and will determine by the frequency at which it rises and falls what the engine speed is. Each time it rises, the computer knows that a piston is approaching top dead center.

While many Hall effect sensors use the concept just described, involving passing iron vanes and windows between a Hall effect crystal and a stationary permanent magnet, there are two other types of Hall effect sensors.

The first variation is to simply move a permanent magnet past a stationary Hall effect sensor (Figure 3–25). One application on GM vehicles mounts the Hall effect sensor through the engine's timing gear cover with the permanent magnet mounted to the camshaft sprocket. As the magnet rotates past the Hall effect

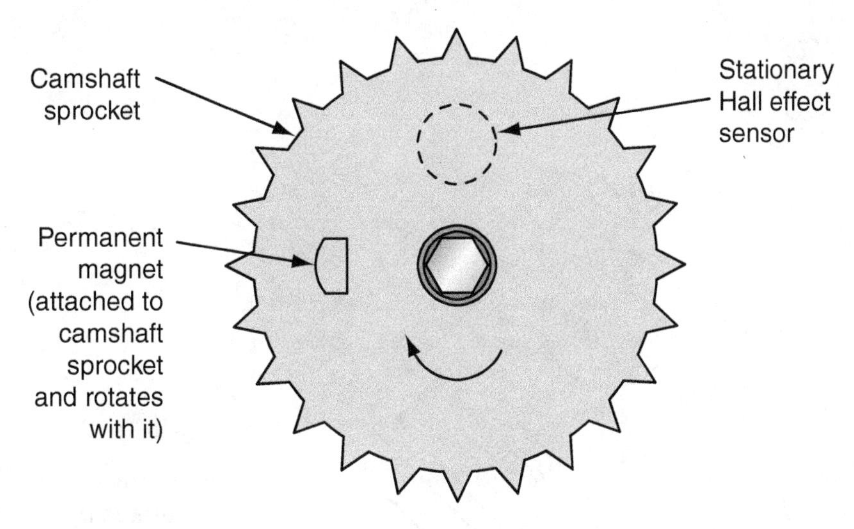

Figure 3–25 Hall effect sensor with rotating magnet.

sensor, a signal is produced that is similar to that of a camshaft sensor, which rotates a vane cup between a Hall effect sensor and a stationary permanent magnet.

The other variation is to install both a permanent magnet and a Hall effect crystal in a plastic body, similar in appearance to many magnetic sensors used today, except that there are three wires connected to the sensor (Figure 3–26). The magnet is placed at the forward end of the sensor's body, and the Hall effect crystal is placed either toward the rear or beside the permanent magnet. Iron vanes and windows that are designed into a flywheel or camshaft sprocket are rotated past the sensor where the magnet is located. When iron is placed near the magnet, its field strength increases enough to envelop the Hall effect crystal placed near it. As the vanes and windows are rotated past the sensor, this action causes the Hall effect crystal to generate a waveform signal that is similar to other Hall effect sensors.

Hall effect sensors are also used as vehicle speed sensors and ignition system pickups. Ultimately, regardless of which style of Hall effect sensor is used, the sensor will have three wires exiting the body (unlike simple magnetic sensors, which only have two). Because a Hall effect sensor is an electronic device, both a voltage supply wire and ground wire are needed for the sensor to power up. The third wire is the signal wire. When testing, check the power and ground wires first. The applied voltage will be as high as the charging system voltage on some applications. On other applications, such as Chrysler vehicles, as little as 8 V is used to power this sensor (this voltage originates at the PCM). Once the power and ground have been verified, check the signal wire for digital DC voltage pulses with the sensor operating (engine cranking). If an oscilloscope is not available, a digital multimeter, or DMM (also known as a digital volt-ohm meter, or DVOM) can be used to test frequency or average DC voltage. Average DC voltage can be measured and should average between about 30 and 70 percent of the applied voltage, indicating that voltage pulses between zero volts and the applied voltage are likely occurring. Some Hall effect sensors are dual sensors, having a total of four wires— one for shared power, one for a shared ground, and two individual signal wires. Both signal wires should be checked for voltage pulses.

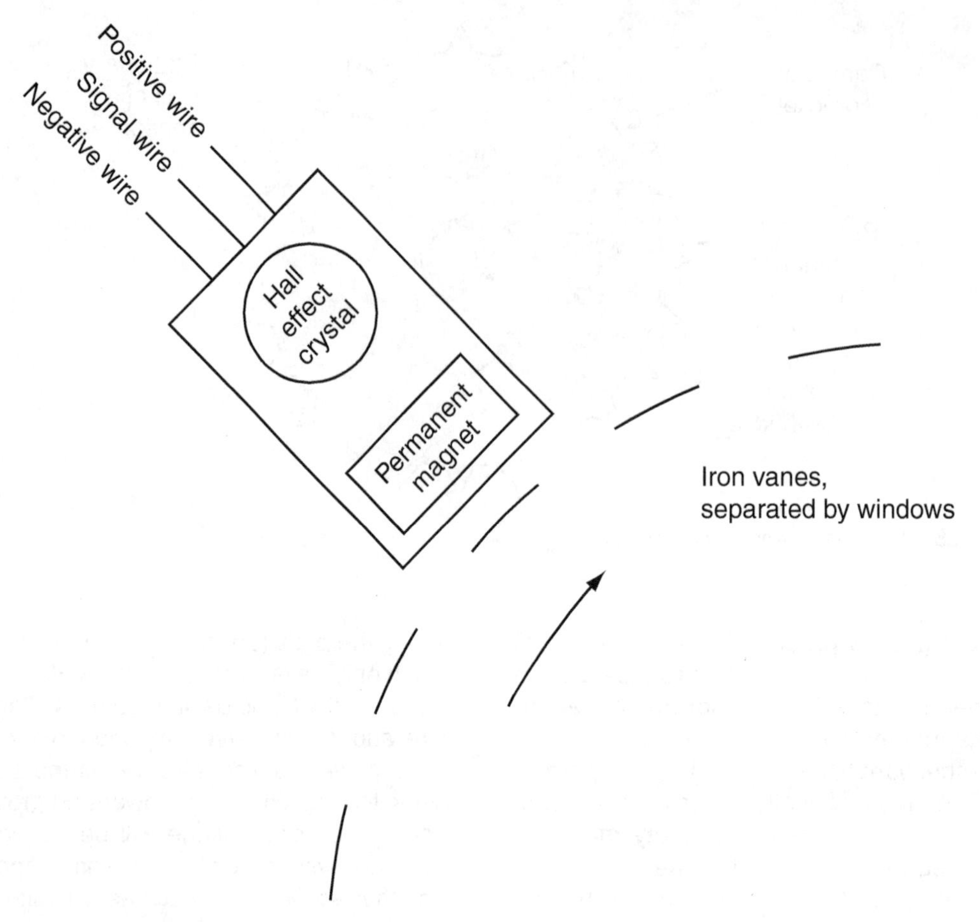

Figure 3–26 Hall effect sensor that contains a permanent magnet.

Magnetic Resistance Element (MRE) Sensors

A **magnetic resistance element (MRE) sensor**, in essence, is a cross between a variable reluctance sensor (permanent magnet sensor) and a Hall effect sensor. This sensor uses a rotating magnetic ring. The rotating magnetic ring contains several magnetic north poles alternating with several magnetic south poles (Figure 3–27). The alternating magnetic fields are monitored by an electronic sensor called a *magnetic resistance element (MRE)*. The rotating magnetic ring induces AC voltage pulses into the MRE sensor as the alternating magnetic fields pass by. The MRE sensor's electronics then convert these AC voltage pulses into crisp digital DC voltage pulses. The final output oscillates in a digital fashion between 0 V and battery voltage (or another voltage, depending upon the sensor's application). Each time a north or south pole passes the MRE sensor, the output voltage signal pulls low to 0 V. In between the north and south poles, the voltage goes back to battery voltage momentarily. The MRE sensor has three wires: battery voltage, signal, and ground. Because it is an electronic device, it must have both power and ground to power up and become functional.

The MRE sensor is manufactured by the Denso Corporation. The Denso Corporation is

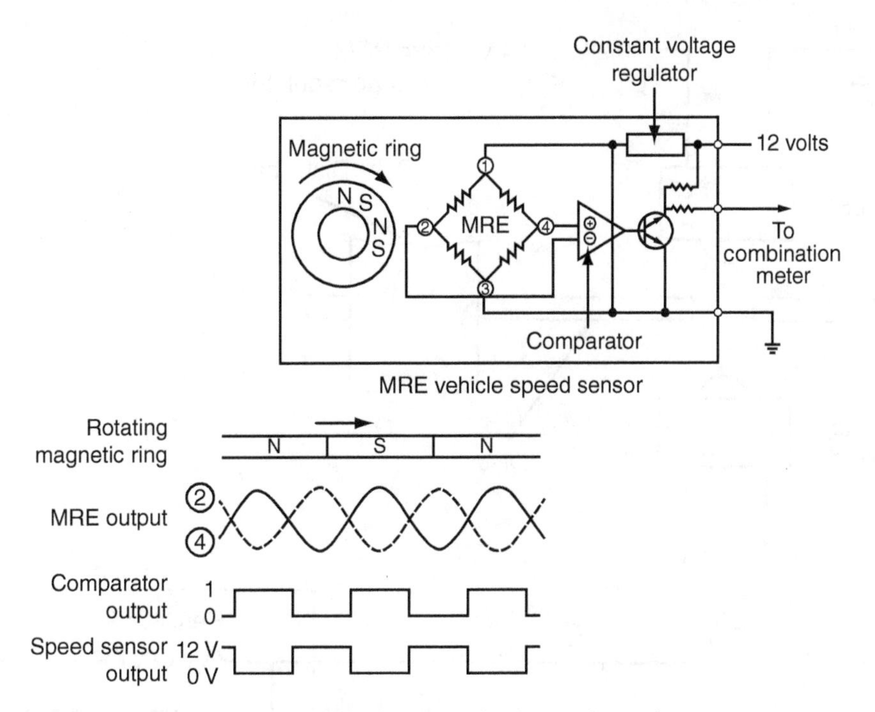

Figure 3–27 Magnetic resistance element (MRE) sensor circuitry and waveforms.

a spin-off of the Toyota Motor Company. When it was part of Toyota, it was known as the *Nippondenso Company*. The Denso Corporation is Japan's leading producer of automobile parts and components. They are also the world's fourth largest automotive parts supplier.

As the manufacturer, the Denso Corporation states that their MRE sensors have a high detection accuracy and are almost 10 times more sensitive than their Hall effect counterparts. The MRE sensor has a high reliability in terms of their construction.

The sensor's electronic IC chip and the cylindrical magnet are integrated together and then covered with polyphenylene sulfide (PPS) resin for greater strength. The PPS resin is highly resistant to chemical substances, including fuel and engine oil.

The Denso Corporation supplies MRE vehicle speed sensors (VSS) and MRE camshaft position (CMP) and crankshaft position (CKP) sensors to various automotive manufacturers, including Toyota and Nissan. Several manufacturers, including U.S. manufacturers, are also using MRE sensors as wheel speed sensors (WSS) in their antilock brake and automatic traction control systems.

Optical Sensors

Several manufacturers use optical sensors to sense crankshaft position and RPM, either in the form of an optical distributor or in the form of an optical crankshaft or camshaft sensor. An optical sensor places a **light-emitting diode (LED)** opposite a photocell with a rotating slotted wheel between them (Figure 3–28). Generally, the LED transmits infrared light and the photocell is sensitive to infrared light, so visible light wavelengths do not affect the sensor. As the slotted wheel rotates, light pulses applied to the photocell cause the photocell to produce enough voltage to forward bias a transistor that then grounds out a monitored circuit, thus producing DC digital voltage pulses. On distributor applications that use an optical sensor,

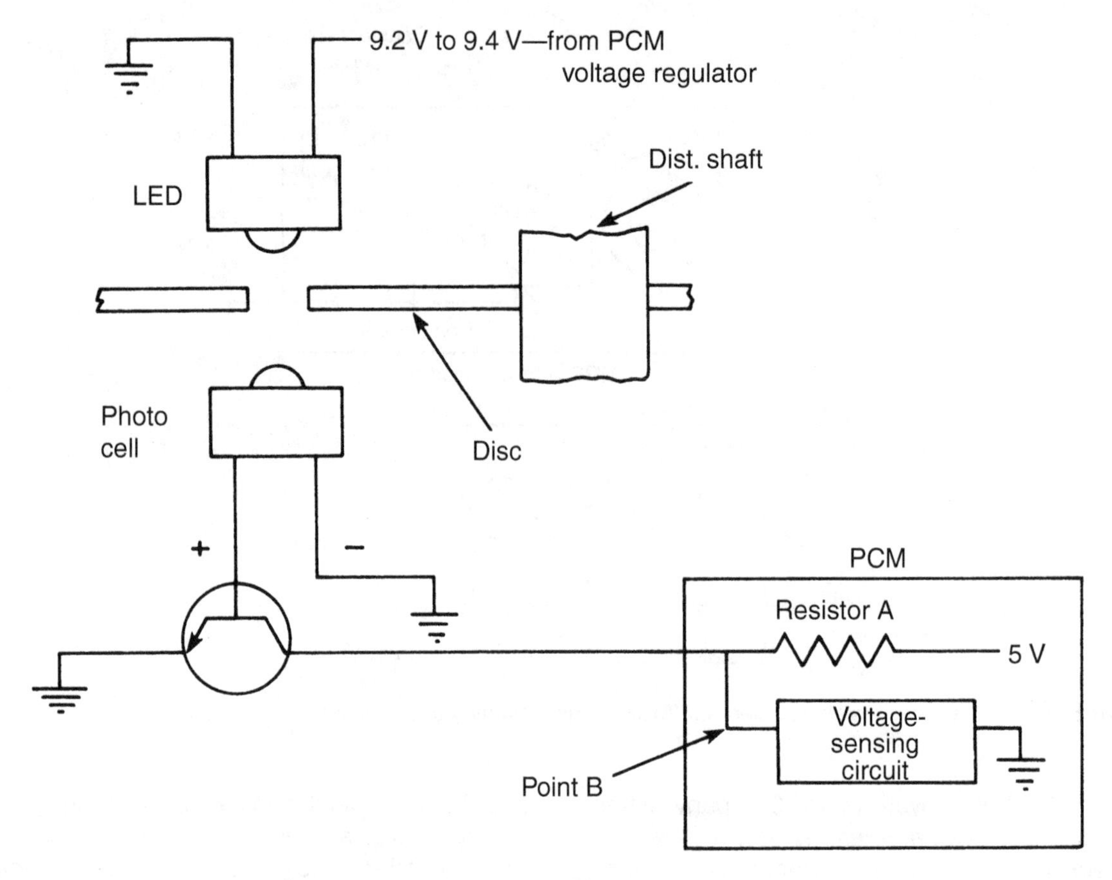

Figure 3–28 Optical sensor in a distributor.

the sensor body contains two LED and photocell assemblies (Figure 3–29). One monitors a low-resolution signal with one slot per engine cylinder and the other monitors a high-resolution signal with one slot per each degree of distributor rotation, or 360 slots (these are actually very small, laser-cut slits). Optical distributors are used by Mitsubishi, and they are used by others as well. One Mitsubishi application uses a dual optical sensor as a camshaft sensor on a distributorless (multiple coil) application. Interestingly, Mitsubishi manufactured many of the optical sensors for General Motors, Nissan, and the Chrysler Mitsubishi engines.

Like a Hall effect sensor, an optical sensor requires a voltage and a ground just to "come alive." The third and fourth wires are individual signal wires for each of the dual optical sensors. When testing, verify proper applied voltage and ground first, then test the signal wires with an oscilloscope or DMM while operating the sensor (cranking the engine). Again, as with Hall effect sensors, average voltage should be between about 30 and 70 percent of the applied voltage.

Detonation Sensors

To optimize performance and fuel economy, ignition timing must be adjusted so that the maximum pressure from combustion will be achieved between 10 degrees and 20 degrees after top dead

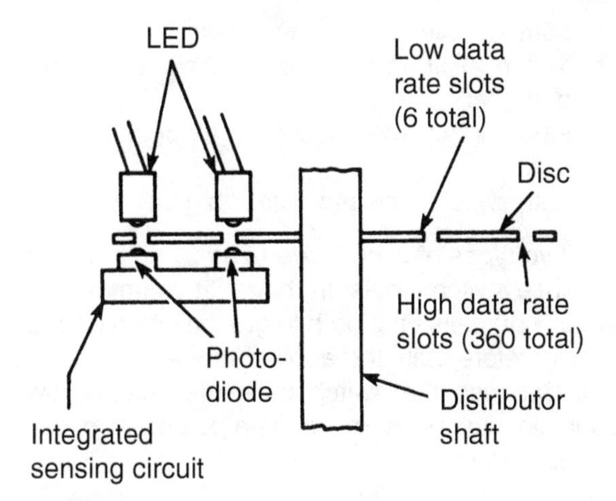

Figure 3–29 Optical distributor pickup with dual sensors.

center (ATDC) during the power stroke. Therefore, because it takes about 3 ms after the spark plug fires to achieve maximum combustion pressure, the spark plug must be fired early enough that maximum combustion pressure will be achieved at the correct time. If the crankshaft is rotating at 1,800 RPM, then it also stands to reason that it is rotating at a speed of 30 rotations per second (1,800 RPM divided by 60 seconds). This equates to 10,800 degrees of rotation per second (30 times 360 degrees) or 10.8 degrees of rotation per ms (10,800 divided by 1,000 ms). Since it takes 3 ms to reach maximum combustion pressure, at this engine speed the spark plug must fire 32.4 degrees ahead of when peak combustion pressure is desired. Therefore, in order to achieve maximum pressure by 10 degrees after top dead center (TDC), the spark plug must fire at 22.4 degrees before top dead center (BTDC). Therefore, engine speed is one of the factors that influences spark timing. Other factors, including engine load, are used by the PCM to determine the ideal spark timing.

If the spark plug fires later than the ideal timing, the maximum combustion pressure will be produced in the cylinder too late, as the piston is already approaching bottom dead center (BDC), and less useful work will be done for the amount of fuel burned. If the spark plug fires too early, detonation (also known as *spark knock*) will occur. Due to variations in fuel quality, cylinder cooling efficiency, machining tolerances, and their effect on compression ratios, a preprogrammed spark advance schedule can result in spark knock under certain driving conditions. In order to provide an aggressive spark advance program and still avoid spark knock, many systems use a detonation sensor, often referred to as a *knock sensor.* It alerts the computer when spark knock occurs so that the timing can be retarded slightly.

The knock sensor, as shown in Figure 3–30, is typically mounted in the side of the engine block or in the intake manifold. This location allows it to be subjected to the high-frequency vibration caused by spark knock. Most knock sensors contain a **piezoelectric** crystal. Piezoelectric refers to a characteristic of certain materials that causes them to produce an electrical signal in response to physical stress, or they experience stress in response to an electrical signal. The most commonly used piezoelectric material produces an oscillating AC voltage signal of about 0.3 V or more in response to pressure or vibration. The oscillations occur at the same frequency as the spark knock (5,000 to 6,000 **Hertz [Hz],** or cycles per second). Spark knock is essentially the ringing of the engine block when struck by the explosion caused by detonation. Although most vibrations within the engine and many elsewhere in the vehicle cause the knock sensor to produce a signal, the computer only responds to signals of the correct frequency.

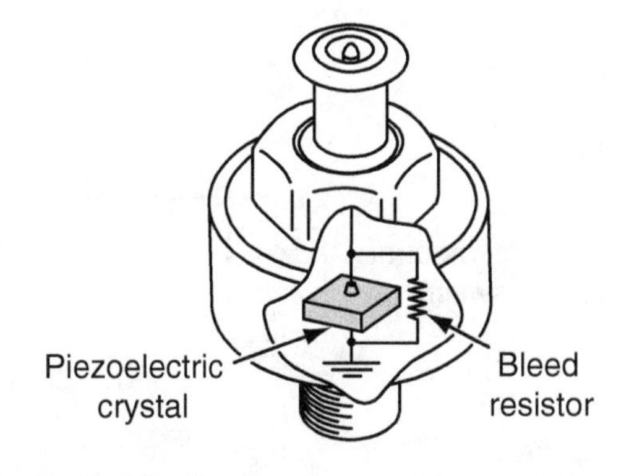

Figure 3–30 Knock sensor.

When testing a knock sensor, tap on the sensor lightly with a metallic object while you measure the resulting output on an oscilloscope or a DMM set to an AC voltage scale.

Switches

Most systems have several simple switches that are used to sense and communicate information to the computer. The information includes whether the transmission is in gear, the air conditioning is on, the brakes are applied, and so on. With an electronic four-speed, automatic transmission, two switches may be used, as follows:

- both switches open = first gear
- switch #1 open and switch #2 closed = second gear
- switch #1 closed and switch #2 open = third gear
- both switches closed = fourth gear

Two types of circuits are used by a computer to sense switch inputs. In the most common, the switch completes the path to ground when closed and therefore pulls the applied reference voltage low. Because the switch pulls the voltage low upon closing, this is known as a *pull-down* circuit (Figure 3–31).

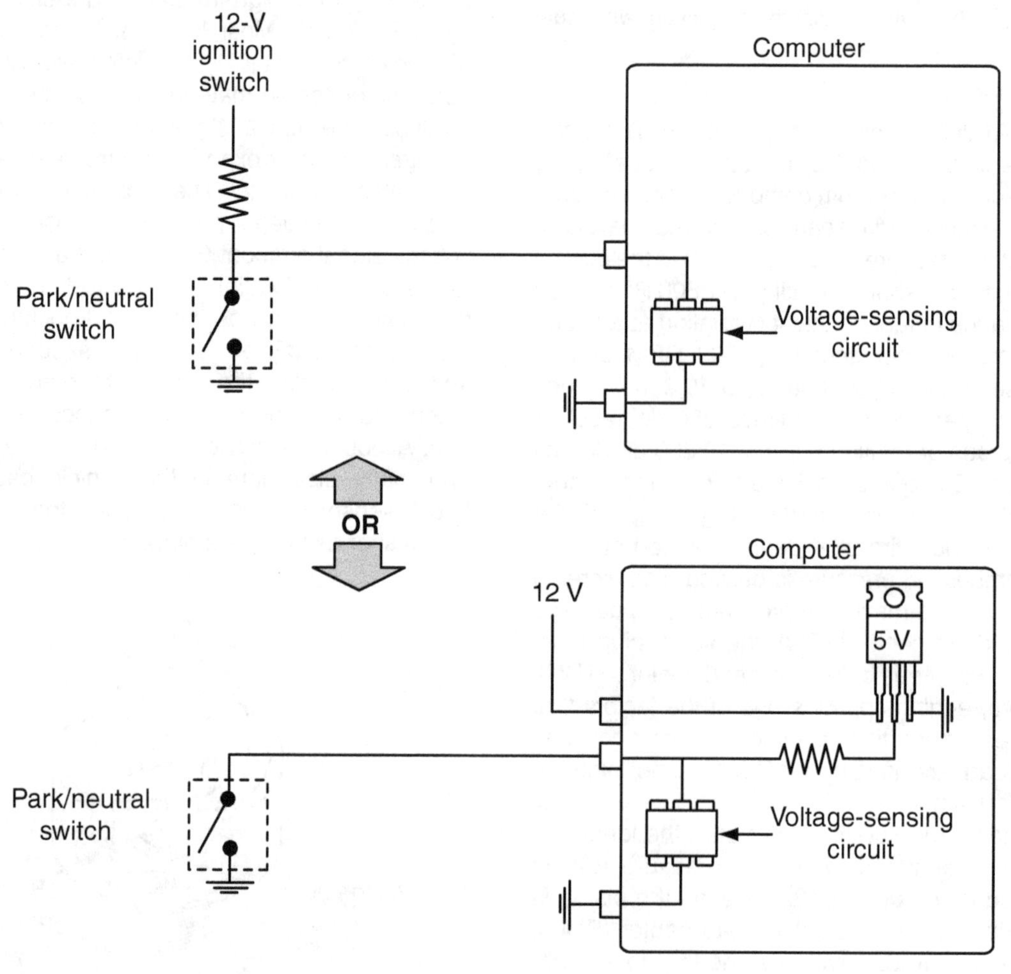

Figure 3–31 Pull-down circuit.

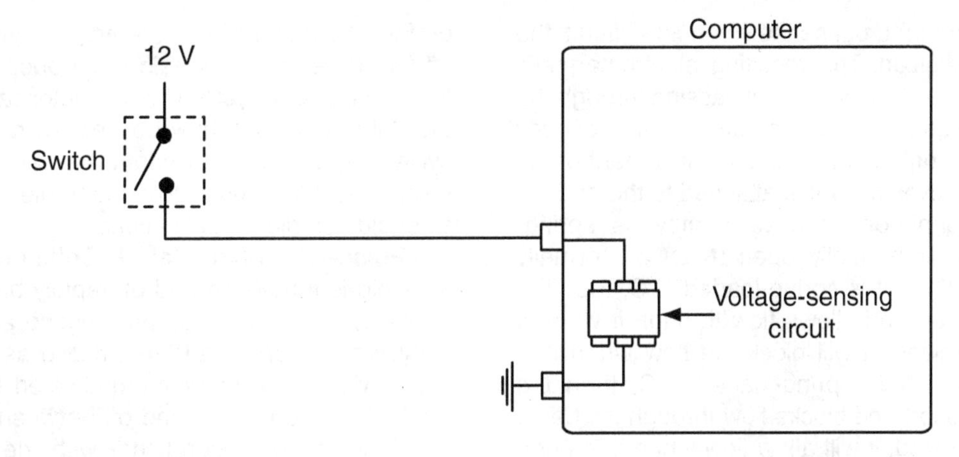

Figure 3–32 Pull-up circuit.

In the other type, a switch applies voltage to a computer-monitored circuit. Because closing the switch pulls voltage high, this is known as a *pull-up circuit* (Figure 3–32).

ACTUATORS

Solenoids

The most common actuator is a valve powered by a solenoid. A solenoid-operated valve is designed to control flow; it may control liquid fuel (fuel injector), fuel vapors (canister purge solenoid), air (idle air control solenoid), vacuum (vacuum solenoid that controls a vacuum signal to a vacuum-operated component such as an EGR valve), or exhaust gases (EGR solenoid that controls exhaust gas flow directly without the use of vacuum). Such valves may also be used to control other liquids such as brake fluid (antilock brake solenoid) and even oil or water.

A solenoid-operated valve consists of a coil of wire around a spring-loaded movable iron core (Figure 3–33). When the output driver (transistor)

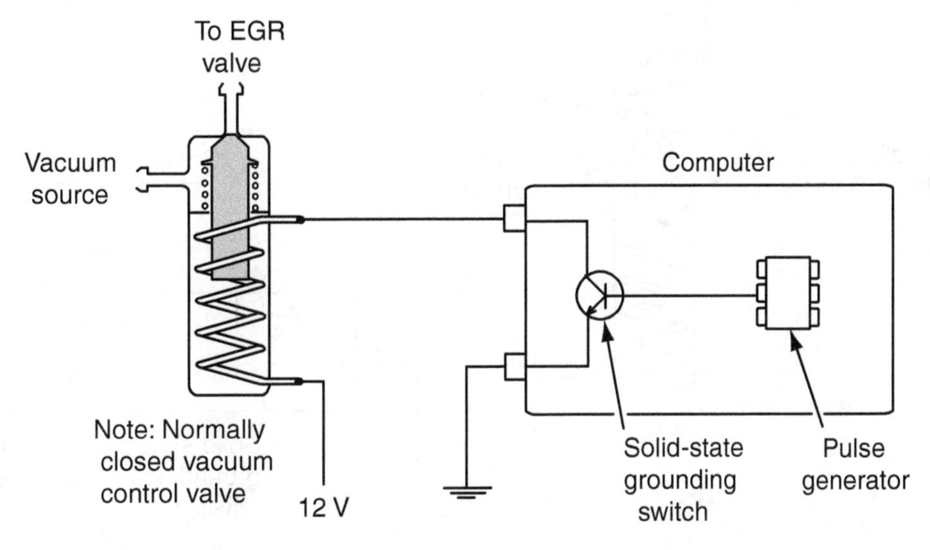

Figure 3–33 Typical solenoid-controlled vacuum valve.

in the computer grounds the solenoid's circuit, the coil is energized. The resulting electromagnetic field developed by the current passing through the coil windings pulls the iron core so as to occupy the entire length of the coil. The movement of the iron core moves the valve attached to the end.

A solenoid-operated valve may be spring-loaded either normally open (N/O) or normally closed (N/C). If it is spring-loaded N/O, then the valve is open and allows flow through it when it is de-energized; it will block the flow when it is energized. If it is spring-loaded N/C, then the valve is closed and blocks flow through it when it is de-energized; it will allow flow when it is energized. Whether the solenoid is designed N/C or N/O is determined many times by safety considerations—that is, consideration is given to the necessary default in the event that the controlling computer becomes incapable of energizing the solenoid.

Duty-Cycled Solenoids. In some cases, the switching circuit that drives the solenoid turns on and off rapidly. If the solenoid circuit is turned on and off (cycled) ten times per second, then the solenoid is duty-cycled. To complete a cycle, it must go from off to on, to off again. If it is turned on for 20 percent of each tenth of a second and off for 80 percent, it is said to be operating on a 20-percent **duty cycle.** The computer can change the duty cycle to achieve a desired result. Duty cycle may be used to control certain emission solenoids; it was used to control idle air control solenoids on older applications.

Pulse-Width-Modulated Solenoids. If a solenoid is turned on and off rapidly but there is no set number of cycles per second at which it is cycled, the on-time is referred to as its *pulse width.* Modern pulse-width-modulated solenoids are typically turned on and off between 500 and 3,500 times per second, not with the intent of opening and closing the valve that many times per second, but with the intent of allowing the on-time, as opposed to the off time, to control the level of average current flowing through the solenoid's winding. This in turn controls the strength of the solenoid's magnetic field, which then *floats* the solenoid to a desired position between closed and open (Figure 3–34). This is sometimes referred to as a high-frequency pulse-width modulation. Pulse-width modulation is used on many modern vehicles to control an idle air control solenoid.

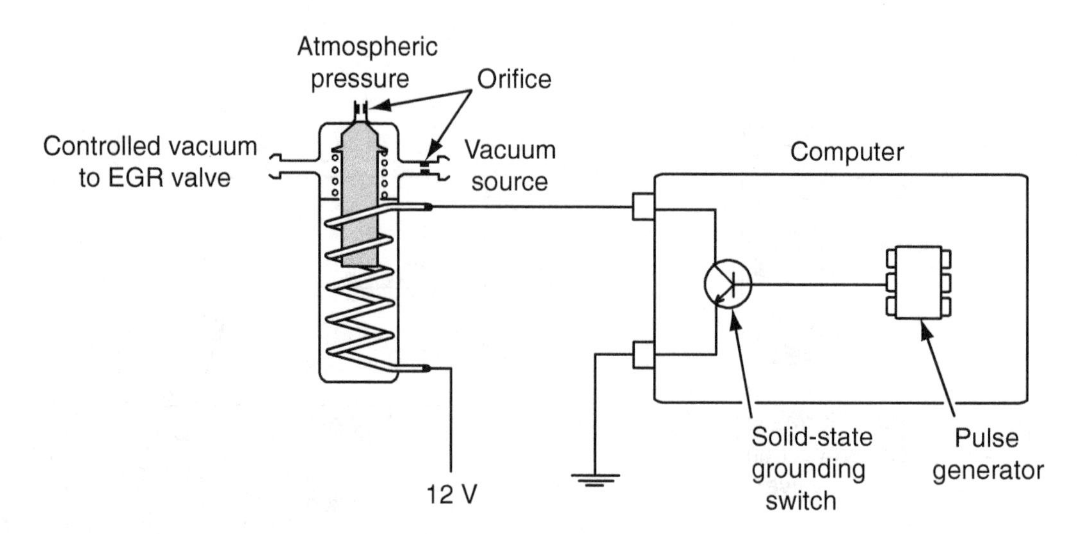

Figure 3–34 Typical pulse-width-controlled vacuum valve.

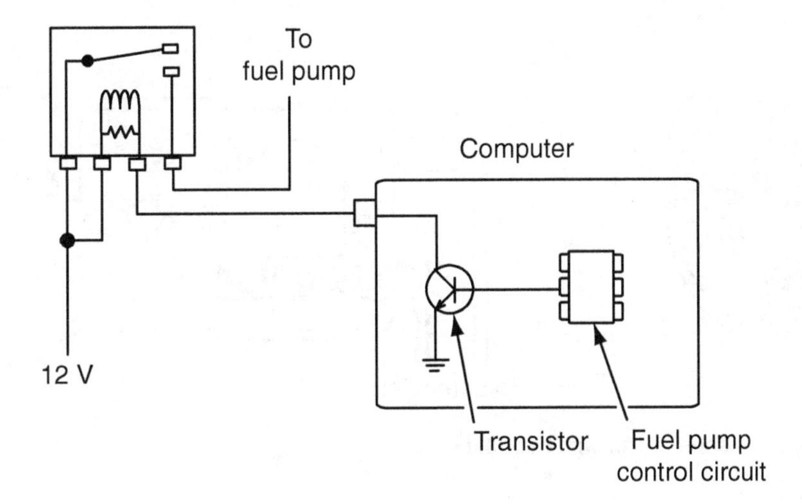

Figure 3–35 Typical fuel pump relay circuit.

Relays

A relay is a remote-control switch that allows a light-duty switch such as an ignition switch, a blower motor switch, a starter switch, or even a driver (transistor) within a computer to control a device that draws a relatively heavy current load. A common example in automotive computer applications is a fuel pump relay (Figure 3–35). When the computer wants the fuel pump on, it turns on its driver. The driver applies a ground to the relay coil, which develops a magnetic field in the core. The magnetic field pulls the armature down and closes the N/O contacts. The contacts complete the circuit from the battery to the fuel pump. Because of safety concerns, the fuel pump relay, at least on some systems, is one of the few instances in which the computer supplies a positive voltage, instead of a ground, to energize a load component.

A relay may be spring-loaded N/O or N/C. If it is spring-loaded N/O, then the switch is open and does not allow current flow across the contacts when de-energized; it will close and allow current flow when it is energized. If it is spring-loaded N/C, then the switch is closed and allows current flow across the contacts when de-energized; it will open the circuit and prohibit current flow when it is energized. Many relays (such as typical Bosch relays) have both N/O and N/C contacts. (In a Bosch relay, terminals 85 and 86 are the positive and negative ends of the electromagnetic coil. Terminal 30 is N/O to terminal 87 and N/C to terminal 87a.)

If this seems confusing when compared to the N/C and N/O solenoids discussed in the previous paragraphs, remember that a *valve* must be *open* to allow flow of whatever it controls, and a *switch* must be *closed* to allow current to flow across its contacts.

Electric Motors

When electric motors are used as actuators, two unique types are used.

Stepper Motors. A stepper motor, as used in these applications, is a small electric motor powered by voltage pulses. It contains a permanent magnet armature with either four or two field coils (Figure 3–36). In the design with four field coils, all coils receive battery voltage and the computer completes the ground paths, one at a time, to pulse current to each coil in sequence. In the design with two field coils, the computer not only pulses current to energize each coil in sequence, but it can also reverse the polarity

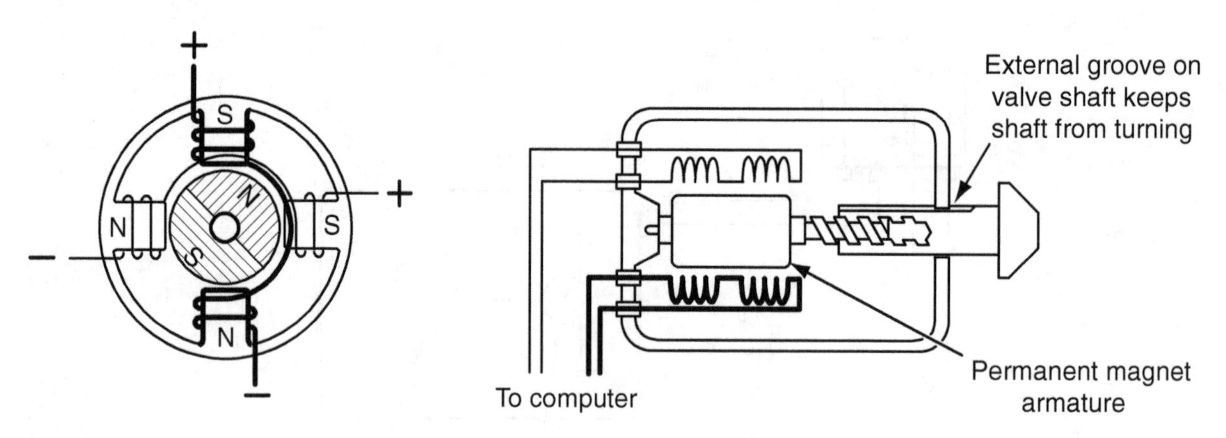

Figure 3–36 Typical stepper motor with two field coils.

applied to each coil as part of the sequencing. Each time the computer pulses current to a field coil, it causes the stepper motor to turn a specific number of degrees. In either design, reversing this sequence causes the stepper motor to turn in the opposite direction. Ultimately, a stepper motor is controlled by the computer to turn a number of specific steps or increments, with the direction being controlled by the sequencing of the current pulses to the windings.

The armature shaft usually has a spiral on one end that connects to whatever the motor is supposed to control. As the motor turns one way, the controlled device (a pintle valve, for instance) is extended. As it turns the other way, the valve is retracted. The computer can apply a series of pulses to the motor's coil windings to move the controlled device to whatever location is desired. The computer can also know exactly what position the valve is in by keeping count of the pulses applied. Stepper motors have been used with feedback carburetors as a method of controlling the air-fuel ratio. They are also used with fuel injection systems as a method of controlling idle speed.

Permanent Magnet Field Motors. Some systems use this type of motor to control idle speed (Figure 3–37). It is a simple, reversible DC motor with a wire-wound armature and a permanent magnet field. The polarity of the voltage applied to the armature winding determines the direction in which the motor spins. The armature shaft drives a tiny gear drive assembly that either extends or retracts a plunger. The plunger contacts the throttle linkage. When it is extended, the throttle blade opening is increased; when it is retracted, the throttle blade opening is decreased. The computer can apply a continuous voltage

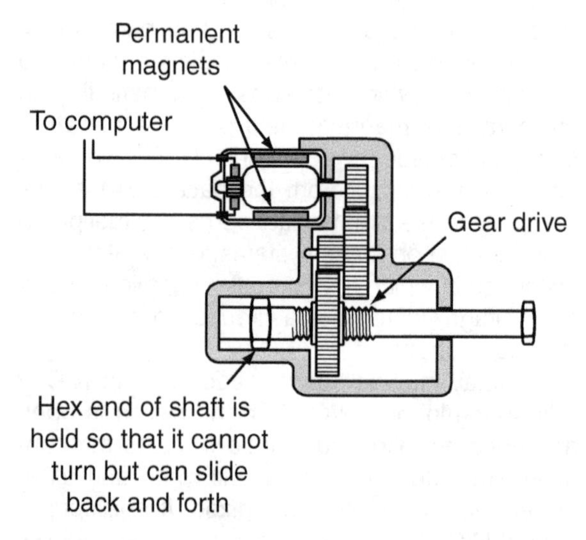

Figure 3–37 Permanent magnet field reversible DC motor.

to the armature until the desired idle speed is reached. On many modern applications, a DC motor is controlled by the computer on a high-frequency pulse-width-modulated signal, which allows the computer to precisely control the speed of the motor.

✔ SYSTEM DIAGNOSIS AND SERVICE

Pre-Checks and Visual Inspection

Many problems that are responsible for a symptom can be found easily if the technician conscientiously and purposefully inspects certain aspects of the problematic system. These pre-checks should include the following four checks:

1. A visual check of all wiring and connectors associated with the system. Look for wires that have rubbed against a component or a bracket and worn through the insulation to the copper. Also check for wires that stretch when you pull them between your fingers—this always indicates broken copper inside the wire, usually due to constant flexing (sometimes due to the movement of the transverse-mounted engine as the vehicle shifts gears).
2. Check all fuses electrically and visually. A visual check may find one leg of a fuse that is corroded where it connects to the fuse socket. Do not rely on a visual check only, however; also perform an electrical check, using a grounded wired or wireless test light with the fuse powered up. Both ends of the fuse should illuminate the grounded test light. Some glass tube fuses will crack under a metal end cap where you may not see it in a visual inspection; this author has seen at least one ATO spade-style fuse that looked okay but did not have electrical continuity through it. Also, most fuses provide power for multiple circuits (usually between three and seven), not just the primary circuit that it is labeled for at the fuse box. Many times this information is provided in the vehicle's owner's manual. So don't just test the one fuse you think is associated with the problem system, but take a minute or two and test all of the fuses.
3. Visually inspect any other components associated with the problem system.
4. Take the time to listen for vacuum leaks in the engine compartment and under the instrument panel with the engine running. While vacuum control is not as prevalent on modern vehicles as it was in the 1970s and 1980s, there are still a few vacuum-operated systems, including the positive crankcase ventilation (PCV) system, the canister purge and evaporative system, and on many vehicles, the EGR system; these systems are discussed further in Chapter 4. Also, some speed control systems continue to be vacuum operated. If a vacuum diaphragm ruptures, it will have multiple effects, as follows:

   - The component with the ruptured vacuum diaphragm will not operate properly.
   - Other components on the same vacuum line will not operate efficiently.
   - The vacuum leak will affect the air-fuel ratio by creating a lean condition.

Vacuum leaks on a speed density system with a MAP sensor will immediately cause a high idle RPM as the PCM will add fuel to the additional intake air. Vacuum leaks downstream from the MAF sensor on a multipoint injection system will initially cause a lean condition, resulting in a poor idle (if the engine idles at all, depending upon the severity of the vacuum leak). In time the PCM's fuel trim program (see Chapter 4) will learn to add fuel to the additional intake air, and a high idle RPM will also result.

These pre-checks will not always identify the problem for you, but on many jobs, as most experienced technicians have learned, just a few minutes of conscientious checks can save an hour or so of diagnostic time.

SUMMARY

We have learned how the computer system's most important sensors work, including variable resistors, NTC thermistors, and signal generators. The zirconium oxygen sensor produces a voltage corresponding to the amount of oxygen remaining in the exhaust gas; temperature sensors modify a reference voltage to match their temperature; and the throttle position sensor varies its return voltage signal through a variable potentiometer, so the computer knows exactly where the throttle is and how quickly it is moving.

We reviewed magnetic sensors, Hall effect sensors, and optical sensors. Permanent magnet sensors use a magnet and a coil to produce an alternating current, the frequency of which corresponds to the speed of a crankshaft or camshaft. The Hall effect sensor may be either a linear Hall effect sensor that outputs an analog voltage or a digital Hall effect switch that outputs a digital DC signal. Optical sensors generate voltage signals in a fashion similar to a Hall effect switch except that they use a photocell opposite an LED.

We have also seen how simple on–off switches can provide information on parameters such as the engagement or disengagement of the air-conditioner compressor, or what gear an electronic automatic transmission is in. Finally, we have seen how the computer controls actuators, including a look at the operation of N/C and N/O solenoid-operated valves, N/C and N/O relays, and motors.

Diagnostic & Service Tip

Oxygen Sensor. When handling or servicing the oxygen sensor, several precautions should be observed:

- A boot is often used to protect the end of the sensor where the wire connects. It is designed to allow air to flow between it and the sensor body. The vent in the boot and the passage leading into the chamber in the oxygen-sensing element must be open and free of grease or other contaminants.
- Sensors with soft boots must not have the boot pushed onto the sensor body so far as to cause it to overheat and melt.
- No attempt should be made to clean the sensor with any type of solvent.
- Do not short the sensor lead to ground, and do not put an ohmmeter across it. Any such attempt can permanently damage the sensor. The single exception is that a digital high-impedance (10 mega-ohm minimum) voltmeter can be placed across the sensor lead to ground (the sensor body).
- If an oxygen sensor is reinstalled, its threads should be inspected and cleaned if necessary with an 18 mm spark plug thread chaser. Before installation, its threads should be coated with an anti-seize compound, preferably liquid graphite and glass beads. Replacement sensors may already have the compound applied.

WARNING: It is often recommended that the oxygen sensor be removed while the engine is hot to reduce the possibility of thread damage. When doing so, wear leather gloves to prevent burns on the hand.

▲ DIAGNOSTIC EXERCISE

A car is brought into the shop with various running and driveability concerns. Among the things the technician notices is a great deal of rust and corrosion throughout the car, including on the connections for the electrical system. What kinds of problems would you anticipate might occur as a result of this kind of excessive rust formation? How would you check circuits to see how much, if at all, they were affected? How would you correct the electrical resistance problems that would occur from the rust?

Review Questions

1. Technician A says that the computer is, most often, designed to control the ground side of an actuator, although there may be an occasional exception. Technician B says that the engine computer is, most often, located in the trunk or luggage compartment of the vehicle. Who is correct?
   A. Technician A only
   B. Technician B only
   C. Both technicians
   D. Neither technician
2. With few exceptions, the reference voltage that engine computers send to the sensors is equal to which of the following?
   A. 1 V
   B. 5 V
   C. 12.6 V
   D. 14.2 V
3. Which of the following best describes the air-fuel range that can be measured with a wide-band oxygen sensor?
   A. 10:1 to straight ambient air
   B. 14.3:1 to 15.1:1
   C. 10:1 to 14.7:1
   D. 14.7:1 to 20:1
4. Zirconium dioxide is a substance that is found in which of the following?
   A. Oxygen sensor
   B. Throttle position sensor
   C. Engine coolant temperature sensor
   D. Manifold absolute pressure sensor
5. During closed-loop operation, the engine computer tries to cause a zirconia oxygen sensor to average which of the following?
   A. 100 mV
   B. 450 mV
   C. 900 mV
   D. 5 V
6. Technician A says that as the temperature of an NTC thermistor increases, its resistance decreases. Technician B says that the engine computer monitors the voltage drop across an NTC thermistor in a temperature-sensing circuit in order to know the measured temperature. Who is correct?
   A. Technician A only
   B. Technician B only
   C. Both technicians
   D. Neither technician
7. Technician A says that a temperature-sensing circuit is a current-divider circuit and that the sensor is an NTC thermistor placed in parallel to a fixed resistance within the computer. Technician B says that a temperature-sensing circuit is a voltage-divider circuit and that the sensor is an NTC thermistor placed in series to a fixed resistance within the computer. Who is correct?
   A. Technician A only
   B. Technician B only
   C. Both technicians
   D. Neither technician
8. A sensor that varies resistance to measure the physical position and motion of a component would be which of the following?
   A. Thermistor
   B. Potentiometer
   C. Piezoresistive silicon diaphragm
   D. Piezoelectric crystal
9. If a temperature-sensing circuit with an NTC thermistor were to develop an electrical open, the computer would see which of the following?
   A. About −40°F
   B. About 0°F
   C. About 150°F
   D. About 212°F
10. If a circuit with a potentiometer were to develop an electrical open in the ground wire, which of the following would be true concerning the voltage that the computer sees?
    A. It would be steady at zero volts.
    B. It would be steady at 2.5 V (one-half of reference voltage).
    C. It would be steady at 5 V (reference voltage).
    D. It would vary normally with the mechanical position of the sensor.

11. The speed density formula is a method of doing which of the following?
    A. Calculating how much air is entering the engine
    B. Measuring how much air is entering the engine
    C. Calculating how much fuel is entering the engine
    D. Measuring how much fuel is entering the engine
12. Technician A says that exhaust gases that are metered into the intake manifold by the EGR valve reduce the amount of ambient air that gets into the cylinders. Technician B says that some engine computers use sensors to determine EGR flow rates, while others assume EGR flow rates by using estimates of EGR flow that are stored in computer memory. Who is correct?
    A. Technician A only
    B. Technician B only
    C. Both technicians
    D. Neither technician
13. Which of the following sensors is used to measure the amount of intake air by measuring the cooling effect of the air that is entering the engine?
    A. Oxygen sensor
    B. NTC thermistor
    C. MAP sensor
    D. MAF sensor
14. Technician A says that a digital Hall effect sensor may be used to sense crankshaft position and RPM. Technician B says that a permanent magnet sensor may be used to sense crankshaft position and RPM. Who is correct?
    A. Technician A only
    B. Technician B only
    C. Both technicians
    D. Neither technician
15. Which of the following is sometimes referred to as a *noncontact* sensor and can be used to measure physical position and mechanical action?
    A. A linear Hall effect sensor
    B. A potentiometer
    C. A thermistor
    D. A piezoelectric crystal
16. What is a detonation sensor (knock sensor)?
    A. A thermistor
    B. A potentiometer
    C. A piezoresistive silicon diaphragm
    D. A piezoelectric crystal
17. Technician A says that when a switch is closed in a pull-down circuit, it will apply battery voltage to a computer input. Technician B says that when a switch is closed in a pull-up circuit, it will apply a ground to a computer input. Who is correct?
    A. Technician A only
    B. Technician B only
    C. Both technicians
    D. Neither technician
18. If a solenoid is turned on and off rapidly but there is no set number of cycles per second, what is its on-time called?
    A. Percent duty cycle
    B. Pulse width
    C. Frequency
    D. Capacitance
19. Technician A says that an N/O solenoid-operated valve must be de-energized to allow flow through it. Technician B says that an N/O relay must be energized to allow current flow through it. Who is correct?
    A. Technician A only
    B. Technician B only
    C. Both technicians
    D. Neither technician
20. Technician A says that a stepper motor can be used to control the air-fuel mixture in a feedback carburetor. Technician B says that idle speed can be controlled with either a stepper motor or a permanent magnet field motor. Who is correct?
    A. Technician A only
    B. Technician B only
    C. Both technicians
    D. Neither technician

Chapter 4

Common Operating Principles for Computerized Engine Control Systems

OBJECTIVES

Upon completion and review of this chapter, you should be able to:

- ❑ Describe the difference between return-type and returnless fuel injection systems.
- ❑ Understand fuel trim and lambda values for diagnosis with a scan tool.
- ❑ Comprehend how the manufacturers are using rear fuel control.
- ❑ Describe the three levels of information available to a powertrain control module (PCM) regarding crankshaft position and camshaft position.
- ❑ Understand the purpose of crankshaft position sensors versus camshaft position sensors.
- ❑ Recognize the purpose and function of various emission-control systems.

KEY TERMS

Camshaft Position (CMP) Sensor
Catalyst
Crankshaft Position (CKP) Sensor
Distributor Ignition (DI)
Electronic Ignition (EI)
Electronic Returnless Fuel Injection
Electronic Throttle Control (ETC)
Flexible Fuel Vehicle (FFV)
Fuel Trim
Gasoline Direct Injection (GDI)
Grounded Shield
Lambda
Long-Term Fuel Trim (LTFT)
Palladium
Platinum
Port Fuel Injection (PFI)
Rear Fuel Control
Reid Vapor Pressure (RVP)
Returnless Fuel Injection
Return-Type Fuel Injection
Rhodium
Sequential Fuel Injection (SFI)
Sequential Multi-Point Injection (SMPI)
Short-Term Fuel Trim (STFT)
Throttle Body Injection (TBI)
Total Fuel Trim (TFT)
Vapor Lock
Variable Valve Timing (VVT)
Volatility
Waste Spark

Understanding the product-specific system designs in the manufacturer-specific chapters of this textbook may seem to be a difficult task, but all these methods are based on a few common operating principles. This chapter covers some of the basic, but often misunderstood, concepts that are used by all manufacturers. Understanding the material in this chapter will help you to understand the manufacturer-specific chapters.

FUEL VOLATILITY AND FUEL OCTANE

There are two fuel qualities that are of particular concern to today's technicians: fuel volatility and fuel octane.

Fuel Volatility

Volatility has to do with a liquid's ability to vaporize, directly affected by its *boiling point*. As the boiling point is decreased, the liquid's volatility becomes greater. Bear in mind that only fuel that is in the form of a vapor will actually burn in the cylinder. Fuel that remains in the form of a liquid as it enters the cylinder will not burn in the cylinder, but will exit the cylinder as unburned fuel. For fuel to be vaporized in time to burn in the cylinder, it must first be finely atomized into millions of tiny liquid droplets. With most modern fuel injection systems these droplets are approximately 100 microns in diameter (100 millionths of a meter), about the same diameter as a human hair. This greatly increases the fuel's surface area, thus increasing the fuel's potential to vaporize readily. A liquid's volatility is dictated by two factors:

- What type of liquid it is
- How much pressure is placed upon the liquid

In terms of the liquid type as it applies to fuel, because alcohol has a lower boiling point than gasoline, it is said to be more volatile than gasoline. Restated, alcohol vaporizes more readily than gasoline. In terms of pressure placed on the liquid, liquids become more volatile under lower pressure such as at high altitude, where the barometric pressure is decreased.

A fuel's volatility is measured by heating the fuel in a confined space to 100°F and then measuring the pressure produced at the point at which it ceases vaporization. For example, if you put a cup of fuel (gasoline or diesel fuel) in a jar, then put a lid with a pressure gauge attached on the jar (closed tightly) and heated it to 100°F, the fuel would begin to vaporize. But because the vapors would have nowhere to go, the jar's internal pressure would increase. At some point, the increased pressure would raise the fuel's boiling point enough so that the fuel would cease vaporization. This pressure is the fuel's **Reid vapor pressure (RVP).** Thus, to put it in perspective, RVP is a direct measurement of a fuel's volatility and is inversely proportional to the fuel's boiling point. Restated, as a fuel's boiling point is decreased, the fuel's volatility and RVP both increase.

A fuel's RVP is increased in the cold weather months to allow easier cold starts and is decreased in hot weather months to keep the fuel from vaporizing prematurely. The Environmental Protection Agency (EPA) dictates that RVP ratings at sea level are not to exceed 9.0 PSI in June, July, and August. RVP ratings in Denver, Colorado (5,280 feet altitude), are not to exceed 7.8 PSI during these summer months. During the winter months, RVP may climb to as high as 14½ to 15 PSI depending upon altitude.

Fuel Octane

A fuel's octane is the ability of the fuel to burn more slowly once ignited so as to resist detonation in the cylinder. A higher octane rating means that the fuel burns more slowly once ignited. While alcohol may vaporize more readily than gasoline, alcohol burns more slowly than gasoline once ignited. Thus, alcohols such as ethanol or methanol can be used to increase a fuel's octane rating.

A fuel's octane rating at sea level tends to be higher than the octane rating of fuel at high altitude such as Denver, Colorado, or Mexico City. This is because barometric pressure is greater at low altitude. This increases the quantity of oxygen molecules that make their way into the cylinder; fuel quantity is increased to match the oxygen content. Thus, hotter combustion will occur at low altitude, such as sea level. This increases the need for a higher-octane fuel.

RVP and Octane Summary

A fuel's volatility, measured as RVP, is a concern *prior to* ignition. A fuel's octane rating is a concern *after* ignition.

If the RVP is too low, cold starts may be difficult. If the RVP is too high, the fuel may vaporize within the fuel management system, resulting in engine stall. This is known as **vapor lock.** Automotive fuel management systems are not designed to deliver enough fuel volume in the form of a vapor to keep an engine running, so the engine will not restart until the fuel recondenses. RVP that is too high can also result in pressurization of a fuel tank, a carburetor's float bowl boiling dry, or excessive HC emissions, particularly in hot weather.

If a fuel's octane ratings are too low for the engine's requirements, the engine may detonate while under load. If the fuel's octane ratings are higher than the engine requires, no disadvantage is experienced, but neither is there an advantage as the commercials would lead you to believe.

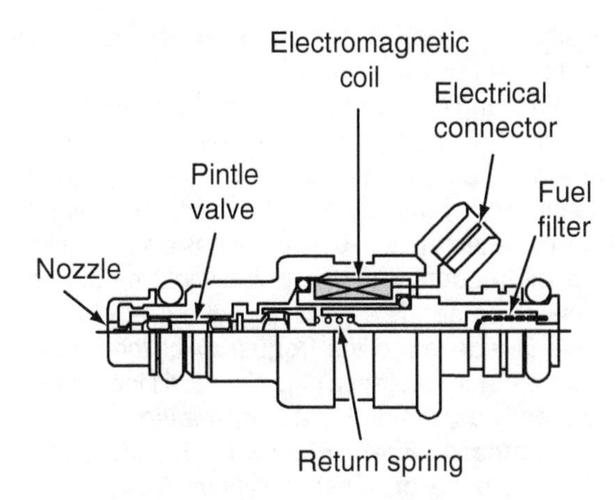

Figure 4–1 Typical high-pressure fuel injector.

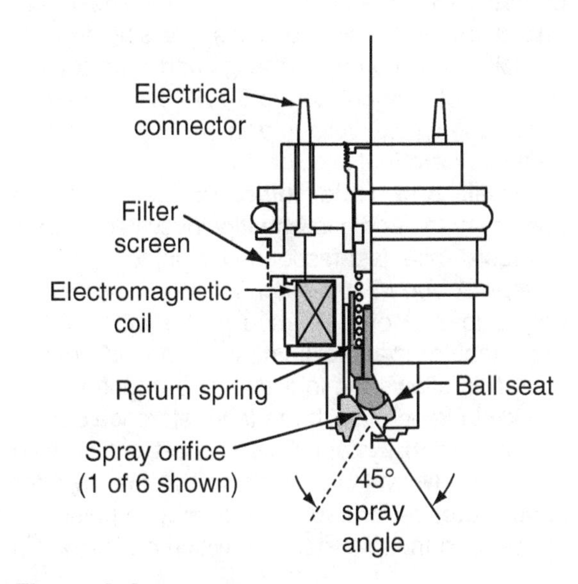

Figure 4–2 Low-pressure fuel injector.

ELECTRONIC FUEL INJECTION SYSTEM OPERATING PRINCIPLES

Fuel Injector Design and Operation

Fuel injectors are nothing more than solenoid-operated valves that are spring-loaded normally closed (N/C) and that spray fuel when electrically energized. Because the injector is under pressure, fuel will spray in a highly atomized pattern each time the injector is energized by the computer. The injector's electromagnetic winding, or coil (Figure 4–1), is also kept immersed in liquid fuel that carries away heat, preventing overheating of the winding; such overheating could result in a short between coils of the winding. The injector has a screen at its inlet to keep any foreign particles from entering the injector. An injector may be designed for use with either high-pressure or low-pressure fuel injection systems (Figure 4–1 and Figure 4–2).

Most fuel injectors are pintle-and-seat designs; the pintle blocks fuel flow when de-energized and protrudes slightly through the seat's orifice. Some low-pressure injectors are designed with a ball valve that seats on six small orifices when de-energized (Figure 4–2). Fuel injectors are rated for flow rate. The required flow rate is determined by engine displacement, volumetric efficiency (VE), and whether the injection system is a throttle body or multipoint system design.

(Multipoint injectors have a lower flow rate than do throttle body injectors.)

The engine computer, or powertrain control module (PCM), controls the fuel injectors on a pulsed signal. The on-time of this signal is known as the injector's *pulse width.* Pulse width is similar to duty cycle, except that duty cycle has a fixed cycling rate or frequency and a pulse-width-modulated signal does not. The frequency of a fuel injector's pulse is indexed to the frequency of the ignition system's tach reference signal and will increase in frequency as engine RPM is increased.

Standard Fuel Injectors. Standard fuel injectors, used on most multipoint fuel injection systems, have an internal winding resistance of between 12 and 22 Ω (typical). These are energized by the computer using a single driver (transistor) to complete the ground path for the duration of the pulse (Figure 4–3). Figure 4–4 shows the typical voltage and current waveforms for a standard fuel injector.

Peak and Hold Injectors. Most throttle body systems and a few multipoint systems use a low-resistance, fast-response fuel injector, known as a *peak and hold injector.* These injectors have only 2 to 3 Ω of resistance in the winding and are capable of approximately 4 amps of current flow. This allows the injector to be opened more quickly to keep up with the tach reference pulses of the ignition system. This is most critical with throttle body systems in which the fuel injector is pulsed once for each tach reference pulse, as opposed to those systems in which an injector is

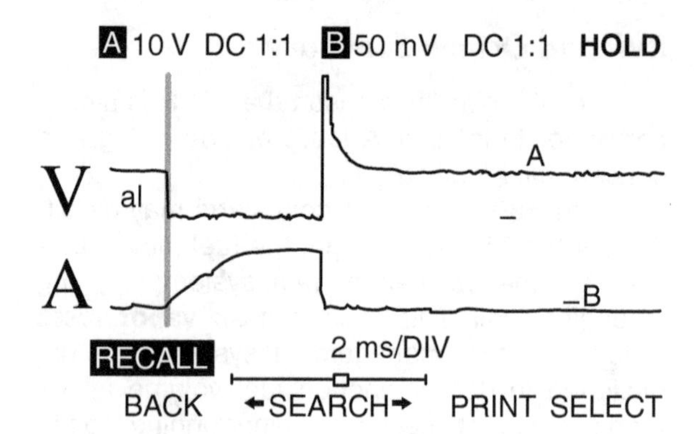

Figure 4–4 Voltage and current waveforms for a standard fuel injector.

pulsed once each one or two crankshaft revolutions. Peak and hold injectors are controlled by two drivers as shown in Figure 4–5: One provides a direct path to ground; the other provides a path to ground through a resistor that is inside the computer. Both drivers are forward biased to initially energize the injector; then the one with the direct path to ground is released, leaving current to continue to flow through the remaining circuit with the resistor. This allows maximum initial current flow through the fuel injector, resulting in good mechanical response. Once the injector has been opened, however, current flow is immediately reduced to approximately 1 amp or less for the duration of the pulse to prevent overheating of the injector's winding. These circuits are known as the injector's "peak" and "hold" circuits.

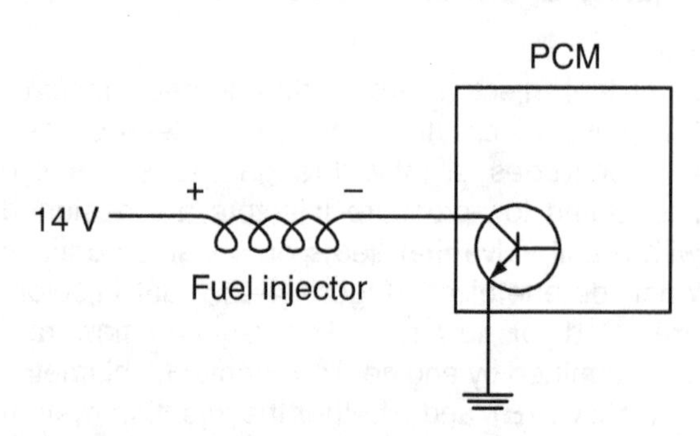

Figure 4–3 A standard injector's driver circuit.

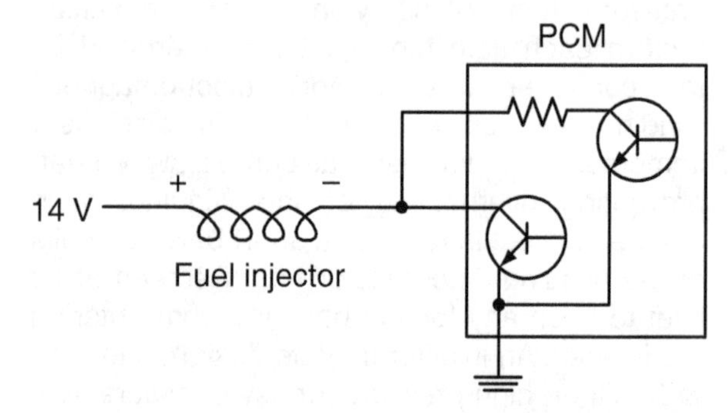

Figure 4–5 A peak and hold injector's driver circuit.

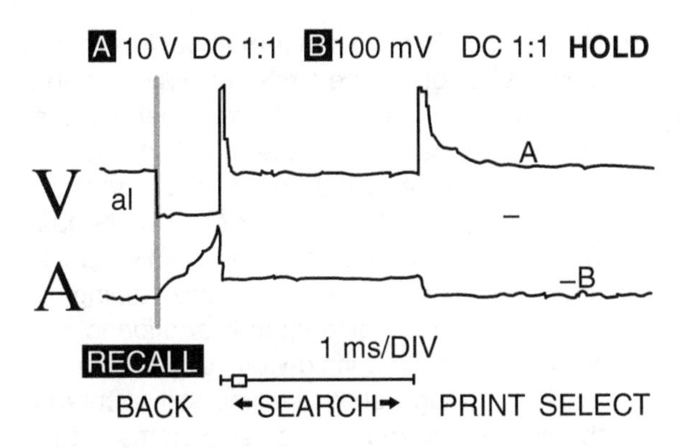

Figure 4–6 Voltage and current waveforms for a peak and hold injector.

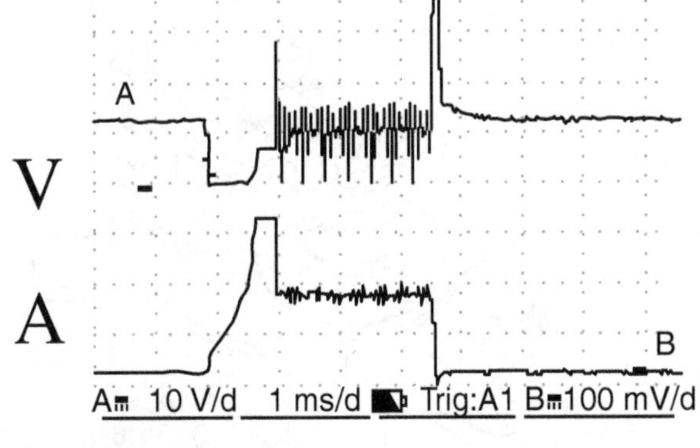

Figure 4–7 Voltage and current waveforms for a current-controlled injector.

Figure 4–6 shows the typical voltage and current waveforms for a peak and hold fuel injector.

Current-Controlled Injectors. Current-controlled injectors are similar in purpose and operation to peak and hold injectors except that a single driver is used to control them. Initially, full current is allowed to flow through the injector's winding for good mechanical response. Once the injector is open, the PCM operates the driver on a high-frequency pulse-width-modulated signal so as to reduce the average current to approximately 1 amp or less to prevent overheating of the winding. This style tends to be found on some Asian throttle body and port fuel-injected applications. Figure 4–7 shows the typical voltage and current waveforms for a current-controlled fuel injector. If the time base on the scope is reduced from 1 ms/div (as shown in Figure 4–7) to 200 µs/div, the modulated portion of the voltage waveform (between the two voltage spikes) appears as a digital square wave.

Injectors with an External Resistor. Some applications design a low-resistance injector in series with an external resistor to limit current flow. Check the electrical schematic to identify systems of this type. When testing any fuel injector for proper resistance in the winding, always use the resistance specifications from the service manual.

Fuel Injection System Types

There are two types of electronic fuel injection systems used with comprehensive computerized engine control systems: single point and multipoint. They each use an intermittently pulsed and timed spray to control fuel quantity.

Throttle Body Injection. Throttle body injection (TBI), also sometimes referred to as single-point injection or central fuel injection (CFI), means that fuel is introduced into the engine from one location. This system uses an intake manifold similar to what was used with a carbureted engine, but the carburetor is replaced with a throttle body unit (Figure 4–8). The throttle body unit contains one or two solenoid-operated injectors that spray fuel directly over the throttle plate (or plates). Fuel under pressure is supplied to the injector. The throttle plate is controlled by the throttle linkage, just as in a carburetor. The computer controls voltage pulses to the solenoid-operated injector, which opens and sprays fuel into the throttle bore. The amount of fuel introduced is controlled by the length of time the solenoid is energized. This is referred to as the injector's pulse width. The amount of air introduced is controlled by the opening of the throttle plate(s); it is also affected by any air that is introduced by the idle air control (IAC) solenoid or stepper motor.

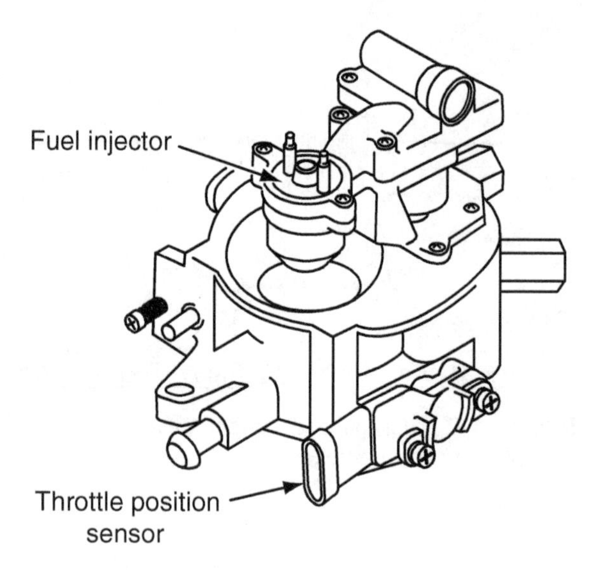

Figure 4–8 TBI unit.

A TBI system is characterized by excellent throttle response and good driveability, especially on smaller engines. Experience has shown, however, that the system is best suited for engines with small cross-sectional area manifold runners that at low speeds will keep the fuel mixture moving at a high velocity. This reduces the tendency for the heavier fuel particles to fall out of the airstream.

Port Fuel Injection. Port fuel injection (PFI), also referred to as a multipoint injection (MPI) system, uses an injector at each intake manifold port. Fuel is sprayed directly onto the back side of the intake valve.

The MPI/PFI system provides the following advantages:

- Spraying precisely the same amount of fuel directly into the intake port of each cylinder eliminates the unequal fuel distribution inherent when already mixed air and fuel are passed through an intake manifold.
- Because there is no concern about fuel condensing (puddling) as it passes through the intake manifold, there is less need to heat the air or the manifold.
- Because there is no concern about fuel molecules falling out of the airstream while moving through the manifold at low speeds, the cross-sectional area of the manifold runners can be larger and thus offer better cylinder-filling ability or VE at higher engine speeds. Some manufacturers have designed variable-geometry intake runners to optimize air ingestion at different engine speeds and loads.
- Most of the manifold-wetting process is avoided, though some wetting still occurs in the port areas and on the valve stems. If fuel is introduced into the intake manifold, some will remain on the manifold floor and walls, especially during cold engine operation and acceleration. Fuel metering has to allow for this fuel to avoid an overly lean condition in the cylinders. It has to be accounted for again during high-vacuum conditions because under such conditions it will begin to evaporate and go into the cylinders.

In general, PFI provides better engine performance and excellent driveability while maintaining

Diagnostic & Service Tip

An unexpected problem with PFI throttle bodies has been fuel deposits on the throttle blades. The source of these fuel vapors is fuel that drifts back from the intake ports when the engine is shut down hot. The heat vaporizes most or all of the residual fuel, and the coolest surface it encounters in the intake path is the throttle blade itself. The problem is that these deposits change the shape and size of the opening in the smallest throttle opening angles. This can adversely affect driveability in unexpected ways, such as hesitation, erratic idle, hard starting, and surging. The throttle blades can be visually inspected for such deposits and cleaned with solvent. Such inspection should be a routine maintenance task on cars with this fuel injection system.

or lowering exhaust emission levels and increasing fuel economy. The major disadvantages are somewhat greater cost and reduced serviceability because of the larger number of components and their relative inaccessibility.

PFI systems use various firing strategies for the fuel injectors that group them together in pairs, banks, or groups with all injectors being pulsed simultaneously.

Sequential Fuel Injection. **Sequential fuel injection (SFI),** also known as **sequential multi-point injection (SMPI),** is a form of a PFI system whereby the PCM will pulse the injectors individually in the engine's firing order, just before the intake valve opens (Figure 4–9). Each pulse, therefore, delivers 100 percent of the cylinder's fuel requirement. However, during the initial engine start on an engine with SFI and a waste spark ignition system (covered later in this chapter), the PCM will pulse all injectors simultaneously to get fuel to the cylinders to allow for a faster start. This is known as a *primer pulse.* Then, once the PCM sees the CMP sensor signal, it switches over to sequential pulsing of the injectors.

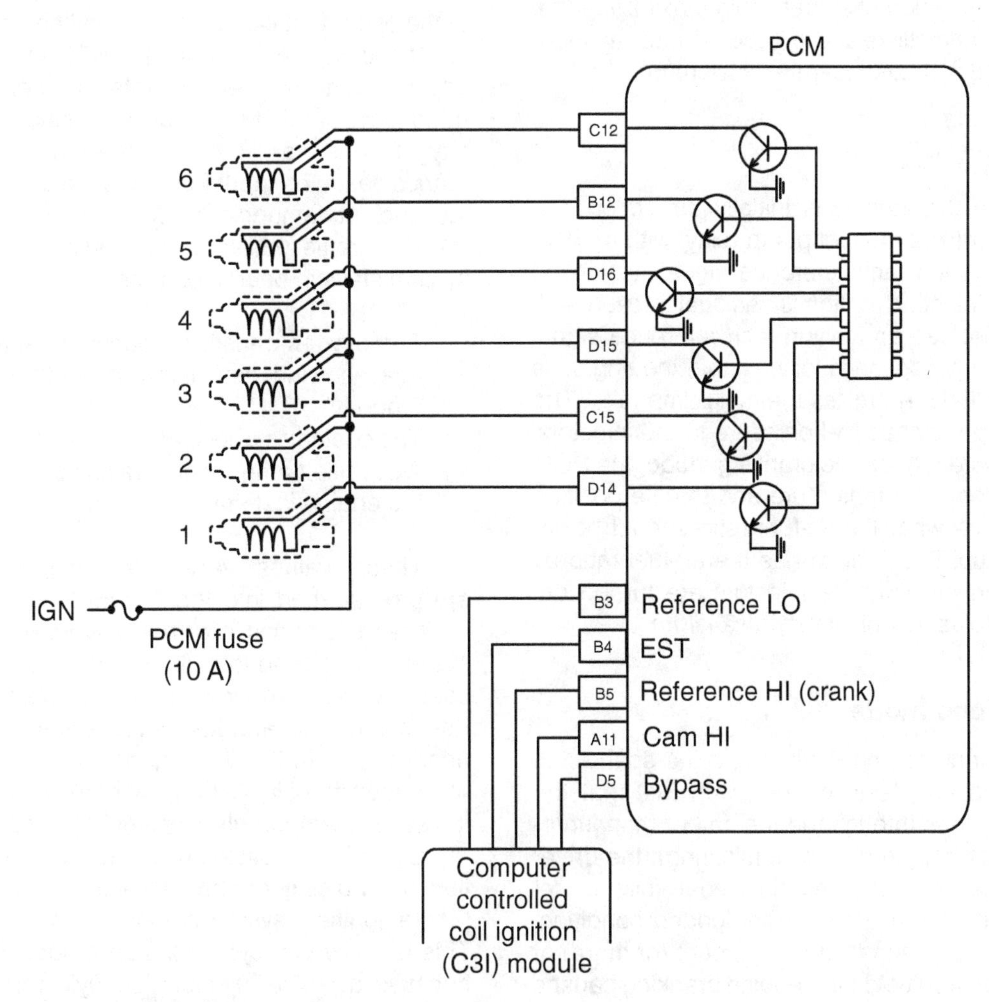

Figure 4–9 Electrical schematic for a typical SFI circuit.

OPERATING MODES OF A FUEL INJECTION SYSTEM

The different operating modes of the TBI and PFI systems control how much fuel is introduced into the intake manifold and include the typical closed-loop and open-loop modes. During open-loop operation, the PCM is programmed to provide an air-fuel ratio to maximize driveability during cold starts and hard accelerations, while minimizing toxic emissions as much as possible. During closed-loop operation, the PCM fine-tunes the air-fuel ratio to minimize the toxic emissions produced in the combustion chamber while maximizing fuel economy and still retaining good driveability under cruise and light-to-moderate acceleration.

Starting Mode

When the ignition is initially turned on, the PCM energizes the fuel pump relay, with or without an ignition tach reference signal, to top off the fuel injection system's residual pressure. If it does not see an ignition tach reference signal occur within 2 seconds telling it that the engine is cranking, it de-energizes the fuel pump relay. The fuel pump provides fuel pressure to the injectors. As the system goes into cranking mode, the PCM checks coolant temperature and throttle position to calculate what the air-fuel ratio should be for this startup. The PCM varies the air-fuel ratio by controlling the time the injectors are turned on, referred to as the injectors' *pulse width*.

Clear Flood Mode

If an engine floods (that is, if the spark plug electrodes are damp enough with fuel that the spark grounds through the fuel rather than jumping the spark plug gap, thus misfiring), the driver should depress the throttle pedal fully to let additional air in to relieve the flooded condition. Depressing the throttle to 80 percent (or more) of wide-open throttle during engine cranking causes the PCM to adopt the clear flood strategy. It will then control the air-fuel ratio to about 20:1. It stays in this mode until the engine starts or until the throttle is closed to less than 80 percent of WOT. If the engine is not flooded and the throttle is held at 80 percent WOT or more, it is unlikely the engine will start with the super lean mixture, especially if the engine is cold. If the TPS ground wire is open, the resulting 5 V signal will automatically cause the PCM to enter clear flood mode during starting and will make for a difficult cold start.

Run Mode

For TBI and PFI applications, run mode consists of the open- and closed-loop operating conditions. When the engine starts and the RPM initially rises above 400, the system operates in open loop. In open loop, the PCM ignores information from the oxygen sensor and determines air-fuel ratio commands based on input from other sensors, including coolant temperature, MAP or mass airflow (MAF), throttle position, and engine speed. The system stays in open loop until:

- The oxygen sensor produces a varying voltage showing that it is hot enough to work properly.
- The coolant is above a specified temperature.
- A specified amount of time has elapsed since the engine last started.

These values vary with application and are programmed into the PROM IC chip. When all three conditions are met, the PCM puts the system into closed loop. In closed loop, the PCM uses oxygen sensor input to calculate air-fuel ratio commands and keeps the air-fuel ratio at a near-perfect 14.7:1. During heavy acceleration, wide-open throttle (WOT), or hard deceleration, the system temporarily drops out of closed loop.

On a TBI application, during normal operation the pulsing of the fuel injector is indexed to the ignition system's tach reference signal. This is known as *synchronized mode*. However, because a single injector (four-cylinder applications) is pulsed with every tach reference pulse (twice per crankshaft rotation), it becomes impossible for the mechanical injector to keep up with this at high engine speeds. On a dual-injector TBI system (on a V6 or V8 application), while the injectors are pulsed alternately, they are still pulsed more often than the PFI counterpart (once every 240 degrees on a V6 and twice per rotation on a V8). Therefore, at high engine RPM the PCM quits pulsing the TBI injectors according to the tach reference signal and begins pulsing them at a fixed frequency. This is known as *nonsynchronized mode* or *asynchronous mode*. This is not necessary on a PFI system where a given injector is either pulsed once per crankshaft rotation (group injection) or once per two crankshaft rotations (SFI). Therefore, a PFI system's injector pulses are always synchronized, in some way, to the ignition system's tach reference signal.

Acceleration Mode

Rapid increases in throttle opening and manifold pressure or the mass airflow rate cause the PCM to enrich the air-fuel mixture. This adds fuel to drive the air-fuel ratio closer to the desired 12:1 needed for hard acceleration and even compensates for the reduced evaporation rate of the gasoline resulting from the higher manifold pressure.

Deceleration Mode

Rapid decreases in throttle opening and manifold pressure or airflow cause the PCM to lean out the air-fuel mixture. If the changes are severe enough, fuel is momentarily cut off completely.

Battery (Charging System) Voltage Correction Mode

When battery voltage drops below a specified value, the PCM:

- Enriches the air-fuel mixture according to a preset formula.
- Increases the throttle bypass air if the engine is idling.
- Increases ignition dwell to compensate for a weakened ignition spark should the coil's magnetic field not build sufficiently to generate a hot secondary spark.

Fuel Cutoff Mode

When the ignition is turned off, the PCM immediately stops pulsing the injectors to prevent dieseling as well as to consume the fuel mixture remaining in the ports. Injection is also stopped any time the distributor reference pulse (REF) stops coming to the PCM.

FUEL INJECTION SYSTEM COMPONENTS AND SYSTEM DESIGNS

Fuel Injector

A fuel injector is a solenoid-operated valve that is spring-loaded normally closed (N/C). The PCM controls the length of time that a given injector is energized in milliseconds in order to control the air-fuel ratio. The injector's on-time is known as its *pulse width*. The typical pulse width for an SFI fuel injector is from 2 or 3 milliseconds (ms) at idle (depending upon coolant temperature) to possibly more than 20 ms under heavy load at WOT.

Electric Fuel Pumps

The electric fuel pumps used with electronic fuel injection systems are most often mounted in the fuel tank (Figure 4–10), which reduces the potential for vapor lock. These pumps are highly capable, being designed to develop more pressure than the system's working pressure (as regulated by the pressure regulator). A few vehicles use more than one fuel pump, locating one in the fuel tank and an additional high-pressure pump in the fuel supply line. The fuel pump contains a check valve, which prevents fuel pressure from pushing fuel already in the supply line back into the fuel tank following engine shutdown. This, combined

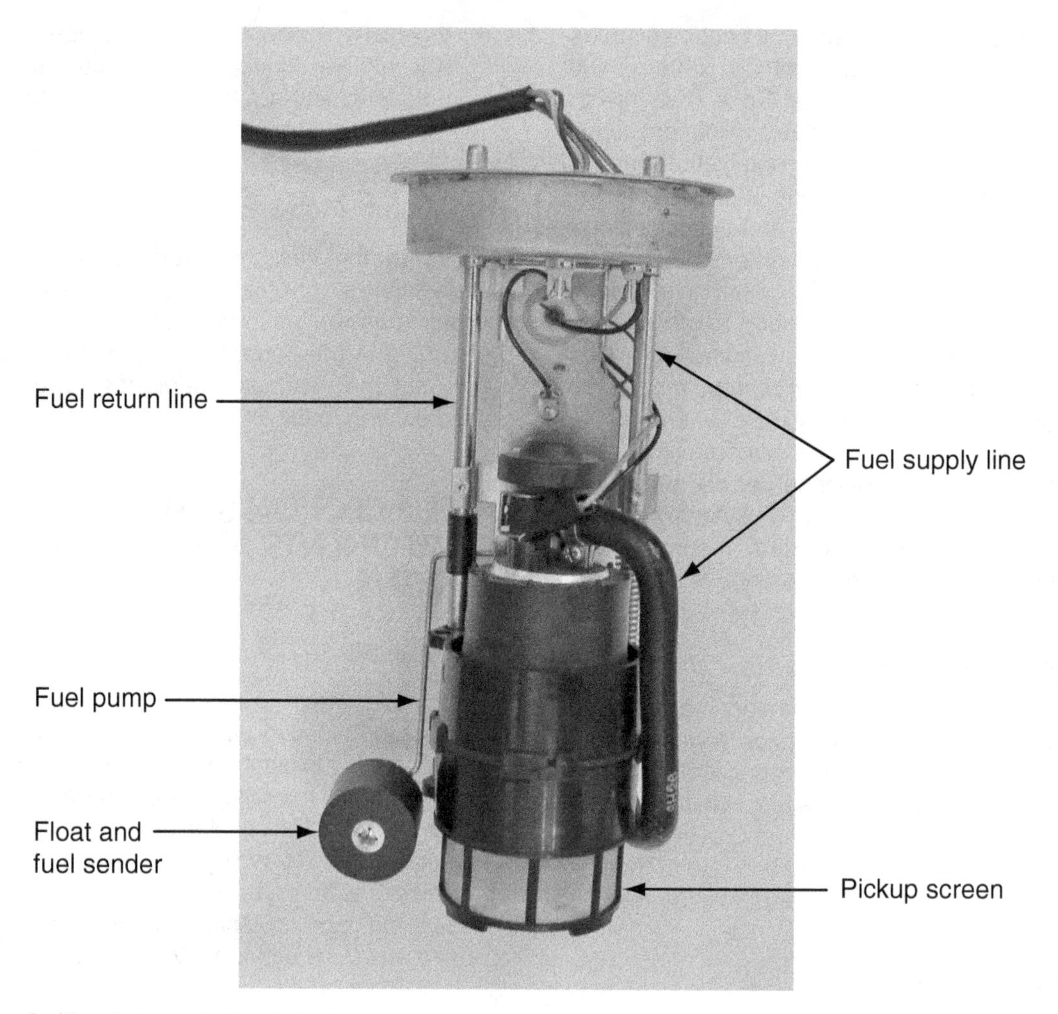

Figure 4–10 An in-tank electric fuel pump.

with the pressure regulator, allows the system to hold residual pressure until the next time the engine is started.

Electric fuel pumps are not directly controlled by the PCM, but rather are controlled by the PCM via a fuel pump relay, which carries the higher current flow. This relay is energized by the PCM whenever the PCM is receiving a tach reference signal from the ignition system. The relay is de-energized if the tach reference signal to the PCM is lost. There is only one exception to this: The fuel pump relay will be energized by the PCM for approximately 2 seconds whenever the ignition switch is first turned on and the PCM is initially powered up, whether or not a tach reference signal is present. (This occurs provided the ignition switch has been turned off long enough to allow the PCM to power down—about 8 seconds on those vehicles that use a motor or stepper motor to control idle speed. This type of actuator must be reset by the PCM after the ignition is turned off.) Manufacturers can thus protect an unconscious driver from a system that might otherwise empty the fuel tank on the ground through a ruptured fuel supply line following an accident: If the fuel supply line ruptures, fuel pressure is immediately lost, the engine immediately stalls, and the tach reference signal to the PCM is lost, resulting in the PCM de-energizing the fuel pump relay.

Fuel Pressure Regulator Design and Operation

A fuel pressure regulator is a simple, spring-loaded, N/C bypass valve, which allows the system to maintain proper fuel pressure at the fuel injector(s). The pressure regulator is located *after* the fuel injector(s) (or, in a returnless system, after the supply line that feeds the injectors) so as to keep the fuel injector(s) pressurized at the inlet screen (Figure 4–11).

Return-Type Fuel Injection Systems

As manufacturers began replacing carburetors with electronic fuel injection systems, system features that had been used with carbureted engines to reduce the potential for vapor lock also made their way into fuel injection system designs. *Vapor lock* is defined as the vaporization of fuel within the confines of the fuel management system. This will typically result in engine stall due to the fact that none of our fuel management systems can supply enough fuel volume in the form of a vapor to keep an engine running.

The primary factor in vapor lock is the fuel's volatility. Fuel that is highly volatile is easily vaporized, having a lower boiling point than fuel that is less volatile. A liquid's volatility is influenced by two factors: what type of liquid it is (for example, gasoline versus alcohol) and how much pressure is placed on it (reducing the pressure placed on a liquid increases its volatility).

Vapor lock is caused when high-temperature fuel is combined with low pressure in the lines, particularly in the fuel pump suction line of a carbureted engine with a mechanical fuel pump located at the engine. If the fuel vaporizes within the lines, the engine will typically stall and will not restart until the hot fuel cools down enough to condense back into a liquid.

Many carbureted systems began using *return lines* from the fuel pump (or from the fuel filter located near the carburetor) back to the fuel tank

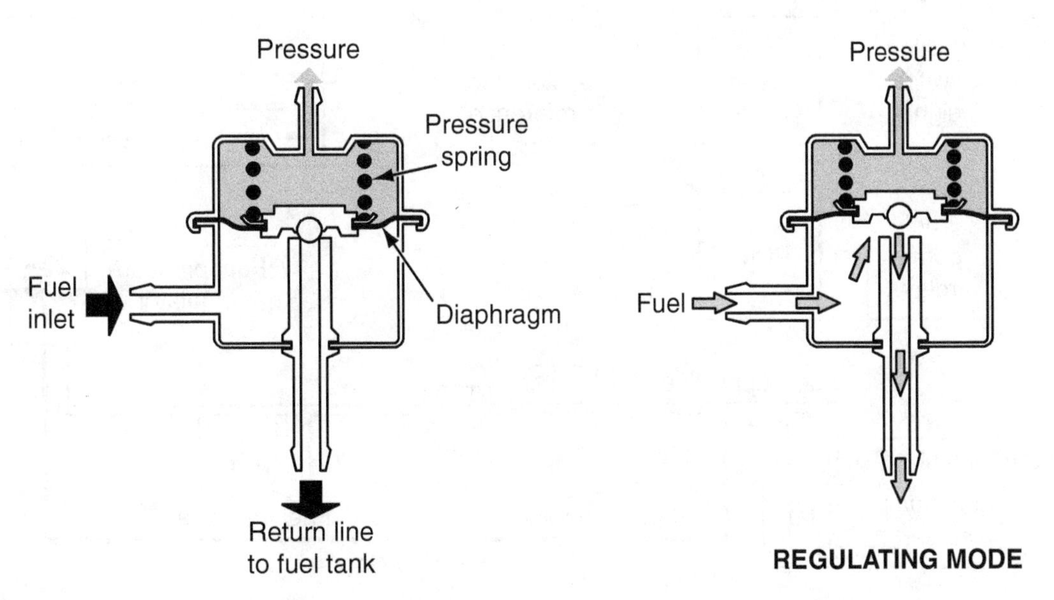

Figure 4–11 Operation of a diaphragm-operated fuel pressure regulator in a return-type system.

as a method of continually drawing cooler fuel from the tank. This helped to keep the temperature of the fuel in the line low enough to reduce the potential for vapor lock, even with low fuel consumption, such as might be experienced in heavy city traffic. These fuel return lines were carried over to most fuel injection systems. Another method of combating vapor lock involves *relocating the fuel pump* from the engine compartment into or near the fuel tank. This replaces the negative pressure (in what used to be the suction line) with positive pressure in the fuel supply line and thereby reduces the volatility of the fuel in the line. A third method is simply *raising the pressure placed on the fuel* in the fuel management system, something that modern fuel injection systems do quite effectively.

In a **return-type fuel injection** system (Figure 4–12), the fuel pump in the tank pushes the fuel through the fuel supply line and primary fuel filter, then past the inlets to the fuel injectors and to the fuel pressure regulator. When enough fuel pressure has developed underneath the regulator's seat to unseat it (against spring tension), fuel is allowed into the fuel return line that exits the pressure regulator, thereby allowing this excess fuel to return to the fuel tank.

Most fuel pressure regulators used on TBI systems use spring tension alone to regulate the pressure of the fuel at the fuel injectors, because the fuel injector is designed to spray fuel above the throttle plate(s) and into an area that is very close to barometric pressure. Since this pressure near the outlet tip of the injector is relatively stable, the line pressure is held stable as well, resulting in a consistent pressure differential across the fuel injector, and therefore a consistent flow rate. A few TBI systems allow atmospheric pressure to the spring side of the regulator, thereby allowing it to adjust fuel pressure slightly to compensate for barometric pressure changes. This increases the consistency of the pressure differential across the injector, regardless of changes in altitude or weather patterns.

In a return-type PFI system, because these injectors must spray fuel below the throttle plate(s) or into existing pressure within the intake manifold, the spring side of the pressure regulator is indexed to intake manifold pressure via a "vacuum hose." Consider how changes in intake

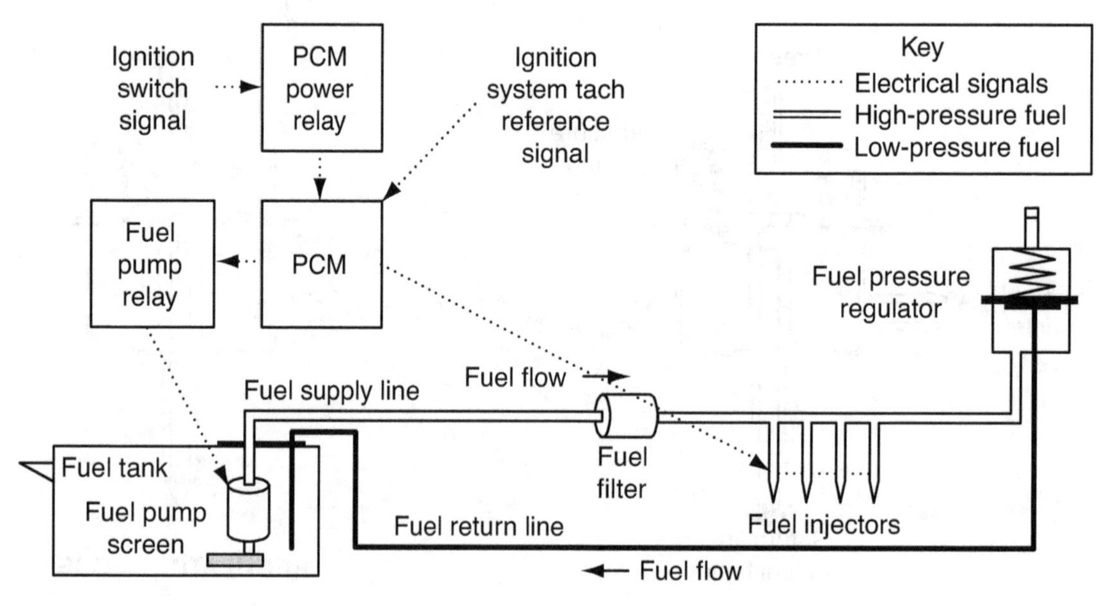

Figure 4–12 Return-type fuel injection system.

manifold pressure would affect fuel delivery without this indexing (that is, with a fixed fuel rail pressure): At idle under light engine load, when there is reduced pressure within the intake manifold, the pressure differential across the injector would be greater, resulting in an increased flow rate when there is less need for fuel. Under heavy load, when intake manifold pressure is greater, the pressure differential across the injector would be reduced, resulting in a decreased flow rate when there is more need for fuel. When the spring side of the pressure regulator is indexed to intake manifold pressure, any change of pressure within the manifold causes an exactly equal change in fuel rail pressure. In actual operation, a low pressure (or vacuum) applied from the intake manifold to the spring side of the regulator results in an equal reduction in fuel rail pressure because the low-pressure signal helps the fuel pressure to unseat the valve. When a higher pressure (loss of vacuum) is applied from the intake manifold to the spring side of the regulator, fuel rail pressure is raised equally. This also holds true when boost pressure might be present within the intake manifold, resulting in an additional equal increase in fuel rail pressure. Therefore, when the spring side of the regulator is indexed to intake manifold pressure, the pressure differential across the fuel injectors is held constant, regardless of engine load. This results in a consistent flow rate whenever the PCM energizes the injector, regardless of engine load. As a result, the only real variable in controlling the quantity of fuel that is delivered to the cylinders is injector on-time—and this is the variable that is under PCM control. As a result, the PCM has complete control over the quantity of fuel that is delivered.

Mechanical Returnless and Semi-Returnless Fuel Injection Systems

Once manufacturers had successfully reduced the potential for vapor lock, their concerns turned to reducing the heat that return-type systems dump into the fuel tank—heat that increases the potential for evaporative hydrocarbon emissions. Most of this heat is picked up in the engine compartment by the excess fuel that is to return to the tank. As a result, in the mid- to late 1990s, many manufacturers began using mechanical **returnless fuel injection** systems that eliminate the return line from the engine compartment to the fuel tank (Figure 4–13). Chrysler Corporation was the first domestic manufacturer to introduce returnless systems in 1995. By 2003 all manufacturers had switched over to some form of returnless system.

On a mechanical returnless fuel injection system, a short return line actually does exist. However, it originates at a pressure regulator either inside the fuel tank (mechanical returnless) or near the fuel tank and internal of the fuel filter (mechanical semi-returnless). So there is no return line from the engine compartment bringing heat back to the fuel tank.

The fuel line from the fuel pump carries fuel to a “tee” intersection. One line from this tee goes to the fuel pressure regulator, which, as in a return-type system, returns excess fuel to the fuel tank. The other line carries fuel through the fuel supply line to the fuel injectors in the engine compartment, after which it dead-ends. The fuel supply line is actually in parallel with the fuel line that goes to the pressure regulator. Therefore, it can be said that the regulator is located after the line that tees off to deliver fuel to the injectors, and as a result it allows the regulator to control the pressure in the supply line. However, the pressure regulator is no longer indexed to manifold pressure as with previous systems that had a pressure regulator at the fuel rail. A vacuum line from the engine compartment to the fuel tank area would be so long that the responsiveness of such a signal would be inadequate. Therefore, the line pressure is fixed and does not vary, but the resulting pressure differential across the fuel injectors does vary. Under light load the pressure differential across the injectors is greater; under heavy load the pressure differential across the injectors is less. Therefore, the PCM must both calculate the pressure differential across the injectors and

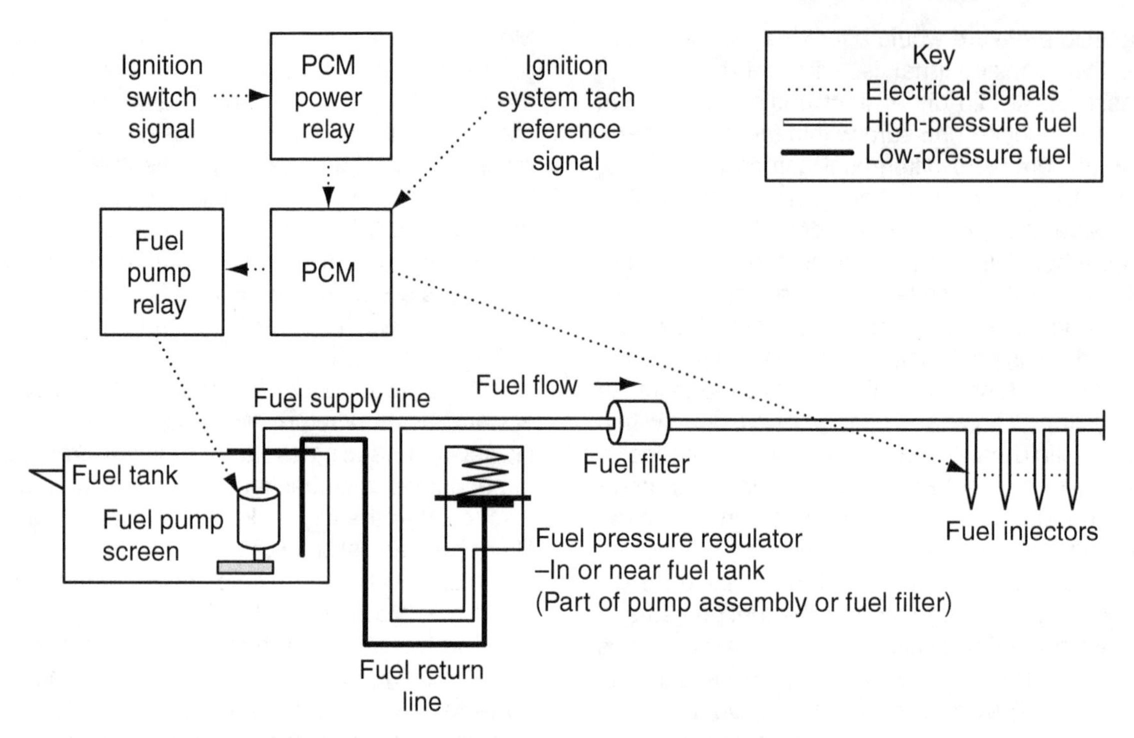

Figure 4–13 Mechanical returnless fuel injection system.

compensate for changes in the differential with an adjustment of the injectors' pulse width.

Electronic Returnless Fuel Injection Systems

A further adaptation of a returnless system is known as an **electronic returnless fuel injection** system (Figure 4–14), which allows the PCM to control the fuel supply line pressure by electronically controlling the speed of the fuel pump motor. A fuel pump relay supplies power to enable a fuel pump driver module, or FPDM (Ford), usually located in the area of the vehicle's trunk. (Many newer Ford products now use the Rear Electronic Module, or REM, to control the fuel pump power and ground side.) The FPDM/REM carries the high current flow for the fuel pump in a manner similar to the fuel pump relay in the traditional system, but because the transistor in the FPDM/REM is solid state and therefore has no moving parts, it can be turned on and off many times faster than a mechanical relay. This characteristic allows the fuel pump to be operated on a high-frequency pulse-width-modulated (PWM) signal. The frequency of this signal can be from 500 Hertz (cycles per second) to 3,500 Hertz. The on-time of this PWM signal controls the average current flowing in the fuel pump circuit and ultimately controls the speed of the fuel pump motor. This allows the PCM to control fuel rail pressure via PWM commands given to the FPDM or REM.

In an electronic returnless system, the PCM receives feedback from a pressure sensor on the fuel rail. This sensor not only measures fuel rail pressure, but also measures manifold pressure through a short vacuum hose attached to it. The fuel rail pressure sensor is a differential (delta) pressure sensor (see Chapter 3); it measures the difference between fuel rail pressure and intake

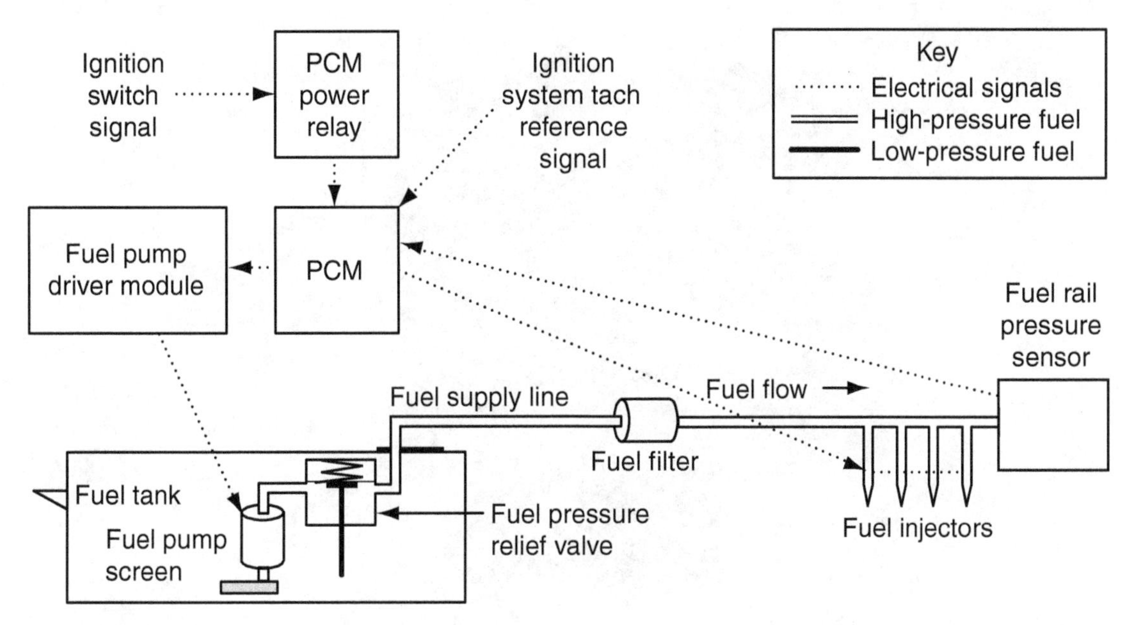

Figure 4–14 Electronic returnless fuel injection system.

manifold pressure. Ultimately, this sensor reports to the PCM the pressure differential across the fuel injectors, which defines the injectors' flow rate when energized. This allows the PCM to accurately calculate the fuel injectors' required pulse width.

Gasoline Direct Injection Systems

A **gasoline direct injection (GDI)** system is a system in which the fuel injectors spray fuel directly into the combustion chamber (Figure 4–15). Robert Bosch Corporation first designed a direct fuel-injected engine for aircraft use in 1937 and then for automotive applications in 1952, so the concept has been around for many years.

In order to spray fuel directly into the cylinder on a modern engine, higher fuel pressures are used, typically from 625 PSI at idle to as high as 2,140 PSI under heavy load. The higher pressure results in a much more finely atomized fuel spray, with the size of the droplets being about one-tenth that of those in a standard PFI engine. The result is more power from less fuel. A GDI engine also uses an overall lean but stratified air-fuel charge during cruise, between 30:1 and 40:1 on vehicles built for U.S. markets and between 40:1 and 60:1 on European vehicles, although the latter also require a specialized NO_x reduction catalytic converter. Also, the majority of modern GDI engines use one or two turbochargers that further increase both performance and fuel economy.

When cruising, the stratified charge mode is accomplished through a combination of the following items:

- Cylinder head and piston dome design
- The placement of the fuel injector
- The timing of fuel injector sprays
- The use of swirl control valves in the intake manifold runner

Swirl Control Valves. By using swirl control valves, also known as *tumble flaps*, in the intake manifold runners to affect the airflow as the air enters the cylinder (Figure 4–16), engineers

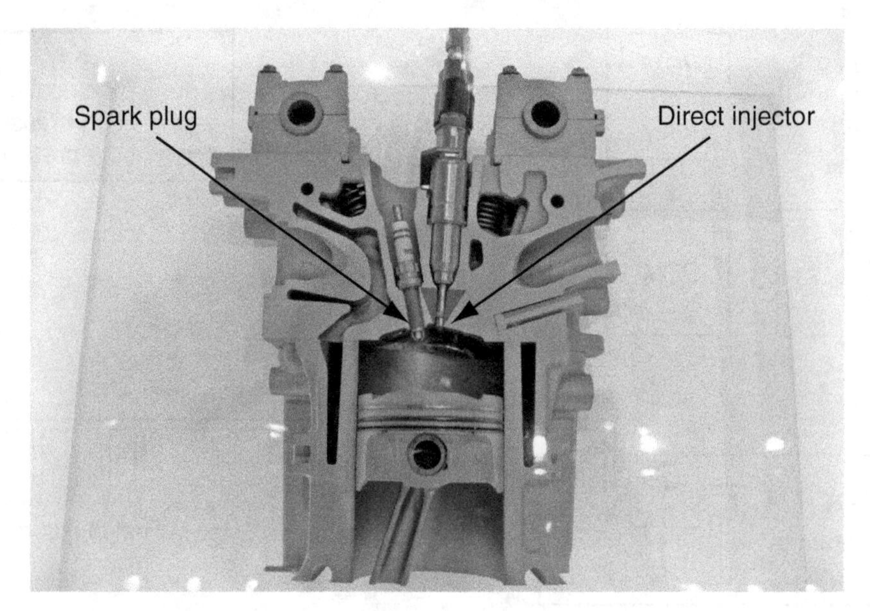

Figure 4–15 GDI cylinder components.

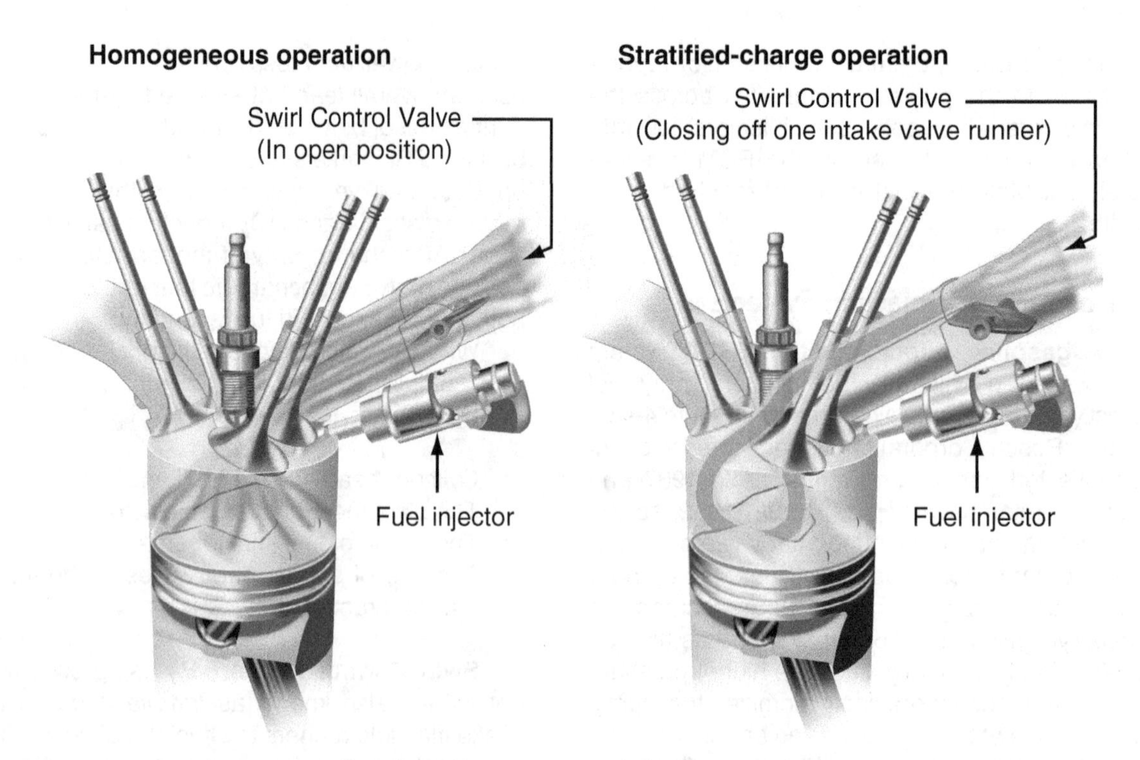

Figure 4–16 GDI cylinder with swirl control valves in the intake manifold runners.

can better control where the fuel sprayed in by the fuel injector ends up in the cylinder. When the swirl control valve is open (both passages now allowing airflow to the two intake valves), the mixing of the air and fuel in the cylinder is more complete, resulting in a homogeneous air-fuel charge that must remain very close to stoichiometric. But when the swirl control valve is closed (now closing up one passage through the intake manifold runner), the airflow's path into the cylinder is changed so as to allow the bulk of the fuel delivered by the fuel injector to remain very close to the spark plug, allowing for an overall-lean stratified air-fuel charge.

When using a stratified air-fuel charge, the engine speed can be controlled through fuel delivery, thus allowing the PCM to hold the throttle plate between 70 and 100 percent open without over-revving the engine. This, in turn, reduces the pumping losses of the crankshaft during cruise conditions because it no longer has to use energy to develop a vacuum in the intake manifold. This further improves fuel economy. However, these engines will typically adjust the throttle during cruise conditions to create around 1 in. Hg of vacuum in the intake manifold so the PCV system can function. And when the PCM must purge the evaporative canister, it will adjust the throttle accordingly in order to create the necessary vacuum, during which time the evaporative canister is purged quickly and efficiently.

Fuel Pumps. A GDI engine uses two fuel pumps, an electric in-tank pump (similar to that on a PFI engine) that supplies fuel to a second, camshaft-driven high-pressure pump (Figure 4–17). The PCM then determines how much pressure the high-pressure pump will deliver to the fuel injectors based on engine load. Most high-pressure pumps are driven by a four-lobe cam, but a few are driven by a three-lobe cam as in Figure 4–17.

Pressure Control Solenoid. A pressure control solenoid is built into the high-pressure pump and must be energized by the PCM to allow fuel pressure to build in the line going to the fuel

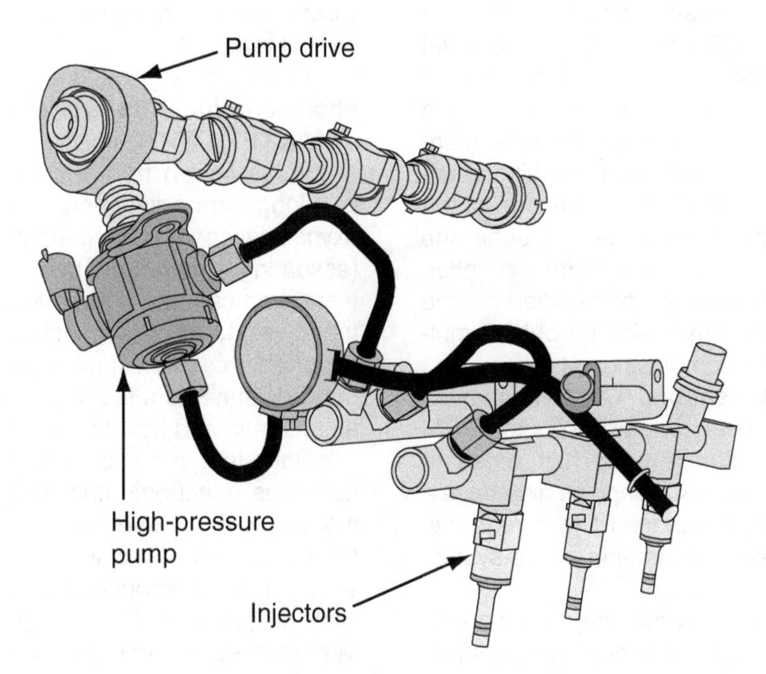

Figure 4–17 Camshaft-driven fuel pump.

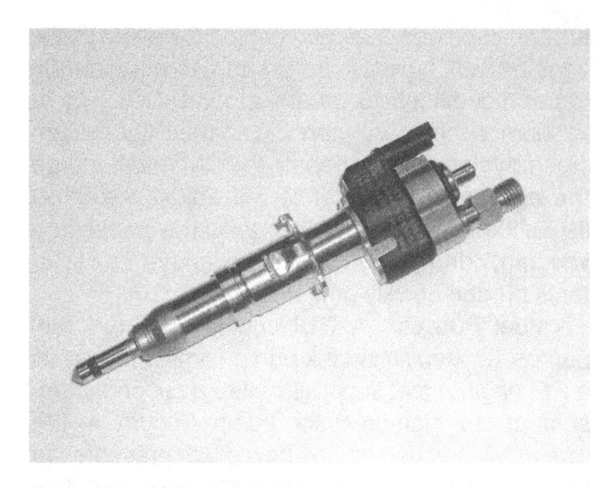

Figure 4–18 GDI fuel injector.

injectors. This solenoid is normally energized by the PCM using a peak and hold strategy, similar to the strategy used to pulse peak and hold fuel injectors.

Fuel Injectors. GDI fuel injectors are fast-response injectors and have a winding with low resistance. A GDI engine uses a transformer and capacitor combination to apply from 80 to 120 V to the GDI fuel injector (Figure 4–18) to initially get the injector open. Then the voltage is decreased to charging system voltage and is either controlled with a peak and hold circuit or is pulse width modulated to achieve the desired on-time while reducing current to a lower level for the duration of the pulse. A GDI injector may be pulsed on either the intake stroke for homogeneous charge operation or on the compression stroke for stratified charge operation. The injector may also be pulsed multiple times during the compression stroke.

Fuel Pressure Sensors. All GDI engines have a fuel pressure sensor that allows the PCM to monitor fuel pressure on the high-pressure side of the system. Most GDI engines also have a pressure sensor that allows the PCM to read fuel pressure on the low-pressure side of the system as well.

Safety Concerns. When diagnosing a GDI engine, use only a scan tool to read fuel pressure on the high-pressure side of the system. The PSI at idle will easily split human skin, and if it is snap-throttled in the service bay, pressure will quickly rise to more than 1,000 PSI. Therefore, if you think you see a possible fuel leak near a fuel injector or fuel rail, DO NOT reach in with your fingers to feel for fuel spray. Many manufacturers also allow technicians to use a scan tool to read pressure on the low-pressure side of the system, including General Motors, Ford, Audi, BMW, and VW.

Because of the high voltage that is initially applied to the fuel injectors, DO NOT EVER back-probe an injector wire/connector on a GDI engine to obtain a voltage waveform. However, you may use a current probe to obtain a current waveform on a lab scope. (See Chapter 5 for additional information on this.)

Fuel Trim

Fuel trim is a function programmed into the PCM on all electronically fuel-injected vehicles, and it affects the PCM's control of the air-fuel ratio. It is designed to allow the PCM to adjust the fuel injectors' pulse width to compensate for the wear and aging of components as they affect the air-fuel ratio.

Fuel trim is divided into two components: **short-term fuel trim (STFT)** and **long-term fuel trim (LTFT).** STFT is the immediate response to the last report from the oxygen sensor. LTFT is a long-term adjustment designed to keep the oxygen sensor averaging around the ideal 450 mV (assuming a homogeneous air-fuel charge). The technician can use a scan tool to look at both of these values as a diagnostic aid. Therefore, it is important that the technician understand how fuel trim adjustments affect the PCM's control of the air-fuel ratio and how to interpret fuel trim data.

Both fuel trims consist of electronic look-up tables (electronic charts) in the PCM's RAM memory chip. These look-up tables are divided (originally) into 16 cells (or blocks) that represent various combinations of engine load and RPM (Figure 4–19, Figure 4–20, and Figure 4–21). Modern systems are likely to use a greater number of cells, not because the range of engine load or

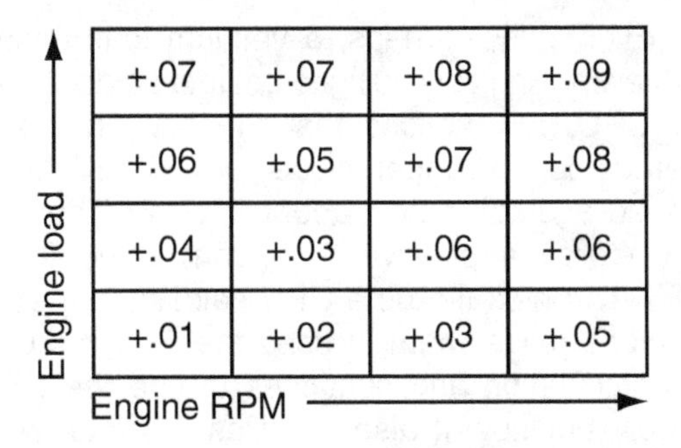

+.07	+.07	+.08	+.09
+.06	+.05	+.07	+.08
+.04	+.03	+.06	+.06
+.01	+.02	+.03	+.05

Figure 4–19 Fuel trim electronic look-up table indicating the PCM's response to an overall slightly lean condition.

Engine load

−.04	−.03	−.03	−.03
−.06	−.05	−.05	−.04
−.08	−.09	−.08	−.06
−.10	−.10	−.09	−.07

Engine RPM

Figure 4–20 Fuel trim electronic look-up table indicating the PCM's response to an overall slightly rich condition.

Engine load

+.04	+.05	+.06	+.09
−.01	+.01	+.05	+.03
−.03	−.02	−.01	+.01
−.05	−.04	−.03	−.01

Engine RPM

Figure 4–21 Fuel trim electronic look-up table indicating the PCM's response to a condition that is rich in the lower range and lean in the upper range.

RPM has been extended, but in order to divide the original cells into smaller cells for more precise control of the air-fuel ratio. On-board diagnostic (OBD) II vehicles use the whole number 1 to represent the default neutral setting. Prior to OBD II, some other values were used, depending upon the manufacturer. For example, pre-OBD II General Motors vehicles used the value 128 as the default neutral setting, derived from the midpoint value of an 8-bit computer (see the discussion of binary code in Chapter 2).

On an OBD II vehicle, the value in a given cell is shown as a percentage of the default value, either positive or negative. As an adjustment is made, this value can move either higher (positive) or lower (negative). Positive percentages indicate that the PCM is adding fuel to the base fuel calculation in response to a "lean report" from the oxygen sensor, and negative percentages indicate that the PCM is subtracting fuel from the base fuel calculation in response to a "rich report" from the oxygen sensor.

When watching these values on a scan tool, the technician will find that STFT values will constantly change (along with oxygen sensor values) if the engine is running in closed loop and the PCM is in proper control of the air-fuel ratio. In fact, not only is the STFT the immediate response to the oxygen sensor's last report, but the STFT adjustment of the fuel injectors' pulse width is what will then cause the oxygen sensor to cross to the other side of the stoichiometric value. That is, the oxygen sensor and the STFT react to each other. It is this interaction that keeps the oxygen sensor cross-counting. LTFT, on the other hand, is a long-term adjustment that is designed to keep a zirconia oxygen sensor averaging around 450 mV (or, at least, between 400 and 500 mV) when STFT is operating in its normal operating range. Therefore, LTFT values, as seen on a scan tool, will tend to be stable, with minor changes occasionally occurring. If, while watching these values, you suddenly rev the engine, you will likely see the LTFT value change. This change in LTFT is not likely the result of an actual change having occurred but is, rather, the result of having entered another cell of the fuel trim table as the engine was revved.

To demonstrate the operation of the fuel trim strategy, let us suppose that a vehicle is traveling at a steady speed, under a consistent load, with the engine fully warmed up and the system in closed loop and with all other inputs remaining constant (set at 60 MPH on cruise control on a long, straight highway with no hills). The central processing unit (CPU) is preparing to turn on a fuel injector. It looks at all the input data that it has recorded in the RAM memory chip. It then looks up this combination of input values in the look-up tables that are programmed into the PROM memory chip (which is tailored to this particular vehicle) and comes up with the base fuel calculation. Perhaps this base fuel calculation is a pulse width of 5.46 ms. The CPU then issues a command of 5.46 ms to the injector's output driver in the input interface.

As the exhaust gases of the resulting air-fuel ratio, having combusted in the combustion chamber, reach the oxygen sensor, the sensor sends a report indicating "slightly rich" to the CPU. The CPU reacts to this by recording an STFT value in the RAM memory chip of minus 3 percent. The next time the CPU turns on the fuel injector, it still sees the same 5.46 ms as the base fuel calculation, but now it is also being instructed by the STFT value to subtract 3 percent, or 0.16 ms. So then the actual pulse width for the next injector firing is 5.30 ms. Of course, now the oxygen sensor sees this as being slightly lean and informs the CPU, which might store an STFT value of positive 3 percent. With the next injector firing, the CPU now takes the base fuel calculation of 5.46 ms and adds 3 percent or 0.16 ms to it and commands a total pulse width of 5.62 ms. As a result, the pulse width varies from 5.30 ms to 5.62 ms, and the oxygen sensor is cross-counting to maintain an average of about 450 mV as the system operates in closed loop. (The variance of the fuel injectors' pulse width can also be seen on an oscilloscope and indicates that the system is in proper control while operating in a closed loop.)

Eventually, perhaps, a vacuum leak occurs. With everything continuing to operate as described, the CPU now notices that the oxygen sensor average is no longer close to 450 mV but is instead averaging only about 300 mV. The CPU now stores a positive 5 percent value in the LTFT. The next time the CPU calculates the fuel injector's pulse width, it uses the 5.46 ms base fuel calculation and continues to use the STFT values, but now it also modifies the total pulse width by the LTFT value, which indicates that 5 percent of the base fuel calculation, or 0.27 ms, must be added. As a result, while operating in a closed loop, the pulse width is no longer varying between 5.30 ms and 5.62 ms but now is varying between 5.57 ms and 5.89 ms. Ultimately, a stoichiometric air-fuel ratio continues to be achieved, in spite of the vacuum leak.

A problem may become so severe that the PCM, while trying to respond to it within the fuel trim program, simply cannot make a large enough adjustment, at which point the fuel trim values are said to have reached their threshold limits. In this scenario, the PCM will no longer be able to maintain a stoichiometric air-fuel ratio and emissions will increase. Driveability and fuel mileage will also likely be affected.

By monitoring the fuel trim values with a scan tool, a technician can get some idea of the types of problems to which the PCM has already responded or is trying to respond. In order to use fuel trim as a diagnostic tool, add the average (median) STFT to the LTFT value in order to obtain a **total fuel trim (TFT)** value. Most fuel trim programs are limited to a total adjustment of around 20 to 25 percent, either positive or negative. Generally, a TFT value that exceeds around 10 percent, either positive or negative, is cause for concern. Many systems that are operating in good closed-loop control will display a TFT value of less than 5 percent. Unfortunately, most aftermarket scan tools can only display the STFT and LTFT values for the RPM/load cell that the engine is currently operating in. Automotive Test Solutions (www.automotivetestsolutions.com) offers a PC-based scan tool (software and hardware) that can display the fuel trim values for all cells simultaneously, as a visual chart, that have been recorded by the PCM and can show in a color-coded format the severity of all fuel trim adjustments.

If an average STFT value is extreme in one direction (either positive or negative) and the LTFT adjustment is extreme in the opposite direction, resulting in a TFT value that is relatively close to 0 percent total adjustment, consider the possibility that someone has recently made a repair that the PCM is still learning to adapt for. As the vehicle continues to be driven, the PCM's LTFT adjustments will gradually move closer to a 0 percent adjustment as it continues to watch STFT, thus allowing the average STFT to move closer to 0 percent adjustment as well.

Lambda and the Air-Fuel Ratio

German engineers have long used the Greek letter **lambda** (λ) to represent a ratio reflecting the air-fuel ratio. The relationship lambda represents is:

> Lambda equals the actual inducted air quantity divided by the theoretical air requirement.

The theoretical air requirement is 14.7 parts, by weight, to each one part of fuel. If the actual air quantity was 14.7 parts to each one part of fuel, then lambda would equal 1, which represents the stoichiometric air-fuel ratio.

Using this method of expression, an air-fuel ratio richer than 14.7:1 is expressed as a number less than 1. For example, if the actual quantity of air was 14.259 parts to each one part of fuel, then lambda would equal 0.97 ($14.259 \div 14.7 = 0.97$). We could then deduce that the air-fuel ratio was 3 percent to the rich side of stoichiometric (the difference between 1.00 and 0.97).

Conversely, an air-fuel ratio leaner than 14.7:1 is expressed as a number greater than 1. For example, if the actual quantity of air was 15.288 parts to each one part of fuel, then lambda would equal 1.04 ($15.288 \div 14.7 = 1.04$). We could then deduce that the air-fuel ratio was 4 percent to the lean side of stoichiometric (the difference between 1.00 and 1.04). Modern engine computers are programmed to keep lambda within 2 percent of lambda 1 (0.98 to 1.02) when delivering a homogeneous air-fuel mixture. This keeps emissions at their lowest levels. (Obviously, when an engine is operating in a stratified charge mode, we expect the lambda value to be potentially much greater than this.)

Incidentally, the stoichiometric air-fuel ratio of 14.7:1 assumes straight gasoline with no alcohol. If you are running 10 percent ethanol, because of the additional oxygen molecules in the fuel, the stoichiometric air-fuel ratio is 14.29:1. And with E85 the stoichiometric air-fuel ratio is 9.87:1. However, because oxygen sensors and air-fuel ratio sensors detect oxygen content in the exhaust stream regardless of its source, whether the oxygen molecules are from the air drawn into the combustion chamber or from the oxygen content of the fuel metered into the cylinder, modern PCMs can monitor lambda accurately regardless of the alcohol content in the fuel. While the alcohol content of the fuel may change the desired air-fuel ratio, the desired lambda value of a homogeneous air-fuel charge is always 1 ± 2 percent, which always represents stoichiometric. The exception is the engines that use a stratified charge coupled with the use of wideband oxygen sensors (discussed in Chapter 3). During cruise conditions, these engines may run a total air-fuel ratio that is much leaner than this.

Summary of Fuel Trim and Lambda

Finally, it is important to note that fuel trim is not lambda and lambda is not fuel trim. Rather, fuel trim is the adjustment that the PCM makes in order to achieve a lambda value of between 0.98 and 1.02 out the exhaust. Therefore, TFT can be showing that an excessive amount of correction is being made, while simultaneously, lambda continues to remain within 2 percent of lambda 1. The only time that lambda will be out of this range while operating in closed loop on a homogeneous engine will be if the fuel trim program has reached its limits of adjustability. This scenario will illuminate the malfunction indicator lamp (MIL) and set a fault code in the PCM's memory if the vehicle is an OBD II vehicle.

Rear Fuel Control. While for many years a PCM would calculate the lambda value from the average of the front (pre-cat) oxygen sensor as it cross-counted in closed loop, in 1988 Toyota and Saab added a rear (post-cat) oxygen sensor to some of their applications to allow the PCM to calculate lambda from a more stable signal. Because a catalytic converter neither creates nor destroys matter, but only changes chemical composition from one set of chemicals to another, and assuming that there are no exhaust leaks ahead of the rear oxygen sensor, whatever lambda value is calculated from the rear oxygen sensor will equate to the average lambda value from the front oxygen sensor. It is simply easier to calculate lambda from the rear sensor's more stable signal. This is referred to as **rear fuel control**.

Between 1994 and 1996 other vehicle manufacturers added the rear oxygen sensor to meet the OBD II catalyst efficiency monitor requirement. By 1997 or 1998 most other manufacturers began reprogramming their PCMs to also use the rear oxygen sensor for rear fuel control since the physical hardware was now already present to meet the catalyst efficiency monitoring requirement.

In a system that uses rear fuel control, the PCM will adjust the pulse width of the fuel injectors to affect the duty cycle (lean time versus rich time) of the front oxygen sensor in order to achieve a correction in the lambda value as reported by the rear oxygen sensor.

As most technicians know, it has always been critical to avoid having any exhaust leaks ahead of the front oxygen sensor due to the ambient air that gets sucked into the exhaust stream. This air leak will cause the front oxygen sensor to report a false lean condition to the PCM while operating in closed loop. However, with a vehicle that uses rear fuel control, it is equally important to avoid having any exhaust leak ahead of the rear oxygen sensor. If there is any ambient air leak into the exhaust ahead of the rear oxygen sensor (including that coming through the body of a cracked oxygen sensor housing), the PCM will drive the air-fuel ratio extremely rich, even possibly to the point of an engine stall, while operating in closed loop.

Because of this, modern secondary air injection systems (using an electric air pump) no longer have a provision to pump air to the oxidation bed of the catalytic converter as older systems did. The pump is turned on during engine warmup only, pumping air to the exhaust manifold. Once the PCM is ready to move into closed loop control of the air-fuel ratio, the pump is turned off.

Another important point to note is this: As wideband oxygen sensors are now being implemented as post-cat sensors, the rear sensor is now being given a greater priority over fuel control.

Flexible Fuel Vehicles

Another variation of a fuel injection system is that of a **flexible fuel vehicle (FFV)**. An FFV is one that allows for the use of alcohol concentrations in excess of 15 percent, such as E85 fuel, which contains 85 percent ethanol. Remember, ethanol contains oxygen. If the driver were to put E85 into a non–flexible fuel vehicle, the first and most prominent problem that would be immediately noticeable would be driveability issues. That's because the fuel trim program on a non–flexible fuel vehicle cannot compensate for how lean the E85 would drive the air-fuel ratio. Remember that the stoichiometric air-fuel ratio for straight gasoline is 14.7:1, for a 10 percent ethanol blend is 14.29:1, and for E85 is 9.87:1. On a non–flexible fuel vehicle, the PCM's fuel trim program can adjust far enough to accommodate a 10 percent ethanol blend and still maintain a stoichiometric air-fuel ratio, but it cannot adjust far enough to accommodate an 85 percent ethanol blend without reaching its limits of adjustability.

An FFV has both enhanced hardware and software that can accommodate the high level of ethanol being used. Hardware changes include enhancements to fuel system components, allowing them to be able to handle the increased ethanol without system damage. Chrysler states that the fuel system components must be upgraded to stainless steel.

But another required change is the additional ability of the PCM to determine the fuel's composition. In other words, the PCM must be able to determine what percentage of fuel in the fuel tank is gasoline and what percentage is ethanol, even if the driver had a partial tank of gasoline that was topped off with E85 or a partial tank of E85 that was topped off with gasoline.

Hardware-Type Fuel Composition Sensor

There are currently several patents held by the major manufacturers for hardware-type fuel composition sensors. These sensors include types such as optical sensors as well as sensors that measure energy vibration that is absorbed by the fuel. These sensors tend to be somewhat pricey, typically around $600, when you need to replace a failed unit. Chrysler used a hardware-type fuel composition sensor on its early FFVs called a "Smart Sensor." While these hardware-type sensors are found on many vehicles currently in use, they have been phased out and replaced with software-type fuel composition sensors.

Software-Type Fuel Composition Sensor

In 1998, Chrysler began replacing its hardware-type fuel composition sensors with software within the PCM, enabling Chrysler engineers to remove the Smart Sensor from the system. This in turn resulted in reductions in the cost of the FFV system. General Motors' engineers also introduced a software-type sensor on the Chevrolet Impala. The PCMs on these systems use input from the oxygen sensor and the fuel trim program along with complex algorithms that involve informing the PCM when the tank is refueled.

Ultimately, when the tank is refueled, the PCM will perform some tests involving its control of the air-fuel ratio in order to determine what percentage of the fuel is gasoline and what percentage is ethanol. The premise of these tests is that ethanol fuel contains oxygen and will cause the oxygen sensors to read leaner than pure gasoline does. You may recall that ethanol has been classified as an oxygenated fuel and has been added to gasoline in smaller concentrations (10 percent) for the purpose of reducing CO emissions for several decades.

Software fuel composition sensors are essentially a program that allows the PCM to remember the vehicle's fuel level during periods that it is shut down. Upon a restart, the PCM does a quick comparison of the last memorized fuel level to the current fuel level. If it is apparent that fuel was added to the tank, then it waits until the engine is operating in closed-loop fuel control and then looks for any changes in rich/lean tendencies. If a sudden rich/lean change is identified, the PCM will attribute this change to a change in the percentage of ethanol that is in the fuel. For example, if the engine is running leaner than the last time the engine was running and the fuel tank has just been filled during the last shutdown, the PCM will attribute this lean condition to an increase in ethanol, rather than to a vacuum leak, and will shift its fuel trim program to one that has been programmed specifically to allow for the use of E85.

The advantages of E85 are that it is less expensive than straight gasoline and 10 percent ethanol blends and that it also can result in more power because of the oxygen content in the fuel. The disadvantage is that E85 does not have the heat energy content of gasoline, so fuel economy is decreased.

Unfortunately, a driver may mistakenly add E85 to the fuel tank of a non-FFV, or may even choose to do so purposely because of the cheaper cost. Either way, his vehicle may get brought into your shop to diagnose engine performance problems that are simply the result of more ethanol in the fuel that the PCM's fuel trim program can compensate for.

Electronic Throttle Control Systems

For many years fuel injection systems allowed the driver to control the volume of intake air through the throttle body and into the engine while the PCM only controlled fuel delivery. Eventually the PCM was given charge of controlling both the fuel delivery and the intake air volume. This is an **electronic throttle control (ETC)** system. Alternative names include *throttle actuator control* (TAC) and *drive-by-wire*.

With an ETC system the physical throttle cable is eliminated and the PCM controls throttle plate position electronically through a reversible DC motor (Figure 4–22 and Figure 4–23). The PCM controls the DC motor using a high-frequency PWM signal. It can reverse polarity as needed through the use of an H-gate. Through this PWM signal the PCM can choose to open or close the throttle plate from very quickly to very slowly.

Before the arrival of ETC systems, the throttle position sensor (TPS) was the sensor that sent the PCM "driver demand" information—that is, whether the driver wished to accelerate, maintain current speed, or decelerate. With an ETC system, the "driver demand" information now originates at the accelerator pedal position sensor (APPS or APP sensor) (Figure 4–23).

Figure 4–22 ETC throttle body.

The TPS is retained for feedback purposes so that the PCM can verify its control of the throttle plate.

On early ETC systems both the APPS and the TPS were potentiometers. One identifiable problem with a potentiometer-type sensor is that it is the only sensor type that has a physical wear factor as the wiper wipes across the length of the resistive carbon material. Beginning in 2004, Toyota began using Hall effect sensors for the APPS on the Prius. By 2010 or 2011 most manufacturers were using Hall effect APP sensors because they are considered a "noncontact" sensor and don't experience the wear problems of a potentiometer. However, there is one difference between the Hall effect sensors used as ignition system pickups and those used as APP sensors—those used as APP sensors do *not* digitize their output voltage. (Remember, the Hall effect sensors used as distributor pickups, crankshaft position sensors, and camshaft position sensors contain the electronics to both amplify and digitize their output signals.) With an APPS, as the driver pushes the throttle pedal down, a magnet rotates within the Hall effect sensor and influences the sensor to change its output voltage in an analog fashion.

With an ETC system that used potentiometers for the APPS, most manufacturers used two potentiometers in the APPS assembly (Figure 4–24). One provided a redundant backup for the other. The same was true for the TPS—two were used within the TPS assembly, with one providing a redundant backup for the other. With many manufacturers, the voltage signal of one would cross the voltage signal of the other as the throttle pedal was depressed or as the throttle plate was opened (one signal moved from low to high as the other signal moved from high to low). Ford Motor Company used three potentiometers within the APPS assembly—as the throttle pedal was depressed, two of the signals moved from low to high (but within a slightly different voltage range) and the third signal moved from high to low.

With a Hall effect APPS, two sensors are also used. One common method is for one sensor

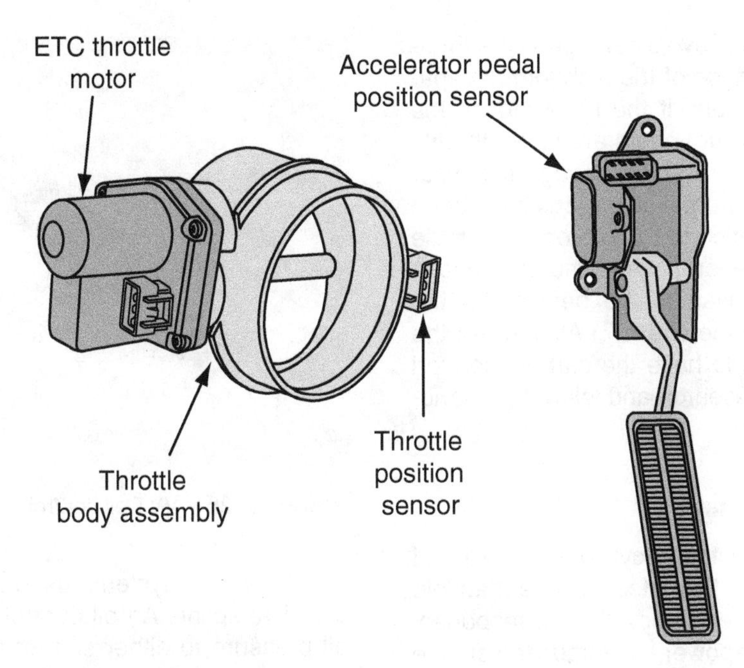

Figure 4–23 ETC throttle body and APP sensor.

to operate in a 0- to 5-volt range while the other sensor operates from 0 V to 2½ V.

ETC System Benefits. The addition of an ETC system offers several advantages. For example, there is no longer a need for a dedicated idle air control actuator to control idle speed. Also, the dedicated speed control servo is now eliminated.

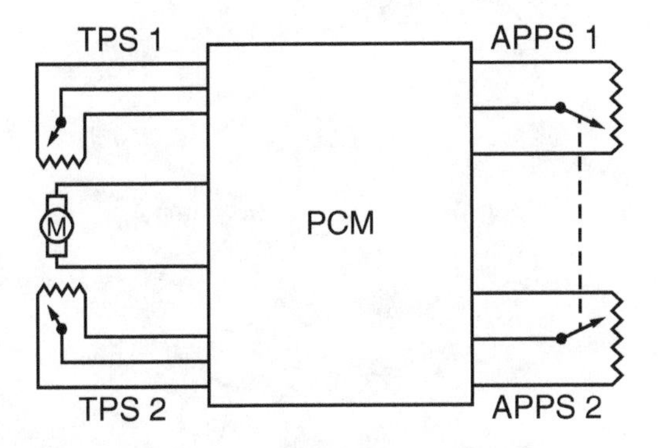

Figure 4–24 Schematic of typical ETC system sensors and throttle control motor.

And an ETC system works in conjunction with the traction control system to assist in reducing engine torque when needed.

ETC System Failure Modes. If one of the two APP sensors were to fail, vehicle operation should be fairly normal. If one of the two TP sensors were to fail, the vehicle's speed would be limited. In either event, the PCM will illuminate the malfunction indicator lamp (MIL) and will set a diagnostic trouble code (DTC) in memory. If both APP sensors were to fail, the PCM would default to an *idle only limp-in mode*. Likewise, if both TP sensors were to fail, the PCM would also default to the idle only limp-in mode.

ETC System Cautions. The throttle body on an ETC system is geared to allow the motor to move the throttle plate with a considerable amount of torque. Because of this, do not ever try to open the throttle plate manually—you may damage the throttle body gears. The throttle plate should only be opened by the PCM. (Note: Some ETC throttle plates are spring-loaded, but caution should still be taken to avoid damage to the throttle body.) Also, because of the high torque capability of these actuators, if the PCM moves the throttle plate while your fingers are in the throttle body area, the torque is high enough that your fingers could be severed. (Technicians have been known to use a wrench to try to prop the throttle plate open before performing a compression test, which, in turn, resulted in the bending of the throttle plate and/or the wrench.) Also, when the throttle body needs to have the carbon cleaned from it, be sure to research and follow the manufacturer's instructions.

Variable Valve Timing

Valve timing affects the cylinders' intake and exhaust flow. In turn, this affects intake manifold vacuum, volumetric efficiency, throttle response, and how much horsepower and torque the engine can develop. Most engines over many years have used a camshaft that has a fixed timing. This fixed timing is, at best, a compromise. Advancing valve timing improves idle quality and low-RPM torque. Retarding valve timing boosts high-end performance.

In an effort to extract more performance from smaller engines while providing good driveability in the low-speed operating range of the engine and keeping vehicle emissions in check, manufacturers implement a **variable valve timing (VVT)** system. Also sometimes referred to as a *variable cam timing* (VCT) system, such a system can vary the timing of the valves' opening and closing relative to the crankshaft's position.

Functional Concept. The PCM can vary valve timing by rotating a camshaft in its relationship to the crankshaft. This is commonly done by using a hydraulic control signal that is under the control of a PCM-controlled solenoid-operated valve (Figure 4–25). The hydraulic signal is applied to a cam phaser to change the position of the camshaft gear on the camshaft. This rotates the camshaft in its relationship to the cam gear and changes the camshaft's relationship to the crankshaft.

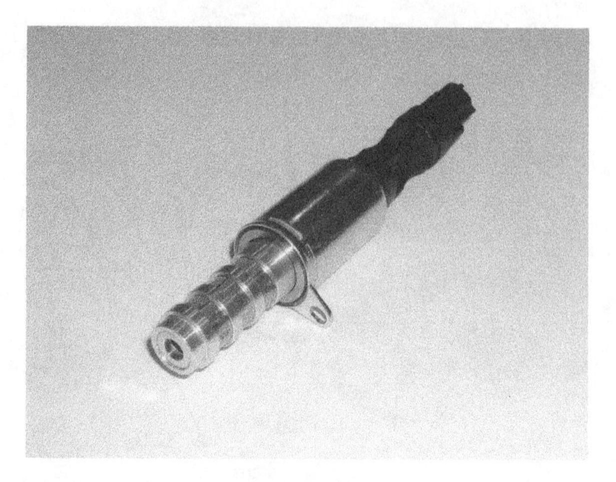

Figure 4–25 VVT oil control solenoid.

Early VVT systems used a cam phaser with a helical spline. An oil control solenoid provided oil pressure to either side of a piston within the cam phaser assembly to rotate the camshaft. Modern VVT systems use a vane-style cam phaser (Figure 4–26). Note the eight oil seals in the cam phaser assembly shown in Figure 4–26. Four are located on the outer legs of the vane

Photo courtesy of William K. Bencini.

Figure 4–26 VVT actuator, or cam phaser.

and four are located on the inner legs of the gear sprocket. Vane-style cam phasers are more responsive than helical gear cam phasers. Either type of cam phaser is under the control of a pulse-width-modulated oil control solenoid valve. This provides the PCM with the ability to precisely control the camshaft's timing.

On most systems the default of the VVT system is for the cam phaser to be spring-loaded to the fully retarded position. The oil control solenoid applies oil pressure to advance cam timing, but it can also apply oil pressure to quickly retard valve timing. General Motors Ecotec engines are one example of a system that functions differently in that the cam phaser is spring-loaded to the fully advanced position.

Concerning which camshafts and valves may be under the control of a VVT system, several variations are used, as follows:

- Exhaust phase shifting (EPS): This design varies the timing of the *exhaust* camshaft only on a dual overhead cam (DOHC) engine. This design allows the PCM to retain some exhaust within the cylinder in order to reduce NO_x emissions. GM and Ford used this approach on some of their early VVT systems.
- Intake phase shifting (IPS): This design varies the timing of the *intake* camshaft only on a DOHC engine. This design allows some improvement in idle quality and low-end response while providing better high-end performance.
- Dual equal phase shifting (DEPS): This design varies the timing of both the intake and exhaust valves equally and simultaneously. This design is used on an engine with one camshaft and allows for moderate improvement in the idle quality and low-end response while providing a moderate improvement in high-end performance. This design was first used by Ford on Triton light truck engines.
- Dual independent phase shifting (DIPS): This design varies the timing of both the intake and exhaust camshafts independently on a DOHC engine. This design provides the ultimate improvement in the idle quality and low-end response while also providing the ultimate improvement in high-end performance. It is commonly used on many applications today.

Both the EPS and the DIPS can be used to help control NO_x emissions by retaining some exhaust gas within the cylinder. However, although an engine with these variable valve timing systems is less likely to have an EGR system, some high-compression engines with these systems continue to also use an EGR system to control NO_x emissions.

The use of variable valve timing also allows a higher-compression engine to use fuel of a lower octane (as opposed to an engine that cannot vary its valve timing) without creating a detonation problem. Ultimately, this system is designed to increase the high-performance capability of the engine while maintaining good driveability and emissions.

While most VVT systems use engine oil pressure to control them hydraulically, Ford Motor Company was the first to introduce an electromagnetic cam phaser. It is controlled by the PCM through a pulse-width-modulated signal to the cam phaser's winding so as to control electromagnetic field strength.

Idle Speed Control

Manufacturers use two primary methods to achieve PCM control of the engine's idle speed. One method is to manipulate the physical position of the throttle plate(s). This concept began with feedback-carbureted systems and carried over into some fuel injection systems. This method worked well with feedback carburetors because as the PCM caused the throttle plate angle to change position, the carburetor automatically compensated with the correct amount of fuel. However, once this concept was applied to fuel injection (usually TBI systems), the fuel compensation was no longer automatic in that the PCM's adjustment of the throttle plate(s) was now only controlling airflow.

The other method is to leave the physical position of the throttle plate(s) at the hard stop screw setting (known as the *minimum airflow rate*) and instead, control the volume of air that bypasses the throttle plate(s), known as *throttle bypass air.* The actuator that the PCM uses to do this is typically either some form of a stepper motor adjusting a pintle's position or some form of N/C solenoid that is controlled by the PCM on either a duty cycle or a high-frequency PWM command. Those solenoids that are operated on a duty cycle are controlled at a relatively low frequency (around 10 Hz), thus allowing the solenoid to completely open and close about 10 times per second while controlling the percentage of duty cycle on-time. Those solenoids that are operated on a PWM command are operated at a frequency typically between 800 Hz and 2,500 Hz while the PCM controls the pulse width of the signal. The solenoid's valve does not actually open and close with each current pulse, but instead will float in a certain physical position according to the average current and magnetic field strength of the solenoid's winding, thus allowing a certain amount of air to bypass the throttle plate(s). The various methods of controlling throttle bypass air were used in most fuel injection systems until ETC systems were introduced.

In an ETC system, the idle speed actuator is eliminated because the PCM now has the ability to control the throttle plate angle directly. In a sense, ETC is a return to the original concept of manipulating the position of the throttle plate(s).

Fuel Injection Systems Summary

While there are many manufacturers, each producing many different models of vehicles, today's vehicles all use electronically controlled fuel injection systems that have common features, whether they are of U.S., European, or Asian design. The commonalities of all electronically pulsed fuel injection systems are these:

- The amount of fuel metered into the engine by the PCM is dependent upon how much air is entering the engine. The PCM knows this value either through a calculation known as the *speed density formula* (identified by the use of a MAP sensor) or through a measurement using a MAF sensor.
- The PCM meters the appropriate amount of fuel into the engine using a fuel injection system of one of the following designs: a mechanical return-type system, a mechanical returnless system, an electronic returnless system, or a gasoline direct injection system.

IGNITION SYSTEM OPERATING PRINCIPLES

Ignition Coil Design and Operation

Because it is critical to understand the operation of an ignition coil before proceeding, a quick description of operation is given here. It is a voltage step-up transformer. It allows the primary winding to induce an intense voltage spike into the secondary winding. This voltage spike in the secondary winding becomes the spark that fires across the spark plug gap. To accomplish this, current flow is turned on and off in the primary winding. When current flow is turned on, the transistor (switch) in the primary circuit completes only the path for current to flow, thus allowing the voltage that is present to begin pushing current. (The transistor itself does not push current; rather, it only completes the circuit. Visualize the *opening* of a solenoid-operated valve that might control the flow of water. It only completes the path.) Thus, current flow builds to full saturation rather slowly (Figure 4–27). But when the transistor is turned off, the transistor itself forcibly stops current flow. (Visualize the *closing* of a solenoid-operated valve that might control the flow of water. Closing the valve forcibly stops the flow.) Therefore, current flow stops much more rapidly than it starts.

Each time current flow is turned on, relative motion exists between the resulting expanding magnetic field and the windings. Each time

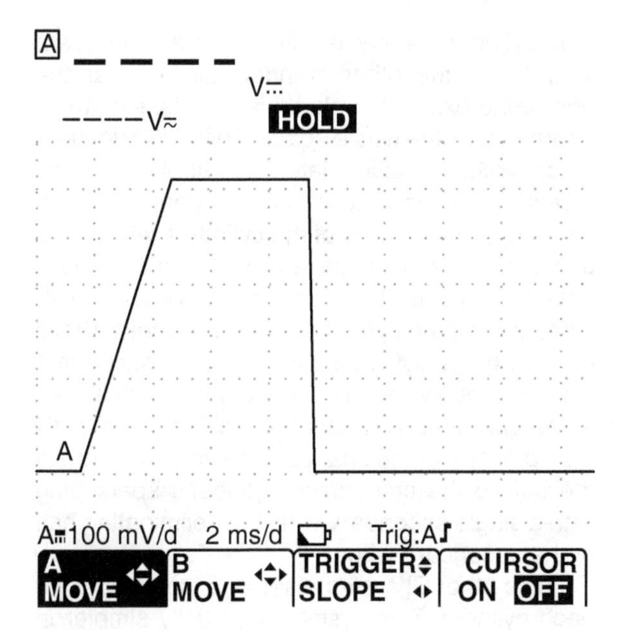

Figure 4–27 Current waveform for a typical ignition coil.

current flow is turned off, relative motion also exists between the collapsing magnetic field and the windings. However, for this relative motion to induce a voltage spike (or voltage imbalance) in the windings, the relative motion must happen quickly. This speed is only achieved when the current flow is turned off, never when it is turned on. Therefore, electromagnetic windings produce voltage spikes only when the current flow is turned off.

In the case of an ignition coil, primary current flow is turned on in anticipation of when it will be turned off. When it is turned off, a voltage imbalance (voltage potential) is induced in both windings. The potential strength of this voltage spike is very low in the primary winding and is dissipated. But in the secondary winding, the potential strength of this voltage spike is much greater and will climb either to the coil's capacity or to the level that it needs to achieve to dissipate. The resistance in the secondary circuit determines this. A jumper wire placed between the secondary winding's negative and positive terminals would allow the excess of electrons (negative) to dissipate to the deficiency of electrons (positive) without much voltage buildup.

In the case of an ignition coil, a complete path must always exist for the secondary spark to flow, allowing electron flow not only from the negative end of the secondary winding but also back to the positive end of the secondary winding. Without this complete path, the voltage potential in the secondary winding would not cause the spark's electrons to flow. It is important at this point to understand that the secondary winding becomes a voltage source each time primary current flow is turned off.

Secondary Ignition System Types

There are three types of secondary ignition system designs. The first uses a single ignition coil to fire all cylinders. The spark from this single coil is distributed mechanically using a physical distributor. Thus this system is known as a **distributor ignition (DI)** system. Both of the other types of ignition systems use multiple ignition coils, in which case the correct coil to be fired must be chosen electronically. Because the spark is said to be distributed electronically, these systems are referred to as **electronic ignition (EI)** systems.

In a DI system, the spark leaves the negative end of the coil primary winding and passes through the distributor and then to the spark plug. After jumping the spark plug gap to engine ground, it returns to the positive end of the secondary winding, either through the battery and positive primary circuitry or, if one is provided, through a more direct circuit.

In the first of the EI (or multiple coil) systems, known as **waste spark** systems, each coil is used to fire two spark plugs located on opposite (or companion) cylinders in the firing order. Each time a coil is fired, the spark leaves the negative end of the secondary winding and fires across one spark plug gap from the center electrode to the side electrode (Figure 4–28). (This plug is said to be firing *negatively* because the center electrode has a negative polarity.) Then the

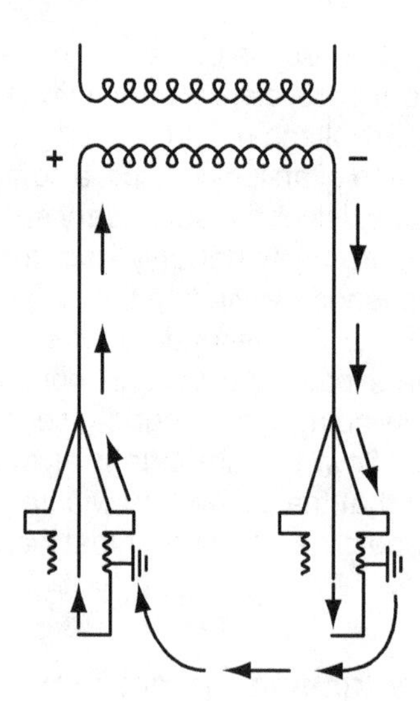

Figure 4–28 Current flow in a waste spark EI system.

spark's electrons move through the metal of the engine block to the second plug, which is wired in series with the first plug. This fires the second plug across the plug gap from the side electrode to the center electrode. (This plug is said to be firing *positively* because the center electrode has a positive polarity.) The spark then continues back to the positive end of the secondary winding, at which point the voltage imbalance of the secondary winding is fully dissipated. In actual operation the two spark plugs fire simultaneously. (Visualize a water pipe that is full of molecules of water, similar to a copper wire that is full of loose electrons. If you force another water molecule into one end, you simultaneously receive a water molecule out the other end. You do not have to wait until the molecule forced into one end is received at the other end in order to have water flow out the other end. Similarly, electron flow throughout an electrical circuit happens simultaneously as the voltage imbalance pushes one electron into one end of the circuit and simultaneously receives an electron at the other end.)

Each time a waste spark ignition coil fires the two plugs, one of the cylinders will be approaching TDC near the end of the compression stroke (this cylinder's spark is referred to as the *event spark*) and the other cylinder will be near the end of the exhaust stroke (this cylinder's spark is referred to as the *waste spark*). Then, 360 degrees of crankshaft rotation later, the strokes will be reversed between these two cylinders; but each time the coil fires, one of the cylinders will always use the spark to ignite an air-fuel charge. Coil polarity stays the same regardless of which cylinder is completing the compression stroke. While most of these systems use two secondary wires to connect a given coil to the two plugs that it fires, some applications mount the ignition coil directly on top of one of the spark plugs and then connect the coil to the companion cylinder's spark plug with a single secondary wire, thus eliminating half of the secondary wires.

The other EI system uses one ignition coil for each cylinder. This system is actually simpler in concept than the waste spark system. Each coil fires only one spark plug from the center electrode to the side electrode (Figure 4–29). Then a direct path is provided for the spark back to the positive end of the secondary winding. These EI systems, generally referred to as *coil-on-plug* (COP), have the ignition coil situated physically on top of the spark plug that it fires. Some applications mount the coil near the spark plug rather than on the plug (due to the close proximity of the exhaust manifold) and then use a short secondary spark plug wire to connect them. This is

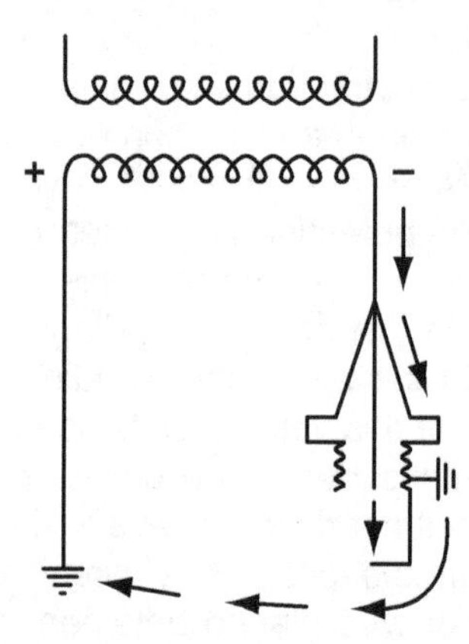

Figure 4–29 Current flow in a COP EI system.

usually referred to as *coil-near-plug* (CNP). Either design may also be referred to as *coil-per-plug* (CPP). An advantage of the COP design is that, because the coil is mounted directly on the plug and no secondary wires are used, it reduces the potential for secondary wiring to induce voltages in other nearby circuits.

It should also be noted that many COP ignition coils contain their own driver (transistor). These simply receive a digital control signal from the PCM to the base leg of the driver.

Primary Ignition System Levels of Information

The information that is provided to the ignition module and/or PCM can be broken down into three distinct levels of information. Understanding these various levels of information will enable you to apply these concepts to any system you might encounter.

The first level of information that might be provided to the ignition module and/or PCM is simply that "*a pair* of cylinders' pistons is approaching TDC." No further information is provided. Systems that would require only this level of information include DI systems in terms of spark management, and feedback-carbureted systems and TBI systems in terms of fuel management. In these systems, it is not necessary for the PCM to know which cylinder is the number one cylinder. (In a DI system, it is the distributor's responsibility, not the PCM's, to deliver the spark, once created, to the proper cylinder's spark plug.)

The second level of information that might be provided to the ignition module and/or PCM is "*which pair* of cylinders' pistons is approaching TDC." This information is needed to correctly choose which coil to fire in a waste spark system. (The PCM does not need to have any additional information because if the correct waste spark coil is fired, both cylinders' spark plugs will fire, and an air-fuel charge will be ignited, regardless of which cylinder was on the compression stroke.)

The third level of information that might be provided to the ignition module and/or PCM is "*which cylinder* of the piston pair approaching TDC is on the compression stroke." This information is needed to correctly choose which coil to fire in a COP system. It is also needed for any sequentially injected PFI or GDI system.

Ultimately, the information that is provided is never more than what is needed for proper system operation, and therefore the level of information provided is always determined by the systems that are used in a given application.

Primary Ignition System Sensors

In a multiple coil system, information of the first level will originate at a **crankshaft position (CKP) sensor.** (Few exceptions to this exist, but one exception was a Mitsubishi multiple coil ignition system produced in the 1990s that used dual optical camshaft sensors only and no crankshaft sensor.) The reason that most multiple coil systems take the first level of information from the crankshaft is that the timing of this signal can be engineered to be permanently fixed, without having to allow for adjustments to compensate for timing chain stretch. This will be the primary tach reference signal. If this signal is lost, no spark plug will fire, no fuel injector will be pulsed, and the fuel pump relay will be de-energized, resulting in an immediate stall of the engine and no restart. Figure 4–30 shows a Hall effect CKP sensor that provides the first level of information

Figure 4–30 Hall effect CKP sensor and harmonic balancer with three vanes and three windows, all equally sized and equally spaced.

only—all vanes are the same size and are equally spaced. When the PCM sees a leading edge of a vane, it only knows that a *pair* of pistons are approaching TDC.

Information of the second level may or may not originate at the CKP sensor. That is, on applications where this level of information is needed, the CKP sensor can be modified to inform the ignition module and/or PCM of this level of information. Figure 4–31 shows a magnetic CKP sensor that provides both the first and second levels of information due to an extra offset notch in the reluctor wheel that is not evenly spaced with the others. This allows the PCM to know the exact position of the crankshaft, thus indicating *which pair* of pistons are approaching TDC.

Information of the third level cannot be obtained from a CKP sensor. No matter how it may be modified, a CKP sensor will always "look the same" on both revolutions that, together, make one full cycle on a four-cycle engine. Rephrased, the PCM will not be able to tell the compression stroke from the exhaust stroke if it only uses a CKP sensor. However, this information can be provided by a **camshaft position (CMP) sensor** due to the fact that a camshaft will only make one revolution to complete all four cycles on a four-cycle engine. Therefore, if the application uses either SFI or a COP ignition system, it will require a CMP signal.

When a CMP sensor is present, the second level of information may come from the CMP sensor simultaneously with the third level of information, leaving only the first level of information to be derived from the CKP sensor. However, if it is used with a waste spark ignition system, the ignition module and/or PCM will not be able to fire a coil until it has substantiated this second level of information. Potentially, it could take up to one full rotation of the camshaft to obtain this level of information, which translates to two full rotations of the crankshaft. If the CKP sensor is modified

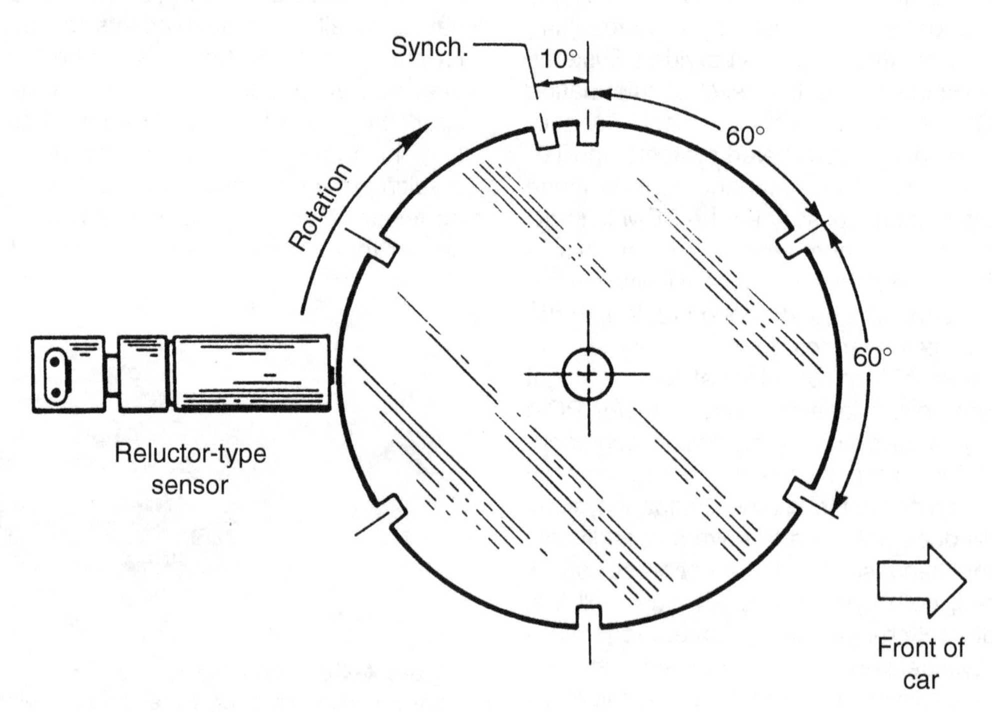

Figure 4–31 Magnetic CKP sensor with an offset notch for identifying the second level of information.

to inform the ignition module and/or PCM of the second level of information, the result is a potential to begin firing the ignition coils earlier during engine start. (A PCM may pulse SFI injectors before obtaining the third level of information if it is programmed to deliver a group injection pulse, called a *primer pulse,* during engine crank. Therefore, starting will not be delayed by reason of the fuel injection system.)

A few applications now create a CMP signal without the use of a physical camshaft sensor. (GM first began using this concept on their Saturn models.) Both the first and second levels of information must be derived from the CKP sensor. On a waste spark application, this can be achieved by commanding a primer pulse at the fuel injectors and then using the CKP information to fire the correct waste spark coil. As the coil is fired, smarter electronics within the coil measure the voltage drop across each of the two spark plugs. The plug with the larger voltage drop is the one that is firing on the cylinder that is on the compression stroke. (Remember that increasing the cylinder pressure increases the electrical resistance across a spark plug gap, therefore requiring a higher voltage to jump the gap.) As a result, the PCM can artificially create a CMP signal and will then begin to pulse the injectors in sequence. Therefore, it is possible that switching the two secondary wires on one of these waste spark ignition coils, even though both wires are on the correct coil and still fire at the correct time, can create a camshaft sensor code on an engine that has no physical camshaft sensor.

A few COP applications create the CMP signal artificially by initially firing two ignition coils on opposite (companion) cylinders after the CKP information is received (Saab, for example). The voltage drop across each plug is measured, the CMP signal is created, and then the PCM begins proper sequencing of the fuel injectors and ignition coils.

With DI systems, a distributor may be designed to provide only the first level of information or all three levels of information, depending upon the type of fuel management system that is used with it.

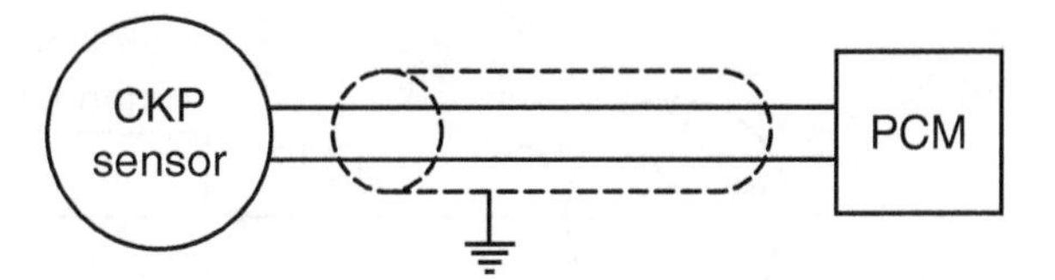

Figure 4–32 Example of a grounded shield.

Regardless of the system design and because of the importance of the primary waveform signals, a **grounded shield** is used with many systems to protect the primary ignition circuitry from the potential induction of a voltage signal caused by a nearby circuit. This grounded shield consists of a bare wire that is grounded at one end only (to ensure that it will not flow any current) and runs the length of the primary circuitry. The bare wire and the primary circuit wires are then wrapped with aluminum foil or aluminum braid. Figure 4–32 shows a typical electrical schematic that identifies a grounded shield. When repairing a primary circuit wire, it is important to avoid compromising the integrity of the grounded shield. A grounded shield may also be found with other circuits/systems, such as around the wiring from a wheel speed sensor in an antilock brake system.

Ignition Module and PCM Overview

Traditionally, many systems have used an ignition module that is separate from the PCM (Figure 4–33). In this design, the ignition module will receive the tach reference signal. This signal may originate either at a CKP sensor or at a distributor pickup and may originate with either a magnetic sensor, a Hall effect sensor, or an optical sensor. The ignition module will use this signal to fire the ignition coil(s) at base timing when necessary (typically during engine crank); it will also pass a digitized version of this signal on to the PCM. (The ignition module will digitize the analog AC voltage pulse that it receives from a magnetic sensor.) The PCM then uses this tach reference signal, along with information from its other sensors, to create a modified digital timing command signal that is sent back to the ignition module.

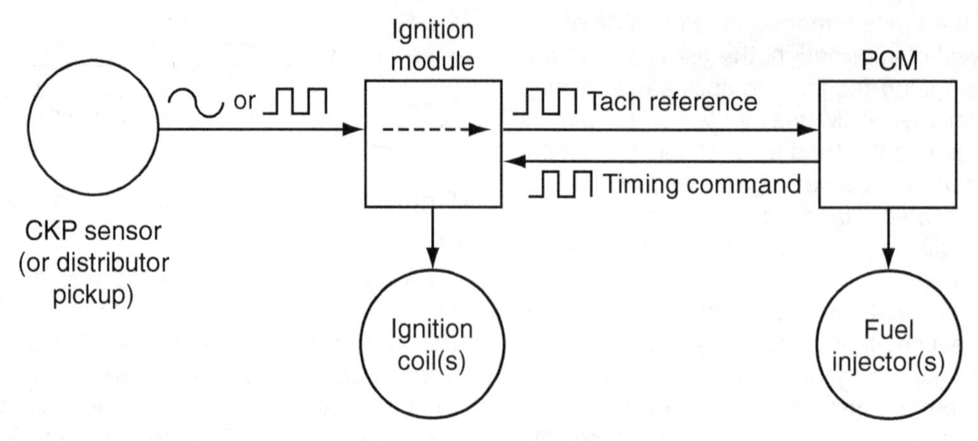

Figure 4–33 System with an ignition module that is separate from the PCM.

This modified timing command signal is actually advanced in time. The PCM can advance this signal because it uses the frequency of the tach reference signal to calculate engine speed—not only revolutions per minute (RPM) but also how many degrees of crankshaft rotation are occurring per millisecond.

If the ignition module is firing the ignition coil(s) from the tach reference signal, the engine is said to be operating at base timing. If the ignition module is firing the ignition coil(s) from the PCM's timing command signal, the engine is said to be operating at computed timing. The strategy used for switching a system between base timing and computed timing will vary among manufacturers.

The tach reference signal is also used by the PCM for injector pulsing and is required if the PCM is to energize the fuel pump relay past the initial 2-second pulse that occurs when the PCM is initially powered up. In a combination EI/SFI system, this tach reference signal will be the one originating at the CKP sensor. The CMP sensor signal is only used by the PCM to determine which injector/ignition coil should be pulsed/fired next.

In another system design, the ignition module is no longer separate from the PCM, but its equivalent is placed inside the PCM (Figure 4–34). In this design, the PCM receives the CKP and CMP signals directly from the sensors and then uses the CKP signal to pulse the injectors and fire the ignition coil(s) directly. Again, the CMP signal is used for injector and ignition coil sequencing. Chrysler Corporation originated this design back in the 1970s, but most other manufacturers have now adopted it. At least one manufacturer states that this design was chosen because it enhances the effectiveness of the antitheft disabling system on the vehicle. A variation of this design uses a PCM to control external switching transistors located in an "igniter" (Asian vehicles) or may control switching transistors that are located within

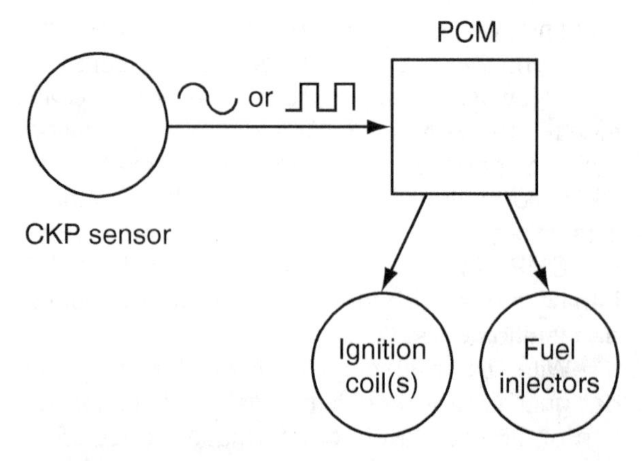

Figure 4–34 System with the ignition module equivalent located internally within the PCM.

the ignition coils (many modern COP ignition systems). The PCM still receives the CKP and CMP signals directly from the sensors, but it does not have to carry the higher current flow that the ignition coils draw.

Diagnosing for the Presence of CKP and CMP Sensor Signals

When chasing a no-start condition on an EI/SFI application, check for the presence of an injector pulse and a primary ignition coil pulse—if either one exists, the PCM is receiving a tach reference pulse from the ignition module, which, in turn, is receiving it from the CKP sensor or distributor pickup and is passing it on to the PCM. If neither one exists, the CKP sensor should be checked, followed by testing of the continuity of the signal's path to the ignition module, through the ignition module, and from the ignition module to the PCM.

In a system with waste spark EI and SFI, if the CKP sensor provides both the first and second levels of information to the PCM, and the PCM is receiving the CKP signal but is not receiving the CMP signal, it will likely guess at the correct injector sequence according to its internal programming, and the correct ignition coil will also be fired. Some PCMs are even programmed to guess at the correct ignition coil firing sequence during engine crank if the second level of information is lost. But if the first level of information is lost, the engine stalls immediately and will not restart.

EMISSION-CONTROL SYSTEMS

Exhaust emissions are dramatically reduced when the air-fuel ratio and spark timing are precisely controlled, when the engine coolant temperature remains between 93°C (200°F) and 104°C (220°F), and when combustion temperature is controlled. However, even these measures do not totally eliminate vehicle emissions. Therefore, other control systems are added to the vehicle with the ultimate goal of further reducing emissions.

Catalytic Converters

Catalytic Converter Overview. The catalytic converter is the single most effective device for controlling exhaust emissions. Located in the exhaust system, the **catalyst** agents cause low-temperature oxidation of HC and CO (oxidizing HC into H_2O and CO_2, and oxidizing CO into some additional CO_2) and reduction of NO_x (into N_2 and O_2). Placing the converter downstream in the exhaust system allows the exhaust gas temperature to drop significantly before the gas enters the converter, thus preventing the production of more NO_x in the oxidation process.

The catalysts that are used most often are **platinum** and **palladium** for the oxidation of HC and CO and **rhodium** for the reduction of NO_x. The catalytic materials are applied over a porous material, usually aluminum oxide in a honeycomb structure. Aluminum oxide's porosity provides a tremendous amount of surface area on which the catalyst material can be accessed to allow maximum exposure of the catalyst to the gases.

A dual-bed, three-way catalytic converter on a modern automobile has a reducing chamber for breaking down NO_x and an oxidizing chamber for oxidizing HC and CO and is referred to as a *three-way cat* (TWC). The exhaust passes through the NO_x reduction chamber first, then through the HC/CO oxidation bed (Figure 4–35). On some older vehicles, additional oxygen is introduced into the oxidation chamber via a secondary air injection system to increase the efficiency of this chamber by ensuring that it is kept up to the ideal operating temperature.

HC/CO Oxidation Bed Operation. The oxidation bed of a catalytic converter uses platinum and palladium to oxidize the HC molecules into water and carbon dioxide and the CO molecules into additional carbon dioxide. To oxidize HC and CO, a lean air-fuel mixture is needed to yield leftover oxygen. During closed-loop operation (when the oxygen sensor is cross counting), the oxidization converter is designed to store oxygen from the lean O_2 sensor pulse (when no CO and little HC are produced) in its extremely porous surfaces

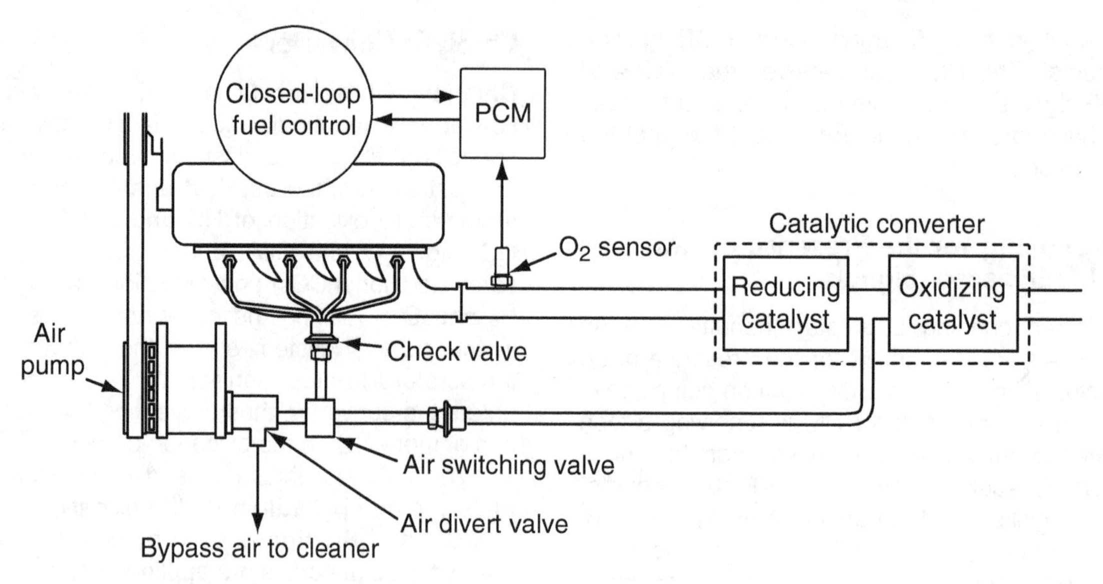

Figure 4–35 Dual-bed, three-way catalytic converter.

and then give up the stored oxygen during the rich O_2 sensor pulse (when CO and the majority of the HC are produced). Thus, the oxygen from the lean exhaust pulses can mix with the HC and CO, allowing all three substances to touch the catalyst metals simultaneously and thus allowing the oxidation of the HC and CO to take place.

NO_x Reduction Bed Operation. The reduction bed of a catalytic converter uses rhodium to break down the NO_x molecules (NO and NO_2, collectively known as NO_x) into nitrogen and oxygen. However, it can only do so if there is a small amount of CO also present simultaneously in the engine's exhaust. The CO molecule is an oxygen-deficient molecule. Therefore, to reduce NO_x, a rich air-fuel ratio is required, resulting in the production of a small amount of CO. When mixed with the NO_x, the CO molecules promote the breakup of the NO_x molecules by attracting an oxygen atom from the NO_x molecule, resulting in the breakup of the entire molecule. Simultaneously, the CO molecule is converted to a CO_2 molecule. With the only oxygen atom extracted from an NO molecule, nitrogen atoms are left to form nitrogen molecules. With one of the oxygen atoms extracted from an NO_2 molecule, some free oxygen atoms are also left to form some additional oxygen molecules.

Initially, NO_x reduction converters only reduced NO_x during the rich O_2 sensor pulse, when the air-fuel mixture was rich enough to produce CO, and never during the lean O_2 sensor pulse, when the majority of the NO_x is normally produced in the combustion chamber. (While the peak combustion temperature is reached slightly to the rich side of stoichiometric, most of the oxygen has been consumed when the air-fuel mixture is rich by the time the peak combustion temperature is reached. Therefore, when the engine is running correctly, most NO_x is produced when the air-fuel mixture is slightly lean due to the leftover oxygen that was not consumed during combustion.) This meant that reduction converters were less than 50 percent efficient in reducing the NO_x produced in the combustion chamber. However, by the late 1990s catalytic converter technological updates gave the NO_x reduction converter the ability to store CO from the rich O_2 sensor pulse and give it off during the lean O_2 sensor pulse when the majority of the NO_x is making its way through the exhaust. Thus, modern NO_x reduction converters are extremely efficient.

Another approach to the reduction of NO_x emissions is to mix the NO_x with urea, a form of ammonia. This approach is used on modern light-duty and heavy-duty diesel engines that require a urea *diesel exhaust fluid* (DEF) to be added to a DEF reservoir at regular intervals. The ammonia reacts with the NO_x to chemically reduce it to water and nitrogen.

Other Emission-Control Systems

Other emission-control systems currently used on most vehicles and those that can be included as part of the computerized engine control system, or added to it, are discussed in this section.

Positive Crankcase Ventilation. The positive crankcase ventilation (PCV) system is designed to recirculate blow-by gases from the crankcase through the intake manifold and into the cylinders to burn them in the combustion chamber. Therefore, the PCV system is designed to reduce HC emissions that were once vented to the atmosphere through a road draft tube.

First introduced on California vehicles in the 1961 model year and in the other 49 states in the 1963 model year as an *open PCV system,* this system drew ambient air through the crankcase to pick up any blow-by gases. In 1968, this system was modified into a *closed PCV system* that draws filtered air in from the air cleaner, then through the crankcase to pick up any blow-by gases (Figure 4–36). The combination of fresh air and blow-by gases is then drawn through a metering device, either a PCV valve or a metering orifice, and into the intake manifold, thus allowing the crankcase hydrocarbons to be burned in the cylinders. This combination of fresh air and blow-by gases tends to lean out the air-fuel mixture, so the PCV system is considered to be a calibrated vacuum leak.

If for any reason the crankcase vapors exceed the PCV system's capacity, either due to a PCV system fault or due to excessive blow-by gases getting past the piston rings (which would indicate worn rings), vapors are routed into the intake airstream through the intended crankcase inlet and will still be drawn through the throttle body into the intake manifold (not true of the early *open PCV system*). However, as a result, the blow-by gases are no longer purged efficiently from the crankcase, and there now exists a greater tendency to form sludge. An incorrect PCV valve can cause as many problems as a worn or plugged one. Therefore, when diagnosing PCV system problems, always begin by replacing the PCV valve with one of the correct part number and verify that proper manifold vacuum is made available to the valve. This is one of the few times that parts replacement is recommended in lieu of initial diagnosis: It would not be wise to spend $100 of diagnostic time due to a worn or incorrect $15 part.

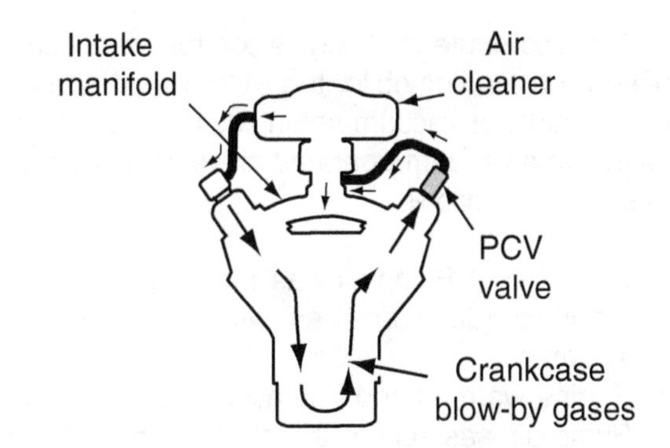

Figure 4–36 PCV system basic operation.

Air Injection Reaction Systems. Throughout the 1970s and 1980s, air injection reaction (AIR) systems used belt-driven air pumps to pump air into the exhaust stream. Known generically as *secondary air injection* systems, these systems used crankshaft horsepower to pump ambient air into the exhaust to help reduce the quantity of toxic emissions exiting the tailpipe.

AIR systems used prior to automobiles having computer controls simply pumped air into the exhaust manifold to help oxidize hydrocarbon emissions that left the cylinder. This reaction also helped to oxidize any carbon monoxide in the exhaust simultaneously.

Once computerized engine controls were introduced on the automobile, the PCM would control air solenoids, or vacuum solenoids that controlled a signal to a vacuum-operated air valve, to control the system as follows:

- Engine cold: Pump air was sent to the exhaust manifold (upstream) as with pre-computer systems.
- Engine warmed and operating in closed loop: Pump air was sent to the oxidizing bed of the two-bed catalytic converter (downstream) to help keep the cat up to operating temperature. It should also be noted that any ambient air ahead of the front oxygen sensor during closed-loop fuel control would cause a false lean signal to the PCM. And any ambient air ahead of the NO_x reduction cat would oxidize the CO that was being used to break up the NO_x.
- Under heavy load such as at wide open throttle (WOT): Pump air was dumped to the atmosphere to get it out of the system. This helped to prevent exhaust backfires and overheating of the catalytic converter during these rich air-fuel conditions.

These systems pumped air into the exhaust through one-way check valves that prevented the heat of the exhaust from reaching the AIR system components. If exhaust heat reaches the components within the AIR system, they will melt and become nonfunctional.

Pulse AIR (PAIR) systems, both pre-computer and computer-controlled, relied on exhaust positive and negative pressure pulses to bring atmospheric air into the exhaust system through one-way check valves. These systems did not use an air pump and were generally used on smaller engines such as four-cylinder engines.

AIR systems are no longer as popular as they once were, but they are still being used on a few modern vehicles with the following differences: Modern AIR systems no longer use a belt-driven air pump. Instead, they use an electric air pump (Figure 4–37). This air pump can be turned off when the ambient air is not desired in the exhaust. Therefore, there is no longer a provision to dump air to the atmosphere. Also, on a modern vehicle, there is no longer a provision to send ambient air downstream to the oxidation catalytic converter due to the influence of the rear oxygen sensor on air-fuel control during closed-loop operation. (The reality of early AIR systems that sent ambient air to the oxidation cat during closed-loop operation was that this did cause the temperature of these cats to remain high enough that the cat's longevity was shortened substantially.) Therefore, modern AIR systems will route ambient air into the exhaust manifold during cold engine warmup only to reduce cold start emissions. Once the engine is warm enough that the PCM can slip into closed-loop fuel control, the PCM shuts the electric air pump off.

Evaporative and Canister Purge Systems. These systems are designed to prevent the release of HC emissions to the atmosphere through evaporation. The fuel tank is vented to a charcoal canister (Figure 4–38). Then, when the engine is running, manifold vacuum is used to purge the HC vapors from the canister to the intake manifold for combustion in the cylinders. On modern systems, the PCM controls a purge solenoid placed directly into the purge vacuum line. OBD II applications use an evaporative system that has become increasingly complex due to the added ability of the engine computer to check the system for leaks to

Diagnostic & Service Tip

Charcoal canister saturation can often be the result of fuel tank overfilling. Slowly filling the tank right to the top of the filler neck causes the expansion area in the fuel tank to fill and thus leaves no room for fuel expansion as it warms. When the fuel does expand, it is pushed up and into the charcoal canister. Canister purging then dumps so many hydrocarbons into the intake manifold that the system may not be able to compensate, even in closed loop. Many driveability problems can be linked to charcoal canister problems; do not overlook that in diagnostic procedures.

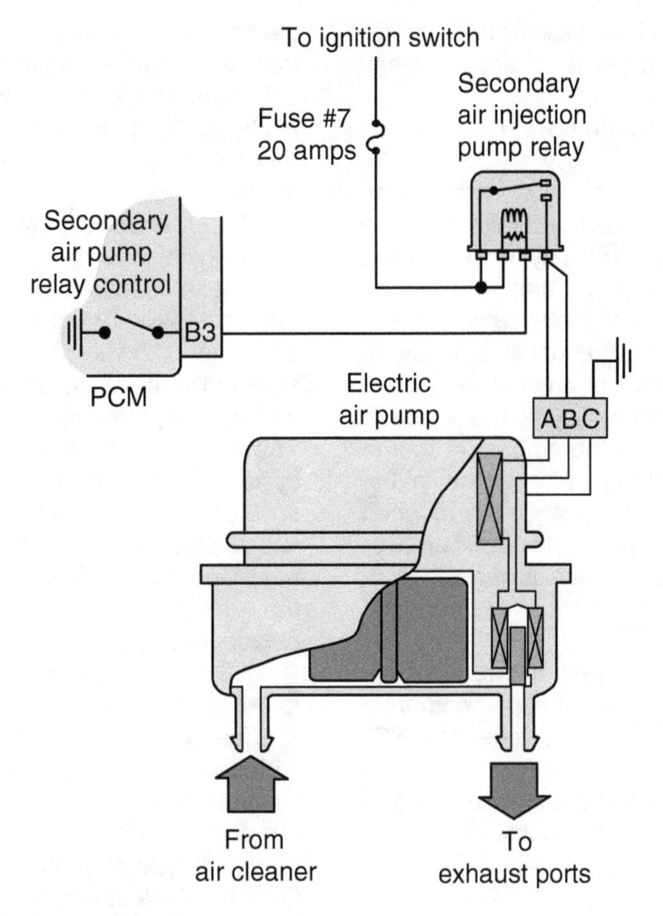

Figure 4–37 An electric air pump circuit.

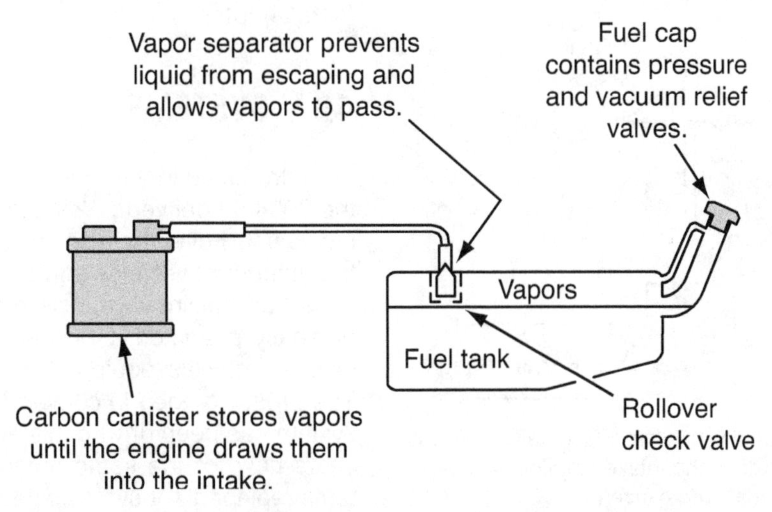

Figure 4–38 In a typical evaporative emission system, the fuel tank is vented to a charcoal canister.

minimize the escape of HC emissions to the atmosphere through a missing gas cap or any other leak (see Chapter 7).

Exhaust Gas Recirculation. Exhaust gas recirculation (EGR) systems use an EGR valve, as shown in Figure 4–39, to allow recirculation of exhaust gases back into the intake manifold to replace incoming air and oxygen with inert exhaust gas. This reduces the amount of ambient air getting into the cylinder. With a carburetor, an automatic reduction in fuel delivery occurs in response to the reduction of ambient air reaching the cylinders. With an electronic fuel injection system, the PCM reduces the amount of fuel to maintain the proper air-fuel ratio whenever the EGR valve is opened. Ultimately, with reduced oxygen and fuel reaching the cylinders, combustion chamber temperatures are lowered, which reduces the amount of NO_x that is produced during combustion.

The use of an EGR system allows the engineer to design an engine with a higher compression ratio or increased VE while still keeping NO_x production at a minimum. An added benefit is that a more aggressive spark advance program can be programmed into the middle RPM and load ranges, resulting in improved engine performance and fuel economy while continuing to control spark knock. As a result, an EGR valve that is physically stuck closed can be responsible for a spark knock complaint under midrange light to moderate acceleration.

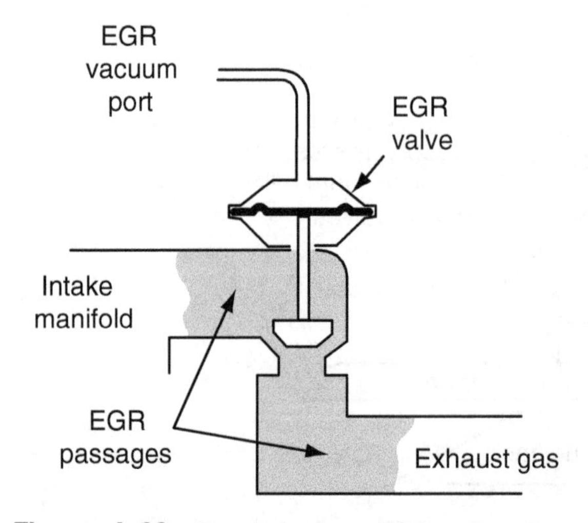

Figure 4–39 Opening of an EGR valve allows exhaust gases to flow into the intake manifold, displacing incoming ambient air and oxygen.

An EGR valve may be a vacuum-operated valve which, in turn, controls a passage between the intake manifold and the exhaust manifold. The PCM will control the opening of a vacuum-operated EGR valve via a solenoid. Typically, this solenoid is normally closed to vacuum application and normally open to an atmospheric vent. The PCM will operate this solenoid on a duty cycle to control the percentage of time that the solenoid is applying vacuum versus the time that it is venting vacuum. Other EGR valves are electrically operated solenoids that control the flow of the exhaust gases directly and without the use of a vacuum signal.

EGR valves are not opened at idle so as not to detract from the idle quality of the engine, nor are they opened at WOT so as not to interfere with all-out engine performance. On some older systems (beginning in 1980), the PCM monitored the EGR valve's position, resulting in a feedback type of control of the EGR system. On OBD II vehicles, the PCM is required to monitor the EGR system for both flow rate and leakage, although several different methods are used (see Chapter 7).

42 V SYSTEMS

Introduced in vehicles in the United States in the 2004 model year, 42 V systems may be used increasingly over the next several years. Although the number of vehicles equipped with a 42 V systems has not increased as dramatically as was originally predicted, GM now has manufactured several vehicles equipped with a 42 V system. The pressing need to convert to a higher-voltage system has been driving this evolution for several years now for the same reasons that 14 V systems replaced 7 V systems decades ago.

The Need for More Power

You may recall that total power, or wattage, is equal to voltage multiplied by amperage. As engineers have continued to add many electrical and electronic systems to the automobile over the past several decades, more and more total power has been required to control and operate these systems. Because power is not free, the manufacturer's reaction to an ever-increasing power requirement traditionally has been to build larger alternators capable of producing more and more amperage. It has been obvious for some time now that eventually we would have to increase the vehicle's operating voltage as a method of meeting these power demands.

Today's vehicles may use electrical power for any combination of the following systems:

- Computerized engine controls
- Electronically controlled automatic transmissions
- Electronically controlled interior and exterior lighting systems
- Electronically controlled gauge systems
- Electronically controlled body systems, including heated front and rear seats as well as seat cooling systems
- Heated windshield wiper systems
- Electronic antilock braking and traction control systems
- Electronically controlled stability control systems
- Electronic active suspension systems, including both load leveling and shock dampening systems
- Electronic passive restraint systems, including frontal air bags, side air bags, head air bags, knee air bags, and seat belt pretensioners
- Variable assist power steering systems
- Electromagnetic steering systems
- Electronically controlled cooling fans
- Heated rear window defrosters
- Heated outside mirrors
- Heated windshield deicing systems
- Heated windshield wipers
- Electrochromic inside and outside rear view mirror systems
- Remote-controlled electrically operated exterior door systems
- Antitheft systems, including both warning and disabling systems
- Smart key systems
- Entertainment systems, including DVD players with relatively high-wattage sound systems
- Integrated voice recognition, navigational, and Bluetooth wireless control systems

42 V as a World Standard

When engineers were looking at what voltage to use as a standard, one of the considerations was the fact that voltage above 50 V can cause a fatal shock. Therefore, the 42 V standard was adopted as a worldwide standard. It is essentially three 12.6 V batteries in series, producing a total of 37.8 V, plus charging system voltage, for a total of 42 V. This triples the kilowatt capability of the electrical system without becoming dangerous to a technician. (Today's hybrids use dangerously high voltage levels, from 144 to 650 V, but the use of 42 V systems creates little potential for a fatal shock.)

42 V System Designs

The automotive world is not likely to transform the entire automobile into a 42 V system overnight. The changes will occur progressively because everything from light bulbs and other electrical parts to battery chargers and other service equipment will have to be upgraded simultaneously. Therefore, today's 42 V systems are used with 14 V subsystems. Retention of a separate 12.6 V battery that is charged through a pulse-width-modulated signal from the 42 V system is generally used, but a 42 V system can use pulse-width modulation to power a 14 V subsystem without the additional battery. The starter motor can be designed as a 12.6 V starter or as a 37.8 V combination starter/generator. The advantages of the 37.8 V combination starter/generator are that reduction gearing is not needed and that regenerative braking can be incorporated. Regardless, these vehicles will continue to use many 14 V loads for some time to come.

Advantages of a 42 V System

With the large increase in voltage, heavy-duty components that have been traditionally belt-driven are now driven by a 42 V motor, including A/C compressors, water pumps, and power steering pumps (in those systems that retain a hydraulic power steering system). Also, with a threefold increase in voltage, the amperage can be cut back proportionally and still maintain a given wattage; therefore, smaller-diameter wiring can be used, thus cutting back on vehicle weight. Engineers are adding 120 V power points to power common devices such as televisions, DVD players, and laptop computers because only a 3× step-up transformer is required.

42 V System Concerns

There are many concerns associated with 42 V system designs, including service equipment, service procedures, and safety issues. Battery chargers and other service and diagnostic equipment must be redesigned. Back-probing of a connector as a diagnostic procedure will no longer be an option. If the wiring's insulation is compromised in a 42 V system, a greatly increased potential for corrosion (even with the system at rest) as well as a greater potential for arcing will exist. In addition, 42 V arcs are more than 50 times hotter than 14 V arcs. The best advice, as these systems are introduced, is to pay particular attention to manufacturer-recommended diagnostic and maintenance procedures.

SUMMARY

This chapter has dealt with many concepts related to the design and operation of modern fuel injection, spark management, and emission-control systems. We concluded our chapter with a discussion on the need for more electrical power on the modern automobile and the reasons some manufacturers have begun to add 42 V systems to the automobile. The intent of this chapter is to create a foundation upon which readers will build as they continue to progress through the remainder of this textbook. We have covered several aspects of each of these systems that are common to most of the vehicles produced today, whether domestic, European, or Asian.

▲ DIAGNOSTIC EXERCISE

A customer brings in a vehicle for a diagnosis. The primary complaint is that the engine has a noticeable loss of performance and that the malfunction indicator light is illuminated. When questioning the customer, you find that the vehicle is far overdue for the engine's oil and filter change. Your discussion also finds that the last time he changed the engine's oil he performed the oil change himself and used a 20W-50 viscosity oil on an engine that requires 5W-20 oil. He also used a "bargain basement" oil filter. How could this information relate to the engine's loss of performance?

Review Questions

1. What is the primary reason that a fuel injector's winding is kept immersed in liquid fuel?
   A. To cool the fuel before it is sprayed from the injector to reduce the potential for vapor lock
   B. To help atomize the fuel before the injector is energized
   C. To help prevent overheating of the injector's winding
   D. To provide electrical continuity through the fuel between the two ends of the winding to reduce the potential for an injector voltage spike

2. What is the purpose of controlling a peak and hold injector with two drivers?
   A. To provide a backup driver in case one driver fails
   B. To provide additional current for the total duration of the injector's pulse
   C. To provide good mechanical response without overheating the injector's windings
   D. To provide a diagnostic circuit for the PCM
3. Why are electric fuel pumps that are associated with fuel injection systems mounted in the fuel tank rather than in the engine compartment?
   A. To make replacement difficult so as to keep the vehicle owner from replacing it with an aftermarket pump
   B. To reduce the cost of replacing the fuel pump
   C. To increase the PCM's ability to diagnose the fuel pump circuits
   D. To reduce the potential for vapor lock within the fuel management system
4. Why is the PCM programmed to de-energize the fuel pump relay if the tach reference signal is lost?
   A. To guard against emptying the fuel tank through a ruptured fuel supply line
   B. To protect the winding in the fuel pump relay from overheating
   C. To protect the fuel pump from overheating
   D. To ensure that the ignition coil(s) will not overheat
5. In a PFI system with a return line, why is the fuel pressure regulator indexed to intake manifold pressure?
   A. To increase the rate of fuel flow through the fuel injectors when the engine is under high load
   B. To maintain a consistent rate of fuel flow through the injectors, regardless of engine load
   C. To reduce the rate of fuel flow through the injectors when the engine is under high load
   D. To ensure that the fuel injectors will not be able to supply fuel if the engine stalls and manifold vacuum is lost
6. Why have manufacturers begun using returnless fuel injection systems?
   A. To reduce the amount of heat that is carried from the engine compartment to the fuel tank
   B. To reduce the potential for evaporative hydrocarbon emissions to be produced within the fuel tank
   C. To reduce the potential for vapor lock within the fuel management system
   D. Both A and B
7. What lambda value does a PCM try to maintain on a modern vehicle?
   A. 128
   B. Between 0 and 1
   C. Between 0 and 5
   D. Between 0.98 and 1.02
8. A technician is using a scan tool and notices that LTFT is at 18 percent (+.08). What is indicated?
   A. The PCM has learned to subtract fuel from the base fuel calculation to compensate for a lean condition.
   B. The PCM has learned to subtract fuel from the base fuel calculation to compensate for a rich condition.
   C. The PCM has learned to add fuel to the base fuel calculation to compensate for a lean condition.
   D. The PCM has learned to add fuel to the base fuel calculation to compensate for a rich condition.
9. *Technician A* says that on modern vehicles, the rear O_2 sensor is used by the PCM to determine the lambda value as it relates to the air-fuel ratio. *Technician B* says that on modern vehicles, any exhaust leak ahead of the rear O_2 sensor can cause the PCM to drive the air-fuel ratio rich, even possibly to the point of an engine stall. Who is correct?
   A. *Technician A* only
   B. *Technician B* only
   C. Both technicians
   D. Neither technician

10. Two technicians are discussing the design of a waste spark ignition system. *Technician A* says that the polarity of a waste spark coil is reversed each time it fires so that the negatively firing spark plug always ignites an air-fuel charge. *Technician B* says that a waste spark coil may be mounted directly on top of one of the spark plugs that it fires. Who is correct?
    A. *Technician A* only
    B. *Technician B* only
    C. Both technicians
    D. Neither technician
11. Which of the following is an advantage of a COP ignition system?
    A. There is less potential for the secondary circuit to induce a voltage in a nearby circuit.
    B. It is easier to connect a secondary ignition oscilloscope to the system for diagnosis.
    C. The ignition coils used in this system no longer contain primary windings.
    D. This system uses a pickup in a distributor, thereby allowing base timing to be adjusted.
12. Which level of information, as described in this chapter, is required for the PCM and/or ignition module to fire the correct ignition coil in a waste spark ignition system?
    A. First level; identifying "a *pair* of cylinders' pistons are approaching TDC"
    B. Second level; identifying *"which pair* of cylinders' pistons are approaching TDC"
    C. Third level; identifying *"which cylinder* of the piston pair approaching TDC is on the compression stroke"
    D. None of the preceding answers provides enough information for the correct ignition coil to be fired.
13. Which level (or levels) of information, as described in this chapter, may originate at a CKP sensor?
    A. The first level only
    B. The first and second levels only
    C. The second and third levels only
    D. The first, second, and third levels
14. Two technicians are discussing the design and use of a grounded shield. *Technician A* says that it is used to protect the primary ignition circuitry from the potential induction of a voltage signal caused by a nearby circuit. *Technician B* says that it consists of a bare wire that is grounded at one end only and then, along with the other wires, is wrapped with aluminum foil or braid. Who is correct?
    A. *Technician A* only
    B. *Technician B* only
    C. Both technicians
    D. Neither technician
15. Which of the following is an advantage of placing the ignition module's equivalent inside the PCM?
    A. It is less expensive to replace the ignition module when it fails.
    B. The ignition module can more efficiently dissipate the heat that it produces.
    C. The effectiveness of the vehicle's antitheft disabling system is enhanced.
    D. The PCM's complexity is reduced.
16. While diagnosing an engine with COP and SFI systems that "will crank, but won't start," the technician finds that the ignition coils are producing spark, but no fuel injector pulse is present while cranking. Which of the following could *not* be the reason the engine fails to start?
    A. The CKP sensor has failed.
    B. The ignition module has failed in its ability to pass the tach reference on to the PCM.
    C. The tach reference wire between the ignition module and the PCM might be open.
    D. The PCM has failed.
17. An evaporative and canister purge system is designed to prevent the release of which of the following emissions from the fuel tank to the atmosphere?
    A. CO
    B. HC
    C. CO_2
    D. NO_x

18. The primary purpose of the EGR system is to reduce which of the following emissions?
    A. CO
    B. HC
    C. CO_2
    D. NO_x
19. *Technician A* says that variable valve timing is designed to increase the high-performance capability of the engine while maintaining good driveability and emissions. *Technician B* says that variable valve timing can reduce the fuel octane requirements needed to control detonation. Who is correct?
    A. *Technician A* only
    B. *Technician B* only
    C. Both technicians
    D. Neither technician
20. All except which of the following are advantages of a 42 V system over a 14 V system?
    A. Heavy-duty components that have traditionally been belt-driven can be driven by an electric motor.
    B. A 42 V system is safer to work on than a 14 V system.
    C. Smaller-diameter wiring can be used.
    D. A 42 V system requires only a 3× step-up transformer in order to equip the vehicle with 120 V power points.

Chapter 5

Introduction to Diagnostic Concepts and Diagnostic Equipment

OBJECTIVES

Upon completion and review of this chapter, you should be able to:

- ❑ Describe the two types of faults associated with automotive computer systems.
- ❑ Describe the two types of diagnostic trouble codes associated with automotive computer systems.
- ❑ Understand the function and purpose of a data stream.
- ❑ Understand the various types of functional tests that may be programmed into an engine computer.
- ❑ Recognize the function of a pinpoint test.
- ❑ Recognize the function of a flowchart.
- ❑ Understand the function and purpose of an electrical schematic.
- ❑ Describe the types of equipment that are used in diagnosing symptoms associated with electronically controlled systems.
- ❑ Understand the functional concepts and procedures associated with each type of diagnostic tool.
- ❑ Interpret the readings associated with each type of diagnostic tool.
- ❑ Understand the advantages, disadvantages, and limitations of each type of diagnostic tool.

KEY TERMS

Active Command
Amplitude
Bidirectional Controls
Breakout Box (BOB)
Current Ramping
Data Link Connector (DLC)
Data Stream
Diagnostic Trouble Code (DTC)
Digital Storage Oscilloscope (DSO)
Digital Multimeter (DMM)
Digital Volt/Ohm Meter (DVOM)
Electrical Schematic
Flowchart
Functional Test
Graphing Multimeter (GMM)
Hard Fault
Logic Probe
Memory DTC
Motor Start-Up In-Rush Current
Multimeter
Non-Powered Test Light
On-Demand DTC
Parameter ID (PID)
Pinpoint Test
Scan Tool
Self-Test
Soft Fault
Technical Service Bulletin (TSB)
Trigger
Voltage Drop Test

Understanding the product-specific diagnostic methods in the manufacturer-specific chapters of this textbook may be difficult and may seem to require a lot of memorization. But all these methods are based on the same fundamental diagnostic concepts. And your noblest efforts to properly diagnose today's electronic systems may also be limited by your knowledge of the diagnostic tools available to you as a technician.

This chapter is designed to cover some of the basic concepts involved in diagnosing modern electronic engine control systems. Mastering the material in this chapter will promote your understanding of the manufacturer-specific chapters. This chapter will also enhance your understanding of today's tools, their purposes, their functions, and their limitations.

WARNING: Modern vehicles are equipped with frontal air bags, side-impact air bags, and/or seat belt pre-tensioners. Any technician working in or around the area of an air bag or around the circuits involved in an air bag system should review the relevant sections in the service manual and disable the air bag systems before performing any service on the vehicle. Failure to do so can result in personal injury from unexpected air bag deployment as well as expensive damage to the vehicle. These systems will be labeled with one or more of the following labels:

- ***Supplemental Restraint System or SRS***
- ***Supplemental Inflatable Restraint or SIR***
- ***Air Bag***

DIAGNOSTIC CONCEPTS

Types of Faults

Hard Faults. A **hard fault** is defined as a fault that is currently present. As a technician, you should always try to verify the symptom as the first step of diagnosis. When you try to verify the symptom associated with a hard fault, the symptom will appear to be present.

Soft Faults. A **soft fault** is commonly known as an *intermittent fault.* If the symptom cannot be re-created, you are likely dealing with a soft fault. As a general rule, hard faults should be diagnosed before attempting to diagnose soft faults.

Diagnostic Trouble Codes

It is common knowledge that most PCMs can communicate to the technician alpha and/or numeric readouts that represent fault codes, but some additional explanation is required about the types of fault codes that can be obtained from a PCM. All manufacturers now refer to this type of fault code as a **diagnostic trouble code**, or **DTC**. The technician should always check the applicable DTCs when trying to diagnose driveability, fuel economy, and emission concerns.

Memory DTCs. A **memory DTC** is a DTC that the PCM has stored in its RAM IC chip at some point in the past. The technician may be able to bring up the memory fault code either through manual code pulling methods or with a scan tool, depending on the make and year of the vehicle (modern vehicles generally require the use of a scan tool). When pulling memory DTCs, current state (or condition) *does not* matter. Rephrased, it does not matter if the engine is properly warmed up when pulling codes from a computer's memory. When pulling memory codes you are not asking the computer to evaluate the condition of system components or the system itself at this time.

Most PCMs are programmed to erase the memory fault on their own if they do not see the fault reoccur after a preset number of restarts of the engine, or on modern vehicles, a preset number of warmup and cool-down cycles of the engine as measured by the ECT sensor. A memory fault code may be either a hard fault or a soft fault. The fact that it is set in the PCM's memory does not necessarily mean that the fault is still present.

If a pre-OBD II vehicle application has a memory code capability only and more than one DTC has set in memory, you should first record and then erase all stored memory codes. Then drive the vehicle on a road test. Any DTCs that do not reset during the road test are soft faults; any DTCs that do reset during the road test are most likely hard faults. After repair of a memory DTC, the technician should again erase the DTC from the PCM's memory to avoid inadvertently chasing an "already-repaired DTC" at a later time.

Then the vehicle should be road tested again to verify that the DTC does not reset.

On-Demand DTCs. An **on-demand DTC** is set as a result of the PCM running a **self-test** (performance test) of the system at the technician's request. A self-test is defined as a test that the computer is programmed to perform on the system that it controls. The technician may be able to initiate a self-test either through manual methods or with a scan tool, depending on the make and year of the vehicle (again, modern vehicles generally require the use of a scan tool). After the technician initiates the self-test, the PCM puts the system through a variety of tests as it checks each circuit's ability to function. The resulting DTCs are hard faults in that the PCM has identified their existence and immediately reported the results to the technician. When initiating a self-test, current state (or condition) *does* matter. For example, if the engine is not properly warmed up, a false DTC may be generated for such components as the ECT sensor, IAT sensor, engine oil temperature sensor, and the oxygen sensors. Or if the air-conditioning system has been left turned on during the self-test when the manufacturer has specifically stated that all accessories should be turned off, a false DTC may be generated. And any other manufacturer-specific instructions should be adhered to before performing the self-test.

A PCM on a given application may be programmed to perform key on, engine off (KOEO), and key on, engine running (KOER) self-tests. In this case, it is important to perform both tests. The engine should be properly warmed up before running this type of test, and all accessories should be turned off. Also, in some cases the technician will be asked to cycle certain switches at some point during the self-test to allow the PCM to check their operation.

The PCM will perform checks of the electrical circuitry during the KOEO self-test and will perform functional tests during the KOER self-test. For example, if the PCM controls the air-injection system through vacuum solenoids, it may energize and de-energize these solenoids during the KOEO self-test while monitoring the voltage drop in these circuits. Then, during the KOER self-test, the PCM will perform a functional test of this system by energizing the solenoid valves and watching the oxygen sensor(s). For example, the PCM may hold the first solenoid valve energized so as to direct pump air to the second solenoid. Then it may energize and de-energize the second solenoid valve so as to direct pump air upstream (to the exhaust manifold) and downstream (to the oxidation converter). Then with the second solenoid valve held energized (upstream), the PCM may energize and de-energize the first solenoid valve, ultimately switching pump air to the upstream position and then dumping it to the atmosphere. Each time the pump air is switched to the upstream position, the PCM expects to see the primary O_2 sensor values go lean. In this way, the PCM can test the secondary air system functionally. Functional tests may also be run during the KOER self-test for other components and systems, including components such as the knock sensor and systems such as the EGR system. Therefore, when these tests are built into the PCM's programming, it is not an option to run one test or the other; both tests should always be run, beginning with the electrical (KOEO) self-test, followed by the functional (KOER) self-test. In fact, repairs should be made for any KOEO hard faults before you perform the KOER functional self-test. Often, a KOER self-test cannot be completed successfully because electrical failures still exist. In such cases, a repair must be sold by the service writer to the customer before testing can be completed. This is not the result of a failing on the part of the technician, but rather is due to the complexity of the modern vehicle and the way that systems interact with each other. The customer understandably will want to know the total dollar amount that the repairs will cost on the first phone call, but often this is simply not practical.

On vehicles manufactured prior to OBD II, an on-demand DTC does not set in the PCM's memory; therefore, you cannot really erase it from memory in the way that you would erase a memory DTC. Instead, you must make the indicated repair and rerun the self-test. If the repair was successful, the DTC will not reset. (Some OBD II PCMs will now retain in memory a failure DTC that is set as the result of a self-test.)

Also, if the application has both self-test and memory code abilities (as with the Ford EEC IV and EEC V systems), a simple comparison of the types of faults received will allow you to know which memory faults are hard and which are soft without having to erase the memory faults and road-test the vehicle. Remember, while a memory DTC may be either soft or hard, a DTC that is set as a result of a self-test is a hard fault. Therefore, on those systems that have both abilities, a memory DTC that also sets following either self-test is a hard fault, but a memory DTC that does not set following the self-tests is likely a soft fault.

Data Streams

A **data stream** is information about a computer system that is delivered via serial data from the PCM (or other control module) to your scan tool (see "Scan Tools" later in this chapter). The PCM believes this information to be true about the system that it controls. There is no manual method to access this type of information, which comes from the PCM's memories in the form of binary code. This information may include memory DTCs, input voltages and/or PCM-interpreted values, output commands, fuel trim numbers (see "Fuel Trim" in Chapter 4), PROM ID numbers, and other types of information. The information accessed in a data stream is identified by the **parameter ID (PID)**. A data stream PID may identify an input value from a sensor, an output command to an actuator, or some other piece of information such as the PID for the long-term or short-term fuel trim values.

Functional Tests

Functional tests are tests that the PCM has been programmed to perform, but the test results are usually not rendered by the PCM. Most often, it is up to the technician to determine the pass/fail result of each functional test. Functional tests are usually used in specific circumstances to aid in diagnosing specific problems and are not used on every vehicle that comes in the door. Many times, a flowchart will guide the technician to a specific functional test.

For example, a common functional test is an **active command.** Through the use of an active command, the technician may be able to command the PCM to energize and de-energize certain actuators or to perform other test procedures, such as an engine power balance test. While most active commands on modern vehicles are quite sophisticated and require the use of a scan tool, some early, low-level active commands could be performed using jumper wires and other manual methods. On modern vehicles, performing an active command involves the use of the scan tool to send commands to the PCM (or other control module); these commands are known as **bidirectional controls.**

There is a strong similarity between an active command and an on-demand self-test in that both will cause the PCM to turn actuators on and off. But with an on-demand self-test, the PCM (or other control module) also determines pass/fail during the test, and it can set a failure code following the test. With an active command, the control module does not make a pass/fail determination; rather, it is up to the technician to determine whether the computer can properly activate the output.

Another advantage of an active command: If the control module can successfully activate the actuator, it can serve as an excellent component locator. For example, if two identical relays are mounted on the fire wall next to each other and you want to know which one is the cooling fan relay, simply use a scan tool to issue a bidirectional command to energize and de-energize the cooling fan relay and feel which relay is "clicking."

Technical Service Bulletins

When manufacturers are alerted to a problem on their vehicles, often they will issue a procedure to help the technician repair the problem properly.

This is known as a **technical service bulletin (TSB)**. TSBs are the manufacturers' method of alerting technicians to problems that they have become aware of as well as their recommended "fix." The manufacturers enable their dealership technicians to report problems that have not already been addressed. The manufacturer will then issue a TSB that addresses the problem and instructs the technician on how to perform a proper repair to correct the problem.

TSBs are issued with regard to many types of problems, including driveability problems, steering and suspension problems, brake problems, wind noise concerns, and so on. TSBs for engine control systems may involve anything from relocating a wiring harness to reduce electromagnetic interference (EMI) to replacing a sensor/actuator with a new part number; they may even involve the reprogramming (reflashing) of the PCM. TSBs are also available to aftermarket technicians through most electronic service manuals and information systems. The technician should always check for applicable TSBs when diagnosing driveability, fuel economy, and emission concerns.

Pinpoint Testing

After a DTC has been retrieved to direct the technician to a general area or circuit, the identified circuit must undergo a **pinpoint test** to determine the exact fault. Bear in mind that connectors and wires can set the same DTCs that a defective input sensor or output actuator can set. Some early systems did not have any self-diagnostic ability programmed into the PCM, thus requiring that the entire computer system be pinpoint tested by the technician when a symptom was encountered. While this can be a huge undertaking, fortunately many of these early systems were not as complex as later systems. Still, this can be quite a daunting task.

Pinpoint testing may include testing the integrity of the PCM's power and ground circuits, testing the circuitry that connects the PCM to each of its sensors and actuators, and testing each of the sensors and actuators for proper operation. The pinpoint test procedure should ultimately determine the exact repair needed so that the technician can identify for the customer what is required to properly repair the vehicle. Often, the pinpoint test procedure may identify a particular fault that the customer must be made aware of before overall testing is complete. Because identified faults should be repaired before continuing with the diagnostic procedure, the technician may have to contact the customer more than once if multiple faults exist.

Method of Performing Pinpoint Tests

When using a voltmeter or an oscilloscope to perform pinpoint tests, some point of electrical contact must be made between the meter's leads and the circuit being tested. It is always recommended that a set of jumper leads be connected between the circuit and the component while taking care not to allow these leads to contact each other.

For example, if a DTC directs you to test the throttle position sensor (TPS) circuit, you can purchase or manufacture a set of three jumpers with the appropriate male and female terminals. Disconnect the sensor's three-wire connector from the sensor and install these jumpers between the harness connector and the sensor. Then, after turning on the ignition switch, simply touch the meter's leads to the jumpers' male/female metal terminals to take measurements of the reference voltage supplied to the sensor, the signal returned by the sensor, or the voltage drop of the ground circuit that completes the sensor's circuit to the battery. When the tests are complete, simply turn off the ignition switch and disconnect and remove the jumpers, then reinstall the harness connector to the sensor. When making voltage drop tests of the computer's power and ground circuits, because of the large number of wires in these connectors, do not use individual jumper wires (inadvertent shorts would occur). Rather, you should use a breakout box (BOB) (discussed later in this chapter) to access these circuits. No connectors or wiring insulation will be damaged or compromised when these procedures are used.

An alternative approach (although often abused) is to use a T-pin (available at the fabric/sewing center of major retail stores) to back-probe the terminal at the sensor or actuator. This is *not* a recommended method, but because it is so commonly used in the industry and will likely continue to be used, the following discussion is included in hopes of minimizing any damage that may result from its use.

When using a T-pin to back-probe an electrical connector, carefully slide the T-pin into the plastic body of the electrical connector beside and *parallel to* the wire coming out of the connector until you feel it touch the metal terminal of the connector (Figure 5–1). Then take the applicable measurement with your test equipment. If done properly and carefully, this approach minimizes the damage done to the wire's copper and insulation as well as to the connector while providing a relatively easy approach to making pinpoint measurements. (While this method does work well with many types of electrical connectors when done properly and carefully, it is not recommended for use with weather-pack connectors because it will compromise the weather-pack seal. With this type of connector, it is best to use the method of connecting temporary jumper wires, as described earlier.)

CAUTION: When back-probing, *do not* poke the T-pin *through* the wire at an angle perpendicular to the wire (Figure 5–1)—this compromises the wire's insulation and will also break copper strands of wire, thus increasing the resistance of the wire at the point of the damage. Many voltage drop problems can be traced to someone who previously worked on the vehicle and damaged the wiring when making pinpoint tests.

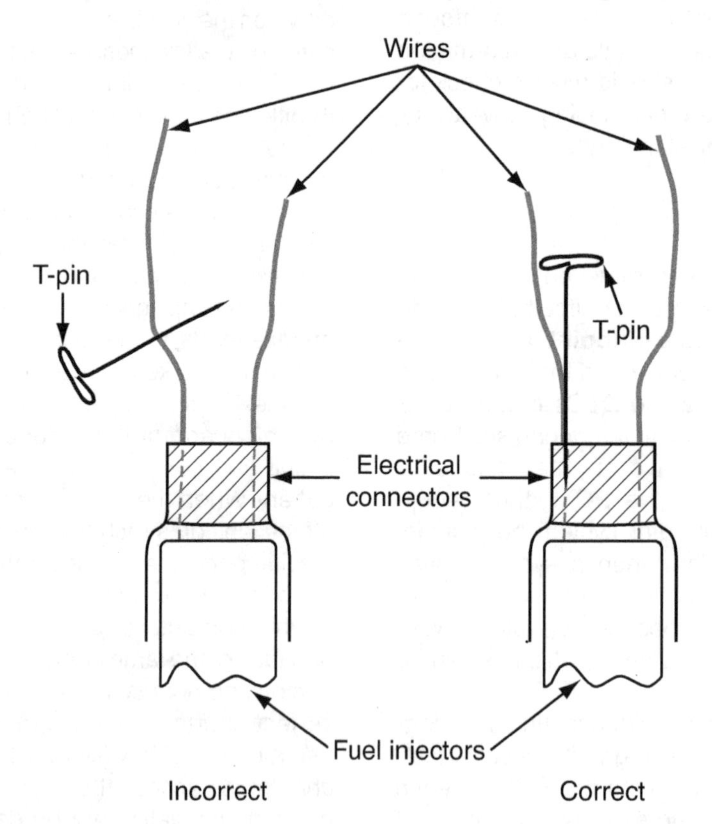

Figure 5–1 Incorrect and correct methods of using a T-pin to back-probe an electrical connector.

CAUTION: When back-probing, *do not* probe directly into a component, such as an O_2 sensor, where the wire exits the sensor. Always perform the back-probing at an electrical connector.

CAUTION: When back-probing, *do not* use more than one T-pin at a time in the same connector—if the T-pins were to touch each other, the load would be bypassed and the computer that controls the load might be damaged. Many damaged control modules can be traced to someone who previously worked on the vehicle using improper procedures.

CAUTION: *Do not ever* use back-probing as a diagnostic method on systems that operate at a voltage above 14 V. Systems that operate at 42 V and higher require other forms of diagnostic tools such as breakout boxes. This includes GDI fuel injectors. Refer to manufacturer recommended procedures for these systems and components.

CAUTION: *Do not* use T-pins to probe the terminals in the data link connector (DLC). You will enlarge these terminals. This will result in the inability to use a scan tool to communicate with the on-board computers in a reliable fashion, causing either intermittent communication or a total loss of communication. Instead, use a breakout box designed for the DLC.

CAUTION: When using a voltmeter or oscilloscope to pinpoint test a circuit, *never* insert the meter's leads directly into a female terminal. You will enlarge it and thus create a faulty connection when it is reconnected.

Pinpoint Testing of a Sensor

When a DTC directs you to a sensor or sensor circuit, always consider that there are two possibilities: (1) the sensor or sensor circuit is faulty and should be replaced or repaired, or (2) the sensor and its circuit are operating correctly but the value that is being measured by the sensor is less than ideal. For example, a coolant temperature gauge might read hot because of a faulty gauge/sender/circuit, but it could also read hot simply because the engine is truly overheating. And so it is with the sensors in a computer system.

An "O_2 always rich" code can be generated because of a faulty O_2 sensor or circuit problem, but the same DTC can also be generated because the engine air-fuel ratio is truly running rich, in which case replacing the sensor would solve nothing. In this case, one method to verify whether there is a problem with the O_2 sensor would be to use an exhaust gas analyzer (discussed in Chapter 6) to determine whether the air-fuel ratio was truly rich.

Similarly, a MAP sensor that reports the engine is under partial load when it is at idle in the service bay could be the result of a faulty MAP sensor. But it could also be the result of low engine vacuum being made available to the MAP sensor due to a restricted or leaking vacuum line, or even low vacuum being produced by the engine, in which case, again, replacing the sensor would not solve the problem. Find out by teeing a vacuum gauge into the line going to the MAP sensor.

Pinpoint Testing of a Load Component

Pinpoint tests of most load components can be broken down into two categories: electrical tests and functional tests. Using a solenoid-operated valve as an example, electrical tests might include the following:

- Test the resistance of the solenoid's coil and compare to specification (checks for both shorted windings as well as open windings).
- Energize the solenoid with jumper wires and note the "click." (This does not constitute a functional test, as it does not test the valve for flow or leakage.)
- Capture a current waveform with an oscilloscope (discussed later in this chapter).

- Some service manuals also advise, with the solenoid's harness connector disconnected from the solenoid, using a jumper wire to jumper battery voltage to one side of the solenoid and then using a voltmeter to measure the voltage at the other terminal (should be source voltage). However, while this tests the winding for an open, it does *not* test the winding for shorts. A resistance test with an ohmmeter is a better test.

These same tests might also be used to test the winding of a relay electrically. For example, with a Bosch relay, simply make a resistance test between terminals 85 and 86. (When applying power and ground to energize a diode-suppressed relay, be sure to carefully observe the correct polarity to avoid burning the diode open.)

Follow the electrical tests with a functional test. The functional test will test whether the load component performs its function properly. The type of test involved depends on the type of load component involved. For example, with a fuel pump motor, you might test the fuel pressure and volume with the motor energized and driving the pump. Or with a TBI fuel injector, you might simply watch it—with the system pressurized and the engine not running, it should not leak; and with the starter cranking the engine, it should spray fuel. With a vacuum solenoid, you might use a vacuum pump opposite a vacuum gauge to test when the solenoid does and does not allow vacuum to be transferred from the pump to the gauge. With a relay, you might use it to power another circuit while testing the voltage drop across the contacts when they are closed. With a Bosch relay, test the voltage drop across terminals 30 and 87 when the relay is energized and being used to power a light bulb (a 12 V test light will work as a load for purposes of the voltage drop test). Also, if applicable, test the volt-age drop across terminals 30 and 87a when the relay is de-energized and being used to power a light bulb.

Flowcharts

In most cases, either a symptom or a DTC can direct the technician to a **flowchart** that can assist the technician in making pinpoint tests efficiently. A flowchart instructs the technician to perform a specific pinpoint test and note the results. Depending on the results, it may direct the technician either to make another specific pinpoint test or to make a specific repair. If all the steps in a flowchart are designed to fit on one page, it is commonly known as a *troubleshooting tree* (Figure 5–2). Figure 5–3 shows a manufacturer-specific flowchart. In most paper service manuals,

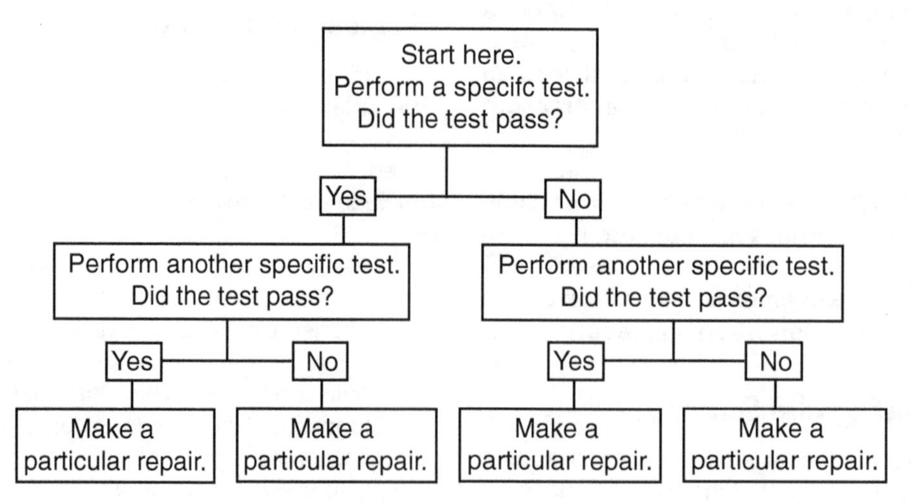

Figure 5–2 Typical form of a troubleshooting tree.

DTC 15
Engine coolant temperature sensor circuit
(Low temperature indicated)

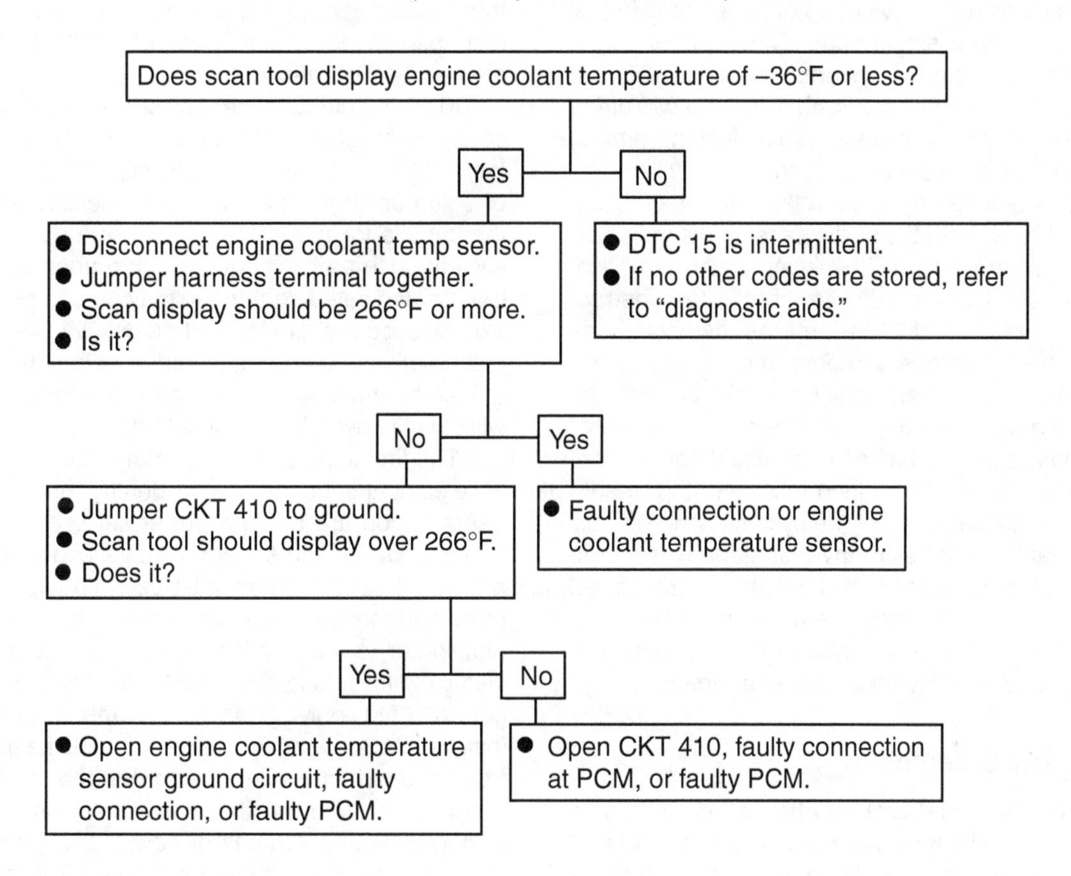

Diagnostic Aid

Temperature vs. Resistance Values (Approximate)		
°C	°F	
100	212	177
80	176	332
60	140	667
40	104	1459
20	68	3520
0	32	9420
–20	–4	28,680
–40	–40	100,700

Figure 5–3 Diagnostic flowchart for locating the cause of an ECT sensor DTC.

the flowchart sequence will typically have the technician flipping back and forth through multiple pages of the service manual as the diagnostic steps are followed, as opposed to fitting all test steps on a single page. With electronic service and information systems, the flowcharts are typically written in paragraph form and will direct you to certain paragraphs, while skipping others, depending upon each test result.

When following a flowchart, the technician should begin with the particular step, as indicated by either the DTC or the symptom. It is critical at this point to read the flowchart's test instructions carefully and to perform the test exactly as described. It is equally important to be sure of the condition of any diagnostic tools used in each test step. If a required jumper wire has no continuity or if a voltmeter lead has an intermittent connection, you may be led to an incorrect test step, resulting in a misdiagnosis of the system. Also, equally important, do not skip any test steps or assume their outcome. Some test steps may seem difficult to perform, but skipping a test step can also result in misdiagnosis of the system or wasted time if you end up starting your tests over again.

Electrical Schematics

An **electrical schematic** is essentially a map of the electrical portion of a system. It is an essential tool that should be used when performing pinpoint tests of a circuit or system. An electrical schematic will show and identify components, wire colors, circuit numbers, connector numbers, and terminal numbers at a connector.

When manufacturers design a flowchart, it is because of the engineers' understanding of how the system is designed to operate that they can tell technicians what tests to perform for each of the possible symptoms and/or DTCs. Likewise, technicians should try to understand how a system with which they are unfamiliar is designed to operate. Part of your ability to do this lies in understanding each of the concepts outlined in this textbook, from simple inputs, processing, and outputs to the more detailed descriptions in various chapters of how each of the systems functions. However, your diagnostic ability will be increased if you can also interpret the electrical schematic for the system you are trying to diagnose. In fact, many flowcharts incorporate an electrical schematic for this purpose.

Many electrical schematics are simple, dedicated schematics. Others are complicated by the fact that an otherwise simple schematic is part of a larger schematic, such as one that shows the circuits for the entire vehicle or for an entire system such as the engine performance system. Even a dedicated schematic may be complicated simply because of the technology involved. As you practice interpreting electrical schematics, be guided by the following recommendations, which will simplify even the most complex schematics.

The first thing that is important in interpreting an electrical schematic is to identify the *components*. If you are using a paper service manual, make a copy of the schematic on a copier; if you are using an electronic service manual, simply print out the electrical schematic. Then use a highlighter to identify the components and specific circuits. Taking the time to identify the components not only gives you some immediate insight into the system, but it will also keep you from inadvertently trying to interpret the outline of a control module or other component as a wire. Second, identify sources of *power* and *grounds*. Then you should identify any *input circuits* to a control module, *output circuits* from a control module, and *data circuits* between control modules on a network.

Certain complex electrical schematics will always be a challenge to any technician, no matter the level of experience. But as you practice interpreting electrical schematics, you will find that they become noticeably easier to read. If you take the time to compare the verified symptom to your understanding of system operation based on the system's electrical schematic, you will be able to narrow down the area where you need to begin testing the system. This is essentially what a flowchart does for you. You should use an electrical schematic *with* a flowchart to enhance your diagnostic abilities. As your level of experience increases you will also find that you may want to use an electrical schematic *in place of* a flowchart in many cases.

In summary, an electrical schematic is used by the technician to turn textbook theory (or classroom theory) into product-specific knowledge that is applicable to the specific vehicle being worked on! In fact, the Automotive Service Excellence (ASE) organization does *not* include any flowcharts in its materials that accompany the L1 Advanced Engine Performance Specialist Certification test. Instead, this diagnostic test provides *electrical schematics* to assist in answering questions regarding the diagnostic process.

Other General Diagnostic Concepts

Not all faults (either soft or hard) will actually generate a DTC, particularly in the early engine computer systems. The PCM was programmed to monitor the system for total failures or those failures that resulted in a voltage value outside the parameters programmed into the PCM. For example, a TPS could actually have an intermittent glitch that would not set a DTC in the PCM's memory because the voltage did not fall below the lower parameter that was programmed into the PCM, but it would still result in a lean hesitation each time the voltage signal dropped. By the late 1980s, the PCM could compare certain sensors and determine if a fault existed even if the voltage was not outside the parameter values. OBD II (discussed in Chapter 7) took this concept further, and the ability of the PCM to monitor the system for efficiency as well as for failure became a standard that applies to all OBD II vehicles. As a result, the PCM on newer vehicles is more often able to alert the technician to a fault through a DTC. In addition, as technology progressed, the PCM was enabled to monitor more functions through additional feedback circuits. Therefore, on a newer vehicle, a DTC can direct the technician to a more specific area than on some of the older systems. However, no matter how specifically a DTC can direct a technician to a fault area, you should always verify any fault through pinpoint testing. Also, bear in mind that multiple DTCs should be repaired in the order specified by the manufacturer.

DIAGNOSTIC EQUIPMENT

Scan Tools

A **scan tool**, also known as a *scanner*, is probably the most familiar tool to anyone working with today's computerized systems. A scan tool is a most important tool in that it provides the technician with a way to communicate with the computers on the modern vehicle. There are essentially four classes of scan tools:

- Manufacturer scan tools: These are the scan tools that are generally found in dealerships and provide the most in-depth features.
- Obsolete manufacturer scan tools: Many times these can be purchased at a lower cost than for the current manufacturer models, and they are fully featured for all but the most recent vehicle models.
- Generic scan tools: These scan tools are generally the most cost effective, but capable, units for most professional technicians working in aftermarket shops. Examples of dedicated generic scan tools are shown in Figure 5–4, Figure 5–5, and Figure 5–6. Professional-level generic scan tool systems that allow the technician to use a Windows PC as a scan tool are available from several manufacturers and include the appropriate software and hardware. These include the *EScan Pro* available from Automotive Test Solutions (www.automotivetestsolutions.com) and the scan tool available from Auto Enginuity (www.autoenginuity.com).
- Generic scan tools designed for the do-it-yourselfer: These hardware and software kits are designed to turn a smartphone, a tablet, or a Windows PC into a scan tool and range from low-end systems that can access DTCs

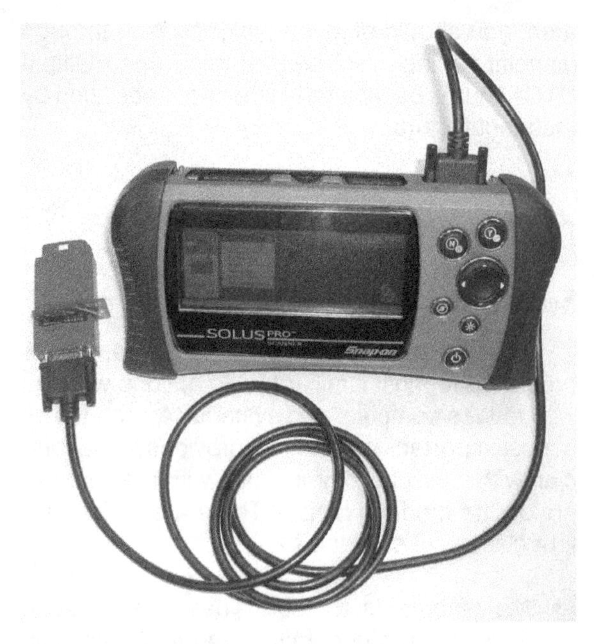

Figure 5–4 A dedicated generic scan tool: A Snap-on Solus Pro.

Figure 5–6 Another example of a dedicated generic scan tool: A Launch Smartbox.

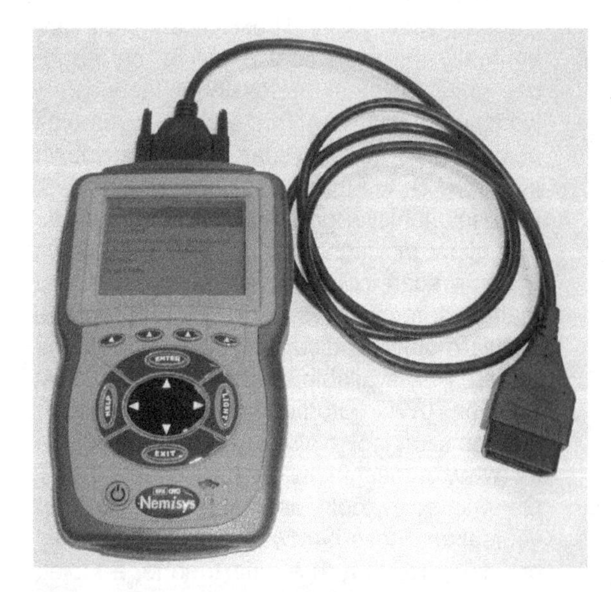

Figure 5–5 Another example of a dedicated generic scan tool: An OTC Nemisys.

and data stream information to high-end systems that approach the abilities of some generic dedicated scan tools. Figure 5–7 shows a wireless OBD II DLC transmitter from OBDLink, and Figure 5–8 shows a less expensive generic Bluetooth DLC transmitter, an ELM327 that can be produced by any of several manufacturers. Depending upon the device, it may be used with an Android smartphone only, or also with an Apple iOS phone (iPhone Operating System) or a Windows PC. Software for these devices ranges from low- to high-end. An example of an upper-end

Figure 5–7 A wireless OBD II DLC transmitter manufactured by OBDLink.

Figure 5–8 A generic Bluetooth OBD II DLC transmitter known as an ELM327.

software program is *ScanXL Professional,* which is available with several software upgrades to expand its abilities. At the lower end of the scale is a software program called *Torque Pro* designed for use with an Android smartphone or tablet only.

Purpose and Functional Concept. Scan tools are designed to communicate with the on-board computers using serial data (binary code). When connected to the **data link connector (DLC),** the scan tool can communicate with all other modules on the vehicle that are also connected to the DLC. It is important to recognize both the distinct abilities of a scan tool and the limitations.

Functional Procedure. A scan tool simply plugs into the DLC after it has been equipped with the correct DLC adaptor. A standardized DLC connector and adaptor is used for all 1996 and newer vehicles. Once powered up, it requires vehicle identification number (VIN) and data input from the technician. (Some OBD II applications will automatically configure the VIN code entry once the scanner is connected to the DLC and the ignition is turned on.) This identifies for the scan tool the communication languages used by the modules connected to the DLC. Then the technician must select the module to be diagnosed, as well as the desired diagnostic path.

Code Pulling. A scan tool's ability to pull codes includes both pulling DTCs from a computer's memory and initiating a computer self-test of the system and displaying the resulting DTCs. In certain early applications, code pulling cannot be done with a scan tool but must be accomplished through manual methods. For these the scan tool may still be able to instruct the technician as to the proper procedure, even though it cannot be connected to the system.

Functional Tests. Scan tools can also be used to perform functional tests that the vehicle manufacturer has programmed into the on-board computer. Some scan tool manufacturers refer to these tests as *special tests.* The primary and best-known functional test is the active command mode, in which the scan tool can use its bidirectional controls to command on-board control modules to perform certain actions, such as energizing a cooling fan relay or a canister purge vent solenoid.

A common use of such a test is to help the technician to quickly narrow a problem down to the input side or the output side of the computer. For example, if the electric cooling fan does not come on during normal vehicle operation, resulting in an over-heating engine, a scan tool may be used to command the PCM to energize the cooling fan relay. If the cooling fan then runs, the output side of the computer has been eliminated as a possible fault and the technician knows that the problem is likely on the input side—either a coolant temperature sensor or its associated wiring. However, if the cooling fan does not run during this test, then the technician knows that the problem is likely on the output side of the computer and can proceed by testing the cooling fan relay, the cooling fan motor, and the associated wiring.

Another purpose of this test is simply to provide an easy method of energizing an actuator when the need arises for the sake of other testing procedures. For example, the technician may use a scan tool to command the PCM to energize the normally open canister purge vent solenoid when using a leak tester to test the evaporative system for leaks. This closes the solenoid and thus enables the system to hold pressure. This eliminates the need to energize the solenoid manually using jumper wires in order to leak-test the system.

Data Stream. Scan tools can read data stream information from the on-board computers. This information includes the *computer's interpretation* of input sensor values, including both the perceived voltage and the interpreted value. For example, the engine coolant temperature (ECT) information may be displayed as a voltage (0.62 V) and as the interpreted temperature that it represents (205°F or 96°C).

The scan data also shows the computer's *intended* output commands. It is always a good idea to check whether the intended commands are actually being carried out.

Scan data will also display other pieces of information from the computer's memory. This includes information such as fuel trim values.

Scan data, whether input data, output commands, or other data, is identified as to subject area by the information's parameter ID (PID). Data stream PIDs include inputs identified as "ECT volts," "ECT temperature," "O_2 sensor volts," and so on. Output PIDs might include commanded "fuel injector pulse width" or commanded "spark advance." Each PID will be displayed with a corresponding value.

Finally, scan data may be customized on many scan tools. This allows technicians to select only the PIDs that they wish to look at. This speeds up the scan tool's update rate because the PCM (or other control module) no longer has to communicate all the available information to the scan tool. As fast as it might be, it still holds true that the binary communication (serial data) between the PCM, or other control module, and the scan tool can only communicate one PID and its value at a time, so reducing the amount of information being communicated increases the update rate of the remaining information.

Data Stream Graphing. Many modern scan tools can take the data stream information and create a graph of it as it receives updated values from the on-board control modules. This gives the technician a visual representation of the updated data over time and can be an immense help in diagnosis.

Additional Features. Many scan tools provide additional features such as a data capture mode (sometimes called *movie mode*) for use in identifying causes of intermittent symptoms. When you are trying to verify an intermittent problem, this mode may be used to capture data while road testing the vehicle, which may then be reviewed later, frame by frame.

Other features may include idle reset and/or idle speed control through the scan tool, octane adjustment, tire size input for vehicle speed sensor (VSS) calibration, and the ability to disable or enable certain features for various systems. Additionally, some scan tools provide an internal source of information (electronic shop manual) for such things as DTC meanings, module terminal pin-outs, and term definitions.

Scan Tool Interpretation. When looking at scan tool data, bear in mind that the scan tool only offers the computer's interpretation of the data. For example, input values are measured at the computer and are subject to errors induced by faulty connections as well as poor computer power and ground connections. Output commands indicate the computer's intentions and cannot compensate for defective drivers, poor connections, and deficient actuators.

For example, a discrepancy between an input voltage value shown on a scan tool and a raw voltage signal when measured at the sensor may occur because of computer power and ground voltage drop problems. Scan data that indicates an actuator is energized simply indicates the computer is trying to forward bias the driver that controls the circuit. Many problems have been overlooked because a technician blindly believed that the intended output command had successfully been carried out simply because the scan tool had indicated it.

Advantages. A scan tool has the advantage that it is easy to connect and can very quickly give the technician a very large amount of information. In addition, the display typically provides a written menu and instructions that make the tool very easy to use.

Disadvantages and Limitations. Although a scan tool may offer a very quick approach to diagnostics, it can only direct you to a problem *area* or circuit. It cannot tell you the exact cause of a symptom. For example, if it shows that a temperature circuit is sending a 5 V signal to the computer (equating to approximately –40°F), it is your understanding of an NTC thermistor circuit that tells you that the circuit is open. This open could be in the sensor, a wire, a connector, or the PCM itself. To identify the exact cause of the open circuit, further pinpoint testing is needed.

As previously stated, when a computer informs you via a scan tool that an action has been commanded, this means only that the computer is trying to forward bias the output driver (transistor) that controls this action. It does not really know (in most cases) whether the action has in fact taken place. The exception would be for those components for which the output function is monitored with a feedback signal. This is common in newer systems, particularly those that are OBD II compliant.

Scan tools are also limited in their ability to identify intermittent problems of short duration. The update speed of a scan tool's data is limited by the PCM's communication speed. Only one PID and its corresponding value can be updated with each binary communication. This increases the probability that a glitch in a TPS circuit, for example, may happen between scan tool updates and therefore may never be seen by the technician who is using a scan tool.

Non-Powered Test Lights

Non-powered test lights (Figure 5–9) continue to have a place in your toolbox for certain quick checks. However, many technicians continue to use these devices in circuits where they can be very dangerous; therefore, it is important to understand a few facts about them.

Purpose and Functional Concept. The test light's only purpose is to serve as a quick-check tool to show if a voltage potential is present. A voltage differential across the test light causes the bulb to glow.

Functional Procedure. Begin by connecting the test light across the battery's terminals to verify the integrity of the test light. Then, while it is grounded, touch the circuit in question and note

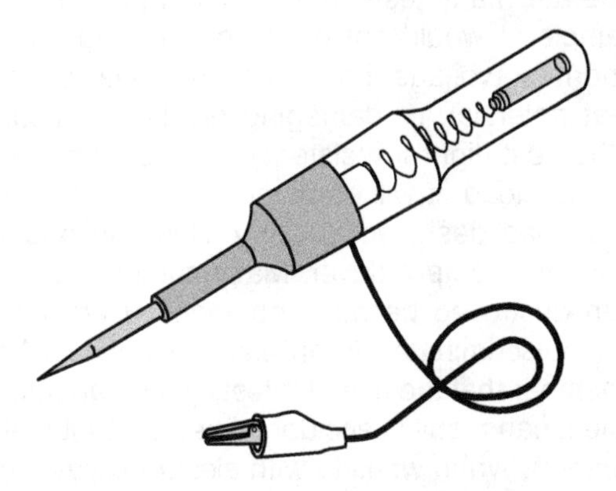

Figure 5–9 A typical non-powered test light.

the light. A test light makes a great tool for testing all of the vehicle's fuses electrically in a short amount of time. It can also be used to test switching of circuits such as ignition coils and fuel injectors as long as the resistance of the test light equals or exceeds the resistance of the intended load. A grounded test light can also be used to provide a path of low resistance for identifying poor insulation in secondary ignition wiring. Secondary ignition wiring that is arcing to ground can be difficult to identify in the shop when cylinder pressures are low and resistance across the plug gap is also low. But many times, even with the engine idling, a secondary wire will arc to a grounded test light to help identify the specific insulation breakdown. For this test, pass the tip of the grounded test light along each of the secondary wires with the engine running.

Interpretation. The fact that a test light illuminates proves only that a voltage potential is present. In situations in which accurate voltage readings are required, you should never try to guess by the bulb's brightness how much voltage is present.

Advantages. The advantages of a test light are that it is easy to use, it is inexpensive, and it takes up very little room in your toolbox. Although it has several limitations, it makes a great quick-check tool for several initial tests as you begin diagnosing certain types of faults.

Disadvantages and Limitations. A test light cannot indicate voltage values; therefore, there are many tests that it cannot perform. For example, it would not be a good tool to use in performing voltage drop tests. Additionally, it has great potential for damaging electronic circuits. If the test light's resistance is less than the intended load, it will draw more current than the circuit was designed to carry. This can destroy a module's output driver. Many computers have been destroyed because someone used a test light indiscriminately in an electronic circuit. Also remember that the use of a test light on an airbag system can result in accidental deployment of the airbag(s). When working with electronic systems, consider a test light to be a jumper wire.

Logic Probes

A **logic probe** (Figure 5–10), is basically a test light that is safe for use on electronic systems because it has high resistance designed into its internal circuitry. In fact, the term *logic probe* implies that it was designed for use with today's computers. Logic probes are generally available from electronics suppliers at a reasonable price. You can also purchase similar devices known as *high-impedance test lights* from tool suppliers.

Purpose and Functional Concept. Like a test light, a logic probe provides the technician with a quick-check tool. Once powered up, a logic probe can be used to check for voltage and ground potential. It is also excellent for testing for the presence of switched signals within the primary ignition system or at the fuel injectors.

Functional Procedure. As an electronic device, a logic probe must first be powered up. To do this, connect the power leads to battery positive and negative. Then, for best results, connect the tip's negative lead to a clean, unpainted ground. (The ground lead in the power lead set contains a protective diode that

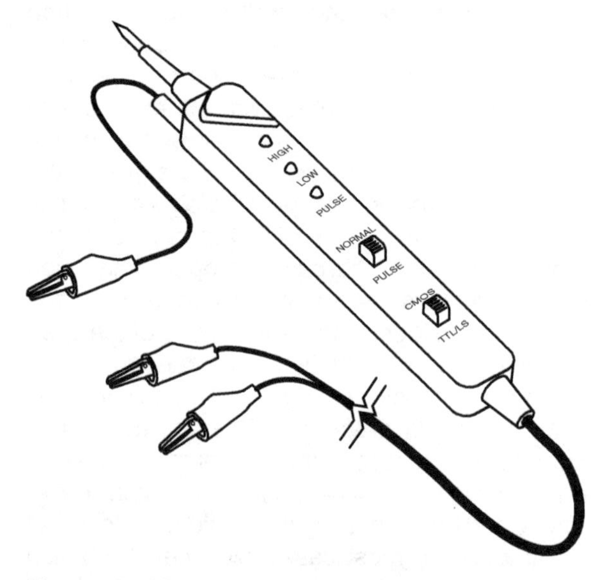

Figure 5–10 A logic probe is an excellent tool for testing for the presence of electrical signals.

influences the ground level if it is the only ground used.) If the logic probe has a CMOS/TTL switch, set it as follows: For working with voltages from 0 to 12 V, set it to CMOS, and for working with voltages from 0 to 5 V, set it to TTL. Then set the NORMAL/PULSE switch as follows: When it is set to NORMAL, the red and green (LEDs) will indicate high- or low-voltage levels within the selected voltage range and a tone generator emits a high or low tone; when it is set to PULSE, a yellow LED is also used to indicate a *pulsed* voltage signal. Unfortunately, the yellow LED is also lit by induced voltage from the secondary ignition wires if you are in close proximity to them on a running engine.

The alligator clips may be cut off and replaced by soldering in a cigarette lighter adaptor (while paying attention to the required polarity). In this configuration, a logic probe can be quickly powered up using a power point in the passenger compartment for the purpose of electrically testing fuses. The plug can still be connected quickly to the vehicle's battery through the use of a cigarette lighter female socket–to–alligator clips adaptor.

Interpretation. With the tip of the probe connected to the circuit in question, note the LEDs. The red LED (and high tone) responds to a higher voltage, whereas the green LED (and low tone) responds to a lower voltage. The brightness of one LED compared to the other can also give you a rough indication of duty-cycle on-time versus off-time. If you have the probe connected to the ground side of a ground-side switched actuator, increasing the on-time increases the brightness of the green LED.

Advantages. A logic probe is an easy-to-use tool best suited for determining the presence (or loss) of primary ignition and fuel injection signals. It also has high internal resistance and is therefore safe for use in diagnosing electronic circuits.

Disadvantages and Limitations. A logic probe, like a standard test light, cannot inform you of exact voltage values or voltage drop readings, and it is therefore limited to certain quick checks.

Multimeters

A **multimeter** may be used to measure voltage, amperage, resistance, and other electrical characteristics. It may be of any of the following designs: analog, digital, or graphing. Analog multimeters are no longer as common in the automotive industry as they once were, so the emphasis will be on the other two designs.

Digital Multimeters

A **digital multimeter (DMM)** is used by most technicians to make pinpoint tests of the computer and its associated circuits and components. Also known as a **digital volt/ohm meter (DVOM)**, a typical DMM is shown in Figure 5–11.

Figure 5–11 A Fluke digital multimeter (DMM).

Purpose and Functional Concept. A DMM is generally the tool of choice for making accurate pinpoint tests of the electrical characteristics of a circuit. When purchasing a DMM, look for the following features: resistance, AC and DC voltage, AC and DC current flow, frequency, duty cycle on-time, pulse width on-time, temperature, and RPM. Also, you might want a bar graph, a min and max function, a hold function, and an average function. If the DMM is auto-scaling, also look at the unit's response time (in the specifications sheet) or how long it takes for the DMM to sense the signal, rescale, and display the reading. This time can range from 2 seconds on some low-end units to 100 ms or faster on high-end units. The DMM should also be rated at no less than 10 MΩ impedance when scaled to a DC voltage scale. This high resistance through the meter ensures that it does not damage electronic circuits, and meter accuracy is increased when measuring DC voltage. Bear in mind that when the DMM is functioning as an ammeter, it effectively acts as a jumper wire.

A DMM is designed to provide in-depth information about the electrical characteristics of a circuit. However, all too often, technicians get careless in their interpretation of what they see on the DMM's display screen and misdiagnose a problem.

Functional Procedure. A DMM has a simple two-lead connection. Connect the black lead to the jack marked "Common" and the red lead to the jack marked for the type of measurement needed. Bear in mind that leads connected into the amp jacks effectively turn the DMM into a jumper wire. Now set the dial to the scale needed. Connect the red and black leads to the circuit or component that you need to test, according to the type of measurement being made (voltage, amperage, or resistance). The red lead should connect to the point that is most positive in the circuit and the black lead to the point that is most negative. If the DMM is a manual-scaling unit, for best accuracy be sure to scale down to the lowest scale that can read the value being measured. Then note the reading on the display and interpret the reading according to the type of scale to which the DMM is currently set. Bear in mind that with each scale that a DMM is scaled upward, the reading becomes less accurate by a factor of 10. Most DMMs are auto-scaling (auto-ranging) in that the DMM will automatically rescale itself to the correct scale to be used as the leads are connected to a circuit or component. On an auto-scaling meter, be sure to verify which type of scale has been selected (K, M, m, or μ) in the DMM's display window.

Even with an auto-scaling DMM, many times it is to your advantage to manually scale the meter, in which case the same rules should be followed as for a manually scaled meter. For example, it would be better to manually scale a DMM to a 20 V scale when making measurements in a circuit such as a TPS circuit that operates within a 0 V to 5 V range as opposed to having an auto-scaling meter rescale itself during the sweep test. This automatic rescaling is sometimes misinterpreted as a fault.

Interpretation. Because a DMM provides very exact measurements of raw data, it makes an excellent pinpoint test tool. But particular attention must be paid to the scale's factor when interpreting the reading on the display. This factor may be shown either on the scale selection dial or on the display, depending on the particular DMM. A "K" (kilo) means that whole numbers represent thousands, and an uppercase "M" (mega) means that whole numbers represent millions. The decimal point should be moved three or six places to the *right,* respectively, to convert these values to numbers that represent single whole units. Likewise, a small "m" (milli) means that whole numbers represent thousandths and a "μ" symbol (micro) means that whole numbers represent millionths. The decimal point should be moved three or six places to the *left,* respectively, to convert these values to numbers that represent single whole units. If a factor value is not shown on the scale selection or the display, read the display as it appears. The ability to use a DMM on any type of scale (ohms, amps, volts, Hertz) depends on the technician's ability to interpret these factors.

PREFIX (Factor)	SYMBOL	RELATIONSHIP TO BASIC UNIT
Mega	M	1,000,000
Kilo	K	1,000
Milli	m	1/1,000
Micro		1/1,000,000

Figure 5–12 The four most common DMM scale factors.

1.234 MΩ = 1,234,000 Ohms
1.234 KΩ = 1,234 Ohms
1,234 mA = 1.234 Amps
1,234 µA = 0.001234 Amps

Figure 5–13 Examples of converting each of the factors to the base unit.

Figure 5–12 shows the four most common factors used with DMMs. Figure 5–13 shows four sample interpretations of these factors.

DC Voltage versus AC Voltage. A DMM that is set to the DC voltage scale will measure the *average* voltage of a rapidly switched voltage. If, with the alternator ripple considered, the voltage that is applied from the alternator to the vehicle's battery with the engine running varies from 14.15 V to 14.25 V, a DC voltmeter will interpret this signal as 14.20 V. Similarly, if a DC voltmeter is used in an attempt to measure the AC voltage produced by a wheel speed sensor in an ABS, no matter how fast you spin the wheel, the measured voltage will always be near zero volts because the average of an AC voltage is zero. Unfortunately, this mistake could lead the technician to falsely believe that the sensor was defective. And if you connect a DC voltmeter to an actuator that is operated on a duty cycle, cycling from zero volts to about 14 V, an average voltage of about 7 V would be indicative of a 50 percent duty cycle.

An AC voltmeter works differently and is much more complex. When the DMM is set to the AC voltage scale, the meter looks at the **amplitude** of the signal. The amplitude is the difference between the highest value and the lowest value. If the meter's leads are connected to an AC wall outlet, the meter will see the voltage go positive by a value of about 170 V and will then see the voltage go negative by a value of about 170 V. Thus, the amplitude of the signal, or peak-to-peak voltage, is about 340 V. The AC voltmeter uses an algorithm to obtain the DC equivalent value, or how much work is getting done when compared to a DC voltage. This is done by multiplying the 170 V value (either positive or negative) by a factor of about 0.707 to obtain a resulting value of about 120 V.

If the AC voltage is a sine wave (known as a *sinusoidal waveform) (*Figure 5–14), the DC equivalent value can be accurately measured and calculated by either a standard AC voltmeter (known as an *average responding* meter) or with a true RMS meter. (RMS stands for *root mean square,* which refers to the algorithm used to obtain the DC equivalent value.) However, if the AC voltage is not a sine wave (known as a *non-sinusoidal waveform*) (Figure 5–15), the true RMS meter will measure and calculate the DC

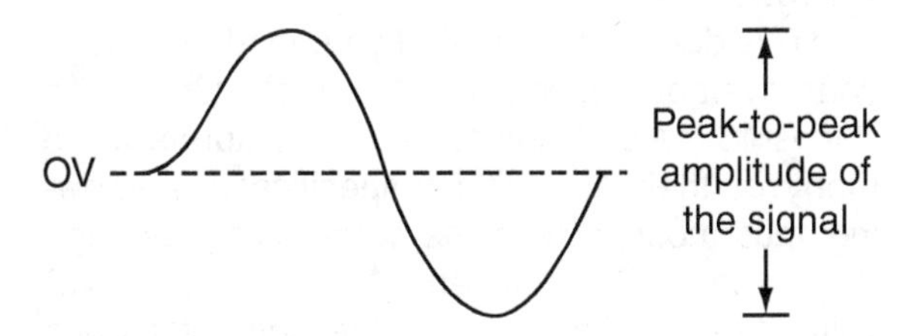

Figure 5–14 Sinusoidal AC voltage.

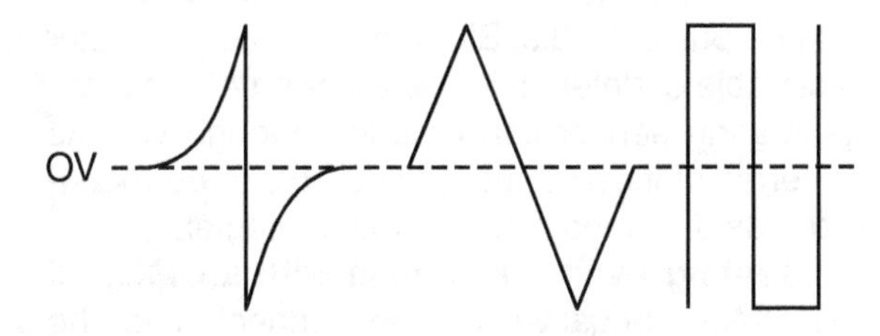

Figure 5–15 Examples of non-sinusoidal AC voltages.

equivalent value of the AC voltage with greater accuracy than an average responding meter is capable of.

The difficulty in measuring an unstable voltage (such as an AC voltage) and calculating its true working value stems from the fact that it is constantly changing and never remains at one voltage value for very long. A true RMS meter can do this by calculating the heat energy loss (measured in wattage) when the current is pushed by the voltage through a fixed resistance within the meter. In essence, the meter compares this energy loss to the energy loss that would be created by a DC voltage pushing current across the same resistance. Thus, the true RMS meter can accurately calculate the DC equivalent value of both sinusoidal and non-sinusoidal AC voltages with equal accuracy. In comparison, an average responding AC voltmeter, while capable of an accurate measurement of a sinusoidal AC voltage, cannot measure a non-sinusoidal AC voltage with the same accuracy that a true RMS AC voltmeter does. As a result, an average responding AC voltmeter will generally display a lesser voltage value than its true RMS counterpart when measuring non-sinusoidal AC voltages.

This does not mean that you need to replace your average responding AC voltmeter with a true RMS meter. Automotive manufacturers wrote many of their AC voltage specifications knowing that most technicians were using average responding AC voltmeters. And with most AC voltage-generating sensors, the computer does not react to how much AC voltage is produced. Instead, what is important to the computer is the frequency of the voltage pulses and when the voltage pulse occurs. Either type of AC voltmeter is capable of determining whether an AC voltage-generating sensor can produce enough voltage (in terms of its amplitude) to be able to be recognized as a voltage pulse by the computer.

Testing Available Voltage with a DMM. If the DMM's negative lead is connected to the negative terminal of the battery, the DMM's positive lead can be used to test for available voltage at any point in a circuit (Figure 5–16). If the positive lead is connected to the positive terminal of the battery, the battery's open circuit voltage (OCV) is measured. While these are easy tests, many technicians limit their testing to this, ignoring the increasingly important art of voltage drop testing.

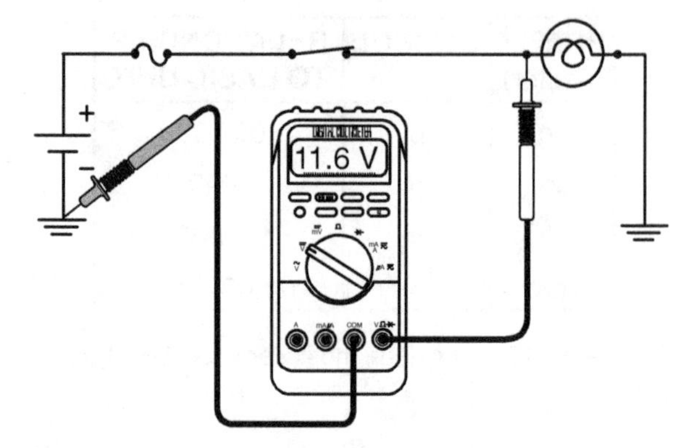

Figure 5–16 Available voltage test.

Voltage Drop Testing with a DMM. A circuit always consists of three parts: the source, the load component, and the connecting circuitry, on both the positive and negative sides of the circuit. When a symptom of poor operation (or even no apparent operation at all) is encountered, the source should always be checked first. This, of course, is the vehicle's battery if the engine is not running. When the engine is running, the alternator (actually, the rear half of the alternator since that is where the rectifier bridge is located) becomes the voltage source.

The second item that should be tested is the connecting circuitry, which is responsible for supplying the load component with as close to the full amount of the source's voltage potential as possible. There are exceptions to this, however. Some examples are:

- The current-limiting resistors used for the low and intermediate speeds of a blower motor control circuit (Figure 5–17A)
- The resistor (ballast resistor) that was used to limit current flow through an ignition coil

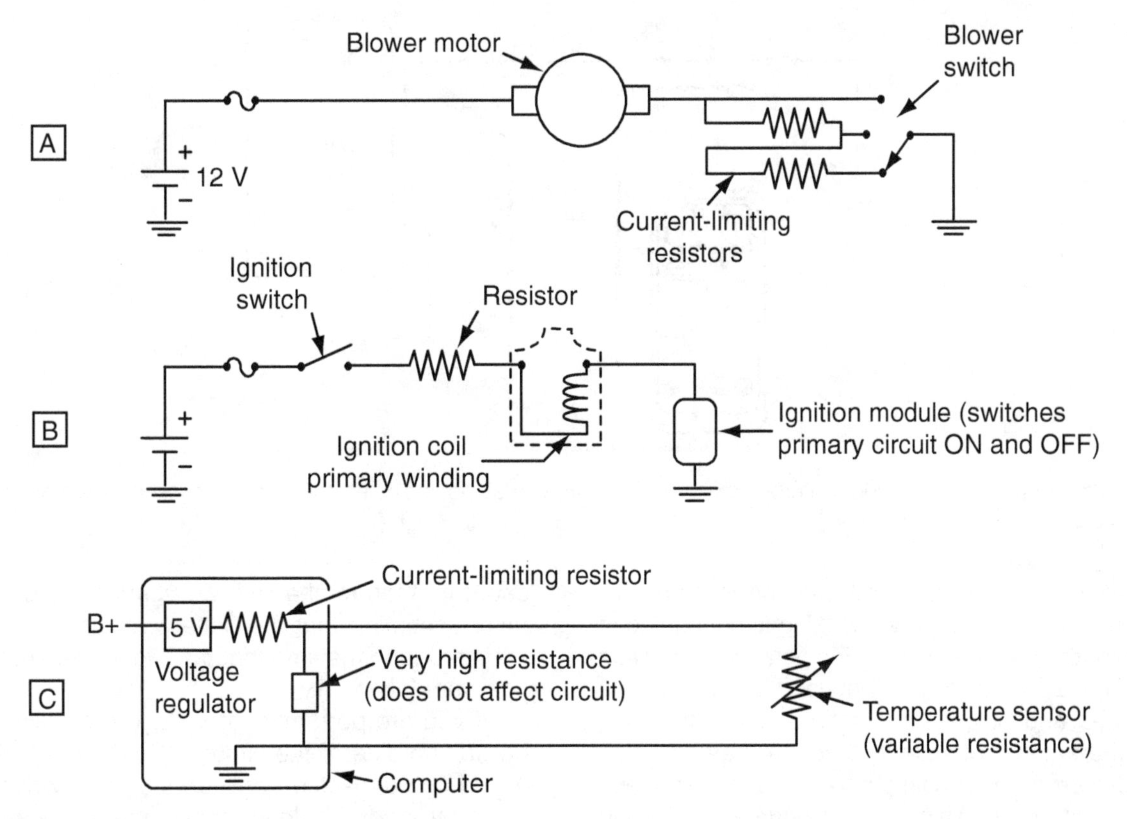

Figure 5–17 Current-limiting resistor circuits.

of many ignition systems a few years ago (Figure 5–17B)
- The current-limiting resistor used in a temperature-sensing circuit (Figure 5–17C)

While testing voltage drop across a circuit, you should remember that:

- In an ideal circuit, most of the voltage drop should take place across the load component.
- In an ideal circuit, little of the voltage drop should take place across the circuitry.
- Voltage drop cannot occur unless the circuit is complete and current is flowing in the circuit.
- All ohms of resistance in a series circuit share equally in the voltage drop occurring in the circuit.
- Voltage that is dropped (used up) is always the result of resistance encountered when current is flowing in the circuit and therefore proves the existence of resistance in the circuit.
- Voltage measurements can then be used to identify the existence of unwanted, excessive resistance in the circuitry, but only if the circuit is complete and current is flowing in the circuit.

To perform a **voltage drop test** of a circuit, the DMM must be connected in *parallel* to the circuitry, either on the positive (insulated) side of the circuit or on the negative (ground) side of the circuit. If you are performing a voltage drop test on the insulated side of the circuit with the KOEO, the DMM should be connected *between* the battery

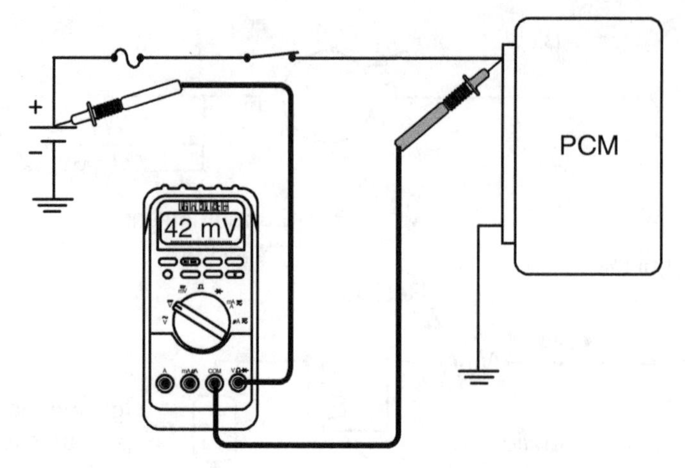

Figure 5–18 Positive-side voltage drop test. A 0.042 V reading indicates that insulated circuit resistance is acceptable.

positive post and the load component's positive terminal with the positive DMM lead connected to the battery (Figure 5–18). This parallel connection compares the difference between the two voltage values and displays the results. This difference is the amount of voltage drop that has occurred between the two points where the DMM's leads are connected. This reading equates to having measured the available voltage at the battery, then having measured the voltage reaching the load, and then mathematically figuring the difference, assuming that the battery voltage did not change during the time that the two voltage readings were taken.

If you are performing a voltage drop test on the ground side of the circuit (KOEO), the DMM should be connected *between* the battery negative post and the load component's negative terminal with the negative DMM lead connected to the battery (Figure 5–19). Again, this parallel

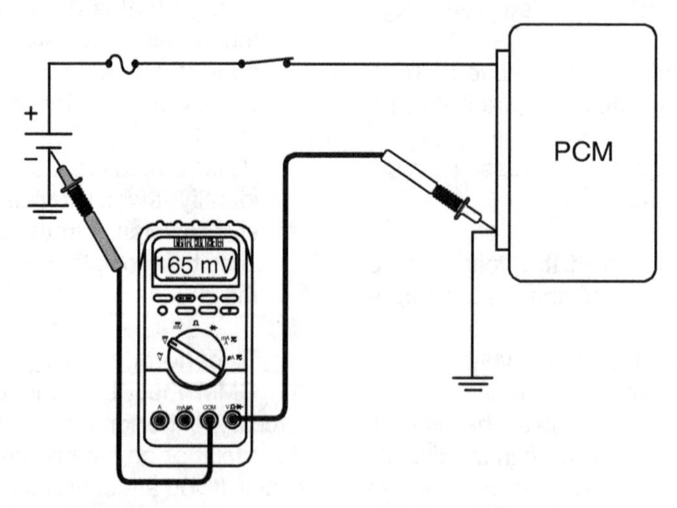

Figure 5–19 Negative-side voltage drop test. A 0.165 V reading indicates excessive resistance in the ground-side circuit and is above acceptable limits.

connection compares the difference in the two voltage values and displays the results. This difference is, again, the amount of voltage drop that has occurred between the two points where the DMM's leads are connected and equates to how much voltage was left in the circuit after the load due to unwanted resistance in the circuit after the load.

Performing voltage drop tests is the most accurate and efficient method to verify the integrity of the circuitry that connects the source to the load; it will alert the technician if excessive unwanted resistance exists in any part of the circuitry, be it in the wiring or connectors, any circuit protection devices, any switches (whether relay-operated or not), and any ground connections (either to engine, body, or chassis ground).

Depending on the type of circuit being tested, the specifications for allowable voltage drop vary, not because we allow more or less resistance for different types of circuits but primarily because more amperage across a given resistance results in more voltage drop across the resistance in the circuitry, and less amperage across a given resistance results in less voltage drop across the resistance in the circuitry. This is due to the fact that all the source voltage will be used up evenly by each ohm of resistance value in a circuit. A high-amp circuit has less resistance, so each ohm of resistance produces more voltage drop; a low-amp circuit has more resistance, so each ohm of resistance produces less voltage drop. As a result, the typical maximum allowable voltage drop specifications are as follows:

- For most circuits, the voltage drop should be less than 0.1 V (100 mV) per connection.
- For high-amp starter circuits, the voltage drop should be less than 0.2 V (200 mV) per connection.

Voltage drop readings less than the maximum allowable specification indicate that all components and connections that make up the circuitry have resistance levels low enough to be acceptable. If an excessive voltage drop reading is encountered, then, with one DMM lead connected to the battery (KOEO), the other lead should be drawn back through the circuit toward the battery (according to the applicable electrical schematic) until the voltage drop reading is reduced to acceptable levels. At this point, you know that the high resistance is located between the point in the circuit where the last unacceptable reading was taken and the point in the circuit where the first acceptable reading was taken.

To perform voltage drop tests on circuits that power load components at the rear of the vehicle or in the interior of the vehicle, a DMM lead of extra length may be manufactured. With this long lead connected to the vehicle's battery using an alligator clip, the DMM may now be carried to the rear of the vehicle or to the interior of the vehicle to perform the voltage drop tests. The wire gauge is not critical except for reasons of its durability because the DMM's high internal resistance does not allow any substantial current flow through the meter when making voltage tests. This extra-length lead should be plugged into the DMM's positive jack and connected to battery positive when performing positive-side voltage drop tests; it should be plugged into the DMM's negative jack and connected to battery negative when performing ground-side voltage drop tests.

Voltage Drop Testing a Computer's Power and Ground Circuits. All too often, technicians overlook the voltage drop tests of the computer's power and ground circuits as shown in Figure 5–18 and Figure 5–19. As with any electronic device, if either the power or ground circuit is faulty, the module cannot function properly. With the ignition key turned on and the engine not running (KOEO), the maximum battery-to-control module voltage drop specifications are as follows:

- The total voltage drop in the positive side of the circuit should never exceed 100 mV (0.100 V).
- The total voltage drop in the negative side of the circuit should never exceed 50 mV (0.050 V).

All automotive computers have internal communication circuits and internal logic gates that perform the decision-making functions of the computer. The computer's internal communication circuits and the logic gates depend upon the 5 V reference voltage that originates at the computer's reference voltage regulator. If the reference voltage regulator quits producing the 5 V signal, the computer will not power up properly and/or will make incorrect decisions. The reference voltage regulator must receive a minimum of 9 V to 9.6 V (minimum voltage that is dropped across the computer itself) to produce the 5 V output signal that is supplied to the internal communication circuits, the logic gates, and to the sensors that require a reference voltage.

If any of the following conditions occurs, the reference voltage regulator quits producing its 5 V output and the computer's internal communication circuits and the logic gates will not function properly:

- If battery voltage falls below 9.6 V.
- If battery voltage is 12.6 V, but excessive voltage drop exists in either the positive-side or negative-side circuitry of the computer so that the voltage drop across the computer is less than 9.6 V.

It should also be noted that operation of the starter motor naturally pulls battery voltage down a couple of volts. Therefore, when starting an engine, as the starter motor pulls battery voltage low, it does not take an enormous amount of unwanted voltage drop in a computer power or ground circuit to keep a computer from functioning properly. This can result in the failure of a PCM to pulse a fuel injector or fire an ignition coil while the engine is cranking over for startup.

When the vehicle's battery is severely discharged or if a power or ground voltage drop problem exists, what is otherwise a good computer may appear to stop functioning properly. Replacing the computer will not repair the problem because it is not the computer that is at fault. As a result, any time a computer seems to be functioning abnormally, or not at all, test the battery's open circuit voltage (OCV) and test the power and ground circuits for excessive voltage drop before deciding to replace the computer. Also test the available voltage at the reference voltage wire of one of the sensors that receive the reference signal, such as the TPS in the engine control system. This will indicate whether the reference voltage regulator within the control module is functioning properly.

Using Voltage Drop Tests to Identify Circuit Intermittent Problems. It should be noted that voltage drop testing can be used to

Diagnostic & Service Tip

For pinpoint testing of a circuit, voltage drop tests are far more reliable than making resistance tests with an ohmmeter. This is particularly true for low-resistance measurements. This is because an ohmmeter cannot pass realistic current through a circuit. In theory, an ohmmeter makes resistance tests using voltage drop and then calculates the equivalent ohms of resistance, but the ohmmeter's amperage output is very limited.

If you were to use a hacksaw to cut through most, but not all, strands of the high-current starter-to-battery cable, this would create an enormous voltage drop when the starter motor tried to pull 140 amps through it, but an ohmmeter would show the circuit as having good continuity. Likewise, if a do-it-yourselfer were to replace a 14- or 16-gauge wire in an exterior lighting circuit with 20- or 22-gauge wire, this would create a voltage drop problem that would result in the bulbs operating dimly. While this can be easily and accurately diagnosed using voltage drop testing procedures, an ohmmeter would see the circuit as having good continuity. As a result, trained technicians use a voltage drop test in preference to a direct resistance measurement when testing the integrity of a circuit. The use of an ohmmeter should be reserved for testing sensors and load components.

identify excessive unwanted resistance in circuits that are experiencing *intermittent* symptoms, even when the symptoms do not appear to be present. This is because as circuit conditions change (cooling off, for example), the symptom may seem to disappear, but in fact the excessive resistance does not totally disappear. It only decreases to a value that allows the circuit to begin operating again. As a result, the excessive resistance can still be identified by a voltage drop test, even with no apparent symptom present. The measured voltage drop will not be as excessive as when the symptom is present, but it will likely still be above the maximum specifications allowed.

Measuring Amperage. An ammeter is used to measure the current flowing in a circuit. There are two types: the conventional type and the inductive type. The conventional type must be put in series as part of the circuit (Figure 5–20). The inductive type has an inductive clamp that clamps around a conductor in the circuit to be tested (Figure 5–21).

The inductive clamp contains a Hall effect sensor that uses the magnetic field around the conductor that it is clamped around to create an electrical signal. The strength of the signal depends on the strength of the magnetic field around the conductor, which in turn depends on the amount of current flowing through it. The DMM displays a voltage reading that corresponds to the strength of the signal. The inductive ammeter is more convenient to use because it eliminates the need to open the circuit.

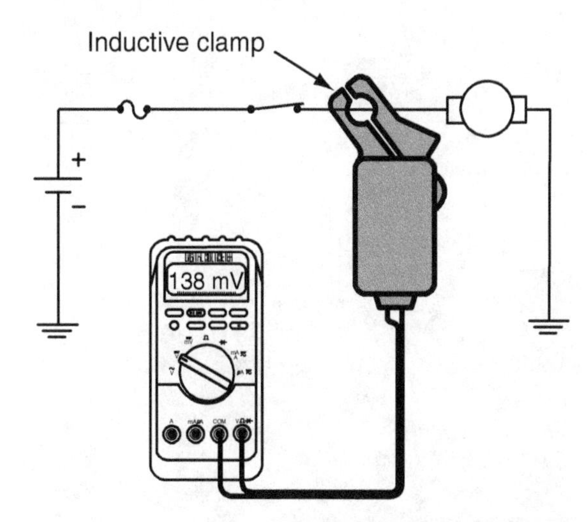

Figure 5–21 DMM used with a current probe to perform a nonintrusive current draw test. If the output of the current probe is labeled as "1 mV/amp," the 138 mV reading equates to 138 amps.

When using a conventional ammeter, it is important to select the correct scale. Passing 5 amps through an ammeter with a 2-amp scale selected can damage the meter. Most meters are fuse protected when on an amperage scale. Figure 5–22 shows a dedicated ammeter.

Measuring Resistance. Resistance is measured with an *ohmmeter.* An ohmmeter is self-powered and has multiple scales that allow testing of resistance values from one-tenth of an ohm (typical) up to 40 MΩ (40,000,000 Ω). When manually scaling a DMM on a resistance scale, it is good practice to begin on the uppermost scale and scale down until the over-range indicator is displayed, then scale back up one scale. If a good reading is obtained on the lowest scale, use the lowest scale to take the measurement.

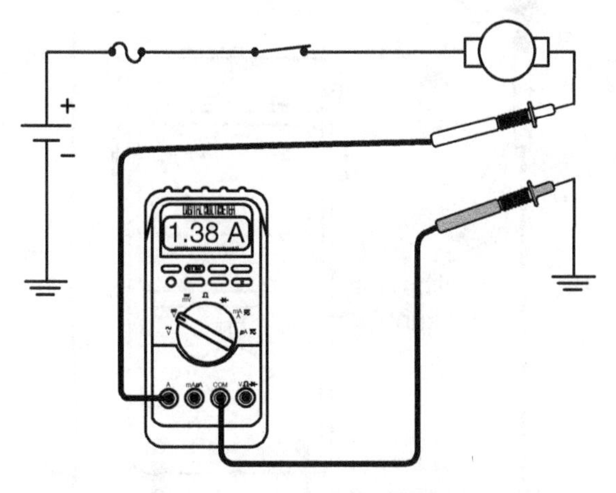

Figure 5–20 Ammeter connected in series as part of a circuit.

Remember that an ohmmeter does have an internal voltage source, and therefore should never be connected to a circuit or component

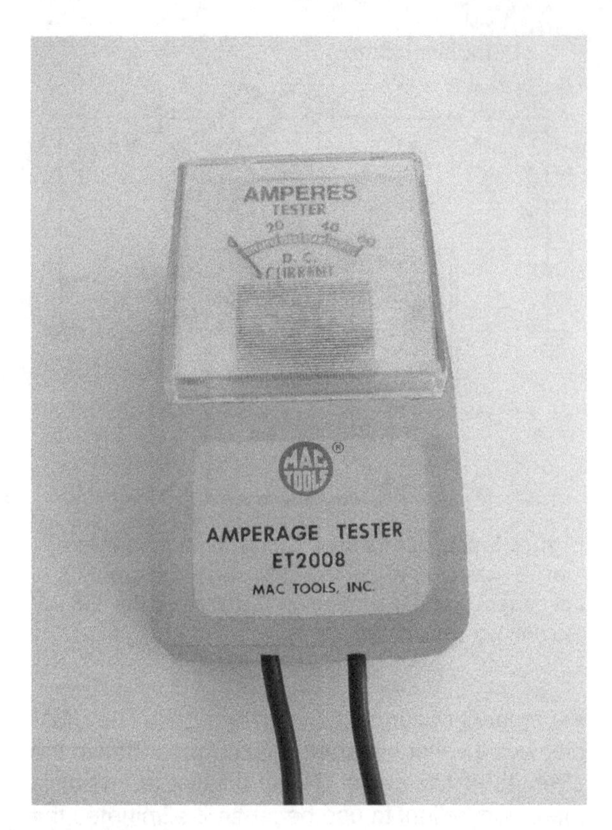

Figure 5–22 A dedicated ammeter.

that has power to it. If such a connection is made, the result could be anything from an inaccurate reading to a damaged meter. Also, because of the internal power source, never connect an ohmmeter to a live air bag or air bag circuit on a supplemental restraint system (SRS)-equipped vehicle.

To help you to understand how the resistance scales of a DMM function, the following text describes how an analog ohmmeter functions. The DMM accomplishes the same things, only using solid-state electronics to achieve them.

When an analog ohmmeter is turned on, its internal battery tries to pass current through the meter coil and the two leads, but unless the leads are electrically connected outside the meter, the circuit is open. If the two leads are touching or if they contact something that has continuity, the circuit is closed. If the leads are connected directly to each other, enough current should flow to cause the meter to swing full scale to the right, which represents zero ohms (Figure 5–23). If the leads are across a conductor, component, or part of a circuit, the resistance of that unit will reduce the current flow through the meter's circuit. The reduced current flow produces a weaker magnetic field, and the meter's needle will not be moved as far to the right. A digital ohmmeter has an electronic circuit that produces a digital readout in response to the amount of current flowing.

Most needle-type (analog) ohmmeters must be calibrated before each use. Calibration adjusts for voltage variation in the meter's battery due to the discharge that occurs during use. This is done by touching the leads together and adjusting the zero knob until the needle reads zero ohms (Figure 5–23). This establishes that the magnetic field produced by the current flowing through the meter's coil and external leads moves the needle to zero. Any additional resistance placed between the external leads will lower the current flow and will be displayed as the amount the needle lacks in reaching the zero line. Analog ohmmeters also must be recalibrated each time a different scale

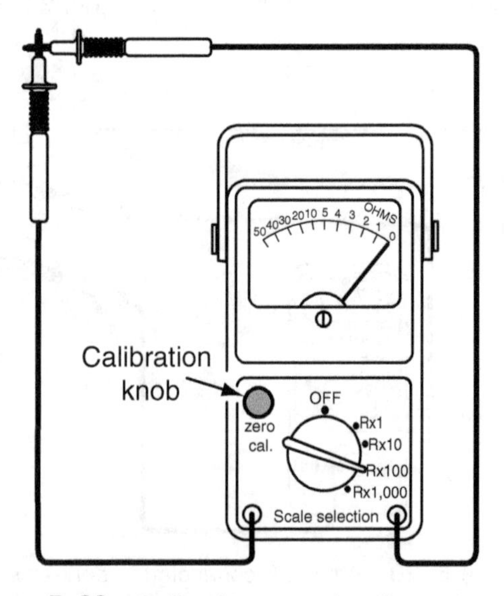

Figure 5–23 Calibrating an analog ohmmeter.

is selected because selecting a different scale changes the meter's internal resistance.

DMMs do not have a zero adjust knob, but the leads should be shorted together while the DMM is on the lowest resistance scale (Figure 5–24). If any value other than zero is displayed, this value should be subtracted from the displayed reading when making resistance measurements of low-resistance components (Figure 5–25). Some high-end DMMs have a relative-resistance feature that allows the DMM to automatically compensate for any resistance in the leads—simply touch the leads together and depress the "relative resistance" button. Remember to change the DMM's internal battery regularly. Many DMMs display a "low battery" indicator when the battery's charge becomes low.

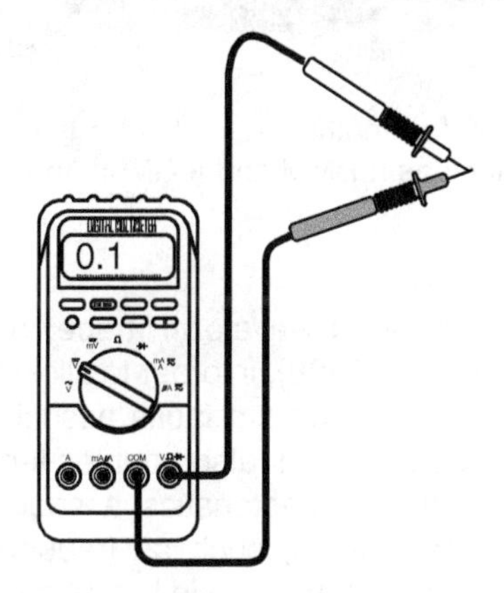

Figure 5–24 Testing the calibration of a DMM's resistance scale.

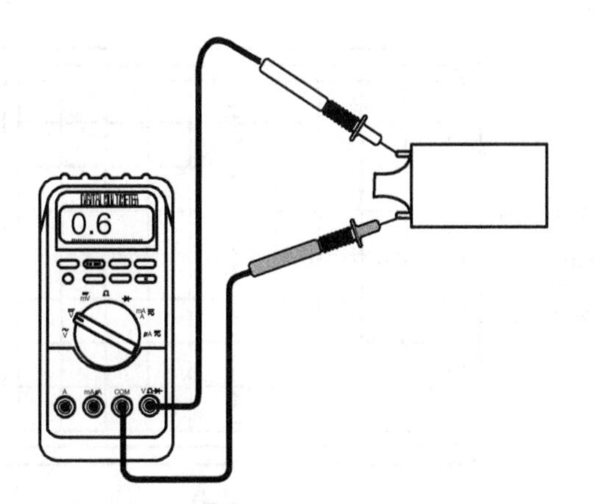

Figure 5–25 When using a DMM to test the resistance of a low-resistance component, the resistance of the leads and meter should be subtracted from the displayed value.

Advantages. A DMM is very easy to connect, read, and interpret. It provides an easy method for accurate pinpoint testing of most components or circuits. Again, you must be careful when interpreting the displayed reading.

Disadvantages and Limitations. A DMM cannot read time and may be ineffective in diagnosing certain component failures in which this is important—for example, an oxygen sensor's response time. If time is a critical measurement, then an oscilloscope should be used.

Graphing Multimeters

A **graphing multimeter (GMM)** is essentially a multimeter that can display any of its measurements as a graph over time. This provides the technician with a visual image that can help detect changes in measurements. A GMM is not to be confused with a lab scope. While the graph of a DC voltage on a GMM may look similar to a trace of the same DC voltage on a scope, there are major differences. For example, a trace of a digital MAP sensor or digital MAF sensor on a scope will appear as a digital waveform that is rapidly switching between 0 and 5 V. But if a GMM is used to graph the frequency (Hertz scale) of this signal, the height of the graph would simply indicate changes in frequency. It would not display the digital voltage pattern that a lab scope displays. This would make it easier during diagnosis to identify a sudden decrease or increase in the frequency caused by a problem with the sensor.

You could equate a GMM's display to that of a graphing scan tool, except that the scan tool is graphing the data stream's parameters that are communicated from an on-board computer in the binary language in order to provide a visual display of this data, whereas a GMM, like a DMM, is used to read raw voltages, resistance

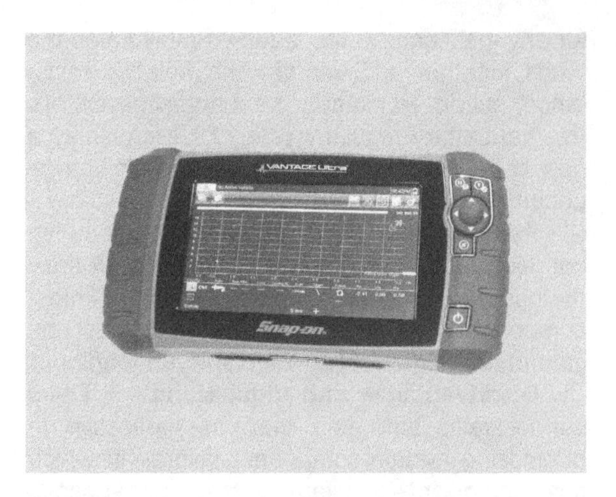

Figure 5–26 An example of a lab scope that also functions as a DMM and a GMM: A Snap-on Vantage Ultra.

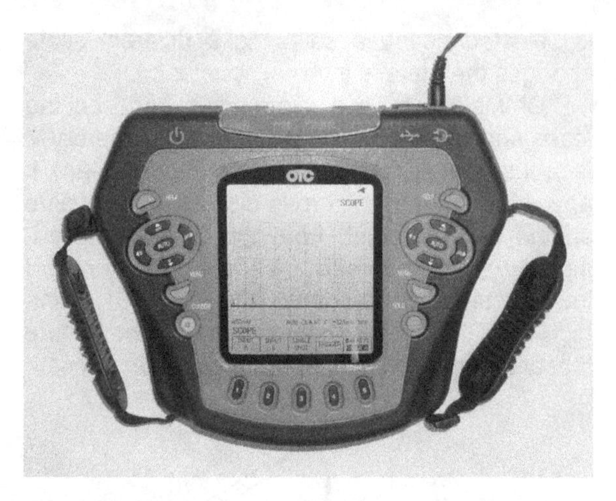

Figure 5–27 Another example of a lab scope that also functions as a DMM and a GMM: An OTC Model 3840.

values, amperage values, frequencies, duty cycle percentages, and pulse width values and then display these values as a graph over time.

Many modern DMMs and DSOs include a GMM capability. For example, the original Snap-on Vantage was a dedicated DMM and GMM. But the modern Snap-on Vantage Ultra, shown in Figure 5–26, is actually a DMM, a GMM, and a lab scope all built into one piece of equipment.

Digital Storage Oscilloscopes

A **digital storage oscilloscope (DSO)**, also sometimes referred to as a *lab scope,* is one of the great forward leaps in diagnosing automotive computer systems. Similar in design to the analog bench-top oscilloscopes of the past that were used in the electronics field, these units also use computer-generated waveforms that can be frozen in time, measured, downloaded to a PC, and printed out. Figure 5–26 and Figure 5–27 show typical DSOs or lab scopes. These DSOs also function as DMMs and GMMs. They also can, with the correct adaptors, function as secondary ignition oscilloscopes.

Purpose and Functional Concept—Voltage and Current Waveforms. A DSO allows the technician to see a waveform (trace) of *voltage over time (*Figure 5–28), in order to allow accurate pinpoint testing of system faults associated with these voltage signals. It also enables technicians to evaluate the characteristics associated with these voltage signals, such as frequency, duty cycle on-time, and pulse width on-time. A DSO can also be adapted to display a waveform of *current over time* to allow the technician to

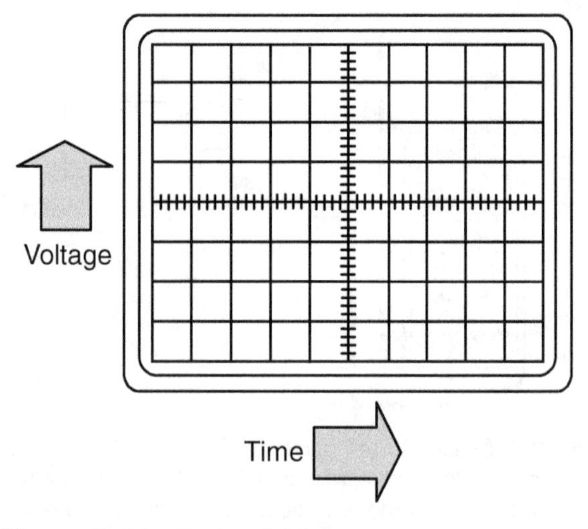

Figure 5–28 Typical DSO screen.

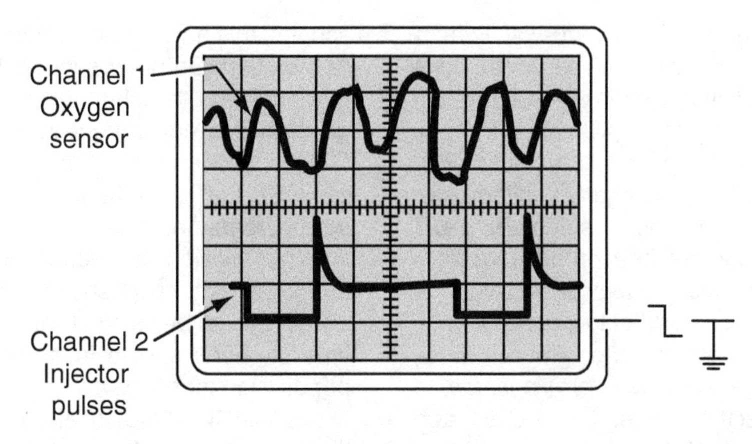

Figure 5–29 Lab scopes may have two or more traces, so you can compare signals—for example, that of an oxygen sensor to that of injector pulse width.

consider a load component's current draw when diagnosing a problem.

A DSO can provide in-depth information about the electrical characteristics of a circuit. Some DSOs provide multiple channel (trace) capability, allowing the technician to capture more than one waveform simultaneously (Figure 5–29). A DSO will commonly have one-, two-, or four-channel capability, although there are exceptions to this. Automotive Test Solutions markets a software and hardware package that is designed to turn a laptop PC into an eight-channel dual–time base lab scope. A DSO has one lead set for each channel. Multiple channels may share a common ground lead. The DSO screen is typically divided into eight voltage divisions and 10 time divisions, although this can vary among manufacturers (Figure 5–28).

Functional Procedure—Voltage Waveforms. A DSO has two leads per channel, one positive lead and one negative lead, making for an easy basic connection. Connect the leads to the DSO as indicated. Connect the negative probe to a clean, unpainted ground. Connect the positive probe to any voltage you wish to capture.

The Eight DSO Adjustments. To adjust the DSO's characteristics to see a proper waveform, several adjustments need to be made. These include four fundamental adjustments and four **trigger** characteristics. The four fundamental adjustments are coupling, ground level, voltage, and time. The four trigger characteristics are trigger source, trigger delay, trigger level, and trigger slope. The four trigger characteristics work together to determine the instant in time that the DSO begins to display a waveform. If the trigger characteristics are misadjusted, the DSO randomly guesses where to begin drawing a waveform, resulting in a waveform that will appear to walk across the screen or will appear randomly. On some DSOs if the trigger is enabled but the trigger level is misadjusted so that the scope does not see a voltage cross the trigger level, no waveform will be displayed.

While scope *"buttonology"* differs from one brand of scope to the next and from one model to the next, these eight adjustments apply to all lab scopes, whether of the digital variety (discussed here as DSOs) or their analog counterparts. Any other adjustment concerns will pertain to the unique buttonology of the oscilloscope and will be covered in the owner's manual that comes with the scope when you purchase it. We will discuss each of these eight adjustments here.

Coupling. Adjust the coupling to the type of voltage signal you will be connected to—AC or DC.

Selecting the AC coupling instructs the scope that the voltage is to be equally displayed both above and below ground (the zero-volt line or ground symbol). If you connect the scope to a rapidly switched DC voltage when the AC coupling is selected, it will display as an AC voltage.

Selecting the DC coupling will actually allow the scope to display both DC and AC voltage signals. When the DC coupling is selected, a DC voltage signal will operate in a range from zero volts to a voltage above ground; an AC voltage signal will go both above and below ground.

Ground Level. Adjust the ground level, also known as the *vertical waveform position,* to the middle of the screen to read an AC voltage and toward the bottom of the screen (ideally, one division from the bottom) to read a DC voltage.

Voltage. This adjustment may be displayed as volts-per-division or volts-per-screen. On some scopes, this adjustment is permanently set to one or the other. On other scopes, the user can use a utilities menu to select the manner in which the scope displays this adjustment. The OTC Pegisys and the Snap-on Verus are examples of DSOs that allow the user to select the preferred setting regarding this adjustment. Generally, the volts-per-division is preferred. This is because if volts-per-screen is shown, often the user must divide the screen's total voltage by the number of voltage divisions shown in order to calculate volts-per-division for diagnostic purposes.

Adjust the voltage to allow you to see the entire amplitude (or range) of the voltage. Adjusting the voltage to a value that is too small may cause your voltage to move off the visible screen and seemingly disappear. Adjusting it to a value that is too large will make the detail of the waveform more difficult to see.

Note that on many scopes a vertical rocker button is used to increase/decrease the voltage. Pressing the upper button decreases the voltage and pressing the lower button increases the voltage. While this may seem illogical at first, note that pressing the upper button increases the height of the waveform and pressing the lower button decreases the height of the waveform.

Time. This adjustment may be displayed as time-per-division or time-per-screen. On some scopes, this adjustment is permanently set to one or the other. On other scopes, the user can use a utilities menu to select which manner the scope displays this adjustment in. As with voltage, the OTC Pegisys and the Snap-on Verus are examples of a DSO that allow the user to select the preferred setting regarding this adjustment. Generally, the time-per-division is preferred. This is because if time-per-screen is shown, often the user must divide the screen's total time by the number of time divisions shown in order to calculate time-per-division for diagnostic purposes.

Adjust the time to allow you to see two or three complete patterns of the waveform. Adjust this value *shorter* to see more *detail* of the waveform and *longer* to see more of the waveform's *trend.* For example, if you were testing an oxygen sensor's ability to produce proper voltage values on an operating engine for the sake of testing the sensor itself, you would ideally adjust this short enough to see more detail—perhaps 100 ms-per-division (ms/div). But if you then wanted to use the oxygen sensor's waveform to verify the ability of the engine computer to properly control the air-fuel ratio while driving the vehicle, you would ideally adjust this to see more of a trend—perhaps 1 s/div. Examples of both an oxygen sensor waveform and a mass air flow (MAF) sensor waveform are shown in Figure 5–30 and Figure 5–31 with the time adjusted to 200 ms/div in both figures.

Trigger Source. Trigger source is used to determine which lead is used to signal the DSO that the trace should begin. In most cases the trigger source is set to the lead you are using to capture the waveform (also known as an *internal trigger).* This keeps the DSO connections simple; just one positive lead and one negative lead must be connected to the circuit. In certain cases you may want to use another lead (also known as an *external trigger)* to trigger the waveform rather than the one being used to capture the waveform. A common example of an oscilloscope that uses an external trigger is a secondary ignition oscilloscope. When using an ignition oscilloscope to capture the secondary waveform of an ignition system, the secondary coil lead captures the waveform for all

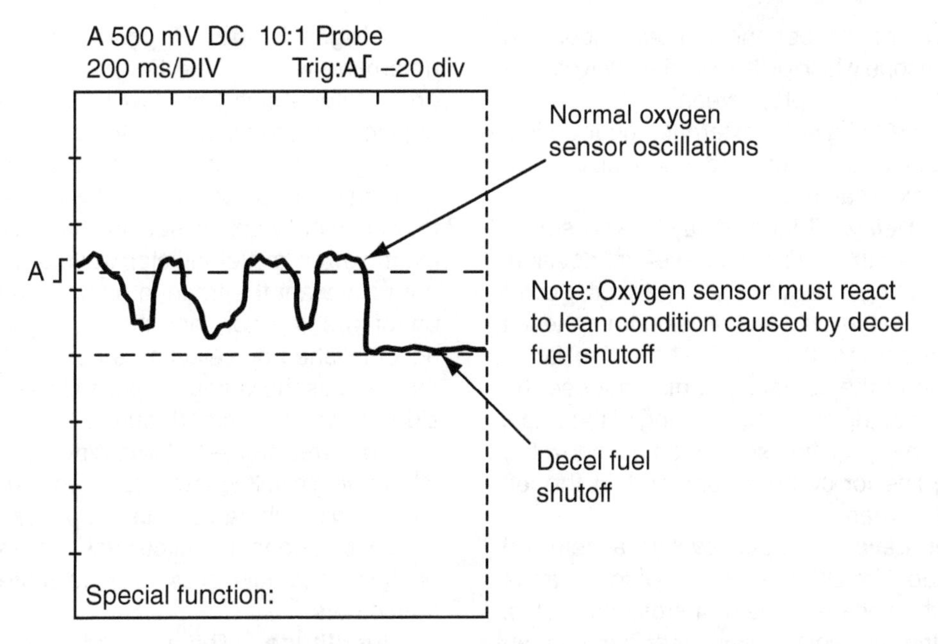

Figure 5–30 A lab scope allows you to see both the quantity and the quality of the signal, as with this O_2 sensor trace.

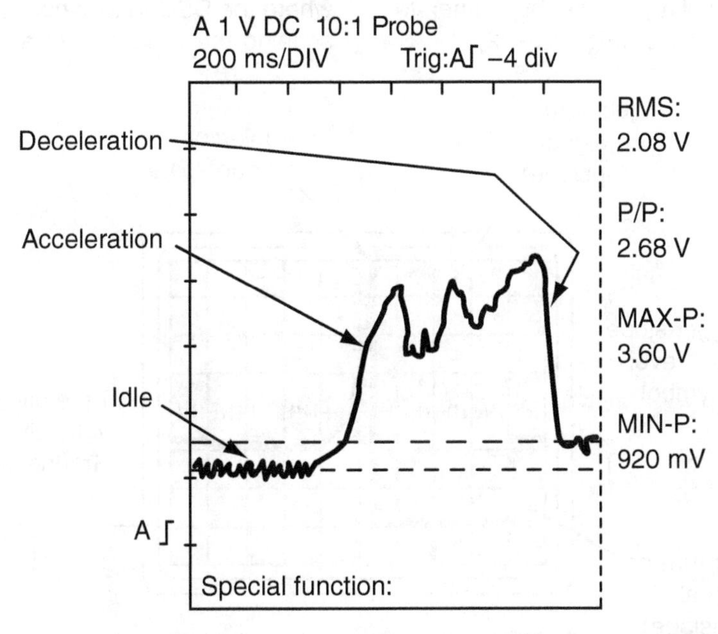

Figure 5–31 All the detail necessary to catch intermittent glitches and fast-occurring faults is present and controllable with a modern lab scope. This figure shows the voltage waveform of an analog MAF sensor.

cylinders. But the number one cylinder trigger lead informs the scope which of the cylinders' waveforms to begin with on the display screen.

Not all DSOs have an external trigger option. In that event another unused channel can be used as an external trigger.

Trigger Delay. Trigger delay is also sometimes known as the horizontal waveform position. Adjust this to begin the trace at a particular point on the screen—typically one division from the left edge of the screen. If you adjust this too far to the left edge of the screen, you may not see the initial event that triggered the drawing of the trace. This is why many ignition scopes do not show the initial firing line for cylinder number 1 at the left edge of the screen.

Trigger Level. Trigger level is a selected voltage value. The DSO begins drawing the trace as it sees the voltage waveform cross this value. Adjust the trigger level to a value that is more than the lowest voltage of the waveform and less than the highest voltage of the waveform. Other factors may help you to determine the proper trigger level as well; for example, you would not want to trigger off a fuel injector's voltage spike but rather to trigger off the turn-on signal (Figure 5–32).

Trigger Slope. Trigger slope is used to determine whether the DSO will begin the trace on a voltage shift from low to high (upslope) or high to low (downslope) as the waveform crosses the trigger level. Adjust this to begin the trace on either an upslope or a downslope according to what you want to see in the waveform. For example, on a fuel injector waveform you would normally want the trace to begin with the turning on of the injector, not the turning off. This is accomplished by selecting a downslope to begin the trace as the voltage is pulled low on a ground-side switched injector (Figure 5–32).

Interpretation—Voltage Waveforms. When interpreting a voltage waveform, you should look at the following characteristics. Depending upon the type of component or circuit you are testing, some of these characteristics may be more important than others.

Amplitude. The amplitude is how high and low the voltage goes. For example, a voltage that does not fall all the way to ground when a ground-side switched fuel injector is turned on indicates that resistance is still in that circuit after the point where the DSO lead is connected to it. Or if when scoping the waveform of a magnetic pickup coil

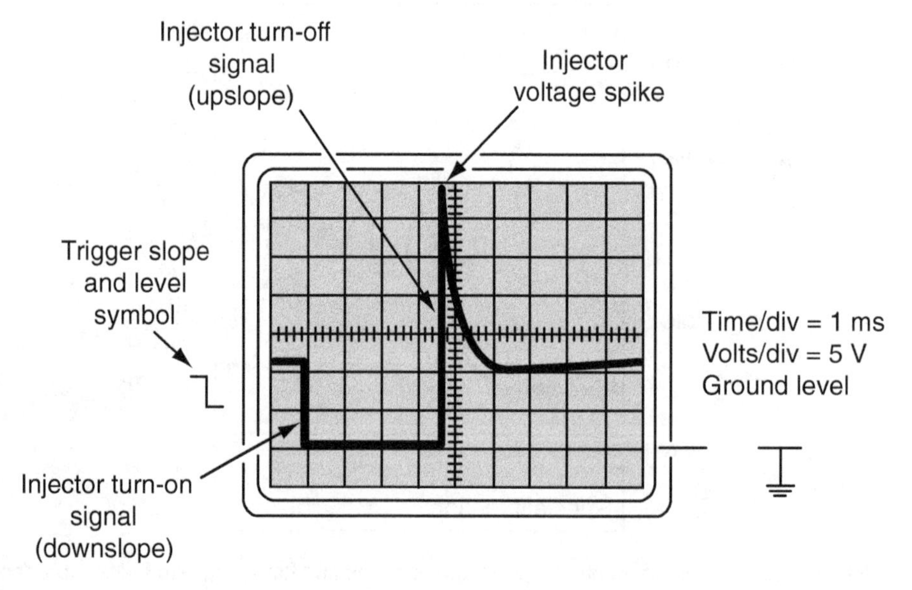

Figure 5–32 A typical fuel injector voltage waveform.

in the distributor of a running engine, you notice that the amplitude varies with each rotation of the distributor at a *steady* engine RPM, worn bushings or a bent distributor shaft is indicated, due to a reluctor-to-pickup air gap that varies with each rotation.

Time. Time in the waveform is important with such things as digital pulse trains (variable frequency, duty cycle, and pulse width), O_2 sensor testing, and signals used to identify speed or RPM. For example, you can see in a DSO's waveform how long it takes for the O_2 sensor to respond to an air-fuel mixture change. This type of failure would indicate a "lazy" O_2 sensor and is difficult to identify without the use of a DSO.

Shape/Symmetry/Sequence of Events. Inspect the waveform for glitches that could cause an improper signal to the PCM—for example, a TPS signal that intermittently drops out. A DSO makes these types of faults much easier to identify.

Cursors. Horizontal and vertical cursors are provided on most DSOs to allow precise measurements of both voltage and time. The horizontal cursors are used to measure a voltage difference; the vertical cursors are used to measure a time difference. These differences may be indicated by the Greek letter *delta* (Δ). For example, ΔV equals delta volts, or the voltage difference between the positions of the two horizontal cursors, and ΔT equals delta time, or the time difference between the positions of the two vertical cursors. Some scopes may show these values as "dV" and "dT," respectively.

Current Ramping. Another diagnostic approach gaining popularity is the use of a DSO to look at a current waveform, rather than a voltage waveform. This procedure is known as **current ramping.** It allows the technician to identify problems with electromagnetic load components that cannot be identified through the use of voltage measurements or voltage waveforms. To do this, you must couple the DSO with a *quality,* low-current, inductive current (amp) probe.

Inductive current probes are often used with a DMM to measure current non-intrusively in that they simply clamp around a wire. They contain a linear Hall effect sensor. They are self-powered and convert the measured magnetic field strength around the wire (indicative of the level of current flow through the wire) to a DC voltage (usually millivolts). The DMM is then set on a DC voltage scale to read the indicated current flow. An adjustment is provided on the current probe to zero out the value on the DMM when the probe is not clamped around a wire or when the circuit is turned off.

A current probe works much the same way when used with a DSO. After connecting the current probe to the DSO, rotate/push the zero dial/button on the current probe to zero the waveform on the DSO to ground level. This can be done before placing the probe around the wire of a load component's circuit. However, due to the magnetic influence of the vehicle, it is potentially most accurate if it is zeroed after placing the probe around the wire, provided that the current flow is turned off in the circuit.

Low-amp current probes are manufactured and distributed by many different tool manufacturers. On the high end is the Fluke 80i-110s AC/DC current clamp. A considerably less expensive, but very common, current probe is a generic current probe that is marketed under the brand names OTC, PDI, and PDA. At a price between the generic current probe and the Fluke current probe is the Snap-on current probe, model EETA308D, which replaces a former Snap-on current probe that had severe problems with lead breakage. The new Snap-on current probe has a beefed-up cable boot where the cable enters the body of the unit. The generic current probe also has problems with lead breakage, but is easy to repair if the lead does break at the point of entry into the unit's body. Some other differences will be noted in the following text. Figure 5–33 shows these three popular current probes.

The Fluke current probe requires you to turn a dial to zero the probe. The generic current probe only requires you to push a button to perform this function. While this may seem advantageous, too often the technician will push this button accidentally after it is connected around a wire with

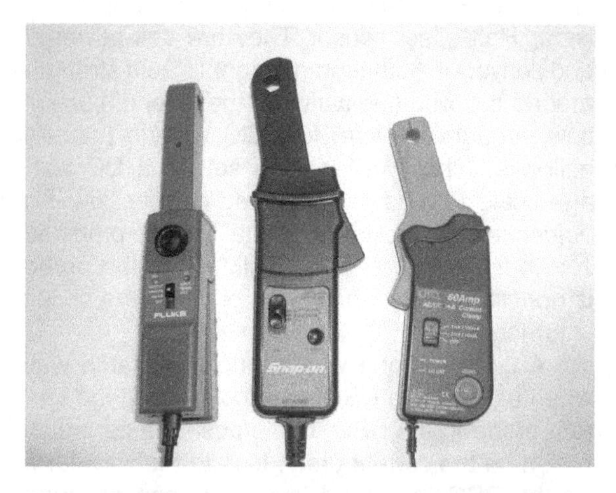

Figure 5–33 Three popular low-amp current probes.

current flowing in the circuit. This will falsely lead the technician to believe that the average current flow is zero amps. On the Fluke probe, you will not accidentally zero it. Snap-on's current probe has a zeroing button like the generic probe, but it is recessed, thus reducing the likelihood of an accidental zeroing.

The probe may be placed around either the positive wire or the negative wire, since current flow is consistent throughout a series circuit. It should also be placed as close to the load component as possible to avoid picking up current for multiple load components. Current ramping is not recommended for input sensor circuits as these circuits usually have high resistance, resulting in very low current levels. Input sensors rely primarily on voltage values to communicate their information to the PCM.

Once the current probe is clamped around a wire and zeroed, the circuit should be energized. If the current waveform appears upside-down and is below ground, the current probe needs to be flipped. The Fluke and Snap-on probes have a polarity indicator on the nose of the probe that points toward negative. The generic probe has a polarity indicator on a sticker inside of the jaw when new, but this sticker typically falls off within a few uses, so if you have one of these units you can apply a more permanent label to it. Once you have a current waveform, the same adjustments already discussed in "Digital Storage Oscilloscopes" should then be performed to obtain the optimum waveform on the visible screen of the DSO.

Most low-amp current probes have a selectable adjustment that determines the millivolts-to-amps relationship. For the current probe's output, the Fluke and Snap-on probes provide a choice of either 10 mV/A or 100 mV/A (Figure 5–34). The 100 mV/A is the more sensitive scale; it enables you to adjust the DSO's volts/division to a less sensitive scale, thus reducing the sensitivity of the DSO to other "noise" on the waveform. The generic probe is scaled differently: 1 mV/100 mA and 1 mV/10 mA (Figure 5–35). Because you need to know how many millivolts represent 1 amp on the DSO screen, you should relabel the scales on the generic probe using a label maker, as follows:

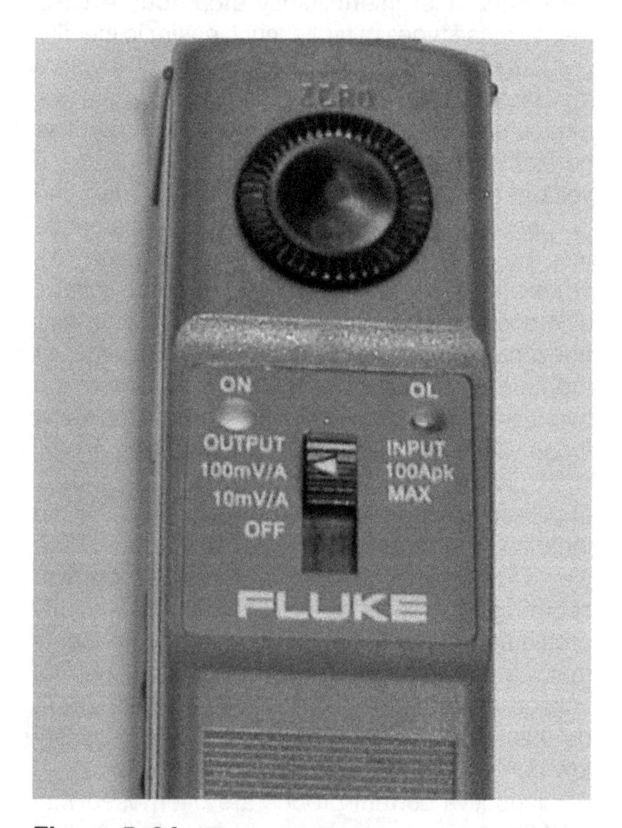

Figure 5–34 The selectable scales on a Fluke or Snap-on current probe are 100 mV/A and 10 mV/A.

Figure 5–35 The selectable scales on a generic current probe are 1 mV/100 mA and 1 mV/10 mA. The 1 mV/100 mA scale is equivalent to a 10 mV/A scale. The 1 mV/10 mA scale is equivalent to a 100 mV/A scale.

The 1 m V/100 mA scale should be relabeled to read "10 mV/A" and the 1 mV/10 mA scale should be relabeled to read "100 mV/A." This is also substantiated under the "output" label on the unit. So the generic probe does have the same two scales as the Fluke and the Snap-on, but you must be careful to select the correct equivalent scale.

Once you have set the probe's scale and have set your volts-per-division on the DSO, you can calculate the amps-per-division on the scope. The formula for calculating amps-per-division is: millivolts per division at the DSO divided by how many millivolts represent 1 amp at the current probe. For example, if the DSO is set to 200 mV/div and the current probe's output is set to 100 mV/A, the DSO's waveform would represent 2 amps per division (200 divided by 100).

Current Ramping a Switched Device. A current waveform of a switched device such as a primary ignition coil winding, a solenoid's winding (a fuel injector, for example), or a relay's winding can show whether the winding is shorted.

When a circuit's switch is turned on, the current waveform *ramps upward* until it is fully saturated; hence the term *current ramping.* However, when the circuit's switch is turned off, this action forces all electron flow in the circuit to stop almost immediately. This concept can be illustrated by a water pipe with water flowing in it and a valve that can be quickly snapped open or closed. If the valve is already open with water flowing in the pipe, and then the valve is suddenly snapped closed, the valve itself will physically and forcibly stop the flow of the water. But the reverse is not true. If the water valve is closed and you suddenly snap it open, the valve will not physically start the flow of water; rather, the open valve simply allows the water pressure that is present to begin pushing water molecules again.

So it is with an electrical circuit—when the transistor (driver) that controls the circuit is switched on, it only allows the voltage to begin pushing electrons again, so the level of current flow ramps upward relatively slowly; but when a transistor is turned off, current flow drops to zero amps very quickly. This is why the turning on of a circuit never produces a voltage spike—the relative motion of the expanding magnetic field to the winding happens too slowly. This also explains why a voltage spike is potentially produced when the current flow is switched off—the current flow stops so quickly that the collapsing magnetic field cuts across the winding quickly enough to induce a voltage spike in the winding of the opposite polarity of the original voltage. Of course, when the circuit is turned on and the current is ramping upward, Ohm's law dictates the level of full saturation.

Figure 5–36 shows a typical voltage waveform and current waveform of a standard fuel injector. When the driver is turned on, the voltage is pulled to ground. It is at this point that the current begins to ramp upward. Then, when the driver is turned off, the current waveform shows a sharp downward line. It is at this point that a voltage spike appears in the voltage waveform.

When using a current waveform to test the primary winding of an ignition coil (Figure 5–37),

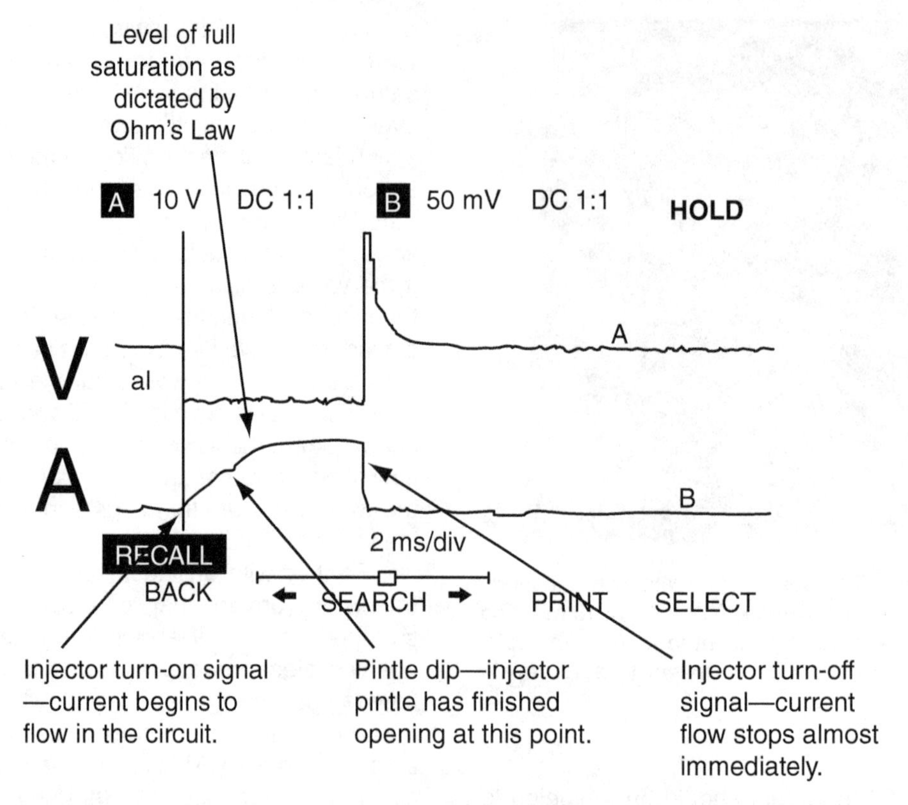

Figure 5–36 Voltage and current (amperage) waveforms for a standard fuel injector.

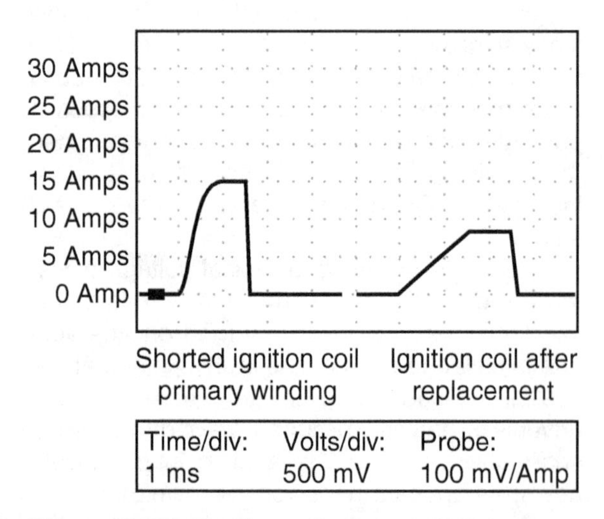

Figure 5–37 Ignition coil primary winding current waveforms, both before and after replacement.

or the winding of a solenoid such as a fuel injector (Figure 5–38), current that builds too quickly indicates that the windings are shorted; this is because the electrons did not have to overcome as much electrical resistance due to multiple coils of wire being shorted together. In most cases, shorted windings cause the current not only to build too quickly, but also to build too high, except in some ignition circuits in which the ignition module or PCM electronically limits the amount of current allowed in the circuit. This prevents a low-resistance primary coil winding from destroying itself.

The Ohmmeter versus Current Ramping. When using an ohmmeter to test a high-energy ignition coil's primary winding for shorts, a good winding may have a resistance of about 0.5 Ω.

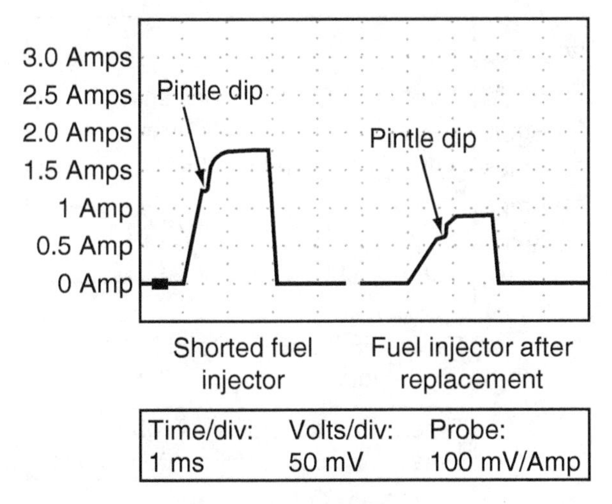

Figure 5–38 Standard fuel injector current waveforms, both before and after replacement.

So if your ohmmeter reads 0.4 Ω, is the winding shorted? And how much of the resistance is in the leads? Actually, high-energy ignition coil primary windings are difficult to diagnose accurately with an ohmmeter due to their ultra-low resistance specifications. But capturing a current waveform of an ignition coil's primary winding can be extremely helpful in determining whether the winding is shorted or not.

Also, note the coil-on-plug schematic in Figure 5–39. As is typical of many such coil-on-plug applications, the driver (transistor) is physically located within the ignition coil assembly. While you could get a voltage waveform of the digital signal that controls the driver, it would be impossible to capture a voltage waveform of the primary winding being switched on and off by the driver. Your DSO voltage lead would have to connect to the ground side (switched side) of the primary winding between the winding and the transistor—this is not possible without cutting the coil assembly physically open and thus destroying the coil. And likewise, in order to use an ohmmeter to test the resistance of the primary winding, you would have to cut the coil physically open in order to test the resistance across the winding itself without testing the resistance across the driver. Therefore, it becomes impossible to test the resistance of this type of primary winding when a shorted primary winding is suspected.

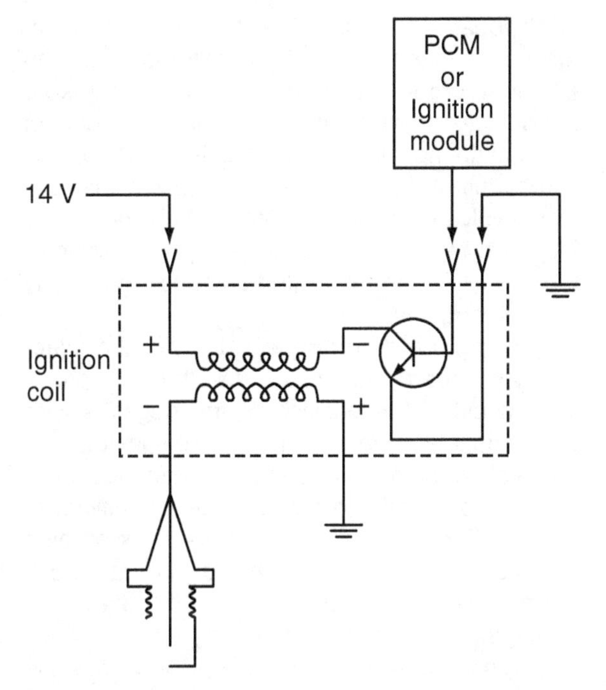

Figure 5–39 Electrical schematic of a typical ignition coil used in a coil-on-plug ignition system.

Because current is consistent throughout a series circuit, a current waveform can be captured by placing the current probe around either the positive wire or the negative wire—both methods provide identical waveforms. In this case, when you cannot access the negative side of the winding, a current probe placed around the positive wire before it enters the coil assembly will provide you with a current waveform of the primary winding. Thus, current waveforms provide you with a way to check for a shorted primary winding in this type of ignition coil without cutting the coil open.

On the other hand, a fuel injector can usually be tested accurately for a shorted winding with an ohmmeter. If the resistance specification for an injector is 14 Ω to 17 Ω and your DMM measures the resistance at 8 Ω, the injector's winding is obviously shorted. But using a current waveform as a diagnostic tool can still make the job easier. In order to use an ohmmeter to measure the injector's resistance, you must disconnect and isolate the injector's winding from the controlling circuit. If performed at the injector, this can also involve trying to get both DMM leads on the injector's terminals simultaneously. Furthermore, fuel injectors are difficult to access on many modern engines.

To capture a current waveform of the injector, you need only place the current probe around either the positive or the negative wire that conducts current to or from the injector. Your connection point can be some distance from the injector if the injector is physically difficult to access. Furthermore, no intrusive back-probing of a connector is needed in order to capture a current waveform, thus reducing the chance of damaging or compromising the circuit.

With sequential fuel injection, if you place the current probe around the positive feed wire for all fuel injectors you can get a current waveform for all injectors as they are pulsed sequentially. This allows you to diagnose the injectors for shorted windings and also to confirm whether each injector is being pulsed by the PCM's driver. It will be obvious if an injector is not firing. A DSO with an external trigger option can be used to determine which injector has a problem. Connect the external trigger lead to the number 1 injector and set the scope's trigger source to *external*. When the engine's cylinder firing order is considered it will be obvious which injector is shorted or is not firing. (If the scope does not have an external trigger feature, one of the unused channels can be used for this purpose.)

One final note on ignition coils: Some coil-on-plug ignition systems use a multiple spark discharge (MSD) when the engine is at or near idle RPM to reduce excessive unburned hydrocarbons that are caused by premature quenching of the flame front. These coils have been problematic on some applications, but current waveforms can quickly identify faulty coils (Figure 5–40). Some Ford MSD coils will actually produce a normal initial current waveform, with the second and third

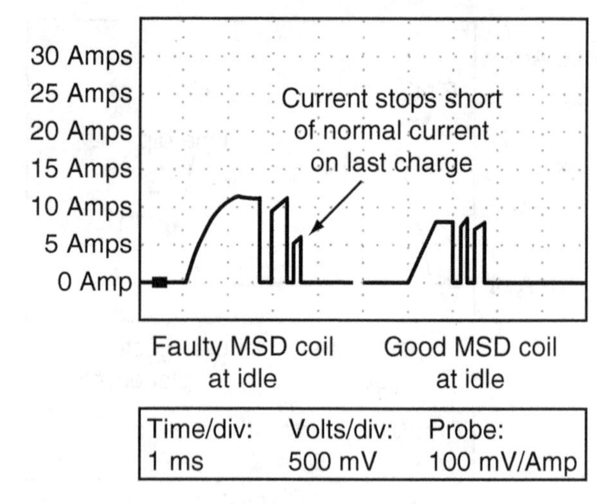

Figure 5–40 Current waveforms of both a faulty ignition coil and a good ignition coil in a Ford COP EDIS ignition system when the engine is at idle.

firings appearing as a negative amperage when the misfire occurs. Simply compare the current waveforms of each of the ignition coils with the engine at idle. These coils will fire just once per cylinder combustion event when the engine RPM is raised from idle.

Current Ramping a Motor. When using current ramping to test a motor, such as a cooling fan motor or a fuel pump motor, the current waveform can tell you quite a lot about the motor's condition. Each time a pair of commutator segments line up with the brushes (representing the two ends of one winding), the current momentarily rises. As the commutator segments slip away from the brushes, the current momentarily falls until the next pair of commutator segments begins to line up. A given winding will line up between the brushes twice per rotation, once every 180 degrees of rotation. Most electric fuel pumps (smaller pumps) contain four windings, resulting in eight commutator segments passing between the brushes in one full rotation. Therefore, eight current pulses on the lab scope would represent one rotation of the fuel pump's motor. Some larger fuel pumps have as many as six windings, resulting in 12 commutator segments passing between the brushes in one rotation and

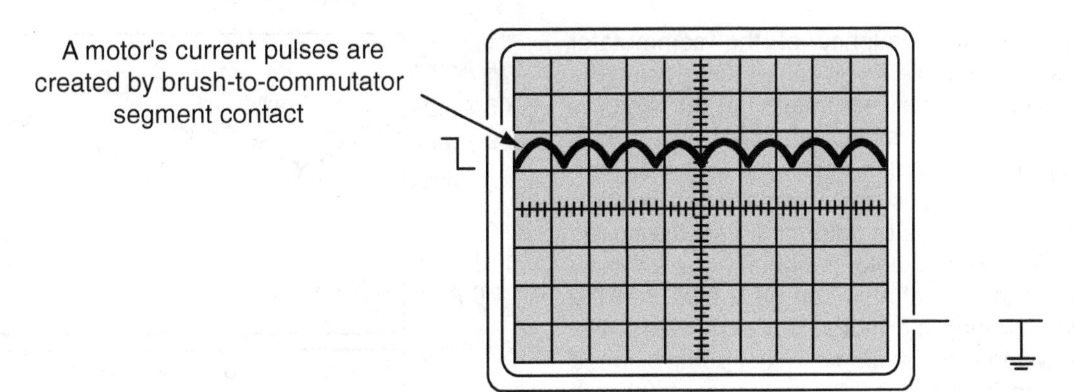

Figure 5–41 A typical DC motor current waveform.

12 current pulses on the lab scope per rotation. See Figure 5–41 for a typical current waveform of a fuel pump motor.

When capturing the current waveform of a motor, consider that the level of current flow is dictated by two things: electrical resistance and mechanical resistance. Of course, electrical resistance affects current flow according to the principles of Ohm's law. To demonstrate the effects of mechanical resistance, take a blower motor without the cage on the shaft and power it up using jumper wires. Then, while watching the current waveform on a lab scope, use a shop rag to physically grab the spinning shaft and forcibly slow it down. (Do not try this without a shop rag as the spinning shaft will burn your fingers!) As you do so, the level of current draw increases. Thus, increasing mechanical resistance increases the current draw of a motor.

In summary:

- A motor's current flow that is excessively high can be caused by either:
  - electrical resistance that is too low (shorted windings within the motor), or
  - mechanical resistance that is too high.
- A motor's current flow that is too low can be caused by either:
  - electrical resistance that is too high (poor brush-to-armature contact within the motor or excessive voltage drop within the circuitry), or
  - mechanical resistance that is too low.

When a motor's circuit is first energized, expect to see a sudden upward spike in the current, which will then settle down to a normal current value as the motor's armature starts rotating (Figure 5–42). This is normal and is due to the fact that the motor's armature is not moving when the circuit is first energized, thus constituting high mechanical resistance initially. This is known as **motor start-up in-rush current.**

If you know the number of windings in a motor, you can calculate the motor's RPM by

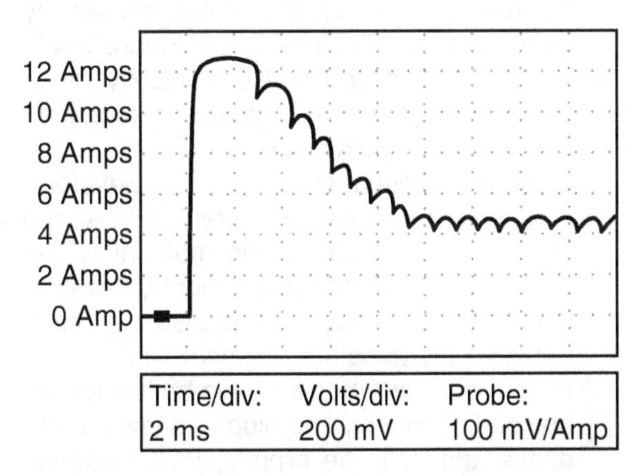

Figure 5–42 Fuel pump motor start-up in-rush current.

looking at the frequency of the commutator pulses. For example, if you can verify by using the scope's cursors that one rotation of the motor is occurring in 10 ms, multiply this by 100 (10 ms is 1/100th of a second) to determine that it is rotating at 100 rotations per second. Then multiply this by 60 seconds to determine that the speed is 6,000 RPM. If a motor is operating at a slower speed than normal, this can be a clue that the circuit powering the motor has excessive resistance within it. For example, a fuel pump's speed should be from 5,000 RPM to the low 6,000s. If the fuel pump is operating at a speed less than this, check the power and ground sides of the circuit using voltage drop tests with the circuit energized.

One huge advantage of current ramping a motor is that you can see faults within the motor assembly even before any symptoms have been experienced, indicating a likely failure in the near future. Current waveforms are therefore also a great way to identify causes of intermittent symptoms due to poor motor condition. While any motor may be tested using current waveforms, fuel pump motors are the most critical in that their failure could potentially leave the vehicle totally stranded.

Some types of faults that may be demonstrated with a motor are as follows:

- Varying current draw due to varying mechanical resistance within the motor, caused by worn bushings (Figure 5–43 and Figure 5–44).
- A current waveform that consistently drops to ground (zero amps) twice per rotation indicates an open winding (Figure 5–45).
- A poorly defined shape of the waveform caused by brush wear (Figure 5–46). As the brushes wear, their carbon material is deposited on the armature segments, decreasing the current flow, but it is also deposited between the armature segments, increasing the current flow when it should be further decreased. Ultimately, the high and low current peaks appear to be clipped, thus flattening the waveform.

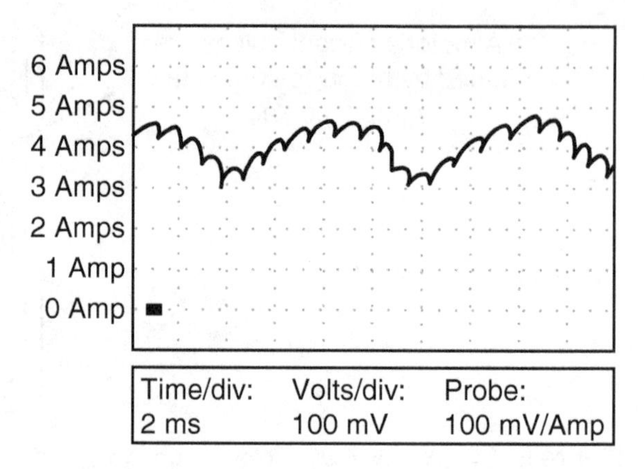

Figure 5–43 Varying current draw due to varying mechanical resistance within a motor caused by worn bushings.

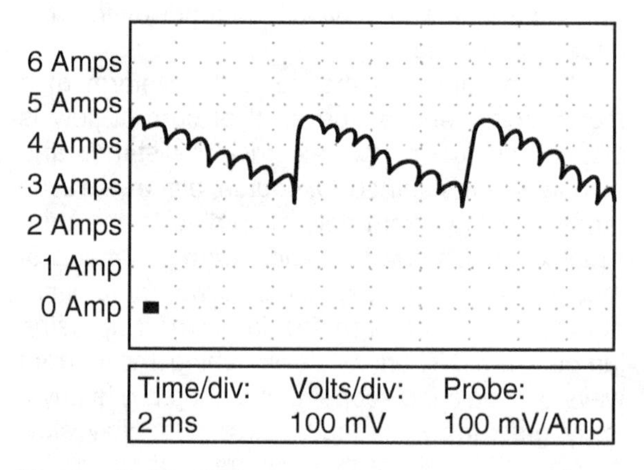

Figure 5–44 Varying current draw due to varying mechanical resistance within a motor caused by worn bushings.

- Noise on the waveform or intermittent dropping to ground (zero amps) indicates poor electrical contact between the brushes and the commutator segments (Figure 5–47). This cannot be an open winding if the armature is seen to rotate more than 180 degrees at any time without dropping to ground.
- Shorted windings that result in a current spike for one winding that rises above the others (Figure 5–48).
- Any combination of the above (Figure 5–49).

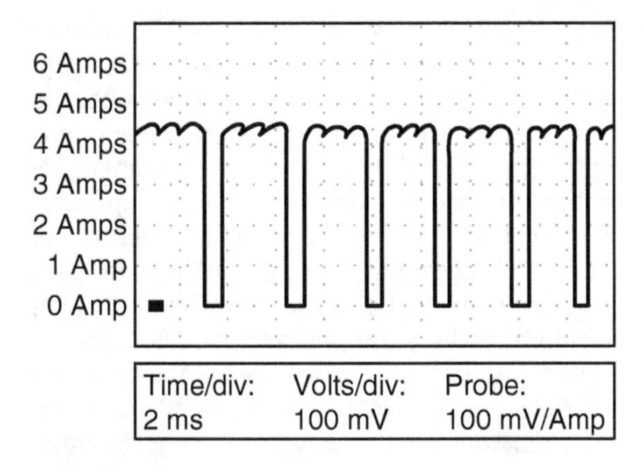

Figure 5–45 An open winding within a motor.

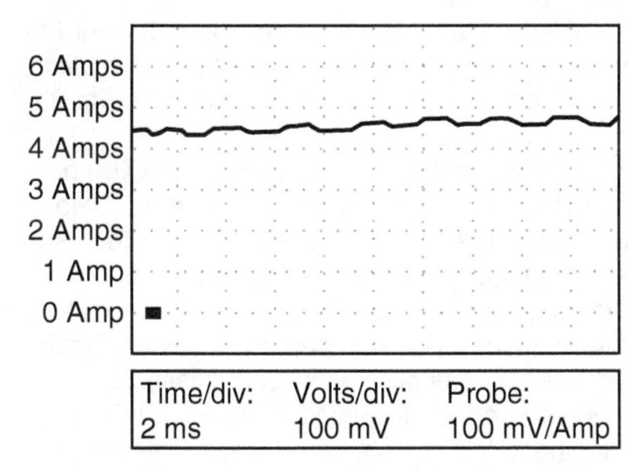

Figure 5–46 A poorly defined shape of a motor's current waveform indicates brush wear.

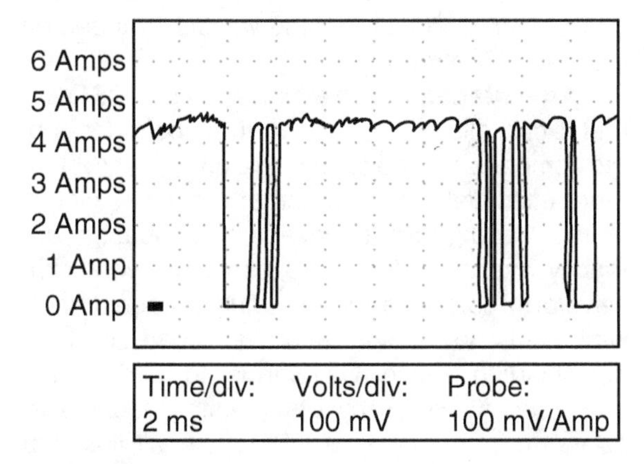

Figure 5–47 Poor brush-to-armature contact within a motor.

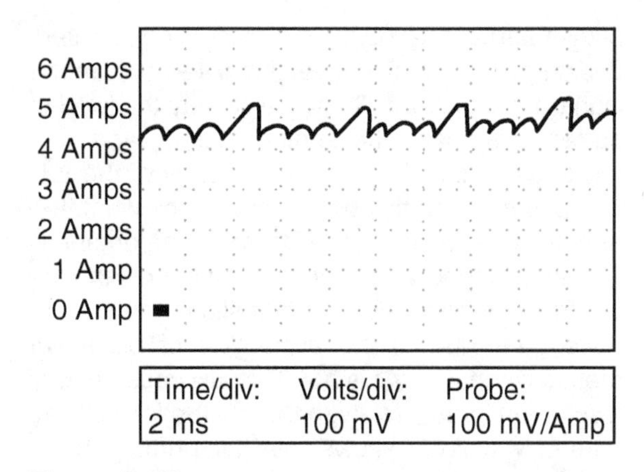

Figure 5–48 A shorted winding within a motor.

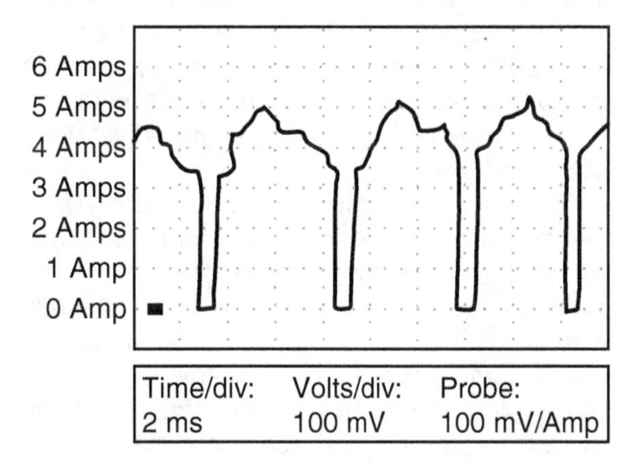

Figure 5–49 A motor with an open winding, a shorted winding, and worn brushes.

Each of these conditions can be responsible for *intermittent* symptoms, particularly those such as "the vehicle stalls on a hot day until it cools down for 20 to 30 minutes." Current ramping can allow you to diagnose intermittent problems with motors more effectively than any other method.

Remember that an electric fuel pump assembly consists of about 80 percent electric motor and 20 percent mechanical pump driven by the motor. If you have a complaint of "stalling on a hot day until cool-down is achieved" and you verify a loss of fuel pressure at a time when the engine is stalled, capture the fuel pump's current waveforms, both when the engine is running and when the engine is stalled. If the waveforms appear okay in terms of everything previously discussed, this only proves that the fuel pump's electric motor is in good condition. Compare closely the level of current draw with the engine running and with the engine stalled. If the current draw of the fuel pump is reduced by about 1 or 2 amps when the engine is stalled compared to when the engine is running, suspect that the high heat conditions are causing the armature's shaft to slip in the pump's drive gear due to the expansion of dissimilar metals when the unit is hot. This will cause a total loss of fuel pressure, even though the lab scope's current waveform shows that the pump's motor is still rotating. Usually, in this event, the amount of fuel pump noise that is heard at the fuel tank's neck is also reduced when the engine is stalled. The speed of the fuel pump, as indicated in the current waveforms, may not necessarily be increased due to the inductance properties of the motor's magnetic field.

Current waveforms of electric cooling fans may also be captured to aid in diagnosis. These look very similar to fuel pump current waveforms. You may also capture current waveforms for many other motors, including wiper motors, windshield washer pumps, power window motors, power seat motors, and starter motors (for a starter motor, use a high-amp current probe connected to the lab scope). When capturing a current waveform from a windshield wiper motor, because mechanical resistance affects current draw, expect these current waveforms to vary as the wipers start and stop on the windshield. Also, because a wiper motor's armature is wound differently due to the use of three brushes in most two-speed motors, the current waveform has an irregular shape compared to most other motors. And a starter motor's current draw is affected by the compression pulses of the cylinders. That makes diagnosis using current waveforms of these motors somewhat more challenging. Even so, such things as open windings can be seen easily.

DSO Waveforms, Databases, and Your Customers. When you have captured a current waveform of a fuel pump that has an open winding (for example), save this waveform to the scope's memory and then download it to a PC. Create a file folder system for "Voltage Waveforms" and another for "Current Waveforms." Under each of these folders, create a folder for each automobile manufacturer (GM, Ford, and so on). Within each of these, create a folder for various components or other tests, and within each of these create folders for both known bad and known good waveforms. Then put your defective fuel pump waveform in the appropriate folder. After replacement of the part, be sure to capture and download a waveform of the new fuel pump to the appropriate folder. In this manner, you can begin to create your own database. Be sure to back up your database regularly.

Also, by downloading a waveform to your PC, you can now print out "before and after replacement" current waveforms. Staple these to the repair order.

One word of caution: Just because you have identified an open winding in the fuel pump of a vehicle that is experiencing intermittent stall symptoms, don't assume that this is the only problem that could cause this symptom. For example, this symptom can also be caused by an ignition module that has outlived its usefulness. However, if you present your printed waveform of the open winding to the customer, you have provided justification for the expense. If replacing the fuel pump does not ultimately address the symptom, you still have made a needed repair for a problem that could have potentially left the vehicle stranded on the side of the road.

Advantages. The advantages of a DSO are that it can allow you to perform pinpoint tests of the components and circuits in a computer system more effectively than any other means available. Voltage and current waveform databases are readily available. Also, you can build your own waveform database by recording voltage/current waveforms both before and after repair and then downloading these waveforms to a personal computer. As with any other testing technique, as you practice using a DSO, you will find that you become increasingly proficient with it and will wonder how you ever did without it.

Disadvantages and Limitations. When a lab scope function is built into a "big box" type of ignition scope, it does not lend itself to procedures such as verifying oxygen sensor waveforms while driving the vehicle under realistic load conditions unless you also have access to a dynamometer in the shop. Also, as a rule, the "big box" versions are not as adjustable as their portable counterparts. Even the more portable versions are typically larger than a DMM, so you may want to use a standard DMM for many tests due to its smaller size, even though most DSOs have a DMM function built into them.

Also note that DSOs and DMMs are not designed to communicate with an on-board computer. This binary communication requires the use of a scan tool.

Safety Considerations When Using a DMM, GMM, or DSO

When using a DMM, GMM, or DSO to test voltage, today's technician must bear in mind several considerations. These considerations concern things to be aware of when purchasing test equipment and even the proper fuse to use when replacing a blown fuse in a DMM.

Electrical Test Equipment for Safely Measuring High Voltages. Some of the concerns that have brought safety issues to the forefront have to do with the vehicles that are now being produced. Vehicles with 42 V systems were introduced in the 2004 model year, and full hybrids with working voltages from 144 to 650 V are being produced by several manufacturers. While there are many other concerns when it comes to measuring high voltages safely, the following information concerns what the technician should be familiar with regarding electrical test equipment.

UL/CSA Listing. To begin, when purchasing a DMM or DSO, look specifically for the following identifications: "UL Listed" indicates that the test equipment is listed by Underwriters Laboratories, and "CSA Listed" indicates that the test equipment is listed by the Canadian Standards Association. Make sure that the UL and/or CSA symbols appear on the test equipment, not simply a statement that says "Meets UL/CSA standards."

Voltage and Category Ratings. When purchasing a DMM or DSO, take note of the voltage and category ratings that are found on the unit. Even the leads should be labeled with the voltage and category for which they are rated. The International Electrotechnical Commission (IEC) sets these standards as used by the UL and CSA listing agencies.

The voltage rating is considered the highest voltage that the unit (or leads) can be used to safely measure continuously. Typical voltage ratings found on quality DMMs and DSOs, as well as their leads, are either 600 V or 1,000 V.

Category ratings have to do with the highest transient voltage (momentarily induced voltage spike) that the test equipment can safely withstand without damage. This could be anything from a voltage spike induced by switching of high-voltage equipment to a nearby lightning strike. Keep in mind that high-voltage transient spikes are dampened by long lengths of wire and/or fuse protection. There are four assigned categories:

1. Category IV (CAT-IV certified) test equipment is designed to be used safely on the power lines that bring power to your home or office building.
2. Category III (CAT-III certified) test equipment can be used safely on power distribution boxes and permanent installations such as central air conditioning units. A CAT-III certified DMM with CAT-III certified leads should be used when diagnosing problems on the high-voltage side of most hybrid electric vehicles.
3. Category II (CAT-II certified) test equipment can be used safely on fuse/circuit breaker–protected wiring within the walls of a home or office building.
4. Category I (CAT-I certified) test equipment can be used safely on fuse-protected electronic devices that are connected to wall outlets.

Fuse Replacement. Many higher-end pieces of test equipment use fuses that are silica-filled. The reason for this is that a voltage above 20 V can sustain an arc even after an air gap is produced. If you replace such a fuse with an inexpensive fuse that does not contain the silica and then make a mistake with a 42 V system (for example, your leads are plugged into the DMM to measure current and you connect the meter between battery positive and negative as if to measure voltage), the fuse will likely burn open, but current will continue to flow across the fuse's air gap and through the meter, thus destroying the DMM. Therefore, always use the manufacturer's recommended fuse when replacing a burned fuse.

Breakout Boxes

A **breakout box (BOB)** (Figure 5–50) is a diagnostic tool introduced in the 1980s, primarily to diagnose Ford computerized engine control systems. Many technicians equated it to a scan tool, but a BOB has a totally different function and purpose than a scan tool, and the technician is best served by access to both diagnostic methods. As a result, today other manufacturers and systems have BOBs designed to help in diagnosis, and aftermarket universal BOBs have become widely available.

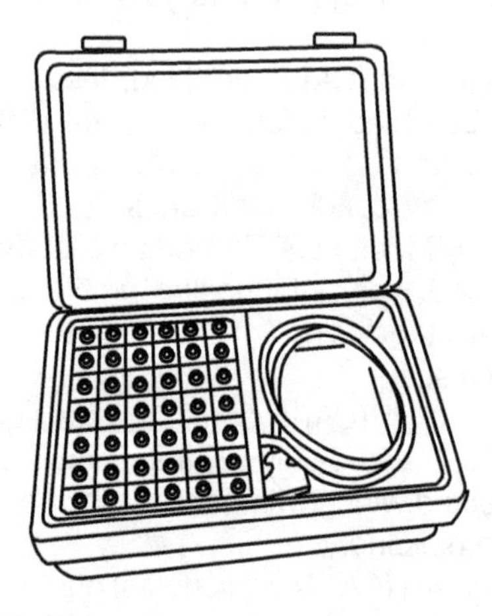

Figure 5–50 A typical breakout box that can be used to pinpoint test a computer's circuits.

Purpose and Functional Concept. A BOB gives the technician a central testing point to make pinpoint tests of input sensors, output actuators, and associated wiring, as well as computer power and ground voltage drop tests—all without back-probing. It connects into the harness between the PCM and the harness connector. As a result, it makes all computer circuits entering or exiting the PCM accessible to your logic probe, DMM, or DSO while maintaining the functionality of the computer system.

Functional Procedure. Carefully disconnect the PCM from the harness connector. Then connect the harness connector to the proper BOB connector. Make only this connection and leave the PCM disconnected if you need to make resistance checks of the circuits or if you will be using jumper wires to test actuators from the box. (Of course, due to their low resistance and high current potential, never use jumper wires to manually energize high-energy ignition coils or low-resistance fuel injectors.) Connect the PCM to the BOB for the system to be fully functional for testing voltage, voltage drop, frequency, duty cycle on-time, or pulse width on-time. Then use a logic probe, DMM, or DSO to probe the designated pins on the BOB, depending on which circuits you need to test.

Interpretation. Tests made with a BOB indicate raw data (voltage, voltage drop, resistance, frequency, duty cycle, or pulse width). Therefore, they verify not only the input sensors or output actuators but also the circuitry that connects these components to the PCM.

Advantages. When used in conjunction with a voltmeter or lab scope, a BOB provides the user with a single location to access all circuits for testing without back-probing. Test results are actual values, not computer-interpreted values. This author fully believes that BOBs are slated to increase in popularity as the automotive world encompasses higher-voltage systems such as the 42 V systems already in use on some vehicles. These systems do not allow for any back-probing whatsoever as a diagnostic method, so the BOB will be required to access the computer power and ground terminals for making voltage drop tests once the computers become powered at 42 V.

Disadvantages and Limitations. Depending on the particular vehicle application, it can be quite time consuming to connect a BOB into the system. Also, a BOB cannot be used to identify things in the computer's memory such as fuel trim values, diagnostic trouble codes, or how an octane adjustment has been set with a scan tool.

Breakout Boxes for the OBD II DLC

Breakout boxes designed for use with the OBD II DLC are now commonly available (Figure 5–51). When such a unit is connected to an OBD II DLC, it not only allows for probing of the 16 terminals with a meter's leads, but also allows for a simultaneous connection of a scan tool to the opposite end of the unit's cable. In this manner, a DMM or DSO may be used to read raw signals while trying to communicate with a scan tool. "Smart" units are also equipped with light-emitting diodes (LEDs) that indicate the availability of power and ground to the appropriate terminals as well as the presence of serial data on the appropriate terminals. OBD II DLC breakout boxes are available from such providers as AESWAVE (http://www.aeswave.com) and Automotive Test Solutions (http://www.automotivetestsolutions.com).

Gas Analyzers

Gas analysis and gas analyzers are more fully explained in Chapter 6 because a full chapter is needed to cover this topic. But in the context of this chapter, the use of a gas analyzer is designed to quickly direct the technician to a problem *area*. That is, gas analysis is not in itself a pinpoint test procedure but instead can quickly direct the technician to problem areas where pinpoint testing should begin.

Combination Tools

Many tools are now available that are combinations of the tools previously discussed in this chapter. These tools are typically viewed primarily as scan tools, but they may also include DSO/DMM capability and gas analyzer capability.

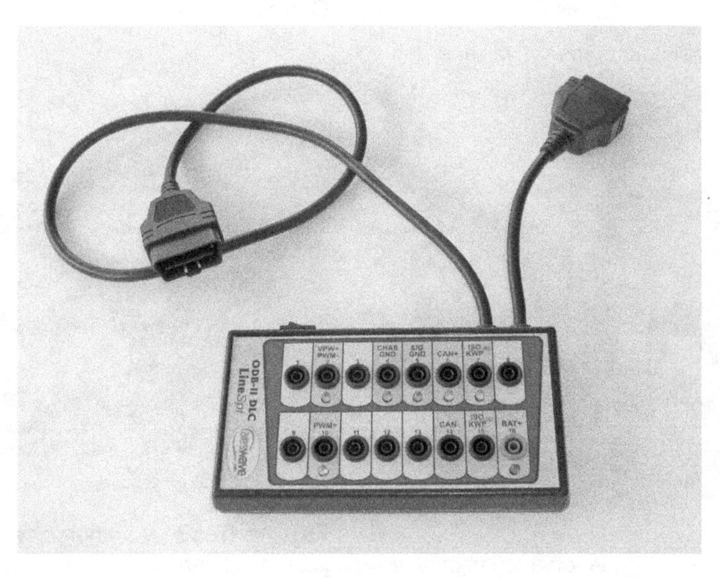

Figure 5–51 DLC breakout box.

Examples of two-in-one combination tools include the OTC Pegisys shown in Figure 5–52 and the Snap-on Verus shown in Figure 5–53. Each of these units can operate as a scan tool or as a two-channel DSO. (As with most DSOs, there is also a DMM function built into the DSO feature.) Another combination scan tool and two-channel DSO is the OTC Genisys Touch. The Genisys Touch is initially a scan tool and a reprogrammer with built-in Internet access. But with an optional pod, it can also function as a DSO.

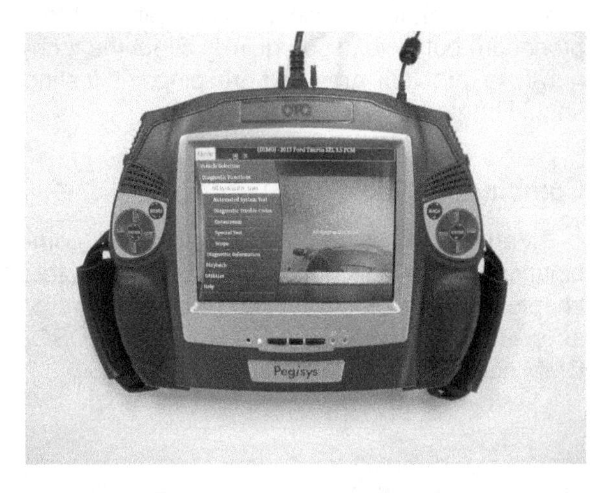

Figure 5–52 A combination scan tool and lab scope: An OTC Pegisys.

Figure 5–53 Another example of a combination scan tool and lab scope: A Snap-on Verus.

Examples of three-in-one combination tools include the OTC Genisys shown in Figure 5–54 and the Snap-on Modis shown in Figure 5–55. Both of these units can operate as scan tools but can also be configured as either DSOs or as exhaust gas analyzers, depending upon the purchased hardware/software options.

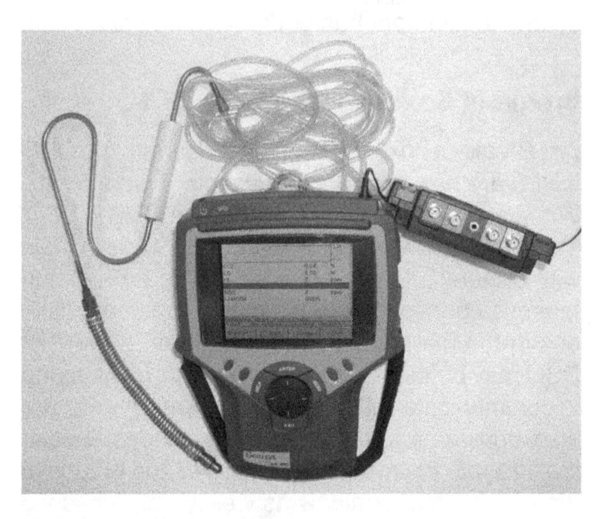

Figure 5–54 An example of a combination scan tool, lab scope, and exhaust gas analyzer: An OTC Genisys.

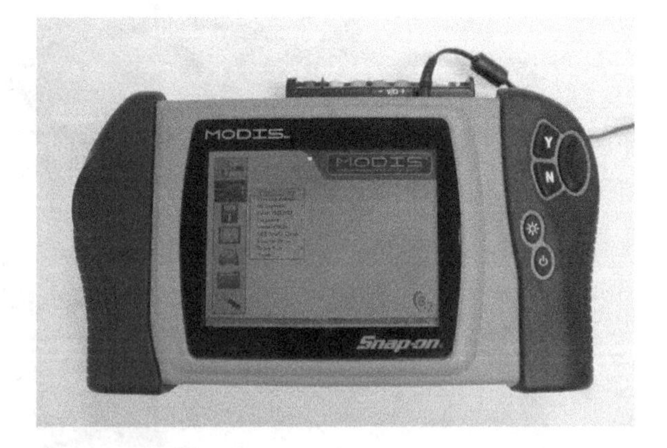

Figure 5–55 Another example of a combination scan tool, lab scope, and exhaust gas analyzer: A Snap-on Modis.

While these combination modular diagnostic tools are popular, there are a few things to consider when purchasing one:

- If the unit malfunctions and has to be sent in for repair, all of these tools become unusable until the unit is returned.
- Anytime that a tool and equipment manufacturer combines multiple tools into one unit, a menu-driven system must be provided to allow the user to navigate to a particular tool feature. This menu may be more or less complex depending upon the tool. With a dedicated DSO, pressing the "on" button puts the user in lab scope mode.
- If the technician wishes to use more than one of the tools simultaneously, this might not be possible. For example, if the technician wished to use a scan tool while also using an exhaust gas analyzer, this combination might not be available, depending upon the particular tool. However, the OTC Genisys *is* capable of displaying scan tool data stream parameters while the unit is operating as an exhaust gas analyzer, which is certainly a plus for this particular tool.

SUMMARY

This chapter presented the basic concepts of diagnosing an engine computer system: hard faults and soft faults, memory code pulling and self-tests, and several other concepts that are used by the manufacturer's systems and that will be discussed throughout the remainder of this textbook. Understanding the concepts that constitute the building blocks of the various diagnostic programs should make it easier for you to understand the manufacturer-specific systems as they are covered.

This chapter also outlined many pieces of diagnostic equipment, along with the advantages and limitations of each. It is designed to give the technician an overall view of when to use each piece of equipment, what types of tests can be performed with each piece of equipment, and what the test results mean to the technician. This chapter also differentiated between diagnostic equipment designed to lead the technician to the general area of a fault quickly, such as a scan tool or exhaust gas analyzer, and equipment suitable for making pinpoint tests of a circuit, such as a DMM or DSO, whether used alone or in conjunction with a breakout box.

Many diagnostic concepts were also covered as the chapter continued into the diagnostic equipment portion, such as voltage drop testing of a circuit.

▲ DIAGNOSTIC EXERCISE

A vehicle is towed into the shop because it "cranks, but won't start." The malfunction indicator lamp (MIL) does not come on, and all attempts to pull DTCs result in "no response." When a scan tool is connected, it cannot seem to communicate with the PCM. What tests should the technician perform?

Review Questions

1. What is a soft fault?
   A. A fault that exists intermittently
   B. A fault that is currently present
   C. A fault that can only be identified through the use of a scan tool
   D. A fault that cannot be properly diagnosed
2. What is a hard fault?
   A. A fault that exists intermittently
   B. A fault that is currently present
   C. A fault that can only be identified through the use of a scan tool
   D. A fault that cannot be properly diagnosed
3. *Technician A* says that memory codes may represent soft faults or hard faults. *Technician B* says on-demand codes represent hard faults. Who is correct?
   A. *Technician A* only
   B. *Technician B* only
   C. Both technicians
   D. Neither technician

4. When manufacturers are alerted to a problem on their vehicles, they will often issue a procedure to help the technician repair this problem properly. This is known as which of the following?
   A. DTC
   B. DLC
   C. TSB
   D. TAB
5. A vehicle is in the shop for diagnosis of a fuel economy complaint. When the memory DTCs are pulled from the PCM, an "O_2 sensor is always lean" code is received. *Technician A* says that this means that the O_2 sensor is faulty and should be replaced without wasting any time on further testing. *Technician B* says that this code does not necessarily mean that the O_2 sensor is faulty. Who is correct?
   A. *Technician A* only
   B. *Technician B* only
   C. Both technicians
   D. Neither technician
6. Two technicians are discussing the proper procedures to test a vacuum solenoid. *Technician A* says that the solenoid may be tested electrically by measuring the winding's resistance or by energizing the solenoid with jumper wires and noting the "click." *Technician B* says that the solenoid may be tested functionally by energizing and de-energizing the solenoid while using a vacuum pump and a vacuum gauge to test the solenoid's ability to allow or block the flow. Who is correct?
   A. *Technician A* only
   B. *Technician B* only
   C. Both technicians
   D. Neither technician
7. Two technicians are discussing a problem vehicle on which the engine will crank, but will not start. *Technician A* says that the voltage signal on the reference voltage wire coming from the PCM to the TPS should be approximately 5 V. *Technician B* says that if the voltage on the reference voltage wire between the PCM and the TPS is too low, the PCM is faulty and should be replaced without further testing. Who is correct?
   A. *Technician A* only
   B. *Technician B* only
   C. Both technicians
   D. Neither technician
8. Which of the following statements concerning a flowchart is *false*?
   A. It is designed to assist the technician in making pinpoint tests to identify the exact fault.
   B. The technician may be directed to a certain step of a flowchart by reason of a DTC.
   C. The technician may be directed to a certain step of a flowchart by reason of a symptom.
   D. It is okay to skip a test step in a flowchart and assume the test result if the test step is too difficult to perform.
9. Which of the following diagnostic tools is designed to help the technician quickly identify the area of a fault?
   A. A scan tool
   B. A DMM
   C. A DSO
   D. Both B and C
10. Which of the following diagnostic tools is designed to allow the technician to access the PCM's data stream through serial data (binary code) communication with the PCM?
   A. A scan tool
   B. A DMM
   C. A DSO
   D. Both B and C

11. Which of the following diagnostic tools is designed to allow the technician to measure time, such as the response time of an oxygen sensor?
    A. A scan tool
    B. A DMM
    C. A DSO
    D. Both B and C
12. *Technician A* says that if a scan tool says that the EGR valve is being commanded to open 50 percent, you can be assured that the valve is working properly and no further testing is necessary on the EGR valve. *Technician B* says that if a scan tool shows that the voltage for the ECT sensor is out of range, then you can be assured that the ECT sensor itself is at fault and no further pinpoint testing is necessary on the ECT sensor circuit. Who is correct?
    A. *Technician A* only
    B. *Technician B* only
    C. Both technicians
    D. Neither technician
13. Which of the following pieces of diagnostic equipment can damage computer circuits if not used properly?
    A. Non-powered test light
    B. Logic probe
    C. High-impedance DMM on a DC volts scale
    D. Scan tool
14. A technician is using a DMM to measure the resistance of a secondary ignition wire. The DMM is set to a 40 kΩ scale. The display shows 7.92. To which of the following does this number equate?
    A. 7.92 Ω
    B. 792 Ω
    C. 7,920 Ω
    D. 316,800 Ω (7.92 × 40,000)
15. A technician is using a DMM to measure the parasitic draw of a vehicle's electrical system on the battery. The DMM is set to a 400 mA scale. The display shows 32. To which of the following does this number equate?
    A. 32 amps
    B. 3.2 amps
    C. 0.32 amps
    D. 0.032 amps
16. A technician has made voltage drop tests on a PCM's power and ground circuits with the ignition switch turned on and the engine not running. The following results were obtained: Positive-side voltage drop = 0.068 V. Negative side voltage drop = 0.124 V. What is indicated?
    A. Both circuits are within specifications.
    B. The positive-side circuit is within specifications; the negative-side circuit has excessive resistance.
    C. The positive-side circuit has excessive resistance; the negative-side circuit is within specifications.
    D. Both circuits have excessive resistance.
17. A customer brings a vehicle into the shop and complains of intermittent stalling requiring a cool-down period before it will restart. During a road test the engine stalls. Testing shows that the fuel system has lost pressure and the fuel pump does not seem to run. After you get the vehicle back to the shop, you make voltage drop tests of the fuel pump's power and ground circuits to test for the intermittent condition, and these tests prove that the circuitry is okay. What is the best test that can be performed to verify whether the fuel pump itself is responsible for the intermittent failure?
    A. Perform a voltage drop test across the fuel pump motor.
    B. Use a DMM to measure the average current that the fuel pump motor draws.
    C. Use a DSO to look at the fuel pump circuit's voltage waveform.
    D. Use a DSO and an inductive current probe to look at the fuel pump motor's current waveform.

18. A DSO is properly connected to a circuit, but the waveform appears to be randomly "walking across the screen." Which of the following adjustments will most likely correct this condition?
    A. Ground level
    B. Volts per division
    C. Time per division
    D. Trigger characteristics
19. When looking at the current waveform of a fuel injector solenoid or an ignition coil primary winding, current that builds too quickly to the saturation level suggests what type of fault?
    A. Open winding
    B. Shorted winding
    C. Circuit has excessive resistance on the insulated side
    D. Circuit has excessive resistance on the ground side
20. When looking at a current waveform of a fuel pump motor, the waveform is seen to go to ground twice for each rotation of the motor. What does this indicate?
    A. Open winding
    B. Shorted winding
    C. Physical binding due to worn bushings
    D. Poor contact between the brushes and the commutator segments

Chapter 6

Exhaust Gas Analysis

OBJECTIVES

Upon completion and review of this chapter, you should be able to:

- ❑ Understand the theory of gas analysis.
- ❑ Describe the exhaust gases measured by a gas analyzer.
- ❑ Recognize the methods used by gas analyzers to sample emission gases.
- ❑ Understand how to use emission gas levels in diagnosing engine performance, fuel economy complaints, and emission failures.

KEY TERMS

Carbon Dioxide (CO_2)
Carbon Monoxide (CO)
Concentration Sampling
Constant Volume Sampling (CVS)
Exhaust Gas Analyzer
Grams per Mile (GPM)
Hydrocarbons (HC)
Nitrogen (N_2)
Oxides of Nitrogen (NO_x)
Oxygen (O_2)
Water (H_2O)

Understanding what produces each of the gases that exits the tailpipe of an automobile is the basis for using these gases to help in diagnosing symptoms of poor engine performance, poor fuel economy, and failure of emission tests. However, bear in mind that gas analysis is used in diagnosis to quickly narrow down the possible problem areas; gas analysis is not a pinpoint test in itself but rather must be followed up by the proper pinpoint tests. Gas analysis can be a very effective diagnostic tool when used properly—a complement to the other diagnostic tools available to today's technician.

THEORY OF GAS ANALYSIS

What Goes In Must Come Out

When the engine is running it is, in effect, operating as an air pump. Everything that comes into the combustion chamber through the intake valve is ultimately expelled in some form through the exhaust valve into the exhaust system. By measuring the gases that exit the tailpipe, we can get a good idea of what happened in the combustion chamber. But first, a good understanding of how each of the gases is formed is in order. Although several exhaust gases are produced in minute amounts during combustion, the discussion here concerns the production of the primary exhaust gases and their implications for the technician (Figure 6–1). An initial description of these gases is also found in Chapter 2, but they are discussed in greater detail in this chapter.

To begin, both atmospheric air and fuel are introduced into the engine. Atmospheric air consists of about 78 percent **nitrogen (N_2)** and between 20 and 21 percent **oxygen (O_2).**

Fuel is made up primarily of **hydrocarbons (HC)**, each molecule containing atoms of hydrogen

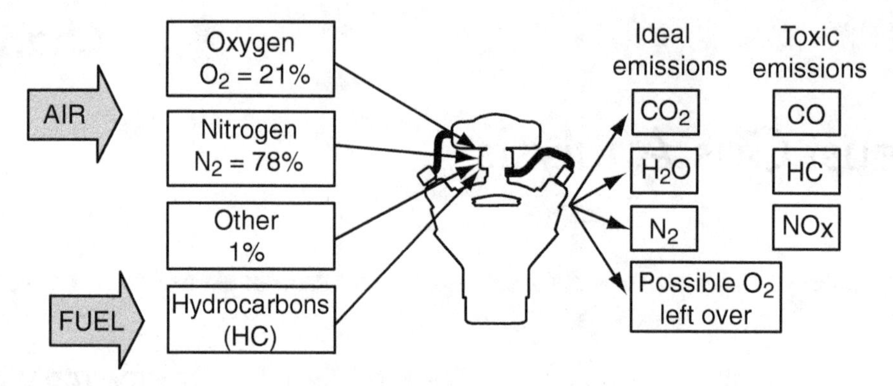

Figure 6–1 Exhaust gases that result from combustion.

and carbon. As the combustion within the cylinder takes place, these components combine in a chemical reaction and then exit the cylinder.

Nontoxic Gases

If an engine has good compression, ignition, and a stoichiometric air-fuel ratio, the combustion process will produce nontoxic gases as follows: During combustion of the hydrocarbons, the carbon separates from the hydrogen. Both of these elements then recombine with oxygen. During combustion, the hydrogen atoms combine with oxygen atoms to form molecules of **water** vapor or **H_2O**, with each molecule consisting of two hydrogen atoms and one oxygen atom. And during combustion, the carbon atoms combine with oxygen atoms to form molecules of **carbon dioxide** or **CO_2**, with each molecule containing one carbon atom and two oxygen atoms. Rephrased, CO_2 is produced when combustion is present with a plentiful supply of oxygen being present as well. Any nitrogen will, for the most part, go through the engine unchanged. And if the air-fuel ratio is slightly lean, a small amount of oxygen may be left over after combustion has taken place.

Toxic Gases

Hydrocarbons. If some of the fuel does not burn during combustion, it exits the engine as unburned hydrocarbons (or **HC** molecules). If a *total misfire* occurs within the cylinder for any reason, HC levels will be *dramatically high.* This includes any misfire due to compression problems, ignition misfire, or a lean misfire, which occurs on a spark-ignition engine when a lean air-fuel ratio causes the fuel molecules to be spread so far apart that the flame front cannot propagate (about a 17:1 air-fuel ratio in most engines, even leaner in some other engines depending upon combustion chamber design). If the air-fuel ratio is too rich, there is not enough oxygen to burn all of the fuel completely. This is sometimes known as a *partial misfire* and results in *moderately high* HC levels.

The low surface temperature of a cylinder wall, which rapidly draws heat from the HC and O_2 molecules that are in contact with it, will prevent the flame front from reaching or maintaining ignition temperature. This causes premature quenching of the flame front and results in excessive HCs in the exhaust. This condition can be caused by low coolant level or low coolant temperature by reason of a stuck-open thermostat. It can also be a result of over-advanced ignition timing or simply the result of low engine speed such as idle. Some engine designs, particularly smaller engines, are more prone to producing excessive HCs at idle than others. Also, when an engine is first started on a cold day, the combustion chamber surfaces will all initially be too low to sustain clean emissions performance.

Carbon Monoxide. Carbon monoxide (CO) is *always* produced as a result of combustion taking place with a shortage of oxygen (rich air-fuel ratio). Misfires do not produce CO. For example, replacing the spark plugs or spark plug wires would not be an effective repair to correct high CO levels. Think of a CO molecule as a molecule that is starved for oxygen. Eventually it will seek out another oxygen atom and be converted to CO_2. In the meantime, if you inhale it, carbon monoxide can cause dizziness, nausea, and even death. You cannot detect the presence of CO in the air through sight, taste, or smell.

Oxides of Nitrogen. Oxides of nitrogen (NO_x) are produced by combining nitrogen and oxygen. Nitrogen combines easily with other elements when placed under enough heat and/or pressure. Remember that atmospheric air is made up primarily of nitrogen (78 percent) and oxygen (20 to 21 percent). NO_x is formed as a result of atmospheric air being placed under enough heat and pressure to cause the oxygen and nitrogen to combine. An NO_x molecule consists of one nitrogen atom combined with one to five oxygen atoms, the most common automotive NO_x emissions being formed by combining the nitrogen atom with just one or two oxygen atoms. When the nitrogen atom is combined with just one oxygen atom, nitric oxide (NO) is formed; NO makes up about 95 percent of automotive NO_x emissions. When the nitrogen atom is combined with two oxygen atoms, nitrogen dioxide (NO_2) is formed, making up less than 5 percent of automotive NO_x emissions. Both types of molecules are collectively referred to as oxides of nitrogen, or NO_x.

In an engine's combustion chamber, while small amounts of NO_x may begin to form at temperatures as low as 2,000°F to 2,100°F, generally NO_x formation does not become critical until about 2,500°F and higher.

As a toxic emission, NO_x has been proven to have a direct negative effect on the human respiratory system. NO_x also reacts with moisture to produce low-level ozone and acid rain. Finally, while some forms of NO_x are colorless, nitrogen dioxide (NO_2) is seen as a reddish-brown layer over urban areas, generally associated with smog, when it is present in heavy concentrations.

MEASURED GASES

Exhaust Gas Analyzers

Of the seven gases previously defined (four nontoxic gases [N_2, O_2, H_2O, and CO_2] and three toxic gases [CO, HC, and NO_x]), up to five of them are generally measured, depending on the particular **exhaust gas analyzer.** Early gas analyzers only measured HC and CO emissions. Unfortunately, on a vehicle that has a catalytic converter in good working order that is up to proper operating temperature, HC and CO emissions may both be effectively oxidized within the converter. If the purpose is to perform an emission test, this does not present a problem. But if the purpose is to analyze the exhaust gases to help diagnose engine performance problems, this reduction of HC and CO can give the impression that the engine is performing more efficiently than it really is.

Today's four-gas analyzers measure not only HC and CO but also CO_2 and O_2. Both CO_2 and O_2 are, of course, nontoxic gases, but they are also valuable for analytical purposes if properly understood.

Modern five-gas analyzers also measure NO_x in addition to HC, CO, CO_2, and O_2. The additional capability to measure NO_x adds very little to our analytical abilities but is a necessary addition if you work in an area where local emission laws require testing for NO_x.

How the Air-Fuel Ratio Affects the Performance of the Gases

Effect of Air-Fuel Ratio on Hydrocarbons. Hydrocarbon levels will increase as a result of any misfire occurring within the cylinder for any reason. If the engine compression and ignition systems are in good working order and valve timing and spark timing are correct, HC levels will

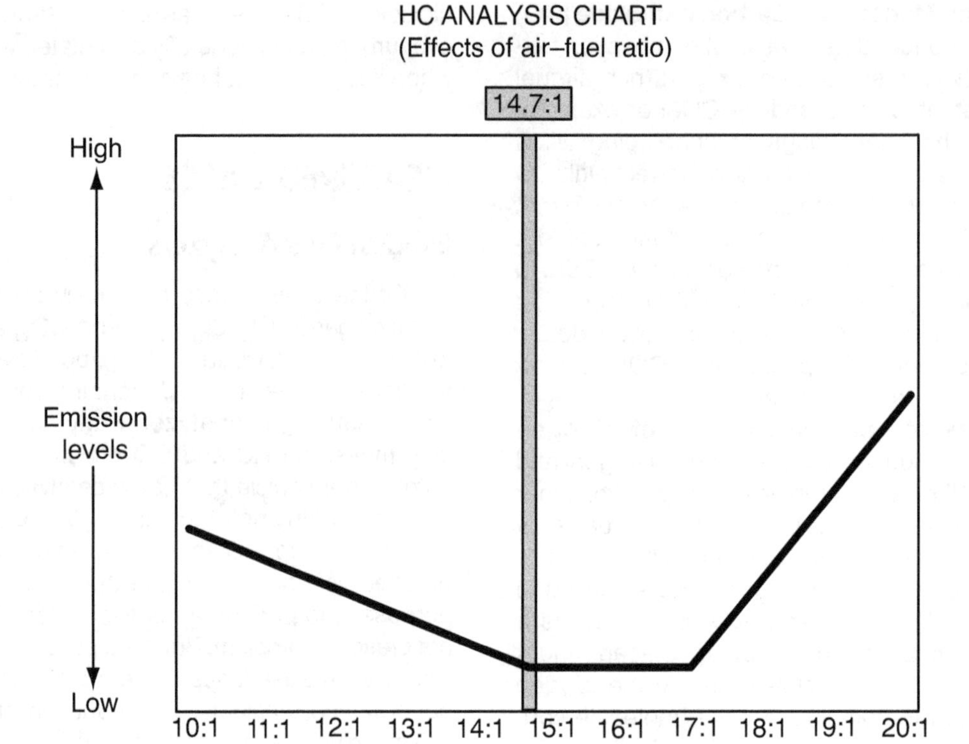

Figure 6–2 Effect of air-fuel ratio on hydrocarbons.

be low at a stoichiometric air-fuel ratio (14.7:1). As the air-fuel ratio moves richer from 14.7:1, HC levels increase moderately as the air needed for complete combustion is proportionately reduced (Figure 6–2). As the air-fuel ratio moves leaner from 14.7:1, HC levels remain low until the point of lean misfire, about 17:1 for many engines. Beyond this point, they tend to increase dramatically.

Effect of Air-Fuel Ratio on Carbon Monoxide. Carbon monoxide levels are not increased because of engine misfire (in fact, they are slightly reduced). They are also low at a stoichiometric air-fuel ratio. As the air-fuel ratio moves richer from 14.7:1, CO levels increase as the air needed for complete combustion is reduced (Figure 6–3). As the air-fuel ratio moves leaner from 14.7:1, the CO level remains low. As a result, you can determine from CO levels how far rich of stoichiometric the air-fuel ratio is (assuming the catalytic converter is not yet up to operating temperature), but CO levels cannot be used to determine how lean of stoichiometric the air-fuel ratio is.

Effect of Air-Fuel Ratio on Oxygen. Oxygen levels are slightly increased because of misfire (a misfire results in a failure to consume both the fuel and the oxygen). When misfire is not occurring, O_2 levels are low at stoichiometric but increase as the air-fuel ratio moves leaner from 14.7:1 due to increased levels of oxygen being "left over" after combustion (Figure 6–4). As the air-fuel ratio moves richer from 14.7:1, O_2 levels stay low. As a result, you can determine from O_2 levels how far lean of stoichiometric the air-fuel ratio is, but O_2 levels cannot be used to determine how rich of stoichiometric the air-fuel ratio is. Of course, all of this assumes that no extra

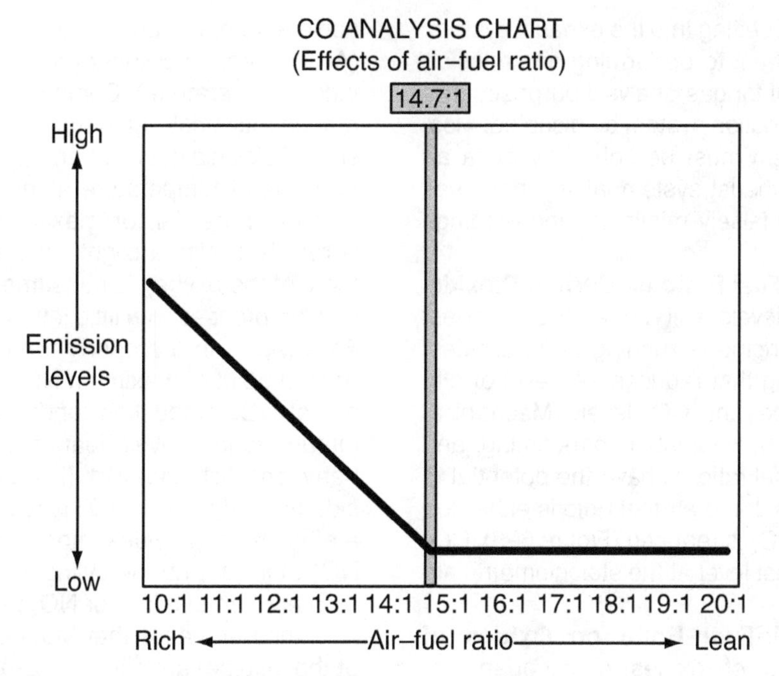

Figure 6–3 Effect of air-fuel ratio on carbon monoxide.

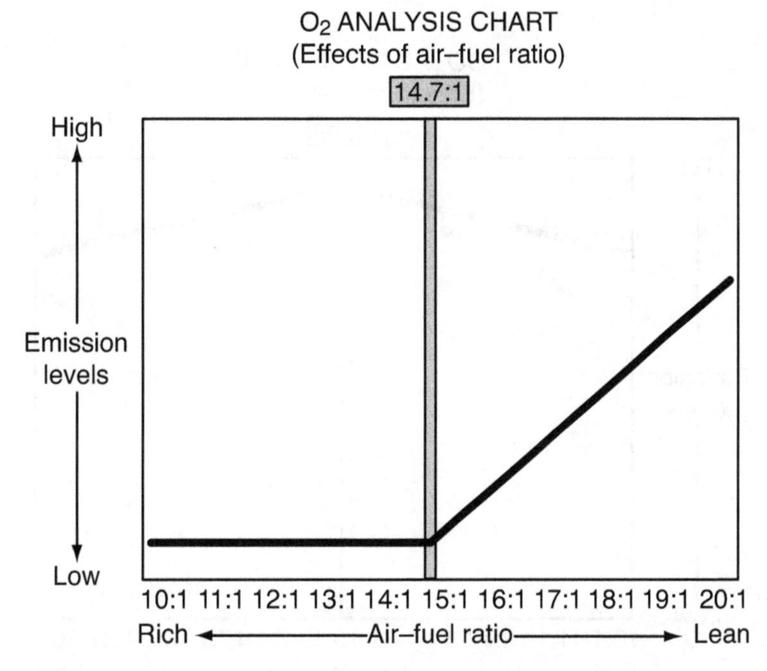

Figure 6–4 Effect of air-fuel ratio on oxygen.

atmospheric air is getting into the exhaust system. Therefore, in contrast to performing an emission test, it is important for gas analysis purposes that any existing air injection system be disabled. Also, the exhaust system must be tight. Any extra air getting into the exhaust system falsely increases the O_2 levels and falsely minimizes the readings of all other gases.

Effect of Air-Fuel Ratio on Carbon Dioxide. Carbon dioxide levels will be at their highest point when the engine is running at its greatest efficiency. Anything that reduces this level of efficiency also reduces the CO_2 levels. Mechanical or ignition misfire, misadjusted spark timing, and an improper air-fuel ratio all have the potential to reduce CO_2 levels. If the air-fuel ratio is either too lean or too rich, CO_2 is reduced (Figure 6–5). CO_2 will be at its highest level at the stoichiometric air-fuel ratio.

Effect of Air-Fuel Ratio on Oxides of Nitrogen. Levels of oxides of nitrogen are elevated dramatically when both sufficient temperature and pressure exist within the cylinder to achieve temperatures of about 2,500°F or more. (An increase in cylinder pressure also raises cylinder temperature.) Combustion in the cylinder reaches its peak temperature slightly to the rich side of stoichiometric, at around 14:1 (Figure 6–6). In terms of temperature alone, it is at this point that the potential for maximum NO_x production is reached. With a slightly rich mixture, however, most of the oxygen is consumed during the combustion process with little left to allow NO_x to form. As a result, in a properly running engine, this is not the point of maximum NO_x production; rather, actual NO_x production tends to increase as the air-fuel ratio moves leaner up to somewhere between 15:1 and 16:1 (just slightly to the lean side of stoichiometric). This is because the leaner air-fuel ratio provides more leftover oxygen for NO_x to form, and the cylinder temperature is still high enough to allow for NO_x production.

Another reason that NO_x levels (as measured at the tailpipe) are seen to increase as the air-fuel ratio becomes leaner beyond 14.7:1 is that a lean air-fuel ratio does not support the operation of the

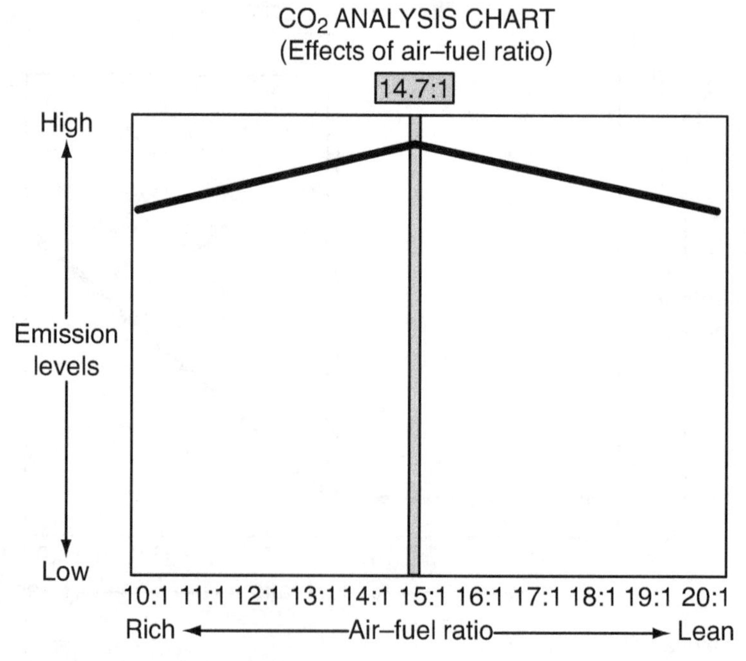

Figure 6–5 Effect of air-fuel ratio on carbon dioxide.

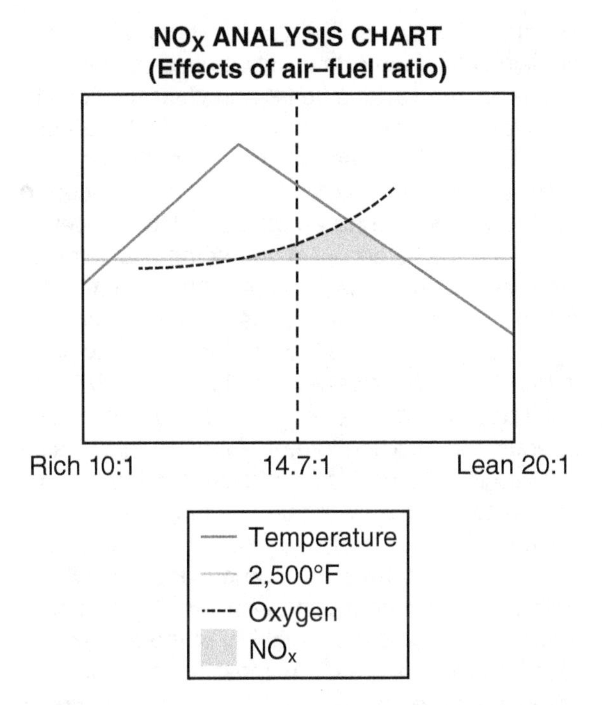

Figure 6–6 Effect of air-fuel ratio on oxides of nitrogen.

NO_x reduction bed of the catalytic converter: the converter actually needs CO, formed only when the air-fuel mixture is rich, to help break up the NO_x molecules. As a result, with the NO_x reduction catalytic converter working properly, NO_x levels after the converter will increase as the air-fuel ratio becomes leaner up to the point at which the NO_x production within the engine subsides.

Spark knock and ping can cause NO_x formation due to the resulting pressure increase. As a result, anything that tends to promote spark knock and ping can raise NO_x levels, including problems with turbocharger intercoolers, thermostatic air cleaners, and the like.

However, a few myths associated with NO_x production must be addressed:

- High engine temperature due to an overheated engine does not directly increase NO_x production. Even a coolant temperature of 300°F cannot directly push cylinder combustion temperature past 2,500°F. (The only exception to this involves the extent to which an overheated engine might contribute to spark knock or ping.)
- Rich air-fuel mixtures do not cool the combustion but rather tend to burn a little hotter than a stoichiometric mixture, particularly if just slightly on the rich side of stoichiometric.
- While high engine loads tend to produce more pressure, resulting in higher cylinder temperatures, they do not actually result in higher levels of NO_x production in an engine that is controlling the air-fuel ratio properly. This is due to the fact that rich air-fuel ratios are required to achieve good engine performance under high load, and, as a result, not much oxygen remains after combustion has taken place. Without the oxygen, NO_x does not form. As a result, most NO_x production takes place in the light to medium load ranges when the air-fuel ratio is not as rich.

If an engine is running abnormally lean during a hard acceleration due to a restricted fuel filter, restricted fuel injectors, and the like, the potential exists for the engine to produce high levels of NO_x. This is because higher cylinder temperature is present due to the high load that the engine is under during hard acceleration, and with an abnormally lean condition plenty of oxygen is available as the combustion temperature peaks out. This combination allows for the formation of excessive NO_x within the combustion chamber.

GAS ANALYZERS

Concentration Sampling

The **concentration sampling** gas analysis machine is the analyzer found in most automotive shops today. Its design may be a big box type (Figure 6–7), or it may be portable (Figure 6–8). Also, the OTC Genisys and the Snap-on Modis, shown in Figure 5–54 and Figure 5–55, respectively, in Chapter 5, can both be configured as concentration-type gas analyzers. A probe, which is simply inserted into the exhaust pipe, allows the analyzer to sample a small percentage of the total exhaust gas volume when the engine

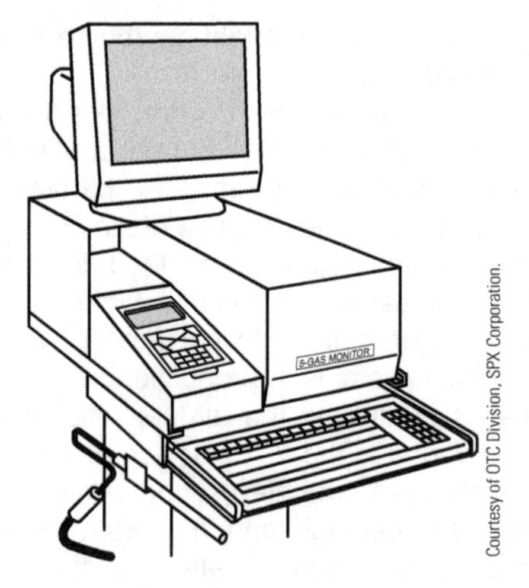

Figure 6–7 Five-gas emissions analyzer.

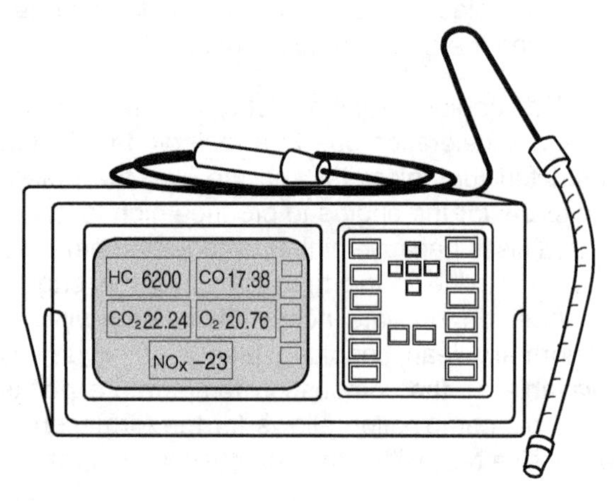

Figure 6–8 A portable exhaust gas analyzer.

is running. Each of the gases is measured as a percentage, or concentration, of the total sample; thus the term *concentration sampling*.

All the measured gases are measured either as a percentage (parts per hundred) or as parts per million (PPM), with PPM simply being a smaller-scale measurement. Because concentration sampling is the method used in most shops, as a technician you must have a thorough understanding of the levels to expect of each of the gases from a well-tuned engine. Keep in mind that these figures are guidelines only, and the results will vary with the year and model of each vehicle. Also, many states issue emission test cut-points for the various years and classifications of vehicles. Such cut-points do not reflect the ideal emission levels for these gases but reflect only the point at which the particular state has chosen to fail the vehicle during an emission test. However, these cut-points can give you some indication of what emission levels to expect if you remember that these cut-points are usually quite generous.

This type of emissions testing is usually done in the shop with the vehicle at idle or at 2,500 RPM, but it does not require the vehicle to be under any operating load. Therefore, this type of test is also referred to as a *no-load emission test.* If you have access to either a dynamometer or a portable gas analyzer that can monitor emission levels while the vehicle is being driven on public roads, you will be able to monitor emissions with the engine under load, which increases the ability of the analyzer to identify problem areas. In most shops, however, the test is still done without the vehicle being under any engine load. Some portable gas analyzers are available that can record and graph the emission readings while the vehicle is being driven. The OTC Genisys is an example of a portable gas analyzer that will record the emission readings, including lambda values, during a road test.

WARNING: Looking away from the road to monitor a portable gas analyzer while driving is extremely dangerous. You should have another technician drive the vehicle for you, allowing you to concentrate on the gas analyzer readings.

Maximum CO Level. A concentration analyzer measures carbon monoxide levels in percent. CO levels should stay below 2.5 to 3 percent if the vehicle does not have a catalytic converter or if the converter is not up to operating temperature. With a converter that is up to operating temperature, CO levels should stay below 0.5 percent, preferably near 0 percent.

Maximum HC Level. A concentration analyzer measures hydrocarbon levels in PPM.

HC levels should stay below 300 to 400 PPM if the vehicle does not have a catalytic converter or if the converter is not up to operating temperature. With a converter that is up to operating temperature, HC levels should stay below 100 PPM, preferably near 0 PPM.

Concerning the measurement of exhaust hydrocarbons, there exist more than 200 different forms of hydrocarbons including oils, lubricants, antifreeze, and even some of the ingredients used to manufacture vinyl products. What exits the tailpipe is primarily nine different forms of hydrocarbons from HC (one hydrogen atom and one carbon atom) to HC_9 (one hydrogen atom and nine carbon atoms). Modern concentration-type exhaust gas analyzers measure the HC_6 molecules (one hydrogen atom and six carbon atoms) or *hexanes* and are therefore not sensitive to many of the hydrocarbons in the exhaust. Only a representative sample is measured.

Maximum O_2 Level. A concentration analyzer measures oxygen levels in percent. O_2 levels should absolutely stay below 2 to 3 percent. With good closed-loop control, the O_2 level should be below 0.5 percent. O_2 levels above this indicate that the air-fuel ratio is too lean, possibly approaching the area of a lean misfire. Any extra air entering the exhaust system artificially increases O_2 levels and artificially minimizes the levels of all other gases. Remember to disable any secondary air injection system by pinching off the air hoses and to check the exhaust system for leaks.

Minimum CO_2 Level. A concentration analyzer measures carbon dioxide levels in percent. CO_2 levels will generally be above 13 percent. CO_2 may reach as high as 15.5 percent. Because engine design parameters (such as valve timing and lift) affect CO_2 levels, it is not possible to set a hard specification for CO_2. Therefore, if the CO_2 is reading close to 15.5 percent, the engine is likely performing well. If the CO_2 is reading at 11 percent or less, the engine likely has performance problems that must be diagnosed. If the CO_2 is between 13 and 14 percent, it could mean that the engine is performing well, but it could also mean that there are performance problems with the engine. For example, an engine that emits 15.5 percent CO_2 when running at its peak efficiency might only emit 13 percent CO_2 when a problem is present. Therefore, the technician must evaluate all of the other gases when making a determination.

Measuring NO_x Levels. A concentration analyzer measures NO_x levels in PPM. Having the ability to measure NO_x is not critical unless you are in an area where vehicles are tested and failed for NO_x; otherwise, the ability to measure NO_x levels is not as important to your diagnosis as the ability to measure the other gases. NO_x is best measured when the vehicle is under normal load because an engine that is not under load may not produce the NO_x levels that would be recorded during an emission test that loaded the vehicle. A dynamometer may be used if you have access to one. Otherwise, drive the vehicle on a road test while recording the results on a portable five-gas analyzer. Make sure your drive cycle is repeatable. Although your concentration-type analyzer shows NO_x in PPM, you can compare the readings obtained following your repairs to your initial readings to see if you made an improvement.

When diagnosing a vehicle that has failed for high NO_x, think in terms of high combustion chamber temperatures. Many engines use EGR and/or variable valve timing to recycle exhaust gases as a method of helping to control these temperatures. Thus, these systems should also be checked. In addition, the engine should be checked for running lean on acceleration due to a restricted fuel filter or restricted fuel injectors. Also, if a failure for high CO is being diagnosed and NO_x is also quite high (whether it fails or not), keep in mind that as the CO problem is corrected and the air-fuel ratio is leaned out toward stoichiometric, additional air will be made available for NO_x formation. If cylinder temperatures are already high, NO_x levels may be increased even as the CO problem is corrected.

Lambda Calculation. Modern exhaust gas analyzers use the lambda formula (Figure 6–9), to calculate the true lambda value from the exhaust

$$\lambda = \frac{[CO_2]+\left[\frac{CO}{2}\right]+[O_2]+\left(\left(\frac{Hcv}{4}\times\frac{3.5}{3.5+\frac{[CO]}{[CO_2]}}\right)-\frac{Ocv}{2}\right)\times([CO_2]+[CO])}{\left(1+\frac{Hcv}{4}-\frac{Ocv}{2}\right)\times([CO_2]+[CO]+K1\times[HC])}$$

Where:

[XX] = Gas concentration in % volume
K1 = Conversion factor for FID measurement to NDIR measurement
Hcv = Atomic ratio of hydrogen to carbon in the fuel
Ocv = Atomic ratio of oxygen to carbon in the fuel

For gasoline, K1 is 6.0, Hcv is about 1,800 (depending on the actual mix), and Ocv is 0.00, except for oxygenated fuels.

Figure 6–9 The lambda formula.

gases. In turn, the analyzer then calculates the air-fuel ratio from the lambda value (assuming straight gasoline is in the fuel tank without any ethanol).

Emission Tests That Use Concentration Sampling

No-Load Emission Test. Some states use a concentration method of exhaust sampling to sample emissions under no load at idle, then at 2,500 RPM, and then again at idle. This type of test may be referred to as a *no-load test* or an *idle test.* The measured emissions at 2,500 RPM may or may not be used as a pass/fail standard. This type of testing is used most often with older vehicles, but it may be used with newer vehicles in less densely populated areas.

Acceleration Simulation Mode Test. Several states have adopted an emissions program referred to as an *acceleration simulation mode* (ASM) test. The ASM test uses a concentration method of exhaust sampling combined with the use of a dynamometer to test CO, HC, and NO_x levels under vehicle load. The ASM test is divided into two parts. During the ASM 5015 part of the test, the vehicle is driven on a dynamometer at 15 MPH. During the ASM 2525 part of the test, the vehicle is driven on a dynamometer at 25 MPH. Both ASM tests require that the vehicle must pass concentration-type emissions standards for at least 10 seconds of a 90-second test, beginning no sooner than 25 seconds into the test. The results of either the idle test or the ASM test are measured in percent and parts per million.

Constant Volume Sampling

Several states have now adopted an emissions-testing program that imitates some characteristics of the federal test procedure (FTP) used to certify new vehicles before they are made legal for sale in the United States. This test, called an *inspection/maintenance test* (I/M test), places the vehicle under realistic load on a dynamometer designed to imitate the vehicle's curb weight, wind resistance, and other rolling resistance. If the vehicle is run for 240 seconds on the dynamometer during the test, the test is referred to as an *IM240 test* (Figure 6–10). An IM240 emission test uses a process called **constant volume sampling (CVS),** which requires that all of the

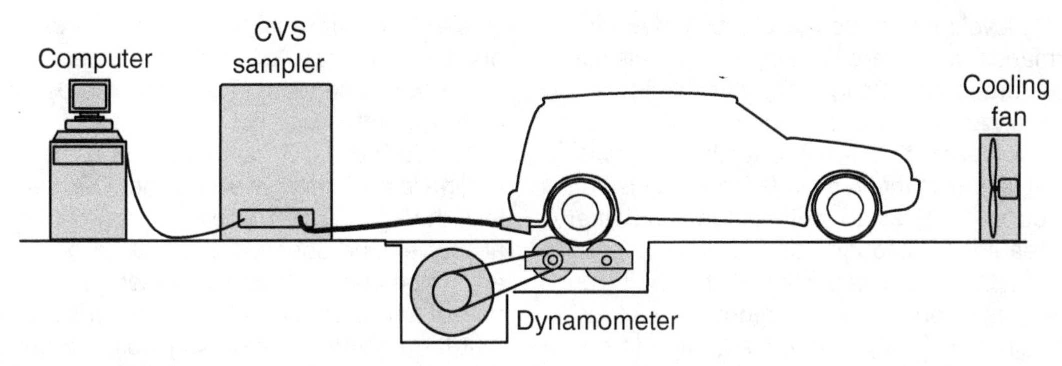

Figure 6–10 Components of a typical IM240 test station.

exhaust be captured and measured along with enough ambient air to create a known volume of gases. Then the density of this known volume is calculated. Finally, the concentration (percentage) of each gas is calculated to figure how many grams of each gas are being expelled from the exhaust system. The analytical computer can also use the dynamometer to determine the distance the vehicle has traveled at any instant in time or over a longer time period. Ultimately, each gas is calculated in **grams per mile (GPM).**

Constant volume sampling generally tests for four gases: the three primary toxic gases (CO, HC, and NO_x) and one nontoxic gas (CO_2). The primary advantage of this type of test is that it tests the vehicle under realistic load conditions. Sometimes the engine computer's strategies are different when the engine is not under load than when the vehicle is driven down the road. Or a secondary spark plug wire with deteriorating insulation may only arc to the engine block when under realistic load because of the increased cylinder pressures when under load. This same problem may not show up at all during a no-load emission test because the spark plug gap may continue to be the path of least resistance when cylinder pressures are low. Ultimately, the best method to test how many emission gases exit the tailpipe when the vehicle is traveling on the road is to test the vehicle under the same conditions.

You should also be aware that most CVS analytical equipment used with IM240 testing uses a process called "flame ionization" to evaluate HC emissions. Unlike the infrared benches used in concentration sampling analyzers, these units are sensitive to hydrocarbons from antifreeze and brake fluid. Because of the manner in which these analyzers draw in additional make-up air near the rear of the vehicle, they also may pick up gasoline hydrocarbons from fuel line leaks near the front of the vehicle or from evaporative system vapor leaks, since all of this is blown toward the rear of the vehicle during the IM240 dynamometer test. Therefore, when diagnosing hydrocarbon failures resulting from an IM240 test, in addition to measuring the hydrocarbon emissions exiting the tailpipe, be sure to perform visual checks for gasoline, brake fluid, and antifreeze leaks, and to conduct a visual inspection of the hoses associated with the evaporative/canister purge system.

DIAGNOSING WITH THE GASES

Concentration Sampling

Because the gas analyzers in most automotive shops are of the concentration sampling type, the emphasis of this book is on this type of testing.

When using gas analysis to aid in diagnosing engine performance problems, fuel economy problems, or emission test failures, the technician should record the readings of all gases at both idle and 2,500 RPM. It is critical to test CO, HC, O_2, and CO_2 levels for the best evaluation of engine performance. If you are working in an emission program area that tests for NO_x, you should have a gas analyzer that also tests for NO_x.

If the reason that you are testing for emission levels is that the vehicle failed an emission test, you should record both before and after repair readings. Although this is always a good idea, it is especially important if the emission test performed on the vehicle produced exhaust gas readings in GPM, as there is no way to reliably convert GPM into percent or PPM. Taking exhaust gas readings before you make any repairs (even before you perform a visual inspection or wiggle any electrical connectors) is known as *baselining.* Baselining the vehicle before you make any repairs allows you to take exhaust gas readings after the repairs have been made and then compare them to the baseline readings to determine whether your repairs have been effective.

Preparing the Gas Analyzer

Before the vehicle is tested, the gas analyzer should be properly warmed up. The analyzer should also be calibrated weekly using a calibration gas and following the manufacturer's specified procedures for the particular analyzer. This test usually includes a leak test of the probe and sample hose.

Before each test, the probe should be left open to sample ambient air while the readings are evaluated. HC, CO, and CO_2 should read very close to zero. O_2 levels should read between 20 and 21 percent. If they do not, the O_2 sensor in the analyzer should be replaced. (It is also a good idea to watch the O_2 reading when the gas calibration is performed. It should show very close to zero during the calibration.)

Evaluating the Gases

This author recommends that the technician first evaluate CO_2 levels, both at idle and 2,500 RPM, to make a quick determination about the efficiency of the engine at both speeds. In theory, CO_2 will *max out* at about 15 percent if CO is about 0.5 percent. (The total of CO_2 and CO added together will not exceed about 15.5 percent.) So if the CO_2 is 15 percent or higher, the engine is performing well. If either CO_2 reading is less than 13 percent, you know a problem exists at that engine speed. If the CO_2 is between 13 and 15 percent, evaluate the other gases to determine if a problem exists. Do the other gases show that the engine is performing well, or should the CO_2 be even higher still? A few engines will give their best performance with the CO_2 at about 13 percent due to cam timing and/or burning of E85. (Ethanol produces less CO_2 than straight gasoline.) Most engines should be closer to 15 percent for CO_2, and a reading of 13 to 14 percent indicates a problem is present.

Once the CO_2 has been evaluated, look at CO and O_2 readings to determine whether the air-fuel ratio is either rich or lean. (The air injection system must be disabled and the exhaust system must be tight for O_2 readings to be meaningful.) If excessive O_2 is present, the engine's air-fuel ratio is lean. If the engine is running lean at idle but is performing well at 2,500 RPM, check for vacuum leaks. Conversely, if the engine is running lean at 2,500 RPM but is performing well at idle, check for a partially plugged fuel filter.

Look at CO levels to check for a rich engine operating condition at either RPM, while keeping in mind that if the catalytic converter is hot, this value may be falsely minimized. (If you are evaluating emission levels to perform an emission test, you want the converter to be up to full operating temperature. But if you are evaluating emission levels for diagnostic purposes, it is best if you avoid getting the converter up to full operating temperature if possible.) If the engine is running rich at idle but is performing well at 2,500 RPM, check for leaking fuel injectors or a small diaphragm tear in the fuel pressure regulator. Of course, larger tears in the regulator's diaphragm will cause overly rich conditions at both engine speeds, although the higher RPM will tend to handle it better.

If excessive amounts of both CO and O_2 are found in the exhaust sample, know that the CO had to be produced in the engine's cylinders as it takes combustion to turn the carbon atom of an HC molecule into either CO or CO_2. Consider the possibility that the O_2 is getting into the exhaust stream by some method other than from the cylinders—either from an exhaust leak or from a secondary air injection system that has not been disabled.

Finally, evaluate HC levels to check for indications of a misfire condition at either RPM, while keeping in mind that if the catalytic converter is hot, this value may be falsely minimized.

If HC levels are *moderately* high and CO levels are also high, the HC levels are high because the air-fuel ratio is rich. For example, 6.50 percent CO can also be responsible for an HC reading of around 650 PPM, give or take a couple hundred. When you correct the overly rich condition to decrease CO levels, HC levels also decrease.

If HC levels are *dramatically* high, a total misfire exists (although it may be intermittent), either due to an ignition misfire, a lean misfire, or mechanical problems. For example, with a CO reading of 6.50 percent and an HC reading of 1,700 PPM, we would know that the engine is experiencing at least two problems. While a CO reading of 6.50 percent indicates a rich air-fuel ratio, 6.50 percent CO cannot be the only cause of an HC reading of 1,700 PPM. In this case, the engine is experiencing a total misfire in addition to the rich air-fuel ratio that is causing the high CO reading.

If HC levels are high at 2,500 RPM only or are high at both engine speeds, you must scope the secondary ignition system. Ignition misfire problems do not decrease with increased RPM or load. That is, if HC levels are high at idle but are normal at 2,500 RPM, the problem is not ignition-related but is due to either a lean misfire or mechanical problems. At this point you may add propane (using a propane enrichment tool) to artificially enrich the air-fuel ratio. If the HC levels now decrease, the problem is air-fuel ratio related. If the HC levels stay high, you should run engine compression tests.

CAUTION: Never spray carburetor cleaner into the throttle body on a running engine as a method of artificial enrichment while a gas analyzer's probe is in the tailpipe. You may damage internal components of the gas analyzer.

Evaluating Lambda

If you are using a modern exhaust gas analyzer that can determine the lambda value from the exhaust gas measurements, make sure that lambda ("λ") is selected in the analyzer's menu as a displayed value. Then note the lambda values both at idle and at 2,500 RPM. Lambda should be between 0.98 and 1.02 at both engine speeds. (Keep in mind that the exhaust system must be tight, as any exhaust leaks will make your lambda values look falsely lean.)

If the lambda value is between 0.98 and 1.02, the air-fuel ratio is correct. If lambda is less than 0.98, the air-fuel ratio is too rich. If lambda is greater than 1.02, the air-fuel ratio is too lean. The advantage of using lambda as opposed to using CO and O_2 readings only is that you can easily calculate the percentage of error when diagnosing rich or lean air-fuel conditions, something that is difficult to do with CO and O_2 readings alone. For example, a lambda reading of 0.87 indicates that the air-fuel ratio is 13 percent to the rich side, and a lambda value of 1.08 indicates that the air-fuel ratio is 8 percent to the lean side. This allows the technician to determine the severity of an air-fuel ratio control problem.

If you are using an older exhaust gas analyzer that does not calculate lambda, a number of Internet websites have automated lambda calculators that allow you to enter the four or five gas readings from your analyzer (with or without NO_x) and then indicate the true lambda value. It will be more accurate with NO_x figured in, but it is fairly accurate without NO_x. One of these websites is the International Automotive Technicians Network at www.iatn.net under *Resources.* There are also lambda formula software programs available for download that can be installed on your computer to allow you to calculate lambda from the gases. Because of the complexity of the lambda algorithm in Figure 6–9, it is suggested that you use an automated lambda calculator.

One point to note: Because there are several forms of hydrocarbons that exit the tailpipe (HC through HC_9), of which your gas analyzer only measures one (HC_6 or hexanes), determining the air-fuel ratio through a lambda calculation is more accurate when hydrocarbon levels are low. In other words, you should correct misfire problems before trying to use lambda values to determine the severity of an air-fuel ratio problem.

Other Tests

A gas analyzer may also be used to perform certain other tests. For example, if you remove the radiator cap from the radiator while the engine is cold and then start the engine, you can use the analyzer probe to check for CO or CO_2 escaping from the radiator. Because most gas analyzers that use concentration sampling use infrared analytical equipment to test for CO, CO_2, and HC emissions and are therefore not sensitive to antifreeze hydrocarbons, you can also monitor the radiator for hydrocarbon emissions. Therefore, if any of these emissions are present at the radiator, a head gasket leak or cracked head is indicated.

CAUTION: Never allow the analyzer probe to suck in liquid coolant because it will damage the gas analyzer.

The ability to use a gas analyzer to evaluate HC levels can allow you to diagnose other sources of unburned fuel, such as leaking mechanical fuel pumps or leaking fuel injectors. To evaluate leaking fuel injectors associated with a loss of residual fuel pressure with a PFI system, start the engine and let it idle several minutes. Then shut it off for five minutes, during which time you should electrically disable the fuel injectors and ignition system. Also, remove all of the spark plugs. After the five minutes are up, crank the engine for about 15 seconds. Then use a gas analyzer to "sniff" each spark plug hole. Normal cylinders will register hydrocarbon levels of approximately 125 to 150 PPM. Cylinders with leaking fuel injectors will show dramatically higher hydrocarbon levels, usually around 5,000 to 10,000 PPM. If leaking injectors are identified, cleaning them will sometimes correct the problem.

Limitations of an Exhaust Gas Analyzer. Treat an exhaust gas analyzer as a "narrow down the problem area" type of tool. It is not a pinpoint test tool. For example, if a high CO reading is obtained, you know that the engine is running rich, due to too much fuel being delivered. And the lambda value can indicate the severity of the rich condition. But the lambda value and high CO reading do not immediately define for you the exact cause of the problem—they only narrow down the possible areas that you will need to pinpoint test.

SUMMARY

This chapter described how each of the exhaust gases is produced and what emission levels mean in terms of diagnostic purposes. We reviewed some basic information on the types of analyzers, what they mean to the technician, and the importance of baselining a vehicle before making any repairs. We also discussed how gas analysis can be used to quickly isolate the general problem area to reduce the amount of pinpoint testing that must be done during diagnosis.

▲ DIAGNOSTIC EXERCISE

A vehicle comes into the shop for a scheduled oil change and normal maintenance. While performing the routine maintenance, the technician notices that the air filter is quite dirty and possibly restricted. The customer has not experienced any noticeable engine performance problems. What effect will this dirty/restricted air filter have on the vehicle's fuel economy?

Review Questions

1. Which of the following best describes the primary makeup of atmospheric air?
   A. 21 percent CO and 78 percent O_2
   B. 21 percent O_2 and 78 percent N_2
   C. 21 percent N_2 and 78 percent O_2
   D. 21 percent O_2 and 78 percent H_2O
2. Which of the following emissions of a gasoline engine does a five-gas analyzer *not* measure?
   A. H_2O
   B. CO
   C. CO_2
   D. O_2
3. Which of the following emission gases should be as high as possible and is an efficiency indicator?
   A. O_2
   B. HC
   C. CO
   D. CO_2
4. Higher-than-normal HC emission levels may be the result of which of the following?
   A. An ignition misfire, a lean misfire, or a mechanical misfire
   B. An overly rich air-fuel ratio
   C. High combustion chamber temperatures
   D. Both A and B
5. Higher-than-normal CO emission levels may be the result of which of the following?
   A. An ignition misfire, a lean misfire, or a mechanical misfire
   B. An overly rich air-fuel ratio
   C. High combustion chamber temperatures
   D. Both A and B
6. Higher-than-normal NO_x emission levels may be the result of which of the following?
   A. An ignition misfire, a lean misfire, or a mechanical misfire
   B. An engine that is slightly overheated
   C. High combustion chamber temperatures
   D. Both B and C
7. *Technician A* says that, on an engine that is running properly, most NO_x production occurs during periods of high engine load. *Technician B* says that if the air-fuel mixture is slightly richer than stoichiometric, NO_x will not form due to the cooling effect of the fuel. Who is correct?
   A. *Technician A* only
   B. *Technician B* only
   C. Both technicians
   D. Neither technician
8. Which of the following describes concentration sampling?
   A. Uses a dynamometer to load the engine to realistic loads
   B. Captures all of the exhaust and displays each gas as GPM
   C. Samples only a small portion of the total exhaust sample, then displays each gas as either percent or PPM
   D. Both A and B
9. Which of the following describes constant volume sampling?
   A. Uses a dynamometer to load the engine to realistic loads
   B. Captures all of the exhaust and displays each gas as GPM
   C. Samples only a small portion of the total exhaust sample, then displays each gas as either percent or PPM
   D. Both A and B
10. When diagnosing engine performance problems, it is important to disable the air injection system and to be sure that the exhaust system does not leak when using a gas analyzer that uses concentration sampling. This is because any extra air getting into the exhaust system will do which of the following?
    A. Falsely decrease O_2 readings
    B. Falsely increase O_2 readings
    C. Falsely decrease the readings of all of the other measured gases (except O_2)
    D. Both B and C

11. If a gas analyzer's probe is placed near a fuel line leak, the analyzer readings will show an increase in which of the following?
    A. H_2O levels
    B. CO levels
    C. HC levels
    D. Both B and C
12. When a gas analyzer is properly warmed up and the probe is sampling ambient air, what should the O_2 level read?
    A. 0 percent
    B. Between 1 and 5 percent
    C. Between 20 and 21 percent
    D. About 78 percent
13. How would an ignition misfire affect the following gases?
    A. O_2 levels would be lower; CO and CO_2 levels would be higher.
    B. CO_2 levels would be lower; O_2 and CO levels would be higher.
    C. CO_2 and CO levels would be lower; O_2 levels would be higher.
    D. CO_2 and O_2 levels would be lower; CO levels would be higher.
14. *Technician A* says that you should take exhaust gas readings before performing any repair work on a vehicle so that when the repair work is complete you have some initial gas readings with which to compare your final gas readings. *Technician B* says that to convert GPM emissions values into percentages (or concentrations), you must multiply the GPM by the atmospheric air pressure and then divide by 78 percent. Who is correct?
    A. *Technician A* only
    B. *Technician B* only
    C. Both technicians
    D. Neither technician
15. The following readings are obtained on a gas analyzer:

	Idle	**2,500 RPM**
HC	1,335 PPM	112 PPM
CO	0.00%	0.48%
CO_2	8.90%	14.80%
O_2	6.80%	0.73%

    What type of problem is most likely indicated?
    A. Insulation breakdown on a secondary ignition wire
    B. A vacuum leak resulting in a lean condition at idle
    C. A plugged fuel filter resulting in a lean condition at 2,500 RPM
    D. A plugged air filter resulting in a rich condition at 2,500 RPM
16. The following readings are obtained on a gas analyzer:

	Idle	**2,500 RPM**
HC	1,585 PPM	1,980 PPM
CO	0.00%	0.00%
CO_2	9.12%	7.98%
O_2	0.80%	0.96%

    What type of problem is most likely indicated?
    A. Insulation breakdown on a secondary ignition wire
    B. A vacuum leak resulting in a lean condition at idle
    C. A plugged fuel filter resulting in a lean condition at 2,500 RPM
    D. A plugged air filter resulting in a rich condition at 2,500 RPM

17. The following readings are obtained on a gas analyzer:

	Idle	2,500 RPM
HC	10 PPM	560 PPM
CO	0.02%	0.00%
CO_2	15.12%	10.79%
O_2	0.70%	4.30%

What type of problem is most likely indicated?
A. Insulation breakdown on a secondary ignition wire
B. A vacuum leak resulting in a lean condition at idle
C. A plugged fuel filter resulting in a lean condition at 2,500 RPM
D. A plugged air filter resulting in a rich condition at 2,500 RPM

18. A vehicle that has failed an emission test for high CO comes into the shop. *Technician A* says that the spark plugs and spark plug wires may be at fault and should be replaced to correct this problem. *Technician B* says that the first thing that should be done is to test the vehicle on the shop's gas analyzer to verify the readings before any repairs are made. Who is correct?
A. *Technician A* only
B. *Technician B* only
C. Both technicians
D. Neither technician

19. Which of the following could cause an IM240 test to fail a vehicle for high HC while a shop's concentration sampling gas analyzer might not be able to identify the source of the hydrocarbons with the probe placed in the tailpipe?
A. Fuel line leak near the front of the vehicle
B. Antifreeze leak near the front of the vehicle
C. Evaporative/canister purge system hoses disconnected
D. All of the above

20. A cylinder power balance test is being run on a vehicle with a V6 engine with port fuel injection while a gas analyzer is sampling the emission gases. With all cylinders enabled, HC levels are low. When the spark for cylinders 1, 2, 4, and 6 is turned off individually, HC levels increase dramatically. When the spark for cylinders 3 and 5 is turned off individually, HC levels stay low. *Technician A* says that this is evidence that cylinders 1, 2, 4, and 6 are getting too much fuel. *Technician B* says that the fuel injectors at cylinders 3 and 5 may be plugged. Who is correct?
A. *Technician A* only
B. *Technician B* only
C. Both technicians
D. Neither technician

Chapter 7

Understanding OBD II

OBJECTIVES

Upon completion and review of this chapter, you should be able to:

- ❑ Define the reasons for the OBD II program.
- ❑ Explain the major aspects of the OBD II program.
- ❑ Describe the features standardized for all manufacturers within the OBD II program.
- ❑ Understand the monitoring conditions that the OBD II PCM requires.
- ❑ Describe the conditions that will cause an OBD II PCM to set a diagnostic trouble code and turn on the malfunction indicator light.
- ❑ Explain the strategies of the diagnostic management software and the monitoring sequences required on all vehicles, both domestic and imported, by the OBD II program.
- ❑ Know how to approach an OBD II–equipped vehicle in terms of diagnostics.

KEY TERMS

Component ID (CID)
Confirmed Code
Continuous Monitor
Data Link Connector (DLC)
Drive Cycle
Efficiency Monitoring
Enable Criteria
Engine Off Natural Vacuum (EONV) Test
Freeze Frame Data
Generic OBD II Mode
Global OBD II Mode
History Code
Intrusive Test
In-Use Performance Tracking
Malfunction Indicator Light (MIL)
Mature Code
Misfire Detection
Monitor
Monitor ID (MID)
Non-Continuous Monitor
OBD II
Pending Code
Protocol
Rationality Testing
Snapshot
Test ID (TID)
Trip
VIN Entry Mode

Through the years, the federal government has been a driving force for change in the automotive industry. Federal regulations apply not only to the domestic product but also to all vehicles marketed in the United States. The second update of on-board diagnostic standards, known as **OBD II,** made foreign and domestic vehicles more similar than dissimilar. This chapter demonstrates those similarities in a broad overview of these vehicles.

CARB/EPA/SAE/OBD BACKGROUND

In the 1970s, Congress recognized that the state of California had a more serious air quality problem than the rest of the United States, and therefore granted California permission as a state to adopt its own regulations. The California Air Resources Board (CARB) is California's state agency charged with adopting regulations to achieve a healthy air quality for people living in that state. However, to avoid having 50 such independent sets of state regulations, the Environmental Protection Agency (EPA), as a branch of the federal government, was charged with achieving a healthy air quality for all people living in the United States.

Another group, the Society of Automotive Engineers (SAE), is a nonprofit organization that consists of more than 90,000 members (engineers, scientists, educators, and students from more than 97 countries) who are dedicated to sharing information and exchanging ideas for advancing the engineering of mobility systems, including designing, building, maintaining, and operating self-propelled vehicles. The EPA and CARB have worked together with SAE to set in place a number of regulations in recent years.

CARB adopted the original on-board diagnostic program (originally OBD, now known as OBD I) in 1985 for 1988 model year vehicles. The OBD I regulation was less than one page in length and required that several simple guidelines be met, many of which were already being met by domestic manufacturers. In fact, General Motors provided most of the ideas that were used to develop OBD I regulations. These regulations only required the PCM to have the ability to recognize a component's total failure and to alert the driver through a **malfunction indicator light (MIL).** There was no standardization associated with OBD I. As a result, the diagnostic connectors and diagnostic procedures varied greatly among manufacturers, making system diagnostics quite difficult for an aftermarket technician.

CARB, having realized how ineffective OBD I was, adopted OBD II regulations in 1989 for 1996 model year vehicles. Simultaneously, the EPA decided to adopt the regulations on a national basis. (Though they were initially slated for the 1994 model year, it was decided to delay compliance requirements until the 1996 model year.) With the 1996 models, many car manufacturers built their first OBD II–compliant vehicles. The OBD II program was intended to standardize the diagnostic procedures associated with emissions and driveability-related problems on all new cars sold in the United States. The required standardized self-diagnostic systems were installed on most cars in the 1996 model year. (Temporary waivers were allowed initially for some manufacturers with unique technical difficulties, but by the 1998 model year all manufacturers were required to be in compliance.)

CARB, the EPA, and SAE continue to work together to update and enhance these regulations to keep pace with new emission controls and technology.

WHY OBD II?

The OBD II standards have been the basis for driveability and emission system diagnostics for many years now and will continue to be for many years to come. These standards make diagnostic tools, trouble codes, and procedures similar, regardless of manufacturer or country of origin. The system was originally crafted to allow plenty of room for growth and the incorporation of additional subsystems. While there are some differences between vehicle manufacturers and between models, the purpose of the OBD II program is to make the diagnosis of emissions and driveability problems simple and uniform. This standardization has reduced the need for product-specific training on the various manufacturers' systems that was so critically important to technicians prior to OBD II.

The chemistry of gasoline combustion, the mechanics of a four-cycle engine, and the emissions control strategies that have proven to be successful are the same for all vehicle manufacturers. These facts, combined with federal law, have made emissions and driveability diagnosis both more successful and easier to learn.

WHAT DOES OBD II DO?

The idea behind OBD II is that any properly trained automotive service technician can effectively diagnose any engine performance, fuel economy, or emissions concern on any vehicle built according to the OBD II standard using standardized diagnostic tools, regardless of vehicle make. As a result, vehicles leave the repair shop running cleaner out the tailpipe, thus improving the overall air quality. OBD II standards have been further enhanced to provide additional manufacturer service information to the repair industry to help aftermarket technicians achieve more effective repairs.

Meanwhile, vehicle manufacturers can introduce special diagnostic tools or procedures for their own systems as long as standardized aftermarket scan tools, digital volt/ohm meters, and oscilloscopes can be used to analyze their systems. These tools can, of course, have additional capacities beyond the designated OBD II–required functions.

Besides standardizing the diagnostic end of the engine control systems, the other major goal of OBD II is to monitor the efficiency of the major emission control systems. To accomplish this, the PCM has been programmed to run specific tests from time to time to test system efficiency. A system test is known as a monitoring sequence or **monitor.** The PCM runs the monitors and determines the results. The PCM is then required to turn on the MIL and store a DTC whenever system efficiency deteriorates to the point where the emission levels reach 1.5 times the allowable standard based on the federal test procedure (FTP) cut-point for a particular vehicle.

DIAGNOSTIC MANAGEMENT SOFTWARE

An OBD II PCM includes diagnostic management software to organize the complex testing procedures. The terms used for this diagnostic management software vary by manufacturer. Ford and General Motors call theirs the “diagnostic executive,” while Chrysler calls it the “task manager.”

OBD II standards require that the engine management system be able to detect faults, turn the MIL on or off, set DTCs in memory, and run drive cycles and trips for each monitored circuit according to very specific sets of operating conditions.

Freeze Frame Data

Besides storing detected DTCs, the diagnostic management software keeps a record of all the relevant engine parameters for a given circuit. If a fault is detected and recorded, that information is stored as a **snapshot.** This data, known as **freeze frame data,** is used by the diagnostic management software for comparison and identification of similar operating conditions when they recur. This data is also available to the diagnostic technician to provide further assistance in determining what might be amiss in the system. The technician may also use the freeze frame data to help in duplicating the symptom during a road test. Freeze frame data can be accessed with a scan tool through the data stream and typically includes the following:

- The DTC involved
- Engine RPM
- Engine load
- Fuel trim (short- and long-term)
- Engine coolant temperature
- MAP and/or MAF values
- Throttle position
- Operating mode (open or closed loop)
- Vehicle speed

While there will be at least one set of freeze frame data available for the first (or higher priority) DTC set in memory, some manufacturers are providing multiple sets of freeze frame data for multiple DTCs set simultaneously (General Motors, for example).

On the basic system, freeze frame data is stored only for the DTC that occurred first, unless a later DTC is of higher priority, such as a severe misfire or fuel system DTC. In that case, the diagnostic management software replaces the stored data from the lower-priority DTC with the freeze frame data related to the misfire or fuel system DTC.

General Motors expands this capacity to include "failure records," which does the same thing as freeze frame but includes any fault stored in the computer's memory, not just those related to emissions component/system failures.

Essentially, freeze frame data is data stream data that has been frozen in time so that it can be accessed by a technician at a later time. Testing has shown that freeze frame data is typically recorded by the PCM about five seconds after the PCM records the DTC in memory. While the driving conditions recorded during freeze frame recording are most often the same as they were when the DTC was recorded (as when at a steady cruise), there is a small potential for inaccuracy if the driver suddenly hit the brakes or mashed the throttle to the floor during this five-second window. Therefore, recorded freeze frame data may not always correspond logically with the recorded DTC.

Diagnostic Trouble Codes (DTCs)

Class A DTCs. A class A code is a DTC that will result in the immediate illumination of the MIL. This type of code sets in response to a gross emission failure. For example, the misfire monitor can store a DTC and start flashing the MIL in response to its first detection of a type A misfire. (A type A misfire is classified as a severe misfire that could result in the overheating of the three-way catalytic converter, resulting in converter damage.)

Class B DTCs. Most DTCs within the engine control system are class B codes. A class B code refers to a fault that does affect the vehicle's emissions. When an emissions-related fault is detected for the first time, a DTC for that fault is stored as a **pending code**. The PCM does not illuminate the MIL at this time. Different manufacturers describe this preliminary code differently. While it is generally referred to as a "pending" code, Chrysler calls it a "maturing" code, and at least one scan tool manufacturer refers to it as a "possible" code. Whatever the name, during the next trip or drive cycle the pending fault code will be erased if the monitoring sequence that first detected the fault is repeated and the same fault does not recur. If the fault does recur on the second trip or drive cycle, the pending code is now stored in memory as a **confirmed code**, also commonly referred to as a **mature code**. It is at this point that the freeze frame data is stored, and the PCM will now illuminate the MIL. (General Motors refers to a confirmed code as a **history code**.)

Class C DTCs. A class C code is a DTC that refers to a fault that does not adversely affect the vehicle's emissions. Depending upon the vehicle, it may result in illumination of the MIL or of a separate "Service Engine Soon" light instead.

Class D DTCs. A class D code is a DTC that refers to a fault that does not adversely affect the vehicle's emissions and does not result in illumination of the MIL. These codes are the least critical of the code types.

OBD II Driving Cycles

Warm-Up Cycle. OBD II standards define a warm-up cycle as a period of vehicle operation after the engine is started in which coolant temperature rises by at least 40°F and reaches at least 160°F. Most OBD II DTCs are erased automatically after 40 warm-up cycles following the PCM turning off the MIL if the failure is not detected again (Figure 7–1).

Monitor	Monitor Type (when it completes)	Number of Malfunctions (on separate drive cycles to set DTC)	Number of Separate Consecutive Drive Cycles (to light MIL and store DLC)	Number and Type of Drive Cycles (with no malfunction to erase pending DTC)	Number and Type of Drive Cycles (with no malfunction to turn MIL off)	Number of Warm-Ups to Erase DTC (after MIL is extinguished)
Catalyst efficiency	Once per drive cycle	1	3	1	3 OBD II drive cycles	40
Misfire type A	Continuous	1	1	1	3 similar conditions	40
Misfire type B/C	Continuous	1	2	1	3 similar conditions	40
Fuel system	Continuous	1	2	1	3 similar conditions	40
Oxygen sensor	Once per trip	1	2	1 trip	3 trips	40
EGR	Once per trip	1	2	1 trip	3 trips	40
Compre-hensive component	Continuous when conditions allow	1	2	1 trip	3 trips	40

Figure 7–1 OBD II DTC/MIL function chart.

Drive Cycle. A **drive cycle** is a series of operating conditions that allows the PCM to test all of the OBD II emissions-related monitors. When all of the driving conditions, known as **enable criteria**, have been met and all of the monitors have been run, the system is said to be inspection/maintenance (I/M) ready. Drive cycle enable criteria vary among manufacturers. A pre-1998 General Motors drive cycle, shown in Figure 7–2, required 12 minutes to run the entire drive cycle. In 1998, General Motors lengthened its drive cycle to a total of 19 minutes.

A drive cycle's enable criteria may be run in any order, but as the PCM sees all of the enable criteria performed that apply to a specific monitor it will check off that monitor as being "complete." A scan tool can be used to determine which monitors have run successfully and which monitors have not yet run and are still needed to complete the full drive cycle.

When using a scan tool to identify the status of each monitor, if all of the monitor's drive cycle enable criteria have been met and the monitor has run, the scan tool will indicate that the monitor is *complete*, *ready*, or *done*. If the monitor's drive cycle enable criteria have not been fully met so that the monitor has not run completely, the scan tool will indicate that the monitor is *incomplete*, *not ready*, or *pending*. If the scan tool indicates *N/A* for a monitor, that monitor does not apply to that particular vehicle.

If the scan tool shows that the monitor has completed, this does not indicate whether the monitor has passed or failed, only that the monitor's tests have been run. If the monitor has run

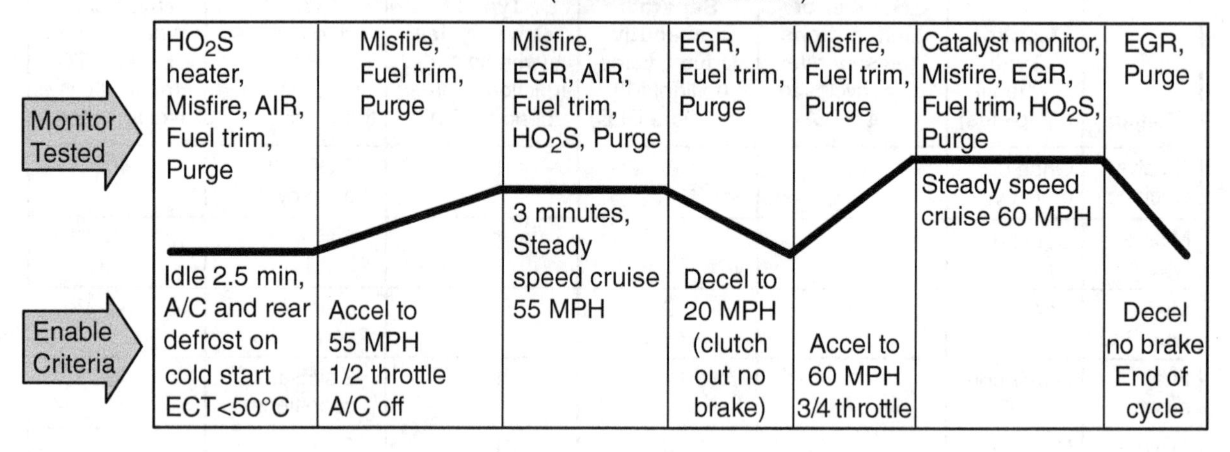

Figure 7–2 General Motors OBD II drive cycle.

and passed, there should be no stored DTCs in memory for that monitor; conversely, if the monitor has run and failed, there should be a DTC stored in memory for that monitor.

Trip

A **trip** is a diagnostic test that is designed to allow the PCM to evaluate a particular fault or DTC. It is generally run when a drive cycle has resulted in the PCM setting a pending code in memory. A trip consists of a key cycle that includes ignition on, engine run, specific enable criteria met that allow the PCM to run a diagnostic test, and ignition off long enough for the PCM to power down (Figure 7–3).

A trip is used by the PCM to confirm a pending code. If the fault is no longer present, the PCM will erase the pending code from its memory. If the fault is still present, the former pending code is turned into a confirmed code and the MIL is illuminated. A trip can also be used by the PCM to confirm a repair after a DTC has been cleared from the PCM's memory with a scan tool.

Because each DTC and its related symptoms are unique, the enable criteria for the various DTCs are also unique to each DTC that the PCM may need to evaluate. Information is readily available that allows the technician to look up the specific enable criteria for any DTC.

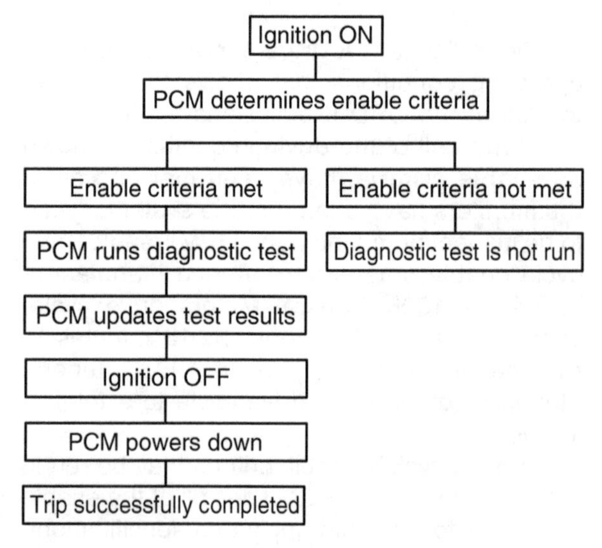

Figure 7–3 An OBD II trip.

Turning Off the MIL

Once the MIL has been turned on for a Class B fault, the vehicle must go through three consecutive trips or drive cycles that include operating conditions similar to those that existed at the time the fault was first detected before the PCM will turn the MIL off. Turning the MIL off, however, does not automatically erase the stored DTC.

Similar conditions mean:

- Engine speed within 375 RPM of the DTC flagged condition
- Engine load within 10 percent of the DTC flagged condition
- Engine temperature similar: cold, warming, or up to operating temperature

Erasing of a DTC

As stated earlier, the PCM will erase most DTCs from memory following 40 warm-up and cool-down cycles of the engine as measured by the ECT sensor without the problem recurring. Most DTCs can also be erased by a technician with the use of a scan tool.

STANDARDIZATION

OBD II standards concerning the diagnostic end of the engine control systems were designed to achieve the following:

- Common communication protocols (SAE Standard J1850)
- Common terms and acronyms (SAE Standard J1930)
- Common DLC shape and pinout (SAE Standard J1962)
- Common DLC location (SAE Standard J1962)
- Common OBD II scan tools (SAE Standard J1979)
- Common diagnostic trouble code format (SAE Standard J2012)
- Common global (generic) diagnostic test modes (SAE Standard J2190)

Common Communication Protocols (SAE Standard J1850)

In computer terminology, a **protocol** is merely an agreed-upon digital binary code, or computer language, that a computer uses for communication with other computers or with a scan tool. OBD II standards require that each manufacturer must use a standardized protocol that allows an aftermarket scan tool diagnostic access to any system that affects the vehicle's emissions. (For more information on this, see Chapter 8.)

Common Terms and Acronyms (SAE Standard J1930)

All vehicle manufacturers must use common terms and acronyms to identify components that perform similar purposes (Figure 7–4). If components perform the same function in the same manner, they must get the same name and acronym. For example, the sensor reporting crankshaft position to the computer will be called a *crankshaft position sensor* by each manufacturer, and its acronym will be a *CKP* sensor. The computer that is in charge of controlling engine performance will be described as the *powertrain control module* or *PCM*. Previously, the PCM might have been referred to as the *electronic control module* (*ECM*), *electronic control assembly* (*ECA*), or *electronic control unit* (*ECU*), depending upon the manufacturer. (The acronym *ECU* is still used generically to refer to a control module of some type without specifically referring to the PCM.) However, if a component performs a different task, or performs a similar task in a different manner, it is allowed a different name and acronym. An example of this is the *Delta Pressure Feedback EGR* sensor or *DPFE* sensor, which is unique to Ford Motor Company. Most manufacturers began using the new terms for their 1993 model year vehicles.

Old Acronyms			New Technology, All Manufacturers	
Chrysler	Ford	GM	Acronyms	Terms
SMEC/SBEC	ECA	ECM	PCM	Powertrain control module
Diag. test connector	Self-test connector	ALDL	DLC	Data link connector
DIS	DIS/EDIS	DIS/IDI/C3I	EIS	Electronic ignition system
CTS	ECT	CTS	ECT	Engine coolant temperature sensor
EVAP	CANP	CANP	EEC	Evaporative emission control solenoid
—	SPOUT	EST	IC	Ignition control
—	Dis module	C3I module	ICM	Ignition control module
CTS	ACT	MAT	IAT	Intake air temperature sensor
KS	KS	DS	KS	Knock sensor
—	PSPS	PS switch	PSP	Power steering pressure switch
PLL/Check engine light	MIL	Check engine/ Service engine	MIL	Malfunction indicator lamp
TPS	TPS	TPS	TPS	Throttle position sensor
—	BP	BARO	BARO	Barometric pressure
Brake switch	BOO	Brake switch	BOO	Brake on/off switch
Sync	CID	Sync	CMP	Camshaft position sensor
REF pickup	CPS	REF	CKP	Crankshaft position sensor
—	PFE		DPFE	Differential pres. feedback EGR
—	TFI-IV	HEI	DI	Distributor ignition
HO_2	HEGO	HO_2	HO_2S	Heated oxygen sensor

Figure 7–4 A partial list of J1930-standardized terminology and acronyms.

Common Data Link Connector Shape and Pin-Out (SAE Standard J1962)

All OBD II vehicles have a standardized **data link connector** or **DLC,** also sometimes referred to as a *diagnostic link connector,* which has a standardized shape, size, and terminal pin-out (Figure 7–5). This allows a scan tool with a standardized adaptor to be used on all OBD II vehicles, regardless of manufacturer. The DLC contains up to 11 terminals that have been standardized according to their purpose when used. The other five terminals are discretionary and may be used by the manufacturer for any purpose. There will be a minimum of four terminals used: terminal 16 is battery positive, terminals 4 and 5 are grounds, and there is at least one additional terminal for communicating with the scan tool, depending on the communication protocol and the number of communication networks that are accessible through the DLC.

The DLC is designed for use with a scan tool only. No jumper connections should be made at the DLC as was sometimes done with pre-OBD II systems. Also, because the DLC contains the scan tool's power and ground circuits, a separate power circuit is no longer required as it was for previous systems.

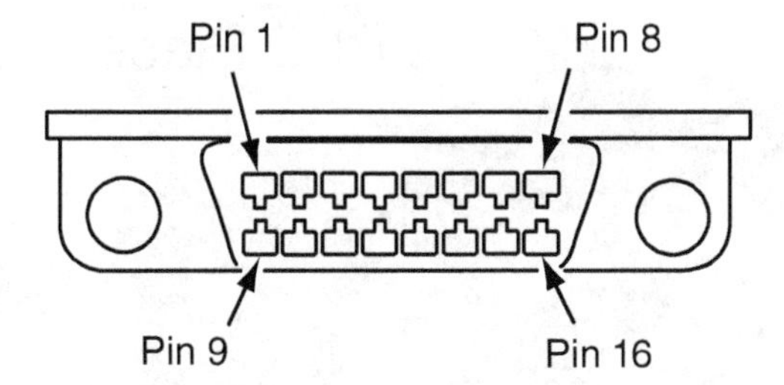

Pin 1: Manufacturer discretionary
Pin 2: J1850 bus positive
Pin 3: MS CAN H
Pin 4: Chassis ground
Pin 5: Signal ground
Pin 6: HS CAN H
Pin 7: ISO 1941-2 "K" line
Pin 8: Manufacturer discretionary

Pin 9: Manufacturer discretionary
Pin 10: J1850 bus negative
Pin 11: MS CAN L
Pin 12: Manufacturer discretionary
Pin 13: Manufacturer discretionary
Pin 14: HS CAN L
Pin 15: ISO 9141-2 "L" line
Pin 16: Battery power

Figure 7–5 Standardized OBD II diagnostic link connector (DLC).

Common DLC Location (SAE Standard J1962)

The DLC also has a standardized location. Initially, the location was required to be in a position between the left side of the interior compartment and a position 300 mm (1 ft.) to the right of the vehicle centerline (Figure 7–6). Unfortunately, this left the manufacturers a lot of latitude in locating the DLC. Some manufacturers even placed the DLC inside of, or to the rear of, the center console or even behind or underneath an ashtray. In 2003 the EPA updated this, now requiring the DLC to be located to the left of the vehicle centerline and from a position below the instrument panel to about halfway up the instrument panel (Figure 7–7). As of the 2003 model year, CARB regulations further restricted this placement, requiring that the DLC be located to the left of the vehicle centerline and underneath the instrument panel, in a position below the lower edge of the steering wheel when adjusted to its lowest position (Figure 7–8). If the vehicle is equipped with a center console, the DLC must be placed to the left of where the console meets the instrument panel. If the manufacturer chooses to place a dust cover over the DLC, the cover must be labeled and the technician must be able to remove it without tools. Furthermore, CARB requires that the DLC must be in a position accessible to a technician entering the vehicle in a crouched position.

Common OBD II Scan Tools (SAE Standard J1979)

OBD II standards mandate that generic, standardized OBD II aftermarket scan tools must be able to access and interpret emission-related DTCs and information, regardless of the vehicle make or model. Most manufacturer-specific and better-quality aftermarket scan tools can communicate with an OBD II PCM in either **VIN Entry Mode** or **Global OBD II Mode** (the latter being known by at least one aftermarket scan tool manufacturer as **Generic OBD II Mode).** These scan tools can also access an abundance of additional information regarding driveability problems, as well as information regarding other vehicle systems.

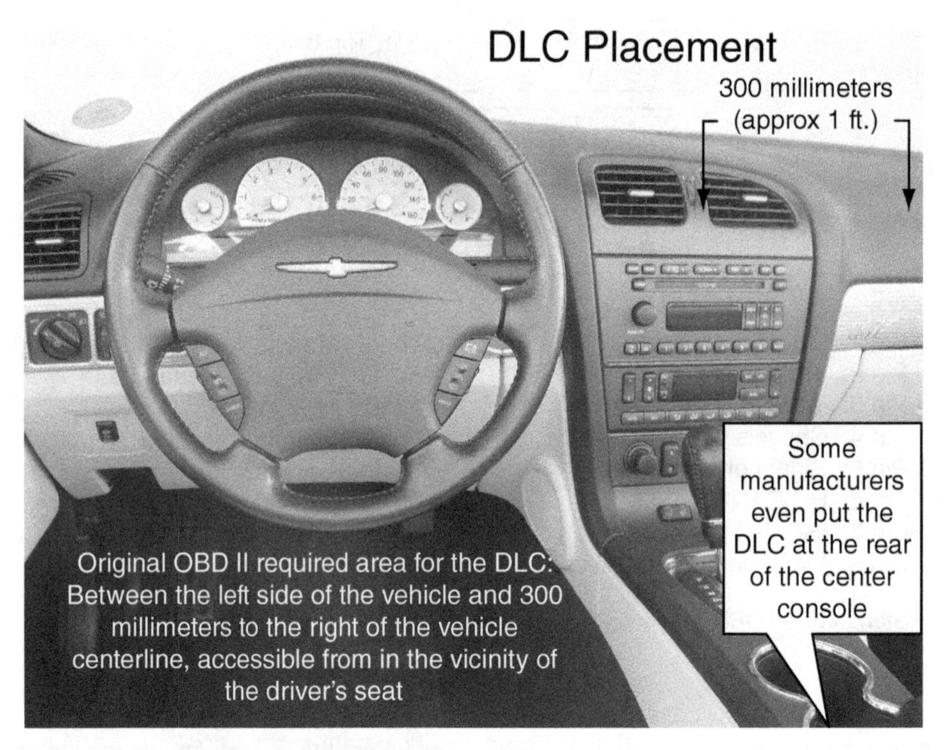

Figure 7–6 Original OBD II DLC location requirement.

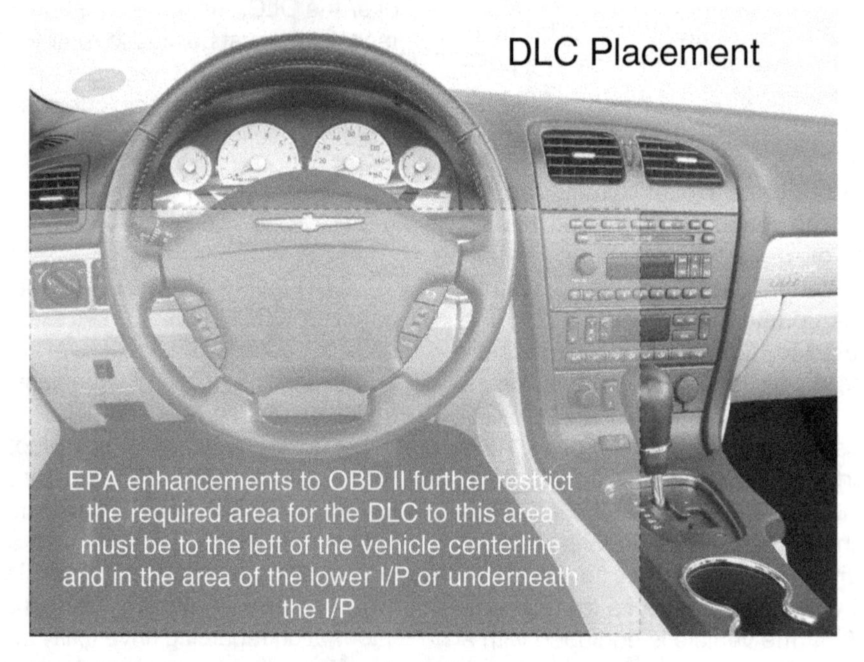

Figure 7–7 OBD II DLC location as enhanced by the EPA.

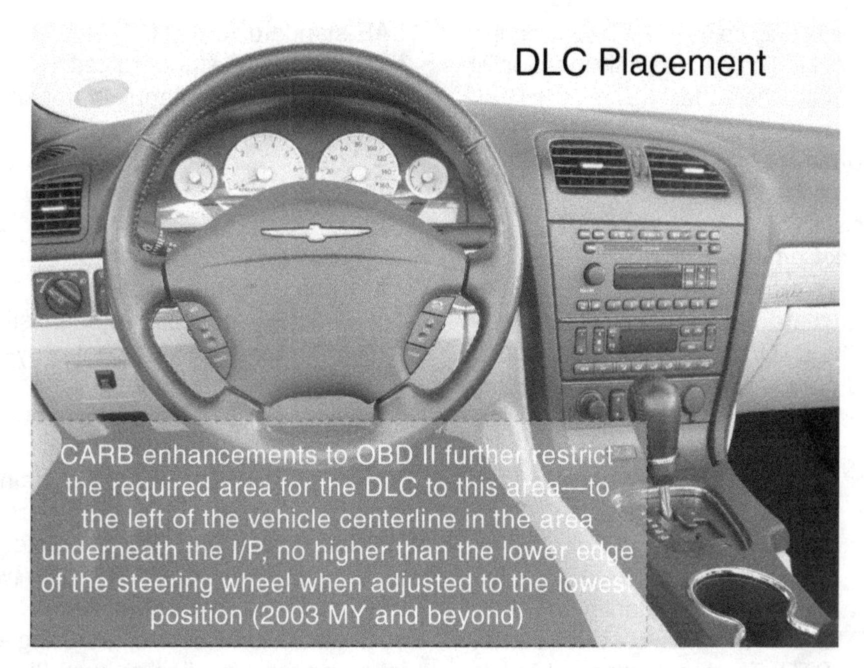

Figure 7–8 OBD II DLC location as enhanced by CARB.

A standardized OBD II aftermarket scan tool will also have a cable and/or an adaptor designed to connect to the OBD II DLC (Figure 7–9).

Depending upon the scan tool, some of the Global OBD II Modes and their related information may be displayed using hexadecimal numbers; alternatively, many scan tools will translate the related information into decimal values.

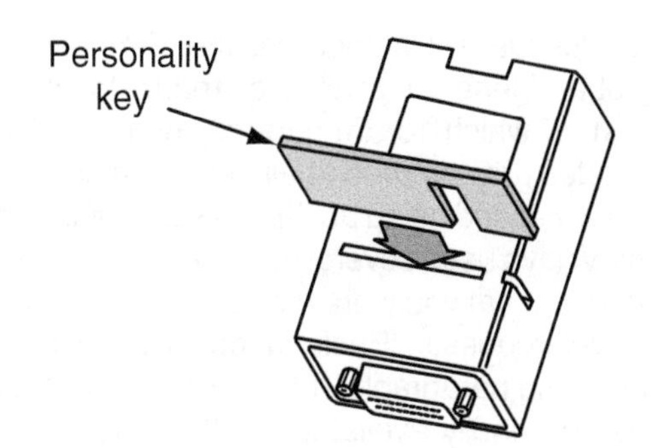

Figure 7–9 The connector of a generic scan tool fits the universal DLC. This aftermarket unit uses plug-in "personality keys" to help take advantage of each manufacturer's unique proprietary diagnostics, i.e., those that go beyond OBD II.

VIN Entry Mode. When you enable the scan tool to communicate with the on-board computers by entering the VIN information of the vehicle, or by identifying the year, make, model, and engine, this is known as VIN Entry Mode. Although most diagnosis is commonly performed in this mode, some of the OBD II information is not available in this mode. Also, 2002 model year and older European vehicles displayed little information in this mode due to the fact that their on-board computer systems were extremely proprietary in design until the EPA forced the issue at a meeting with the manufacturers in June 2001.

Common Diagnostic Trouble Codes (SAE Standard J2012)

SAE J2012 mandates a standardized diagnostic trouble code format, a five-character alphanumeric code in which each character has a specific meaning (Figure 7–10). The first character is a letter that indicates the basic function of the computer that set the code:

P = Powertrain
B = Body
C = Chassis
U = Network Communication

The second character is a number that indicates whether the DTC to follow is an SAE code (a standardized code, having the same meaning regardless of manufacturer) or one specific to the manufacturer:

0, 2, or 3 = SAE
1 or 3 = Manufacturer

Originally, a 0 was used to identify an SAE-standardized DTC and a 1 was used to identify a manufacturer DTC. Then a 2 was added to identify an SAE-standardized DTC and a 3 was added to identify a manufacturer DTC. However, because the EPA and CARB are pressuring the manufacturers to use fewer manufacturer-specific DTCs and more SAE-standardized DTCs, and because a 4 as the second digit is not considered an option, a 3 is also used to identify an SAE-standardized DTC. In fact, the majority of P3 codes are SAE-standardized DTCs, and only a few P3 codes are manufacturer-specific as originally intended.

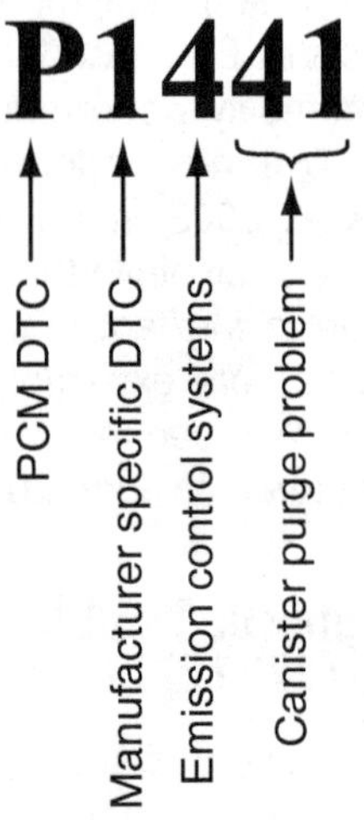

The third character of a powertrain DTC (one beginning with *P*) indicates the system subgroup:

0 = Total system
1 and 2 = Fuel-air control
3 = Ignition system/misfire
4 = Emission control systems
5 = Idle/speed control
6 = PCM and inputs/outputs
7 = Transmission
8 = Non-EEC powertrain

The fourth and fifth characters identify the specific fault area. While these two characters are most often numbers, some manufacturers, such as General Motors, are using a hexadecimal letter for the fifth digit as well (fifth digit may be 0 through 9 or A through F).

In addition, some manufacturers are using a letter value following the fifth digit to increase the number of codes available to the PCM. In fact, the Toyota hybrids use up to three letters following the fifth digit to further define the code.

P1441
PCM DTC
Manufacturer specific DTC
Emission control systems
Canister purge problem

Figure 7–10 OBD II DTC format.

Common Global (Generic) Diagnostic Test Modes (SAE Standard J2190)

Under OBD II standards, the EPA reserved 15 global (generic) scan tool requests (or test modes), of which 10 are currently defined. These test modes are common to all OBD II vehicles and can be accessed using an OBD II scan tool. In fact, you may have used several of these modes without recognizing that you were initializing global OBD II scan tool requests. Each mode is described in the following paragraphs. Many times, information that is not readily available in VIN Entry Mode is available in Global OBD II Mode.

Think of the PCM as having two rooms: a VIN Entry room for manufacturer-specific scan data and a Global OBD II room for generic scan data (Figure 7–11). Think of the Global OBD II Mode as

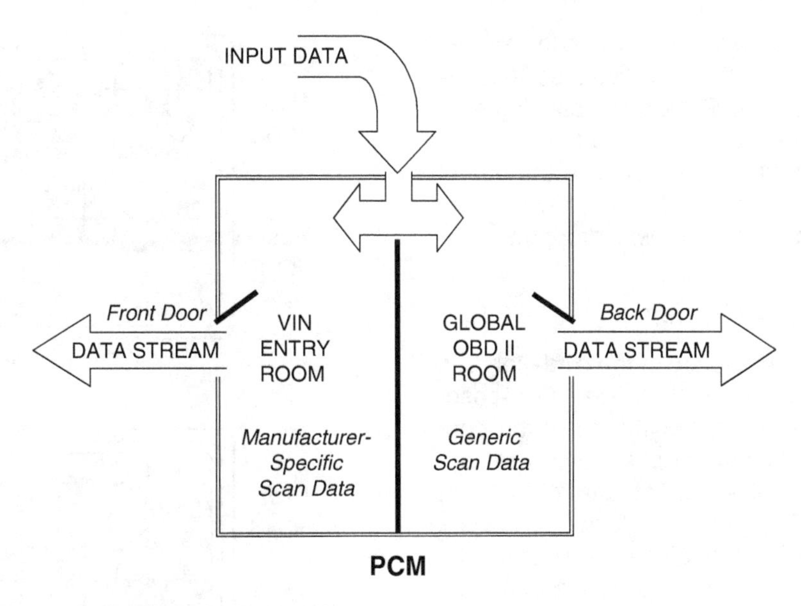

Figure 7–11 The two rooms of an OBD II PCM.

the "back door access" for data that is not always available in the VIN Entry Mode. Whenever you cannot seem to find the information you are looking for in the VIN Entry Mode, do not forget about the back door. Many times the needed information is available in Global OBD II Mode. For 2002 model year and older European vehicles, this mode was sometimes the best an aftermarket scan tool had to offer.

OBD II MONITORS

As mentioned previously, a monitoring sequence, generally called simply a *monitor,* is an operating strategy the computer uses to check the operation of a specific circuit, system, function, or component. One type of monitor is one that is monitored continuously whenever the vehicle is operating and is therefore known as a **continuous monitor.** Another type, known as a **non-continuous monitor,** requires the PCM to energize/de-energize specific output actuators, not because of engine/vehicle requirements, but specifically to test the monitor so that the resulting change in conditions can be measured. This is commonly known as an **intrusive test.** An intrusive test is a system/component test initiated by the PCM, although, in concept, it is similar to a self-test (described in Chapter 5) that may be initiated by a technician. The driver may even feel an intrusive test as the PCM performs it.

Sometimes the diagnostic management software may refuse to perform a test because of an additional problem with significant connection to the test circuit. For example, if the computer already knows the oxygen sensor is not working, it will not conduct a monitoring sequence test of the catalytic converter or the evaporative system because the results would be meaningless. That test will be postponed until the oxygen sensor circuit problem is corrected and the correction is recognized by the computer.

Chlorofluorocarbon Leakage Monitor

While OBD II standards required the monitoring of chlorofluorocarbons (CFCs), other federal laws prohibited the production of the product, so no vehicles continue to employ this type of refrigerant.

Had vehicles continued to use R-12 in their air conditioning systems, the EPA required that, as of the 1996 model year, R-12 leak detection sensors would have to have been used to sense any leaks of the refrigerant. But, because the use of R-12 was discontinued by 1996, the requirement for the leak detection monitor was dropped.

Catalyst Efficiency Monitor

To monitor the efficiency of the oxidation bed of the catalytic converter, a heated O_2 sensor was added just slightly downstream of the converter. Known as the rear, or post-cat, O_2 sensor, the PCM compares this signal to that of the front O_2 sensor to determine converter efficiency. Of course, on modern engines the post-cat O_2 sensor is not only used for monitoring converter efficiency, but also serves to allow the PCM to control exhaust gas lambda values through *rear fuel control* while operating in closed loop (see Chapter 4).

As the engine's fuel system cycles from slightly lean to slightly rich in response to the oxygen sensor signal, an effective TWC stores some oxygen during the lean cycles and uses that oxygen to oxidize (burn) the excess hydrocarbons and oxidize the carbon monoxide during rich periods. Because of this phenomenon, the percentage of oxygen in exhaust discharged from the converter tends to be nearly constant despite fluctuations in the amount entering the converter. An effective converter will even this out, but one that has begun to lose its capacity to store oxygen will not be able to do so as efficiently; therefore, its downstream oxygen signal will begin to fluctuate just as the upstream sensor does (Figure 7–12).

Catalytic converter effectiveness is tested once per OBD II drive cycle, and this test is conducted by comparing the readings of the downstream O_2 sensor (sometimes called the *catalyst monitor sensor* or *CMS*) with the signals from the upstream O_2 sensor(s). As the PCM commands the fuel mixture to be rich and lean across the stoichiometric value, the upstream oxygen sensors should produce a rising and falling voltage

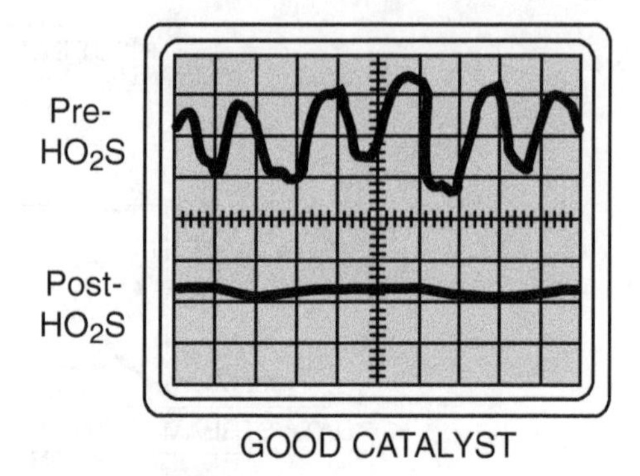

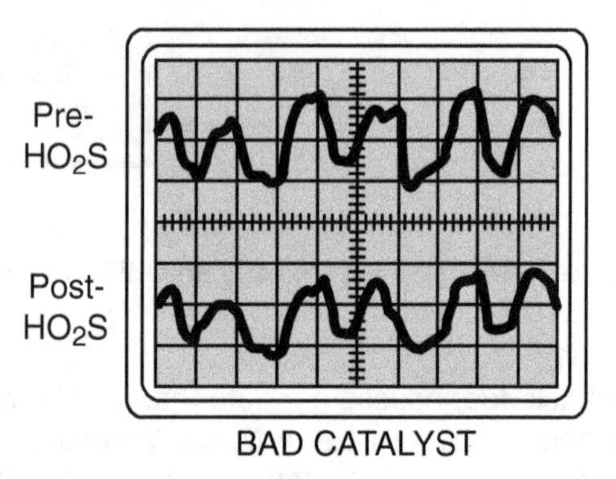

Figure 7–12 There is a difference between the voltage signals of the two oxygen sensors if the catalytic converter is operating efficiently, but the signal from the rear O_2 sensor will start to fluctuate along with the front sensor if the catalyst is deteriorating.

signal in response to the residual oxygen leaving the combustion chambers. The downstream sensor should produce a signal of much lower frequency and amplitude, as shown in Figure 7–12. To prevent mix-ups, the downstream sensor has a different harness connector than the pre-cat sensor, although operation of the two sensors is identical.

In reality, while the previous description is technically correct for proper understanding, the difference in the waveforms of the two sensors when comparing an efficient converter to one that is just inefficient enough to fail OBD II standards (resulting in a "cat efficiency" DTC) may be minuscule. It is therefore not advisable to try to use these waveforms to test converter efficiency directly. The PCM is programmed to look for particular differences in the waveforms that might be misinterpreted by a technician. Additionally, the technician is not likely to be testing the catalytic converter under the same enable criteria that the PCM is programmed to use.

Manufacturers are also using pre-cat and post-cat O_2 sensors to monitor the efficiency of NO_x reduction catalytic converters. The NO_x reduction converter uses CO to break up NO_x, thus releasing oxygen. The effect of this chemical reaction on exhaust oxygen content has been programmed into the PCM so that the PCM can determine the efficiency of the NO_x reduction bed as well as that of the CO/HC oxidation bed.

Misfire Monitor

Any time a cylinder misfires, the raw fuel and air are pumped from that cylinder into the exhaust and through the catalytic converter. With this much oxygen and fuel dumped into and burned in the catalytic converter, it gets very hot. The ceramic honeycomb may begin to melt into a solid mass, thus damaging its ability to convert toxic emission gases and plugging the exhaust with a molten lump. Emissions performance and driveability both suffer dramatically.

Misfires are therefore monitored continually by measuring the contribution of each cylinder to engine speed. This is referred to as **misfire detection.** As the engine runs, the crankshaft speed is not actually constant but changes slightly as each cylinder delivers torque during its power stroke. Between power strokes, the crankshaft actually slows down by a few RPM, only to reaccelerate with the next cylinder's power stroke. By using a high data rate crankshaft position (CKP) sensor, manufacturers can provide the computer with a means to track these slight variations in engine RPM. When it does not detect the appropriate acceleration for a given cylinder (identified by the CKP sensor, in some cases comparing the data with that from the CMP sensor), the computer knows that cylinder has misfired.

Even in engines that are running perfectly, combustion is not perfect, and there are occasional misfires. Misfires that are severe enough to raise catalytic converter temperatures to more than 1,800°F are regarded as type A misfires. Type B misfires are those that could cause emissions to increase to more than 1.5 times the FTP standard, while type C misfires could cause a vehicle to fail a state emissions test. Most OBD II PCMs allow a misfire rate of somewhere between 1 and 3 percent at idle before a type B or C misfire fault is stored, but they may allow a misfire rate as high as 40 percent at idle before a type A misfire fault is stored. However, under high load and RPM conditions, a type A misfire fault will be stored if the misfire rate exceeds 4 percent. This is because, as the load and RPM are increased, it takes a lower misfire rate to create the potential for overheating the catalytic converter.

The OBD II misfire monitoring sequence includes an adaptive feature compensating for variations in engine characteristics caused by manufacturing tolerances and component wear, as well as the adaptive capacity to allow for vibration at different engine speeds and loads. When an individual cylinder's contribution to engine speed falls below a certain threshold, however, the misfire monitoring sequence calculates the vibration, tolerance, and load factors before setting a misfire DTC. Also, on some applications, the PCM receives input from the antilock brake system's wheel speed sensors to determine when perceived cylinder misfire is simply the result of drivetrain influences due to a rough road surface.

When cylinder misfire is present, you should check to see whether two cylinders are involved. If there are, look for something shared by both cylinders, such as a common waste spark coil or exposure to the same vacuum leak.

Type A Misfire Monitoring Sequence. If, during 200 revolutions of the crankshaft, the misfire monitoring sequence, shown in Figure 7–13, detects a misfire rate that would cause the catalytic converter temperature to reach 1,800°F or above, the MIL begins to flash and the system

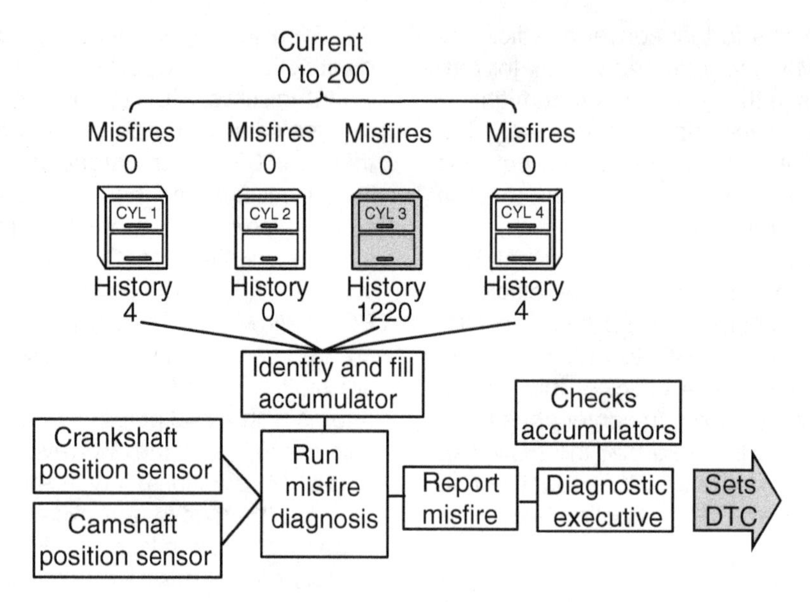

Figure 7–13 Misfire counters are similar to files kept on each cylinder. Current and historical misfire counters are maintained, and the diagnostic executive reviews this information before setting a DTC.

defaults to open loop to prevent the fuel control system from commanding a greater pulse width (because of the extra oxygen in the exhaust stream from the misfired cylinder). Once the engine is out of the operating range in which the high misfire occurs, the MIL stops flashing but stays on constantly. A DTC is immediately set. Many manufacturers whose products use sequential fuel injection have programmed the PCM to turn off one or two injectors associated with the faulty cylinders, to keep from pumping fuel into the exhaust. As a safety exception, however, if the engine is under load, as when passing or climbing a hill, the PCM does not deny fuel to the misfiring cylinder or cylinders.

Type B/C Misfire Monitoring Sequence. If, during 1,000 revolutions of the crankshaft, the misfire monitoring sequence detects a misfire rate of 1 to 3 percent, it sets a pending DTC and records all the operating conditions at the time as freeze frame data. If the same pattern is repeated on the next drive cycle, the diagnostic management software turns on the MIL.

EGR System Monitor

Different manufacturers use different methods to obtain EGR system feedback, to confirm that the system is working under engine conditions in which it is needed. One of the simplest methods is the one used by Chrysler.

In normal operation, when the EGR valve opens, the pulse width to the injectors is somewhat reduced to compensate for the oxygen displaced by the inert exhaust gas. Chrysler's strategy is to select an operating condition that meets the test criteria, including the criteria that the EGR valve should open. Then, without any other change in the pulse-width command to the injectors, the EGR valve is disabled and closed. This should cause the air-fuel ratio to go lean, since there is, in effect, more ambient oxygen in the air-fuel charge now. Monitoring the oxygen sensor signal and the resulting STFT values thus indicates the proper (or improper) function of the EGR valve.

Ford Motor Company uses a differential (Delta) pressure sensor to measure the pressure drop across a restriction placed in the EGR passage in order to know true EGR flow rates. (This is somewhat akin to a voltmeter measuring the voltage drop across a fixed resistance—the more current flow that is occurring, the more voltage drop will be measured.) The restriction may be on either the exhaust side of the EGR valve or on the intake side, depending upon year, model, and engine. Another method is to use a MAP sensor to monitor the change in intake manifold pressure as the EGR valve is suddenly opened or closed. The difference in pressure can inform the PCM as to whether the EGR valve is actually allowing flow when open and whether it is leaking when closed. Several manufacturers use a MAP sensor to evaluate EGR operation, including General Motors, Ford Motor Company/ Mazda (2.0L and 2.3L engines), and Toyota. These strategies are covered in more detail in the appropriate manufacturer-specific chapters of this textbook.

Fuel Trim Monitor

The fuel trim monitor is one of the highest-priority monitors, but it is also one of the simplest ones. Whenever the system is running in closed loop, the fuel system monitoring sequence continuously watches short-term fuel trim (STFT) and long-term fuel trim (LTFT).

If a problem such as a severe vacuum leak, plugged or leaking fuel injectors, restricted fuel filter, or incorrect fuel pressure causes the adaptive fuel control to make changes exceeding a predetermined limit in the STFT and LTFT (thus pushing the fuel trims to their threshold limits), the fuel system monitoring sequence reports a failure, and a pending DTC is set. On the next drive cycle, if the failure does not reappear, the pending code is erased; if it does appear again, a confirmed DTC is set and the MIL is turned on. How fuel trim functions did not change with OBD II; the only change is that the PCM must illuminate the MIL and store a DTC in memory if the fuel trim limits are reached—something that pre-OBD II PCMs did not do.

When the average (median) STFT is added to the LTFT, the total fuel trim (TFT) should be within a window of plus or minus 10 percent. Fuel trim values that exceed plus or minus approximately 20 percent, 25 percent, or in some cases, even 30 percent, will usually set a DTC. This is because as the fuel trim adjustment becomes too great, the PCM loses its ability to respond to the rich or lean air-fuel condition, thus allowing exhaust emissions to increase.

Oxygen Sensor Monitor

Prior to OBD II implementation, a PCM had to assume the O_2 sensors were accurate when it made an adjustment in the fuel trim values. The obvious problem with this is that if the O_2 sensor voltage was not averaging stoichiometric due to a fault with the sensor, when the PCM, not being aware of the faulty sensor, made a fuel trim adjustment, it could actually adjust a correct air-fuel ratio to an improper ratio and cause an increase in emission levels. When a zirconia O_2 sensor could not produce a high voltage to indicate a rich condition, it was said to be biased lean. In response, a pre-OBD II PCM would adjust the air-fuel ratio too rich when attempting to maintain a 450 mV average. If the O_2 sensor was contaminated with a substance such as lead, oil, antifreeze, silicone, or carbon, thus not allowing the oxygen in the exhaust to react with it, it could produce a higher voltage but not a lower voltage and was said to be biased rich. In response, a pre-OBD II PCM would adjust the air-fuel ratio too lean when attempting to maintain a 450 mV average. And if an O_2 sensor was slightly contaminated, it might not respond very quickly to the PCM's fuel trim adjustments. For example, if the mixture is lean and the sensor is reporting this as a 100 mV signal, the PCM responds to this instantly. But if the O_2 sensor voltage, while beginning to climb, is still less than 450 mV 200 ms later, the PCM will drive the air-fuel mixture even richer. Thus, a lazy O_2 sensor would cause the air-fuel ratio to vary too widely on a pre-OBD II system. Because of this, OBD II PCMs are required to be able to test and verify the accuracy of the O_2 sensors so that when a fuel trim adjustment is made it will unquestionably bring an improper air-fuel mixture back to the correct lambda value.

The O_2 monitor tests upstream and downstream oxygen sensors separately, testing each once per drive cycle. Once the diagnostic management software identifies the correct engine operating conditions, the computer pulses the injectors at a fixed pulse width to drive the O_2 sensors both rich and lean. The oxygen sensor monitoring sequence checks the frequency of the oxygen sensor signal to see that it produces a signal corresponding to the pulse width of the fuel injectors. The frequency is high enough that a slow-responding oxygen sensor cannot keep up and exhibits a reduced amplitude sign as well. To pass the oxygen sensor monitoring sequence, a zirconia O_2 sensor must generate a voltage output greater than 800 mV when the air-fuel ratio is rich, fall below 175 mV when the air-fuel ratio is lean, switch across 450 mV a minimum number of times during a 120-second period, and demonstrate a rapid voltage rise and fall, known as *response time.* An O_2 sensor's response time must show that the sensor can switch between 300 mV and 600 mV in each direction in less than 100 ms.

Because of the way the catalytic converter works, the rear oxygen sensor should normally produce a low-amplitude and a fairly low-frequency signal. During O_2 sensor testing, the computer forces the air-fuel ratio from rich to lean to force the rear O_2 sensor to produce higher amplitudes as well. If a rich air-fuel condition is momentarily sustained going into the combustion chambers, available stored oxygen in the catalytic converter is consumed, and the converter will contain less oxygen. As a result, the rear O_2 sensor's voltage should go higher. If a lean condition is then momentarily sustained, the converter gets saturated with oxygen and the rear O_2 sensor's voltage should go lower.

If the system is equipped with a pre-cat wideband air-fuel ratio sensor, or with both pre-cat and post-cat wideband air-fuel ratio sensors, the PCM's diagnostic program is designed to accurately test these sensors as well. In fact, it is good advice, when working with air-fuel ratio sensors, to let the PCM do the testing due to the small variances in the current that the PCM is programmed to look for. A wideband air-fuel ratio sensor should never be tested with a DMM or lab scope. If you want to check the operation of a wide-band air-fuel ratio sensor and the PCM's ability to respond to it, use a scan tool only. However, it is always a good idea to perform a visual check of an O_2 sensor of any type to verify that the sensor is not physically cracked or broken.

Oxygen Sensor Heater Monitor

The PCM's programmed procedures to monitor the heaters (and their controlling circuits) associated with the exhaust gas oxygen sensors (including both the primary O_2 sensors ahead of the converters and the post-cat O_2 sensors) differ among manufacturers.

Ford, for example, uses the PCM to actuate the heater circuits. In this case it is easy to monitor whether the circuit is working by monitoring the circuit voltage.

General Motors handles the testing differently. In these vehicles, the PCM feeds a bias voltage of approximately 0.45 V to the heated oxygen sensor (HO_2S) signal terminal. When the oxygen sensor reaches operating temperature and starts to generate a signal, the bias voltage is turned off and the system works normally. When the ignition is turned on, it feeds battery voltage to the front oxygen sensor heater circuit, which has a fixed ground. After a cold start the PCM measures how long it takes for the forward oxygen sensor to start generating signals. The sensor, of course, reaches operating temperature faster with the heater circuit energized. If the PCM determines from its memory that it took too long to start generating mixture feedback signals, a pending DTC is set. The amount of time allowed for the sensor to start producing signals depends on the ECT and IAT temperature signals.

On their SBEC III and SBEC IIIa systems, Chrysler used yet another strategy to test the front oxygen sensor heater circuit. The heater is powered directly by the automatic shutdown (ASD) relay, controlled by the PCM. After the ignition is shut off, the PCM uses the battery temperature sensor to sense ambient temperature. It then waits for a specific time, based on the ambient temperature, for the oxygen sensor to cool down long enough to stop generating any signal. After that time, the PCM energizes the ASD relay. If the heater brings the oxygen sensor back to operating temperature, it resumes generating a low-voltage signal, even though the engine is not running. If the sensor does not produce a signal within a predetermined amount of time, a pending DTC is set. It should be noted that Chrysler discontinued the use of this strategy with the introduction of the NGC systems.

Additionally, many manufacturers program the PCM to measure the amperage in the heater circuits. This information allows the PCM to calculate the resistance of the heater element.

Comprehensive Component Monitor

The remaining inputs and outputs affecting emissions may not be individually tested by a monitoring sequence. They are instead checked by the comprehensive component monitor (CCM). In many cases, monitoring these components is done in the same way it was on earlier systems (Figure 7–14). Inputs are checked for open circuits and shorts, and their information is compared against each other. For example, the PCM may compare the TPS value against engine load as reported by the MAP or MAF sensor. Or it may compare the ECT value against that of the IAT. This is known as **rationality testing** or **efficiency monitoring.** Another rationality test involves energizing an actuator while watching a pertinent input. That is, the PCM may use the IAC solenoid valve or stepper motor to adjust the engine's idle speed while it watches the tach reference signal for an equivalent change in frequency. Or, with the fuel pump relay commanded to be energized, it may watch for source voltage to be returned on the fuel pump monitor (FPM) circuit.

Important: Not all vehicles have these components.

Components Intended to Illuminate MIL
Automatic transmission temperature sensor
Engine coolant temperature (ECT) sensor
Evaporative emission canister purge
Evaporative emission purge vacuum switch
Idle air control (IAC) coil
Ignition control (IC) system
Ignition sensor (cam sync, Diag)
Ignition sensor Hi res (7X)
Intake air temperature (IAT) sensor
Knock sensor (KS)
Manifold absolute pressure (MAP) sensor
Mass airflow (MAF) sensor
Throttle position (TP) sensor A and B
Transmission 3/2 shift solenoid
Transmission range (TR) mode pressure switch
Transmission shift solenoid A
Transmission shift solenoid B
Transmission TCC enable solenoid
Transmission torque converter clutch (TCC) control solenoid
Transmission turbine speed sensor (HI/LO)
Transmission vehicle speed sensor (HI/LO)

Figure 7–14 By the CARB regulations, comprehensive component monitoring will illuminate the MIL if there is a failure in any of these components. Important: Not all vehicles have these components.

Digital and frequency input signals are checked by plausibility. Again, this is done by using other sensor values and calculations to determine whether a given sensor's reading is approximately what would be expected for the existing conditions. For example, the diagnostic management software compares the CKP signal to the CMP signal. Some examples of these interrelated signals are:

- crankshaft position (CKP)
- camshaft position (CMP)
- vehicle speed sensor (VSS)
- transmission output shaft speed (OSS)

Output State Monitor. The output state monitor, or OSM, is actually a part of the CCM. It tests other outputs and actuators for opens and short circuits by monitoring the voltage in the actuator's driver circuit. A circuit is added within the PCM that tests the voltage potential between the actuator and the driver (Figure 7–15). Whenever the actuator is not energized, the voltage on the ground side of the actuator, but before the driver, should be at the charging system voltage. If not, the PCM knows that there is an open in the circuit ahead of that point. If this test passes, the PCM will energize the actuator by forward biasing the driver, thus completing the circuit to ground. This pulls the voltage to near zero volts. If this test fails, the PCM knows that the driver is faulty. The following are actuators typically monitored by OBD II systems:

- Idle air control coil
- EVAP canister purge vacuum switch
- Fan control (high speed)
- Heated oxygen sensor heater (HO_2S)
- Catalytic converter monitoring oxygen sensor (CMS)
- Wide-open throttle air conditioning cutout (WAC)
- Electronic pressure control (EPC) solenoid
- Shift solenoid 1 (SS1)
- Shift solenoid 2 (SS2)
- Torque converter clutch (TCC)

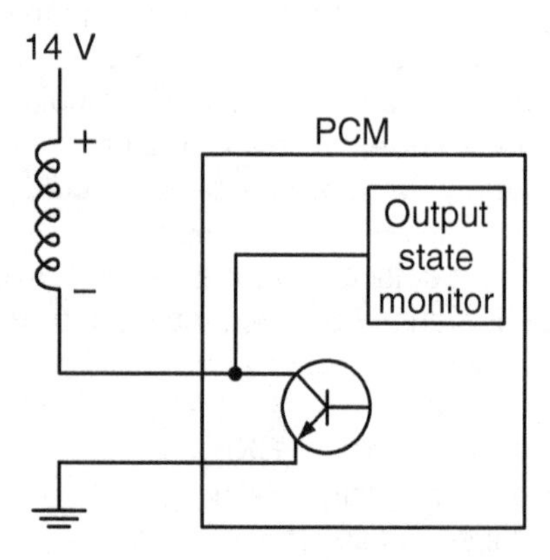

Figure 7–15 OSM voltage-sensing circuit.

Faults in the last four of these actuator circuits most often result in the PCM turning on the transmission control indicator light (TCIL) rather than the MIL, if there is such a light on the vehicle.

Canister Purge and Evaporative System Monitor

The PCM uses this monitor to test two things: whether the canister purge system can properly purge the hydrocarbons from the charcoal canister and whether the evaporative system and fuel tank can keep hydrocarbons from being released into the atmosphere.

Initially, OBD II standards required that the PCM have the capacity to test the purging potential of the canister purge system. Two methods were used. One method involved using a purge flow sensor in the purge line immediately after the purge solenoid. Ford Motor Company used an NTC thermistor that measured the cooling effect of the flow of the hydrocarbon vapors as the PCM commanded the purge solenoid to open. The other method used no additional sensors but rather allowed the PCM to wait until the oxygen sensors were active and then commanded the purge solenoid to open. The PCM would then watch the STFT values to see if the supposed purging of the charcoal canister was actually causing the air-fuel ratio to go rich.

Early on in the OBD II program, the standards were enhanced to require the PCM to monitor the evaporative system and the fuel tank for leaks that could allow unburned hydrocarbons to reach the atmosphere. Known as *enhanced OBD II*, this enhancement was introduced on some models as early as 1996 and was required to be on all vehicles sold in the United States by the 1999 model year. The enhanced system adds a PCM-controlled vent solenoid (normally open) to the atmospheric side of the charcoal canister, in addition to the normally closed canister purge solenoid in use on most systems (Figure 7–16). A pressure sensor is mounted at the fuel tank (usually part of the sender and fuel pump assembly).

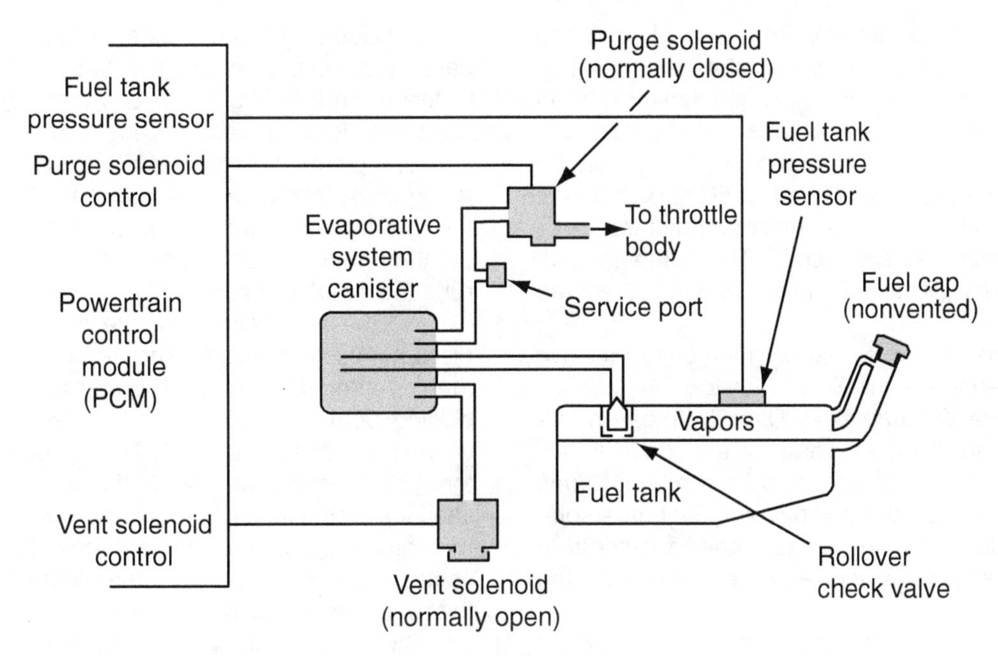

Figure 7–16 Evaporative system components.

From 1996 through model year 2000, the PCM had to be able to verify that no leaks were present that exceeded 0.040 in. Beginning in 2001, the PCM was required to identify any leaks larger than 0.020 in.

Testing the OBD II EVAP system begins when the charcoal canister is charged as the vehicle is parked. The concrete, asphalt, gravel, or dirt the vehicle is parked on has absorbed heat from the sun's rays. The warm surface of the parking pad causes heat to rise to the bottom of the fuel tank and warms the fuel inside. The close proximity of the exhaust pipe adds to this heat radiation. Fuel absorbs heat during the evaporation process through a process similar to air conditioning heat transfer—that is, when heat is transferred into a liquid, the liquid changes into a gas. The gas then contains the heat. There is only one way the expanded gas can leave the tank en route to the atmosphere—it must go through the charcoal canister and the normally open computer-controlled vent valve. Fuel vapors travel through the activated charcoal of the canister, where they are absorbed. This means that only heated air is permitted to vent to the atmosphere. This cycle completes the charging process.

Charging does not occur on extremely cold days. The computer knows the outside temperature and compensates for it. The test schedule is modified by the computer, or may not be completed at all, if the fuel tank level is either full or empty. The tank sender unit provides fuel level information to the computer. Depending upon the manufacturer, typical enable criteria include:

- Ambient temperature above 40°F to 45°F
- Fuel level between 15 and 85 percent, or on some vehicles between 26 and 74 percent

To begin the test, turn the ignition on but do not start the engine. This puts the normally open vent valve in the de-energized position. The computer should be able to sense atmospheric pressure in the fuel tank. The computer checks the fuel tank pressure sensor against the intake MAP sensor (one reason that MAP sensors have become popular again). Both should read the same atmospheric value. If the values differ, the computer must decide which sensor has an incorrect reading. It does this through comparison. The two readings are compared against the computer's memory of the last 50 engine starts. Deviations from these memory parameters cause the computer to set a code. The computer can install a default value for intake manifold pressure sensor failure.

There are three different methods that have been used to run the EVAP monitor.

Active Vacuum Test. Used by most manufacturers in the early years of enhanced EVAP systems, the PCM performs both a *gross leak test* and a *small leak test* by using the purge solenoid to actively apply intake manifold vacuum to the evaporative system and fuel tank when the engine is running.

The computer energizes the normally closed EVAP solenoid when the engine reaches normal operating temperature. The opened EVAP solenoid allows engine vacuum to pull fuel vapors from the charcoal. The computer uses the oxygen sensor to monitor the exhaust. *Cross-counts* are the number of times the oxygen sensor switches from a rich air-fuel mixture to a lean air-fuel mixture within a specific period of time. The cross-counts should indicate a sudden richness in the air-fuel mixture. If this richness is not indicated, the computer knows that either the charcoal canister was not charged properly or the charcoal lost its charge. Charcoal charge may be lost when the fuel tank cap is left off. It may also be lost if there is a leak in one of the evaporative system hoses. The computer only runs further tests if it senses that the charcoal canister was not sufficiently charged.

The next test has a relatively short cycle. It is called a *gross* or *large leak test.* The computer energizes the normally open vent solenoid valve to close it. Then it simply energizes the normally closed EVAP purge solenoid momentarily while watching the fuel tank pressure sensor to determine if it can make a change in pressure inside the fuel tank. (If the computer detects engine vacuum inside the fuel tank, it must act rapidly because if this vacuum draw continues for too long, it will cause the fuel tank to collapse. Vacuum is limited to less than 2 in. Hg or 1 PSI.) If the computer does not detect a vacuum, it suspects a gross leak or a pinched purge hose leading to the tank. If the computer detects a vacuum, it assumes the charcoal canister charge must have leaked out through a very small hole. The computer then runs the *small leak test*.

For the *small leak test*, the computer energizes both solenoids as in the gross leak test and then de-energizes the EVAP purge solenoid, thus closing both valves. This traps intake manifold vacuum inside the fuel tank. The computer monitors how long the tank holds this vacuum. If the vacuum does not hold over a predetermined time, the computer has discovered a small leak. If the vacuum does hold, the computer determines that the system is operating properly.

The most common problem found with this system is caused by the owner of the vehicle, who might fail to tighten the fuel filler cap properly. Although the computer does not detect this at the time of refueling, when the vehicle is started the next morning, the computer discovers a leak, turns on the check engine light in the dashboard instrument cluster, and stores a code. The owner then returns the vehicle to the dealership for service. Many modern vehicles now have a *gas cap loose* light on the instrument panel. If this light illuminates, the owner's manual instructs the owner to verify that the gas cap is tight. If the *gas cap loose* light is ignored, expect the MIL to illuminate within two or three days.

Leak Detection Pump. Used by Chrysler and some European manufacturers in the early years of enhanced EVAP systems, the PCM energizes an electric pump known as a *leak detection pump* (LDP) in order to pressurize the evaporative system and fuel tank to test for leaks. This method was discontinued after the 2001 model year.

Engine Off Natural Vacuum Test. An **engine off natural vacuum (EONV) test** replaces the active vacuum test and requires no hardware changes, with only minor electrical changes and a reprogramming of the PCM. The EONV test may also be referred to as a *natural vacuum leak detection (NVLD)* test. This test uses a natural vacuum to test the evaporative system and fuel tank for leaks. The natural vacuum develops within the fuel tank and evaporative system as the fuel in the tank cools down after a drive cycle.

During a drive cycle, the engine compartment and exhaust system will transfer some heat to the fuel in the fuel tank, causing the fuel to expand. To perform the EONV test, the PCM is kept alive for up to 40 minutes after the driver shuts off the ignition. This is done through a timer IC chip contained in the relay used to power the PCM (Figure 7–17). During the 40 minutes, the PCM keeps the normally open EVAP vent solenoid energized, thus keeping it closed. Since the EVAP purge solenoid is normally closed, the result is that both solenoid-operated valves remain closed, during which time the fuel in the tank cools and contracts, thus causing a slight natural vacuum of about 3 to 5 inches of water to be drawn on the fuel tank and evaporative system. The PCM monitors the fuel tank pressure sensor to determine whether the system is capable of passing the leak test.

The EONV test was introduced by Chrysler Corporation on their 2002 model year vehicles. General Motors implemented the EONV test on their light trucks in the 2003 model year and had the system in place on most of their vehicles by the 2005 model year. Ford Motor Company introduced the EONV test on their vehicles in the 2005 model year. Today, the EONV test has largely replaced the former EVAP system tests on most vehicles being produced.

Secondary Air Pump (SAP) Monitor

Manufacturers are permitted to use a variety of methods to monitor the secondary air injection system, provided they detect increases of 1.5 times the FTP emissions cut-point. With modern electric air pumps there is no longer an atmospheric dump mode as the electric air pump can simply be turned off. Neither is there any longer a downstream airflow routing due to the fact that modern PCMs use rear fuel control to control the exhaust lambda values to within 2 percent of lambda 1. With a rear fuel control system, ambient air cannot be permitted to get into the

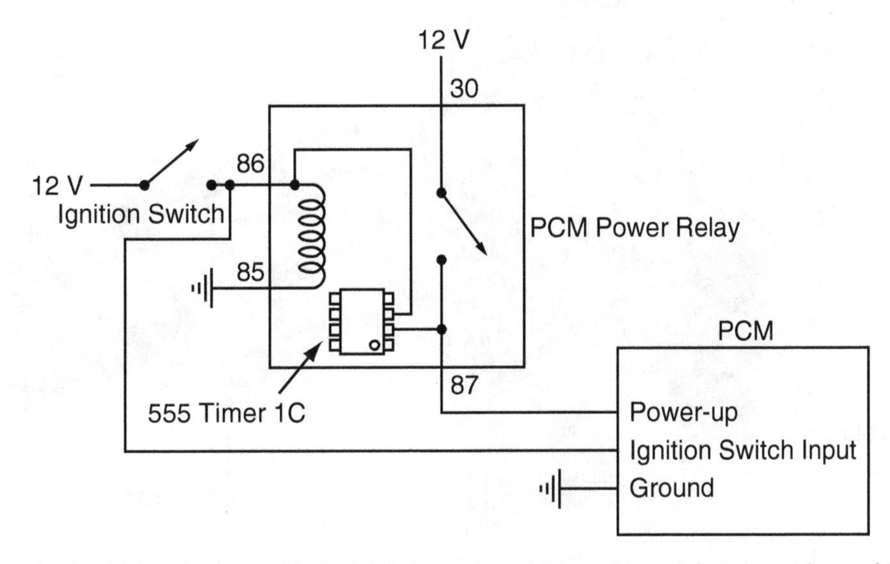

Figure 7–17 A relay with a timer IC chip built into it, designed to provide a delayed shutdown of the device that it powers.

exhaust ahead of the rear O_2 sensor during closed-loop fuel control. Additionally, because the oxidation of HC and CO emissions does not occur efficiently in the exhaust unless the temperature is high enough, a modern secondary air injection system is now required to inject ambient air to within 20 to 25 millimeters of the exhaust valve seat in order to achieve maximum efficiency. On older systems that pumped air downstream to the catalytic converter during closed-loop fuel control, the additional oxidation created by the injected air was dependent upon the heat produced by the converter. Thus, this downstream oxidation was less efficient than injecting air near the exhaust valve seats. Because, on a modern vehicle, most CO is produced during cold engine operation, in normal operation the PCM will energize the pump during engine warmup only and then turn it off once the PCM is ready to enter closed-loop control of the air-fuel ratio.

One method of monitoring the AIR system for proper operation is for the PCM to wait until the system is operating in closed-loop fuel control and then simply cycle the pump on and off while monitoring the O_2 sensors and STFT.

The secondary air system monitor was further enhanced beginning in 2006 due to the possibility that some system faults might go undetected. For example, if a solenoid were to malfunction only when the engine was cold, waiting until the engine was at normal operating temperature and in closed-loop fuel control to test the system would likely not be able to detect the fault. Because of this, some additional sensors have been added. One method to test the system's operation when the engine is cold is to use a secondary mass air flow (SMAF) sensor (Figure 7–18). The PCM uses the SMAF sensor to monitor the airflow rate whenever the pump is energized. An added pressure sensor or pressure switch may also be used.

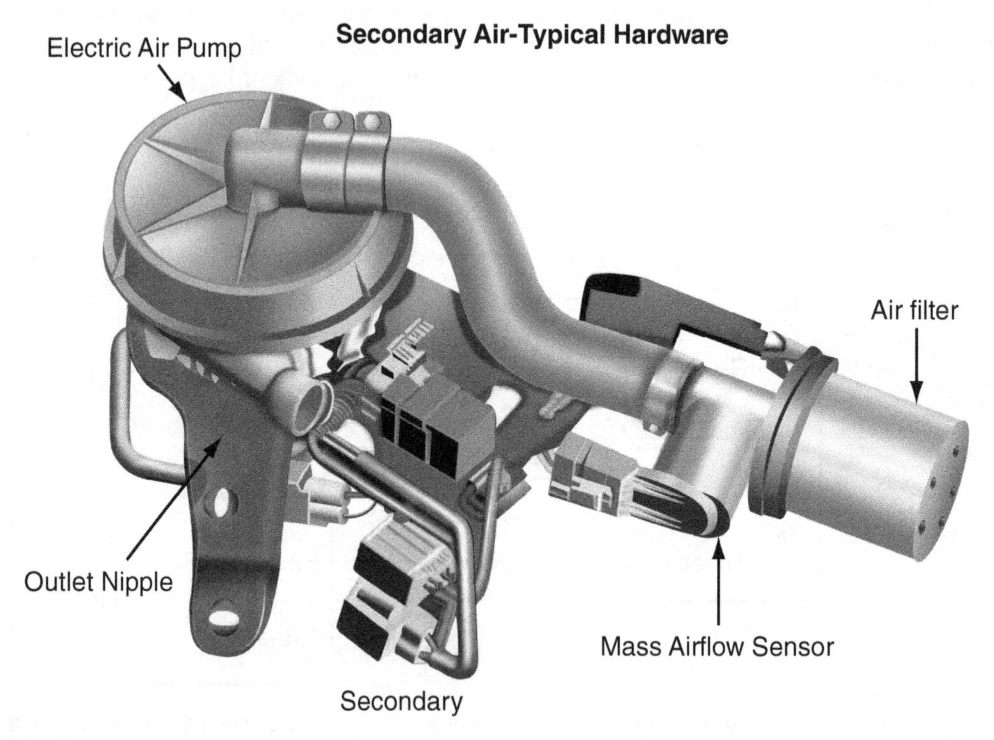

Figure 7–18 A modern secondary air injection system using a secondary mass airflow (SMAF) sensor.

And a wideband oxygen sensor can be used to sense the ambient air that is pushed into the exhaust system, even before the engine enters closed-loop fuel control.

Cylinder Imbalance Monitor

The *cylinder imbalance monitor* is the newest monitor, just having been added in the 2008 model year. The EPA required this feature to be on 25 percent of U.S. vehicles by the 2011 model year, 50 percent by the 2012 model year, 75 percent by the 2013 model year, and 100 percent by the 2014 model year.

Think of the cylinder imbalance monitor as a more rigidly defined misfire monitor. The misfire monitor is designed to allow the PCM to determine total misfire only, whether intermittent or not, occurring within a cylinder. The cylinder imbalance monitor is designed to detect how much each cylinder is contributing to engine performance. There are several monitoring methods being used.

BMW incorporates a strategy that allows the PCM to use the CKP sensor to more acutely determine each cylinder's contribution to the rotation of the crankshaft than the misfire monitor does. General Motors and Ford Motor Company use a strategy that allows the PCM to look at O_2 sensor "noise." If a misfire occurs, neither the fuel nor the oxygen in the cylinder is consumed. The unused puff of oxygen can be read by the O_2 sensor as a momentary lean pulse and occurs much more often than the normal cross-counts (Figure 7–19). If, instead of a total misfire, the cylinder's entire air-fuel charge is not completely consumed, the cylinder's contribution to the rotation of the crankshaft is reduced. In this case, the portion of the oxygen not consumed in the cylinder's combustion can be read by monitoring an O_2 sensor. To accomplish this, however, the physical location of the O_2 sensor, as it is relative to each cylinder, must be reengineered so that all cylinders can be equally monitored. In some cases, additional upstream O_2 sensors are added to do this, as in an inline four-cylinder engine with dual upstream O_2 sensors (bank 1 and bank 2). Ford PCMs look at the frequency of the O_2 sensor's waveform to determine cylinder imbalance. GM PCMs look at the *locus length* of the O_2 sensor's waveform. *Locus length* is the length of the waveform over a predetermined period of time. Imagine that the O_2 sensor waveforms in Figure 7–19 were pieces of string—the one that shows not all of the oxygen in the cylinder is being consumed would be longer over the same time period.

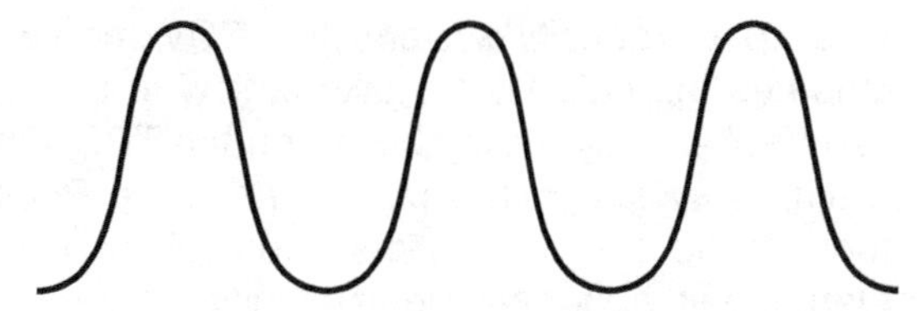

Normal oxygen sensor waveform with the engine operating in closed-loop control

Oxygen sensor waveform with the engine operating in closed-loop control, but the sensor waveform shows indication of extra oxygen pulses in the exhaust due to either cylinder misfire or inefficient cylinder operation, resulting in a cylinder imbalance problem

Figure 7–19 Oxygen sensor waveforms.

PCV Monitor

The PCV monitor requires the PCM to identify a large vacuum leak, as might occur with a disconnected PCV valve and/or hose to the PCV valve. It is used on those engines that use a PCV valve that is external of the valve cover.

On a speed density system, the PCV monitor compares the calculated intake airflow to calculated throttle body airflow based on the TPS. On an airflow system with a MAF sensor, the PCM monitors the oxygen sensors and fuel trims to determine an abnormally lean condition.

NO_x Catalyst Monitor

The NO_x catalyst monitor is an enhanced catalytic converter efficiency monitor that requires the PCM to use enhanced algorithms to determine the efficiency of the NO_x reduction bed of the converter as well as the CO/HC oxidation bed. On LEV-II vehicles the PCM was required to determine when NOx emissions had exceeded 3.5 times the FTP limit for the 2005 and 2006 model years, and 1.75 times the FTP limit for the 2007 model year. On SULEV applications the PCM must determine when NOx emissions exceed 2.5 times the FTP standard. (As with the initial catalytic converter efficiency monitor, the hydrocarbon threshold remains at 1.5 times the FTP standard.)

Thermostat Monitor

Also known as the *cooling system* monitor, the thermostat monitor requires the PCM to determine when engine warm-up time becomes excessive. The PCM will compare actually warm-up time to preprogrammed values to determine when the cooling system's thermostat is malfunctioning.

Variable Cam Timing/Variable Valve Timing Monitor

The variable cam timing (VCT) and variable valve timing (VVT) monitor requires the PCM to vary the timing of the camshaft while monitoring the respective camshaft position (CMP) sensors for feedback. If the PCM finds that it cannot properly control the position of a camshaft, it will set a DTC in memory.

THE TEN GLOBAL MODES OF OBD II

Mode $01—Data Stream and Monitor Readiness Status

Mode $01 is a scan tool request to access both the PCM's data stream and the readiness status of the OBD II monitors. At least one aftermarket scan tool, the OTC Genisys, divides this global mode into two separate menu items: *Mode $01 Data Stream* and *Mode $01 Readiness Status*.

The data stream allows access to specified data, each identified by a *parameter identification* or PID, as well as the associated values for each PID. PIDs include inputs, outputs, and other data in the PCM's memory. Input PIDs include the TPS, ECT sensor, IAT sensor, MAP or MAF sensor, O_2 sensor, and VSS. The associated values for these PIDs may include both the raw voltage or frequency values for the sensor and the interpreted values. For example, if the data stream PID is identified as the ECT sensor, the associated values may include both the voltage that the PCM sees returned from the sensor (example: 0.62 V) and the interpreted value (example: 215°F). Output PIDs include commanded spark timing, commanded fuel injector pulse width, and commands to emission system actuators. The data associated with these PIDs may show commanded spark timing in degrees of crankshaft rotation, fuel injector pulse width in ms of injector on-time, and information about other actuators concerning whether they are commanded to be energized or de-energized by the PCM. PIDs identifying information contained in the PCM's memory would include short-term fuel trim (STFT) and long-term fuel trim (LTFT) as well as their relative values.

References to PID values exist throughout the automotive service literature. Some of the PID references are from a generic OBD II PID list that all scan tools must be able to reference. Other PID references are for PIDs that can be accessed in the VIN Entry Mode. Generally, a greater number of PIDs and their respective values can be accessed from VIN Entry Mode than from the global mode, but the global mode ensures that certain PIDs can be accessed, regardless of the vehicle manufacturer, when using an OBD II aftermarket scan tool.

This mode also allows an OBD II aftermarket scan tool to check the monitor readiness status of all of the applicable OBD II monitors. Depending upon the scan tool, the results may be displayed as "complete," "done," or "ready" for monitors that have successfully run and "incomplete," "pending," or "not ready" for monitors that have not yet run since the last time codes were erased or the battery was disconnected. If a monitor is identified as "complete," this does not indicate whether the monitor has passed or failed—look for DTCs in the PCM's memory to determine pass or fail. Ford EEC-V PCMs are unique in that they set a DTC P1000 in continuous memory whenever there is at least one monitor that has not run.

Mode $02—Freeze Frame Data Access

Mode $02 is a scan tool request to access emission-related *freeze frame data* values from specific generic PIDs. As stated earlier, these values represent the operating conditions at the time the fault was recognized and logged into memory as a DTC.

When a scan tool is used to erase a DTC, it automatically erases all freeze frame data associated with that DTC event. Therefore, make sure to record the freeze frame data when you record the DTCs and before you erase DTCs.

Mode $03—Confirmed Fault Codes

Mode $03 is a scan tool request to obtain confirmed DTCs stored in memory (also commonly known as *mature* DTCs or by General Motors as *history* DTCs). Generally, the scan tool will display both the DTC and its descriptive text. The specific menus and access techniques to obtain emissions-related DTCs are left up to the scan tool manufacturer, but such data should be relatively simple to extract. Most aftermarket scan tools can access and interpret the manufacturer-specific codes as well as the SAE standardized codes. Complete lists of SAE and manufacturers' DTCs are available in both manufacturer and aftermarket service manuals.

Mode $04—Erase DTCs and Clear Diagnostic History

Mode $04 is a scan tool request to erase all DTCs and to clear the diagnostic history. It is actually a PCM reset mode that allows the scan tool to clear all emissions-related diagnostic information from its memory, including DTCs, freeze frame data, and monitor readiness status. Once the PCM has been reset, the PCM may store an inspection maintenance readiness code (for example, Ford's P1000) until all the OBD II system monitors have successfully run again. Be sure to record all DTCs, freeze frame data, and monitor readiness status before performing a mode $04 request.

Mode $05—$O_2$ Sensor Monitor Test Results

Mode $05 is a scan tool request for monitor test results concerning only the oxygen sensors. Some of the tests performed include *minimum oxygen sensor voltage, maximum oxygen sensor voltage, rich-to-lean oxygen sensor switch time,* and *lean-to-rich oxygen sensor switch time.* The available information includes each test's fault limits as well as the results of each test. The PCM's ability to test all of the oxygen sensors is critical because the PCM uses the oxygen sensors to determine the test results for many of the other monitors. You should be aware that the oxygen sensor data described under mode $05 can also be found under mode $06 many times along with data from the other monitors.

Mode $06—On-Board Monitor Test Results

Mode $06 is a scan tool request to obtain the test results of noncontinuous monitors. CAUTION: The monitor drive cycle must be complete BEFORE you try to use mode $06 data, or the data will have no value.

Mode $06 data includes information concerning the monitors associated with the oxygen sensors, the oxygen sensor heaters, the catalytic converters, the EGR system, and the evaporative emission system. It can also include information about the engine misfire monitor. A technician can use mode $06 to accurately interpret information as the computer interprets it. This mode can help the technician to verify a failure or repair, to understand why the PCM may have blocked a monitor or test due to the failure of a sensor needed for the test, and to identify monitors that may be just within the PCM's set programmed limits but have the potential to fail soon.

Most manufacturers supply mode $06 charts that provide information to allow the technician to interpret mode $06 data. Under the CAN protocol (see Chapter 8) a specific monitor is identified as a **monitor ID (MID)** and a specific test is identified as a **test ID (TID).** For systems that predate the use of the CAN protocol, a specific monitor is identified as a **test ID (TID)** and a specific test is identified as a **component ID (CID).** The newer MID/TID terminology more accurately defines the type of information that is available than the TID/CID terminology did.

The MID and TID values, or the TID and CID values, are hexadecimal values, but the manufacturers' charts provide information about what monitors and tests these values relate to. These charts also show whether the MID/TID or TID/CID refer to a test that uses a minimum failure value or a test that uses a maximum failure value. The charts then show what that min/max value is and what range of values the PCM operates within.

Many manufacturers' mode $06 charts have been available on the manufacturers' websites in the past. Some were available for free and some required a small subscription fee. However, legislation enacted in June 2009 now requires that all manufacturers provide this information free of charge. The National Automotive Service Task Force website, http://www.nastf.org, has links to most manufacturers' websites that provide mode $06 information. Some manufacturers' charts are also available on the International Automotive Technicians Network (IATN) website (http://www.iatn.net). Still others are now found in aftermarket electronic service manuals such as *Mitchell On-Demand* and *Pro-Demand*, and *AllData*. In Mitchell, click on *Engine Performance* and then look for a folder in the submenu labeled *Mode $06*. In AllData, select *Powertrain Management*, then *Computers and Control Systems*, then *Mode $06 Data*.

If a scan tool is used to enter global mode $06 directly and without taking the time to identify the vehicle being worked on, the MID/TID or TID/CID values will usually not display a description for what they represent. If the scan tool does not display the monitor/test description, then the manufacturer mode $06 charts must be used to determine what monitor/test the MID/TID or TID/CID values represent. But if the technician begins by identifying the vehicle being worked on for the scan tool and before entering mode $06, the scan tool is more likely to give a text description of the MID/TID or TID/CID values.

The scan tool will also display whether the test is a min-type failure or a max-type failure, what the actual measured value is, and whether the test passes or fails. The actual measured value identified on the scan tool is an average of any drive cycles run for the test since the last time codes were erased or the battery was disconnected. For pre–CAN systems, be careful to observe any footnotes in the manufacturers' charts that may contain information about a multiplier (or factor) that should be applied to the scan tool values to get relevant information.

If the scan tool identifies a *minimum* failure value for a test, it means that, as the system degrades, the actual measured value *decreases.* Likewise, if the scan tool identifies a *maximum* failure value for a test, it means that, as the system degrades, the actual measured value *increases.* In some cases the scan tool will identify both *minimum* and *maximum* values for the same test. Because both cannot apply to the same test, in this case it is important to use the manufacturer's mode $06 charts to determine whether the test is a min-type failure or a max-type failure.

Generally speaking, if the mode $06 data identifies a failure, there should also be a relevant DTC set in memory. However, if a mode $06 failure is identified and there are no relevant DTCs, check the manufacturer mode $06 charts to determine if the MID/TID or TID/CID values are actually applicable to a specific vehicle. Many times a scan tool will show a failure for tests that do not apply to the vehicle that it is connected to.

If the mode $06 values shown on a scan tool for a particular MID/TID or TID/CID are binary column values (see Chapter 2) such as *64*, *128*, *256*, *512*, *1024*, and so on, then the monitor has not likely been run. Drive the vehicle in the appropriate drive cycle so as to get the monitor to run, then recheck the mode $06 values.

If your scan tool displays any of these values in the hexadecimal format, use a hex-to-decimal calculator such as the one available with Microsoft's PC Windows operating systems.

The unfortunate thing about mode $06 on many vehicles is that it can be quite challenging for an entry-level technician to use and interpret the results. The best advice is to begin to use mode $06 and to practice interpreting the results. In many situations, the technician can make a simple comparison of mode $06 data in order to determine a failure. For example, if using mode $06 data to determine which cylinder is the one responsible for the engine's misfire, a quick comparison of mode $06 data among all cylinders will easily identify the cylinder responsible for the misfire.

Mode $07—Pending Fault Codes

Mode $07 allows a technician to use a scan tool to request pending DTCs. Pending DTCs are those codes that have only been identified by the PCM during one drive cycle and have not yet been identified as confirmed codes. Pending DTCs do not result in the illumination of the MIL.

Mode $08—Active Command Mode

Mode $08 allows a technician to use a scan tool to activate and deactivate the system's actuators on command. This is sometimes referred to as *active commands* or *bidirectional controls*. With only one EPA-defined test so far, most mode $08 tests are left to the manufacturer's discretion. As a result, most manufacturers allow a generic OBD II scan tool access to relatively few of these active commands as compared to a dealership scan tool.

As an active command mode, this mode allows the technician to quickly narrow down a problem to either the input side or the output side of the computer by commanding and testing the computer's ability to activate the output on command. For example, if the output actuator does not energize properly during normal driving, but does respond properly during the active command mode, then the technician can rule out the computer and the output as a possible fault, thus narrowing the problem down to the input side of the computer (either a sensor or the sensor's circuitry).

Another function of this mode is that it can make a great component locator. It allows the technician to locate and identify a particular actuator, providing the computer can energize it successfully when the command is given. For example, if the cooling fan relay is selected for activation from the scan tool's menu and then one of several relays can be heard (or felt) to be clicking, the technician can easily identify exactly which relay is responding to the computer.

Currently, the EPA has defined one mode $08 test. It is Test ID (TID) $01, which mandates that an aftermarket scan tool be able to energize the evaporative system's normally open vent solenoid in order to assist in performing a leak test of this system with a smoke machine.

Mode $09—Vehicle-Related Information

Mode $09 is designed to allow a technician to request information about the specific vehicle.

This information can include such items as the vehicle identification number (VIN), a module calibration identifier (CALID), and a calibration verification number (CVN).

In-Use Performance Tracking. Beginning in 2005, CARB required an added ability of global mode $09 that records how often the conditions are present for a monitor to run and how many times that monitor has actually run during real-world driving conditions. **In-use performance tracking** counters are required for the following monitors: catalyst, EGR, O_2 sensors, evaporative system, and secondary air system.

Mode $0A—Permanent Fault Codes

The EPA and CARB have now defined the tenth OBD II global mode, mode $0A. This mode is designed to allow a technician to identify a fault code that might have been erased previously by a scan tool or through the disconnection of the vehicle's battery. The EPA and CARB have mandated that mode $0A was to begin phase-in in the 2010 model year and was mandated to be on 100 percent of vehicles sold in the United States by the 2012 model year.

When a pending code in mode $07 is identified by the PCM a second time under similar driving conditions, it is then stored as a confirmed code in mode $03, freeze frame data is stored in mode $02, and the MIL is illuminated. At this time, the same fault code is set in mode $0A as well, using the same *Pxxxx* (or *Uxxxx* or *Cxxxx*) format as modes $07 and $03 use. But a DTC set in mode $0A is stored in a nonvolatile RAM (NVRAM) memory IC chip and can only be erased by the PCM. It cannot be erased by a scan tool command or by the disconnection of the vehicle's battery.

If the PCM detects that the fault is no longer present three times under similar conditions during further trips, the MIL is turned off, but the mode $03 DTC (confirmed code) and the mode $0A DTC (permanent code) remain in memory. The PCM will erase both the mode $03 DTC and the mode $0A DTC following 40 warm-up and cool-down cycles of the engine with the system sensed as being clean.*

But if a scan tool is used to erase the mode $03 confirmed code following a repair, or if the battery has been disconnected long enough to erase this code, while the PCM will turn off the MIL and the monitor readiness status will be set to *not ready/incomplete,* the mode $0A permanent code is retained in memory. Therefore, a scan tool can still be used to retrieve the fault code through a mode $0A request, even though the MIL has been turned off.

Following the erasure of the mode $03 confirmed code with a scan tool (or a battery disconnect), the PCM will then run a trip designed to identify whether the mode $0A fault is still present. If it finds the fault is still present, the PCM will store the mode $03 confirmed code again and will again illuminate the MIL.

In summary, if a previous technician has unsuccessfully attempted a repair followed by the use of a scan tool to erase both the mature codes and the pending codes, and the customer now brings the vehicle to your shop complaining that the symptom is still present, the permanent codes still recorded in mode $0A provide you with a method to look back at past recorded faults, even if the MIL is turned off. And if a vehicle owner has disconnected the vehicle's battery in an attempt to pass an emission test with a fault present, mode $0A gives the technician a way to retrieve the fault after the fact and to define the fault without having to run a drive cycle in order to test the monitor.

COLD START EMISSIONS REDUCTION

During warm-up from a cold start, an engine will normally emit a much larger percentage of toxic emissions than it does once it is warm and

Source: Mike McCarthy, California Air Resources Board (CARB).

in closed loop. To reduce cold start emissions, a couple of different strategies are currently used.

Ignition Timing Retard

Beginning in the late 1990s, manufacturers began programming their PCMs to retard ignition timing during engine warmup from a cold start. By beginning combustion slightly later, more waste heat is generated, thereby increasing the temperature of the engine more quickly. The amount of spark timing retard is from as little as 2 degrees retarded (General Motors) to as much as 20, or even 25, degrees.

Strategy Changes for Entering Closed Loop

Another strategy to reduce cold start emissions is to get the engine into closed loop as quickly as possible. You might recall that the manufacturers' initial criteria for entering closed loop were as follows:

- The oxygen sensors must reach operating temperature; some of these used heaters and some did not.
- The ECT sensor must indicate that the engine is warmed up to at least 150°F. (This was a typical value, but some manufacturers even had some late 1980s PCMs that required 180°F to 190°F to achieve closed-loop operation; they lowered the specification in later years.)
- A certain time must have elapsed since the last engine restart, typically as long as 1 or 2 minutes.

During the late 1990s and early 2000s, manufacturers began changing this strategy to achieve closed-loop operation more quickly. The required elapsed time on modern vehicles is typically 10 to 20 seconds; oxygen sensors use heaters to get them up to operating temperature more quickly, and PCMs are now programmed to require ECT sensors to achieve only 40°F to 50°F (some as low as 32°F on a few applications). Basically, for many cold starts, depending upon weather and also whether the vehicle is garaged, the ECT sensor has effectively been taken out of the equation. This allows the PCM to get into closed-loop control more quickly and results in further reduced toxic emissions during most of the engine warm-up process. (Chrysler Corporation actually began implementing these changes in 1987 to achieve a *level one* closed-loop mode at 32°F whereby the STFT values could respond, but the LTFT values were prohibited from responding. When the ECT reached about 150°F, then a *level two* closed-loop mode allowed the LTFT values to begin to respond.)

REDUCING THE ESCAPE OF HC EMISSIONS DURING REFUELING

Another federal mandate requires the capture of any fuel vapors that would find their way to the atmosphere during refueling. If a 20-gallon fuel tank has only 3 gallons of liquid fuel in it, then it stands to reason that it also contains 17 gallons of hydrocarbon vapors. In years past, when the tank was then filled with 17 gallons of liquid fuel, the 17 gallons of HC vapors were forced out of the filler neck into the atmosphere. There are now a couple of methods in use to capture the HC vapors.

On-Board Refueling Vapor Recovery (ORVR) System

The refueling process begins with a spring-loaded, closed, anti-spit-back valve, which is located at the bottom of the fuel tank filler tube. The filler tube is designed like the venturi area of a carburetor. The filler pipe diameter is reduced to 24 mm, or approximately 1 inch. When fuel travels down this small pipe, it produces a liquid seal that prevents the escape of fuel vapors from the tank during refueling. Vented fuel vapors in the vapor recirculation line are drawn back

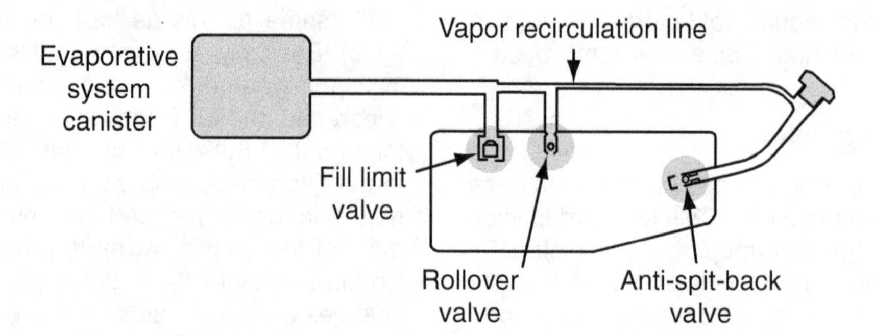

Figure 7–20 ORVR system components.

down the filler pipe by the venturi action of the pipe (Figure 7–20).

As the fuel tank fills, the fill limit valve rises. This causes the passageway to the charcoal canister to close and prevents liquid from overcharging the canister. Liquid in the tank rises up the filler neck and shuts off the automatic gas pump nozzle.

Fuel Tank Bladder

The Toyota Prius uses a rubber membrane, called a *bladder*, that encapsulates the fuel pump and sender assembly. The bladder also contains the liquid fuel. As fuel is used, this bladder shrinks so as to reduce the volume of fuel vapors that are contained within it. As the fuel tank is refueled, the bladder expands to contain the additional liquid. Because this method is so effective at reducing the release of HC vapors to the atmosphere during refueling, other manufacturers are also beginning to adopt the concept. The unfortunate trade-off is that if the fuel pump or sender fails, the entire fuel tank assembly will need to be replaced.

OBD III

The PCM in an OBD II vehicle knows whether the vehicle is capable of passing emission standards. OBD III is simply an OBD II system whereby the PCM has been given the ability to communicate the status of the system to a government agency through an added transponder of some type.

Some of the types of technology that can be used to detect and relay data pertaining to emission system malfunctions include roadside readers, cell phones, or satellites. CARB began testing the roadside reader in 1994. It is capable of reading eight lanes of bumper-to-bumper traffic at 100 miles per hour. If a fault is detected by a reader unit, it can send the VIN and the DTCs to a program regulator.

One manufacturer, General Motors, is already monitoring PCM data and is informing its vehicle owners of problems with the system as a customer courtesy. The implementation of OBD III could eliminate the need to visit a formal emission program station. Additionally, a fault affecting the vehicle's emissions would be reported closer to when it actually occurred, rather than simply testing the vehicle once every 2 years as many emission programs do currently. This would reduce the length of time that the fault was allowed to pollute the atmosphere. At least, that is the official position on OBD III. Regardless of how consumers feel about the technology, the reality is that this technology is already present on production vehicles, but the communication ability has not been enabled. The manufacturer's scan tool is generally required to enable the communication ability. More than a decade ago Colorado legislators toyed with the idea of enabling the communication ability on a voluntary basis in order to allow the vehicle's owner to be taken out of the formal emission program, but they later scrapped the idea.

SUMMARY

In this chapter we have covered the why and the how of OBD II and have shown how the OBD II program has made diagnosis much more standardized for all vehicles. This standardization has assisted today's technicians in diagnosing vehicle problems more accurately, resulting in cleaner exhaust emissions and, ultimately, cleaner air.

▲ DIAGNOSTIC EXERCISE

1. An OBD II-equipped vehicle has an engine with an oil-fouled spark plug on one cylinder. How will the system prevent fuel from passing through the cylinder and overheating the catalytic converter?
2. What advantage comes from reading the sensors' information and testing actuators through the OBD II connector rather than directly testing them on the car with the harness disconnected?

Review Questions

1. What advantage does OBD II provide for the independent technician?
   A. It is designed to standardize diagnosis of vehicle emissions and driveability-related problems.
   B. It is designed to force all vehicle manufacturers to use exactly the same engine computer system, including sensors and actuators.
   C. It is designed to force vehicle manufacturers to use new, totally redesigned computer systems in place of the systems that were already in use.
   D. It is designed to standardize the wiring colors used by all manufacturers.
2. The standardized self-diagnostic computer systems known as OBD II were required on all cars as of what year?
   A. 1986
   B. 1991
   C. 1996
   D. 2000
3. *Technician A* says that the reason OBD II was adopted was to make the diagnosis of a driveability complaint more uniform. *Technician B* says that OBD II was adopted due to concern about automotive emissions and air pollution. Who is correct?
   A. *Technician A* only
   B. *Technician B* only
   C. Both technicians
   D. Neither technician
4. OBD II standardizes all *except* which of the following?
   A. DLC shape, size, and location
   B. Terms and acronyms
   C. Wiring colors used throughout the vehicle
   D. DTC format
5. *Technician A* says that OBD II represents a revolutionary change in the way that computers control engines to maximize performance, economy, and emissions. *Technician B* says that the snapshot (freeze frame) feature of the OBD II self-diagnostics program is the ability of the PCM to record relevant engine parameters when a fault is recorded. Who is correct?
   A. *Technician A* only
   B. *Technician B* only
   C. Both technicians
   D. Neither technician
6. *Technician A* says that for class B faults, the first time that a fault is detected, a "pending" code will be set. *Technician B* says that for class B faults, the second time a fault is detected, the PCM will set a confirmed DTC in memory and will turn on the MIL. Who is correct?
   A. *Technician A* only
   B. *Technician B* only
   C. Both technicians
   D. Neither technician

7. *Technician A* says that an OBD II DTC that begins with P1 is for the powertrain and is specific to the vehicle manufacturer. *Technician B* says that an OBD II DTC that begins with P0 is for the powertrain and is an SAE standardized code. Who is correct?
   A. *Technician A* only
   B. *Technician B* only
   C. Both technicians
   D. Neither technician
8. *Technician A* says that an OBD II drive cycle is a specific set of driving conditions designed to allow the PCM to test all of the OBD II monitors. *Technician B* says that the PCM performs all of the tests required by OBD II instantaneously as soon as the engine is started. Who is correct?
   A. *Technician A* only
   B. *Technician B* only
   C. Both technicians
   D. Neither technician
9. What is the purpose of the heated oxygen sensor placed after the catalytic converter?
   A. To allow the PCM to monitor the catalytic converter's efficiency in storing and using oxygen to oxidize CO and HC emissions
   B. To help the PCM control the air-fuel ratio while operating in closed loop
   C. To allow the PCM to test the oxygen sensor placed before the catalytic converter
   D. Both A and B
10. How does an OBD II PCM detect cylinder misfire?
   A. Through input from the vehicle speed sensor (VSS)
   B. Through input from a high-data-rate crankshaft position (CKP) sensor
   C. Through input from the knock sensor
   D. Through input from the manifold absolute pressure (MAP) sensor
11. If an OBD II PCM detects a type A misfire, it will default to open-loop operation and shut down the fuel injectors on the misfiring cylinders to protect which of the following?
   A. Engine block
   B. Ignition system
   C. Oxygen sensors
   D. Catalytic converter
12. *Technician A* says that if a misfire of sufficient severity to cause the catalytic converter to reach 1,800°F is detected, the PCM will flash the MIL. *Technician B* says that the PCM will not turn on the MIL until the third detection of a type A misfire. Who is correct?
   A. *Technician A* only
   B. *Technician B* only
   C. Both technicians
   D. Neither technician
13. To evaluate the EGR system, Chrysler's OBD II PCM selects an operating mode in which the EGR is held open. Then it closes the EGR valve and maintains the current pulse-width command to the fuel injectors while monitoring which of the following?
   A. MAP sensor
   B. Oxygen sensor
   C. Crankshaft RPM
   D. Vehicle speed sensor
14. For vehicles with an enhanced evaporative system that were built after 2001, the PCM is required to recognize any leak greater than which of the following?
   A. 1.000 in.
   B. 0.500 in.
   C. 0.040 in.
   D. 0.020 in.
15. By which year was the cylinder imbalance monitor required to be fully phased in on 100 percent of U.S. vehicles?
   A. 2008
   B. 2010
   C. 2014
   D. 2016

16. How many global test modes (scan tool requests) are reserved under OBD II standards?
    A. 7, of which 5 are defined today
    B. 12, of which 7 are defined today
    C. 15, of which 10 are defined today
    D. 21, of which 15 are defined today
17. Where are manufacturer global mode $06 charts available?
    A. On the manufacturers' websites
    B. On the International Automotive Technicians Network (IATN) website
    C. In aftermarket electronic service manuals such as *Mitchell On-Demand* and *AllData*
    D. All of the above
18. Which OBD II global mode allows a scan tool to retrieve permanent codes stored in nonvolatile RAM memory that cannot be erased by a scan tool or by disconnecting the vehicle's battery?
    A. Mode $01
    B. Mode $03
    C. Mode $07
    D. Mode $0A
19. What is the purpose of encapsulating the fuel pump and sender assembly within a rubber bladder inside the fuel tank?
    A. Reduce the audible noise from the fuel pump
    B. Reduce the cost of replacing the fuel pump if it were to fail
    C. Reduce the escape of HC emissions to the atmosphere during refueling
    D. Reduce the escape of CO emissions to the atmosphere during refueling
20. Technician *A* says that an OBD III system is an OBD II system in which the PCM can communicate the status of the system to a government agency through an added transponder of some type. *Technician B* says that a benefit of an OBD III system would be that a fault affecting the vehicle's emissions would be reported closer to when it actually occurred, as opposed to testing the vehicle's emissions once every 2 years as with many current emission programs. Who is correct?
    A. *Technician A* only
    B. *Technician B* only
    C. Both technicians
    D. Neither technician

Chapter 8

Automotive Multiplexing and Networking of Computers

OBJECTIVES

Upon completion and review of this chapter, you should be able to:

- ❑ Recognize why multiplexing is reducing hard wiring on many electronic systems, both automotive and nonautomotive.
- ❑ Describe the differences between the two types of circuitry commonly used to form a data bus.
- ❑ Understand the methods used to minimize the effects of an induced voltage on the data bus.
- ❑ Describe the electrical configurations of a data bus and their advantages and disadvantages.
- ❑ Describe how a control module communicates on an SAE-standardized J1850 data bus.
- ❑ Understand the abilities of a scan tool and a lab scope when using them to perform any diagnosis of computers that communicate on a data bus or to diagnose problems with the data bus itself.

KEY TERMS

Active Voltage
Arbitration
Asynchronous
Binary Code
Data Bus Network
Demultiplexor (DEMUX)
Hard Wiring
Intelligent Junction Box
Kilobyte
Linear Topology
Loop Topology
Master Node
Megabyte
Multiplexing (MUX)
Multiplexor
Node
Passive Voltage
Pulse-Width Modulated (PWM)
Serial Data
Slave Node
Star Topology
Termination Resistor
Variable Pulse Width (VPW)

For many years, when an automotive manufacturer decided to add another feature, it involved simply adding more wiring to complete the needed circuits. Even with computer-controlled systems, adding more features meant adding more computers, more actuators, and more wiring to connect them. Many of these computers required inputs from the same sensors or from multiple similar sensors. For example, many older vehicles had three coolant temperature sensors or switches: one to control the operation of the electric cooling fan, one to control the coolant temperature gauge or warning light, and one that acted as an input to the PCM. What would be the results if we could provide sensor information to just one computer and then have that computer communicate what it knew about these sensors to the other computers that needed the same information? This concept is known as **multiplexing (MUX).**

MULTIPLEXING OVERVIEW

Hard Wiring

Before the advent of multiplexing, each sensor was wired to each of the computers that needed that sensor's input with **hard wiring.** Hard wiring is any copper wire that performs only one function 100 percent of the time. This includes simple electrical circuits that operate electrical devices, input circuits from sensors to a computer, output circuits through which a computer controls an actuator, and circuits that provide power and ground to a computer.

Elimination of Hard Wiring

A multiplexing circuit is a circuit that can be used to communicate more than one message, thereby eliminating the need for multiple wires to perform the same work. By this simple definition, Ford Motor Company has been using multiplexing in its speed control systems since the mid-1960s. *The speed control command switches* (SCCS), mounted in the steering wheel, consisted of either four or five switches. As the driver operated each switch, various voltage or ground signals, modified by resistances, were communicated to the transducer or control module over a single circuit. This was important because any additional wires would have created the need for additional slip rings in the steering column and steering wheel assembly (prior to the use of clock springs that are used with air bags). This is known as *resistive multiplexing*.

Of course, today's modern multiplexing systems, as they are used with digital computers, do not simply use resistances to modify the message but instead communicate with digitized code, known as **binary code** or **serial data.** Circuits over which serial data is communicated are known as **data bus networks** (Figure 8–1). Any computer that communicates on a data bus is called a **node.** If a node contains a **multiplexor (MUX)** IC chip, it can send messages on the data bus. If a node contains a **demultiplexor (DEMUX)** IC chip, it can receive and decipher messages on the data bus. Most major nodes contain both a multiplexor and a demultiplexor and thus have the ability to both send and receive messages on the bus. The data bus circuits do not fall into the category of hard wiring because they are not limited to one job 100 percent of the time but instead may communicate one message in one instant and a totally different message in the next.

Some of the computers that may constitute nodes on the data bus include the PCM, the body control module (BCM), the instrument panel controller (IPC), the electronic brake control module (EBCM) or antilock brake control module (ABCM), the automatic transfer case module (ATCM), and several other computers, including trip computers, message centers, antitheft warning system computers, antitheft disabling system computers, electronic suspension system computers, and the computer that controls the heating, ventilation, and air conditioning system. Instead of hard wiring connecting each of the sensors to each of the computers, sensor information is shared on the bus. Additionally, the data link connector (DLC) will typically be connected to one or more of the data buses on the vehicle. This allows a scan tool to become a node on these buses for diagnostic purposes.

Typically, one computer (often the BCM or PCM), known as the **master node**, is the dominant node on the bus and controls the data bus. A dependent node on the data bus is known as a **slave node**. Some manufacturers prefer to refer to these nodes as *primary* and *secondary* nodes. The master node may also be referred to as the *dominant* module.

Advantages of Multiplexing

Ultimately, multiplexing eliminates dedicated hard wiring and wiring connectors, thereby reducing vehicle weight while improving dependability. Multiplexing reduces manufacturing costs and increases the computers' diagnostic abilities. It also allows the addition of many more features while potentially using fewer computers to perform

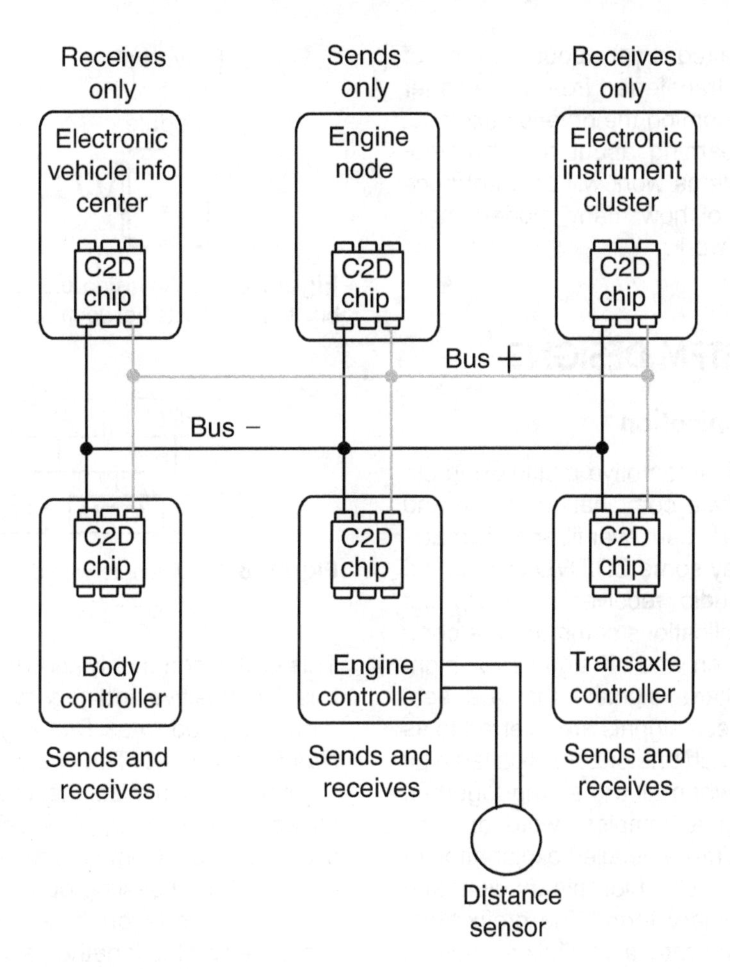

Figure 8–1 Chrysler data bus network schematic.

them. In fact, on a fully multiplexed vehicle, the manufacturer could conceivably add a feature to an existing vehicle by simply providing a software update over the multiplex network to a particular control module. This would install the added feature without having to add control modules or wiring.

The Popularity of Multiplexing

Multiplexing is employed in many of today's technologically advanced components and systems, both automotive and nonautomotive. Telephone systems employ multiplexing extensively. Your personal computer uses the same principles in communicating to and from the monitor, keyboard, printer, scanner, and other computers on a network. For years, popular video cameras have been equipped with a multiplex port that can be linked with a patch cable to a video editor. The patch cable provides a two-way data bus for communication purposes between the components that it connects. Home security systems use a data bus to connect the keypads to the primary controller, which typically has the ability to communicate serial data over the phone lines to a central monitoring station. And a modern clothes washing machine may even communicate information to the paired dryer about the type of clothing that will be transferred from the washer to the dryer, thus becoming the dryer's next load. As you can see, learning about how automotive multiplexing systems work will also enhance your understanding of how many modern non-automotive systems work.

MULTIPLEX SYSTEM DESIGNS

Multiplex Communication

In certain cases, automotive multiplexing circuits may communicate combinations of on and off light signals that are sent over fiber optic material (similar to the way some CD/DVD players are connected to an audio receiver/amplifier), but most automotive applications communicate combinations of voltage on and off signals (or high-voltage and low-voltage signals) that are sent over copper wire. These signals are referred to as "ones" and "zeros." Each one or zero is called a bit of information and, when all are strung together, they eventually form a complete word or byte. (See Chapter 2 for a more detailed explanation of bits and bytes.) "Bit" is short for "binary digit" and "byte" is short for "binary term." The prefix "kilo" (normally used to indicate a multiplier value of 1,000), when used with "byte" to form **kilobyte,** indicates a value of 1,024 bytes of information. (1,024 is equal to the number of *combinations* of zeros and ones available when a base-2 numbering system is carried out to 10 places. It is also the base-10 value of the eleventh column in a base-2 numbering system.) Similarly, a **megabyte** indicates a value of 1,048,576 ($1{,}024^2$) bytes of information. And a gigabyte indicates a value of 1,073,741,824 ($1{,}024^3$), while a terabyte indicates a value of 1,099,511,627,776 ($1{,}024^4$).

Two-Wire versus Single-Wire Data Buses

Many multiplexed circuits use two wires to complete the data bus circuitry between all nodes, one for bus positive and the other for bus negative.

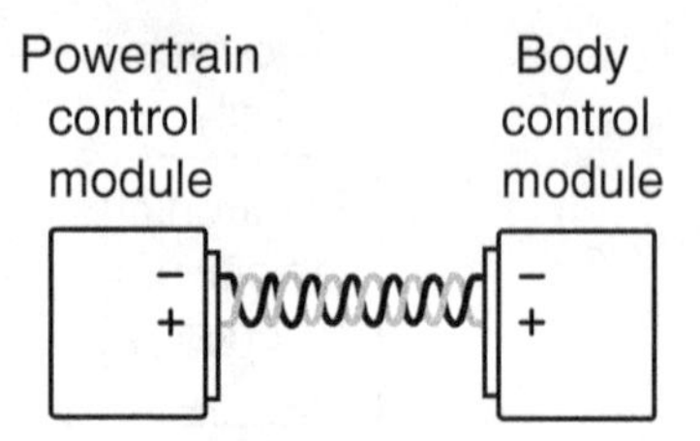

Figure 8–2 A twisted pair of wires forms the data bus that connects between nodes.

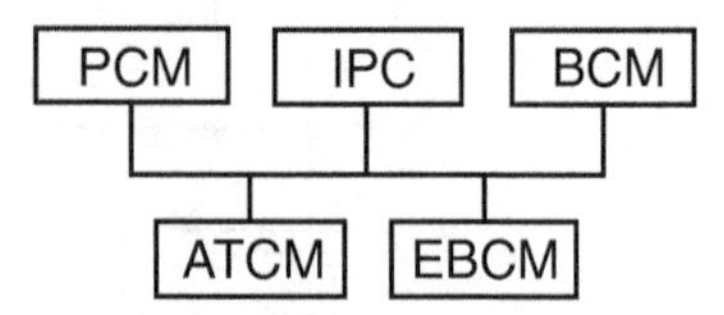

Figure 8–3 A single-wire data bus.

This is also commonly known as a dual-wire (DW) bus. The positive and negative bus wires are commonly referred to as *Bus High* and *Bus Low*, or *Bus H* and *Bus L*. The two wires are then twisted together to minimize the effects of an induced voltage (Figure 8–2). This data bus is known as a *twisted pair.* Comparison of the wires to each other, not to chassis ground, is used to decipher the binary code on these wires. Because the wires are twisted together, a voltage that might be induced on either wire by a nearby circuit is likely to be induced equally on both wires, and thus will not affect the voltage difference between the two wires. Therefore, the binary code is unaffected.

To reduce wiring complexity even further, many data buses use a single-wire (SW) bus to connect all of the nodes together (Figure 8–3).

Pulse-Width Modulated versus Variable Pulse Width

Typically, most DW buses and some SW buses use binary code that has a fixed pulse width. That is, all bits (zeros and ones) will be the same length (Figure 8–4). The serial data on these data buses is said to be **pulse-width modulated (PWM).** With this type of binary code, a binary

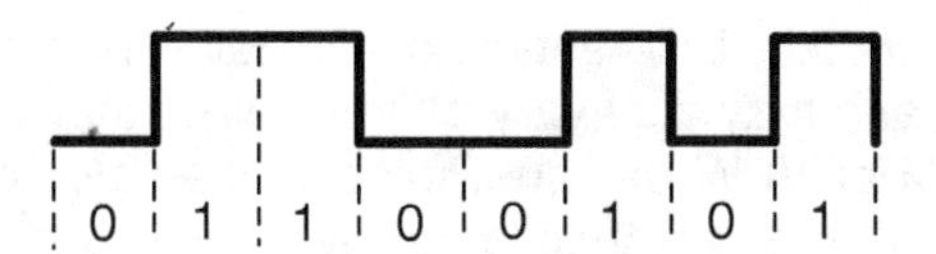

Figure 8–4 Pulse-width-modulated serial data is formed from binary code in which all bits of information are equal in length.

one is usually a high-voltage pulse (which may be $2^1/_2$ V, 5 V, or 6 to 8 V) and a binary zero is a low-voltage pulse (zero volts or even a negative voltage).

However, many SW buses use a **variable pulse width (VPW)** design. On these buses, the binary ones and zeros are not only a function of the voltage value but are also a function of bit length. With VPW data, a binary one may be represented as a short, high-voltage pulse, but it may also be represented as a long, low-voltage pulse. Conversely, a binary zero may be represented as a short, low-voltage pulse or a long, high-voltage pulse. This allows the voltage to switch between high and low with each bit of information (Figure 8–5). On many SW buses that use VPW data, the digital bits are slightly trapezoidal in shape (Figure 8–6). If you were to look at this serial data with a lab scope, you would see that the waveform is not truly vertical at the bits' edges, but, rather, the edges are slanted. This allows the

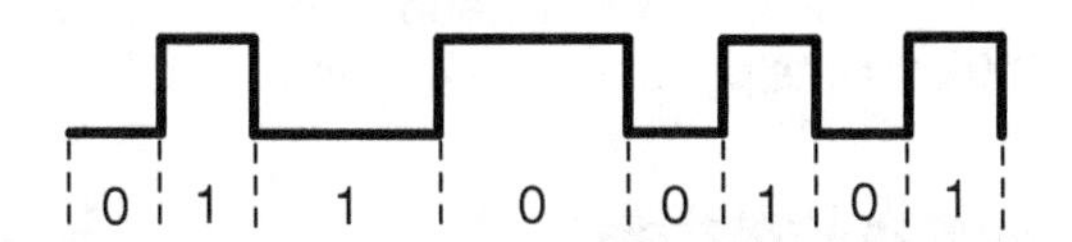

Figure 8–5 Variable-pulse-width serial data is formed from binary code in which bits of information are of different lengths.

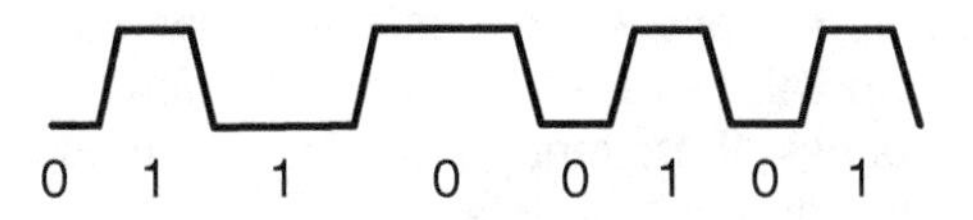

Figure 8–6 VPW serial data on a single-wire data bus has a slightly trapezoidal shape to its waveform.

nodes on the data bus to distinguish between serial data and an induced voltage from a nearby circuit, which would tend to be more vertical at the edges.

Data Bus Configuration

Early automotive data buses were wired from one node to the next in a series circuit or **loop topology** (Figure 8–7). Unfortunately, if one node on the bus were to lose power or ground, it could interrupt bus communication, affecting all nodes on the bus. Newer configurations tend to wire the nodes to the data bus in a parallel fashion, known as either a **linear topology as in** Figure 8–3, or **star topology as in** Figure 8–8. The star topology may have a node at the center hub of the star where all the buses come together. The center node is usually the master node that controls the

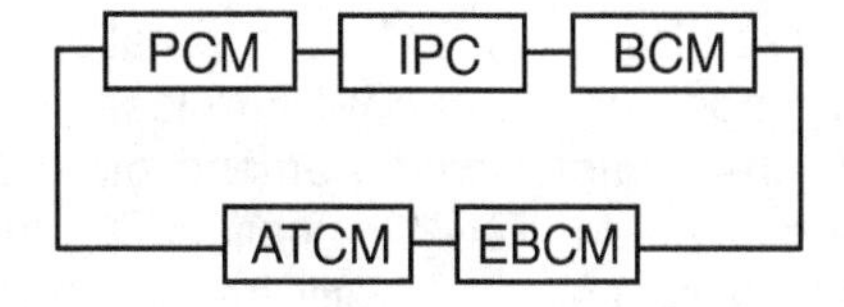

Figure 8–7 Loop design data bus, with all nodes wired in series to one another.

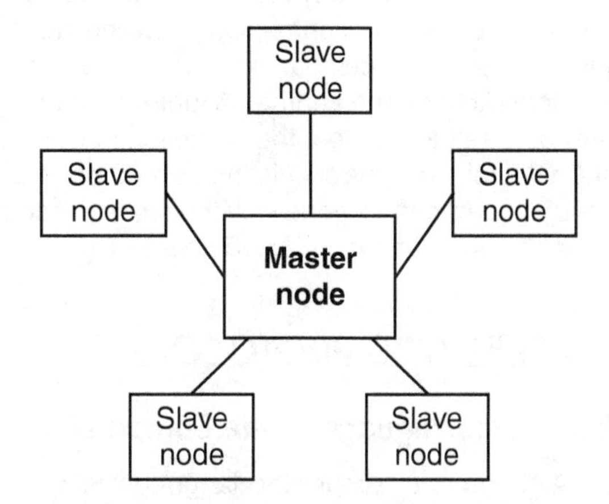

Figure 8–8 Star topology data bus with a master node in the center.

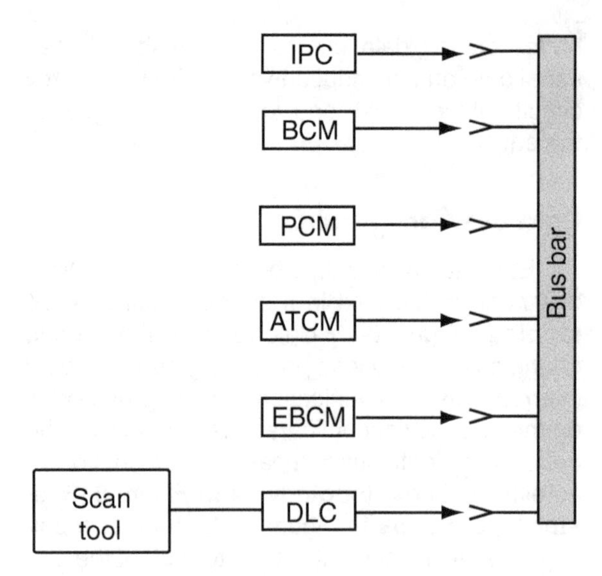

Figure 8–9 Linear design data bus with a bus bar present for diagnostic purposes.

data bus, while the remaining nodes are usually dependent slave nodes. In the parallel topologies, if a node on the bus were to lose power or ground, the communication among other nodes on the bus would likely be unaffected, although the data bus itself could still be affected by a short to power or ground. Another advantage is that, during diagnosis, the system can be modified to allow scan tool connectivity to only one node or to a selected few. A bus bar may be used to complete the bus connection between nodes (Figure 8–9). If the bus bar is removed, all nodes are isolated from one another. A jumper wire may then be used to connect the bus circuit from the DLC to a selected bus circuit. In this method, scan tool diagnosis can be done with the scan tool connected to one or more selected nodes.

MULTIPLEXING PROTOCOLS

Protocols: Language of the Computer

A *protocol* is simply the language in which computers communicate with one another over a data bus. In other words, it is a standardized binary code. Different protocols may vary in baud rate and may also vary as to whether they are a PWM or VPW design. Protocols may also differ according to what voltage level equals a one or a zero. And, while nodes on some older applications communicated to a scan tool continuously with an unending, steady stream of information, nodes on modern networks communicate intermittently and only as needed. The latter is known as an **asynchronous** protocol.

Protocol Classes

The Society of Automotive Engineers (SAE) has defined four different classes of protocols according to their speed of communication or baud rate. A class A protocol is a low-speed protocol that has a baud rate of less than 10,000 bits per second (10 Kb/s), as was used with older systems. A class B protocol is a medium-speed protocol that has a baud rate of 10,000 bits per second (10 Kb/s) to 125,000 bits per second (125 Kb/s); this is commonly used on modern systems to control most systems except real-time control systems. A class C protocol is a high-speed protocol that has a baud rate of 125,000 bits per second (125 Kb/s) to 1,000,000 bits per second (1 Mb/s); this is used for some real-time controls, such as “drive-by-wire” and “brake-by-wire” systems. Finally, a class D protocol is an ultra-high-speed protocol that is also used for some real-time control systems and has a baud rate in excess of 1,000,000 bits per second (1 Mb/s).

Common Protocols

Early protocols used on automotive applications include GM’s UART (universal asynchronous receiver-transmitter) protocol and Chrysler’s C2D (Chrysler Collision Detection) protocol. These are class A protocols that handle PWM data. Chrysler’s term *collision detection* refers to the process of preventing two or more messages from colliding on the data bus.

While many protocols have been developed and used by the different manufacturers over time, in the interest of maintaining standardized communications with aftermarket scan tools, OBD II standards initially defined four protocols that could be used on U.S. automobiles. The accepted protocols were defined by the SAE and the ISO (International Standards Organization). These standards apply to the protocols that are to be used between the nodes on the bus and an OBD II–standardized scan tool (as per the SAE J1978 standard) for diagnostic purposes. For example, Chrysler, Ford, and General Motors engineers, working closely with each other to meet the OBD II regulations, developed the SAE J1850 protocol.

As these protocols were mandated under OBD II standards, the manufacturers were given a certain time frame within which to meet each standard. The four protocols that were initially accepted under OBD II standards were phased in by most manufacturers between the 1996 model year and the 2000 model year, as follows:

1. *SAE J1850 VPW protocol:* This protocol is a class B protocol at 10.4 Kb/s and is used as an SW bus by General Motors as its Class 2 network and by Chrysler as its Programmable Communication Interface (PCI) network. This protocol was no longer allowed to be used after the 2007 model year as the primary communication protocol between an aftermarket scan tool and those computers whose systems affect the vehicle's emissions.
2. *SAE J1850 PWM protocol:* This protocol is a class B protocol at 41.6 Kb/s and is used as a DW, twisted-pair bus by Ford in its Standard Corporate Protocol (SCP) and Audio Corporate Protocol (ACP) networks. This protocol was also no longer allowed to be used after the 2007 model year as the primary communication protocol between an aftermarket scan tool and those computers whose systems affect the vehicle's emissions.
3. *ISO 9141/ISO 9141-2:* This protocol is used primarily on European vehicles. However, it is also used in some domestic applications. In fact, CARB initially adopted the ISO 9141 protocol for all vehicles sold in California with feedback fuel management systems. The ISO 9141 protocol uses an SW bus called the *K-line.* The K-line allows for bidirectional communication between a scan tool and the nodes on the bus. An optional *L-line* is sometimes used to allow one-way communication from the scan tool to the nodes on the bus (Figure 8–10). The original ISO 9141 protocol was enhanced in 1994 to meet CARB requirements and became the ISO 9141-2 protocol. This protocol was no longer allowed to be used after the 2006 model year as the primary communication protocol between an aftermarket scan tool and those computers whose systems affect the vehicle's emissions.
4. *ISO 14230-4,* also known as *KWP 2000* (Keyword Protocol 2000): This protocol is a newer

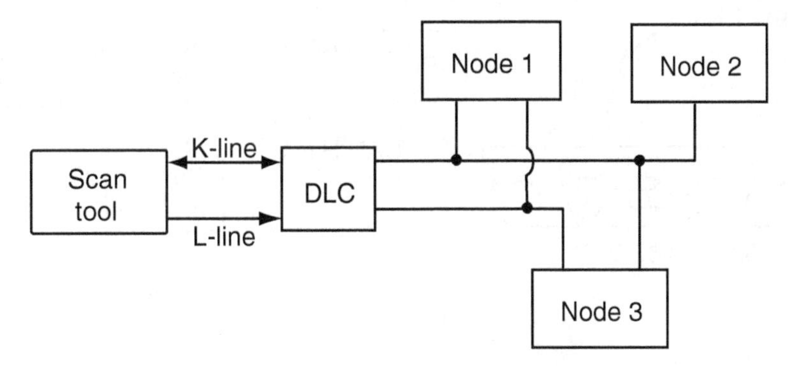

Figure 8–10 ISO 9141 data bus.

version of the ISO 9141-2 protocol and was introduced in 2000. This protocol was also no longer allowed to be used after the 2007 model year as the primary communication protocol between an aftermarket scan tool and those computers whose systems affect the vehicle's emissions.

A fifth protocol, known as the ISO 15765-4 Controller Area Network (CAN) protocol, was added to the list of OBD II protocols. The CAN protocol was originally developed by Robert Bosch in the 1980s and was first used on the 1991 Mercedes S-Class to provide an intra-vehicle communication network between the nodes on a bus. Then one particular node, the PCM for example, would serve as a gateway and translate the CAN protocol into the ISO 9141 protocol for communication with a scan tool connected to the DLC (Figure 8–11).

To meet the OBD II mandate, manufacturers began phasing in the ISO 15765-4 CAN protocol on a few 2003 model year vehicles. Full implementation was required by the 2008 model year, requiring that the CAN protocol replace all other existing protocols for using an aftermarket generic scan tool in diagnosing those systems that can affect the vehicle's emissions. However, additional protocols are also permitted as long as the CAN protocol requirement is met for emissions system diagnosis. Rephrased, there is no restriction on using manufacturer-specific protocols in addition to the CAN protocol once the requirement is met. Another point to note is the fact that one of the factors resulting in CAN implementation prior to the 2008 model year is the addition of real-time control systems, such as drive-by-wire, which benefit from the higher communication speed, or baud rate, provided by the HS CAN protocol.

The CAN protocol mandate is broken down into two categories: *Legislated CAN,* known as the ISO 15765-4 CAN protocol, is required to provide functionality to a generic scan tool connected to the DLC concerning emissions system diagnostics, while *OEM Enhanced CAN* is allowed to conform to OEM proprietary specifications for providing manufacturer scan tool functionality when diagnosing powertrain, body, or chassis systems that do not affect the vehicle's emissions.

Note: Intel has been one of the primary companies involved in the design of automotive multiplexing networks and has designed products that incorporate both J1850 and CAN functionality.

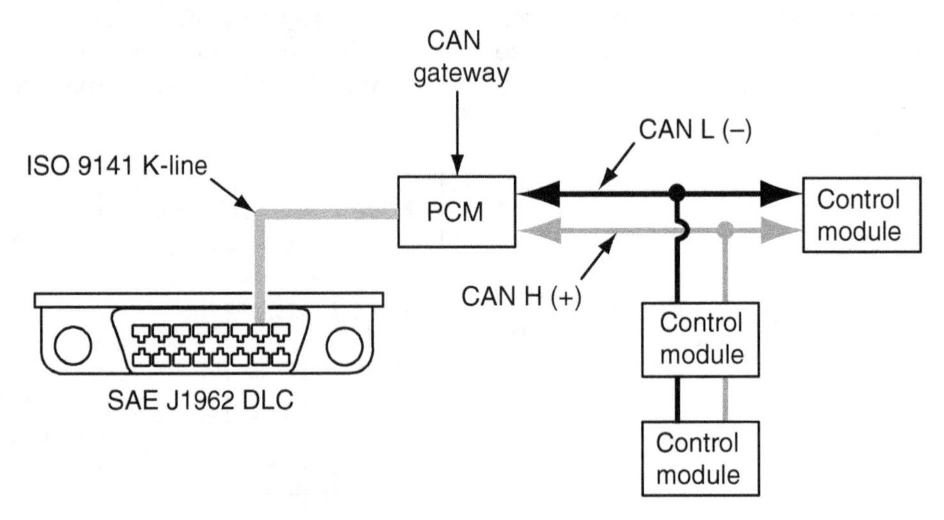

Figure 8–11 Early use of the CAN protocol on European vehicles showing the PCM as a gateway to the DLC.

COMMUNICATION ON A J1850 VPW DATA BUS

Overview of the J1850 VPW Protocol

Because the VPW version of the SAE J1850 protocol was used on a large number of General Motors and Chrysler vehicles, it is presented in some detail here. GM named this protocol *Class 2* and Chrysler called it *Programmable Communication Interface (PCI)*. However, many similarities exist between this protocol and other protocols, including the ISO 15765-4 CAN protocol. Although the ISO 15765-4 CAN protocol may use a much faster baud rate, the CAN message contains many of the same parts as an SAE J1850 VPW message (simply adding some segments to it) and also uses the same method of **arbitration** as the SAE J1850 VPW protocol. Therefore, the following breakdown of the SAE J1850 VPW protocol can be taken as a typical example of modern protocols used throughout the automotive industry.

Passive versus Active Bits

A **passive voltage** is a voltage that is at rest. An **active voltage** is a voltage that is actively pulled either high or low by a computer circuit. On a J1850 data bus, a passive voltage is a low voltage that is close to ground (between zero volts and 3.5 V). The J1850 standard defines an active voltage as one that is pulled high (between 4.25 V and 20 V). To allow the voltage to switch with each bit of binary code, a one may be either a short, active voltage pulse (a high-voltage pulse that lasts 64 μs) or a long, passive voltage pulse (a low-voltage pulse that lasts 128 μs). Conversely, a zero may be either a short, passive voltage pulse (a low-voltage pulse that lasts 64 μs) or a long, active voltage pulse (a high-voltage pulse that lasts 128 μs) (Figure 8–12).

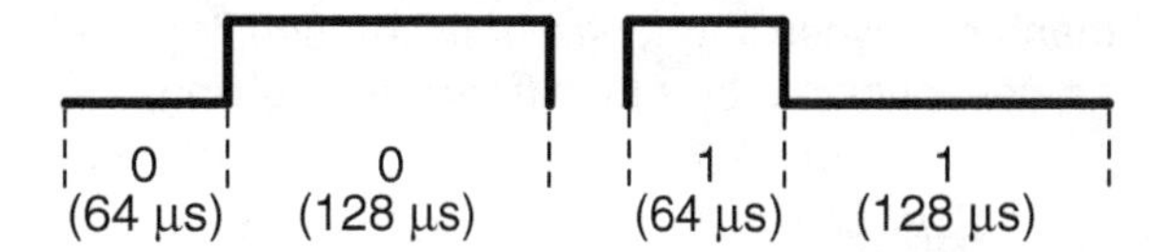

Figure 8–12 SAE J1850 VPW protocol bit length.

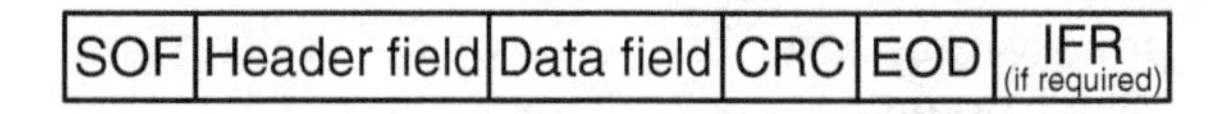

Figure 8–13 The SAE J1850 VPW bus message.

The J1850 Message

The J1850 message begins with a *start-of-frame* (SOF) signal, an active voltage pulse that lasts for 200 μs (Figure 8–13). This is followed immediately by the next portion of the message, called the *header.* The header's first byte designates message priority, whether the header is one or three bytes long, whether an *in-frame response* (IFR) is required from the receiving node, an address identification that identifies both the sender and the intended receiver, and the intended message type.

The next portion of the message is the data field, which contains the "meat" of the message and may be up to 11 bits in length. This is immediately followed by a *cyclic redundancy check* (CRC) byte. The CRC and the data field together are considered one "word" and do not exceed 12 bits in length. For example, if the data field is 11 bits, the CRC is one bit. If the data field is less than 11 bits, the CRC can be more than one bit as long as the total does not exceed 12 bits when added to the data field. The purpose of the CRC bit is to guard against message error. The receiving nodes add the CRC value to the value of the data field to calculate whether message errors have occurred.

The CRC is then followed by an *end-of-data* (EOD) pulse. This is a passive voltage that lasts 200 μs. If requested in the header field, the receiving node must issue an IFR within the next 80 μs. This allows the transmitting node to be certain that the intended receiver received the message. Think of it as a handshake between the transmitting node and the receiving node, similar in concept to the "handshake" that occurs between fax machines when the tones from the sending unit are responded to by a tone from the receiving unit.

Arbitration

A node on the data bus will "listen" to make sure that the bus is "quiet" before it begins transmitting data on the bus. But because the J1850 protocol supports peer-to-peer networking that allows equal access to all nodes on the data bus, occasionally, more than one node may begin to transmit data simultaneously. Arbitration is the process of determining which node can continue to transmit data on the bus when two or more nodes begin transmitting data at the same time. The SAE J1850 protocol uses bit-by-bit arbitration to compare active and passive bits on the bus. When a node that is transmitting a passive, low-voltage pulse sees the data bus circuit voltage pulled actively high by another node on the bus, it quits transmitting. That is, when a node transmits a passive voltage and sees another node's active voltage, it is said to lose arbitration. The nodes with the active voltage continue to arbitrate until only one node is left transmitting data. In reality, a *zero* is the dominant bit and always wins the arbitration when compared with a *one* bit. Therefore, this design is such that the node that initially transmits the most zeros wins arbitration and can continue to transmit data on the bus (Figure 8–14). The nodes that lost arbitration for the current message can try transmitting again after the winning message is complete.

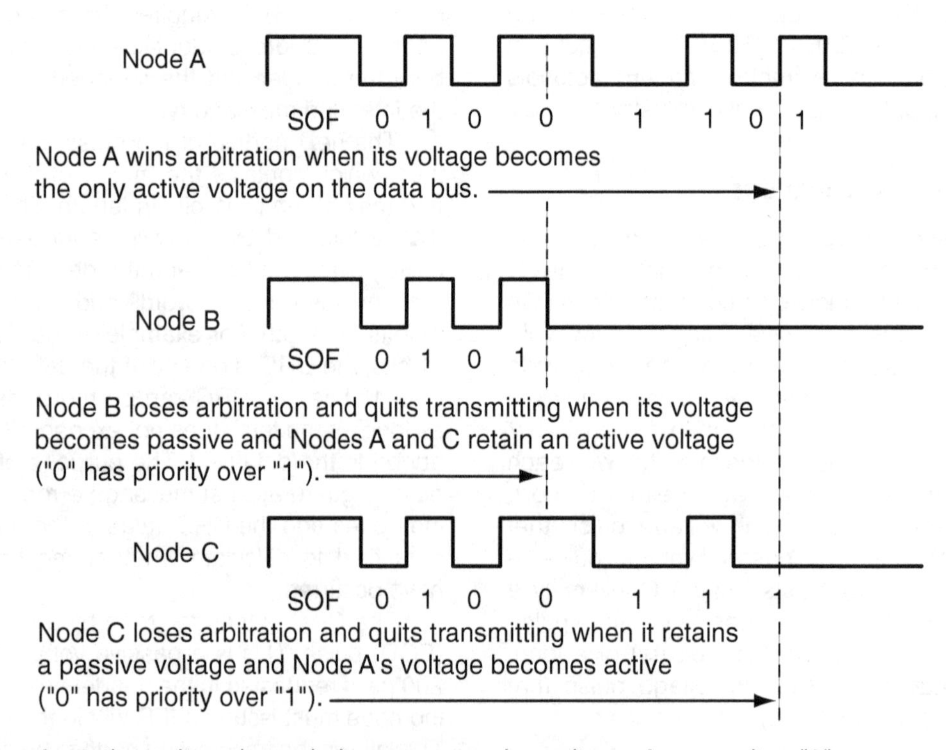

Figure 8–14 SAE J1850 VPW bus arbitration.

COMMUNICATION ON A CAN DATA BUS

Overview of the CAN Protocol

When Robert Bosch developed the CAN protocol, it was designed with an open architecture that purposefully allowed for some product-specific changes to be implemented while leaving its basic architecture unchanged. Therefore, variations of the CAN protocol are used, and yet they are still defined as the CAN protocol. For example, the CAN protocol can be designed at different baud rates from 1 Kb/s to 1 Mb/s, that is, as a class A, class B, or class C protocol. The most common CAN configurations are medium speed (class B) and high speed (class C). CAN-B is commonly found at baud rates between 33.3 Kb/s and 83.3 Kb/s. CAN-C operates at 500 Kb/s. The CAN protocol can also be configured as a single wire (SW) bus or as a dual wire (DW) bus. It should be noted that General Motors refers to their CAN bus as LAN (Local Area Network).

The CAN protocol has several similarities with the J1850 VPW protocol, but there are also some differences. The CAN protocol does not use a message that defines the intended receiving nodes by addresses, but rather uses a message-oriented protocol. This allows all nodes on the CAN bus to simply accept or deny a broadcast message according to its content.

Recessive versus Dominant Bits

Bosch refers to a passive bit as *recessive* and to an active bit as *dominant.* However, on an SW CAN bus the recessive voltage is actually a bias voltage that is supplied by one of the nodes, thus allowing the natural state of the bus to have a high voltage present when none of the nodes are attempting to communicate. A dominant value is created when one of the nodes pulls the voltage low by grounding out the bus momentarily. During active communication, the recessive voltage (high value) equates to a binary one and the dominant voltage (low value) equates to a binary zero.

The CAN Message

A base frame format (standard format) is used with the CAN 2.0 A specification (Figure 8–15). The first part of this message is the SOF, during which the bus voltage is pulled low to the dominant value (unlike the J1850 VPW bus, which at this point pulls the voltage from low to high). The SOF is followed by an 11-bit *identifier* that is used for arbitration.

Next is the *remote transmission request,* or RTR, which is a single dominant bit (0), followed by an *identifier extension,* or IDE, which is also a single dominant bit (0). A single *reserve bit,* or r(0), comes next. This is then followed by the meat of the binary message, beginning with a *data length code,* or DLC, which consists of four bits and indicates the number of data bytes that are to immediately follow (up to eight bytes of data may be transmitted).

Finally, a 15-bit CRC follows, which is used to check for message corruption. This is followed by a single recessive (1) bit known as the *CRC delimiter.* An *acknowledgement,* or ACK, field follows, during which the transmitting node will allow a recessive (1) bit. Any receiving node can respond by pulling this value low to the dominant value (0) as an acknowledgment to the transmitting node that the message was received intact. Once the receiving node releases the bus from ground, the voltage returns to the recessive value, which is now referred to as the *ACK delimiter.* The ACK field equates to the IFR in the J1850 VPW

SOF	Identifier (A)	RTR	IDE	r(0)	DLC	Data Bytes	CRC	CRC Delimiter	ACK	ACK Delimiter	EOF
1 bit	*11 bits*	*1 bit*	*1 bit*	*1 bit*	*4 bits*	*Up to 8 bytes*	*15 bits*	*1 bit*	*1 bit*	*1 bit*	*7 bits*

Figure 8–15 The CAN 2.0 A bus message.

protocol. The final part of the message is the EOD byte, which consists of seven recessive (1) bits in a row.

An *extended frame format,* known as the CAN 2.0 B specification, is also used. The primary difference is that the base frame format (CAN 2.0 A) has an identifier of 11 bits, while the extended frame format (CAN 2.0 B) follows this with a second identifier of 18 bits; the two identifiers, when added together, contain a total of 29 bits.

Arbitration

When two or more nodes begin to transmit a data message at the same instant in time, a dominant (0) bit in the identifier wins arbitration over a recessive (1) bit. Therefore, the lowest binary value in the identifier will always win arbitration. In reality, when one of the nodes in arbitration sees the bus voltage pulled low to a dominant (0) value at a time when it is, itself, allowing a recessive voltage, it knows that another node with a higher-priority message is communicating. It will then quit transmitting and will wait until it has seen a minimum of seven recessive (1) bit pulses in a row before it tries to communicate its message again.

There is one restriction on the identifier: The first seven bits cannot all be recessive (1) bits, since this would cause other nodes on the bus to believe that no communication is currently in progress. This is also why the EOD byte *does* consist of seven recessive bits: This serves to communicate to other nodes that the bus is now idle.

DW CAN Bus Operation

On a DW CAN-B or DW CAN-C data bus, the passive voltage is 2.5 V for both bus-H and bus-L. When a node begins transmitting data on the bus, a binary one is represented by an active voltage on bus-H increasing to 3.5 V and an active voltage on bus-L being pulled low to 1.5 V. A zero is represented by both bus-H and bus-L returning to their passive voltage of 2.5 V. Therefore, a 2-volt difference represents a binary one and no voltage difference represents a binary zero.

As with a DW J1850 data bus, the two wires are twisted together to reduce the effects of EMI. If a nearby circuit were to induce a voltage on this circuit, it would likely induce an equal voltage on both wires of the twisted pair, thus leaving the *voltage difference* and *no voltage difference* values unaffected.

CAN B Data Bus Fault Tolerance

On a DW HS CAN (CAN-C) or MS CAN (CAN-B) data bus, the digital binary data on the bus-L wire displays as a mirror image of the data on the bus-H wire when viewed with a dual-channel lab scope (Figure 8–16). On a CAN-B

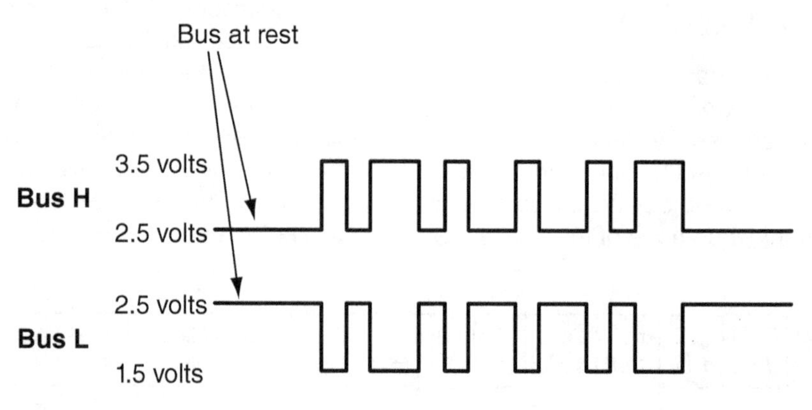

Figure 8–16 CAN DW protocol binary voltages.

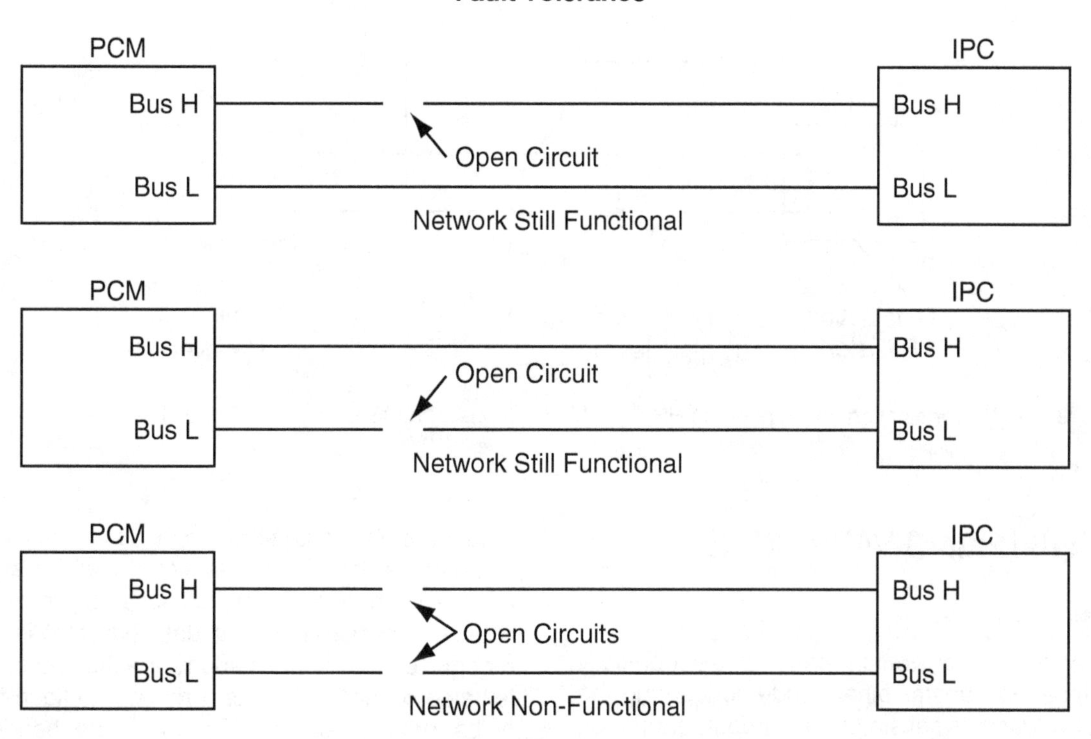

Figure 8–17 Both the DW CAN-B bus and the DW J1850 bus are fault tolerant if only one wire is open or shorted to ground.

data bus, if one data bus wire is open or shorted to ground, the nodes on the network can still communicate over the other bus wire, and the data bus is still functional (Figure 8–17). This condition, however, should cause a code that begins with the letter "U" to be set in memory. These are commonly referred to as "U-codes." The only time that the data bus communication is truly interrupted is if both data bus wires are open or shorted to ground. A DW HS CAN (CAN-C) bus does not have this fault tolerance. If an open or short to ground develops on an HS CAN bus, the bus becomes nonfunctional. An interesting footnote is that, from this aspect, the Ford J1850 DW data bus (used through 2007) has a fault tolerance similar to that of a DW CAN-B data bus.

CAN Bus Termination Resistors

A DW CAN data bus, both CAN-B and CAN-C, has a 120 Ω **termination resistor** located at each end of the data bus (Figure 8–18). With two 120 Ω termination resistors connected in parallel, the total resistance between bus-H and bus-L is 60 Ω. This is part of the CAN standard. The termination resistors may be located within a node or externally, sometimes in a harness end-cap type electrical connector. (See Chapter 1 for calculating the resistance of a parallel circuit when it contains two resistors in parallel.) Understanding the termination resistors can be useful when diagnosing problems associated with a DW CAN data bus.

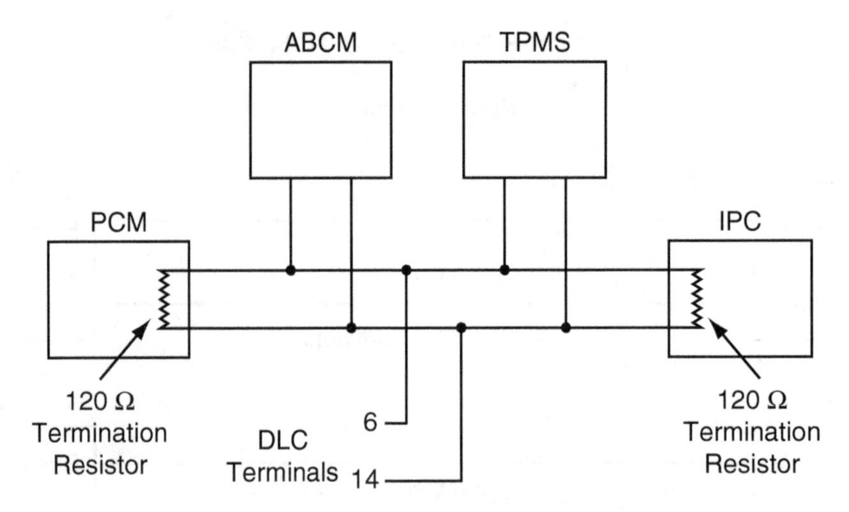

Figure 8–18 Termination resistors on an HS CAN (CAN-C) bus. The MS CAN (CAN-B) bus is similar.

MULTIPLEXING VARIATIONS

Smart Devices

A sensor or switch can be electronically equipped to transmit binary code concerning the input values it is sensing to other components on a data bus. Conversely, an actuator can also be electronically equipped to receive, translate, and react to binary code that it receives from other components on a data bus. These components are said to be electronically "smart."

For example, many of today's vehicles have driver control switches mounted on the steering wheel assembly (either front, side, or back) that are used to control the heating, ventilation, and air conditioning system, as well as the sound system. These switches send serial data over a data bus that connects both to the climate control module and to the radio. If a command is sent to increase the radio's volume, the binary code is received and acted upon by the radio, but it is ignored by the climate control module.

Another example that demonstrates the use of smart devices is an exterior lighting system in which the driver-controlled switches are smart input devices and the exterior bulbs' sockets are actually smart actuators on a data bus. If the driver uses the directional signal stalk to signal a left lane change, the switch assembly transmits a binary message on the data bus that is only recognized by the front and rear actuators on the left side of the vehicle as a request to flash their bulbs' bright filaments. Or, when the headlight switch is turned on, all exterior actuators react to the binary request to illuminate their dim filaments continuously, including even the license plate illumination. In this manner, you can control all of the bulbs at the rear of the vehicle simply by connecting all of them with a single- or dual-data bus, thus eliminating numerous hard-wired circuits. Even the deck lid solenoid can be connected to the same data bus.

Multiple Body Computers

Whereas some vehicles have used a body control module (BCM) since the mid-1980s to control interior and exterior lighting systems and other body functions of the vehicle, some manufacturers have recently divided this control module into multiple computers. For example, some modern Ford vehicles now have a *front electronic module,* a *rear electronic module,* and a *driver's door module.* With such a system design, input devices and actuators are only hard-wired to the closest module, whether or not the particular system is under control of that module. Then the pertinent information is shared between modules on the data bus. This allows for a further reduction of hard wiring.

For example, if the driver depresses the *deck lid release* button on the driver's door, the closing of the switch completes a hard-wired circuit to the driver's door module. This module then communicates the request over the data bus to the rear electronic module, which, in turn, energizes the deck lid release solenoid through a hard-wired circuit. As a result, a hard-wired circuit does not have to be run all the way from the driver's door switch to the deck lid release solenoid. Since the request is communicated for much of this distance over a data bus and because this bus is also used for data communication concerning many other devices as well, hard wiring is reduced.

Intelligent Junction Boxes

Many modern junction boxes (i.e., the main fuse and relay box, usually located near the vehicle's battery) now function as nodes on a data bus and are therefore labeled **intelligent junction boxes**. Instead of running hard wiring from the various controls and computers to the relays in the junction box, the junction box now receives binary requests from these controls and computers to energize or de-energize the various relays that it contains. In this way, a large amount of additional hard wiring can be eliminated.

Many modern control modules are designed to sense current levels in the output circuits they control. If an output circuit's current level increases beyond programmed limits, the module may either de-energize just that circuit or may shut itself entirely down in a self-protective mode. Intelligent junction boxes are also designed with this ability. If an intelligent junction box receives a binary request to energize an output and then finds that the current level is too high in the output circuit, it will de-energize the output. For example, if one of the relays in the junction box has a shorted winding, when the intelligent junction box tries to energize that relay, it will see the high current flow and immediately de-energize the relay. There is a preprogrammed limit to the number of times that the intelligent junction box will do this in an individual output circuit. When that limit is reached, the intelligent junction box will set a DTC that indicates junction box replacement is now necessary due to damaged drivers.

Multiplexing and Driver Control

With the continuing explosion in the use of electronics on the modern vehicle and the expansion in intra-vehicle communication networks, the driver is becoming less and less in control of the vehicle's systems. In fact, on many modern vehicles, the driver actually controls little or nothing. For example, when the driver depressed the horn switch on a 1980 vehicle, the operation of the switch actually completed the ground needed to energize the horn relay over a hard-wired circuit. But rarely is this true today. On many modern vehicles, depressing the horn switch only makes a request of a control module. Then if the module has no reason to disagree, it will complete the hard-wired circuit to energize the horn relay. Therefore, the driver does not actually control the horn relay, but only makes a request. The computer is providing the actual control.

For example, let us say that the driver tries to operate a convertible top while moving down the road at 20 MPH. Operating the convertible top's switch simply makes a request of the appropriate control module. But this module also sees that the vehicle's transmission is not in Park and also sees from the VSS that the vehicle is moving. In obedience to its internal program, it would refuse to comply with the driver's request.

Another advantage of this type of engineering is the ability to take full advantage of another feature that is in its early stages: vehicle-to-vehicle communication. SAE is currently working with several manufacturers in developing this technology. Once in place, the vehicle ahead of yours could hit a four-inch-deep puddle of water and hydroplane. Sensing what had happened, its on-board computers would communicate the danger to your on-board computers and, even though you have your foot pressing the throttle pedal fully, your drive-by-wire system would refuse to comply with your request due to the sensed danger ahead. Such a communication system will also be used to reduce road junction accidents due to the ability of the vehicles to communicate with each other. In short, today's drivers only *think* that they are controlling their vehicles.

FIBER OPTIC PROTOCOLS

Fiber optic protocols are being used on some vehicles in addition to copper wire data buses. For example, BMW uses five data buses on the E65 model, three of which are copper wire data buses and two of which are fiber optic data buses. There are a total of 54 modules (nodes) communicating over the five data buses. The fiber optic data buses use a visible red light wavelength to create the binary messages (similar to the fiber optic communication used with the digital audio in a home theater system). One of the fiber optic data buses used by BMW is known as *MOST*, an acronym that stands for *Media Oriented Systems Transport*. It serves the *infotainment* systems providing communication between the radio, the CD changer, the navigational system, and several other modules. This fiber optic bus was added to their vehicles with an initial baud rate of 22.5 Mb/s with plans to increase the baud rate to 50 Mb/s and eventually to 150 Mb/s (yes, that's correct—150 million bits per second!).

OTHER BOSCH PROTOCOLS

Overview

Bosch, a corporation that employs approximately 9,000 engineers, has continued to develop other protocols designed for mobile vehicle use. Two other protocols have been developed for manufacturers' proprietary use and/or for use with systems not affecting the vehicle's emissions and are described here.

Local Interconnect Network

The first version of the Local Interconnect Network, or LIN, protocol was released in 1999 by Robert Bosch Corporation as version 1.1. The second version, version 2.1, is now being used. LIN exists, not as a protocol to replace CAN, but rather as a protocol to complement and work as a subsystem of CAN. It is a low-cost protocol that uses no arbitration, but instead uses a single master node that does all of the transmitting. Up to 16 dependent slave nodes only listen except for an occasional response. The dependent slave nodes may be linked to a CAN bus through the LIN bus master node. The LIN protocol is a single-wire, low-speed protocol with possible baud rate up to 20,000 bits per second (20 Kb/s).

FlexRay

The FlexRay protocol is now being used on some European vehicles. Developed by Robert Bosch Corporation, FlexRay is not a fiber optic protocol, but uses a copper wire data bus. In spite of this fact, it is a high-speed protocol, transmitting data at a baud rate of up to 10,000,000 bits per second (10 Mb/s). It was developed for use with real-time control systems with its initial use aimed at brake-by-wire and steer-by-wire systems.

The FlexRay protocol uses a timed communication as its form of arbitration. That is, each node on the data bus is programmed to communicate in a timed cycle according to a "bus clock," with each communication taking only a few microseconds. As long as each node "takes its turn" as specified by its program, more than one node should never try to communicate at the same time.

BMW added the FlexRay protocol to the E70 model in December 2006. FlexRay, while not as fast as the MOST fiber optic protocol used by

BMW, is still considered fast enough to perform any task needed. BMW suggests that the FlexRay copper wire protocol may eventually replace all of their other protocols on their vehicles.

DIAGNOSIS OF MULTIPLEXED CIRCUITS

Scan Tools

Multiplexing allows several computers to be accessed through the DLC for diagnostic purposes. Simply connect an OBD II–standardized scan tool (SAE J1978 standard) to the DLC and then, after the VIN information has been entered, you can choose the computer you want to communicate with from the scan tool's menu. Once you have selected a computer, you will be able to pull DTCs, access the computer's data stream, perform functional tests, and perform any other functions the particular computer has been programmed to allow.

Lab Scopes

If your scan tool does not seem capable of communicating with any computer on the network, a DSO may be required to evaluate whether the computer is capable of transmitting the binary code that the scan tool must see. However, do not forget that common reasons for a scan tool's failure to communicate are improper entry of the VIN information or the scan tool failing to power up properly. On OBD II systems, do not forget to check the fuse that provides battery voltage to terminal 16 of the DLC. Also, in certain cases, another scan tool may communicate where one scan tool has failed.

If you cannot get a scan tool to communicate on the data bus, use a lab scope in conjunction with a breakout box designed for use at the DLC to determine whether serial data is present on the bus. With an SW data bus, connect the lab scope between the bus wire and terminal 4 or 5 of the DLC. With a DW data bus, while you can connect the lab scope between the two wires (bus high and bus low), you can also evaluate each wire's communication individually using a dual trace scope.

Once you have connected the lab scope and turned on the ignition switch, you are simply looking for the presence (or lack) of serial data. You should not expect to be able to interpret the serial data with a lab scope, as a scan tool does. In a few older systems, you should open the vehicle's doors, turn on some of the accessories such as the A/C system, and/or connect the scan tool to the DLC breakout box while you try to verify the presence of data on the bus with a DSO. With most modern systems, serial data will be present at the DLC just by virtue of turning on the ignition switch. Be sure to properly set the trigger characteristics of the DSO to reliably display the quick, intermittent serial data as it occurs. If the lab scope shows that serial data is present at the DLC but the scan tool fails to communicate, look for a terminal at the DLC that has been inadvertently damaged, resulting in a loose connection with the scan tool.

When performing this type of diagnosis, a DLC breakout box should be used in order to avoid further damage to the terminals of the DLC. The modern "smart" DLC breakout boxes also have LEDs to show whether proper power (red LED) and grounds (green LEDs) are present at the appropriate terminals and also have yellow LEDs that will flash with each binary message received. However, while these DLC breakout boxes do generally have yellow LEDs for the terminals used for the J1850, ISO/KWP, and CAN-C data buses, they do not generally have LEDs for the terminals used by CAN-B data buses and other manufacturer-specific data buses. But these breakout boxes can make great quick-check tools (available from Automotive Test Solutions and AESWave).

Also, most systems design in a wiring hub that allows the technician to separate all the nodes on the network for diagnostic purposes. GM uses a splice bar connector, sometimes referred to as a comb (Figure 8–9), and Chrysler uses a diagnostic junction port (DJP) or a connector at a specific node as the wiring hub of the system. Once the connector is disconnected, a single jumper wire can be used to connect the wire coming from the DLC to one node at a time while retesting with a DSO or scan tool. On a DW bus, two jumper wires may be used. Be sure to follow manufacturer recommendations when performing this test.

A DMM may be used in place of a lab scope to test for the presence of serial data if the passive and active voltage values are known. The DC voltage scale will only display the average voltage. For example, on a GM/Chrysler J1850 SW bus (Class 2 and PCI respectively), the passive voltage is 0 V and the active voltage is between 6 V and 8 V. The average voltage is also lowered by the fact that the data bus generally spends more time at rest (at the passive voltage) than it does communicating. Therefore, a DMM would typically display the average voltage as less than 3 to 4 V, but greater than 0 V.

Diagnosis of a DW CAN Bus

If you cannot communicate on a DW CAN data bus or if U-codes are present, the following steps can be taken with a DMM with the aid of a DLC breakout box to help determine the cause:

- With the ignition turned on, the average DC voltage between bus-H and ground should be around 2.6 V.
- With the ignition turned on, the average DC voltage between bus-L and ground should be around 2.4 V.
- With the ignition switch turned off long enough for any binary communication to stop, an ohmmeter should show that the total resistance between bus-H and bus-L is close to 60 Ω. CAN-C resistance is measured between terminals 6 and 14. CAN-B resistance is measured between terminals 3 and 11. If the ohmmeter shows that the resistance is 120 Ω, an open exists on one of the bus circuits.

Diagnostic Summary

The key to diagnosing problems associated with an automotive data bus is:

- Understand the system you are working on.
- Use a DLC breakout box in conjunction with a lab scope or DMM to test whether or not there is any data on the data bus.
- Use the manufacturers' service manuals to perform pinpoint tests that are specific to the vehicle being worked on.

SUMMARY

This chapter describes revolutionary electrical systems that had a limited introduction into the automotive world in the 1980s; by the end of the 1990s these systems were being used extensively on most makes of vehicles. Such systems will continue to expand in use. These electrical systems will be part of what the technician needs to diagnose when a vehicle comes into the shop for anything from a minor problem such as an inoperative trip computer to something as major as a no-start condition. We discussed how multiplexing is used to reduce hard wiring, thereby reducing weight and increasing dependability. We also saw how multiplexing allows additional features to be added without increasing the number of control modules or wires in the vehicle. Then we looked at how the protocols have been standardized, and we saw details of two of the most popular standardized protocols currently in use—the J1850 VPW protocol and the CAN protocol. Finally, we discussed the ways both a scan tool and a lab scope aid the technician in diagnosing problems with a multiplexed network.

▲ DIAGNOSTIC EXERCISE

A technician is diagnosing a vehicle that was towed into the shop with a "cranks, but won't start" complaint. After verifying the symptom, he connects a scan tool to the DLC and turns the ignition switch on. The scan tool displays a "No communication" message. At that point, he disconnects the bus bar to isolate all the nodes on the data bus (linear topology). He then uses a jumper wire to connect just the PCM to the DLC. With the ignition switched on, the scan tool begins communicating with the PCM. What type of problem is indicated, and how can the technician isolate the problem?

Review Questions

1. *Technician A* says that a multiplexed circuit can be used to communicate multiple messages. *Technician B* says that modern multiplexed circuits transmit serial data to allow computers to communicate with each other. Who is correct?
   A. *Technician A* only
   B. *Technician B* only
   C. Both technicians
   D. Neither technician
2. Advantages of multiplexing include all *except* which of the following?
   A. Dedicated hard wiring and vehicle weight are both reduced.
   B. All functions on the vehicle are under the control of one physical computer.
   C. Dependability is increased.
   D. Computer diagnostic ability is increased.
3. A megabyte of information is equal to which of the following?
   A. 1,000 bytes of information
   B. 1,024 bytes of information
   C. 1,000,000 bytes of information
   D. 1,048,576 bytes of information
4. *Technician A* says that some multiplexing systems use a data bus that consists of two wires that are twisted together, known as a *twisted pair.* *Technician B* says that some multiplexing systems use a data bus that consists of only one wire. Who is correct?
   A. *Technician A* only
   B. *Technician B* only
   C. Both technicians
   D. Neither technician
5. Why are the wires twisted together on a data bus that uses two wires to communicate?
   A. To provide greater physical strength of the wires
   B. To increase the likelihood that if one wire is severed, they both will be severed (for safety purposes)
   C. Because the data bus circuitry flows high levels of current that could affect other nearby circuits if the data bus wires were not twisted together
   D. To minimize the effects of an induced voltage on the data bus
6. *Technician A* says that a data bus that connects the computers in a series circuit is known as a *star configuration.* *Technician B* says that a data bus that connects the computers in a parallel circuit is known as a *loop configuration.* Who is correct?
   A. *Technician A* only
   B. *Technician B* only
   C. Both technicians
   D. Neither technician
7. What is a standardized binary code (or computer language) known as?
   A. Protocol
   B. Node
   C. Byte
   D. Data bus

8. *Technician A* says that multiplexing is unique to the automotive industry and is only used on automotive applications. *Technician B* says that multiplexing allows features to be added to a vehicle while at the same time reducing hard wiring. Who is correct?
   A. *Technician A* only
   B. *Technician B* only
   C. Both technicians
   D. Neither technician
9. How are the effects of an induced voltage minimized on many single-wire data buses?
   A. The wire is routed inside a grounded shield.
   B. The edges of each bit of information are slightly slanted, resulting in a slightly trapezoidal waveform shape.
   C. Multiple messages are transmitted simultaneously to ensure that the other computers on the data bus understand the message.
   D. Capacitors are connected to the data bus to minimize any voltage change.
10. *Technician A* says that PWM serial data has a fixed pulse width with all zeros and ones being the same length. *Technician B* says that VPW serial data has bits of different lengths. Who is correct?
   A. *Technician A* only
   B. *Technician B* only
   C. Both technicians
   D. Neither technician
11. What is a computer that can communicate on a data bus known as?
   A. Protocol
   B. Node
   C. Byte
   D. Control module
12. What is the term for the process of determining which of two computers on a data bus can continue to transmit data when they begin transmitting at the same time?
   A. Arbitration
   B. Qualification
   C. Annotation
   D. Defragmentation
13. On an SAE J1850 VPW data bus, if multiple nodes on the bus tried to transmit data simultaneously, which of the following messages would win arbitration?
   A. 00101001
   B. 00100111
   C. 00100100
   D. 00100011
14. *Technician A* says that the computer that controls a data bus is known as the *master*. *Technician B* says that a dependent computer on a data bus is known as a *slave*. Who is correct?
   A. *Technician A* only
   B. *Technician B* only
   C. Both technicians
   D. Neither technician
15. Which of the following protocols was mandated under OBD II standards to replace all other protocols by the 2008 model year for aftermarket scan tool communication with systems that affect the vehicle's emissions?
   A. ISO 9141-2
   B. ISO 15765-4 CAN
   C. SAE J1850 VPW
   D. ISO 14230-4 (KWP 2000)
16. *Technician A* says that most multiplexing networks may allow you to access the computers for diagnostic purposes by connecting a scan tool to the DLC. *Technician B* says that some multiplexing networks may allow you to access the computers for diagnostic purposes through the climate control and/or audio system control head. Who is correct?
   A. *Technician A* only
   B. *Technician B* only
   C. Both technicians
   D. Neither technician

17. *Technician A* says that when you connect a scan tool to the DLC, it becomes a node on the network to which it is connected. *Technician B* says that either a scan tool or a lab scope can be used by the technician to interpret the serial data on a data bus. Who is correct?
    A. *Technician A* only
    B. *Technician B* only
    C. Both technicians
    D. Neither technician
18. Which of the following conditions can keep a scan tool from communicating on a data bus when it is connected to the DLC?
    A. The scan tool is not properly powered up.
    B. The VIN is improperly entered into the scan tool.
    C. Serial data does not exist on the data bus.
    D. All of the above.
19. Which of the following is the best tool to check for the presence of serial data on a data bus?
    A. Scan tool
    B. Lab scope
    C. Test light
    D. Short finder
20. Which of the following is the best tool to retrieve information from or issue commands to computers on a data bus?
    A. Scan tool
    B. Lab scope
    C. Test light
    D. Short finder

Chapter 9

Hybrid and Electric Vehicles

OBJECTIVES

Upon completion and review of this chapter, you should be able to:

- ❑ Understand the difference between a series-hybrid-drive and a parallel-hybrid-drive vehicle.
- ❑ Define the components and system features of Honda gasoline/electric hybrid-drive systems.
- ❑ Define the components and system features of Toyota gasoline/electric hybrid-drive systems.
- ❑ Define the components and system features of Ford gasoline/electric hybrid-drive systems.
- ❑ Explain the basic principles and understand the advantages associated with a fuel cell vehicle.

KEY TERMS

AC Synchronous Motor
Belt Alternator Starter (BAS)
Continuously Variable Transmission (CVT)
Electric Motor/Generator (EMG)
Flywheel Alternator Starter (FAS)
Fuel Cell Electric Vehicle (FCEV)
Grade Logic Control
High-Voltage (HV) Battery
Home Energy Station (HES)
Parallel Hybrid
Polymer Electrolyte Membrane (PEM)
Proton Exchange Membrane
Regenerative Braking
Series Hybrid

In an effort to dramatically increase fuel mileage and reduce CO_2 emissions, most manufacturers now offer straight electric vehicles, gasoline/electric hybrid vehicles, and plug-in hybrid vehicles. Fuel cell vehicles will soon be released to hit the showroom floors. This chapter is intended to help you understand how these technologies operate by describing the components and concepts that make up these systems. Several of the early mainstream manufacturers' vehicles are used to demonstrate these concepts and components.

WARNING: The high-voltage systems discussed in this chapter are only to be worked on by qualified service personnel who have been trained in the specific manufacturer training programs that pertain to these systems. Any attempt to diagnose or repair a high-voltage system without the proper manufacturer training puts the technician at risk of receiving a fatal electric shock. Also, at the high voltages that are present in these systems, there is a very real threat of an electrical short creating either an arc flash, resulting in the release of tremendous thermal energy, or an arc blast, resulting in the release of tremendous pressure. (An arc blast results from the instantaneous vaporization of copper, creating a rapid pressure expansion.) The discussion of the systems in this chapter is for informational purposes only and is not intended

to provide the training that is required to work safely on these systems.

COMMON ACRONYMS

Following is a list of acronyms that are common to the vehicles discussed in this chapter:

- BEV: battery electric vehicle (also referred to as an *EV*)
- EV: electric vehicle (also referred to as a *BEV*)
- FCEV: fuel cell electric vehicle (also referred to as a *FCV*)
- FCV: fuel cell vehicle (also referred to as a *FCEV*)
- HEV: hybrid electric vehicle
- PHEV: plug-in hybrid electric vehicle
- EREV: extended range electric vehicle

COMMON COMPONENTS

Electric Motor/Generator

An **electric motor/generator (EMG)** is defined as a component that can (a) turn electrical energy into mechanical energy (as a motor does) and (b) turn mechanical energy into electrical energy (as a generator does). An EMG can be used in combination with a gasoline engine to accelerate the vehicle. In city driving, an electric motor has an advantage over a gasoline engine in that it can develop its maximum torque beginning just above zero RPM. A gasoline engine must reach several thousand RPM to generate its maximum torque. An EMG is also known as an *electric machine* or simply a *motor/generator (MG)*. The EMG on a modern gasoline/electric hybrid vehicle uses three-phase AC voltage.

High-Voltage Battery

A **high-voltage (HV) battery**, as shown in Figure 9–1, consists of multiple cells wired together in series. Its total voltage output may range from 37.8 V DC to 144 V DC on a mild HEV. (A 42 V system has three 12.6 V batteries in series for a total of 37.8 V plus charging system voltage.) A strong HEV, sometimes referred to as a *full HEV*, may have voltage at the HV battery of between 200 and 300 VDC. This voltage may also be up-converted to a substantially higher voltage. It is important to note that the voltage at the HV battery is DC voltage. It is electronically converted to AC voltage for use by the EMGs. The wire harnesses/cables that are used to route this high voltage from the HV battery to other components are colored as follows:

Figure 9–1 The HV battery in a Toyota Camry hybrid.

- A blue cable or harness has 42 V present.
- An orange cable or harness has more than 42 V present—this may be from 144 to 650 V.

Do not confuse these with the yellow harnesses used to identify wiring associated with passive restraint systems.

It should also be noted that manufacturers have experimented with using super-capacitors in place of the HV battery. It is still considered possible that these capacitors could someday replace battery technology. Capacitors will charge and discharge much more quickly than batteries, are extremely lightweight compared to batteries, and will not likely wear out like batteries do.

Inverter

An inverter, as shown in Figure 9–2, is used to convert the HV battery's DC voltage to three-phase AC voltage for use by the EMGs. The inverter typically uses *insulated gate bipolar transistors* (*IGBTs*) to perform this function (Figure 9–3).

Internal Combustion Engine

The internal combustion engine (ICE) used on HEVs is usually smaller in displacement than in gasoline-powered vehicles because of the fact that it is used in conjunction with an EMG. Also, many of these engines use an Atkinson cycle cam grind that reduces the engine's power while increasing its fuel economy. This reduction in power is not generally noticeable to the driver because the EMG compensates for it.

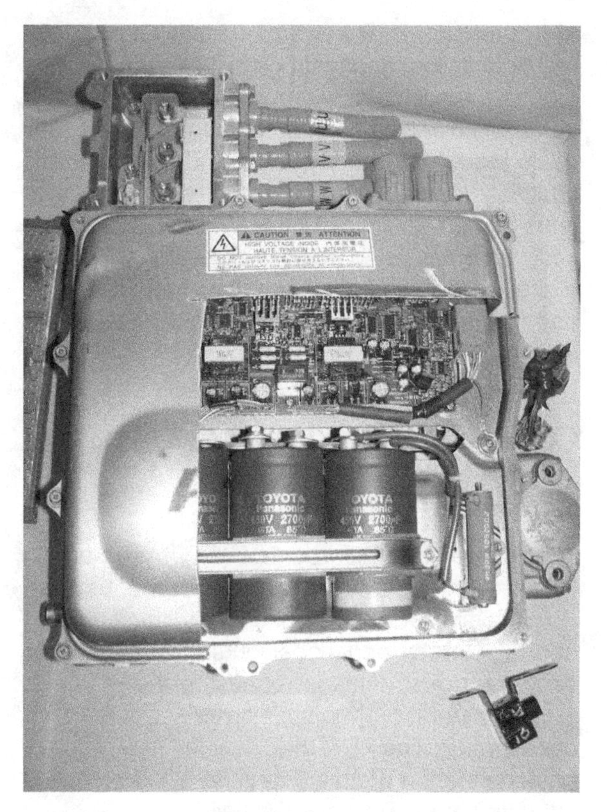

Figure 9–2 A cutaway view of a typical inverter.

Auxiliary 12.6-Volt Battery

Most HEVs use an auxiliary 12.6 V battery to power many of the vehicle's systems, circuits,

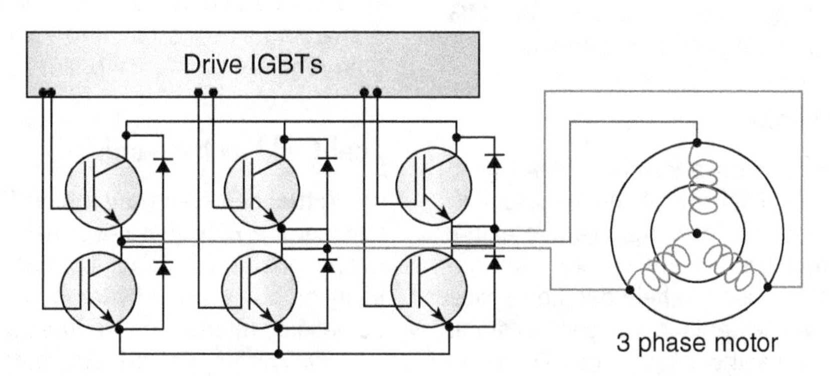

Figure 9–3 Insulated gate bipolar transistors (IGBTs) used to convert DC voltage to three-phase AC voltage.

and components. For example, these vehicles still use light bulbs, gauges, power seat and window motors, windshield wiper motors and washer pumps, and electric fuel pumps that are still designed to operate at 12.6 V. And the electronic control modules that control the various systems still power up at 12.6 V. While this could be accomplished using a PWM signal to reduce the higher voltage from the HV battery to a lower voltage for use by these systems, most HEVs retain an auxiliary battery for this purpose.

COMMON CONCEPTS

Electronic Throttle Control

Electronic throttle control (ETC) is used on HEVs. With two power plants on board, both an ICE and an EMG, the driver uses the throttle pedal and accelerator pedal position sensor (APPS) to send the driver demand request to the PCM, be it for acceleration, maintaining cruise speed, or deceleration. The PCM then decides how much of an acceleration or cruise request is to be carried out using the ICE versus the EMG.

Auto Start/Stop

Both mild HEVs and strong HEVs use an auto start/stop feature with the ICE. When decelerating or even coming to a full stop, most HEVs can shut the ICE totally off to save fuel. When the driver presses the throttle pedal again, the PCM can seamlessly restart the ICE to help the EMG to get the vehicle moving again as needed.

Regenerative Braking

Most HEVs use **regenerative braking** to translate the forward inertia of the vehicle into electrical energy during deceleration and braking. This is used to recharge the HV battery. Then the same energy can be used to help get the vehicle moving again. (With a traditional braking system, vehicle braking turns the forward inertia of the vehicle into wasted heat energy.)

While hybrid vehicles do also have a hydraulic braking system, a control module determines how much slowing is to be done using regenerative braking versus hydraulic braking. Because an electronic control module makes this determination, this is actually a form of a *brake-by-wire* system. Because the regenerative braking system is used first to slow the vehicle during coasting or braking and then the hydraulic brakes are only used if additional braking force is needed, the longevity of the hydraulic brakes is greatly increased.

HYBRID AND ELECTRIC VEHICLE DESIGNS

Battery Electric Vehicle

A battery electric vehicle (BEV), sometimes also referred to as an *electric vehicle* (EV), is a vehicle that uses an HV battery and an EMG to accelerate the vehicle and maintain cruise. It does not have an ICE on-board.

A BEV uses regenerative braking to recharge the HV battery during deceleration and braking. However, a BEV has a range, top speed, and acceleration ability that is limited by the quantity of cells that make up the HV battery, as well as the technology used in battery construction. Therefore, a BEV is usually only suitable for commuter driving and must be recharged at regular intervals, usually overnight. It is not likely that a BEV would become popular for road trips, even if charging stations became common, due to the time required for the HV battery to recharge.

Fuel Cell Electric Vehicle

A **fuel cell electric vehicle (FCEV)** is another version of a BEV that never has to be plugged in to be recharged. Instead, it generates its own electricity on board using hydrogen fuel cells. It is also sometimes referred to as a *fuel cell vehicle* (*FCV*).

Sir William Robert Grove (1811–1896) first invented fuel cell technology in 1839. He determined that if the charging of a battery (sending electrical current through water) could turn the water in the electrolyte into oxygen and hydrogen, then there must be a way to perform this function in reverse—that is, through the combining of oxygen and hydrogen, water and electron current flow could be produced. In the 1960s, the technology was used to provide electrical power on spacecraft in the Gemini and Apollo programs. During the 1980s, automobile manufacturers began working on applying fuel cell technology to automobiles. In fact, a *hydrogen highway* has already been set up along the West Coast of the United States, with hydrogen refueling stations from Mexico to Canada.

Fuel Cells. Several types of fuel cells (distinguished by the type of electrolyte used) have been created. Honda, Ford, Toyota, and General Motors are using **polymer electrolyte membrane (PEM)** fuel cells. These fuel cells are also commonly known as **proton exchange membrane** fuel cells (Figure 9–4). PEM fuel cells were first used in the Gemini space program. PEM fuel cells use solid polymer membrane (a thin plastic film) as an electrolyte. PEM fuel cells combine hydrogen (H_2) with oxygen (O_2) from atmospheric air to create electrical current flow. A hydrogen atom has a single electron orbiting around a single proton in its nucleus and contains no neutrons. If hydrogen is brought to one side of the membrane and oxygen is brought to the other side, the hydrogen is attracted to the oxygen. However, the membrane will allow only the hydrogen atoms' protons to pass through it (thus the name proton exchange membrane). The hydrogen atoms' electrons cannot pass through the membrane but must be directed through another path to get to the other side, thus generating electrical current flow.

PEM fuel cells are capable of generating a very powerful electric current, ranging from 1 W to 250 kW, depending on their physical size. Fuel cells are grouped together to create fuel cell stacks. PEM fuel cells are capable of operating at relatively low temperatures (less than 212°F or 100°C). The resulting emissions of an operating

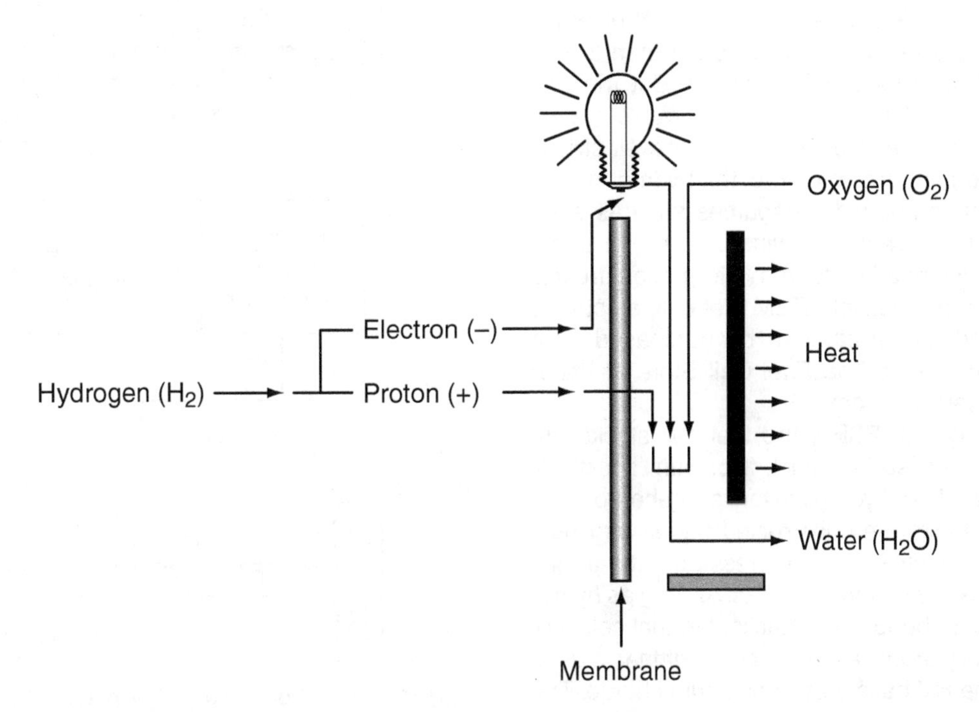

Figure 9–4 PEM fuel cell.

PEM fuel cell are only water and heat. No toxic emissions are produced. Nor does a PEM fuel cell produce any CO_2 emissions.

Another advantage of powering a vehicle by using hydrogen to generate electricity via a fuel cell is the high energy efficiency that is realized (about 83 percent efficiency) when compared to the efficiency of a gasoline engine (typically considered to be between 30 and 40 percent).

The hydrogen consumed by a PEM fuel cell can be obtained from several different sources including water, gasoline, methanol, and magnesium. For example, a *reversible fuel cell* can be used to turn liquid water into separated oxygen and hydrogen by applying a small voltage to it. Then this same fuel cell can be used to turn the same oxygen and hydrogen back into water, creating electrical power in the process. Unfortunately, when the total wattage (power) is considered, there is a net power loss in this process. But there are still some advantages: If electrical power generated from coal can be used to obtain the hydrogen, the dependence on oil is still lessened; and it is easier to control emissions from a stationary power plant than it is to control them on a moving vehicle. But the real advantage can be realized if a renewable energy resource can be used to create the electricity that is used, in turn, to extract the hydrogen from its original compound—resources such as solar power, hydro power, and wind power. (For some experiments relating to the information in this paragraph, a reversible PEM fuel cell, electronic board, and model car can be purchased from a company called The Fuel Cell Store at http://www.fuelcellstore.com.)

On a typical FCEV, the fuel cell stacks are contained in a six-inch-thick floor pan. There will be one or two hydrogen tanks on-board. The hydrogen tanks are constructed under very rigid guidelines, similar to compressed natural gas (CNG) tanks and propane tanks. As long as hydrogen is fed to the fuel cell stacks, the fuel cells will continuously produce electrical power that is used to keep the HV battery at its optimum charge. The optimum charge is considered to be a little less than 100 percent, usually around 80 to 90 percent, to allow the HV battery to take advantage of regenerative braking. Because the EMG can draw power from the HV battery and is not restricted to the wattage being produced by the fuel cell stacks, good acceleration ability is maintained.

Hybrid Electric Vehicle

A hybrid electric vehicle (HEV) is a vehicle that uses both an ICE and an EMG to accelerate the vehicle and maintain cruise. There exist different versions of an HEV: a parallel hybrid, a series hybrid, and a series-parallel hybrid.

Parallel Hybrid Electric Vehicle. A **parallel HEV** has the ICE and EMG arranged in such a manner that both the ICE and the EMG can contribute power to accelerate the vehicle and maintain cruise (Figure 9–5).

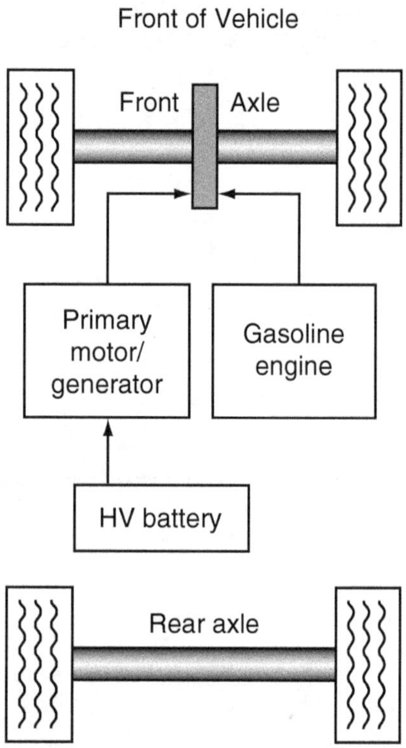

Front of Vehicle
Front Axle
Primary motor/ generator
Gasoline engine
HV battery
Rear axle

Figure 9–5 Power flow of a parallel hybrid drive vehicle.

Series Hybrid Electric Vehicle. A **series HEV** has the ICE and EMG arranged in such a manner that only the primary EMG can contribute power to accelerate the vehicle and maintain cruise. The ICE is strictly used for operating a secondary EMG as a generator for the purpose of providing electrical power to the primary EMG and the HV battery (Figure 9–6). The Chevrolet Volt is an example of a series HEV.

Series-Parallel Hybrid. A series-parallel HEV has the ICE and EMG arranged in such a manner that it can operate both as a parallel HEV and a series HEV. This arrangement makes up the vast majority of HEVs.

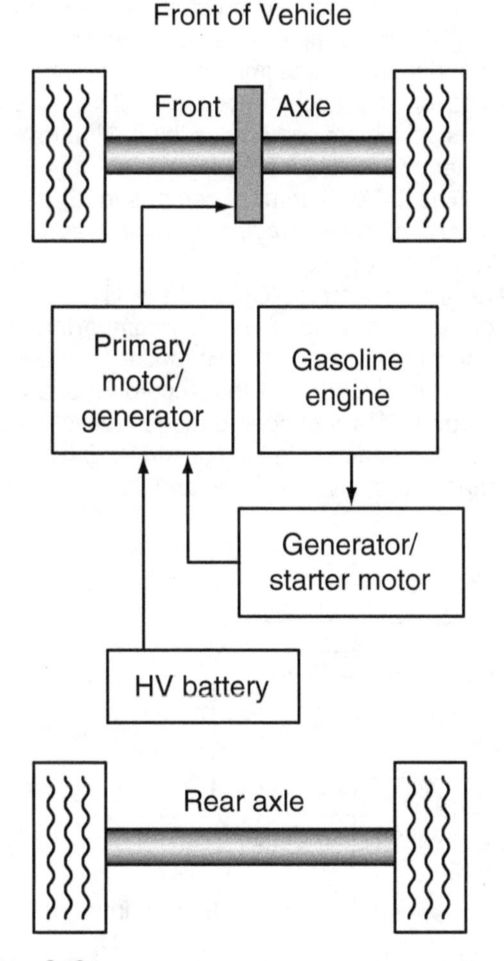

Figure 9–6 Power flow of a series hybrid drive vehicle.

Plug-In Hybrid Electric Vehicle

A plug-in hybrid electric vehicle (PHEV) is an HEV that can also be recharged off the electric grid by connecting its power cord to 120 VAC or 240 VAC, similar to a straight EV. Electric grid power is considerably less expensive per mile traveled than gasoline. One major difference between a standard HEV and a PHEV is that the electronic control unit (ECU) that controls the charging of the HV battery on a standard HEV strives to maintain a state-of-charge (SOC) that is just a little less than 100 percent, around 80 to 90 percent. This allows the vehicle's charging system to be able to take advantage of regenerative braking. However, on a PHEV, the HV battery ECU allows the battery's SOC to discharge down to around 30 percent to allow the HV battery to be recharged using less expensive grid power. The 30 percent SOC may vary depending on how much reserve is desired to allow for hard acceleration.

A PHEV can operate as a BEV with the HV battery powering a primary EMG to accelerate the vehicle and maintain cruise until the 30 percent SOC is reached. Then the ICE will start and operate a secondary EMG to keep the HV battery's SOC from falling lower than 30 percent and to provide electrical power to the primary EMG. In this mode, the PHEV is operating as a series HEV. The distance that the PHEV can operate as a BEV depends upon the quantity of cells that make up the HV battery, as well as the technology used in battery construction. The driver may be able to select when the vehicle will operate as a BEV. A PHEV can also be called an *extended range electric vehicle* (*EREV*) in that it operates initially as a BEV, and then the operation of the ICE can extend this range beyond what the HV battery is capable of. Therefore, this is a modification of a BEV that can be used for cross-country road trips and is not limited to city commutes.

HONDA HYBRIDS

Integrated Motor Assist

Honda gasoline/electric hybrid-drive vehicles use a system called *Integrated Motor Assist (IMA).* This IMA system was first used on the Honda Insight (a vehicle that was manufactured as a hybrid only). The IMA system was then added to two mainstream Honda vehicles: the Honda Civic and the Honda Accord. The Honda Civic Hybrid and the Honda Accord Hybrid are offered alongside the standard gasoline engine versions of these vehicles.

The IMA system uses a single electric motor/generator (EMG) coupled with a gasoline engine to provide acceleration and cruising for the vehicle, as well as regenerative braking. Honda offers a choice of two transmissions—either a manual transmission or a **continuously variable transmission (CVT).** The CVT transfers optimum torque either from the gasoline engine, from the electric motor, or from both to be applied to the front-drive wheels regardless of vehicle speed or load. As a result, the driver never has to "shift gears" as with normal multispeed transmissions.

Honda's IMA system is considered to be a mild hybrid in that the gasoline engine is the primary power source to accelerate the drive wheels, as shown in Figure 9–7, and will always run except when the vehicle is stopped. The IMA system's single electric motor/generator (EMG) runs to assist the gasoline engine as the need arises. During acceleration and under heavy engine load, the EMG contributes considerable torque, resulting in both lower fuel consumption and powerful acceleration. At cruising speeds, when engine load is low, the EMG shuts down, allowing the gasoline engine to be the sole source of power. During deceleration and braking, the EMG converts the vehicle's kinetic energy into electricity to recharge the high-voltage battery (referred to as *regenerative braking).* The EMG also functions as a high-RPM starter motor to start the gasoline engine as needed. When stopping—as at a traffic light, for example—the gasoline engine shuts off and then restarts immediately when the driver steps on the accelerator pedal. This automatic idle-stop system contributes to both improved fuel efficiency and lower emissions.

The IMA system is composed of an ultra-thin DC brushless motor, a nickel-metal hydride (NiMH) battery, and a power control unit (PCU). The IMA system's PCU and the HV battery are both located under the cargo compartment floor behind the rear seat. The HV battery in the Insight, the Civic Hybrid, and the Accord Hybrid all use 120 cylindrically shaped batteries (cells) that produce 1.2 V each for a total of 144 VDC. These batteries are relatively smaller and lighter than those

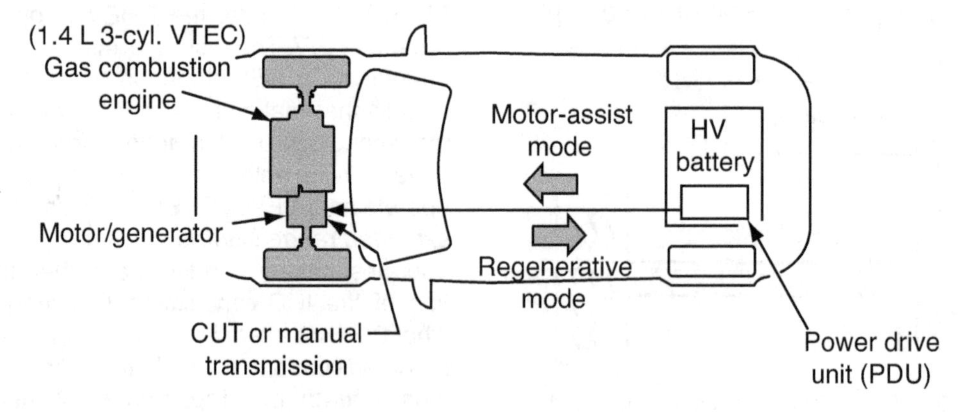

Figure 9–7 Honda IMA system (Insight).

in most other hybrid vehicles, a technology that Honda pioneered in its EV PLUS electric car. The high-voltage cables are bright orange to make them easily identifiable.

Insight

The Honda Insight was first introduced in the United States in early 2000 and is billed as the first gasoline/electric hybrid-drive car offered in the United States. With EPA fuel economy ratings of 61 MPG city/70 MPG highway, this vehicle earns the highest fuel economy rating for any passenger car. However, in the interest of achieving these figures, the Insight is designed as a small two-seater passenger car weighing less than 1,900 pounds. In addition to the IMA system's motor/generator, the Insight uses a tiny 1.0L three-cylinder VTEC gasoline engine that weighs only 124 pounds and can produce up to 67 horsepower at 5,700 RPM. It also uses Honda's lean-burn technology to maximize fuel economy.

In 2014, the Honda Insight was still offered as an HEV, although it is a somewhat larger vehicle than its predecessor. For 2014 the Insight was offered with a 1.3L, eight-valve, SOHC i-VTEC four-cylinder engine. The combined engine and electric motor horsepower is rated at 98 HP.

Civic Hybrid

The Honda Civic Hybrid was introduced in March 2002, marking the first time an existing nameplate had been offered with a gasoline/electric hybrid-drive system. The Civic Hybrid qualifies as an Advanced Technology–Partial Zero Emission Vehicle (AT-PZEV). The EPA fuel economy ratings are 48 MPG city/47 MPG highway for an automatic CVT Civic Hybrid and 46 MPG city/51 MPG highway for the manual transmission version. In addition to the IMA system's motor/generator, the Civic Hybrid uses a small, newly developed 1.3L four-cylinder VTEC gasoline engine that can produce up to 86 HP. The new engine incorporates i-DSI lean-burn combustion technology combined with a VTEC *Variable Cylinder Management (VCM)* system.

The 1.3L i-DSI lean-burn combustion technology uses an engine design that promotes rapid combustion with the use of two spark plugs per cylinder. This allows the fuel-air ratio to run quite lean without lean misfire occurring. Thus, fuel economy is further improved.

The VTEC VCM system uses rocker arms that are similar to those used in the Honda VTEC engines, except that the synchronizing piston allows a rocker arm to engage an adjacent rocker arm to provide "valve-lift mode" (normal operation), or it may be retracted into itself to provide a "rocker arm idle mode." The latter mode effectively fails to open the intake and exhaust valves, thus leaving them closed, allowing the deactivated cylinders' pistons to act as air springs. This reduces the normal pumping losses. The fuel injector for the deactivated cylinders is also deenergized. The 1.3L i-DSI engine's VCM function can deactivate as many as three of its four cylinders during times of low power requirements and can thus achieve a reduction of engine friction of 50 percent during deceleration (as compared with the previous IMA system). This improves the vehicle's electrical regenerative efficiency and thereby increases fuel mileage even further.

Accord Hybrid

The Honda Accord Hybrid was introduced as a 2005 model, marking the first time a V6 was used in a hybrid in conjunction with an electric motor, thus giving the vehicle V8-like performance and exceptional fuel mileage. This hybrid utilizes Honda's third generation of IMA technology. With EPA fuel economy ratings of 29 MPG city/37 MPG highway, the Accord Hybrid's 3.0L i-VTEC V6 and IMA system can produce a combined net peak horsepower of 255 while achieving near-peak torque across most of its operating range. In addition to the IMA hybrid system, the V6 in the Accord Hybrid also uses the VCM cylinder deactivation system when the need for sustained power is low. The VCM system deactivates the three cylinders on the rear bank. When operating in three-cylinder mode, engine vibration is reduced through the use of an "active control engine mount," which compresses/extends an actuator in same-phase, same-period motion to dampen the engine's vibration. Similarly, an audio speaker creates an opposite phase sound (called "Active Noise Control") to provide a frequency-canceling effect. The Active Noise Control uses microphones that are placed in the front and rear of the passenger compartment to monitor the low frequency "booming" noise created by three-cylinder engine operation. It then generates a counter-frequency sound wave through the vehicle's audio system that effectively cancels the booming noise. This makes for a quieter interior, thus leaving the driver unaware of changes in cylinder activation.

The Accord Hybrid also uses a control module strategy called **Grade Logic Control** to reduce the need for the driver to compensate constantly at the throttle pedal for transmission upshifts and downshifts when driving in mountainous terrain.

For example, most vehicles use input only from the TPS and VSS to determine upshift and downshift points. As a result, on certain uphill grades the driver must depress the throttle pedal farther as the vehicle starts to lose speed in order to downshift and bring the vehicle back to the desired speed. However, the vehicle is then likely over-accelerating, requiring the driver to back off the throttle position to avoid an over-speed condition, at which point the transmission upshifts again and the vehicle speed begins to decrease again. In the case of a downhill grade, the driver is constantly downshifting to reduce vehicle speed and then upshifting again to avoid loss of too much vehicle speed. This is commonly known as "gear hunting."

Honda's Grade Logic Control strategy uses input from the accelerator pedal position (APP) sensor, vehicle speed sensor (VSS), and the brake applied switch. The control module also uses vehicle speed to determine rate of acceleration or rate of deceleration. It even monitors barometric pressure. As a result, the control module can figure out when the vehicle is on an uphill slope or a downhill slope. The control module then adjusts the engine speed and/or EMG output accordingly and also makes related adjustments within the CVT. This reduces the need for the driver to keep compensating with the throttle pedal when driving in mountainous terrain.

TOYOTA HYBRIDS

Toyota has earned a reputation for being a major player in the manufacturing of hybrid electric vehicles (HEVs). Today, Toyota offers several versions of an HEV as follows:

- Prius Hybrid
- Prius c Hybrid
- Prius v Hybrid
- Camry Hybrid
- Avalon Hybrid
- Highlander Hybrid
- Prius Plug-in Hybrid Electric Vehicle

While Toyota currently offers several models that use the Toyota Hybrid System, the emphasis in this part of the chapter is on the original Prius, the Highlander four-wheel drive hybrid, and the Prius as a plug-in hybrid electric vehicle (PHEV).

Toyota Hybrid System (THS)

The Toyota Prius made its debut in Japan at the end of 1997 and was first offered in the United States and Canada as a 2001 model. The Prius is a front-wheel-drive vehicle that uses the Toyota Hybrid System (THS). A second generation of THS, known as THS-II, was introduced in the 2004 model year. With several improvements having been made, the 2004 Prius with THS-II runs cleaner than the original Prius, earning both SULEV (Super Ultra-Low Emissions Vehicle) and AT-PZEV ratings. At 2,900 pounds, the Prius has interior and trunk space that is similar to that of a Corolla. With an 11.9-gallon fuel tank, the Prius, with EPA ratings of 60 MPG in the city and 51

MPG on the highway, can exceed 500 miles on a tank of fuel. (The quoted mileage estimates were EPA estimates through 2007. In 2008, the EPA recalculated fuel mileage estimates to be more realistic. This revision in the method of calculating fuel economy seemingly hurt the high-mileage vehicles the most, revising small hybrid cars downward more than low-mileage vehicles. The EPA estimates the fuel mileage of a 2010 Toyota Prius at 51 MPG city and 48 MPG highway. For more information, go to www.fueleconomy.gov.)

It is also worth noting that Toyota has licensed its hybrid system to Nissan, which introduced the technology on the 2006 Nissan Altima.

Major Components and Operation of the Toyota Hybrid System

High-Voltage Battery. A high-voltage (HV) battery is located in the forward area of the trunk, just behind the rear seat (Figure 9–1 and Figure 9–8). The HV battery provides electrical energy to the drivetrain as required. The HV battery consists of multiple nickel-metal hydride (NiMH) cells that are connected in series.

In the original THS design (through the 2003 model year), there are 38 modules, each consisting of six cells (a total of 228 cells) at 1.2 V each for a total of 273.6 VDC. When the HV battery is being charged, the total voltage approaches 300 VDC. The cells are connected by a single contact connecting one cell to the next. While the cost of replacing the entire HV battery can be quite high, Toyota provides a 100,000-mile warranty on it.

The HV battery has an airflow cooling system. Fans within the HV battery assembly blow air across the battery's cells. The vent for this system is located externally on the left rear of the Prius. It looks like a trim decoration just forward of the rear window and just to the rear of the driver's door.

Auxiliary Battery. A 12.6 V auxiliary battery is located on the left side of the trunk. This battery provides electrical energy to systems other than the drivetrain—lighting and electronic control systems, for example. If the auxiliary battery has been drained (for example, by leaving the headlights turned on), it may be jump-started using the traditional method from a 12.6 V donor vehicle.

MG2. Unlike Honda's IMA system, the Prius uses a three-phase AC electric motor/generator (MG2), as in Figure 9–8 and Figure 9–9, to provide most of the power for light to moderate acceleration and for cruising up to 40 MPH. MG2 uses electrical energy from the HV battery to create mechanical energy to accelerate the vehicle. MG2 also uses mechanical energy from the front-drive wheels during coasting and braking to create electrical energy to recharge the HV

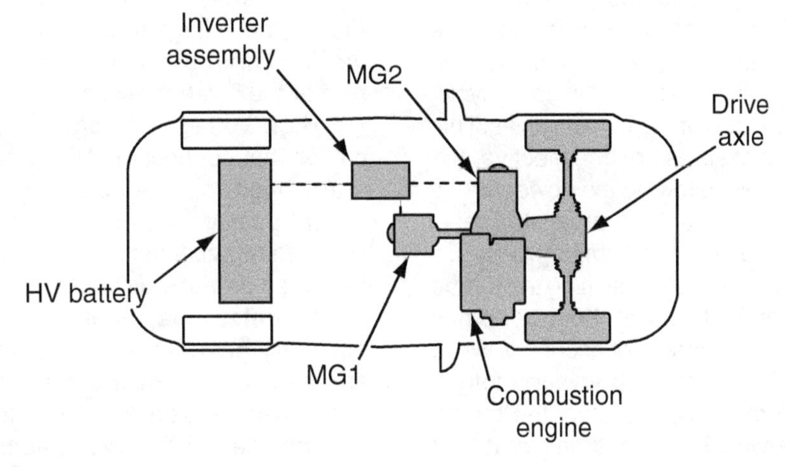

Figure 9–8 Toyota Prius components.

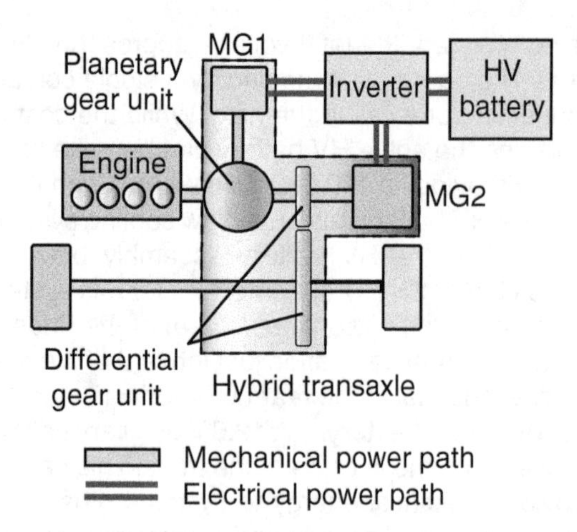

Figure 9–9 The Toyota Hybrid System operates as both a parallel hybrid and as a series hybrid.

battery. With the original THS design, MG2 was rated at 30 kW or 44 HP and 258 ft.-lb of torque, with the maximum torque being achieved from 0 to 400 RPM.

INZ-FXE 1.5L Gasoline Engine with VVT-i. The Prius uses the INZ-FXE 1.5L four-cylinder gasoline engine with the variable valve timing with intelligence (VVT-i) system. The engine is mounted transversely in the engine compartment on the passenger side of the vehicle (Figure 9–8). Through the 2003 model year, it was rated at 70 HP and 82 ft.-lb of torque. The engine is limited to 5,000 RPM, thereby allowing the use of lighter components designed to improve efficiency. The engine also uses the Atkinson cycle, which retards the closing of the intake valves. This reduces cylinder pressures to an effective 9:1 compression ratio while allowing the power stroke to take full advantage of a 14:1 expansion ratio. This also reduces the engine's pumping losses.

The Atkinson cycle results in shifting the torque curve into a higher RPM range than a normal engine, resulting in increased fuel economy. While this design does not have much low-end torque for an initial acceleration from a stop, the motor/generator known as MG2 is more than capable of providing this initial torque.

Unlike Honda's IMA system, the Toyota gasoline engine is not the primary source of power for accelerating at low speeds, but rather is a secondary source of power used to assist MG2 when the acceleration load requires it or when the HV battery needs to be recharged.

MG1. A second motor/generator known as MG1, shown in Figure 9–8 and Figure 9–9, is used to start the gasoline engine when needed. Once the engine is running, MG1 acts as a generator to produce electrical energy. In other words, MG1 uses electrical energy from the HV battery to create mechanical energy to start the engine. MG1 also uses mechanical energy from the engine when it is running to create electrical energy, which is used to recharge the HV battery and to provide additional electrical power for MG2.

Inverter. The inverter is used to convert the HV battery's DC voltage to three-phase AC voltage to power either MG1 (to start the engine) or MG2 (to accelerate the vehicle). It also supplies MG2 with the AC voltage produced by MG1 during vehicle acceleration. The inverter is also used to convert AC voltage produced by MG1 (when the engine is running) or produced by MG2 (during coasting or braking) to DC voltage to charge the HV battery. The inverter is located above MG2 in the engine compartment. A separate cooling system and radiator are used to carry away excess heat from the inverter.

Converter. A voltage converter is used to reduce the voltage from the HV battery to charge the auxiliary battery. The converter is physically part of the inverter assembly.

High-Voltage Cables. The high-voltage cables that connect the HV battery to the inverter are a bright orange color. These cables should never be handled, except by properly trained service personnel, and then only after the HV battery has been properly disabled.

Planetary Gear Set. Due to an ingenious planetary gear set, which operates as a "power split device," the Prius can operate as a parallel hybrid vehicle, as a series hybrid vehicle, or both. As a parallel hybrid drive system, MG2, the gasoline engine, or both can be used to accelerate the vehicle. This allows the vehicle to take advantage of the characteristics of both power plants. When the gasoline engine runs, however, MG1 generates electrical energy, which can be used in addition to the HV battery's electrical energy to operate MG2, thus allowing the system to act as a series hybrid. In fact, under heavy load the HV battery provides electrical energy to MG2 to drive the vehicle, the gasoline engine runs (causing MG1 to provide additional electrical energy to MG2 to drive the vehicle), and the gasoline engine also provides torque directly to the front-drive wheels, thus allowing the drivetrain to act simultaneously as a series hybrid and a parallel hybrid.

The planetary gear set, shown in Figure 9–9 and Figure 9–10, sometimes referred to as a clutchless single-speed transmission, actually acts as a CVT, allowing optimum torque either from the gasoline engine, from MG2, or from both to be applied to the front drive wheels regardless of vehicle speed or load. As a result, the vehicle never has to "shift gears" as with normal multi-speed transmissions.

Hybrid Vehicle Electronic Control Unit. The hybrid vehicle electronic control unit (HV ECU) is shown in Figure 9–11. The HV ECU monitors and controls all of the power flow among all of the key hybrid system components, including the gasoline engine, MG1, MG2, and the HV battery. It is programmed to achieve optimum performance, efficiency, fuel economy, and emissions.

Initial Starting. After the ignition key is turned to *Run*, a READY indicator is displayed on the instrument panel. At this point, the gasoline engine may or may not begin running, depending on the HV battery's state of charge. In any case, once the READY indicator is displayed, the vehicle is ready to begin driving.

WARNING: When the ignition is turned on and the READY indicator is displayed, the engine will automatically begin running unexpectedly from time to time. If the hood is open, keep hands out of the area of the engine and verify that the READY indicator is not illuminated when preparing to change the engine oil.

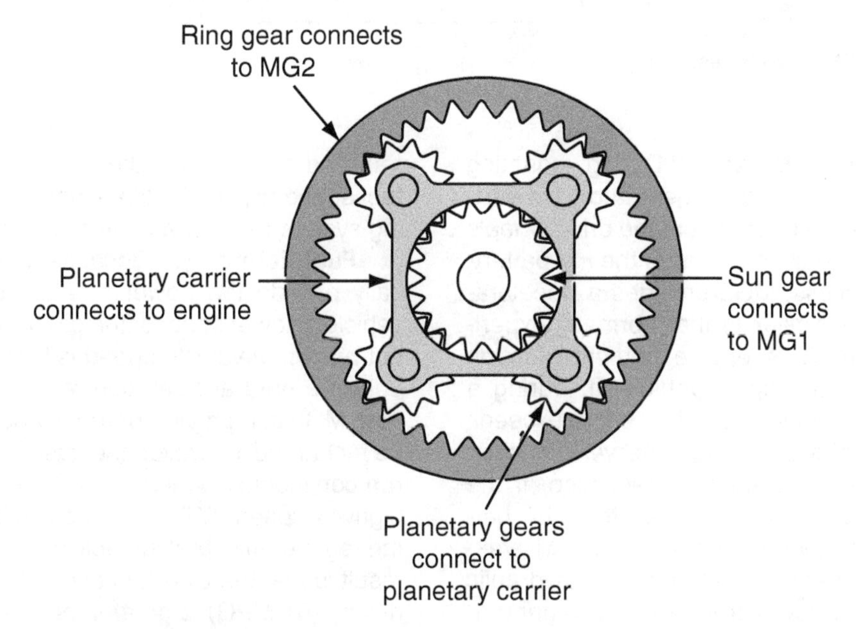

Figure 9–10 THS planetary gear set.

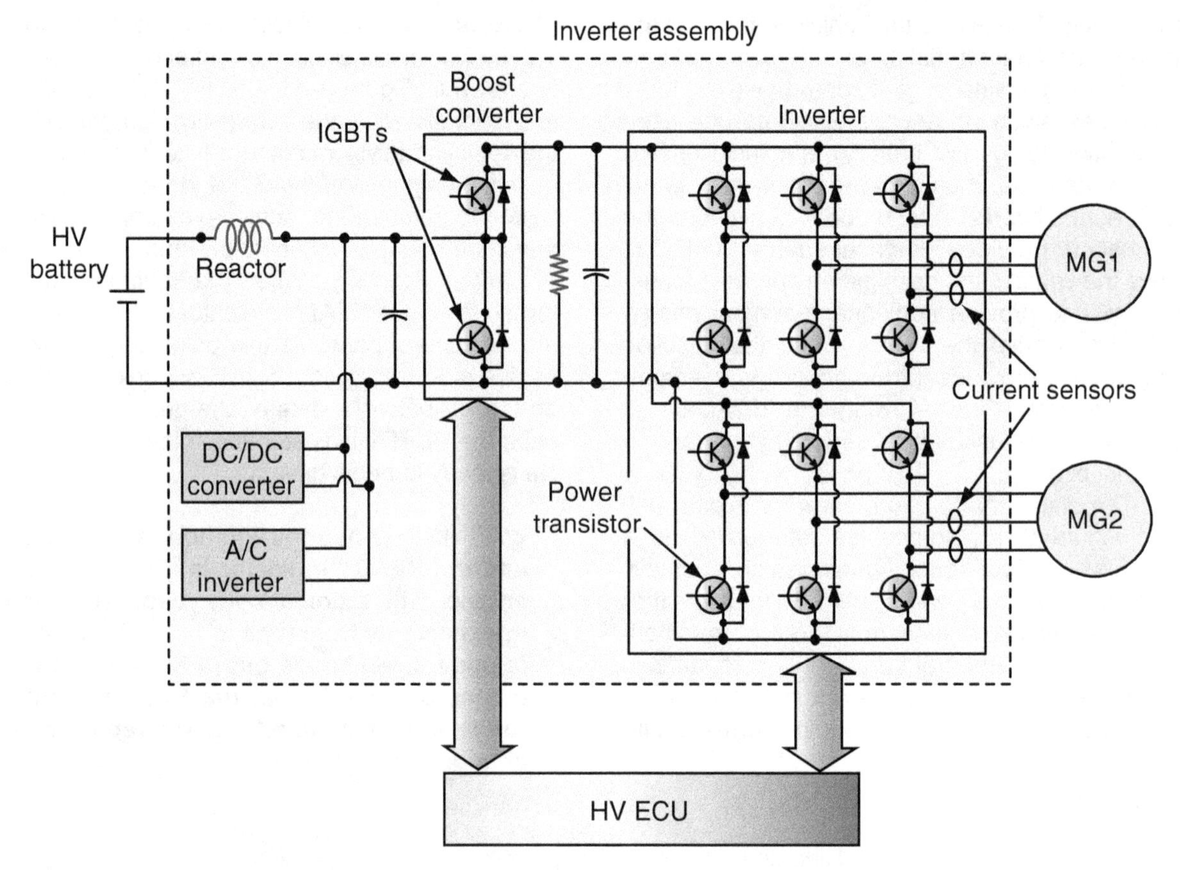

Figure 9–11 THS inverter assembly.

Regenerative Braking. During coasting or braking, MG2 acts as a generator, converting the mechanical energy from the drive wheels into electrical energy to recharge the HV battery. During this process, MG2 provides a deceleration rate that is similar to that normally experienced as a result of engine compression on a normal gasoline engine vehicle. If, during a hard stop, the required braking effort exceeds the capability of MG2 to slow the vehicle, then the hydraulic brakes are actuated. Also, if the control module determines that the HV battery is already fully charged, the required braking energy is dissipated through the hydraulic braking system. However, due to the regenerative braking design, not only is braking energy returned to the HV battery, thus improving city fuel economy, but the life expectancy of the braking system hardware is greatly increased.

Fuel Economy. Because MG2 can electrically provide the power needed to operate the vehicle at lower speeds, the gasoline engine does not run when vehicle speed is less than 40 MPH unless a hard acceleration requires more power than MG2 can provide or the HV battery needs to be recharged. However, the gasoline engine must run continuously when the vehicle is operated at highway speeds. When this design is coupled with the regenerative braking ability of this vehicle, the result is that the EPA fuel economy rating for city driving (51 MPG) is greater than that for highway driving (48 MPG).

Electronic Throttle Control. The Prius uses an electronic throttle control (ETC) system similar to that used by other manufacturers. In fact, because the Prius uses multiple power plants, it is necessary to use an ETC system. When the driver operates the throttle pedal, an APP sensor sends "driver demand" information to the control module, which then determines whether to operate the gasoline engine in addition to MG2.

Fuel Tank. With normal fuel tanks, putting 10 gallons of liquid fuel into the tank during refueling results in displacing 10 gallons of hydrocarbon vapors, most of which are pushed from the tank into the atmosphere during the refueling process. The Prius fuel tank incorporates an expandable bladder, which contains the supply of liquid gasoline used to operate the engine. As fuel is used, the bladder contracts within the tank assembly. As the tank is refueled, the bladder expands. This avoids storing excessive hydrocarbon vapors in the fuel tank, thereby greatly reducing the escape of hydrocarbon vapors to the atmosphere during refueling.

Toyota Hybrid System Second-Generation (THS-II)

The second generation of THS (THS-II) was introduced on the 2004 Prius. Also known as Toyota's Hybrid Synergy Drive system, THS-II includes several improvements. The HV battery consists of only 28 modules, each consisting of six cells (total of 168 cells) at 1.2 V each for a total of 201.6 VDC. The cells are connected by dual contacts connecting one cell to the next to reduce the HV battery's internal resistance. The THS-II design also uses a DC-to-DC boost converter to convert the 201.6 VDC to 500 VDC, which the inverter then changes to AC to power MG1 or MG2. This DC-to-DC boost converter is physically located within the inverter assembly. Also, because the auxiliary battery is located in the trunk, Toyota added a jump-start terminal to the junction block in the engine compartment to simplify the connection of a battery charger or a set of jumper cables to start the system from a 12.6 V donor vehicle.

With THS-II, MG2's rated horsepower is increased from 44 to 67 HP due to a stronger permanent magnet rotor with the motor/generator assembly and a higher voltage applied to the windings. This gives an increase in horsepower of about 50 percent. Also, while THS-II uses the same gasoline engine that the original THS used, its horsepower has been increased from 70 to 76 HP. The ability of MG1 to generate electrical current for use by MG2 and to recharge the HV battery is also increased due to an increase in its maximum RPM from 6,500 to 10,000. Ultimately, the 2004 Toyota Prius with THS-II experienced a substantial increase in its "off-the-line" performance.

Other changes to the 2004 Prius include redesign of the radiator for the inverter's cooling system to make it part of the radiator used in the engine's cooling system, thereby increasing its effectiveness and compactness. A coolant heat storage capability also was added. A double-wall stainless steel container that can keep 3 liters of coolant warm for up to 3 days is used to prewarm the intake ports in the cylinder head prior to operating the gasoline engine during normal cold-start operating conditions (unless the engine must operate immediately following a cold start to provide additional power or to recharge the HV battery). This is designed to dramatically reduce cold-start emissions from the gasoline engine. The heat storage container contains both an electric heat pump that is under computer control and a temperature sensor.

A smart key system was added as an option in the 2004 model year. Antennas are placed in each door and in the trunk, as well as within the instrument panel. If the vehicle key is within 24 inches of a door, the door will automatically unlock and allow the driver to operate the door handle to open the door, without having either to put the key into the lock cylinder or to push a button on a transmitter. The trunk can be opened similarly: If the vehicle key is within 24 inches of the deck lid's lock cylinder and antenna, the trunk release button will be enabled, even if the key is still in a pocket. If the key is in a pocket or in a purse on the front seat, the start button can be operated and the READY indicator will display, thus allowing the vehicle to be driven without having to put the key into the ignition switch.

Toyota Hybrid System Third-Generation (THS-III)

Toyota introduced the third generation of the Toyota Hybrid System, known as THS-III, on the 2010 Prius. Some of the changes include the following:

- THS-III uses a 1.8L DOHC Atkinson cycle engine with 16 valves and VVT-i that provides quicker performance and greater fuel economy, most notably on the highway, than the 1.5L it replaces.
- THS-III uses a new battery pack has been reduced in size and repositioned to increase cargo space. (The battery is warranted for 10 years or 150,000 miles.)
- THS-III uses the DC-to-DC boost converter to convert the 201.6 VDC at the HV battery to 650 VDC, which the inverter then changes to AC to power MG1 or MG2.
- THS-III adds another driving mode, called EV Mode, which helps it stay in electric mode for up to a mile when driving less than 25 MPH.

Toyota Highlander Hybrid

Billed as the world's first seven-passenger electric/gasoline hybrid SUV, the Highlander Hybrid was introduced by Toyota as a 2006 model. Like the Honda Civic and the Honda Accord, the Toyota Highlander was an existing nameplate that was converted to hybrid-drive technology. This SUV incorporates the THS-II system (Toyota's Hybrid Synergy Drive system) used on the 2004 and newer Prius, with a few minor changes, earning a SULEV designation. The Highlander uses a 3.3L V6 gasoline engine (instead of the 1.5L four-cylinder engine that the Prius used). When combined with the power of MG2 (and the high voltage provided to MG2 by the boost converter), this powertrain provides "off-the-line" performance similar to a V8. Toyota's internal tests indicate a zero-to-sixty MPH time of only 7.3 seconds on the four-wheel-drive with intelligence (4WD-i) model.

The 4WD-i version adds one more motor/generator at the rear drive axle, shown in Figure 9–12, known as Motor/Generator Rear (MGR). MGR does for the rear-drive wheels what MG2 does for the front-drive wheels. MGR uses power from the HV battery to accelerate the rear wheels but also uses regenerative braking from the rear wheels to charge the HV battery. MGR generates 96 ft.-lb of additional torque.

The HV battery is located under the second-row seat (the third-row seat is standard equipment). It consists of 240 NiMH cells at 1.2 V each for a total of 288 VDC. This voltage is supplied to all three motor/generators through the boost converter and inverter assembly. The boost converter boosts this voltage to 650 VDC and then sends it to the inverter, which then converts it to AC. This voltage is also used to power an electric air conditioning compressor.

Toyota Prius Plug-in Hybrid Electric Vehicle

Toyota introduced the Prius as a plug-in hybrid electric vehicle (PHEV) in 2012. Currently, the PHEV version of the Prius continues to use a 1.8L I-4 gasoline engine with a compression ratio of 13:1. The traction EMG is a 60 kW permanent magnet **AC synchronous motor**. This motor's speed is controlled by the frequency of the AC voltage that provides its power.

The HV battery is a lithium-ion (Li-ion) battery. As a PHEV, the Prius is rated to be able to travel up to 11 miles in electric vehicle (EV) mode. Unlike earlier versions, the PHEV version can travel at speeds up to 62 MPH while operating in EV mode. The driver can select when EV mode is desired as opposed to HEV mode. The EPA mileage ratings for the Prius PHEV are as follows:

- Mileage while operating in EV mode: 95 MPGe
- Combined mileage: 50 MPG

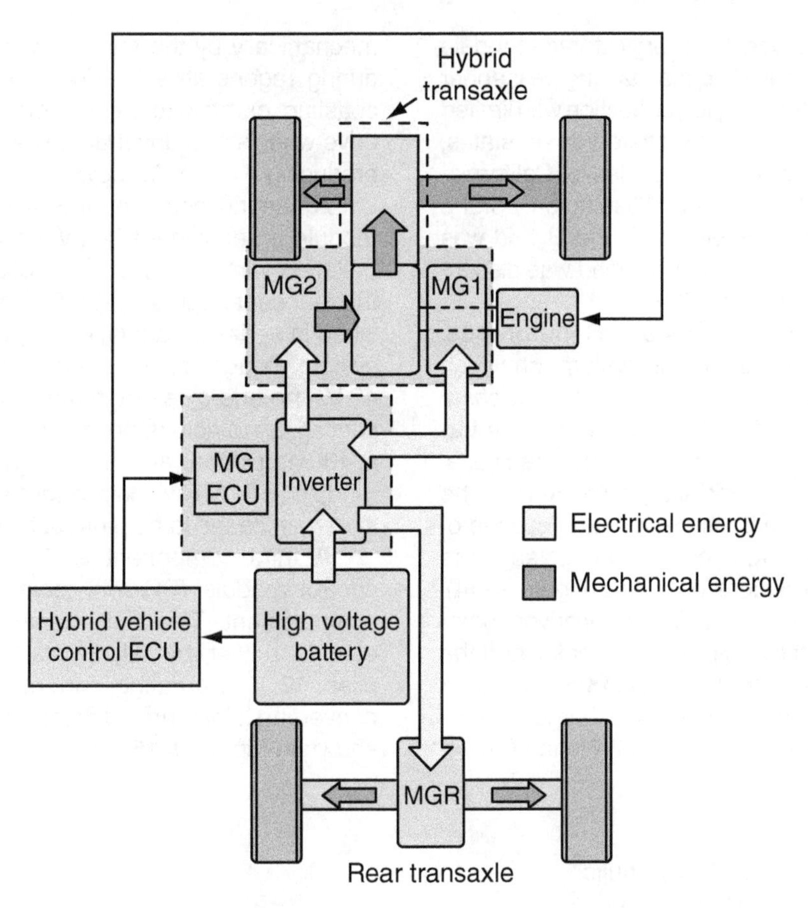

Figure 9–12 Toyota Highlander hybrid drivetrain.

GENERAL MOTORS HYBRID VEHICLES

Flywheel Alternator Starter System

GM 42 V 2004-2007 Chevrolet Silverado and GMC Sierra Hybrid. In the 2004 model year, General Motors introduced its first vehicle with a 42 V system, a special edition Chevrolet Silverado/GMC Sierra Hybrid with a Vortec 5.3L V8. This vehicle is a *dual voltage* vehicle in that it uses neither a straight 42 V system nor a straight 14 V system but rather a combination system that uses both voltages. The Silverado/Sierra Hybrid uses what GM calls the **Flywheel Alternator Starter (FAS)** system.

While this vehicle does not technically use an EMG to assist the ICE in accelerating the vehicle and maintaining cruise, it does have some fuel-saving features in common with strong HEVs—regenerative braking and a transparent engine auto stop/start feature. Thus, this vehicle is considered a mild HEV.

The Silverado/Sierra Hybrid was the first production full-size pickup to use 42 V/14 V hybrid technology. Its initial availability in 2004 and 2005 was only as a 1,500 extended cab model. This is because the extended cab configuration provided a great way to package the energy storage module beneath the rear seat while maintaining passenger comfort. As a 2005 model, its production was limited and, as a result, it was only offered in six states, based upon consumer demand: Alaska, California, Florida, Nevada, Oregon, and Washington. For the 2006 model year the Silverado/Sierra Hybrid was offered in all 50 states. This mild hybrid was discontinued during the spring of 2007.

Silverado/Sierra Hybrid Components. The primary component in this system, on which everything else depends, is the *electric machine* (*EM*), a form of an EMG. It replaces both the traditional alternator (generator) and the traditional starter motor. It is located just to the rear of the engine, ahead of the transmission, and outside of a smaller 258 mm torque converter coupled to the engine's flywheel (Figure 9–13 and Figure 9–14). The EM can be used as an electrically driven motor to begin rotating the engine's crankshaft when the engine is to be started. It can also be used as a mechanically driven generator to recharge the system's batteries when the engine is running (driven mechanically by the rotation of the crankshaft) or during regenerative braking when the vehicle is coasting or braking (driven mechanically by the drive wheels and drivetrain). The EM is capable of producing 7 kW of AC power.

A second component is the energy storage module, a set of three 12.6 V "advanced" lead acid batteries wired in series and located underneath the rear seat (Figure 9–15). The three batteries are wired in series for a total of 37.8 V. When charging system voltage is added, the voltage total is about 42 V. The energy storage module also contains a cooling fan to help remove heat from the batteries, a 400-amp fuse, and an energy storage control module (ESCM), which provides automatic shutoff in certain cases to help protect the system.

A third component is the starter/generator control module (SGCM), located in the engine compartment. This unit manages the vehicle's electric power by taking starter/generator-supplied 42 V alternating current (AC) power and converting it into three different forms of power as shown in Figure 9–16:

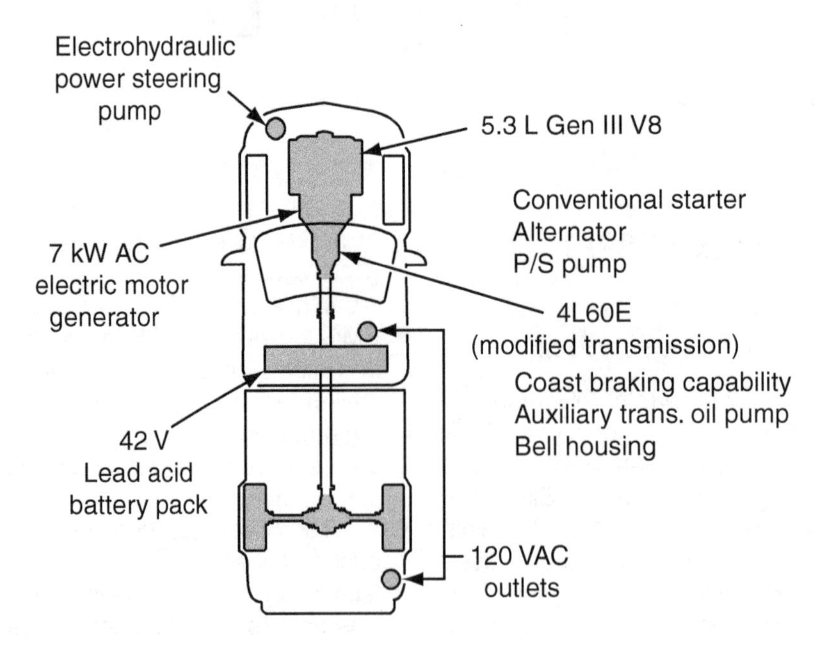

Figure 9–13 Silverado 42 V hybrid vehicle overview.

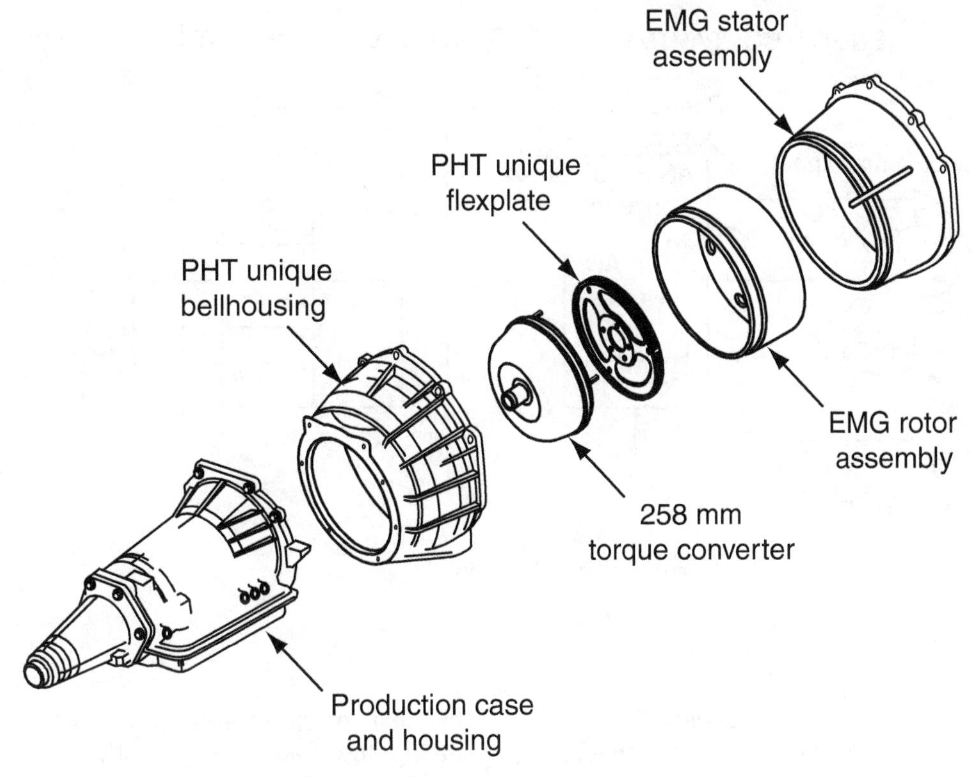

Figure 9–14 Transmission and electric machine on a Silverado 42 V hybrid.

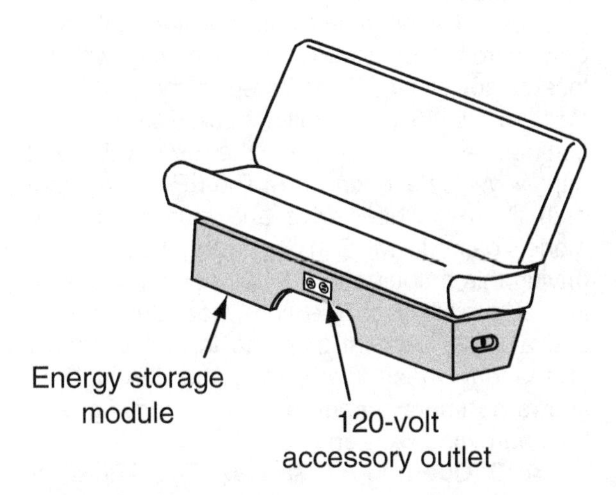

Figure 9–15 Rear seat showing location of the energy storage module and the 120 V accessory power outlet (APO).

1. 42 V direct current (DC) to charge the energy storage module.
2. 14 VDC to charge a 14 V battery and operate the vehicle's 14 V subsystems.
3. 120 VAC to power four 120 VAC power points, each referred to as an *accessory power outlet* (*APO*). Two of these APOs (a normal household double wall outlet) are located in the right rear of the truck's bed. Two additional APOs are located underneath, but to the forward side, of the rear seat (Figure 9–15). All four APOs can collectively supply 2,400 watts of 120 VAC power.

Increased Fuel Mileage. The Silverado/Sierra Hybrid combines intelligent computers with regenerative braking to give this vehicle a

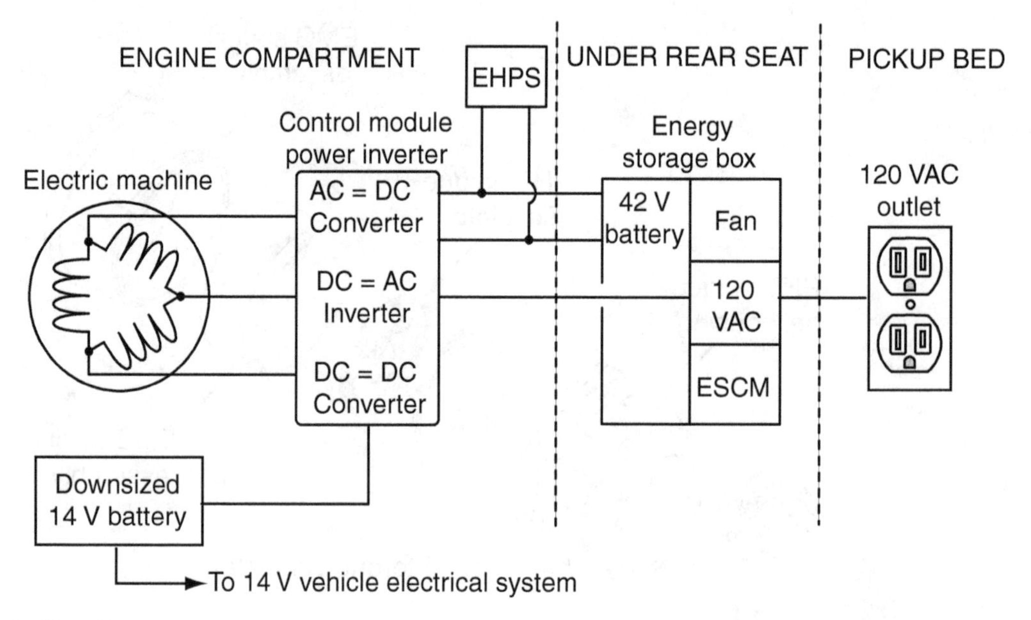

Figure 9–16 Silverado 42 V hybrid system overview.

substantial increase in fuel economy. The regenerative braking ability of the EM means that braking energy (which is turned into waste heat on traditional vehicles) is now translated into electrical energy to recharge the system's 42 V energy storage module.

This vehicle also features an engine start/stop system that automatically shuts off the engine to save fuel when the driver is braking at speeds below 13 MPH. The engine also shuts off when the vehicle is not moving. When the driver releases the brake pedal and pushes on the throttle to reaccelerate from a stop or deceleration (or if the Tow/Haul button has been pressed), the engine begins to run again. Because the EM uses no gears (due to the fact that it is designed to operate at 42 V), the EM can begin turning the engine's crankshaft without the audible gear whine that would be associated with a traditional starter motor. As a result, the restart of the engine is almost transparent to the driver.

Another energy-saving feature is the electro-hydraulic power-assist steering system. This system uses a 42 V electric motor to drive the hydraulic power steering (P/S) pump, thus eliminating the traditional belt-driven pump. As a result, the power assist can be reduced, or even turned off, as driver demand dictates, thereby saving energy compared to a belt-driven pump that requires continual energy usage.

The collective result is an average fuel economy increase of 2 MPG in city driving with no loss of power when compared to the traditional Vortec 5.3L V8 in a similar truck. This information has been updated using the EPA's website at http://www.fueleconomy.gov. The EPA shows that both the two-wheel drive and four-wheel drive models gained 2 MPG in city driving. Thus, for city driving this amounts to a 14.28 percent increase in fuel economy on the two-wheel drive model and a 15.38 percent increase in fuel economy on the four-wheel drive model. The EPA website shows no improvement for highway driving over the non-hybrid 5.3L engine.

APO Operational Modes. The APOs, located under the rear seat and also at the right rear of the truck's bed, can provide a combined 2,400 watts of 120 VAC power. Each outlet is protected by a 20-amp fuse and has ground fault protection. An APO switch mounted on the instrument panel is used to activate the outlets.

The APOs operate in two modes. Normal mode is used while the vehicle is being driven. The driver simply presses the APO switch and waits for an APO indicator to light. The power outlets can then be used until the APO switch is pressed again or until the ignition switch is turned off. Continuous mode operation allows the 42 V energy storage module, combined with the 5.3L engine and the EMG, to function as a generator when the vehicle is stationary. The transmission must be in Park for this mode to operate. (The key does not need to be in the ignition, however.) The engine will automatically run only as needed to help the 42 V energy storage module provide the 120 V of AC power. This mode activates the horn relay if the level of fuel in the tank gets too low. If the low-fuel warning goes unheeded for 5 minutes, the APO system automatically shuts down to avoid completely emptying the fuel tank.

Belt Alternator Starter System

GM 42 V 2007 Saturn VUE and 2008 Chevrolet Malibu. GM introduced another hybrid technology, known as the **Belt Alternator Starter (BAS)** system, on the 2007 Saturn VUE and on the 2008 Chevrolet Malibu. This system, like the FAS used on the Silverado/Sierra Hybrid, uses a 42 V electrical system in combination with regenerative braking and a transparent engine auto stop/start feature. The BAS system is also used in conjunction with an ETC system.

What makes this system different is that the EMG that constitutes both the starter and the alternator is not designed as part of the transmission's torque converter as on the Silverado/Sierra Hybrid. Rather, this 36 V motor and alternator assembly mounts in line with the crankshaft's pulley, similar to a traditional alternator (Figure 9–17). It has a belt that allows it to either be belt driven by the crankshaft or to drive the crankshaft.

With the gasoline engine running, the EMG operates as a belt-driven alternator, producing

Figure 9–17 The EMG on a GM BAS engine.

42 V to charge the 36 V battery. And, following an auto-stop function during deceleration or braking, the EMG acts as a 36 V starter, using the belt to turn the crankshaft until the engine is running again. As a result, the incorporation of this design allowed the engineers to add the BAS system to four-cylinder and six-cylinder engines with minimal changes to the engine and transmission, unlike the FAS system used on the Silverado/Sierra Hybrid, which required an extensive redesign of the transmission's torque converter.

The auto start/stop feature will not function when the transmission is in Park. Neither will it function after an initial engine start but before the vehicle has been driven. Therefore, when starting the engine with the transmission in Park, a traditional 12 V starter is used to turn the crankshaft. But once the engine is running, the transmission has been shifted into Drive, and the vehicle has been driven, the BAS system performs the following functions:

- Auto stop/start feature: During deceleration and braking, the BAS system cuts the fuel to the engine, resulting in engine stall. As the driver operates the accelerator pedal position (APP) sensor so as to accelerate again, the computer responds by using the alternator/starter to start the engine again.
- Regenerative braking: The BAS system uses regenerative braking during deceleration or braking to cause the alternator/starter to use the vehicle's mechanical energy to charge the 36 V battery.
- Electrical power assist: The BAS system allows the alternator/starter to provide some electrical power assist during periods of high engine load. While the BAS system operates at a low voltage compared to a full hybrid vehicle, the vehicles that it is used on are also smaller than the Silverado/Sierra Hybrid, thus allowing the 36 V motor to be more effective in its ability to assist the gasoline engine than on a heavier vehicle.

The vehicles with the BAS system are designed to achieve a fuel mileage increase of about 16.5 percent, which breaks down as follows:

- Auto stop/start: 4 percent
- Deceleration fuel cutoff: 3.5 percent
- Regenerative braking: 1 percent
- Increased efficiency of engine and transmission: 4.5 percent
- Improved aerodynamics: 0.5 percent
- Ultra-low rolling resistance tires: 3 percent

Ultimately, this mild hybrid offers a fair gain in fuel economy at an attractive price, thus making these vehicles an excellent value.

WARNING: When working on or around any hybrid, always remove all jewelry, including rings, watches, and necklaces. This does include mild hybrids with batteries rated at 37.8 V and a charging system that operates at 42 V. A short to ground in a 42 V system can produce an arc that is between 50 and 100 times hotter than the same short to ground in a 14 V system.

Chevrolet Volt

The Chevrolet Volt (Figure 9–18) is a series HEV and is also a plug-in hybrid electric vehicle (PHEV). It was first shown as a prototype at the North American International Auto Show (NAIAS) in Detroit in January 2007. GM began production of about 10,000 units in late summer of 2010 as a 2011 model. These early ones were distributed in the fourth quarter of 2010 to an initial market that included California, Michigan, and Washington, DC. A complete national rollout was in place by the end of 2011. The vehicle's initial styling, as the vehicle was displayed in 2007, had extremely poor aerodynamics. Therefore, the body was restyled for the production car so as to produce a wind drag coefficient of 0.28, slightly lower than that of a 2010 Corvette at 0.29. The restyling was designed to optimize a balance between aerodynamics, performance, and styling.

The Cadillac ELR (Figure 9–19) is the luxury version of the Chevrolet Volt and was first introduced as a 2014 model. The following discussion

Figure 9–18 A Chevrolet Volt.

Figure 9–19 A Cadillac ELR.

is based on the Chevrolet Volt, but the ELR shares many of the same features.

The Volt is designed with a seating capacity of four people, can accelerate from 0 to 60 MPH in less than 9 seconds, and can reach a top speed of 100 MPH. It is propelled using electrical power only. The first 40 miles uses electrical power derived from recharging the batteries, and thereafter the electrical power is generated through the operation of a gasoline engine. Fuel tank capacity is designed to provide a total range of over 300 miles.

Operating Concepts. The Volt is a series hybrid in that solely electrical energy is used to accelerate the vehicle and maintain cruise using an EMG. The initial electrical energy is provided by Li-ion batteries that can be recharged by plugging them into an electrical outlet in your home or at the office. A 1.4L gasoline engine is used to drive a generator to provide electrical power in addition to the power provided by the vehicle's batteries. As a result, the vehicle's range is not limited to the power that the HV batteries can provide. The Volt is therefore considered an extended range electric vehicle (EREV).

The Volt has two modes of operation: all-electric vehicle (EV) mode and extended range (ER) mode. Beginning with a full charge, the vehicle operates in EV mode up to about 40 miles. Then, when the HV batteries reach about 30 percent charge, the vehicle switches over to ER mode and the gasoline engine begins operating to provide additional electrical power from the generator for the purpose of continuing to power the drive wheels. During ER mode operation, the generator is not used to recharge the HV batteries above about the 30 percent charge level. Rather, the HV batteries must be recharged back to full by connecting the vehicle to the power grid.

Economy. General Motors states that when the Volt is operating in EV mode, the cost of the electric power to achieve 40 miles of driving is a just a fraction of the cost required to run a similar-sized vehicle on gasoline. In 2008, the average residential retail price of electricity in the United States was about 9.74 cents per kilowatt-hour (kWh). At this price, GM says that the Volt should operate in EV mode for about 80 cents for a 40-mile charge, or for about 2 cents per mile. At 11 cents per kWh, the calculated cost rises to about 2¼ cents per mile. Even at 12 cents per kWh, the calculated cost rises to just under 2½ cents per mile. At a cost of $3.00 per gallon for gasoline, a gasoline engine vehicle that achieves 30 MPG will cost 10 cents per mile to operate. Recharging the batteries on the Volt overnight should use an amount of electrical power comparable to leaving a personal computer turned on for 24 hours. Figure 9–20 shows an EPA sticker for a

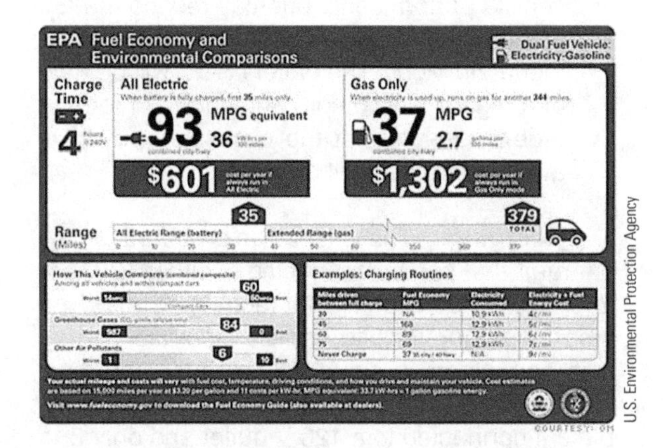

Figure 9–20 An EPA sticker for a Chevrolet Volt.

Chevrolet Volt. MPGe is the MPG *equivalent* measure of gasoline fuel efficiency for electric mode operation. The EPA created the MPGe standard to provide a way to make this comparison.

Why 40 Miles? Adequate batteries are provided to achieve about 40 miles in EV mode with a few comfort accessories operating. Of course, as additional accessories are turned on, this 40-mile EV range is reduced. The engineers' goal was to provide all-electric power for about the first 40 miles of a daily commute because the daily commute of 78 percent of the workforce is 40 miles or less. Therefore, 78 percent of commuters will only use electric power on a daily basis and will never need to use any gasoline. Providing enough batteries to achieve more than 40 miles in EV mode would not be cost effective when the cost of the additional batteries is considered against the number of commuters who would make use of them. Even then, a person who only drives a 10-mile round trip on a daily basis is actually purchasing more battery power than is needed.

Driver-Selectable Modes of Operation. The Volt incorporates three driver-selectable modes of operation as follows:

- Economy Mode: The battery will be discharged to a lower state of charge (SOC) before entering ER mode, and gasoline engine on/off cycling is optimized for economy.
- Sport Mode: This mode offers an increase in vehicle performance, but may reduce the EV mode by 15 to 20 percent.
- Mountain Mode: The battery SOC will be held at a higher level before entering ER mode in order to insure that the necessary power is available for long hill climbs.

Charging. The Volt has an on-board "smart charger" that can be connected to either 120 V or 240 V, either 50 or 60 Hz, for recharging. Also, the smart charger inhibits vehicle operation when it is connected to an AC outlet for recharging. Recharging should take 8 to 10 hours to complete if connected to a 120 V outlet and only 3 to 4 hours to complete when connected to 240 V.

HV Battery. The HV battery is rated at 16 kWh and contains 288 cells at 3.6 V each. While conventional wisdom would suggest that the resulting voltage would be 1,036.8 V (288 × 3.6), this is not the case. A battery electronic control module (ECU) can instantly re-circuit the cells of the HV battery either in parallel or in series with each other (and in different combinations) in order to maintain a voltage of 345.6 V, although actual operating voltage applied to the EMG is much higher. (You might recall that adding battery cells in series increases voltage while leaving the amperage potential unaffected, but adding battery cells in parallel increases the amperage potential without increasing voltage.) By recombining the cells in different series and parallel combinations, the computer can electronically control the output voltage to a stable 345.6 V as the battery is discharged while operating in EV mode.

The HV battery is physically contained in a sturdy T-shaped structure that is located underneath the body and in the area of where the normal drivetrain's tunnel forms a T with the rear axle. This design allowed the engineers to further solidify the "living space" of the vehicle's interior to further protect passengers during an accident while also protecting the HV battery. The battery control module is mounted on top of the T structure.

Driver Convenience Features. The Volt includes a driver-configurable instrument display, two 7-inch touch screen navigation and vehicle information displays. This includes touch-sensitive climate and *infotainment* controls and an optional navigation system with an on-board hard drive for maps and music storage. A Bluetooth wireless connection is standard for cell phones and for music streaming.

GM Two-Mode Hybrids

Allison Transmissions, a division of General Motors, has developed a two-mode hybrid drive vehicle that borrows some technology from their Allison hybrid buses and incorporates it into pickups and SUVs on a smaller scale. A joint venture between GM, Daimler, BMW, and Chrysler, this hybrid technology was introduced in July 2008 on the Chevrolet Tahoe, GMC Yukon, and Cadillac Escalade and in the fall of 2008 on the 2009 Chevrolet Silverado and GMC Sierra. This technology was also intended to be used on the Dodge Durango until Chrysler lost its contract with Daimler.

Most vehicles using hybrid technology are designed to get their best fuel economy in city driving and really have little advantage, if any, when driving on the highway. In fact, with highway driving, the gas engine provides most, if not all, of the power and the additional weight of the HV batteries is a burden. The GM/Allison two-mode hybrid system is designed to give a vehicle an advantage at highway speeds as well as in city driving.

Functional Concept. The system uses a 300 V NiMH HV battery and a two-mode transmission. The transmission, shown in Figure 9–21, houses two 60 kW EMGs referred to as *MG-A* and *MG-B*. The transmission is designed with four initial gear ratios, but the EMGs are used in conjunction with planetary gear sets and multi-disc friction clutches to provide a continuously variable gear ratio at the final output shaft (Figure 9–22).

Figure 9–21 The motors and clutches inside a two-mode transmission.

For acceleration and sustaining cruise, the two-mode hybrid vehicle can operate solely on electric power, operate solely using the ICE, or operate using a combination of both. As a rule,

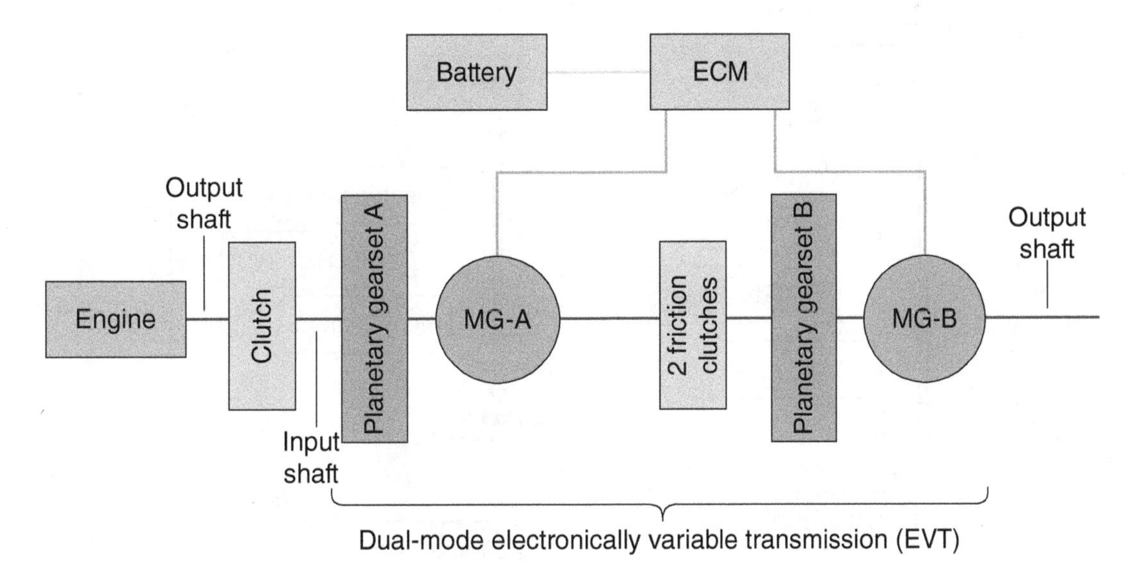

Figure 9–22 The two-mode system uses two motors and two planetary gear sets to move the vehicle or to assist the ICE in moving the vehicle.

if MG-A is not providing power to the output shaft, it is operating as a generator that is driven by the engine and providing electrical power to the HV battery. Likewise, if MG-B is not providing power to the output shaft, it is operating as a generator that is driven either by the engine or by the drive axle during regenerative braking and also providing electrical power to the HV battery.

This vehicle also uses GM's Active Fuel Management (AFM) system discussed in the General Motors chapter of this textbook. The PCM can disable one-half of the cylinders, three on a V6 and four on a V8, when cruising and the additional power is not needed.

The GM/Allison hybrid system operates in two distinct modes of operation, referred to as the *input split mode* and the *compound split mode* (Figure 9–23). An electronic control module (ECM) controls the operational modes.

The input split mode of operation is used for low-speed cruise and light to moderate acceleration. While operating in this mode, the vehicle may operate using battery power, engine power, or both. When the module determines that battery power is sufficient for current driving conditions, the ICE will either be shut down or will operate in the AFM mode with half of its cylinders disabled. During this time one EMG is working as a traction motor to sustain cruise and the other EMG is working as a generator. If the engine is commanded to begin running, the EMG operating as a traction motor may be converted to operate as a generator as needed.

The compound mode of operation is used for cruising at highway speeds and during heavy acceleration. The ICE will operate and one or both EMGs will be used to assist the ICE. However, during high-speed cruise the ICE may still operate in the AFM mode of operation and disable one-half of its cylinders. All cylinders will be enabled for hard acceleration.

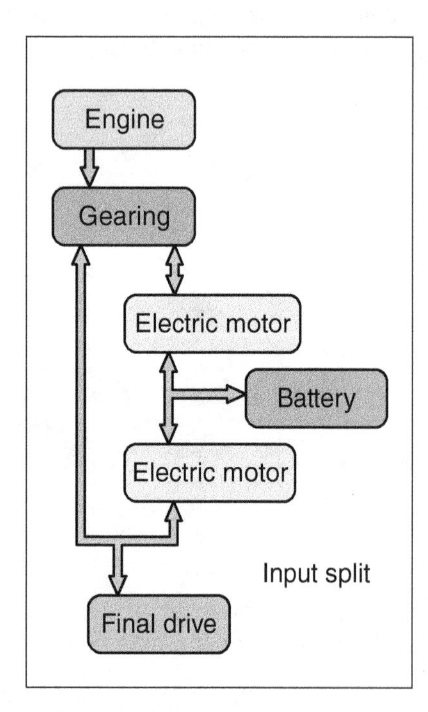

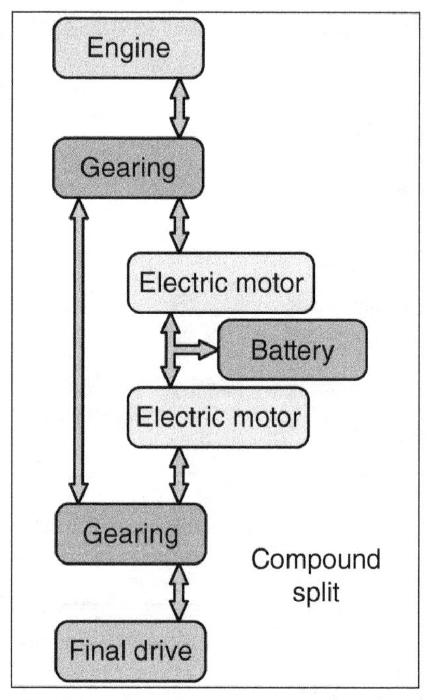

Figure 9–23 The flow of energy in each of the operating modes.

General Motors' Approach to Fuel Savings Technology

While many other manufacturers are applying their hybrid drive technology, fuel cell technology, and other technologies designed to increase fuel mileage to smaller vehicles such as the Honda Civic and the Toyota Prius, GM's engineers are instead applying the "fuel-saving" technologies to their larger vehicles with the V6 and V8 engines. These are the vehicles that many consumers want to purchase and drive—the people-mover SUVs and minivans, the workhorse pickup trucks, and even the larger passenger cars. Certainly one idea here is to increase the fuel economy on the vehicles that ordinary people drive. But another view of fuel economy involves the "Chart of Diminishing Returns" (Figure 9–24).

If you study this chart, you will see that at a cost of $3.00 per gallon applied to a vehicle that gets 5 MPG, the vehicle's fuel cost per mile is 60 cents. As technology is applied to increase the vehicle's fuel mileage, each time you double the vehicle's fuel mileage you cut the fuel cost per mile in half. But consider the application of this statement.

If you were to double this vehicle's fuel mileage to 10 MPG (an increase of just 5 MPG), you would cut the cost in half to 30 cents per mile and therefore save 30 cents per mile. You can never again save another 30 cents per mile, as that would reduce the cost to zero.

If you were to again double this vehicle's fuel mileage to 20 MPG (an increase of 10 MPG), you would cut the cost in half to 15 cents per mile and therefore save 15 cents per mile. And if you again double this mileage to 40 MPG (an increase required this time of 20 MPG), you cut the 15 cents per mile cost in half to 7.5 cents per mile—a savings this time of only 7.5 cents per mile. And you would have to double this again to 80 MPG (an increase of 40 MPG—not easy to accomplish) to save another 3.75 cents per mile.

Thus there are substantial savings for every small increase in fuel mileage when the mileage is less than 18 to 20 MPG, but very little increase

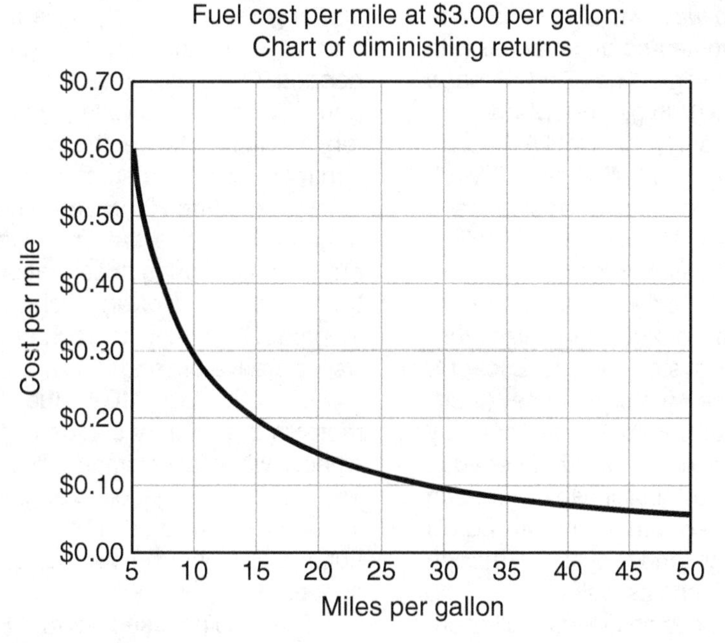

Figure 9–24 Fuel costs per mile.

in savings when the mileage is already above 30 MPG. The chart also works, showing the same curve, for any cost per gallon, whether it is $1.00 per gallon, $2.00 per gallon, $3.00 per gallon, $4.00 per gallon, or more.

Therefore, if a vehicle is achieving less than around 15 to 18 MPG, it becomes quite profitable to achieve an additional one or two MPG increase out of the vehicle as new technologies are applied. The same mileage increase applied to a small car that already achieves 32 to 34 MPG is not nearly as cost effective. It is no wonder GM is targeting the larger vehicles with their newest technologies. These are the vehicles where the application of these technologies will make the largest difference in the consumer's wallet.

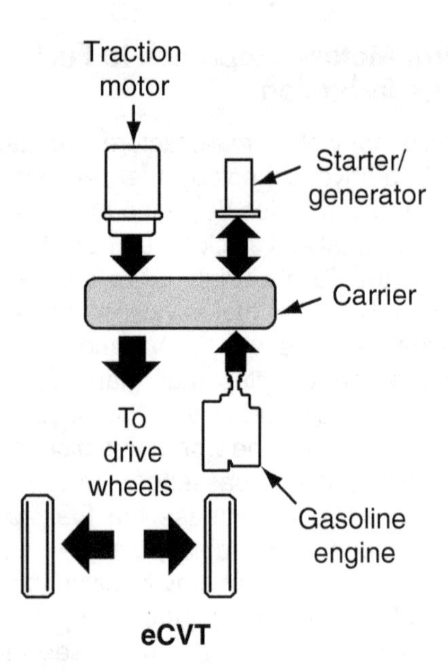

Figure 9–25 Ford Escape Hybrid primary components.

FORD HYBRIDS

Ford Escape Hybrid Electric Vehicle

Ford applied hybrid-drive technology to the Escape beginning in the 2005 model year. Like the Honda Civic, the Honda Accord, and the Toyota Highlander, the Escape was available either with a standard gasoline engine and drivetrain or with the hybrid-drive technology. The Ford Escape became the first Ford SUV to get the hybrid-drive technology. The Escape Hybrid (4WD) has an EPA fuel economy rating of 30 MPG city/27 MPG highway (2010 estimates) and is labeled as a Partial Zero Emissions Vehicle (PZEV). The Escape Hybrid was available as a standard front-wheel-drive vehicle or with an intelligent four-wheel-drive system. For 2006, Ford also introduced the hybrid-drive system on the Escape's Mercury counterpart, the Mercury Mariner (introduced in the fall of 2005). Ford Motor Company developed Ford's hybrid technology, but because Ford and Toyota share certain licenses with each other, several similarities are found in Ford's hybrid-drive technology and Toyota's, though there are several differences as well.

The Escape Hybrid uses an electronically controlled Continuously Variable Transmission (eCVT) that consists of a planetary gear designed to split the power, as in the Toyota system (Figure 9–25). The *starter/generator* is equivalent to Toyota's MG1 and is used to start the gasoline engine as needed. Once the engine is running, the starter/generator produces voltage to charge the HV battery and to help power the *traction motor.* The traction motor is equivalent to Toyota's MG2 and uses voltage from the HV battery and from the starter/generator to drive the drive wheels through the output shaft. During deceleration and braking, the traction motor uses the vehicle's kinetic energy to produce voltage to recharge the HV battery (regenerative braking).

Like Toyota's MG2, the Escape's traction motor is the primary source of power to the drive wheels when accelerating and when cruising at low speeds. The gasoline engine runs as needed to assist the traction motor, to recharge the HV battery, or when the vehicle is cruising at higher speeds.

The engine used in the Escape is a 2.3L four-cylinder that uses the Atkinson cycle. By delaying the closing of the intake valves to about 90 degrees ABDC (instead of the 40 degrees ABDC designed into many engines), the engine can shift the torque curve higher to increase fuel economy. The compression stroke is effectively a short stroke while the power stroke takes advantage of the engine's full compression ratio, becoming effectively a long stroke. While this type of camshaft does not do well for quick starts "off the line," the traction motor provides the needed initial torque for such acceleration.

The PCM will lean out the fuel injectors' pulse width for two or three crankshaft rotations just prior to shutting down the gasoline engine. This is referred to as a "lean engine shutdown" and is done to ensure that the Atkinson cycle engine does not release excessive hydrocarbons into the atmosphere by way of the intake manifold due to the late-closing intake valves. During this lean engine shutdown, the engine will stumble slightly. The engine is equipped with active powertrain mounts that reduce the vibration felt in the passenger compartment during engine shutdown.

The HV battery consists of 250 1.2 V NiMH batteries connected in series to produce 300 VDC. An inverter converts this voltage to AC for use by the traction motor or the starter/generator. The HV battery assembly is located below the floor of the rear cargo area. A power disconnect switch on top of the HV battery assembly is rotated to electrically open the high-voltage circuit by dividing the battery electrically at its midpoint (Figure 9–26). As on the Toyota and Honda hybrids, all of the high-voltage cables are bright orange. The following are some of the differences between the Escape and Toyota systems:

- The Escape's HV connectors incorporate a third wire that signals the battery control module to open the HV battery at its midpoint if a high-voltage connector is disconnected before using the proper procedure to disable the HV circuit.
- The Escape uses two inertia switches—one located at the front of the vehicle and one located at the rear of the vehicle—that open upon impact, as during a vehicle accident. The rear inertia switch is an input directly to the battery control module. The front inertia switch is an input to the PCM. If the PCM sees that the front inertia switch has opened, it relays this information to the battery control module. If an accident causes either inertia switch to open, the battery control module opens the HV circuit by electrically dividing the HV battery at its midpoint.
- While the Escape also uses two cooling fans to blow air through the HV battery assembly for cooling, it also uses "opportunistic cooling" to provide refrigerated air from the climate control system to cool the HV battery assembly any time the driver has requested A/C operation for passenger compartment cooling. This allows the HV battery to be cooled below its normal operating temperature, which may delay having to run the engine again just to cool the HV battery.

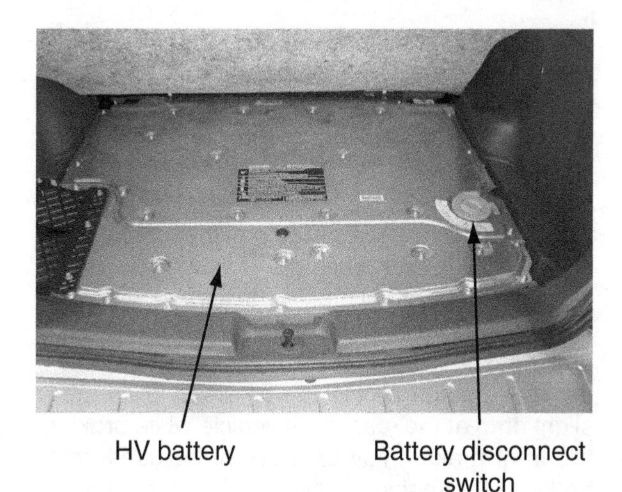

Figure 9–26 Ford Escape battery disconnect switch.

The Ford Escape Hybrid also uses Electronic Power Assist Steering (EPAS). This system varies the level of power steering assist electronically; it can draw up to 80 A at parking lot speeds when maximum assist is required.

The Ford Escape's Electro-Hydraulic Brake (EHB) system allows a control module to interpret how much deceleration is to be achieved through the regenerative braking system and how much is to be achieved through the hydraulic braking system. The brake pedal has a "pedal feel" emulator that allows the feel of the brake pedal to seem normal to a driver while the computer uses the electronics to control braking. In the event of electronic failure, the vehicle reverts to full hydraulic braking. The brake control module always applies the rear brakes slightly on all stops to create a slight drag at the rear of the vehicle while braking; the front brakes may or may not be used while braking, depending on the severity of the braking situation and the ability of the regenerative braking system to slow the vehicle. As a result, it is likely that the rear brakes on the Escape will wear out long before the front ones do—just the opposite of a normal vehicle.

Warning: The brake control module will apply braking pressure to the front calipers 40 seconds after a vehicle door has been closed. Make sure to follow Ford Motor Company cautions when working on this braking system. A mistake here could cost a technician his fingers or hand if performing a brake job improperly.

Ford Fusion and Ford C-Max Hybrid Electric Vehicles

The Ford Escape is no longer offered as a hybrid, but instead is offered with a gasoline direct injection (GDI) turbocharged engine. Ford Motor Company now offers its hybrid technology on the Ford Fusion and the Ford C-Max. Both the Fusion and the C-Max are offered as an HEV with a 2.0L Atkinson-cycle I-4 hybrid engine and also as a PHEV with the same 2.0L Atkinson-cycle I-4 hybrid engine. The Ford Fusion is also offered as a non-HEV with a GDI engine.

Both non-plug-in HEVs are equipped with a 1.4 kWh Li-ion HV battery. Both PHEVs are equipped with a 7.6 kWh Li-ion HV battery. The recharge time for the PHEVs is 7 hours at 120 VAC or 2.5 hours at 240 VAC. Both HEVs and both PHEVs are equipped with a traction EMG, which is an 88 kW permanent magnet AC synchronous motor. And both HEVs and both PHEVs are rated to be able to travel at speeds up to 85 MPH in electric vehicle (EV) mode.

The EV range of both PHEVs is rated as being able to operate in EV mode up to 19 miles before switching over to gasoline operation. The driver may choose when to operate in EV mode, when to operate in gasoline-only mode, and when to operate in a combination gasoline/electric mode. The driver may also choose whether to operate in *charge sustain mode* or *charge depletion mode.* When used for daily commuting the vehicle should be operated in charge depletion mode if the driver wants to take advantage of cheaper electric grid power with an overnight recharge of the HV batteries.

The EPA estimated fuel mileage ratings are as follows:

- Fusion HEV MPG: 44 city/41 highway
- C-Max HEV MPG: 42 city/37 highway
- Fusion PHEV MPGe: 95 city/81 highway
- C-Max PHEV MPGe: 95 city/81 highway

The Fusion and the C-Max, as well as other non-hybrid Ford offerings such as the Ford Escape, are now equipped with Electric Power-Assisted Steering. This power steering system uses no hydraulics for steering assist and therefore no longer has a power steering fluid reservoir that needs to be checked or refilled.

Ford Focus

It should also be noted that Ford Motor Company is offering the Ford Focus as a battery electric vehicle (BEV). The Focus is rated as being able to travel up to 76 miles between charges. The Focus is equipped with a 107 kW EMG and is rated at 110 MPGe in city driving.

FUEL CELL VEHICLES

Honda FCX

The Honda FCX (Fuel Cell Power) is an electric vehicle that uses PEM fuel cell stacks to power the EMG that accelerates the drive wheels. Honda has already leased or sold several of these to cities for use in their fleets on U.S. roads. Because the resulting output of the PEM fuel cell stack is only heat and water, the Honda FCX is rated as a Zero Emission Vehicle (ZEV). The EPA mileage ratings on the FCX, in "miles-per-kilogram of hydrogen consumed," are 51 city and 48 highway. The maximum speed of the FCX is limited electronically to 95 MPH.

Honda first introduced two FCV prototypes in 1999: the FCX-V1 and the FCX-V2. In 2000, Honda introduced a third FCV prototype (FCX-V3) at the California Fuel Cell Partnership in Sacramento. In 2001, Honda introduced the fourth FCV prototype (FCX-V4) and also designed the first solar-powered hydrogen production and refueling station, which was built in southern California. In 2002, the 2003 Honda FCX became the world's first FCV to be certified by the EPA and CARB. Also in 2002, the city of Los Angeles leased two Honda FCX vehicles. In 2003, Honda began building its PEM fuel cell stacks using a Honda proprietary membrane that allows it to begin operating at subzero temperatures. Also in 2003, the city of Los Angeles leased three additional FCX vehicles.

Honda also began working on designing a hydrogen **Home Energy Station (HES)** in 2003. In 2004, Honda began experimental operation of the first HES. This came about because automotive manufacturers discovered, as a side note, that people who owned the early straight electric cars enjoyed being able to charge them at home and as a result, never had to visit a refueling station. Today the HES has been reduced to a size that can be hung on a garage wall and can use tap water with electricity from a wall outlet to create the hydrogen needed to refill a vehicle's hydrogen tanks overnight.

In 2004, the city and county of San Francisco took delivery of two Honda FCX vehicles, and two additional FCX vehicles went to the State of New York. In February 2005, Honda announced that the city of Las Vegas, Nevada, also was leasing two FCX vehicles. As of February 2005, the number of Honda FCX vehicles on U.S. roads stood at 14.

The Honda FCX uses an EMG to accelerate the front wheels and to provide regenerative braking capability. Honda also experimented with using an *ultra-capacitor,* a unit that consists of many smaller, high-efficiency capacitors that provide a high-energy storage capability (Figure 9–27). The fuel cell stacks, capable of producing 78,000 W of power, are located below the main floorboard of the vehicle. Two high-pressure hydrogen storage tanks are located just beneath and behind the rear seat. A liquid cooling system is provided for the fuel cell stacks, using one large radiator. (Two additional smaller radiators are placed one on either side for cooling of drivetrain components.) A power control unit (PCU) controls the operation of the fuel cell stacks to keep the ultra-capacitor properly charged. During cruising and mild acceleration, output to the EMG is provided directly (and only) from the fuel cell stacks. During startup and moderate to heavy acceleration, the ultra-capacitor provides the EMG with additional electrical power to assist the fuel cell stacks, resulting in crisp acceleration. During braking or coasting, the EMG uses regenerative braking to recharge the ultra-capacitor. To conserve hydrogen fuel, the PCU shuts down the fuel stacks when the vehicle is stopped. When the vehicle is stopped, electrical energy to operate the electric A/C compressor and other vehicle electrical systems is provided by the ultra-capacitor.

This is Honda's second version of the ultra-capacitor. The output of the THS/THS-II systems' NiMH HV battery is limited to approximately 900 W per kilogram (W/kg) due to the effects of heat loss. The ultra-capacitor's low resistance enables it to handle much higher output. The ultra-capacitor used on FCX vehicles further improves the performance of the previous model, achieving an output density of 1,750 W/kg or more.

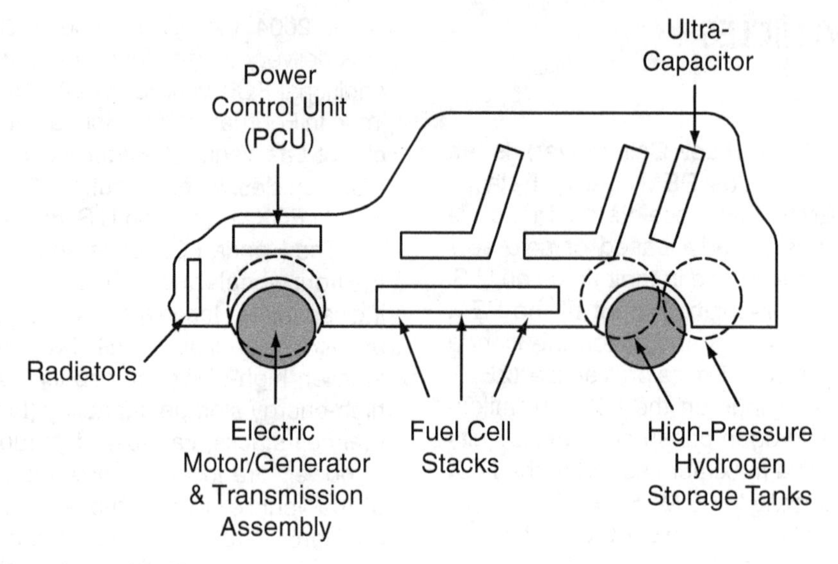

Figure 9–27 Components of the Honda FCX.

Ford Focus FCV

Ford Motor Company has adapted the Ford Focus to become a fuel cell vehicle prototype, known as the Ford Focus FCV. Like the Honda FCX, the Ford Focus FCV uses PEM fuel cell technology. The system supplies up to 330 A at between 250 and 400 V with a nominal voltage of 315 V. Similar in concept to the Honda FCX, the Ford Focus FCV has also been awarded a ZEV rating. The maximum speed of the Focus FCV is electronically limited to 80 MPH.

General Motors Fuel Cell Vehicles

General Motors has a couple of fuel cell prototype vehicles—the HydroGen3 and the Hy-wire. Both vehicles demonstrate GM's expertise in fuel cell technology. These vehicles use a fuel cell stack that can produce 94 kW of available power and are capable of a top speed of 99 MPH.

The GM Hy-wire, known as the *Autonomy,* is the world's first vehicle to combine fuel cell technology (to power the drivetrain) with control-by-wire technology. The Autonomy was displayed at the 2002 North American International Auto Show in Detroit, Michigan. The fuel cell stacks, hydrogen tanks, batteries, control modules, and four EMGs—one at each wheel—are contained within an 11-in.-thick chassis that has a skateboard-like appearance. In fact, the Autonomy has become known as the *skateboard.* The Autonomy's design means that there is no gasoline engine, and interior space is maximized. Ultimately, it can accommodate five occupants as well as some cargo depending upon the body that is attached to it.

The Autonomy, because it uses drive-by-wire, steer-by-wire, and brake-by-wire systems, can be electronically programmed to accommodate the characteristics of any type of vehicle body that is attached to its skateboard chassis. Even the suspension can be electronically tuned. The driver has no pedals to operate. The driver's controls are contained in a single module, called the *driver control unit,* that can be easily configured to either a left-hand drive (as in the United States and Canada) or a right-hand drive (as in Australia) position. The driver control unit allows drivers to control the throttle, the steering, and the braking of the vehicle and may allow them to control other vehicle systems electronically as well.

Fuel Cell Hybrid Vehicle Update

Fuel cell vehicles are now beginning to hit the showroom floors and be made available for lease or sale to the general public, primarily in southern California due to the hydrogen refueling infrastructure that has already begun to be set in place there. As of early 2015, and according to the California Fuel Cell Partnership, there were just eight hydrogen refueling stations in metro Los Angeles and one in the San Francisco Bay Area that could serve the public. However, the California legislature intends to provide $20 million annually over several years for the construction of additional locations. By the end of 2016 a network of more than 50 hydrogen stations should be operational in the San Francisco and Los Angeles areas. Some of these will be stations that provide both gasoline and hydrogen fuel and some of these will be dedicated to providing hydrogen fuel only. Manufacturers suggest that a system of Home Energy Stations (HES) is not being considered at this time.

Hyundai, Toyota, and Honda all appear to be the frontrunners to bring a fuel cell hybrid vehicle to market in 2015 and 2016.

Hyundai Tucson. Hyundai released its hydrogen-powered Tucson SUV in southern California in 2015. It was initially offered on a 3-year lease, but the lease included maintenance and unlimited hydrogen fuel for the term of the lease. As a fuel cell hybrid vehicle, Hyundai states that the Tucson will be capable of traveling up to 300 miles between refueling stops and suggests that refueling will take less than 10 minutes, as compared to the several hours that is required to recharge a PHEV. Hyundai builds the Tucson at its production facility in Ulsan, South Korea, and says it plans to produce a minimum of 1,000 of the hydrogen fuel cell vehicles annually through 2016.

Toyota Mirai. Toyota first began working on the development of the hydrogen fuel cell vehicle in 1992. In December 2002, Toyota began limited marketing of Toyota's fuel cell vehicle, known as the *Fuel Cell Hybrid Vehicle*, or *FCHV*, in the United States and Japan. Toyota billed the FCHV as the "closest thing yet to the ultimate eco-car." Toyota has also used the Toyota fuel cell stack in a wide variety of applications including city buses, small cars, and even fuel cells for applications in the home.

In 2015 the Toyota FCHV became a reality that was made available to the general public, first in Japan, then in the United States, and then in Europe. And, for the release to the general public, the Toyota FCHV was renamed the *Mirai*, the Japanese word for *future*.

The Mirai was unveiled at the November 2014 Los Angeles Auto Show. Toyota released the Mirai in the United States in mid-2015 and in Europe in September 2015. Toyota planned to build just 700 of these vehicles during 2015. Toyota also stated that its lease would include unlimited hydrogen refueling during the lease period, similar to the Hyundai Tucson.

Honda FCX/FCV Update. The current Honda fuel cell vehicle, called the *FCX Clarity*, has been leased to just a few users who live in southern California. Honda has scheduled the release of an exotic-looking FCV, called the *FCV Concept*, for introduction in the United States in 2016. Honda says the FCV Concept will have a 300-mile range and will require only 3 minutes to refuel the tank with hydrogen pressurized to 10,000 PSI. Today, this is considered to be a typical pressure for hydrogen storage and dispensing, twice the pressure that was being used only a few years ago.

Also, Honda has stated that an "all-new Honda BEV" and an "all-new Honda PHEV" will both be released by 2018.

Fuel Cell Hybrid Vehicle Summary

In the future, expect to see an increase in fuel cell vehicles as the technology progresses and prices become more affordable. As more manufacturers begin producing fuel cell vehicles, the number of available models will increase. And look for government programs to offer energy tax incentives on these vehicles.

✔ SYSTEM DIAGNOSIS AND SERVICE

WARNING: As stated previously, the high-voltage systems discussed in this chapter are to be worked on only by qualified service personnel who have been trained in the specific manufacturer training programs that pertain to these systems. Any attempt to diagnose or repair a high-voltage system without the proper manufacturer training puts the technician at risk of receiving a fatal electric shock. Also, at the high voltages that are present in these systems, there is a very real threat of an electrical short creating either an arc flash, resulting in the release of tremendous thermal energy, or an arc blast, resulting in the release of tremendous pressure. (An arc blast results from the instantaneous vaporization of copper, creating a rapid pressure expansion.) The discussion of the systems in this chapter is for informational purposes only and is not intended to provide the training that is required to work safely on these systems.

Required Equipment and Procedures for Servicing a Hybrid Electric Vehicle

High-Voltage Battery Disconnect. One of the first steps in performing any service work on these vehicles is to follow the manufacturer's recommendations for disconnecting the HV battery (or ultra-capacitor). These procedures typically include the following steps:

1. Wear an electrically insulating glove on the hand that is used to disconnect power.
2. Keep the other hand in your pocket or in your belt, not on the vehicle.
3. Use the hand with the electrically insulating glove to pull the "power disconnect plug" or to rotate the "power disconnect switch" (depending upon manufacturer) that is located in the trunk area.

Also, a Generation I Toyota Prius has three capacitors on-board, each of which maintains a 450 V charge, for a total of 1,350 V. After removing the high-voltage disconnect, these capacitors should discharge within about 50 seconds, but wait at least 5 minutes before trying to work on the high-voltage system.

Lineman's Gloves. When working on the high-voltage system of a hybrid electric vehicle, be sure to wear high-voltage linesman's gloves. The gloves must be certified for at least 1,000 V and must also be certified as class "0" or class "00" by the American National Standards Institute (ANSI) and the American Society for Testing and Materials (ASTM). Protective leather gloves must also be worn over the high-voltage insulating gloves to prevent them from being pierced by sharp objects, which would render them useless. The high-voltage insulating gloves also have an expiration date that must be adhered to. These gloves may be rolled up from the cuff toward the fingers to identify whether any pinholes exist. They can also be checked by certified professionals to verify whether any electrical leakage exists within a glove.

Nonconductive Shepherd's Hook. A nonconductive shepherd's hook must be readily available when a technician in the shop is working on a high-voltage system. This tool allows a stricken technician, following electrocution, to be pulled away from the high-voltage source without endangering another person.

Using a DMM on a Hybrid Electric Vehicle. Following the disconnect of the HV battery, a DMM should be used to verify whether power has been successfully removed from the HV cables (colored orange) before working on the high-voltage system. The DMM and voltage leads should be certified as follows when diagnosing problems on the high-voltage side of a hybrid electric vehicle (see Chapter 5):

- Generation I Toyota Prius: CAT-II, 1,000 V
- Most newer HEVs except the GM/Allison Dual Mode HEV: CAT-III, 600 V
- GM/Allison Dual Mode HEV: CAT-III, 1,000 V
- If in doubt, use a DMM and lead set that is CAT-III certified at 1,000 V

When using a DMM in this manner, a *LIVE-DEAD-LIVE* test sequence must be performed. What this means is that the technician must complete the following test procedure before working on the high-voltage system:

- LIVE test: Begin by using the DMM to measure voltage at the 12.6 V auxiliary battery. If the DMM successfully reads close to 12.6 VDC, only then should you proceed to the next test.
- DEAD test: Connect the voltage leads of the DMM between the orange high-voltage cables. The DMM should read close to zero volts DC. Also connect the DMM between each orange high-voltage cable and ground. In each test here, the DMM should also read close to zero volts DC.
- LIVE test: Complete this test sequence by, once again, connecting the DMM to the auxiliary 12.6 V battery. Again, the DMM must read close to 12.6 VDC.

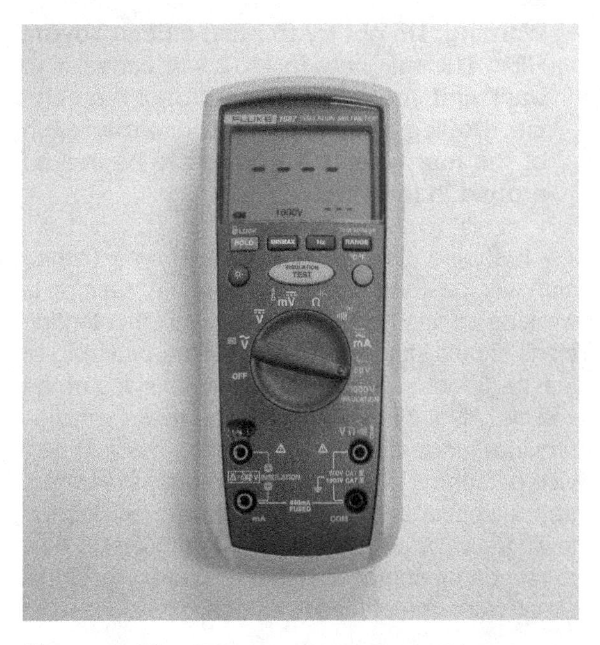

Figure 9–28 Fluke 1587 insulation multimeter.

The preceding test sequence verifies the DMM and the test leads both *BEFORE* and *AFTER* the DMM is used to identify that voltage has been removed from the high-voltage cables. By following this test sequence, you ensure that you have not overlooked the potential presence of high voltage by reason of a faulty meter, a faulty meter setting, or a faulty test lead.

Using an Insulation Tester on a Hybrid Electric Vehicle. Use an insulation tester, such as the *Fluke 1587 Insulation Multimeter* as shown in Figure 9–28, to determine whether there is an electrical path between the HV battery/HV cables and the body or frame of the vehicle. This tester can be set to a rated output of the meter of 50, 100, 250, 500, or 1,000 V. Connect the tester between a high-voltage lead and the body or frame of the vehicle. Select a voltage that is at least twice the rated DC voltage of the HV battery. When the Insulation Test button is pressed, the meter will output the selected voltage and determine both current flow and resistance between the two points where the leads are connected. Refer to manufacturer's specifications to determine pass or fail. For example, Toyota states that an insulation test on their HEVs must show at least 10 mega-ohms of resistance between the high-voltage system and the vehicle's body and frame. On many HEVs, this resistance could be compromised by a technician using an incorrect oil in the air-conditioning system. A proper insulation tester, capable of producing the required voltage output, must be used for this resistance test. The typical ohmmeter is *NOT* a suitable tool for this test.

Hoisting a Hybrid Electric Vehicle. When preparing to lift a hybrid electric vehicle on a shop hoist, research and use the proper lift points. Be very careful to verify that the pads are not placed under the orange high-voltage cables, as a mistake here could weld the vehicle to the lift and endanger the people in the immediate area.

Warning: Do not try to push a dead Toyota HEV. The magnets in MG2 will cause it to buck and you may lose control of the vehicle. Dollies should be placed under each of the four wheels if it needs to be moved around in the shop.

It should also be noted that when a technician who has had proper manufacturer training is working on one of these vehicles, all other technicians should stay clear of the vehicle, providing at least a 3-foot safety zone around all sides of the vehicle. Also, all manufacturers of these vehicles provide information for emergency medical service personnel regarding how to safely extricate a trapped occupant from one of these vehicles following a collision, as well as information on how to respond if someone has received an electrical shock from one of these vehicles.

SUMMARY

Most manufacturers now offer hybrid electric vehicles, plug-in hybrid electric vehicles, and battery electric vehicles. And fuel cell vehicles are now available in dealership showrooms. In this chapter we have discussed a sampling of several early mainstream vehicles that used some of these technologies, including:

- Hybrid-drive technology that uses a gasoline engine as the primary power source and an EMG to assist it as required
- Hybrid-drive technology that uses an EMG as the primary power source at low speeds and a gasoline engine to assist it as required
- Electric-drive vehicles that use fuel cell technology to provide the needed electric power

We have also taken a look at several technologically advanced concepts, including auto start/stop of a gasoline engine, regenerative braking, and Honda's Grade Logic Control. All the systems discussed in this chapter are designed to substantially increase fuel mileage and reduce CO_2 emissions while continuing to provide acceptable performance. In the near future, you should expect to see dramatic increases in the following:

- The number of vehicles on the road that are equipped with the technologies discussed in this chapter
- The number of vehicles that will come into your service and repair facility that are equipped with the technologies discussed in this chapter

The vehicles discussed in this chapter are just the beginning of a technological revolution that is changing the way technicians must approach a vehicle that has come in for service. For the technician who understands that upgrade training classes, textbooks, and other training materials are becoming increasingly important to his or her success in the industry, this technological revolution is not something to shy away from, but rather provides the potential to pull ahead of the competition.

▲ DIAGNOSTIC EXERCISE

A customer has had his Toyota Prius towed to your service and repair facility with the Output Control Warning Light illuminated. Your service advisor asks you to take a look at it. What is the best course of action for you to take?

Review Questions

1. What is the advantage of an electric motor over a gasoline engine when driving in city traffic?
   A. An electric motor can develop higher horsepower than a gasoline engine can be designed to develop.
   B. An electric motor can develop its maximum torque beginning just above zero RPM.
   C. An electric motor can develop its maximum torque beginning at about 1,500 RPM.
   D. An electric motor can develop its maximum torque beginning at about 2,500 RPM.

2. Which of the following can occur if a technician does not follow proper procedures when working on the systems discussed in this chapter?
   A. Fatal shock
   B. Arc flash
   C. Arc blast
   D. All of the above
3. *Technician A* says that the information contained in this chapter is not sufficient to allow readers to work on the high-voltage systems discussed. *Technician B* says that the information contained in this chapter is for informational purposes only and that a technician should only work on the high-voltage systems discussed in this chapter after completing the appropriate manufacturer training. Who is correct?
   A. *Technician A* only
   B. *Technician B* only
   C. Both technicians
   D. Neither technician
4. How many cylindrically shaped batteries (cells) are used in the Insight, the Civic Hybrid, and the Accord Hybrid to form the HV battery?
   A. 1
   B. 3
   C. 120
   D. 168
5. On the Honda IMA system, which of the following is the primary power source to power the drive wheels?
   A. EMG
   B. CVT
   C. PCU
   D. Gasoline engine
6. Which of the following gasoline engines is used in the Honda Insight?
   A. 1.0L VTEC
   B. 1.3L i-DSI
   C. INZ-FXE 1.5L
   D. 3.0L i-VTEC
7. The Honda Civic Hybrid qualifies as which of the following?
   A. ULEV
   B. SULEV
   C. ZEV
   D. AT-PZEV
8. On the Honda Civic Hybrid, how does the VCM system deactivate a cylinder?
   A. It only disables the fuel injector and allows the valves to continue to operate.
   B. It only disables the ability of the valves to open and allows the fuel injector to continue to operate.
   C. It disables the fuel injector and also disables the ability of the valves to open.
   D. It only disables the ignition system's spark to the cylinder.
9. Which of the following is used in the Honda Accord Hybrid to lessen the awareness of the driver to the deactivation of cylinders by the VCM system?
   A. An active control engine mount
   B. An "Active Noise Control" system
   C. A 4-inch-thick fire wall
   D. Both A and B
10. *Technician A* says that the EPA fuel economy rating on a Toyota Prius is greater for city driving than for highway driving. *Technician B* says that a Toyota Prius uses a throttle cable to allow the throttle pedal to physically control the throttle body on the gasoline engine. Who is correct?
    A. *Technician A* only
    B. *Technician B* only
    C. Both technicians
    D. Neither technician
11. What is the name of the electric motor/generator that drives the front-drive wheels on a Toyota Prius?
    A. MG1
    B. MG2
    C. MGF
    D. MGR

12. What is the primary purpose of the coolant heat storage capability on the Toyota Prius?
    A. To reduce vehicle emissions following a cold engine start
    B. To bring the climate control system up to operating temperature more quickly following a cold engine start
    C. To provide extra heat to the HV battery following a cold start
    D. To provide extra heat to MG2 as the vehicle is initially accelerated following a cold engine start
13. *Technician A* says that on the THS-II system as it is used on the Toyota Prius, the voltage that is applied to MG1 or MG2 is 201.6 V. *Technician B* says that on a Toyota Prius equipped with the Smart Key system, the driver can enter the vehicle while locked and drive it away without having to physically put the key into the door's lock cylinder or the ignition switch. Who is correct?
    A. *Technician A* only
    B. *Technician B* only
    C. Both technicians
    D. Neither technician
14. On the 2006 Toyota Highlander Hybrid, what is the voltage output of the boost converter/inverter?
    A. 201.6 VAC
    B. 273.6 VAC
    C. 500 VAC
    D. 650 VAC
15. Which of the following vehicles is a series PHEV that is rated to be able to travel up to 40 miles in EV mode?
    A. Toyota Prius PHEV
    B. Ford Fusion PHEV
    C. Chevrolet Volt PHEV
    D. All of the above
16. Which of the following HEVs uses the Allison two-mode hybrid system?
    A. Toyota Prius HEV
    B. Honda Insight HEV
    C. Chevrolet Tahoe HEV
    D. Ford C-Max HEV
17. The Honda FCX emits which of the following emissions?
    A. CO_2 only
    B. NO_x only
    C. CO_2 and NO_x
    D. Heat and water
18. Which of the following technologies was being used on the Nissan Altima beginning with the 2006 model year?
    A. Honda hybrid technology
    B. Toyota hybrid technology
    C. Ford hybrid technology
    D. Fuel cell technology
19. Which of the following manufacturers' hybrid-drive vehicles uses the gasoline engine as the primary power source and uses an electric motor/generator to assist it only when the load requires?
    A. Honda
    B. Toyota
    C. Nissan
    D. Ford
20. Which of the following hybrid vehicles uses a control module strategy called Grade Logic Control to reduce the driver's need for constantly having to compensate at the throttle pedal for transmission upshifts and downshifts when driving in mountainous terrain?
    A. Honda Accord
    B. Toyota Prius
    C. Toyota Highlander
    D. Ford Escape

Chapter 10

Modern Systems that Interact with the Engine Control System

OBJECTIVES

Upon completion and review of this chapter, you should be able to:

- ❑ Understand body control modules and the systems they may control.
- ❑ Explain how an anti-theft warning system differs from an anti-theft disabling system and how both operate.
- ❑ Understand how smart key and remote start systems operate.
- ❑ Explain the operation of a menu-driven information and control system.
- ❑ Understand navigational systems, including their mapping databases, their features, and the signals they rely on.
- ❑ Understand voice recognition systems, including their purpose and their features.
- ❑ Explain the various systems that were derived from an antilock brake system, including automatic traction control, electronic throttle control, collision avoidance systems, adaptive cruise control, stability systems, lane departure and lane sway warning systems, and traffic collision avoidance systems that use vehicle-to-vehicle communication.
- ❑ Explain the operation of driver alert systems, including back-up cameras, back-up alert, cross traffic alert, blind spot alert, and park assist systems.

KEY TERMS

Capacitive-Touch Screen
Micro-Electronic Mechanical System (MEMS)
Resistive-Touch Screen
Telematics
Yaw Rate

Modern vehicles incorporate many electronically controlled systems that interact with the engine control system in various ways. These systems are designed to provide the driver with additional convenience and safety features. Many of these systems have been formed through the modification of antilock braking systems.

BODY CONTROL MODULES

A body control module, commonly referred to as the BCM, is used to control the vehicle's bodily functions. These functions commonly include systems that were once under the control of small dedicated computers. A BCM typically provides control of the following features:

- Electronic automatic temperature control (EATC) systems. This includes control of the mode doors, blend air door, and the blower motor. The blower motor is generally controlled on a high-frequency pulse-width-modulated signal as a method of controlling the motor's speed without using resistors.
- Cooling fan operation. The cooling fan motor may also be controlled on a high-frequency pulse-width-modulated signal.
- Exterior lights, including headlights, both high and low beam operation, parking and taillights, side marker lights, directional signals, back-up lights, and the brake lights.
- Interior lights, including the courtesy lamp circuit, glove box light, under-hood and trunk lights, and even some of the warning lamps on the instrument panel. The BCM typically controls the illuminated entry feature, initiated through a wireless key transmitter (key fob), a keypad on the driver's door, or the operation of an outside door handle on a front door.
- Horn circuit.
- Memory seat systems, including memorized seating positions and outside mirror adjustments.
- Heated and cooled seats.
- Heated steering wheels.
- Electronic back lights, or EBL (also known as the rear window defogger).
- Heated outside mirrors. These are usually activated simultaneously with the operation of the rear window defogger.
- Front and rear wiper motors and washer pumps.
- Heated windshield wipers. The physical wiper blades are not heated, but rather wipe across a small heating grid at the base of the windshield. In this way enough heat is transferred to them to prevent wiper blade icing. The system is typically energized either by turning on the front/rear defrost systems or when exterior ambient temperature is sensed to drop below a predefined temperature.
- Power windows, including one-touch down and one-touch up features. This may include the driver's window only or all windows that are controlled from the driver's switch and front seat passenger switch.
- Power door lock and unlock relays.

Many vehicles use multiple body control modules. For example, a front body module may be used to control exterior lighting and other functions at the front of the vehicle while a rear body module may control the exterior lighting and other functions at the rear of the vehicle.

ANTI-THEFT SYSTEMS

Two types of anti-theft systems are used on modern automobiles: anti-theft warning systems and anti-theft disabling systems.

Anti-Theft Warning Systems

An anti-theft warning system will typically energize the horn circuit intermittently and flash the lights if an unauthorized person opens a door or hood. The purpose of this system is to simply draw attention to the vehicle. This type of system is usually armed 30 seconds after the doors are locked, provided all doors are properly closed. It can also be armed using a wireless key transmitter. On many vehicles if the driver tries to use the wireless key transmitter to arm the system, the system may activate the horn circuit in an unusual manner to alert the driver in the event that a door is still ajar, thus preventing the system from arming.

It may be disarmed by using the wireless key transmitter, by unlocking the door with the door key, or by using a key to turn the ignition switch on. The wireless transmitter can also be used to activate a panic alarm to draw attention.

When replacing the wireless transmitter, or when adding another transmitter, follow manufacturer-recommended procedures. These procedures may involve the use of a scan tool, a transmitter that is already programmed, or a simple routine such as cycling the ignition switch on, off, and on a predetermined number of times within a period of a few seconds.

Anti-Theft Disabling Systems

An anti-theft disabling system prevents the engine from running if an improper key is used in an attempt to start the engine. An electronic transponder is placed in the key head. When the ignition key is inserted in the ignition lock cylinder and turned to Start, the PCM will initially start the engine. At this time an oscillating voltage is created within a coil of wire located around the lock cylinder, just behind the plastic trim panel, and in close proximity to the key. The coil's pulsing electromagnetic field induces a voltage into the key head, allowing the transponder IC chip to power up. The transponder in the key head then sends a radio frequency (RF) code to a decoder module located in close proximity. The decoder module compares the transmitted code with a preprogrammed code located in a specific node on the data bus. If the code is a match, within 1 second of engine start the PCM is instructed to allow the engine to continue to run. If the code is not a match, the PCM is instructed to shut down the engine's fuel injection and ignition systems for 15 minutes (typically). This system can use any of about three trillion different key codes as a recognized code for a particular key. The system can be programmed to recognize a dozen or more key codes.

A modern anti-theft disabling system is extremely effective at preventing vehicle theft when a programmed key is not available. In fact, this system is so effective that vehicles that incorporate this system as standard equipment will typically have the ignition switch located in the instrument panel. There is no longer a need to locate the ignition switch lock cylinder in the steering column to combat theft.

SMART KEY AND REMOTE START SYSTEMS

Smart Key Systems

A smart key system uses an electronic key (Figure 10–1), which transmits an RF signal to receiving antennas on the vehicle. Typically, these antennas are located within each front door, the trunk or cargo area, and the instrument panel.

Figure 10–1 A wireless key transmitter. Pressing the lock button once followed by pressing the top button twice sends a request to the PCM to remotely start the engine.

If the wireless key transmitter is within 1 meter (typical) of the lock cylinder on either front door, a touch pad on the door handle allows the driver/passenger to unlock the doors without having to push any buttons on the transmitter or insert a key. If the wireless key transmitter is within 1 meter of the rear liftgate on a sport utility vehicle (SUV), the driver may press a button that will release the liftgate or activate an electrically controlled liftgate. Some SUVs now allow the driver to simply pass a foot underneath the rear bumper to activate the electrically controlled liftgate, provided the smart key is nearby (such as in a pocket). This allows the liftgate to be operated hands-free.

Similarly, a smart ignition switch allows the driver to start the vehicle without having to insert the key in an ignition switch lock cylinder. With the vehicle in Park and the brake pedal depressed, simply push the ignition Start button on the instrument panel. When the smart key electronic control unit (ECU) recognizes that the wireless key transmitter is inside the vehicle, it sends a request to the PCM to energize the starter circuit and start the engine.

In practical use, a smart key system allows the driver to enter the vehicle, start the engine, and drive away with the key still in a pocket or purse.

Remote Start Systems

A remote start system is often used in conjunction with a smart key system. Used with automatic transmission vehicles, the remote start system allows the driver to use the wireless key transmitter to remotely make a request of the PCM to energize the starter circuit and start the engine. This allows the driver to start the engine from outside the vehicle, as when getting ready to punch out at work and leave the office. By the time the driver gets to the vehicle, the engine coolant has already begun to warm up.

In order to make a remote start request, the driver must begin by pressing the lock button on the wireless key transmitter followed by pressing the remote start button twice (typically), as shown in Figure 10–1. This procedure reduces the potential to remotely start the vehicle unintentionally. This procedure also insures that the doors are locked before starting the engine. The smart key ECU places the vehicle in *remote start mode*, which will not allow the vehicle to be driven until the wireless key transmitter is inside the vehicle. Once the driver uses the smart key system to enter the vehicle, the Start button must still be pressed with a foot on the brake pedal to exit the *remote start mode* and enter *driving mode* before the vehicle can be shifted out of Park. The remote start will keep the engine running in this mode for 5 or 10 minutes maximum (typically), at which point the PCM will shut the engine off if the driver has not yet arrived. The maximum amount of time that the engine is allowed to run in *remote start mode* may be adjustable by the driver.

Ford's recently-announced *Sync Connect*™ (to make its debut on the 2017 Ford Escape) will allow the driver to program engine remote start times on a smart phone using a Ford app. If you leave your workplace at 3:30 PM every afternoon, you can program the engine to start remotely at 3:20 PM, thus allowing the heating or air conditioning system to adjust the vehicle's interior temperature. The Sync Connect™ app will also allow you to access vehicle status information including tire pressures, fuel level, and the vehicle's location.

MENU-DRIVEN INFORMATION AND CONTROL SYSTEM

A menu-driven information and control system uses a touch-sensitive screen to allow the driver to obtain information about the vehicle's systems and to allow the driver to control some systems such as the climate control system and the sound system. Therefore, this display screen provides two-way communication between the driver and the vehicle. The display screen is also a node on a multiplexed data network.

General Motors was the first manufacturer to introduce such a system on the 1986 Buick

Riviera and the 1986 Cadillac Eldorado. The early GM system used a two-tone cathode ray tube (CRT) and was considered to be far ahead of its time. It was produced for only 3 years, with the final production year being 1988.

Modern systems use a **resistive-touch screen** or **capacitive-touch screen**, either one being a full-color LED display. In addition to providing information to the driver and allowing the driver to control the climate control and sound systems, these touch-sensitive screens also provide a method to display the map that is an integral part of a modern navigational system, a feature that forced the comeback of such a display screen.

NAVIGATIONAL SYSTEMS

A navigational system is primarily an electronic map with some added features. The original electronic maps simply allowed the user to look up an address, similar to using a paper map. The next feature to be added was a "trips" feature, which allowed the user to look up two addresses and then determine the best route between them. A navigational system has an added feature that allows the mapping program to determine the user's current location on the mapping database automatically. With this feature, the user has only to look up his or her destination. Then the mapping database will determine the best (fastest or shortest) route to the selected destination.

To determine the user's location automatically, several different methods were experimented with in the late 1980s, including land-based transmitters, satellite-based transmitters, placing active signposts at every intersection that would update the on-board computer with the vehicle's location, and even systems that required an initial entry of the vehicle's location and then would monitor the distance and direction the vehicle was driven. Beginning in the 1990s, consumer-level navigational system receivers began receiving and using signals from the U.S. military's Global Positioning System (GPS) of satellites.

Global Positioning System

The GPS consists of 24 satellites orbiting about 10,900 miles above the Earth. The system is owned by the U.S. Department of Defense. The United States began working on the GPS project in 1973. It became fully operational in 1995.

The system actually has additional satellites, a total of about 32. But a satellite may be temporarily labeled as *unhealthy* while it is repositioned and then be labeled *healthy* again. Generally this system will have 24 *healthy* satellites at any time. The satellites are spaced in their orbits so that, at that altitude, a minimum of four satellites will be above the horizon and will have a direct line of view to any point on Earth. The satellites receive updates from the control centers. The GPS control segment consists of a global network of ground facilities that track the GPS satellites, monitor their transmissions, perform analyses, and send commands and data to the satellite network.

The current operational control segment includes a master control station, an alternate master control station, 12 command and control antennas, and 16 monitoring sites. The master control station is Schriever Air Force Base (Schriever AFB) located 10 miles east of Peterson AFB near Colorado Springs, Colorado.

GPS Signals

Each satellite has an on-board atomic clock, a computer, and a radio transmitter. The computer transmits radio signals that include the atomic time and the satellite's position. There exist two signals that are transmitted from the GPS satellites:

- Precise Positioning Service (PPS): This is an encrypted signal designed for use by the U.S. military. Consumer-level receivers cannot decrypt this signal.
- Standard Positioning Service (SPS): This signal is not encrypted and is designed for use by consumer-level receivers. However, in times of war, satellites in the part of the world we are at war with may use algorithms

to transmit these signals that will be purposefully made to be in error.

Initially, the SPS signal was purposefully downgraded to be ten times less accurate than the PPS signal. Originally, the PPS signal was designed to be accurate within 3 meters (about 10 feet) 90 percent of the time, while the SPS signal was intended to be accurate within about 30 meters (about 100 feet) 90 percent of the time. A California university did a study in the late 1990s and concluded that the SPS signal was about three times more accurate than the U.S. military had intended. Today, the SPS signal is generally accurate within about 5 meters (about 16 feet). However, with enhanced algorithms, the SPS signal has been known to gain an increase in accuracy down to as little as two centimeters.

The Automotive Navigational System Receiver

The navigational system receiver uses complex algorithms and signals from a minimum of four satellites to calculate latitude, longitude, and altitude. To do this, the receiver must calculate the position of each of the four satellites first, known as *satellite geometry*. Then the receiver calculates how long it took for each of the signals to travel through the Earth's atmosphere from the satellite to the receiver. Finally, the GPS receiver uses these signals to calculate changing latitude, longitude, and altitude over time and then determines the vehicle's speed and direction of travel.

Additional features on many automotive navigational systems include such things as point-of-interest (POI) entry, estimated time of arrival (ETA), and the ability to display speed limits on many roads and highways. While many vehicles have navigational systems built in, there are many dedicated add-on systems also, including *Garmin*, *Magellan*, *TomTom*, and others. There are also automotive mounts available for these units that allow the user to mount the unit to an A/C vent or a power point.

For travelers, there are also many apps for smartphones, such as *Waze* and *Scout GPS Maps*. These apps also use user input to warn the driver of a stalled vehicle or an accident ahead.

VOICE RECOGNITION SYSTEMS

A voice recognition system is designed to allow the driver to control many of the vehicle's systems through voice commands. In turn, this allows the driver to keep his or her eyes on the road and decreases the level of distraction while controlling these systems. Systems that can commonly be controlled through the voice recognition system include the climate control system, the sound system, and the vehicle's navigational system. Voice recognition also allows the user to initiate and receive phone calls through a wireless connection with a Bluetooth phone within the vehicle. Other features include access to satellite radio and all of its features, including the use of voice recognition to request traffic reports, weather reports, and movie theater listings. The operation of some systems such as control of the power windows and the power sun roof may also be supported on some models.

The voice recognition ECU is a node on a data bus. The driver simply operates the *voice recognition* button and then verbalizes a command. The ECU recognizes a much larger vocabulary on modern systems than on the early systems. And the ECU speaks to the driver over the vehicle's sound system speakers using a voice that is much more human-like than that on the early systems. Because the sound system speakers are part of a voice recognition system, the radio or CD player will automatically mute as needed for the system to operate.

Voice recognition systems were introduced on the automobile in 2003 and found a larger market on luxury vehicles beginning in 2005. Today, the systems are also available on many SUVs, minivans, light trucks, and mid-class to low-end passenger cars.

ANTILOCK BRAKING SYSTEMS AND RELATED ELECTRONIC SYSTEMS

Antilock Brake Systems

Ford was the first automotive manufacturer to experiment with antilock braking systems (ABS) in 1954 when they fitted a 1954 Lincoln with an English aircraft anti-skid system called the *Maxaret* system. Throughout the 1960s and into the early 1970s, several manufacturers experimented with ABS, during which time analog computers were used to control vacuum solenoids. The vacuum actuation was quite slow, capable of only about four cycles per second. During the middle 1970s Robert Bosch Corporation continued to pursue ABS technology. In this period analog computers gave way to digital computers and vacuum actuation gave way to hydraulic actuation, which is much faster. Robert Bosch Corporation introduced to first modern ABS system using a digital computer and hydraulic actuation on the 1978 Mercedes Benz.

An antilock brake control module (ABCM), also sometimes referred to as the electronic brake control module (EBCM), monitors a magnetic wheel speed sensor (WSS) at each wheel. The WSS monitors a reluctor ring with iron teeth that are closely spaced, resulting in a high-resolution signal. The ABCM monitors the frequency of this signal to known wheel speed. The ABCM also calculates overall vehicle speed from the multiple WSSs in order to know allowable rates of deceleration.

An ABS system operates using three aggressive modes of operation: *apply*, *maintain*, and *release*. (The *maintain* mode is also known as a *hold* mode and the *release* mode is also known as a *dump* mode.) When the driver begins to *apply* braking force to the brakes, the ABCM monitors the frequency of each WSS. When a WSS indicates a wheel is decelerating so quickly that it is likely to lock up, the ABCM isolates that wheel's caliper from the master cylinder, therefore putting that particular wheel into the *maintain* mode for a few milliseconds. This will keep the driver from applying additional braking force to that wheel's caliper. If the WSS is still indicating that wheel lockup is likely after the predetermined hold period, the ABCM will *release* hydraulic pressure at that wheel's caliper while watching the WSS to determine the re-acceleration of the wheel. Once the wheel has re-accelerated and regained traction with the road surface, the ABCM will put that wheel's hydraulic channel back into the *apply* mode and allow the driver to apply additional pressure to the wheel's caliper. All of this takes place in just a few ms.

The *apply* mode is engineered to be the default mode. If an ABS system fails electrically, it defaults to the *apply* mode. That does not mean that the brakes are applied, only that the default positioning of the hydraulic solenoids when they are de-energized gives the driver the ability to apply hydraulic pressure to the wheels as needed. Rephrased, in the event of an electrical or electronic failure of the system, the system will default to normal hydraulic braking.

The cycling rate through theses three modes of operation is determined according to the demand. If a tire with good tread is on a dry road surface, it may only take two or three cycles per second to prevent wheel lockup. As the friction coefficient between the tires and the road surface becomes more slippery, a higher cycling rate is required to prevent wheel lockup. Under slippery condition such as gravel or snow, this system may cycle through these three modes of operation 10 or 15 times per second. And there are limits to the system's abilities—on wet ice the tire will not likely regain traction with the road surface during the *release* mode. The result of ABS operation is that the tire rotates in a jerky rotation—slowing, accelerating, slowing, accelerating, and so on. Engineers have determined that to achieve optimum braking, wheel slip needs to be held between 8 and 35 percent (Figure 10–2). As a result, it is normal to hear momentary squeals from the tires during ABS functioning.

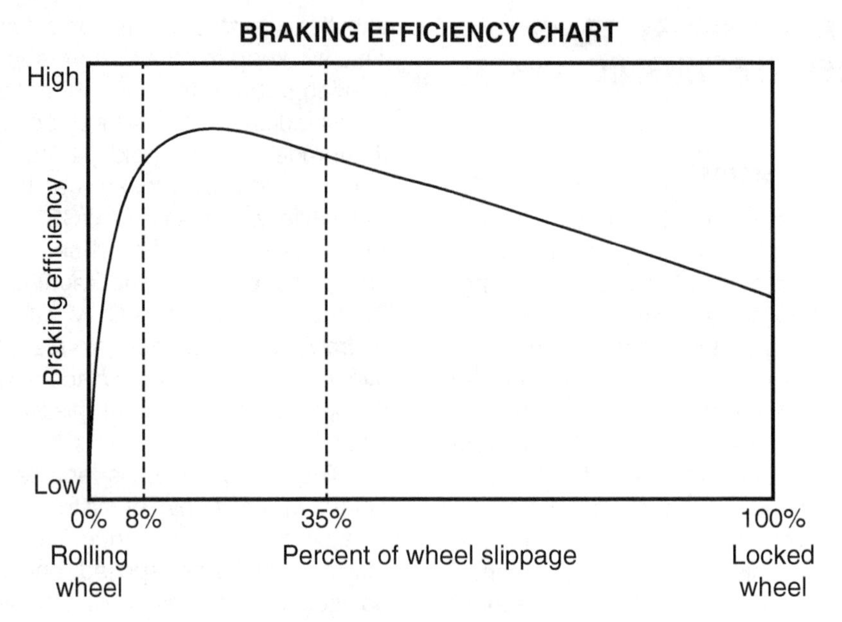

Figure 10–2 With an ABS system, the ABCM is programmed to achieve between 8 and 35 percent wheel slippage.

There are several benefits to be gained from antilock system functioning:

- Shorter stopping distances when compared to allowing the tire to totally break traction with the road surface and skid.
- Controllability of the vehicle during hard braking, also known as *directional stability*.
- Steerability of the vehicle while braking hard.
- Reduced potential for flat-spotting of tires during a hard braking.

Pertaining to shorter stopping distances, it is important to note that an ABS system does not take the optimum stopping distance and shorten it up even further. Rather, it brings the vehicle back toward the optimum stopping distance. The driver may be able to achieve the optimum stopping distance by applying the optimum pressure to the brake pedal during a hard braking situation but is not like to do this during a panic stop. Too little force applied to the brake pedal results in less than optimum braking, and too much force applied to the brake pedal can result in wheel lockup, also resulting in less than optimum braking. And if the right front wheel is on gravel while the left front wheel is on dry pavement, the ABCM can modulate pressure to these two wheels independently to obtain maximum slowing potential while preventing lockup at each wheel.

With regard to controllability of the vehicle (also known as *directional stability*), an ABS system can prevent the vehicle from getting sideways during a hard braking situation. This happens with vehicles that don't have ABS when all four tires are allowed to lock up and skid for a period of time.

Regarding steerability, an ABS system allows the driver to have steering control while braking hard. Prior to ABS systems, if both front wheels were locked up and skidding, the driver did not have an option to steer the vehicle around an object. In this situation, the driver would have to release the brake pedal before steering control could be regained. With an ABS system, steering control is retained while braking hard due to the continuing jerky rotation of the tires.

Automatic Traction Control

An automatic traction control (ATC) system, also referred to as a traction control (TC) system, uses the same wheel speed sensors as the ABS system uses. Some early ATC systems had a separate ECU to control this function, but with many modern systems the ATC function is also under control of the ABCM. For ATC functioning, the ABCM looks for too rapid an acceleration of the drive wheels, particularly at speeds less than 20 MPH, indicating that a drive wheel has lost traction with the road surface during acceleration. If this happens, the ABCM will apply hydraulic brake pressure to that wheel's caliper which, in turn, transfers existing engine torque to the opposite drive wheel. If the driver is mashing the throttle so hard that the second drive wheel loses traction with the road's surface, the ABCM will make a request of the PCM over the data bus to reduce engine torque.

Some early ATC systems such as Corvette and Infiniti used two throttle plates in series. One was cable-operated and controlled by the driver and the other was electronically controlled by the PCM. If the driver was holding the primary throttle plate wide open, the PCM would close the secondary plate until the engine torque at the drive wheels matched the friction coefficient between the tires and the road surface. All of this would take place in a few milliseconds. Of course, the closing of the secondary throttle plate only reduced airflow. The PCM would match it with a reduction in fuel delivery as well.

On modern ATC systems, engine torque may be reduced by the PCM reducing fuel delivery and retarding spark timing. The PCM or TCM (transmission control module) may also upshift an automatic transmission to reduce available engine torque at the drive wheels.

You should note that an ATC system was the first electronic system in which a control module could apply hydraulic brake pressure at a wheel's caliper without the driver pressing the brake pedal. It was also the first electronic system that could reduce engine torque even though the driver was pressing the throttle pedal to wide open.

Electronic Throttle Control

An electronic throttle control ETC system, sometimes also referred to as a *throttle actuator control* (*TAC*) system, was derived from the early ATC systems that used two throttle plates in series. Essentially, the cable-operated throttle plate was deleted and the electronically controlled throttle plate was retained. Modern ATC systems use a modern ETC system to assist in controlling engine torque during ATC functioning and also in cooperation with some of the following systems' functions. Additional information about ETC systems can be found in Chapter 4.

Pre-Collision Throttle Management

On some newer vehicles, the ETC system includes an active pre-collision throttle management system that will reduce engine torque when an obstacle is detected in front of the vehicle, even if the driver is pressing hard on the throttle pedal. This could help reduce the potential for running into a garage door or into another vehicle parked in front of the primary vehicle if the driver were to unintentionally place the transmission in **Drive** when intending to back out of a parking space. In this situation, it can reduce engine torque enough to prevent the vehicle from climbing over a parking block placed ahead of the vehicle's front tires. The pre-collision throttle management system also works with the active collision avoidance system described later in this chapter in order to avoid a collision or to reduce the level of severity of a collision.

Brake Assist Systems

Many times during a panic stop situation, the driver will not apply enough pressure to the brake pedal to avoid a collision. In this event the ABCM senses that the driver is in a panic stop situation and will apply additional hydraulic braking force to all brake calipers while using its ABS to prevent wheel lockup.

Collision Avoidance Warning Systems

A collision avoidance warning system is designed to detect the potential for a collision to occur and alert the driver, both visually and audibly. Statistics have shown that given an extra one-half second of warning time, a driver can avoid up to 60 percent of all rear-end collisions, 50 percent of accidents occurring at an intersection, and 30 percent of all head-on collisions.

The ECU in a collision avoidance warning system monitors vehicle speed, the distance to an obstacle ahead, and whether the vehicle is being steered at an obstacle ahead. The reason that the vehicle's speed is important is that each time a driver doubles the vehicle's speed:

- The driver reaction time, measured in distance traveled, also doubles (as you might expect).
- Once the brakes are applied, the actual braking distance, measured in distance traveled, is quadrupled (times four).

Figure 10–3 shows that if the vehicle's speed is doubled from 20 to 40 MPH, the actual average braking distance increases from 20 feet to 80 feet. If the vehicle speed is doubled from 30 to 60 MPH, the actual average braking distance increases from 45 feet to 180 feet. And if the vehicle speed is doubled from 40 to 80 MPH, the actual average braking distance increases from 80 feet to 320 feet. Therefore, it is important that the ECU be able to monitor the vehicle's speed in order to calculate stopping distances.

Radar sensors are used to monitor the distance to an obstacle ahead (Figure 10–4). These sensors are intended to be capable in all weather conditions. (While modern vehicles can also incorporate laser sensors and cameras, these devices are not considered to be all-weather and can be adversely affected by rain, snow, fog, or blowing dust.) The ECU uses the measured distance to an obstacle ahead over time to calculate the closing speed to the obstacle ahead.

The ECU uses input from a steering angle sensor (SAS) to determine the vehicle's heading. The SAS may be an optical sensor that produces digital voltage pulses as the steering wheel is turned, or it may be an analog sensor that can output positive and negative voltages depending upon which direction the steering wheel is turned from center steer. Most often, the SAS assembly will contain two sensors for redundancy. Some SAS modules are a node on a CAN bus and simply communicate the steering angle to other nodes on the data bus in the form of the binary language.

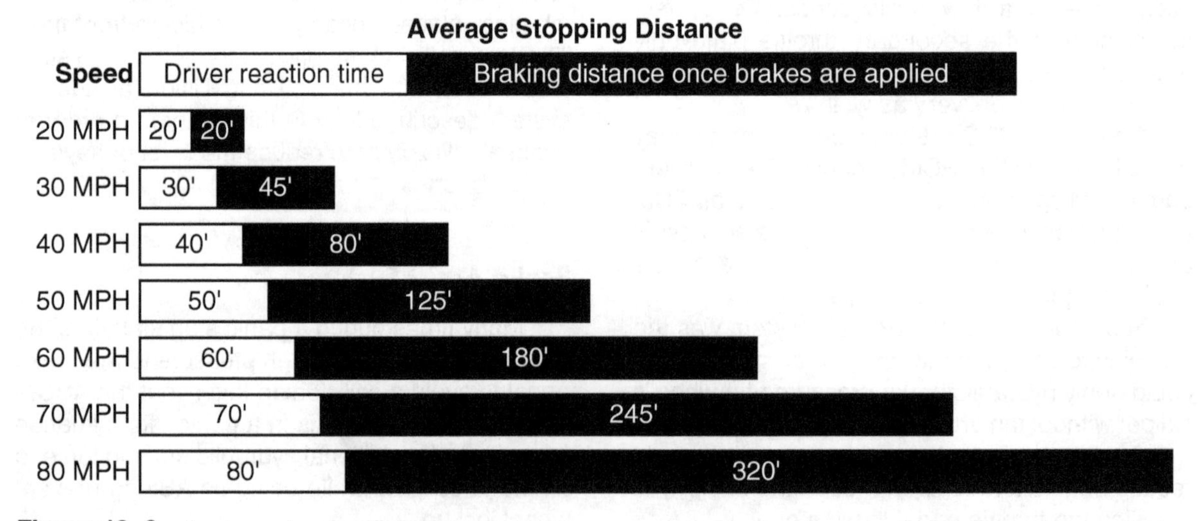

Figure 10–3 Average stopping distances.

Figure 10–4 A typical radar sensor.

The SAS must be correctly set whenever a wheel alignment is performed on the vehicle. Follow the manufacturer's instructions to perform this task correctly. Generally, the best method to test the sensor or to interpret what the sensor is communicating to the modules that need the information is through the use of a scan tool.

Active Collision Avoidance Systems

An active collision avoidance system is one that uses an ECU to apply hydraulic pressure to the brakes at all four wheels and reduce engine torque in order to avoid a collision or, at least, reduce the severity of a collision. This system uses the same sensors as the warning system, but it can actively override the driver to stop the vehicle quickly even if the driver is not pressing the brake pedal and even if the driver is holding the throttle pedal to the floor.

Active collision avoidance systems were first introduced in the United States in the 2007 model year on European automobiles, and they are becoming increasingly popular on many other makes.

Adaptive Cruise Control Systems

An adaptive cruise control system is a modification of a conventional cruise control system that allows the ECU to automatically adjust the vehicle's speed in response to the flow and speed of the traffic ahead of the vehicle. To an extent, the driver can select a preferred following distance. Like the collision avoidance systems (both warning and active), the PCM uses radar sensors to determine distance to the vehicle ahead. This system will not bring the vehicle to a complete stop to avoid a collision. The driver must remain ready to brake when driving conditions require it. Adaptive cruise control systems were first introduced in the United States in the 2004 model year.

Electronic Stability Control Systems

Manufacturers are now using a feature on many vehicles that is designed to reduce the potential for a collision by aiding the driver's control of the vehicle if the vehicle's tires lose their grip with the road surface during cornering. Known generically as an electronic stability control (ESC) system, this feature helps the driver to maintain vehicle directional stability control, providing both oversteer and understeer correction to maintain the vehicle's behavior on the road surface. General Motors calls this system *StabiliTrak*, Ford Motor Company calls it *AdvanceTrac*, and Chrysler calls it their *Electronic Stability Program* (*ESP*).

In a cornering situation where the tires have lost traction with the road surface, the vehicle's **yaw rate** is changed abnormally, resulting either in oversteer or understeer. (Yaw rate is a determination of how far left or right a vehicle has moved from its intended course.) If the vehicle's back end begins to fishtail (oversteer) during a turn, the ECU detects this through a "yaw rate and lateral acceleration sensor" and applies a braking pulse at the outside front wheel to help the driver stabilize the car (Figure 10–5). This action allows the side of the vehicle on the inside of the turn to begin to move ahead of the other side, thus reducing the oversteer. If the front end of the vehicle begins to drift to the outside of a turn (understeer), the ECU detects this and applies a braking pulse to the inside rear wheel. This action places a drag on the side of the vehicle at the inside of the turn and allows the side of the vehicle on the outside of the turn to move ahead, thus reducing the understeer. In both situations, it can also reduce engine torque as needed.

A gyrometer-type sensor is also used to sense if the vehicle were to tip during high-speed cornering, indicating that it is beginning to roll over. When tipping of the vehicle is sensed in a high-speed cornering situation, the controlling ECU will reduce engine torque and apply hydraulic braking pressure to the wheels so as to reduce the potential for rollover. One manufacturer states that the ECU will reduce the vehicle's speed by 10 MPH within 1 second to reduce the potential for vehicle rollover.

The sensors that are used to detect yaw rate and rollover are a **micro-electronic mechanical system (MEMS)** design, similar to the modern

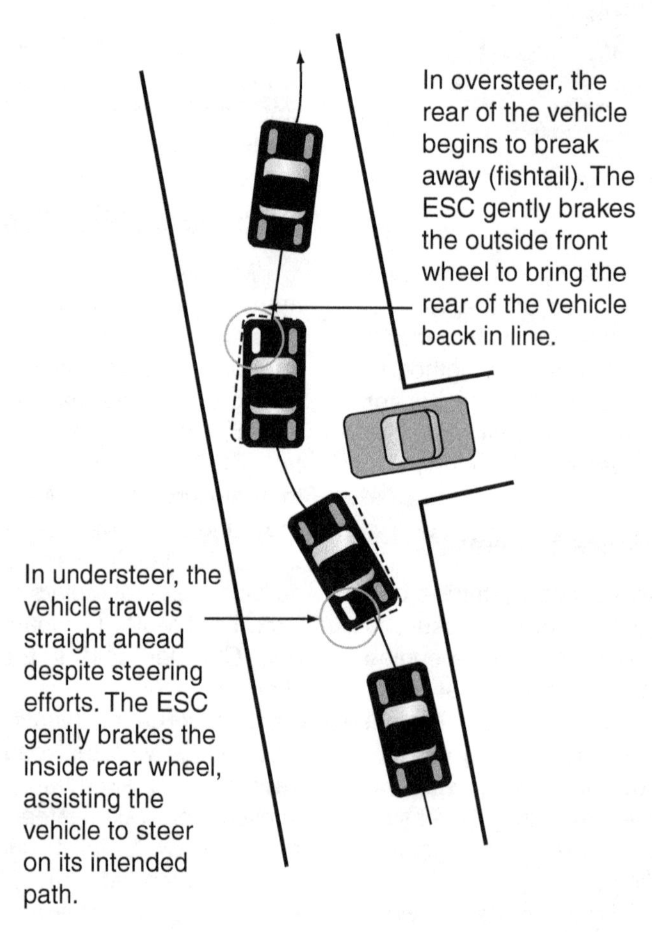

Figure 10–5 With an electronic stability control system, the ECU reacts to correct understeer or oversteer.

G-force sensors (crash sensors) used with passive restraint systems. Also known as *nano-technology*, these sensors are built on a silicone substrate and can be designed to detect movements such as yaw rate, rollover, G-force, and many other conditions. These sensors are very similar to the ones in modern smartphones that allow the phone to detect movement and tilt.

Lane Change Departure and Lane Sway Warning Systems

Some manufacturers are now using a camera mounted ahead of the rearview mirror and facing forward out the windshield to allow an ECU to monitor the white and yellow lane markings. If the vehicle crosses a lane marking without the directional signal activated, the ECU will sound a chime to alert the driver. If the lane change is completed, the chime will be turned off. The intent of this system is to alert the driver when a lane marking has been crossed unintentionally.

The warning system can be turned off because sometimes the lane marking are faded and unrecognizable to the ECU. Also, if a strip of new pavement has been added alongside and in parallel to old pavement, the place where the new pavement meets the old pavement may look like a lane marking, and the ECU can become confused.

Subaru Eyesight® System. Subaru uses a version of this system known as *Eyesight®*. Introduced in the 2013 model year on - Subaru Legacy and Outback models, by 2015 the Eyesight® system was also available on Subaru Forester, Impreza, and XV Crosstrek models. The Eyesight® system uses two cameras mounted behind the windshield and facing forward (Figure 10–6). The cameras are mounted 34 cm apart (about 13½ inches). The camera views cross each other. This triangulation allows the ECU to establish depth perception similar to human eyes. The Subaru Eyesight® system can be turned off, but the default is set to *ON*.

The Subaru Eyesight® system includes a *lane sway warning* feature that alerts the driver to the vehicle swaying within a lane, even though no lane markings have been crossed. This could

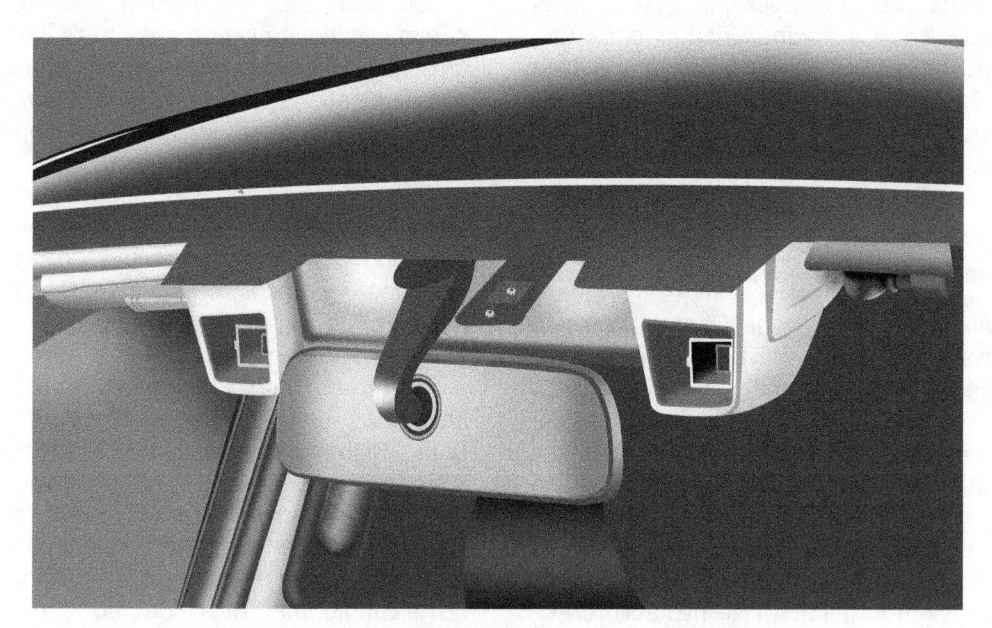

Figure 10–6 The cameras of a Subaru Eyesight® system.

be consistent with the behavior of a driver who is becoming drowsy. The ECU will audibly alert the driver to "stop, revive, and survive."

The Subaru Eyesight® system includes the following features that have already been discussed in this chapter:

- Pre-collision throttle management
- Pre-collision braking (brake assist)
- Adaptive cruise control
- Lane departure warning
- Lane sway warning

An additional feature of the Subaru Eyesight® system that has not yet been discussed is the *lead vehicle start alert* feature. In heavy stop-and-go traffic, this feature alerts an inattentive driver that the traffic in front of the vehicle has begun moving again. An audible alert is given if the vehicle in front of the Eyesight® vehicle has moved 3 meters or more (about 10 feet) and the Eyesight® vehicle has not yet begun to move.

The Eyesight® system is not used on black vehicles due to reflection issues in the vehicle's hood, nor is it used on vehicles with manual transmissions. Also, the Eyesight® system does ***not*** include an active lane change departure prevention feature, as is discussed in the next section. On an Eyesight® vehicle it is the driver's responsibility to take corrective action if the lane change departure or lane sway warning is given.

Active Lane Change Departure Prevention Systems

Infiniti introduced a Lane Change Warning (LDW) system and an active Lane Departure Prevention (LDP) system on its 2008 model year vehicles. The LDW system works similarly to the systems discussed in the preceding section, but it uses only one camera mounted ahead of the rearview mirror. The LDP system uses the same camera, but if a lane marking is crossed without the directional signal activated, the ECU uses the ABS system to gently pulse hydraulic brake pressure to the front and rear wheel calipers on the opposite side of the vehicle. This action, while slowing the vehicle a little, will also gently bring the vehicle back into its own lane. Either system can be turned off.

Traffic Collision Avoidance Systems (TCAS)

A traffic collision avoidance system (TCAS) involves vehicle-to-vehicle communication, a form of **telematics**. In essence, *telematics* is the ability of the vehicle, or something within the vehicle, to communicate with an outside entity. Telematics may include something as commonplace as the driver using a linked Bluetooth phone to communicate with other people through the vehicle's microphone and speaker system, or something such as GM's *On-Star* system that allows the driver to seek assistance. Telematics also includes the ability of the PCM to report the vehicle's emissions status to a roadside transponder if this communication ability has been enabled with the manufacturer's scan tool (OBD III).

A TCAS system uses telematics to communicate with other vehicles. Much as an on-board multiplexed data networking system involves communication between a vehicle's control modules, this system extends the on-board computers' communication ability to be able to communicate with on-board computers on other vehicles that are located in close proximity.

Bear this in mind: On modern vehicles, the driver controls nothing except the door handle, the hood latch, and maybe the vehicle's brakes and steering. For all other operations the driver makes a request of an electronic control module. If the driver wants to honk the horn, the horn switch is no longer hardwired to ground the horn relay directly. Instead, the operation of the horn switch makes a request of a computer. Then the computer energizes the horn circuit. Because drivers in modern vehicles only make requests of computers, these computers can, in some cases, be programmed to deny the driver's request. For example, the driver does not directly control engine speed on a fuel-injected engine. Instead, the pressing of the throttle pedal simply uses the APPS (or TPS) to request vehicle acceleration. With a traffic collision avoidance system, the PCM has programmed permission to deny this request when it recognizes that to honor the request would endanger the driver or others.

Two potential features of a traffic collision avoidance system include:

- The ability to warn computers on nearby vehicles of a road hazard in the immediate area.
- The ability to reduce the potential for accidents, particularly at road intersections.

In the first case, if a vehicle with this ability were accelerating up a hill a half mile behind another vehicle that also had this ability, and the vehicle in the lead were to encounter a hazardous road condition after cresting the top of the hill, the danger would be communicated to the vehicle in the rear. Even though this vehicle's driver is requesting hard acceleration at the throttle pedal, the PCM would deny the request and work with the collision avoidance and ABS systems to slow the vehicle or even bring the vehicle to a complete stop in time to avoid the danger.

As for reducing the potential for accidents at road intersections, each vehicle would receive the same GPS signals used by navigational systems. This would allow a computer on each vehicle to know the vehicle's latitude, longitude, and altitude. The computer would then calculate changing latitude, longitude, and altitude over time to determine the vehicle's speed and direction of travel. The computer would then wirelessly and continually communicate this information to all vehicles in the immediate area.

If a vehicle, approaching a stoplight intersection, was still six car lengths back when the driver saw the red light turn green with no stopped traffic ahead in his lane, the driver might mash the throttle pedal fully to the floor. But, in this example, what he failed to notice was that another vehicle was running a red light through the same intersection. The on-board computers would already have been made aware of the other vehicle's latitude, longitude, altitude, speed, and direction of travel. Therefore the PCM on the first vehicle would deny the driver's request for a hard acceleration and would work with the collision avoidance and ABS systems to bring the vehicle to a complete stop in time to avoid broadsiding the second vehicle. It is true that the first vehicle had the legal right-of-way and that the second vehicle was clearly in the wrong, but an accident would be avoided that could have resulted in injury and death.

DRIVER WARNING AND ALERT SYSTEMS

Radar sensors and rearview cameras are now being used to alert the driver to various conditions in order to reduce the potential for accidents resulting in injury and death. These sensors also reduce the potential for damage to occur to the vehicle.

Rearview Cameras

Many modern vehicles use a rearview camera mounted just above the rear license plate in conjunction with the display screen that is also used for the menu-driven information and control system. When the driver shifts the transmission into Reverse, the display screen automatically links to the rearview camera and allows the driver to view what is immediately behind the vehicle.

Some years ago the U.S. government had considered making rearview cameras required equipment by 2014, but then backed off. However, the National Highway Traffic and Safety Administration (NHTSA) has now ruled that rearview cameras will become mandatory equipment on all new vehicles sold in the United States that weigh less than 10,000 pounds. The rules will phase in over several years. Manufacturers will be required to have compliant rearview camera systems in 10 percent of the vehicles they build from May 1, 2016 to May 1, 2017. That share rises to 40 percent for the next year and to 100 percent starting on May 1, 2018.

About 200 people are killed each year in the United States because a vehicle backed over them. Most of these are children and, unfortunately, it is often the child's parent who backs the vehicle over the child, having not seen their child in the rearview mirror. It is hoped that the increased use of rearview cameras will prevent many of these unfortunate accidents.

Back-Up Alert Systems

A back-up alert system uses radar sensors mounted in or near the rear bumper to detect distances to objects behind the vehicle. When the transmission is in Reverse, the ECU warns the driver with multiple audio tones when an object is behind the vehicle. As the distance to the object decreases, the frequency of the audio tones increases until, at about 12 inches distance, the tone becomes continuous.

Cross Traffic Alert Systems

A cross traffic alert system uses radar sensors at the rear of the vehicle to detect when traffic is approaching from either side of the vehicle. If the driver is backing out of a parking space, the ECU will audibly warn the driver when cross traffic is approaching. A display on the instrument panel indicates whether the traffic is coming from the left or the right.

Blind Spot Alert Systems

A blind spot alert system uses radar sensors to detect when another vehicle is traveling in the driver's blind spot on either side of the vehicle. The ECU uses a visual indicator in the left and right outside mirrors to alert the driver when another vehicle is alongside. The indicators in the left and right outside mirrors are specific to traffic on each side of the vehicle.

Park Assist Systems

A park assist system uses radar sensors located at the front and rear of the vehicle and on the sides of the vehicle to audibly warn the driver of objects that are close to and around the entire vehicle. This aids the driver in parking in tight spaces or when entering a garage. As the vehicle becomes close to an object, a series of pulsed audio tones are emitted. As the distance decreases, the frequency of the tones increases until, at about 12 inches distance, the tone becomes continuous.

Active Park Assist Systems

Another version of a park assist system is an *active* park assist system. With this system, the ECU will control the steering when the driver wishes to parallel park. It is still the driver's responsibility to control the throttle, brakes, and transmission gear appropriately. Figure 10–7, Figure 10–8, and Figure 10–9 show the radar sensors and typical locations that are used with active park assist systems.

An active park assist system can parallel park the vehicle on either the right side or the left side of the street. The ECU knows which side of the street to park the vehicle on based on which directional signal is activated. If no directional signal is activated, the ECU will default to the right side of the street.

Figure 10–7 A radar sensor that is mounted in the front bumper.

Figure 10–8 A radar sensor that is mounted in the trim over a front wheel well.

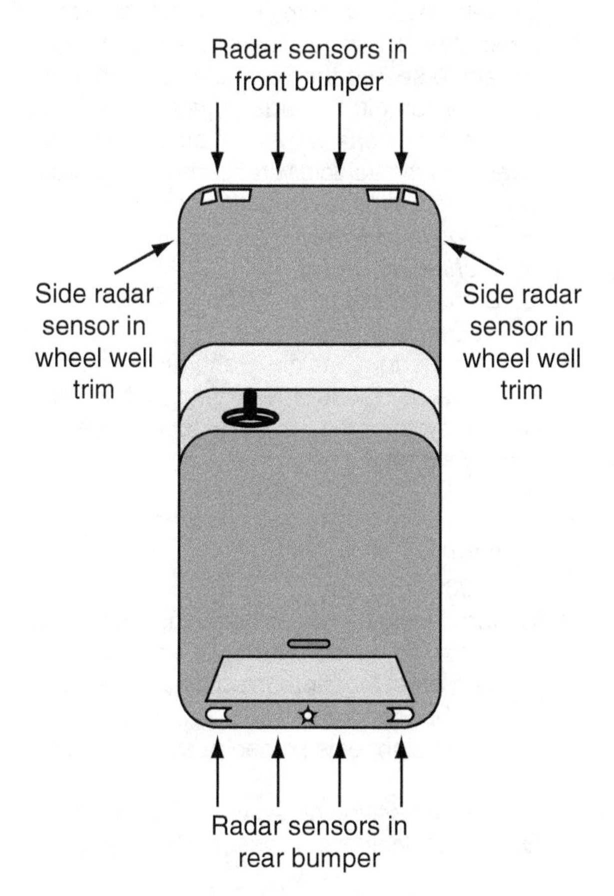

Figure 10–9 Typical radar sensor locations.

When approaching a parking space, the driver presses a *park assist* button on the instrument panel or center console. Then, at a speed less than 20 MPH, the ECU uses radar sensors to look for an appropriate parking space, which it also measures the length of. The ECU will then alert the driver to stop, shift the transmission into Reverse, and take his or her hands off the steering wheel. The driver then releases the brake pedal and allows the vehicle to back up, during which time the ECU will turn the steering wheel first in the direction toward that side of the street and then in the opposite direction. The ECU will then alert the driver when to stop and then will instruct the driver to shift the transmission into Drive. With the brake pedal again released and the vehicle moving forward, the ECU will turn the steering wheel to straighten the vehicle in the parking space. If the driver touches the steering wheel during this parking maneuver, the ECU will quit controlling the steering wheel and allow the driver to take over control of the steering again. In essence, the ECU is telling the driver at this point "Okay, you think you can do better, go for it!"

Toyota first introduced its Intelligent Parking Assist System (IPAS) on the Toyota Prius hybrid in Japan in 2003. The Toyota added it to the U.S. version of the Lexus LX in 2006 with several upgrades. Ford Motor Company added its Active Park Assist system to the Ford Escape in 2013.

SUMMARY

In this chapter we have discussed many systems that interact with the engine control system to provide the driver with additional convenience features and/or to increase safety. These systems represent some of the modern technologies that require the technician to fully understand the fundamentals and operating principles of electronics and computerized engine controls. As technology becomes increasingly complex, technicians can use these advances in technology to their advantage if they actively and continually pursue upgrade training. With continued education, these advances in technology can become your advantage over the other technicians.

▲ DIAGNOSTIC EXERCISE

A vehicle owner has a 2003 Lincoln LS towed into the shop with a complaint of "Every time I step on the brake, the engine stalls." Further questioning of the vehicle's owner reveals that he had just replaced the left rear stop lamp bulb before experiencing this symptom. What diagnostic steps should be taken to determine the nature of the problem?

Review Questions

1. *Technician A* says that body control modules control many functions that were once under the control of a small, dedicated computer. *Technician* B says that many vehicles use multiple body control modules. Who is correct?
   A. *Technician A* only
   B. *Technician B* only
   C. Both technicians
   D. Neither technician
2. *Technician A* says that an anti-theft warning system will typically energize the horn circuit intermittently and flash the lights if an unauthorized person opens a door or hood. *Technician B* says that an anti-theft disabling system prevents the engine from running if an improper key is used in an attempt to start the engine. Who is correct?
   A. *Technician A* only
   B. *Technician B* only
   C. Both technicians
   D. Neither technician
3. What is the purpose of a smart key system?
   A. It allows the driver to enter the vehicle, start the engine, and drive away with the key still in a pocket or purse.
   B. It allows the driver to enter the vehicle, start the engine, and drive away if he or she has lost the vehicle's key.
   C. It allows the vehicle owner to loan the vehicle to a friend or family member without having to give a key to that person.
   D. It allows the driver to use the key as an Internet wi-fi hot spot.
4. *Technician A* says that a driver should never activate the remote start system unless he or she is in sight of the vehicle because once the engine is running it would be very easy for another person to enter and drive away with the vehicle. *Technician B* says that the use of a remote start system uses the door locking system to secure the vehicle and, even if another person did manage to gain entry to the vehicle, he or she would not be able to drive away with the vehicle without having a proper ignition key. Who is correct?
   A. *Technician A* only
   B. *Technician B* only
   C. Both technicians
   D. Neither technician
5. In what model year did General Motors first introduce a menu-driven information and control system that used a touch-sensitive display screen?
   A. 1986
   B. 1996
   C. 2001
   D. 2006
6. Modern navigational systems use which of the following to automatically track the vehicle's current location on an electronic map?
   A. Land-based transmitters
   B. Active signposts placed at every intersection
   C. Global positioning system PPS signal
   D. Global positioning system SPS signal

7. Which of the following systems can be controlled through a voice recognition system?
   A. Climate control system
   B. Sound system
   C. Navigation system
   D. All of the above
8. If an ABS systems fails electrically, which operational mode does it default to?
   A. Apply
   B. Maintain
   C. Release
   D. Hold
9. If one drive wheel loses traction with the road surface during an acceleration, how does an automatic traction control system ECU transfer engine torque to the opposite drive wheel?
   A. It sends a binary command to the drive wheel that lost traction to tell it to slow down.
   B. It applies hydraulic brake pressure to the caliper of the wheel that lost traction.
   C. It closes up the throttle plate.
   D. It upshifts the automatic transmission and retards spark timing.
10. *Technician A* says that a pre-collision throttle management system will warn the driver when an object is sensed to be in front of the vehicle, but the driver must then release the throttle pedal in order to avoid a collision. *Technician B* says that a pre-collision throttle management system will reduce engine torque when an object is sensed to be in front of the vehicle even if the driver is pressing hard on the throttle pedal. Who is correct?
    A. *Technician A* only
    B. *Technician B* only
    C. Both technicians
    D. Neither technician
11. Which of the following sensors is considered to be an *all-weather* sensor and is least affected by rain snow, fog, or blowing dust?
    A. Camera
    B. Laser sensor
    C. Radar sensor
    D. Both B and C
12. A driver is travelling at 30 MPH. From this speed, in addition to the driver reaction time, it would take about 45 feet to stop the vehicle once the brakes have been applied. If the driver doubled the vehicle's speed to 60 MPH, how far, in addition to the driver reaction time, would it take to stop the vehicle once the brakes have been applied?
    A. 45 feet
    B. 90 feet
    C. 135 feet
    D. 180 feet
13. A collision avoidance warning system or an active collision avoidance system uses input from all except which of the following sensors?
    A. Yaw rate sensor
    B. Vehicle speed sensor
    C. Radar sensor
    D. Steering angle sensor
14. *Technician A* says that an active collision avoidance system will not override the driver if the driver has the throttle pedal depressed fully to the wide-open throttle position. *Technician B* says that an adaptive cruise control system will automatically adjust the vehicle's speed in response to the flow and speed of the traffic ahead of the vehicle. Who is correct?
    A. *Technician A* only
    B. *Technician B* only
    C. Both technicians
    D. Neither technician
15. What is meant by the term *yaw rate*?
    A. How far left or right a vehicle has moved from its intended course
    B. How far the vehicle can lean before it actually rolls over
    C. How quickly the vehicle can stop with the ABS system deactivated
    D. How quickly the PCM can respond to a driver request from the APPS
16. All of the following systems use radar sensors to detect distance to an object **EXCEPT**:
    A. Blind spot alert system.
    B. Subaru Eyesight® system.
    C. Active collision avoidance system.
    D. Cross traffic alert system.

17. Under which of the following conditions is a *lane sway warning* feature designed to alert the driver?
    A. If the vehicle crosses a lane marking at least three times
    B. If the vehicle crosses a lane marking at least four times
    C. If the vehicle comes within 12 inches of a lane marking
    D. If the vehicle begins to sway within a lane, systematic with a driver becoming drowsy, even though no lane markings have been crossed
18. *Technician A* says that a traffic collision avoidance system (TCAS) uses vehicle-to-vehicle communication. *Technician* B says that a traffic collision avoidance system (TCAS) can bring a vehicle to a complete stop in time to avoid an accident, even though the driver is pressing the throttle pedal to the floor. Who is correct?
    A. *Technician A* only
    B. *Technician B* only
    C. Both technicians
    D. Neither technician
19. The NHTSA has mandated that rearview camera systems will be mandatory on 100 percent of new vehicles weighing less than 10,000 pounds that are sold in the United States by what date?
    A. May 1, 2016
    B. May 1, 2017
    C. May 1, 2018
    D. May 1, 2025
20. Which of the following systems is designed to alert the driver that traffic is approaching from either side of the vehicle when the driver is backing the vehicle out of a parking space?
    A. Back-up alert system
    B. Blind spot alert system
    C. Park assist system
    D. Cross traffic alert system

Chapter 11

Approach to Diagnostics

OBJECTIVES

Upon completion and review of this chapter, you should be able to:

- ❑ Understand how to narrow down the problem area before beginning pinpoint testing.
- ❑ Describe what steps are involved in chasing an intermittent symptom.
- ❑ Define the steps involved in diagnosing an emission test failure.
- ❑ Understand how to use a smoke machine in diagnosing an EVAP system DTC.
- ❑ Identify the important steps in using a flowchart to pinpoint test a DTC or symptom.
- ❑ Understand how to use an electrical schematic in reducing your dependence on a flowchart.
- ❑ Define the three essential tools of electronic system diagnostics.

KEY TERMS

Baseline
Cranking Compression Test
Non-Responsive Unit (NRU)
Off-Board Reprogramming
On-Board Reprogramming
Pattern Failures
Running Compression Test

If you have read the previous chapters in this book, you might be wondering how best to apply the information when a vehicle is brought into your service bay for diagnosis and repair. Sometimes it can be difficult to put theory into action. This chapter is designed to help you to understand how to convert the information in this book into hands-on diagnostics.

NARROWING DOWN THE PROBLEM AREA

When a vehicle pulls into your service bay, there could be as many as a thousand possibilities as to what the specific problem could be that is causing the customer complaint. If someone asked you to guess a number between 0 and 1,000 in a minimum number of guesses with a promise to tell you whether each guess was too high or too low, you would do well to begin guessing with the number 500. This cuts the possibilities in half. If your number is too high, your next best guess is 250, cutting the possibilities in half again. So it is with diagnosing complex automotive electronic systems.

Your initial goal in diagnosing a fault of any type, be it a hard fault or a soft fault, should be to narrow down the problem area associated with the symptom to as small an area as possible. You may take several approaches to do this. Failure to reduce the possibilities can result in extreme frustration, as most technicians have experienced at one time or another.

Use the following tips to help you narrow down the area of the problem before you begin pinpoint testing. You can thereby minimize your frustration and increase your diagnostic effectiveness in spite of the complexity of modern automotive systems.

Talking to the Customer

First, when the vehicle is brought into your shop, gather information from the customer and record it on the repair order. Include as much information as possible about the conditions under which the symptom appears. It is also important to find out if the vehicle has been taken to a shop for the same symptom prior to this service. If you listen carefully, what the customer says about the problem may provide clues to the type of problem the vehicle has, such as, "This problem has been happening *ever since. . . .*" That "ever since" might be just the clue you need.

You might also ask the customer to show you the problem that is being experienced. For example, this author once received a repair order identifying the symptom as "headlights will not stay turned on" on a vehicle that had been purchased new 3 days earlier. When I was unable to verify the symptom, I asked the customer, who was in the waiting room, to come back to my service bay and show me the problem she was experiencing. She literally reached through the open driver's door window, pulled back on the "flash-to-pass" switch, and said, "See, when I let go, the headlights turn off." I then explained to her that, though this new vehicle was new technology in her eyes, it still had a traditional push/pull headlight switch in the dash.

Symptom Verification

The first real step in diagnosis is to try to verify the *exact* symptom. If the symptom can be verified, you are likely dealing with a hard fault. If the symptom cannot be verified, you are likely dealing with a soft fault. Hard faults should always be diagnosed before trying to diagnose any soft faults. While there are some differences between diagnosing a hard fault and diagnosing a soft fault, most of the following information applies to faults of either type.

First, it is important to understand how the system is designed to operate. Secondly, you must verify the exact symptom—not what the customer said is wrong (such as, "My exterior lights are acting strangely"), but rather the *exact* symptom. In order to understand how the system is designed to operate, you must have:

- A working knowledge of electrical laws
- A working knowledge of electronics—computer inputs, processing, and outputs
- An understanding of system strategies
- The ability to read and interpret an electrical schematic

Along with your understanding of how the system is designed to operate, symptom verification can help you narrow down the area in which you need to perform pinpoint tests (Figure 11–1). In fact, what you are really doing is mentally creating a flowchart.

For example, if the customer complains that there seems to be a short in the exterior lighting circuits because other circuits seem to energize the running lights (park, tail, and side marker lights), the first step would be to gain an understanding of how the system is designed to operate. If it is an unfamiliar system, study the electrical schematics associated with the system. The next step would be to verify the exact symptom. This can be accomplished by operating each of the exterior lighting systems and noting the results.

For example, if the exterior running lights illuminate when the left directional signal is activated but not with the right directional signal, it is likely that the short is located on the left side of the vehicle. If the exterior running lights illuminate when the brake pedal is depressed, the short is at the rear of the vehicle. Thus, verification of the exact symptom can be used to narrow

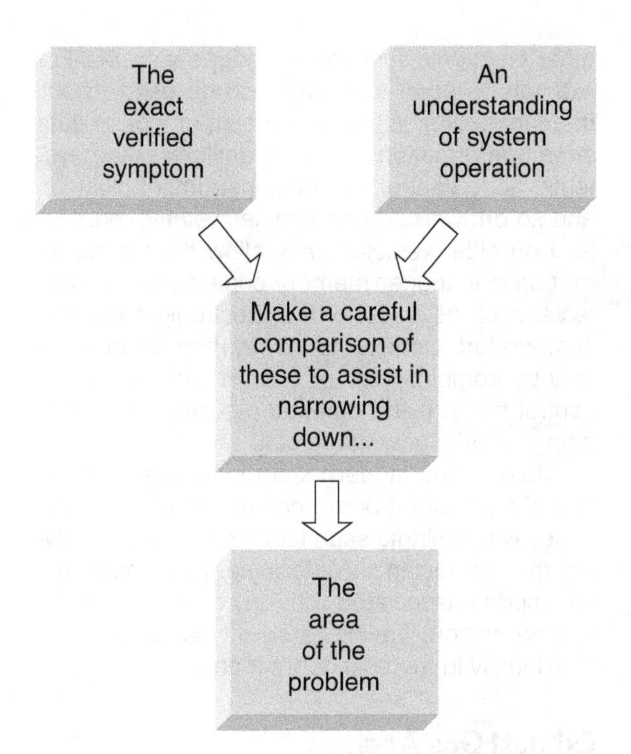

Figure 11–1 Basic approach to narrowing down a problem area.

down the problem area, in this case to the left rear corner of the vehicle. Next perform a visual check at that corner of the vehicle, beginning with the bulbs—you may find a dual-filament bulb that has a filament that has broken loose and is physically touching the other filament. Because you narrowed the problem down to one corner of the vehicle by using the *exact verified symptoms,* you do not have to perform visual checks and pinpoint tests at all four corners of the vehicle—only at the one corner identified by the symptoms. This type of approach ultimately results in less frustration in that it is easier to perform pinpoint tests of a particular area than to open up harnesses and pinpoint test components throughout the vehicle.

Visual Inspection

A thorough visual inspection of the system(s) associated with the symptom should be performed. If this is done properly and conscientiously, problems can often be identified at this time. Additional visual inspections may need to be performed as the area of the fault is further narrowed down through diagnostic processes such as code pulling and gas analysis.

Assistance from the On-Board Computers

An important aid to helping the technician narrow down the problem area is to take advantage of the abilities of on-board computers; they are programmed to offer you their assistance.

Diagnostic Trouble Codes. Modern on-board computers are programmed, at the very least, to allow the technician to pull memory diagnostic trouble codes (DTCs). (OBD I standards required a memory code capability as of 1988; many manufacturers had programmed their PCMs with this ability well before this.) Others will also allow the technician to perform a self-test of the system. Both memory codes and self-test codes can quickly identify a problem area or circuit for the technician. (Remember that a memory DTC can relate to either a hard fault or a soft fault, whereas a self-test DTC identifies a hard fault.) But, for example, if a DTC is set as the result of an electrical open in a specific circuit, the technician must still perform pinpoint tests to identify the exact location and cause of the open. The DTC only narrows down the area of the problem.

One precaution worth remembering: With the considerably increased number of diagnostic trouble codes (DTCs) in the OBD II system, technicians are often startled to find a large number of DTCs stored. Check to see what circuits are involved: Often all the recorded codes are for a single circuit or a set of closely related circuits. Also check for excessive AC ripple coming off the alternator when the engine is running and several loads are turned on. AC ripple in excess of 200 mV can cause a computer to generate and store multiple false DTCs in memory.

Data Stream. All OBD II systems and most pre–OBD II systems allow the technician to look at data stream parameters (PIDs) and related information concerning sensor values and actuator commands. This is extremely effective in helping the technician to narrow down the problem to a specific area or circuit. But suppose, for example, the data stream information identifies an NTC thermistor circuit as reporting a temperature of −40°F; while the technician now knows that there is an open in that circuit, further pinpoint tests must be performed to identify the exact location and cause of the open. The data stream only narrows down the area of the problem.

Other Scan Tool Functions. Most computers will also perform other functional tests, either using manual methods to trigger these functions or with the help of a scan tool. For example, the technician may be able to use a scan tool to command the computer to energize a specific actuator (known as *active commands* or *bidirectional controls*). This helps the technician to quickly narrow down a problem associated with that system to either the input side or the output side of the computer. If an actuator that does not operate under normal circumstances now operates under the scan tool's direction, we know that the computer can successfully energize the actuator, thus eliminating the output side of the computer as a possible fault; the technician would check the sensors, switches, and other inputs next. But if the actuator will not operate when the scan tool commands it to, the actuator and its related wiring/connectors should be checked.

Other functional tests may also be available, depending upon the scan tool and the vehicle itself. For example, on many systems a scan tool can command the PCM to conduct a cylinder power balance test (this may be built into the PCM's program or the scan tool's program). The cylinder power balance test would narrow down the problem area to a specific cylinder, but the technician would still need to conduct further pinpoint tests to determine if the fault is ignition related, air-fuel ratio related, or volumetric efficiency related.

PCMs versus Other Control Modules. These scan tool diagnostic features are not limited to engine control systems; they also apply to other electronic systems on today's vehicles. For example, a scan tool may also be used to pull memory codes, initiate a self-test, or look at data stream information on many antilock brake systems, body control systems, instrument clusters, and so on. Certain non-engine systems, particularly on older vehicles, may allow the technician to manually trigger memory code pulling or self-tests, such as a test of the Electronic Automatic Temperature Control (EATC) system, by pushing various combinations of buttons on the climate control head. If the computer is programmed to help you, take advantage of it.

Also, many modern scan tools offer a mode that checks all on-board computers for any fault codes with a single scan tool request, thus allowing the technician to determine very easily any fault codes associated with any computer on the vehicle without having to select each computer individually in the scan tool's menu.

Exhaust Gas Analysis

Another approach to narrowing down the problem area when diagnosing an engine system for poor performance, poor fuel mileage, or a failed emission test is to use an exhaust gas analyzer to capture exhaust readings at both idle and 2,500 RPM, and then to use these readings to give you an idea of the type of problem that you are trying to diagnose. For example, a high CO reading would indicate that you should look for problems that could result in a rich air-fuel ratio (too much fuel). But the technician would still need to perform pinpoint tests to determine the exact cause. Analyzing the exhaust gases of an engine can help a technician quickly narrow down the problem area, even though it cannot identify the exact problem. An exhaust gas analyzer becomes an invaluable tool in minimizing the frustration factor.

When using an exhaust gas analyzer as a diagnostic tool, use the gases to determine the actual lambda and air-fuel ratio values. Newer exhaust gas analyzers perform this calculation for you from the measured gases. If you are using an older analyzer that does not perform this task, use an automated "lambda and air-fuel ratio" calculator that is available on the Internet, such as the one found at the International Automotive Technicians' Network (IATN) (http://www.iatn.net). Simply enter your four or five measured gases (with or without the NO_x value) in the automated calculator to calculate the lambda and air-fuel ratio values. It is very helpful to know what these values are when trying to narrow down the problem area. This will also serve as an early indicator that will help you determine whether or not the PCM is in proper control of the air-fuel ratio.

PINPOINT TESTING

The primary tools used for pinpoint testing include a DMM and a lab scope (DSO). Other tools can also be classified as pinpoint test tools, including breakout boxes, battery load testers, and even test lights (used for the purpose of testing fuses electrically). Pinpoint testing should not begin until the problem area has been narrowed down to as specific an area as possible.

Basic Circuit Testing

To isolate circuit faults, visualize the electrical circuit in three parts: the source (battery or alternator), the circuitry, and the load component (Figure 11–2).

Testing the Source. The source is the first component that should be verified. For example, if the starter cranks slowly, check the battery's open-circuit voltage and then perform a battery load test. It is a good idea to test the battery's open-circuit voltage for symptoms associated with smaller load components, even if the starter cranks normally, due to the effects of low open-circuit voltage on computer operation. However, if the starter cranks okay but the washer pump isn't strong enough to spray water up on the windshield more than a couple of inches, a battery load test is unnecessary—if the battery is strong enough to crank the starter, it is not the problem that is causing the washer pump to perform poorly.

With the alternator, do not overlook the effects of excessive alternator ripple on computer-controlled systems. To test this, connect a DMM between the alternator positive and negative terminals while scaled to an AC voltage scale. With the engine running and some vehicle loads turned on, AC voltage should stay well below 200 mV, preferably less than 100 mV. Excessive alternator ripple can, of course, also be seen with a DSO (lab scope).

Testing the Circuitry. To test the circuitry, use a DMM on a DC voltage scale to perform voltage drop tests on both the positive and negative sides of the circuit. This verifies the integrity of the circuitry, including all fuses, switches, relay-operated switches, terminal connections, and grounds used to connect the source to the load component (Figure 11–2). While 100 mV per connection

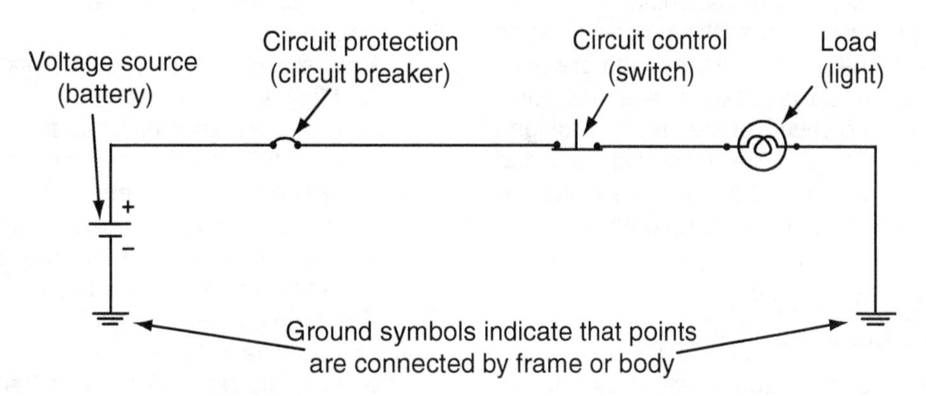

Figure 11–2 Circuit components.

is allowed in most circuits (and even 200 mV per connection in starter circuits due to the high-current, low-resistance designs of these circuits), a power or ground circuit that is used to power up a PCM or other ECU should not exceed 100 mV *total* voltage drop on the positive side of the circuit and 50 mV *total* voltage drop on the negative side of the circuit with the key on, engine off (KOEO). This tight specification is due to the high-resistance, low-current flow designs of these circuits. Many good PCMs are replaced simply because of a poor power or ground circuit. Therefore, after verifying each of the input and output circuits, but before replacing the PCM, you should get in the habit of using an electrical schematic to identify each of the PCM's power and ground terminals. Then with the ignition on, perform voltage drop tests between the battery and the PCM.

Testing the Load. Ultimately, the load component can be tested by the following methods: If the load component is an electronic control module, simple elimination of possible faults elsewhere in the circuitry through voltage drop testing is used most often to narrow down the problem to the control module itself. Other load components can be tested electrically and functionally. Electrical tests involve the use of an ohmmeter to measure a solenoid's (or relay's) resistance in the windings, or jumper wires to energize the component. Functional testing can be done using other methods, depending upon the type of actuator being tested. For example, use a vacuum pump opposite a vacuum gauge to test a vacuum solenoid.

Other tools that can make pinpoint testing easier include DSOs, logic probes, and breakout boxes. DSOs can show voltage waveforms, logic probes can quickly check for the presence of ignition primary signals and fuel injector signals, and breakout boxes can be used to improve computer circuit accessibility without back-probing.

Diagnosing a Loss of the Computer's Reference Voltage

If the reference voltage (sometimes referred to as the *VREF signal*) from an ECU is lost, that ECU will not function properly. This is because not only is this signal sent from the ECU to the sensors that need it, but the ECU's internal logic circuits depend on it for both processing information and communication within the module.

The VREF signal should be between about 4.8 and 5.2 V when measured at a sensor. If the computer's ground is lost or has excessive resistance, this signal can be too high. If the battery's OCV is less than 9.6 V or the computer's power circuit has excessive resistance, this signal can be too low. And if a sensor shorts internally, the defective sensor can pull this signal to near zero volts. This was common in years past with Ford *pressure feedback EGR* (PFE) sensors and, more recently, with Chrysler CMP and CKP sensors.

In the engine control system, if the PCM's VREF signal is not produced within the PCM or is shorted to ground external of the PCM, the engine will not start and the vehicle will likely be towed in. A voltage check at the TPS (or other sensor that receives the VREF signal) will show less than about 1 V. In order to determine whether the signal is not being produced at the PCM or is being shorted to ground external of the PCM, perform the following test:

- Use a current probe coupled to a DSO or graphing multimeter (GMM) to monitor the current in the VREF wire at the PCM end of the circuit.
- While watching the current level, switch the ignition on.
- A normally operating VREF signal should read less than 100 mA of current flow due to the high resistance built into the sensors.
- If the VREF signal is being produced within the PCM and is then being shorted to ground externally by either a shorted sensor or even a wiring fault, the current flow will spike upward following the turning on of the ignition switch, possibly to more than 2 amps. The PCM will then ramp the current down to around 400 mA in a self-protective mode.

If the current never exceeds 100 mA following the turning on of the ignition switch, the PCM is not producing the VREF signal. Verify the PCM's power and ground circuits using voltage drop tests. If either voltage drop test is excessive, use voltage drop testing procedures to narrow down the excessive resistance and repair. Then retest for the 5 V VREF signal. (If excessive voltage drop is identified on the positive side of the PCM, try replacing the PCM's power relay and retest. This is a common source of excessive resistance in the positive side circuit that is used to power up the PCM.)

If the voltage drop tests prove the integrity of the power and ground circuits to be okay and the PCM is not producing the VREF signal, replace the PCM.

If the preceding test shows that the PCM is producing the VREF signal and that it is being shorted to ground externally of the PCM, disconnect each of the sensors one at a time while repeating the testing procedures until the circuit is identified that is shorting the VREF signal to ground. Also, carefully inspect the wiring harnesses at this time for wiring insulation faults.

While the PCM has been used in the example here, this test procedure can be used with any ECU that is not powering up properly.

Pinpoint Testing a DTC

Monitor Readiness Status Verification. The first step in beginning DTC diagnosis on any OBD II vehicle is to use a scan tool to evaluate the *monitor readiness status* of all OBD II monitors. If a monitor has not run since the last time DTCs were erased or the battery was disconnected, a DTC will not be set for a potential failure.

To verify monitor readiness status, use a scan tool. You may either go in through VIN Entry Mode or you may select Global Mode $01. If all of the monitors are complete, proceed to retrieve the memory DTCs. If a monitor that *could* apply to the symptom is incomplete, road-test the vehicle in order to get the appropriate monitor to run. (Many aftermarket scan tools will give instructions on how to drive the vehicle to get a specific monitor to run or to complete a full drive cycle that allows all of the monitors to run.) Then use a scan tool to verify that it has run and proceed to retrieve memory DTCs. If a monitor is displayed on the scan tool as *N/A*, it does not apply to the vehicle you are working on.

Retrieving Memory DTCs. A scan tool should be used to perform any applicable self-tests (such as on Ford Motor Company vehicles) before retrieving memory DTCs because these tests identify hard faults. Then use the scan tool to retrieve confirmed DTCs (also known as *mature DTCs* or *history DTCs*) through VIN Entry Mode or Global Mode $03. If any confirmed DTCs are present, use either VIN Entry Mode or Global Mode $02 to access the freeze frame data for the DTC(s). Study the freeze frame data to help narrow down the problem area. Freeze frame data can also be used to help you confirm a symptom during a road test. If no confirmed DTCs are found, use VIN Entry Mode or Global Mode $07 to retrieve pending DTCs. On some vehicles beginning in the 2010 model year you may also use Global Mode $0A to retrieve permanent DTCs. This is particularly useful if you have reason to believe that the vehicle has been worked on recently at some other shop.

Data Stream. After you retrieve DTCs, a scan tool should be used to access the data in the data stream. This can be done either through VIN Entry Mode or through Global Mode $01. Many scan tools will allow you to select only the parameters (PIDs) that are of concern to the symptom or DTC that you are diagnosing. Selecting only a few PIDs will speed up the scan tool's update rate since only one PID's value can be communicated from the PCM to the scan tool at a time.

Flowcharts. Flowcharts are designed to help the technician to make appropriate pinpoint tests that are applicable to the DTC or symptom. After retrieving DTCs, use a *DTC chart* (or *symptom chart*) to determine which flowchart test step you should begin with. Each flowchart test step will give an instruction on how to make a specific test, ask you about the results of the test you made, and take you to another appropriate test step depending upon your test results. At some point, the flowchart will ask you to make a specific repair. When using a flowchart, be sure to do the following:

- Read and follow the test instructions carefully.
- Do not skip any test steps or assume their outcome.
- Be sure of your equipment: Does your DMM or DSO read battery voltage when the leads are placed across the battery? Or does your jumper wire have continuity?
- Make the repair as indicated in the final step—you can be confident of what it is asking you to do if you have carefully followed all test steps and have not skipped any.

Reducing Your Dependency on Flowcharts. While they are a necessary diagnostic tool, flowcharts may be restrictive for the following reasons:

- Their content is limited to the engineer's ability to imagine all potential problems that could develop with the system.
- They will occasionally contain an error.
- They are time consuming—it can take you longer to read and understand what the flowchart is asking you to test than the flat rate for the job pays.
- If someone has altered the OEM design in the aftermarket, the flowchart may no longer apply.

An approach that will reduce your dependency on flowcharts is to diagnose the system through your own understanding of how the system is designed to operate. As stated earlier in this chapter, this requires a working knowledge of electrical laws and electronics, an understanding of system strategies, and the ability to read and interpret an electrical schematic.

Interpreting an Electrical Schematic. An electrical schematic is used to turn classroom theory (or textbook theory) into product-specific knowledge. You can sit in a classroom for many hours to learn how today's complex electronic systems operate. You can also read a textbook, such as this one, to learn about the system strategies they use. But when a specific vehicle is in your service bay, the electrical schematic becomes the primary tool to help you turn all of that intensive theory into product-specific knowledge that pertains strictly to the vehicle you are working on. Following are a few ideas that will help you interpret a complex electrical schematic. It helps to begin by copying or printing out the electrical schematic so that you can draw on it.

- *First: Identify the components.* Highlight the components with a yellow highlighter if necessary. This quickly simplifies a seemingly complex schematic. It also keeps you from inadvertently trying to interpret the outline of a control module (or other component) as a circuit.
- *Second: Identify the powers and grounds.* You might use red or orange for power and green for grounds.
- *Third: Identify the input and output circuits.* You might use blue for inputs and purple for outputs. Bear in mind that sometimes an ignition-switched voltage operates as an input to a control module, as with an illuminated entry system where the system must be kept electrically alive even when the key is out of the ignition. In this system, battery voltage is used to power up the module. The ignition voltage only acts as an input to let the module know that it may turn off the interior lights once the ignition switch is turned on.
- *Fourth: Identify any data buses used for communication between computers.* You might use brown for data network buses that connect between the various nodes, of which most will also connect to the DLC. There may be multiple data buses on the same vehicle; some may be single-wire (SW) buses, some may be dual-wire (DW) buses, and still others may be fiber optic buses.

Finally, flowcharts and electrical schematics are invaluable when it comes to pinpoint testing. A flowchart is written according to the design engineer's understanding of system operation. An electrical schematic helps you as a technician to understand how the system operates.

Basic Engine Function Testing

Do Not Overlook the Basics. When warranted, the basic engine systems—compression, ignition, and air-fuel delivery—must not be forgotten. A vacuum gauge can be used to quickly give the technician ideas about possible fault areas that should be tested further. A cylinder power balance test, performed through a functional test programmed into the PCM (as on Ford EEC-IV systems) or with a scan tool (under bidirectional controls) can quickly identify cylinders that may be lacking in compression, volumetric efficiency, spark, or fuel delivery.

It is not recommended that you disable spark as a method of performing a cylinder power balance test (as the old ignition scopes used to do) on a vehicle that has a catalytic converter—1,500 RPM for 5 minutes with unburned fuel being delivered can severely overheat a converter. Ford's automated cylinder power balance test (EEC-IV only) and the method of using a scan tool to disable a cylinder through bidirectional controls involve the disabling of the fuel injectors, not the ignition coils.

If one or more cylinders are identified as being weak, the following areas should be looked at.

Compression and Volumetric Efficiency Tests. Performing a compression test on each of the cylinders while the starter is cranking the engine and the throttle is held open, known as a **cranking compression test**, tests for cylinder seal. For example, a burned valve could easily cause a particular cylinder to fail this test due to the leak it creates. If a cylinder fails the cranking compression test, a leak-down tester should be used to verify the severity of the problem as well as the cause.

However, if all cylinders pass the cranking compression test and spark and fuel are verified with an exhaust gas analyzer, a **running compression test** should be performed. This test will evaluate each cylinder's volumetric efficiency (the cylinder's ability to breathe in and out efficiently). To perform this test, all spark plugs must be reinstalled except the one for the cylinder in which the compression gauge is installed. Then the engine is started. Once the engine is running, momentarily release the compression gauge pressure. Now note the compression pressure at idle, at 1,500 RPM, and on a snap throttle. At idle and 1,500 RPM, the pressure reading will be about one-half of cranking compression due to the closed throttle and the faster piston speed. Because of the faster piston speed, this is a very poor test to check cylinder seal, but it is an excellent test to check and compare each cylinder's volumetric efficiency. It is difficult to apply a hard specification to this test because of different camshaft grinds; instead, the cylinders should simply be compared. One or two cylinders showing lower readings than the rest indicates an inability to *breathe in* because of an improper opening of the intake valve, and one or two cylinders showing higher readings than the rest indicates an inability to *breathe out* because of an improper opening of the exhaust valve. This inability of the engine to breathe properly is caused by problems in the valve train such as loose rocker arms, worn pushrods/lifters, or a worn cam lobe. Because these valves still seal when closed, the cranking compression test does not identify this type of problem. One point to note: Have some spare Schrader valves on hand for your compression gauge when performing running compression tests.

Ignition. If the engine cranks but does not start, both secondary ignition and primary ignition should be checked. If the tach reference signal is not present, either from a distributor or from a crankshaft sensor, neither secondary spark nor fuel injection pulses will be present. If there is a spark being produced in an ignition coil or if the fuel injectors are being pulsed, the distributor pickup/crankshaft sensor is functioning.

If the engine starts and runs but has performance problems or misfire present, the secondary ignition waveform should be checked on an ignition oscilloscope. And if a shorted primary coil winding is a possibility, use a DSO coupled with a current probe to test each primary coil winding through its current waveform.

Also, if the engine uses a DI system, be sure to verify that ignition timing is correct with computer control of timing disabled and the engine at base timing. See the service manual for each manufacturer's procedure to disable the PCM's control of ignition timing.

Air-Fuel. This area includes the common maintenance items such as air and fuel filters and the PCV valve and crankcase ventilation filter. Also, if the engine has failed a cylinder power balance test, measure the exhaust emissions with an exhaust gas analyzer at both idle and 1,500 RPM. Then turn off the engine and disable the spark for the failed cylinder. On distributor and waste spark systems, simply use a jumper wire to short one spark plug wire to or from engine ground. On coil-on-plug (COP) systems, simply disconnect the primary connector from the coil for the failed cylinder. Now restart the engine. If unburned hydrocarbons increase dramatically from their previous values, the failed cylinder has both spark and (at least some) fuel present. If this is the case, then cranking and running compression tests should be performed (if they have not already been performed).

Another concern is the quality of the fuel. For these tests, use a fuel test kit (such as one marketed by BG Products, Inc., part number 995) to test for both excessive alcohol in the fuel and the fuel's density. To test for excessive alcohol, mix a sample of the vehicle's fuel with some water in a graduated cylinder. Because it is less dense than water, the gasoline will float above the water, creating a gasoline/water line between them. Shake the cylinder vigorously for 30 seconds and then let it settle. After settling, if the height of the gasoline/water line has increased, it is because of alcohol content in the gasoline sample—the alcohol separates from the gasoline and combines with the water during this test. Use the increase in the height of the gasoline/water line to calculate the percentage of alcohol content. Except for E85-capable vehicles, any alcohol content in excess of 10 percent can create fuel volatility problems.

Finally, use the fuel test kit to measure the specific gravity of a sample of the vehicle's gasoline (similar to specific gravity tests of a lead-acid battery's electrolyte when the cell caps were removable). Although it does not directly measure the fuel's Reid vapor pressure (RVP), the fuel's specific gravity will be indicative of the fuel's volatility. Fuel that is too volatile can be responsible for vapor lock problems and problems involving excessive pressure building within a fuel tank. Fuel that is not volatile enough can be responsible for difficult cold starts. The specific gravity of the fuel should also quickly alert you to any diesel fuel that might have been put into a gasoline engine's fuel tank inadvertently.

Technical Service Bulletins

Another test step that should be performed is checking the manufacturer-issued technical service bulletins (TSBs). These are usually available either through electronic information systems or directly from the manufacturer. A TSB is a manufacturer-suggested repair for a problem the manufacturer is already aware of. It may involve a simple fix or it may involve something as in-depth as replacing a PCM, replacing a PROM chip in the PCM, or reflashing the PCM. Make sure that you use the latest TSB issued for the particular symptom that you are trying to diagnose and repair, as multiple TSBs have sometimes been issued in trying to correct a single problem. Ultimately, access to TSBs is critical to a technician trying to diagnose engine performance problems, and it can lead to increased diagnostic effectiveness.

DIAGNOSING INTERMITTENT SYMPTOMS

This question is often asked: How can I pinpoint test a fault when the symptom does not appear to be present? There are several answers to this question. Certainly, past experience on like models with similar problems, known as **pattern failures**, counts for something. Unfortunately, this approach often results in improper repairs and unnecessary parts replacement. A better approach is suggested in the following paragraphs.

Memory DTCs

The first step in diagnosing an intermittent fault, also known as a soft fault, is to determine whether the PCM has stored any DTCs in memory that could direct you to a probable problem area. For example, a symptom of "stalling on a hot day and then restarting after a 20-minute cool-down period" can be the result of a faulty ignition module. But it can also be the result of a faulty electric fuel pump or even a faulty fuel pump relay. Beginning in the mid- to late 1980s, feedback circuits were typically added to the PCM to assist the technician in identifying problem areas. At the very least, a memory DTC can help you determine if the problem is the likely result of the ignition system or the fuel system. However, additional tests can also be performed to pinpoint-test the exact cause, even when the symptom does not appear to be present.

Scan Tool Movie Mode/Record Mode

Most scan tools have a record capability, known by some as *movie mode,* that allows the scan tool to record data from the data stream during a road test. This mode allows the technician to play back recorded data frame by frame in the shop environment where it can be studied. Use this mode to diagnose intermittent glitches with a sensor or in a sensor's circuit.

Voltage Drop Tests

If the soft fault is in the circuitry, performing voltage drop tests of the circuitry can locate the fault, even when the circuit appears to be functioning normally. Remember, voltage drop testing is used to identify excessive unwanted resistance in the circuitry. This excessive unwanted resistance increases with an increase in temperature to the point of creating a symptom. Likewise, this resistance decreases with a decrease in temperature, to the point that the circuit again becomes operational. The excessive unwanted resistance does not totally disappear, even when the circuit appears to be operational and the symptom does not appear to be present. Thus, if a soft fault is in the circuitry, voltage drop testing can identify the circuit problem even without the symptom being present. While the voltage drop reading will not be as excessive as when the symptom is present, it will still likely be above acceptable limits.

Current Ramping

If the soft fault is in the load component, using a DSO (lab scope) and a low-current amp probe to sample the current waveform of the load component can identify faults in the load component that can cause intermittent operation. This is particularly true of motors. The current waveform can identify an open or shorted winding within the motor as well as some other faults that will weaken the motor's strength and result in intermittent symptoms associated with the motor. In this way, a load component can be proven to be defective, even though it may still be operational while in your shop. This technology allows you to show your customer a printed waveform that proves that a load component is not being replaced needlessly. It can even allow you to identify a load component that is ready to fail, even before symptoms have been encountered.

Electronic Control Modules

If the intermittent fault is in an electronic module, such as an ignition module, use a choke tester to heat and/or cool the module while watching for the symptom to resurface. This technique may help you identify the fault. This method also helps in diagnosing problems with other electronic components as well, such as Hall effect sensors or MAF sensors. You may also tap lightly on electronic devices in an attempt to verify the problem.

TESTING CATALYTIC CONVERTERS

Catalytic Converter Backpressure Tests

Most manufacturers have experienced some problems with catalytic converters becoming restricted and have developed backpressure tests to test for this condition. Depending on the severity of the restriction, the symptoms vary from slight loss of power after several minutes of driving to the engine being unable to start at all.

Always begin by tapping on the converter with a rubber mallet, listening for anything loose inside. One quick test that can be indicative of excessive exhaust backpressure is to measure intake manifold vacuum at idle and then again at 2,500 RPM. As the RPM is increased, a loss of more than about 2 in. Hg of vacuum is indicative of excessive exhaust backpressure. (As vacuum is the inverse of pressure, this amounts to an increase in pressure of about 1 PSI.) A severe exhaust restriction that affects the vehicle's driveability can also sometimes be identified by removing the precat O_2 sensor(s) and checking for a difference in driveability.

To test for excessive exhaust backpressure directly, tap into the exhaust ahead of the catalytic converter with an exhaust pressure gauge that can read from 0 to 8 PSI. (A standard pressure gauge will work for this test, but an exhaust backpressure gauge that is designed for this purpose is calibrated with more detail and is therefore easier to use when determining the results.) Then, with the engine running at 2,500 RPM in the service bay (no load), backpressure should generally not exceed about 1½ PSI. Backpressure tests can be performed on all converters, regardless of whether they are oxidation converters or reduction converters and whether they are OBD II or pre–OBD II converters.

Testing Oxidation Catalytic Converters on Pre–OBD II Vehicles

HC-to-CO_2 Conversion Test. One chemical test that has been developed can test an oxidation converter's efficiency by testing for the converter's ability to convert hydrocarbons to carbon dioxide. However, this test should only be used on vehicles manufactured prior to OBD II and only on oxidation converters. Remember, NO_x reduction converters were not that efficient prior to OBD II. And OBD II PCMs test their own converters with much more accuracy than chemical tests are capable of, so this test should also be restricted to pre–OBD II converters.

To perform this test, begin by getting the converter up to normal operating temperature. This involves running the engine until the upper radiator hose is hot and pressurized, then running it for an additional 2 minutes at 2,000 RPM. (An easier method is to take the vehicle on a short road test.) Once this is accomplished, insert a concentration-type gas analyzer's probe into the tailpipe. Turn the engine off, and then disable the secondary ignition system without disabling the fuel injectors. Then crank the engine for 10 seconds. Follow this by using the gas analyzer to measure the HC and CO_2 concentrations exiting the tailpipe for the next 30 or 40 seconds. HC levels that remain below about 500 ppm or CO_2 levels that go above 12 percent indicate that the converter is working quite efficiently. Because we cannot absolutely control the quantity of fuel injected into the engine, an "either/or" pass specification is allowed. With a smaller quantity of fuel, the converter passes if it can hold the hydrocarbons below 500 ppm even though CO_2 does not go very high (Figure 11–3). With a greater quantity of fuel, the converter passes if the carbon dioxide rises above 12 percent even though it cannot hold hydrocarbons low (Figure 11–4). The failure is when hydrocarbons go above 500 ppm but CO_2 does not reach 12 percent (Figure 11–5). Then the question becomes, "If there was that much unburned fuel available at the converter, why wasn't it able to convert more of the hydrocarbons into CO_2?" Use this test simply to give you an indication as to whether the converter's oxidation bed is functioning efficiently.

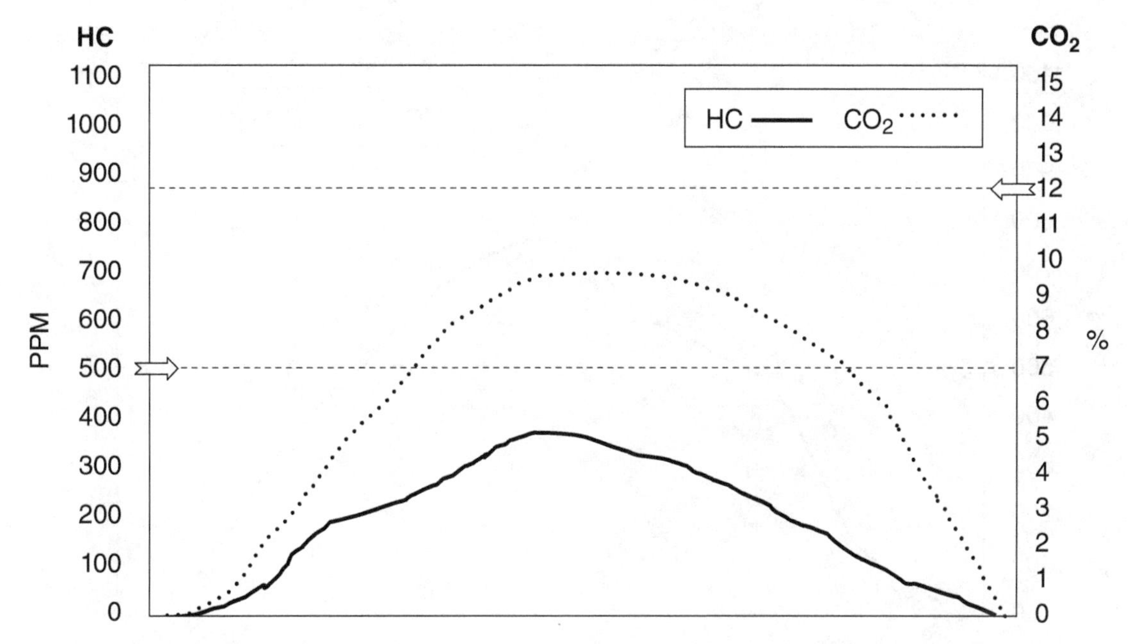

Figure 11–3 Pre–OBD II oxidation catalytic converter that passes the HC-to-CO_2 conversion test by reason of HC that stayed below 500 ppm.

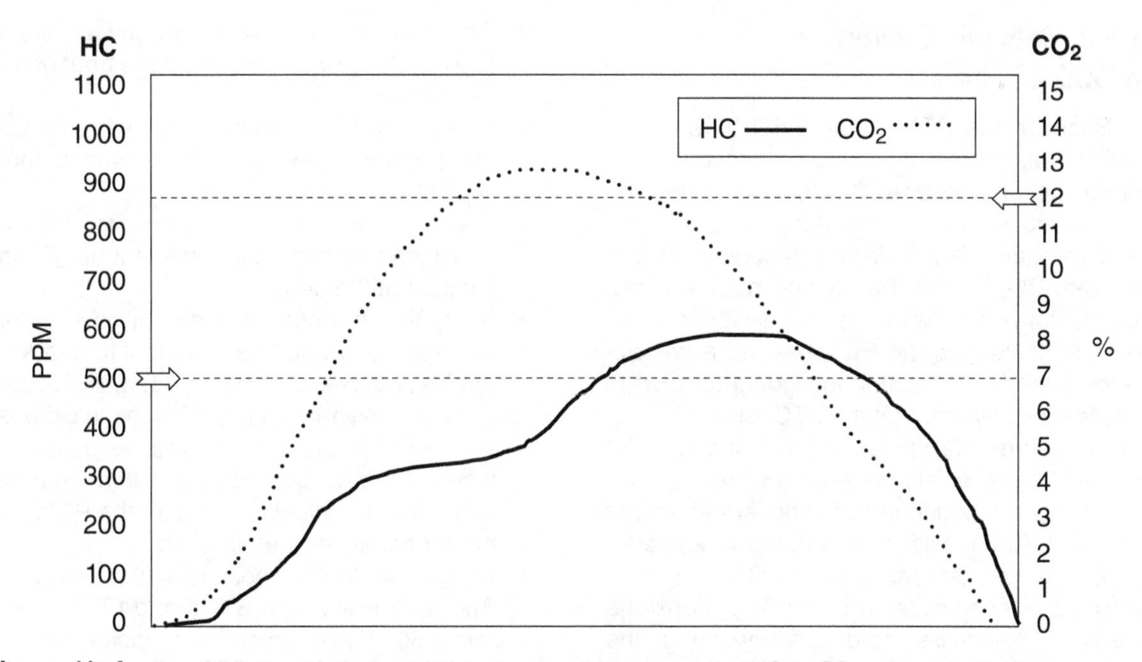

Figure 11–4 Pre–OBD II oxidation catalytic converter that passes the HC-to-CO_2 conversion test by reason of CO_2 that went above 12 percent.

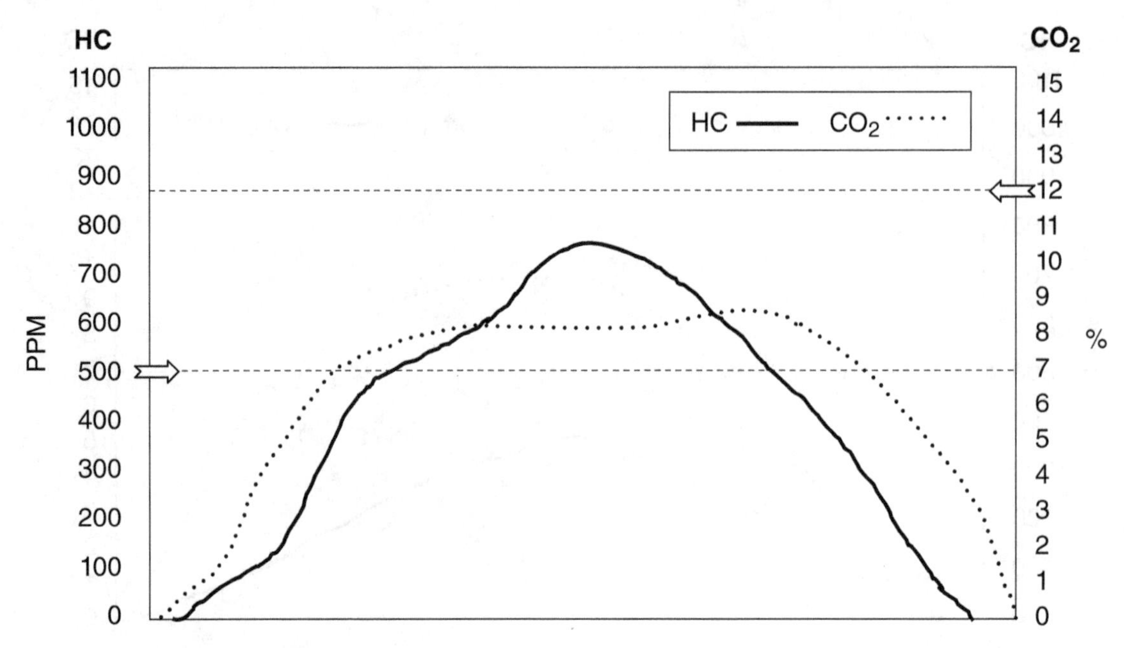

Figure 11–5 Pre–OBD II oxidation catalytic converter that fails—the converter could not boost CO_2 above 12 percent and neither could it hold HC below 500 ppm.

Testing Catalytic Converters on OBD II Vehicles

Because the PCM on an OBD II vehicle is programmed to test the catalytic converters on the vehicle, and because the PCM has been programmed to test these converters with a lot more accuracy than a technician can accomplish with an HC-to-CO_2 conversion test, do not try to test OBD II converters with the HC-to-CO_2 conversion test. Instead, if the PCM has set a *cat efficiency code,* you are probably going to have to replace the converter that the DTC refers to. However, there are a few prequalifiers that should be checked before replacing the converter.

Excessive oxygen in the exhaust can cause an OBD II PCM to falsely set a catalytic converter efficiency code. So if an OBD II PCM has set a catalyst efficiency code, you should perform the following pre-checks before condemning the converter:

- Check the exhaust system for any leaks, particularly those ahead of either the front or rear O_2 sensors.
- Use a scan tool to check for O_2 sensor codes and to determine whether the O_2 sensor monitor has run.
- Visually check the O_2 sensors for cracked housings that might leak ambient air into the exhaust at the sensor.
- Verify that the engine does not have excessive misfire, and use a scan tool to check for misfire codes—misfiring cylinders do not consume the oxygen but expel it into the exhaust.
- Use an exhaust gas analyzer to determine whether the lambda value is within 2 percent of lambda 1, indicating whether the PCM is in proper control of the air-fuel ratio.
- Use a scan tool to check total fuel trim (TFT). Add up the average STFT and LTFT at idle and 2,500 RPM, which also indicate whether the PCM is in proper control of the air-fuel

ratio. It is also a good idea to check TFT during a road test.

- Check for excessive alcohol (ethanol or methanol) in the fuel. Non–flexible fuel vehicles are not designed to run on fuel with greater than 10 percent alcohol concentrations. Because alcohol contains oxygen, high alcohol concentrations can falsely set a cat efficiency DTC.
- Check for TSBs concerning this issue; some vehicles have TSBs that recommended reflashing the PCM when 10 percent alcohol is used.

If all of the preceding checks are okay, then simply replace the converter that the DTC refers to. No further tests are required.

EVAPORATIVE SYSTEM TESTS

Using a Smoke Machine to Test an EVAP System

Smoke machines have been around for years. There are many uses for such machines. But for many years, shop owners have held the view that there are other ways to find such leaks as vacuum leaks, exhaust system leaks, and wind and water leaks without having to invest in a smoke machine. However, with modern evaporative systems, a smoke machine is a must when diagnosing these potentially tiny leaks (Figure 11–6). The EPA has set precise standards for EVAP system leaks that the PCM has to be able to detect: through model year 2000, the PCM must be able to detect any leaks larger than 0.040 in.; and for model year 2001 and newer, the PCM must be able to detect any leaks larger than .020 in. These tiny leak standards have been the driving force behind the sales of smoke machines in recent years.

Two major manufacturers produce most of the smoke machines available today, although they may be marketed under different brand names. Both manufacturers use similar controls. The following steps describe the basic functions of a smoke machine.

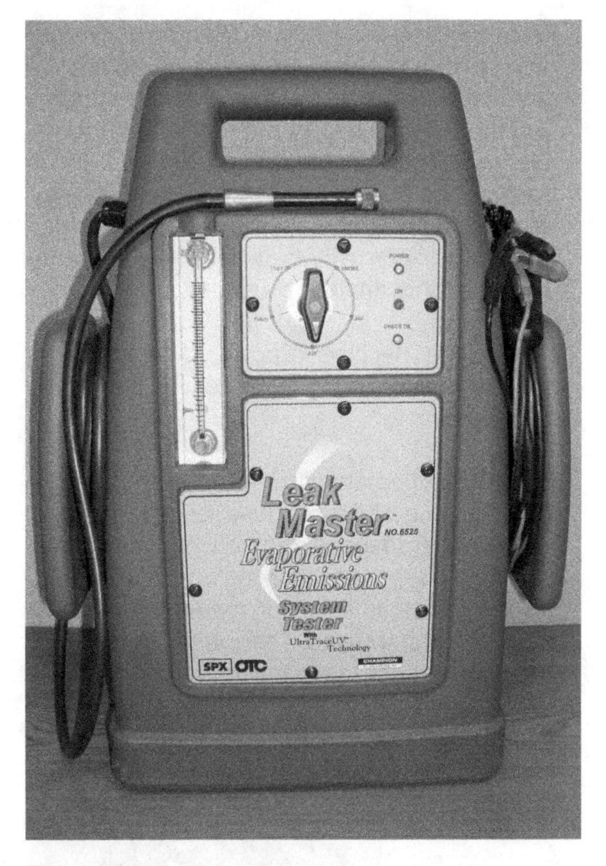

Figure 11–6 Typical modern smoke machine.

First, connect the smoke machine to power and turn on the power switch. Most machines can be powered from the vehicle's 12 V battery. Then, connect the smoke machine to a bottle of an inert gas that does not support combustion—either nitrogen (most common) or carbon dioxide.

Warning: Never use shop air for EVAP system testing. The ambient air found in a shop's compressor contains 21 percent oxygen, which supports combustion. While shop air can be used for other leak tests such as vacuum leaks and exhaust leaks, you should never use ambient air to pressurize the EVAP system and fuel tank. It will displace the fuel vapors in the tank. Although the fuel vapors

from fuels that contain alcohol already contain some oxygen, introducing additional ambient oxygen to the fuel tank can create a potential fire hazard in the event that a loose electrical connector within the tank were to produce a spark or the fuel pump motor is energized (which can normally produce small arcs internally between the commutator segments and the brushes).

Once the machine is powered on and connected to the bottle of inert gas, follow these three steps:

1. **Set the acceptable flow rate:** Set the controls on the face of the machine to either .040 in. or .020 in., depending upon the year of the vehicle that you are working on (Figure 11–7). There is a third selection marked *Future* that is not currently used. Then, with the inert gas flowing through the machine, set the arrow on the flow meter to the acceptable flow rate.
2. **Leak-test the EVAP system:** Connect the smoke machine to the *EVAP system service port* using the correct Schrader valve connector supplied with the machine. This port is usually located in the engine compartment and is identified by a green dust cap. Energize the EVAP *vent* solenoid. This solenoid is on the atmospheric side of the charcoal canister and is spring-loaded normally open. The EVAP system cannot hold pressure unless this solenoid is energized. In most cases, particularly on 2005 model year vehicles and newer, a scan tool may be used to command the PCM to energize this solenoid. On those vehicles where this active command is not an option, turn on the ignition switch (KOEO) and then use a jumper wire to ground the negative-side control wire. (Use an electrical schematic to verify that the computer controls the ground side before doing this.) Select Test on the machine's controls. Give the system some time to pressurize. If the flow meter's ball comes down to or below the acceptable flow rate, the system passes. If it stays above the acceptable flow rate, proceed to the next test.

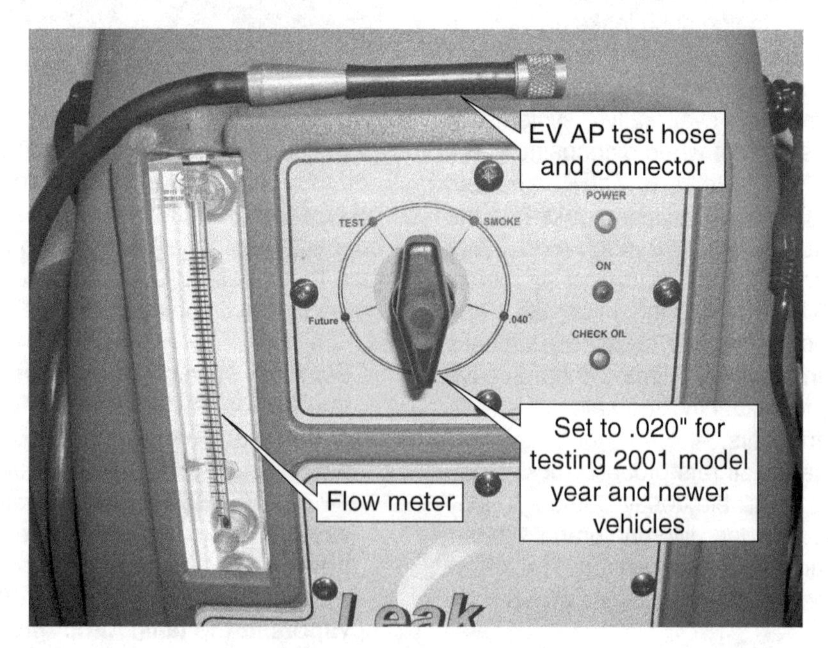

Figure 11–7 Control panel on a typical modern smoke machine.

3. **Use oil smoke to identify the leak:** Once it has been determined that the EVAP system has a leak that exceeds the acceptable flow rate, select Smoke on the machine's controls. The machine will begin to generate oil smoke and to fill the EVAP system and fuel tank with this smoke. Allow a few minutes (patience is key here) and then begin to inspect the system for visible smoke escaping from the EVAP system and fuel tank. Begin in the area of the gas cap, then the fuel tank, the charcoal canister, and the EVAP system lines. Also closely inspect the canister purge solenoid and the EVAP vent solenoid.

Caution: Always use the oil specified by the smoke machine's manufacturer to refill the machine's oil reservoir. Never substitute oil that is not recommended. Use of incorrect oil may shorten the machine's life and/or may also affect the operation of the vehicle's purge solenoid and EVAP system vent solenoid.

DIAGNOSING AIR-FUEL RATIO PROBLEMS

Overview

When diagnosing a complaint of poor engine performance, poor fuel economy, or an emissions test failure, you should perform the following steps to verify whether the PCM does or does not have proper control of the air-fuel ratio. Then, if the PCM is shown not to have proper control of the air-fuel ratio, the additional steps shown here can help you quickly narrow down the type of concern to either an electronic control system fault or a mechanical fault.

Verifying the Oxygen Sensors

The oxygen sensor should be verified before you try to verify the PCM's ability to properly control the air-fuel ratio. How this should be done depends upon whether the vehicle is an OBD II vehicle or a pre–OBD II vehicle.

Verifying an O_2 Sensor on an OBD II Vehicle. Because an OBD II PCM is programmed to test its own O_2 sensors, and because air-fuel ratio sensors are now also being used on some systems, the best advice is to let the OBD II PCM determine pass or fail for the O_2 sensors in the system. Begin by using a scan tool to verify whether the O_2 monitor has run. If it has not run, drive the vehicle in order to allow the O_2 monitor to run. Once it has run, check for O_2 sensor codes. Also perform a visual check of the O_2 sensors to make sure that their housings are not cracked or otherwise physically damaged. If O_2 sensor DTCs are present, or if any visible damage is present, replace the O_2 sensors as applicable. It should be noted that if both banks of a V-type engine have pre-cat O_2 sensors present, and a DTC is set for an O_2 sensor on one bank only, both pre-cat O_2 sensors should be replaced. Experience has shown that replacing just one O_2 sensor on an OBD II vehicle can result in the PCM quickly setting a DTC for the remaining O_2 sensor due to the manner in which the PCM's diagnostic management software is programmed to compare the O_2 sensors. It should also be noted that some four-cylinder engines have bank 1 and bank 2 O_2 sensors, with each one monitoring a pair of cylinders.

Verifying an O_2 Sensor on a Pre–OBD II Vehicle. To verify the O_2 sensor on a pre–OBD II vehicle, use a DSO to monitor the oxygen sensor's waveform while forcing the air-fuel ratio both rich and lean. This can be done on most vehicles by simply throttling the engine in the service bay. On a fuel-injected engine, suddenly opening the throttle (acceleration) results in a rich air-fuel mixture, and allowing the throttle to close (deceleration) results in a lean air-fuel mixture. If needed, a propane enrichment tool can be used to force the mixture rich. Then suddenly turning off the propane will force the mixture lean.

A zirconia-type oxygen sensor should be capable of exceeding 800 mV while the air-fuel ratio is rich, then descending to less than 175 mV when the air-fuel ratio is lean. It should be able to switch between 300 mV and 600 mV in either direction in less than 100 ms. It should never show a negative value. Any failure in any of these areas is cause for replacement. If you replace the oxygen sensor, drive the vehicle about 10 miles to break in the new oxygen sensor before you try to test it.

Verifying PCM Control of the Air-Fuel Ratio

Once the O_2 sensors have been verified, sample the O_2 sensor waveform with either a DSO or a graphing scan tool while driving the vehicle on a road test. Be sure to have another technician drive the vehicle so that you can safely monitor the DSO/scan tool. A DSO or scan tool with a recording capability may also be used. (A DSO displays the raw voltage; a scan tool displays the PCM's interpreted value communicated to the scan tool in the form of binary code. As a result, a graphing scan tool will update more slowly than a DSO.) The O_2 sensor should show a consistent cross-counting to both sides of stoichiometric. A zirconia O_2 sensor should also average between 400 and 500 mV. Additionally, fuel trim values should be monitored with a scan tool during the road test. The short-term fuel trim should be reacting to the oxygen sensor's voltage. And if the vehicle has electronic fuel injection (either TBI or PFI), the oxygen sensor's voltage should indicate a lean condition during vehicle deceleration. All of this indicates that the PCM is in proper control of the air-fuel ratio. In this case, no further tests are required.

If the oxygen sensor's voltage indicates that the air-fuel ratio is pegged either rich or lean, or is averaging outside of the 400 to 500 mV range (zirconia-type O_2 sensor), the PCM is not in proper control of the air-fuel ratio and the following steps should be taken to narrow down the area of the problem to an *electronic control system fault* or to a *mechanical fault.*

Narrowing Down the Fault Area

If the oxygen sensor's voltage is pegged either rich or lean, the key to effective diagnosis is to compare the air-fuel *command* to the air-fuel *condition* (Figure 11–8).

COMMAND \ **CONDITION**	*Rich*	*Lean*
Rich	**Electronic Control System Problem**	**Mechanical Problem** (or outside of the PCM's ability to control it)
Lean	**Mechanical Problem** (or outside of the PCM's ability to control it)	**Electronic Control System Problem**

Figure 11–8 Air-fuel ratio diagnostics: Command versus condition chart.

Air-Fuel Condition. The air-fuel condition can be verified either with the O_2 sensor or with an exhaust gas analyzer. When using the O_2 sensor, use a scan tool to read whether the PCM sees it as rich or lean. When using an exhaust gas analyzer, it is critical to have a tight exhaust system, and any secondary air system must be disabled or de-energized. Then check the lambda value on the analyzer. If the lambda value is within 2 percent of lambda 1, the PCM is likely in proper control of the air-fuel ratio. If lambda is above 1.02, the condition is lean. If lambda is below 0.98, the condition is rich.

Air-Fuel Command. The air-fuel command can be verified by the pulse-width command to the fuel injectors or by using a scan tool to verify total fuel trim (TFT). When using the fuel injector's pulse width, use a scan tool, DSO, or DMM with an "ms" scale to identify the on-time of the pulse-width command. Unfortunately, the specification is very vague due to the different fuel system designs. With the vehicle's engine running in the service bay under light load, if the pulse width is less than 2 to 3 ms, a lean command is likely indicated. If the pulse width is greater than 2 to 3 ms, a rich command is likely indicated. When using TFT, use a scan tool to read both the STFT and the LTFT. Then add the average STFT to the LTFT value to obtain a TFT value. The TFT value indicates whether the PCM is trying to add fuel to or subtract fuel from the base fuel calculation.

Compare the Command to the Condition. If the air-fuel condition is pegged lean and the air-fuel command indicates that the PCM is trying to enrich the air-fuel ratio, we can rule out the electronics (sensors and computer) as being responsible for the fault—the PCM is responding properly to the air-fuel condition. This indicates that the fault is mechanical and is outside of the computer's ability to control it. Likewise, if the air-fuel condition is pegged rich and the air-fuel command indicates that the PCM is trying to lean out the air-fuel ratio, we can again rule out the electronics as being responsible for the fault. Again, the fault is mechanical. When the fault is identified as being a mechanical fault, look for the following types of problems:

- Vacuum leaks
- Plugged or leaking fuel injectors
- Shorted or open injector windings
- Plugged fuel filter
- Ruptured fuel pressure regulator diaphragm (return-type system)

If the air-fuel condition is pegged lean and the air-fuel command indicates that the PCM is trying to further lean out the air-fuel ratio, the fault is one of electronic control—the PCM is not responding properly to the lean air-fuel condition. Likewise, if the air-fuel condition is pegged rich and the air-fuel command indicates that the PCM is trying to further enrich the air-fuel ratio, the PCM is not responding properly to the rich air-fuel condition. Again, the fault is one of electronic control. When the fault is identified as being an electronic control system fault, look for the following types of problems:

- Battery open circuit voltage (OCV) that is less than 12.4 V
- Sensors and other input circuits reporting improper or out-of-range values to the PCM
- 5 V reference voltage to the sensors (TPS, for example) is too low or too high—indicative of the voltage that is also applied to the computer's internal logic gates
- Excessive resistance in the PCM's power or ground circuits (voltage drop should not exceed 100 mV total on the power side or 50 mV total on the ground side)

If all above tests are okay, replace the PCM.

As a rule, the tests described are an excellent method for a technician to quickly narrow down an air-fuel control problem to either an electronic control system fault or a mechanical fault. However, there is one scenario where using the TFT values can be misleading. When using a scan tool to look at fuel trim values to identify the condition, try to verify that the PCM is operating in the correct fuel trim cell for the current engine operating condition. If an engine load sensor (MAP or MAF sensor) is incorrectly reporting that the engine is under more load than it really is, the PCM will operate in an incorrect fuel trim cell in an upper load area; as a result, the PCM will be locked into a window of longer fuel injector pulse widths. If you simply note the fuel trim values without noting the fuel trim cell number, these values may indicate that the PCM is trying to subtract fuel—and yet the total pulse width will be too long for the actual low-load condition. Therefore, it is always a good idea to back up the TFT value with a pulse-width measurement of the fuel injectors' on-time.

Because of this, when using an aftermarket scan tool that does not indicate which fuel trim cell the PCM is operating in, it is always a good idea to begin these tests by using a scan tool to evaluate the accuracy of the engine load sensor. When the engine is equipped with an analog MAP sensor, the voltage should follow pressure in a zero-to-five-volt range (low voltage at idle and high voltage at WOT). When the engine is equipped with a MAF sensor, the scan tool should indicate about 1 gram of air per second (G/S) for each 1 liter of engine displacement when the engine is operating at idle. For example, the MAF sensor on a 4.6L engine should indicate airflow of about 4.6 G/S during an idle condition. And when the engine is under heavy load, such as when operating at WOT, the G/S should be 40 times the engine's rated displacement in liters.

A helpful tool in this area is a scan tool that shows the values of all fuel trim cells, such as the PC-based scan tool from Automotive Test Solutions (http://www.automotivetestsolutions.com).

DIAGNOSING AN EMISSION TEST FAILURE

Because many states have incorporated emission testing programs in their major metropolitan areas, many technicians are finding themselves with the task of diagnosing emission test failures. The following discussion is designed to provide some tips for efficiently diagnosing an emission test failure.

Emission Testing Programs

Emission testing programs are divided into two basic groups: those that sample tailpipe exhaust gases and those that simply query an OBD II PCM about whether any emission problems are known to exist.

Emission programs that sample tailpipe exhaust gases include no-load idle/off idle testing, ASM testing, and IM240 testing. No-load idle/off idle tests generally evaluate HC and CO levels to determine pass/fail and are generally used with older vehicles or vehicles that for some reason cannot be tested on an ASM or IM240 dynamometer. ASM and IM240 tests also evaluate NO_x levels as well as CO and HC levels to determine pass/fail. Many of these emission programs also require visual checks for the existence of such components as catalytic converters, O_2 sensors, and other emission systems that were originally installed on the vehicle. And some emission programs require either the fuel tank and EVAP system and/or gas cap to be pressure tested. In addition, most states also look for an MIL that illuminates properly during KOEO conditions but goes out once the engine is running.

Emission programs that query an OBD II PCM about whether any emission problems are known to exist are known as *OBD II testing (or simply OBD testing)*. These emission programs essentially use a scan tool to verify that the OBD II monitors are complete and that there are no DTCs present. Monitor readiness status is critical here in that if one or more OBD II monitors have not run since the last time codes were erased or the battery was disconnected, the lack of DTCs is meaningless. Most states that use OBD II testing allow up to two monitors to be incomplete on older vehicles (typical: 2000 model year or older), but they only allow one monitor to be incomplete on newer vehicles. As a result, if the vehicle's owner can't seem to get enough of the OBD II monitors to run, he or she may bring it to your shop with a request for you to drive the appropriate drive cycles in order to get these monitors to run. Some states that use OBD II testing as their emission testing program have also considered setting up kiosks that allow the vehicle owner to perform the emissions test by connecting a state-monitored scan tool to the DLC and then paying by credit card.

Diagnostic Steps

Failure Evaluation. The first step in diagnosing an emission test failure is to evaluate the reason(s) for the failure. With a no-load idle/off idle test failure, simply determine what gases failed. With an ASM test failure, determine what gases failed and during which speed test. With an IM240 test failure, determine which gases failed and under what driving conditions: light acceleration, moderate acceleration, low-speed cruise, and/or high-speed cruise. And with all of these tests, determine how much each gas failed by. Keep in mind that the cut-points for these gases are usually quite liberal—emissions programs usually set the limits high in an effort to identify only the dirtiest vehicles, not every vehicle that has a problem.

With an OBD II test failure, determine if the vehicle was rejected because not enough of the OBD II monitors had run or if the vehicle failed due to DTCs that were present. And note any visual failures, gas cap or fuel tank/EVAP system failures, and MIL failures.

Baseline the Exhaust Gas Levels. Before you make any repairs, you should **baseline** the exhaust gas levels. This means that before you make any repairs, and before you even wiggle any electrical connectors, use whatever method is available to you in your shop to sample the vehicle's exhaust gases. Then, once the vehicle is repaired, you will use the same method to sample the vehicle's exhaust gases again, at which time you will compare the final readings to your baseline readings to determine the effectiveness of your repairs. Your method may be to use a large exhaust gas analyzer to sample the exhaust gases with the vehicle in the service bay at both idle and 2,500 RPM, but under no load. (You should note that an NO_x problem may not be as apparent under a no-load test.) It may involve running the vehicle on a dynamometer so that the gases can be tested under realistic load. Or it may involve using a portable gas analyzer to sample and record the exhaust gases during a road test. (Caution: Be sure to have another technician drive the vehicle during the road test if your intent is to monitor the exhaust gas readings. Some portable gas analyzers include a Record function that allows you to record the exhaust gas readings during the road test so that you don't have to monitor the analyzer. Then the recording can be played back once you are back at the shop.)

Whatever the method, make sure it is one that you can repeat once repairs are complete. In addition to recording the exhaust gases, also record the lambda value. Or use an automated lambda calculator to determine the lambda value from the recorded gas readings if you are using an older machine that doesn't have the lambda feature. Do not forget to record your baseline readings on the repair order for review later.

Visual Inspection. Once the exhaust gas readings have been baselined, perform a conscientious visual check of the fuel system, spark management system, emission systems, and associated electrical harness and connectors. Look for chafing of wires against metal brackets, loose connectors, or connector terminals that may have been damaged through abuse, heat, or age. Also, test all fuses in all fuse blocks, both visually and electrically. And listen for vacuum leaks with the engine running. Remember to visually check the O_2 sensors for physical damage and cracks. If done purposefully, this visual check can sometimes save you a tremendous amount of time.

Monitor Readiness Status. Now it is time to connect your scan tool to the DLC. Begin by checking monitor readiness status (Global OBD II Mode $01). If one or more OBD II monitors have not run, the absence of DTCs is meaningless. In this case, you will need to drive the vehicle in order to run the appropriate drive cycles. This will allow the incomplete monitors to run.

Road Test and Drive Cycles. At this point you should road-test the vehicle to identify driveability issues such as poor performance, hesitation on acceleration, surging, misfire, detonation, and exhaust smoke. During this road test you should also drive the vehicle through the appropriate drive cycles in order to allow the PCM to run all of the OBD II monitors that were incomplete during the previous test. Many scan tools can display the required enable criteria for each monitor and the required enable criteria to complete a full drive cycle.

Diagnostic Trouble Codes. Use a scan tool to verify that all of the OBD II monitors are complete. Once they are complete, use the scan tool to check for confirmed DTCs (Global OBD II Mode $03) and pending DTCs (Global OBD II Mode $07). Pay particular attention to DTCs that represent faults with any of the O_2 sensors. Because several OBD II monitors can be dependent on the O_2 sensor, if an O_2 sensor fault is present you may not be able to get some of the monitors to run. In this case you will have to replace the respective O_2 sensors and then road-test the vehicle again in order to complete the appropriate drive cycles.

If you are working on a Ford Motor Company product, remember that you should also run both the KOEO and KOER self-tests, which can identify faults currently present. And if you are working on a 2010 or newer vehicle, remember that Global OBD II Mode $0A can display permanent DTCs that cannot be erased by a scan tool or by disconnecting the vehicle's battery.

Freeze Frame Data. Use a scan tool to check the OBD II freeze frame data that set for the first (or higher priority) confirmed DTC set in the PCM's memory. Some PCMs will record multiple sets of freeze frame data for multiple confirmed DTCs. This information can provide some important clues as to what the operating conditions were at the time the DTC was set.

Data Stream. Use a scan tool to look at PIDs that might apply to the failure that you are concerned with. This might include such things as O_2 sensor data, LTFT, STFT, commanded fuel injector pulse width, ECT, and IAT. Use the scan tool's Record function to record the data while driving the vehicle under conditions similar to those that existed when the confirmed DTCs were recorded, as indicated by the freeze frame data.

Misfire/Hydrocarbon Failures. If the vehicle failed for high hydrocarbons, if the engine runs rough or misfire was felt during the road test, or if a misfire DTC is present, begin with a cylinder power balance test to confirm the misfire. Either the misfire DTC or the cylinder power balance test can identify which cylinder is responsible for the misfire condition. Many scan tools can now help you perform this test by allowing you to command the PCM to deactivate a cylinder's fuel injector through the active command mode while you monitor the reduction in engine RPM. Remember that some vehicles, beginning in 2008, can set a cylinder imbalance DTC as well.

If a cylinder is determined to be misfiring or weak in its contribution to the rotation of the crankshaft, you will need to test ignition, fuel delivery, cranking compression, and running compression to determine the nature of the fault.

Fuel Management: CO or NO_x Failures. Air-fuel ratio problems can cause excessive CO when the mixture is rich and excessive NO_x production when the mixture is lean, particularly if the mixture goes lean on a hard acceleration. As a result, if you are diagnosing a CO failure and NO_x levels are more than three-quarters of their limit, as you correct the overly rich condition, thus leaning out the air-fuel ratio back toward stoichiometric, NO_x levels will naturally increase. Therefore, even with a rich mixture, you need to concern yourself with the NO_x levels as well to avoid turning a CO failure into an NO_x failure.

Were any pending or confirmed DTCs recorded earlier that relate to air-fuel control problems? These DTCs may give you the clue you need to help diagnose air-fuel control faults. And be sure to use the lambda value from your baseline readings to determine how far lean or rich the actual air-fuel ratio is. In terms of air-fuel ratio concerns, also refer to the discussion earlier in this chapter titled "Diagnosing Air-Fuel Ratio Problems."

For NO_x failures, this is also the time to take a look at any pending or confirmed DTCs that relate to the EGR system. To see *engine out NO_x*, use a scan tool's bidirectional controls to command the PCM to lean out the air-fuel ratio until your gas analyzer shows that the lambda value is slightly leaner than 1.02. This lean air-fuel ratio effectively shuts down the NO_x reduction catalytic converter due to the lack of CO production.

Exhaust Backpressure. Another concern is the possibility of a plugged exhaust system, a problem most commonly associated with a plugged or restricted catalytic converter. As stated earlier, a vacuum gauge can provide a quick indication of excessive exhaust backpressure, but an exhaust backpressure gauge should be used to pinpoint test this concern.

Catalytic Converter. With the previous concerns now diagnosed and repaired, and with the PCM now in proper control of the fuel management and spark management systems, there is still one more critical concern: Is the catalytic converter operating efficiently? Because the catalytic converter is the single most efficient emissions device, its operation can reduce emission levels to well below the liberal cut-points.

Were any pending or confirmed DTCs recorded earlier that relate to catalytic converter efficiency concerns? As discussed earlier, with an OBD II vehicle, the PCM will test the converters and set a DTC if a concern is present. With a pre–OBD II vehicle, proceed at this point to test the converter yourself, using the tests discussed earlier in this chapter.

Secondary Air System. Check the emission decal to determine if a secondary air injection system is present. They largely disappeared in the 1990s, but some newer vehicles are using them again with electric air pumps and without any downstream injection point (due to rear fuel control). The modern ones should turn on during engine warmup only to pump air upstream to the exhaust manifold. Once the PCM enters closed loop, they should be turned off. Check the system for proper operation.

Final Exhaust Gas Evaluation. Use your exhaust gas analyzer to retest the exhaust gas levels again, using the same method as you used earlier. Compare these readings to your baseline readings recorded earlier. Did the readings show improvement?

Erase DTCs and Clear Diagnostic History. Now that your repairs have been completed, use a scan tool to erase the OBD II data (Global OBD II Mode $04). This mode clears the DTCs, erases freeze frame data, resets the monitors to *incomplete,* and resets the fuel trim values.

Final Road Test. Road-test the vehicle again. How does this road test compare to the initial road test in terms of vehicle performance? This final road test is also used to allow the PCM to begin the relearn process if repairs were made, if OBD II data was erased, or if the vehicle's battery was disconnected. You should also drive the vehicle to allow the PCM to complete a full drive cycle, which sets the monitors to Complete again.

Final DTC Check. One last test: Use a scan tool to verify that the monitors have completed and to determine if any of the DTCs returned during the final road test and drive cycle.

REPROGRAMMING A COMPUTER

Overview

As computer-controlled systems replaced mechanically controlled systems, the complexity of these systems moved into the computer itself. Early engine control computers generally had few terminals, and these were used for powers, grounds, inputs, and outputs. Considering Ford systems as an example, an EEC-II PCM in 1979 had a maximum of 32 terminals, not all of which were used. EEC-IV increased the number of potential PCM terminals to 60, and EEC-V further increased the number of potential PCM terminals to 104 and then to 150. Modern computers monitor many more inputs and control many more outputs than their predecessors. As the computer is required to take on additional functions, the computer's internal programs become more complex. At times, this complexity can result in unintended problems created by the engineers themselves. It can be very difficult for the manufacturers' research and development departments to stay abreast of these problems and to create solutions to resolve them. (And how much more difficult this might be for an aftermarket parts manufacturer. This is why OE parts are recommended when making repairs on today's systems.)

With some early systems, when the manufacturer determined that an internal program of the computer was causing a problem, a newly designed part number PROM chip would be issued as a correction. Ford Motor Company called this IC chip an "engine calibration assembly" and made it a replaceable chip from 1978 through 1984, attached externally to the PCM; General Motors referred to this chip as the PROM chip or CALPAK and made it a replaceable chip throughout the 1980s, with the chip accessible underneath a small access cover on the PCM. With later systems, such as Ford EEC-IV, the chip was permanently soldered into the PCM, and repair of such a problem required that the entire PCM be replaced with a new part number module. In the 1990s, manufacturers began making these control modules reprogrammable. Reprogramming an automotive computer is also known as *reflashing*.

Reprogramming an Automotive Computer

When a manufacturer creates a reflash file, it is created as a specific fix to a specific problem. This file will be posted to the manufacturer's technician/service website. In order to reprogram a control module, the technician must download this file to a PC or laptop and then install it electronically into the specific control module that it is intended for.

Downloading the Reflash File. To download the file, you must connect to the manufacturer's website via the Internet. (The National Automotive Service Task Force lists the OEM service websites on their website: www.NASTF.org.) The EPA requires that the manufacturers make those files intended for on-board computers that affect the vehicle's emissions available to the aftermarket repair industry. Most manufacturers make files for other control modules available as well, such as reflash files for body computers. However, it is not required that access to these files be free of charge. As a result, most manufacturers charge a subscription fee. Generally these subscriptions can be purchased as an annual subscription, a six-month subscription, a monthly subscription, or a daily subscription. The fee for a daily subscription is usually around $20 to $25. One manufacturer charges this low fee for a 72-hour subscription. For an aftermarket technician, it is recommended that the daily subscription be purchased and that the fee then be passed on to the retail consumer. Once the subscription has been purchased, you may download as many reflash files as you want during the subscription period. Download the needed files and store them in a specific folder on your computer.

Installing the Reflash File. To install the reflash file, the PC that contains the reflash file must be connected to the appropriate control module via a secondary module that is used as an interpreter. This secondary module is known as a J2534 tool (Figure 11–9). J2534 reflash tools are available in many different brands from many different automotive tool and equipment companies. (Be sure to comparison shop before purchasing one.) It is through this tool that the reflash file will be used to reprogram the appropriate control module. Of course, when reprogramming a module, the module must be properly powered up before beginning the reflash procedure and must remain powered up throughout the procedure, except when instructed to cycle the ignition off and back on at specific points in the procedure.

There are two primary methods for reprogramming an automotive computer: on-board reprogramming and off-board reprogramming.

On-Board Reprogramming. On-board reprogramming refers to loading the reflash file to the appropriate control module via the vehicle's DLC. With this method, there is no need to remove the control module from the vehicle. The PC or laptop is connected to the vehicle's DLC via

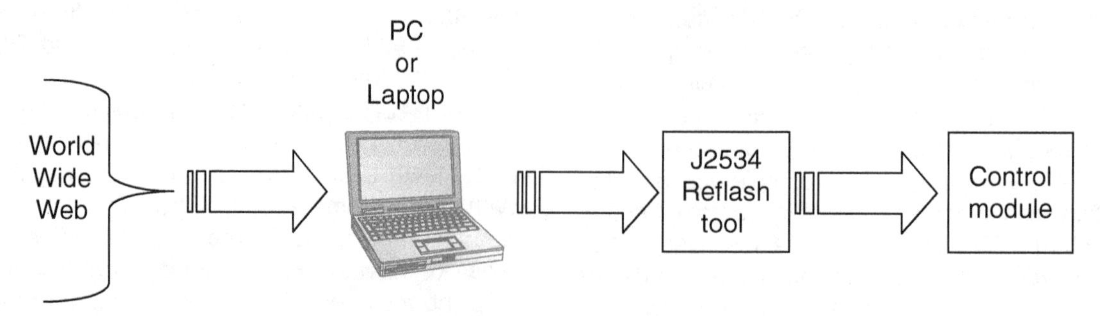

Figure 11–9 Using a J2534 tool to reprogram a control module.

the J2534 tool. Then simply turn on the ignition switch to the Run position to power up the module for the reflash procedure.

Off-Board Reprogramming. **Off-board reprogramming** refers to removal of the control module from the vehicle and then installing the reflash file using the J2534 tool. The J2534 tool will also power up the module during the procedure.

Determining When to Reprogram an Automotive Computer

Many times a TSB will direct the technician to reprogram a control module in response to a specific problem. However, to further complicate the issue, many reflash files also exist for symptoms for which no TSB has been issued. Therefore, reprogramming should be considered only after all of the basics have been checked and normal diagnostic and repair procedures have been fully implemented. Only then should the download and installation of a reflash file be considered. This is because *problems associated with a control module can also be created due to the installation of a reflash file!!!* (Have you ever created a problem with a PC or laptop *because* you installed a new program? This is why Microsoft provides a *restore* feature.) Therefore, reprogramming on-board computers should never be done indiscriminately! Do not reprogram a control module simply because the manufacturer has an updated reflash file available for the control module. Reprogramming should be reserved for correcting real problems with the module.

Concerns with Reprogramming of an Automotive Computer

If a technician has begun the installation of a reflash file and communication with the control module being reprogrammed is interrupted, the module may become a **non-responsive unit (NRU)**. Recovery of an NRU is risky at best. When an NRU is encountered, recovery should be attempted by beginning the reprogramming procedure again. While it is possible that the NRU may be recovered on the first recovery attempt, it is also possible that the module may still exist as an NRU many attempts later—in which case the module is considered junk and will have to be replaced.

There are several reasons why a control module may inadvertently become an NRU during a reprogramming attempt. Keep in mind that the method used to reprogram a control module is through your PC or laptop communicating with the module in binary code. So, one reason is if another control module on the data bus were to try to *talk* (via binary code) to the module being reprogrammed during the reprogramming procedure. If you connect a DSO to the DLC terminals so that the binary code is displayed, you will find that on some vehicles, the communication may continue to be present on a network data bus for 30 seconds to several minutes after the ignition switch is turned off, the ignition key is removed, and all of the vehicle's doors are closed. And then simply the act of opening a door causes the binary communication to begin again. What if you are in the middle of a reprogramming procedure and another technician or service writer/service manager opens a door on the vehicle, maybe to get a personal item from the vehicle that the customer has requested? There is a good chance that the control module you were reprogramming just became an NRU. Obviously, this is one advantage of off-board reprogramming.

Another reason a control module might become an NRU is if it were to lose power during the reprogramming. With on-board reprogramming, care should be taken to ensure that the vehicle's battery voltage does not deteriorate substantially during the reprogramming procedure. This is not a huge problem if the reprogramming only takes about 5 minutes as is typical with many PCMs and BCMs. However, at least one European manufacturer requires all on-board control modules to be reprogrammed simultaneously due to the manner in which the modules work together. (This requirement deletes the possibility of performing an off-board reprogramming procedure.) In this case, the reflash file is quite large and may take aftermarket equipment up to

20 hours to successfully perform the reprogramming procedure. Therefore, the vehicle's battery should be kept alive using a specialty battery charger. Do not use the standard battery charger found in most automotive shops; they produce a large amount of AC ripple (you can see this on a DSO—try looking at a fuel injector's voltage waveform with the charger turned off and then turned on) and may cause interference in the computer communication. Specialized chargers that are designed for reprogramming are available, although they are usually quite expensive. These chargers are designed to output a steady voltage without the ripple.

Also, with on-board or off-board reprogramming, make sure that your PC is powered through an uninterruptable power supply (UPS) or that your laptop's batteries are charged and that the laptop is plugged into an AC outlet using the proper adaptor. This will allow your reprogramming procedure to survive any momentary AC outages that might cause the module to become an NRU.

If a module does become non-responsive, then begin the reprogramming procedure again, taking care to avoid these problems.

DIAGNOSIS OF EXCESSIVE PARASITIC DRAW ON A MODERN VEHICLE

Because a modern vehicle has multiple ECUs on board, the methods for an accurate diagnosis of excessive parasitic draw have changed when compared to traditional methods. The traditional method was to place an ammeter in series between the battery negative cable and the negative post and measure the parasitic draw. However, this method requires disconnecting the battery from the vehicle long enough to reconnect the ammeter in series. Upon reconnection, many ECUs will reset their internal timers and can require from 20 minutes to an hour to go to sleep, during which time any parasitic draw measurements will be artificially excessive and could result in a technician trying to diagnose a nonexistent problem.

Because of this, many technicians have learned to use a low-amp current probe connected to a DMM (or DSO) to determine initially whether or not an excessive parasitic draw exists. With the vehicle parked (with the ignition off and the doors closed) for an ample amount of time, the technician zeroes the current probe and then places it around a battery cable. If an excessive parasitic draw is identified, the alternator should be temporarily disconnected in order to identify an excess current drain caused by a shorted positive diode within the rectifier bridge. Following the testing of the alternator, the standard procedure in the industry has always been to begin removing fuses one at a time, beginning with the primary junction box, and then reinstalling each fuse and removing the next fuse, until the circuit that is responsible for the excessive current drain on the battery is identified. The problem with this approach is that when a fuse that provides power to an ECU is removed and reinstalled, that ECU has now been triggered to power up again, and its internal timers are reset. Thus, any further testing during the next 20 minutes to an hour can produce inaccurate results.

A new approach to narrowing down excessive parasitic draw to the responsible circuit(s) has been developed that requires absolutely no disconnections. Two original equipment manufacturers (OEMs), Chrysler and Volkswagen, were instrumental in developing this new technique and Ford Motor Company now also recognizes it. This technique was also tested and validated by CARQUEST Technical Institute, which also produced its own charts for the technique (Figure 11–10, Figure 11–11, and Figure 11–12).

This technique is based on a little-known fact that the fusible element within a fuse will have a very tiny amount of resistance across it due to the type of material that is used in its construction. Other concepts that are important to note here are:

- If current is flowing in a circuit, all of the voltage is dropped by all of the resistance.
- In a series circuit, if current is flowing, all ohms of resistance share equally in the

A

Voltage Drop Across a Mini Fuse

Measurement mV	Mini 5 Amp	Mini 7.5 Amp	Mini 10 Amp	Mini 15 Amp	Mini 20 Amp	Mini 25 Amp	Mini 30 Amp
0.1	6	10	14	22	29	40	50
0.2	12	20	28	44	57	80	100
0.3	18	30	43	67	86	120	150
0.4	24	40	57	89	114	160	200
0.5	30	50	71	111	143	200	250
0.6	36	60	85	133	171	240	300
0.7	42	70	99	156	200	280	350
0.8	48	80	114	178	229	320	400
0.9	54	90	128	200	257	360	450
1	60	100	142	222	286	400	500
1.1	66	110	156	244	314	440	550
1.2	72	120	171	267	343	480	600
1.3	78	130	185	289	371	520	650
1.4	84	140	199	311	400	560	700
1.5	90	150	213	333	429	600	750
1.6	96	160	227	356	457	640	800
1.7	102	169	242	378	486	680	850
1.8	108	179	256	400	514	720	900
1.9	114	189	270	422	543	760	950
2	120	199	284	444	571	800	1000

B

Voltage Drop Across a Mini Fuse

Measurement mV	Mini 5 Amp	Mini 7.5 Amp	Mini 10 Amp	Mini 15 Amp	Mini 20 Amp	Mini 25 Amp	Mini 30 Amp
2.1	126	209	296	467	600	840	1050
2.2	132	219	313	489	629	880	1100
2.3	138	229	327	511	657	920	1150
2.4	144	239	341	533	686	960	1200
2.5	150	249	355	556	714	1000	1250
2.6	156	259	369	578	743	1040	1300
2.7	162	269	384	600	771	1080	1350
2.8	168	279	398	622	800	1120	1400
2.9	174	289	412	644	829	1160	1450
3	180	299	426	667	857	1200	1500
3.1	186	309	441	689	886	1240	1550
3.2	192	319	455	711	914	1280	1600
3.3	198	329	469	733	943	1320	1650
3.4	204	339	483	756	971	1360	1700
3.5	210	349	497	778	1000	1400	1750
3.6	216	359	512	800	1026	1440	1800
3.7	222	369	526	822	1057	1480	1850
3.8	228	379	540	844	1086	1520	1900
3.9	234	389	554	867	1114	1560	1950
4	240	399	568	889	1143	1600	2000

Figure 11–10 Voltage drop across a mini fuse.

C

Voltage Drop Across a Mini Fuse

Measurement mV	Mini 5 Amp	Mini 7.5 Amp	Mini 10 Amp	Mini 15 Amp	Mini 20 Amp	Mini 25 Amp	Mini 30 Amp
4.1	246	409	583	911	1171	1640	2050
4.2	252	419	597	933	1200	1680	2100
4.3	258	429	611	956	1229	1720	2150
4.4	264	439	625	978	1257	1760	2200
4.5	270	449	639	1000	1286	1800	2250
4.6	276	459	654	1022	1314	1840	2300
4.7	282	469	668	1044	1343	1880	2350
4.8	288	479	682	1067	1371	1920	2400
4.9	294	488	696	1089	1400	1960	2450
5	300	498	711	1111	1429	2000	2500
5.1	306	508	725	1133	1457	2040	2550
5.2	312	518	739	1156	1486	2080	2600
5.3	318	528	753	1178	1514	2120	2650
5.4	324	538	767	1200	1543	2160	2700
5.5	330	548	782	1222	1571	2200	2750
5.6	336	558	796	1244	1600	2240	2800
5.7	342	568	810	1267	1629	2280	2850
5.8	248	578	824	1289	1657	2320	2900
5.9	354	588	838	1311	1686	2360	2950
6	360	598	853	1333	1714	2400	3000

Courtesy of CARQUEST Technical Institute a Division of Advance Auto Parts.

D

Voltage Drop Across a Mini Fuse

Measurement mV	Mini 5 Amp	Mini 7.5 Amp	Mini 10 Amp	Mini 15 Amp	Mini 20 Amp	Mini 25 Amp	Mini 30 Amp
6.1	366	608	867	1356	1743	2440	3050
6.2	372	618	881	1378	1771	2480	3100
6.3	378	628	895	1400	1800	2520	3150
6.4	384	638	909	1422	1829	2560	3200
6.5	390	648	924	1444	1857	2600	3250
6.6	396	658	938	1467	1886	2640	3300
6.7	402	668	952	1489	1914	2680	3350
6.8	408	678	966	1511	1943	2720	3400
6.9	414	688	981	1533	1971	2760	3450
7	420	698	995	1556	2000	2800	3500
7.1	426	708	1009	1578	2029	2840	3550
7.2	432	718	1023	1600	2057	2880	3600
7.3	438	728	1037	1622	2086	2920	3650
7.4	444	738	1052	1644	2114	2960	3700
7.5	450	748	1066	1667	2143	3000	3750
7.6	456	758	1080	1689	2171	3040	3800
7.7	462	768	1094	1711	2200	3080	3850
7.8	468	778	1108	1733	2229	3120	3900
7.9	474	788	1123	1756	2257	3160	3950
8	480	798	1137	1778	2286	3200	4000

Courtesy of CARQUEST Technical Institute a Division of Advance Auto Parts.

Figure 11–10 Voltage drop across a mini fuse. (*continued*)

E

Voltage Drop Across a Mini Fuse

Measurement mV	Mini 5 Amp	Mini 7.5 Amp	Mini 10 Amp	Mini 15 Amp	Mini 20 Amp	Mini 25 Amp	Mini 30 Amp
8.1	486	807	1151	1800	2314	3240	4050
8.2	492	817	1165	1822	2343	3280	4100
8.3	498	827	1179	1844	2371	3320	4150
8.4	504	837	1194	1867	2400	3360	4200
8.5	510	847	1208	1889	2429	3400	4250
8.6	516	857	1222	1911	2457	3440	4300
8.7	522	867	1236	1933	2486	3480	4350
8.8	528	877	1251	1956	2514	3520	4400
8.9	534	888	1265	1978	2543	3560	4450
9	540	897	1279	2000	2571	3600	4500
9.1	546	907	1293	2022	2600	3640	4550
9.2	552	917	1307	2044	2629	3680	4600
9.3	558	927	1322	2067	2657	3720	4650
9.4	564	937	1336	2089	2686	3760	4700
9.5	570	947	1350	2111	2714	3800	4750
9.6	576	957	1364	2133	2743	3840	4800
9.7	582	967	1378	2156	2771	3880	4850
9.8	588	977	1393	2178	2800	3920	4900
9.9	594	987	1407	2200	2829	3960	4950
10	600	997	1421	2222	2857	4000	5000

Figure 11–10 Voltage drop across a mini fuse. (*continued*)

A

Voltage Drop Across a Standard Fuse

Measurement mV	Standard 5 Amp	Standard 10 Amp	Standard 15 Amp	Standard 20 Amp	Standard 25 Amp	Standard 30 Amp
0.1	7	13	23	30	47	62
0.2	13	27	45	61	94	123
0.3	20	40	68	91	141	185
0.4	26	54	91	122	188	246
0.5	33	67	113	152	235	308
0.6	40	80	136	183	281	370
0.7	46	94	158	213	328	431
0.8	53	107	181	244	375	493
0.9	59	120	204	274	422	554
1	66	134	226	305	469	616
1.1	73	147	249	335	516	677
1.2	79	161	272	366	563	739
1.3	86	174	294	396	610	801
1.4	92	187	317	427	657	862
1.5	99	201	340	457	704	924
1.6	106	214	362	487	751	985
1.7	112	228	385	518	797	1047
1.8	119	241	407	548	844	1109
1.9	125	254	430	579	891	1170
2	132	268	453	609	938	1232

Figure 11–11 Voltage drop across a standard fuse.

B

Voltage Drop Across a Standard Fuse

Measurement mV	Standard 5 Amp	Standard 10 Amp	Standard 15 Amp	Standard 20Amp	Standard 25 Amp	Standard 30 Amp
2.1	139	281	475	640	985	1293
2.2	145	294	498	670	1032	1355
2.3	152	308	521	701	1079	1417
2.4	158	321	543	731	1126	1478
2.5	165	335	566	762	1173	1540
2.6	172	348	589	792	1220	1601
2.7	178	361	611	823	1267	1663
2.8	185	375	634	853	1313	1725
2.9	192	388	656	884	1360	1786
3	198	401	679	914	1407	1848
3.1	205	415	702	944	1454	1909
3.2	211	428	724	975	1501	1971
3.3	218	442	747	1005	1548	2023
3.4	225	455	770	1036	1595	2094
3.5	231	468	792	1066	1642	2156
3.6	238	482	815	1097	1689	2217
3.7	244	495	837	1127	1736	2279
3.8	251	509	860	1158	1782	2340
3.9	258	522	883	1188	1829	2402
4	264	535	905	1219	1876	2464

C

Voltage Drop Across a Standard Fuse

Measurement mV	Standard 5	Standard 10	Standard 15	Standard 20	Standard 25	Standard 30
4.1	271	549	928	1249	1923	2525
4.2	277	562	951	1280	1970	2587
4.3	284	575	973	1310	2017	2648
4.4	291	589	996	1341	2064	2710
4.5	297	602	1019	1371	2111	2772
4.6	304	616	1041	1401	2158	2833
4.7	310	629	1064	1432	2205	2895
4.8	317	642	1086	1462	2252	2956
4.9	324	656	1109	1493	2298	3018
5	330	669	1132	1523	2345	3080
5.1	337	683	1154	1554	2392	3141
5.2	343	696	1177	1584	2439	3203
5.3	350	709	1200	1615	2486	3264
5.4	357	723	1222	1645	2533	3326
5.5	363	736	1245	1676	2580	3387
5.6	370	749	1268	1706	2627	3449
5.7	376	763	1290	1737	2674	3511
5.8	383	776	1313	1767	2721	3572
5.9	390	790	1335	1798	2768	3634
6	396	803	1358	1828	2814	3695

Figure 11–11 Voltage drop across a standard fuse. (*continued*)

D

Voltage Drop Across a Standard Fuse

Measurement mV	Standard 5	Standard 10	Standard 15	Standard 20	Standard 25	Standard 30
6.1	403	816	1381	1858	2861	3757
6.2	409	830	1403	1889	2908	3819
6.3	416	843	1426	1919	2955	3880
6.4	423	857	1449	1950	3002	3942
6.5	429	870	1471	1980	3049	4003
6.6	436	883	1494	2011	3096	4065
6.7	442	897	1517	2041	3143	4127
6.8	449	910	1539	2072	3190	4188
6.9	456	923	1562	2102	3237	4250
7	462	937	1584	2133	3284	4311
7.1	469	950	1607	2163	3330	4373
7.2	475	964	1630	2194	3377	4434
7.3	482	977	1652	2224	3424	4496
7.4	489	990	1675	2255	3471	4558
7.5	495	1004	1698	2285	3518	4619
7.6	502	1017	1720	2315	3565	4681
7.7	508	1030	1743	2346	3612	4742
7.8	515	1044	1766	2376	3659	4904
7.9	522	1057	1788	2407	3706	4866
8	528	1071	1811	2437	3753	4927

Courtesy of CARQUEST Technical Institute a Division of Advance Auto Parts.

E

Voltage Drop Across a Standard Fuse

Measurement mV	Standard 5	Standard 10	Standard 15	Standard 20	Standard 25	Standard 30
8.1	535	1084	1833	2468	3800	4989
8.2	541	1097	1856	2498	3846	5050
8.3	548	1111	1879	2529	3893	5112
8.4	555	1124	1901	2559	3940	5174
8.5	561	1138	1924	2590	3987	5235
8.6	568	1151	1947	2620	4034	5297
8.7	575	1164	1969	2651	4081	5358
8.8	581	1178	1992	2681	4128	5420
8.9	588	1191	2015	2712	4175	5482
9	594	1204	2037	2742	4222	5543
9.1	601	1218	2060	2772	4269	5605
9.2	608	1231	2082	2803	4316	5666
9.3	614	1245	2105	2833	4362	5728
9.4	621	1258	2128	2864	4409	5789
9.5	627	1271	2150	2894	4456	5851
9.6	634	1285	2173	2925	4503	5913
9.7	641	1298	2196	2955	4550	5974
9.8	647	1312	2218	2986	4597	6036
9.9	654	1325	2241	3016	4644	6097
10	660	1338	2263	3047	4691	6159

Courtesy of CARQUEST Technical Institute a Division of Advance Auto Parts.

Figure 11–11 Voltage drop across a standard fuse. (*continued*)

Voltage Drop Across a Cartridge Fuse

Measurement mV	Cartridge 20 Amp	Cartridge 30 Amp	Cartridge 40 Amp	Cartridge 50 Amp
0.1	100	67	100	200
0.2	200	133	200	400
0.3	300	200	300	600
0.4	400	267	400	800
0.5	500	333	500	1000
0.6	600	400	600	1200
0.7	700	467	700	1400
0.8	800	533	800	1600
0.9	900	600	900	1800
1	1000	667	1000	2000
1.1	1100	733	1100	2200
1.2	1200	800	1200	2400
1.3	1300	867	1300	2600
1.4	1400	933	1400	2800
1.5	1500	1000	1500	3000
1.6	1600	1067	1600	3200
1.7	1700	1133	1700	3400
1.8	1800	1200	1800	3600
1.9	1900	1267	1900	3800
2	2000	1333	2000	4000

A

Voltage Drop Across a Cartridge Fuse

Measurement mV	Cartridge 20 Amp	Cartridge 30 Amp	Cartridge 40 Amp	Cartridge 50 Amp
2.1	2100	1400	2100	4200
2.2	2200	1467	2200	4400
2.3	2300	1533	2300	4600
2.4	2400	1600	2400	4800
2.5	2500	1667	2500	5000
2.6	2600	1733	2600	5200
2.7	2700	1800	2700	5400
2.8	2800	1867	2800	5600
2.9	2900	1933	2900	5800
3	3000	2000	3000	6000
3.1	3100	2067	3100	6200
3.2	3200	2133	3200	6400
3.3	3300	2200	3300	6600
3.4	3400	2267	3400	6800
3.5	3500	2333	3500	7000
3.6	3600	2400	3600	7200
3.7	3700	2467	3700	7400
3.8	3800	2533	3800	7600
3.9	3900	2600	3900	7800
4	4000	2667	4000	8000

B

Voltage Drop Across a Cartridge Fuse

Measurement mV	Cartridge 20 Amp	Cartridge 30 Amp	Cartridge 40 Amp	Cartridge 50 Amp
4.1	4100	2733	4100	8200
4.2	4200	2800	4200	8400
4.3	4300	2867	4300	8600
4.4	4400	2933	4400	8800
4.5	4500	3000	4500	9000
4.6	4600	3067	4600	9200
4.7	4700	3133	4700	9400
4.8	4800	3200	4800	9600
4.9	4900	3267	4900	9800
5	5000	3333	5000	10000
5.1	5100	3400	5100	10200
5.2	5200	3467	5200	10400
5.3	5300	3533	5300	10600
5.4	5400	3600	5400	10800
5.5	5500	3667	5500	11000
5.6	5600	3733	5600	11200
5.7	5700	3800	5700	11400
5.8	5800	3867	5800	11600
5.9	5900	3933	5900	11800
6	6000	4000	6000	12000

C

Figure 11–12 Voltage drop across a cartridge fuse.

Voltage Drop Across a Cartridge Fuse

Measurement mV	Cartridge 20 Amp	Cartridge 30 Amp	Cartridge 40 Amp	Cartridge 50 Amp
6.1	6100	4067	6100	12200
6.2	6200	4133	6200	12400
6.3	6300	4200	6300	12600
6.4	6400	4267	6400	12800
6.5	6500	4333	6500	13000
6.6	6600	4400	6600	13200
6.7	6700	4467	6700	13400
6.8	6800	4533	6800	13600
6.9	6900	4600	6900	13800
7	7000	4667	7000	14000
7.1	7100	4733	7100	14200
7.2	7200	4800	7200	14400
7.3	7300	4867	7300	14600
7.4	7400	4933	7400	14800
7.5	7500	5000	7500	15000
7.6	7600	5067	7600	15200
7.7	7700	5133	7700	15400
7.8	7800	5200	7800	15600
7.9	7900	5267	7900	15800
8	8000	5333	8000	16000

Courtesy of CARQUEST Technical Institute a Division of Advance Auto Parts.

D

Voltage Drop Across a Cartridge Fuse

Measurement mV	Cartridge 20 Amp	Cartridge 30 Amp	Cartridge 40 Amp	Cartridge 50 Amp
8.1	8100	5400	8100	16200
8.2	8200	5467	8200	16400
8.3	8300	5533	8300	16600
8.4	8400	5600	8400	16800
8.5	8500	5667	8500	17000
8.6	8600	5733	8600	17200
8.7	8700	5800	8700	17400
8.8	8800	5867	8800	17600
8.9	8900	5933	8900	17800
9	9000	6000	9000	18000
9.1	9100	6067	9100	18200
9.2	9200	6133	9200	18400
9.3	9300	6200	9300	18600
9.4	9400	6267	9400	18800
9.5	9500	6333	9500	19000
9.6	9600	6400	9600	19200
9.7	9700	6467	9700	19400
9.8	9800	6533	9800	19600
9.9	9900	6600	9900	19800
10	10000	6667	10000	20000

Courtesy of CARQUEST Technical Institute a Division of Advance Auto Parts.

E

Figure 11–12 Voltage drop across a cartridge fuse. (*continued*)

voltage drop. This includes any resistance that is built into the fuse in the circuit.

- As current flow is increased in a circuit, the voltage drop across a fixed resistance (in this case, the fuse) is also increased. This is due to the fact that the increase in current flow is always the result of less total resistance in the circuit to share in the voltage drop.
- If current is not flowing in a circuit, resistance does not drop voltage.

With the vehicle parked (with the ignition off and the doors closed) for an ample amount of time, measure the voltage drop across each fuse. To perform this task with cartridge fuses, the plastic cover must first be removed. If current flow exists through a fuse, there will be a very small amount of voltage drop across the fuse. This voltage drop is measured in millivolts (mV). If no current is flowing through the fuse, the voltage drop will be zero mV.

The charts in Figure 11–10, Figure 11–11, and Figure 11–12 are provided courtesy of CARQUEST Technical Institute and will provide the technician with a calculated current flow value without actually having to measure current flow. Note that the current flow is in milliamps (mA). Figure 11–13 has been added to show some voltage drop and current flow relationships with a maxi fuse. With any of the charts, if the measured voltage drop across a fuse is in excess of what is shown on the chart, you can use the existing values on the chart to calculate a greater current flow with a fair amount of accuracy. If there is no voltage drop across a fuse, as indicated by a reading on the DMM of 0.0 mV, there is no current flowing in the applicable circuit.

To determine the allowable parasitic draw for a specific vehicle, refer to the specification decal on the vehicle's battery and note the battery's *reserve capacity*. Reserve capacity is the amount

Voltage Drop Across a Maxi Fuse

Measurement mV	Maxi 20 Amp	Maxi 30 Amp	Maxi 40 Amp	Maxi 50 Amp	Maxi 60 Amp
0.1	30	50	80	100	160
0.2	65	100	150	190	270
0.3	95	155	220	280	380
0.4	130	205	290	370	490
0.5	160	255	360	460	600
0.6	195	305	430	550	710
0.7	225	360	500	640	820
0.8	260	410	570	730	910
0.9	290	460	640	820	1020
1.0	325	510	710	910	1130
1.1	355	565	780	1000	1240
1.2	390	615	850	1090	1350
1.3	420	665	920	1180	1460
1.4	455	715	990	1270	1570
1.5	485	770	1060	1360	1685
1.6	515	820	1130	1450	1800
1.7	550	870	1200	1540	1910
1.8	580	925	1270	1630	2020
1.9	615	975	1340	1725	2135
2.0	645	1025	1410	1820	2250

Current in milliamps (mA)

Figure 11–13 Voltage drop across a maxi fuse.

of time in minutes that a fully charged battery (at 80°F) can sustain a current draw of 25 amps without falling below 10.5 V.

Divide the battery's reserve capacity by 4 to obtain an approximate maximum allowable parasitic draw in milliamps. For example, if the battery's rated reserve capacity is 120 minutes, the allowable parasitic draw would be 30 mA (120/4 = 30).

A very few vehicles will have a higher parasitic draw than that which is reflected in their battery's reserve capacity. For example, a few late 1980s Lincolns had a normal parasitic draw of approximately 200 mA. And, more recently, some Chrysler vehicles were designed with an *Ignition Off Draw (IOD) fuse* that is to be removed by the vehicle's owner if the vehicle is to sit for more than a day or two. This is due to the fact that these vehicles may have a normal parasitic draw of up to 500 mA with the vehicle at rest.

THE THREE ESSENTIAL TOOLS OF ELECTRONIC SYSTEM DIAGNOSIS

There are three major tools that are essential for efficient driveability and electronic systems diagnosis. They are:

1. **Scan tool.** A scan tool is the most critical tool for electronic systems diagnosis. In today's world of electronic controls, you must be able to communicate with the on-board computers. However, you should recognize that a scan tool is used to quickly narrow down a problem area—it is not a pinpoint test tool.
2. **Exhaust gas analyzer.** An exhaust gas analyzer is a requirement for diagnosing concerns involving engine performance, fuel economy, and emission test failures. However, like a scan tool, it is not a pinpoint test tool. It is only used to help the technician quickly narrow down a fault area.
3. **DSO with a DMM capability and a current probe.** A lab scope with a built-in DMM capability, or a lab scope used with a separate DMM, is required to accurately pinpoint test concerns with modern electronic systems. Most technicians prefer to have a separate DMM. Many jobs do not require a lab scope, only a DMM for pinpoint testing. So, for most jobs, leave that pricey lab scope locked up securely in your toolbox and use the DMM. However, there are some pinpoint tests that absolutely require a lab scope, or at the very least, a graphing multimeter (GMM).

It should also be noted that, in addition to these three major pieces of diagnostic equipment, a low-amp current probe is also an important tool required for many diagnostics and is used in conjunction with either a DSO or a DMM. It can also be used in conjunction with a GMM.

Three-in-One Diagnostic Tools

Many tool companies, including Snap-on and OTC, market test equipment that combines three pieces of diagnostic equipment into one tool. The Snap-on Modis and the OTC Genisys can be configured to include a scan tool, a DSO, and an exhaust gas analyzer. While many technicians speak highly of these three-in-one tools, there are some points to consider before purchasing one:

- A three-in-one tool uses a *buttonlogy* that tends to be more menu-driven than dedicated stand-alone tools.
- What if the technician wishes to use more than one of these tools at the same time (such as analyzing exhaust gases while using a scan tool's bidirectional controls to command the PCM to enrich or lean out the fuel injectors)?
- All three tools are temporarily out of order if the tool breaks down and needs to be sent in for repair.

Regardless, if you have these three tools in your toolbox and understand the concepts covered in this textbook, you will excel when it comes to diagnosing electronic systems on the modern automobile.

OTHER DIAGNOSTIC RESOURCES

In the most difficult diagnostic situations, the odds are that someone else, somewhere else, has already encountered and solved the problem that you are striving to solve. This is particularly true for pattern failures. Following are some examples of resources that can potentially link you to a solution already discovered by another technician.

International Automotive Technicians' Network

The International Automotive Technicians' Network (IATN) can be a valuable resource in helping the technician to solve difficult problems. Accessible at http://www.iatn.net, this website offers many valuable tools, from technician interaction to manufacturer-provided information and other technical resources. IATN offers both a free membership and a sponsoring membership.

Identifix

An updated version of IATN, Identifix offers an online diagnostic information service designed to help the automotive service technician in diagnosing difficult problems. Accessible at http://www.identifix.com, this website is administered by Automotive Information Systems (AIS), Inc. (founded in 1987). AIS employs more than 30 factory-certified ASE master technicians.

World Wide Web

Finally, the World Wide Web (WWW) can sometimes be helpful in solving a difficult problem. Simply perform a Google search for a specific symptom on a specific make, model, and year of vehicle and you may find a helpful repair hint.

SUMMARY

This chapter is designed to help you put some of the information contained in the previous chapters of this book into perspective. No single chapter is going to provide a quick repair guide to all the problems encountered on today's complex vehicles. Rather, it takes an understanding of the concepts covered in the first chapters, in combination with the product-specific information covered in the later chapters, to effectively diagnose and repair today's modern systems. Even then, in spite of this author's best efforts, none of us will ever fully retain all of the system and product-specific information necessary for effective repair and diagnosis. Therefore, a source of product-specific information is as necessary as any other tool that you might use in your business. These service manuals may be paper or electronic, but they are very necessary.

Good technicians today cannot know it all. Rather, they must strive to understand the operating and diagnostic concepts associated with today's modern systems and then know where to look up the product-specific information when an unfamiliar system is encountered. If this textbook has provided the reader with an understanding of the underlying concepts and the knowledge required to comprehend the paper/electronic service manual literature, then our goal has been achieved.

▲ DIAGNOSTIC EXERCISE

A vehicle is brought in with complaints of surging and erratic idle. When the technician uses a scan tool to access the confirmed DTCs, the PCM displays DTCs pointing to the TPS circuit. Describe the steps a technician should follow between identifying the DTCs and deciding to replace the TPS.

Review Questions

1. *Technician A* says that the most efficient method to diagnose a fuel economy complaint is to begin by making pinpoint tests of the fuel delivery system, including fuel pressure and volume and the operation of the fuel pressure regulator. *Technician B* says that diagnosis of a fuel economy complaint should begin with taking the necessary steps to narrow down the area of the problem before beginning pinpoint test procedures. Who is correct?
   A. *Technician A* only
   B. *Technician B* only
   C. Both technicians
   D. Neither technician
2. Which of the following is the *most efficient* procedure to help you in narrowing down a problem area?
   A. Compare the symptom as described by the customer to your knowledge of how most similar systems work.
   B. Research the vehicle records in your shop from a similar vehicle that had a similar symptom.
   C. Begin by replacing parts until the symptom appears to be no longer present.
   D. Use the electrical schematic for the specific vehicle/system to help you understand the system; then compare the exact verified symptom to your understanding of system operation.

3. *Technician A* says that after you have attempted to verify the symptom, you should begin diagnosis by performing a thorough visual inspection of the system(s) associated with the symptom. *Technician B* says that a scan tool can be used to narrow down a problem area through DTCs and/or the data stream. Who is correct?
   A. *Technician A* only
   B. *Technician B* only
   C. Both technicians
   D. Neither technician
4. Which of the following is ***not*** a pinpoint test tool?
   A. Exhaust gas analyzer
   B. DMM
   C. Lab scope
   D. Breakout box
5. Which component in a circuit should be verified first when beginning circuit diagnostics?
   A. Source (battery)
   B. Circuitry
   C. Load
   D. Grounds
6. Which of the following can be used to reduce a technician's dependence on flowcharts?
   A. Breakout box
   B. TSB
   C. Electrical schematic
   D. DSO
7. Which of the following tests can be used to evaluate each cylinder's volumetric efficiency?
   A. Cranking compression test
   B. Running compression test
   C. Leak-down test
   D. All of the above
8. Which of the following can a fuel test kit be used to test?
   A. For excessive alcohol in the fuel
   B. The specific gravity of the fuel (indicative of the fuel's RVP)
   C. The fuel's octane
   D. Both A and B
9. *Technician A* says that voltage drop tests can be used to identify an intermittent fault within the circuitry, even when the symptom does not appear to be present. *Technician B* says that current ramping can be used to identify an intermittent fault within the load component, even when the symptom does not appear to be present. Who is correct?
   A. *Technician A* only
   B. *Technician B* only
   C. Both technicians
   D. Neither technician
10. A pre–OBD II oxidation catalytic converter is being tested using the HC-to-CO_2 conversion test. The following results are obtained following the cranking of the starter:
    Peak HC = 1,120 PPM
    Peak CO_2 = 13.1 percent
    What test result is indicated?
    A. Pass
    B. Fail, because the HC reading went too high
    C. Fail, because the CO_2 reading went too high
    D. Fail, because the HC reading did not go high enough
11. *Technician A* says that an exhaust leak ahead of an O_2 sensor can cause an OBD II PCM to falsely set a catalytic converter efficiency DTC. *Technician B* says that if an OBD II PCM has set a catalytic converter efficiency DTC, the technician should use the HC-to-CO_2 conversion test to test the converter before deciding if it should be replaced. Who is correct?
    A. *Technician A* only
    B. *Technician B* only
    C. Both technicians
    D. Neither technician
12. What specification should a smoke machine be set to when testing the EVAP system on a 2007 model year vehicle?
    A. .100 in.
    B. .040 in.
    C. .020 in.
    D. .010 in.

13. When narrowing down the area of an air-fuel ratio problem, which of the following can be used to verify the air-fuel *condition*?
    A. O_2 sensor
    B. Lambda, as measured with an exhaust gas analyzer
    C. Total fuel trim
    D. Both A and B
14. When narrowing down the area of an air-fuel ratio problem, which of the following can be used to verify the air-fuel *command*?
    A. O_2 sensor
    B. Fuel injector pulse width
    C. Total fuel trim
    D. Both B and C
15. When diagnosing an air-fuel ratio problem, a technician finds that the problem is one of electronic control. This means that the problem could be any of the following except which?
    A. Restricted fuel filter
    B. Faulty sensor or input value
    C. PCM power or ground fault
    D. PCM itself
16. When diagnosing an emission test failure, you should *baseline the exhaust gas levels*. What does this mean?
    A. Calibrate the exhaust gas analyzer.
    B. Warm up the exhaust gas analyzer and record the gas levels while it is sampling ambient air.
    C. Record the exhaust gas readings from the vehicle before making any repairs.
    D. Record the exhaust gas readings from the vehicle after the repairs are complete.
17. As a technician makes the required repairs to correct a CO emission failure, which of the following emission gases may be elevated?
    A. NO_x
    B. HC
    C. N_2
    D. All of the above
18. To see *engine out NO_x* when diagnosing an NO_x failure, you should use a scan tool's bidirectional controls to command the PCM to lean out the air-fuel ratio until the exhaust gas analyzer displays a lambda value that is at least as lean as which of the following?
    A. 0.95
    B. 0.98
    C. 1.02
    D. 1.05
19. All except which of the following are essential tools in diagnosing modern electronic systems?
    A. DSO with a DMM capability and a current probe
    B. Test light
    C. Scan tool
    D. Exhaust gas analyzer
20. *Technician A* says that when diagnosing a difficult problem on a modern automobile, resources such as the International Automotive Technicians' Network (IATN) and Identifix may be helpful in solving the problem. *Technician B* says that when diagnosing a difficult problem on a modern automobile, simply researching the problem on the World Wide Web may offer a solution. Who is correct?
    A. *Technician A* only
    B. *Technician B* only
    C. Both technicians
    D. Neither technician

SECTION TWO

MANUFACTURER-SPECIFIC CHAPTERS

Chapter 12

General Motors Computerized Engine Controls

OBJECTIVES

Upon completion and review of this chapter, you should be able to:

- ❑ Understand the inputs used with GM computerized engine controls.
- ❑ Describe the operation of GM fuel injection and idle speed control systems.
- ❑ Explain the operation of GM's throttle actuator control and active fuel management systems.
- ❑ Describe the operation of the GM spark management system.
- ❑ Understand the GM emissions control systems.
- ❑ Explain how GM's variable valve timing system functions.
- ❑ Explain the operation of a turbocharger and the turbocharger wastegate.
- ❑ Explain how GM's torque converter clutch lockup functions.
- ❑ Be familiar with the protocols that GM uses on its vehicles.
- ❑ Describe some of the newer electronic features found on GM vehicles.

KEY TERMS

Active Fuel Management (AFM)
Background Noise
Central Multiport Fuel Injection (CMFI)
Central Sequential Fuel Injection (CSFI)
Close-Coupled Catalytic Converter
Displacement on Demand (DOD)
EEPROM
Engine Metal Overtemp Mode
Normally Aspirated Engine
Torque Management
Wastegate

General Motors (GM) introduced an electronic engine control system in 1980 that was known as *Computer Command Control,* also called the *CCC system* or *C3 system*. Within the realm of the C3 system, GM had four different fuel management systems: feedback carburetion, throttle body injection (TBI), port fuel injection (PFI), and sequential fuel injection (SFI). The C3 system also encompassed two spark management systems: distributor ignition (DI) and waste spark electronic ignition (EI) systems. All C3 systems, regardless of the fuel and spark management systems used, provided similar scan tool functionality and flash code diagnostics, including initially a four- or five-pin, or, later on, a standardized 12-pin data link connector (DLC) under the instrument panel, originally known by General Motors as the *assembly line communications link (ALCL)* or *assembly line data link (ALDL)*. The 12-pin DLC was used in the C3 system from 1982 through 1995 when OBD II mandates standardized diagnostic protocol, marking the end of the C3 system.

Another GM system was a dual-injector TBI system used by Cadillac known as the *Digital Fuel Injection (DFI)* system. The DFI system allowed for scan tool diagnostics and also allowed for diagnostics to be performed through the driver's Electronic Climate Control Panel (ECCP).

Like all manufacturers, GM continues to develop engines and computerized engine control systems that use more advanced levels of technology. Modern GM vehicles incorporate sequential fuel injection (SFI) and gasoline direct injection (GDI) fuel management systems and waste spark, coil-near-plug (CNP), and coil-on-plug (COP) ignition systems.

POWERTRAIN CONTROL MODULE

The heart of the GM engine control computer system is the PCM (Figure 12–1). It is a much faster and more powerful computer than those previously used to monitor and control engines for emissions control, fuel economy, and driveability. In addition, the PCM is responsible for all the diagnostic functions required for compliance under OBD II regulations.

The PCM may be located above or near the glove compartment, under the instrument panel, behind the passenger kick panel, or even inside of a fender on older vehicles. On some Corvettes, it is in the battery compartment behind the driver's seat. On most modern GM vehicles it is located in the engine compartment.

Among the engine and vehicle functions the PCM controls are:

- Fuel mixture control
- Ignition control
- Knock sensor (KS) system (detonation control)
- Automatic transaxle shift functions
- Cruise control
- Generator (alternator) output
- Evaporative emission (EVAP) purge (canister purge)
- Exhaust gas recirculation (EGR)
- Air injection reaction (AIR)
- A/C clutch engagement control
- Radiator cooling fan control
- Traction control

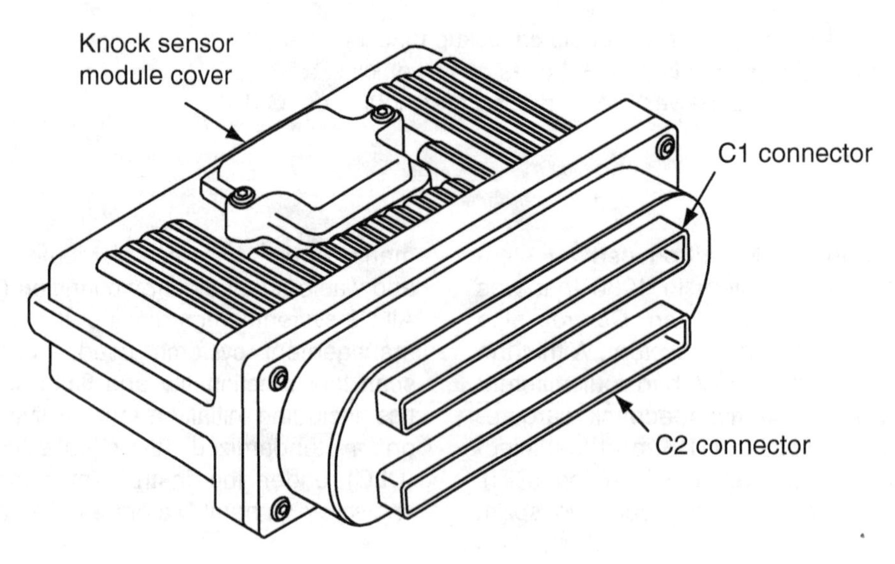

Figure 12–1 PCM.

PCM Learning

The PCM has a much more advanced learning capability than previous engine control computers. But if erroneous data has somehow become recorded in memory, the technician may no longer be able to merely disconnect the battery for a few minutes while the computer's volatile memory (keep-alive memory) deletes. Changes to the memory data may require the use of a dedicated scan tool to make the deletions. More detailed discrimination learning comes as the system performs various tests and experiments, while the vehicle is driven at part throttle with moderate acceleration. Modern PCMs also allow reprogramming of the control module with or without removal from the vehicle, a process called *flashing.*

Fuel Trim

Fuel trim is an adaptive fuel control strategy that concerns the control of the air-fuel ratio. Fuel trim is divided into two functions: Short-term fuel trim (STFT) and long-term fuel trim (LTFT). Originally, GM referred to the STFT as *integrator* and the LTFT as *block learn*. Fuel trim is discussed in greater depth in Chapter 4.

PCM Memory and the EEPROM

Instead of the fixed, unalterable PROM of previous years, the PCM has an electrically erasable programmable read-only memory (**EEPROM**) it uses to store a large body of information. While the EEPROM is soldered into the PCM and is not service-removable, it can store data, like the PROM, without a continuing source of electrical power. The PCM uses it as a storage bank for its throttle position/learned idle control tables, its transaxle adapt values, its transmission oil life information, its cruise control learning, and more. The EEPROM includes several areas available to store this data, and the PCM moves the data to a good location if there is damage to an earlier place. It will store a code if internal damage prevents the EEPROM from storing the information. The same data is stored in the keep-alive memory, so even if there is an EEPROM failure, the driver will probably not notice any difference. If the EEPROM defect code is set, the PCM must be replaced since the EEPROM is not removable.

PCM Self-Test and Memory Test

Like computers in earlier vehicles, this computer checks its internal circuits continuously for integrity. Likewise, it checks its EEPROM for the accuracy of its data. In computer terms, it tests the *checksum* of its files against what they are supposed to be and sets a DTC if they are different. Besides the hard-wired memory, the PCM also monitors its volatile keep-alive memory. If that has been improperly changed or deleted, the PCM sets a DTC. *Note:* Such a code will be set when the battery is disconnected for more than a brief instant.

Vehicle Identification IC Chips

Vehicle identification (VID) integrated circuit chips were introduced on GM vehicles in the late 1990s and are now commonly found in most major computers on a GM vehicle. Just as body panels such as fenders and hoods have had VIN labeling for some years now (to signify the vehicle for which they were originally manufactured), major computers on the vehicle are now labeled electronically as to the VIN number of the vehicle they belong to. This is designed to reduce the potential for "chop shops" to successfully sell or use the major computers from a stolen vehicle.

The VID chip is a "one-time-write" IC chip; once the vehicle's VIN information has been programmed into the chip, it cannot be erased or overwritten with new VIN information. When a module with a VID chip is "remanufactured," the original VID chip must be replaced with an electronically blank chip in order to allow the module to be used again.

If a module with incorrect VIN information is installed in a vehicle, the instrument panel may falsely illuminate most or all of the warning lamps or, at the very least, a DTC for an incorrect module will be set in memory. Unfortunately, this also means that a replacement module from a junkyard vehicle is not recommended for use as a replacement module.

Sensor Reference Voltage

The PCM also provides a reference voltage to the various sensors and switches that it uses to collect information on how the engine is running. To protect both the PCM and the sensor, this voltage is "buffered" in the PCM before it is sent to the switches and sensors. Many voltmeters, particularly analog voltmeters, may not give an accurate reading of this reference voltage because their impedance is too low. In order to make accurate measurements of reference voltage, or any other voltage throughout a modern system, use only digital voltmeters with a minimum of 10 MΩ input impedance—specification *J 39200*.

Torque Management

Besides the functions normally provided by an engine control computer, this PCM also works to control **torque management** through a reduction of engine torque under certain circumstances. Torque reduction is performed for three reasons:

1. To prevent overstress of powertrain components.
2. To limit engine power when the brakes are applied.
3. To prevent damage to the vehicle or powertrain from certain abusive driving maneuvers.

To calculate whether to employ its torque-limiting functions, the PCM analyzes manifold pressure, intake air temperature (IAT), spark advance, engine speed, engine coolant temperature (ECT), the EGR status, and the engagement or disengagement of the A/C clutch through its control circuit (Figure 12–2). It also considers

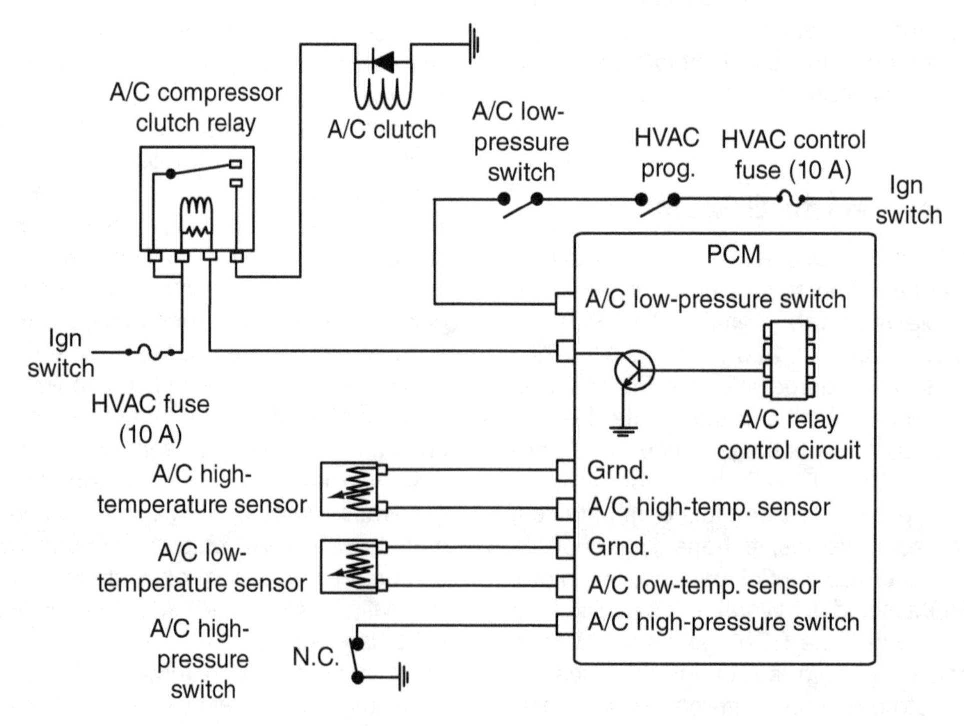

Figure 12–2 A/C compressor clutch control circuit.

whether the torque converter is locked up, what gear the transmission is in, and whether the brakes are applied. When the numbers add up to torque reduction, the PCM's first strategy is to reduce spark advance to lower engine torque output. In some more extreme cases, the PCM can also shut off fuel to one or more cylinders to further reduce engine power.

There are several occasions when **torque management** is likely to come into play:

- During transaxle upshifts and downshifts
- When there is hard acceleration from a standstill (traction control)
- If the brakes are applied simultaneously with moderate to heavy throttle
- If the driver initiates abusive or stress-inducing actions, such as shifting gears at high open throttle angles

In the first two cases, the operation of the torque management will probably not be noticeable.

Traction Control

The PCM is also responsible for the vehicle's traction control system. If the computer learns from the wheel speed sensors that the drive wheels are slipping during acceleration, it can individually apply the appropriate brake to prevent spin. In extreme cases, the torque management system can reduce engine torque as well, disabling as many as seven of the fuel injectors to do so. This strategy will not be adopted, however, if any of the following conditions occur:

- ECT is below 40°F.
- Engine coolant level sensor indicates the coolant's level is low.
- The engine is at a speed below 600 RPM.

If the traction control system is disabled for any of the preceding reasons or for a system failure, the TRACTION OFF light will come on in the driver's information panel.

The torque management system can sometimes have surprising consequences. If a car is running under cruise control and encounters slippery pavement—perhaps slick ice—the wheel speed sensors will report extremely rapid acceleration. The traction control, among its other measures, will shut off the cruise control. Not only will it shut off the cruise control, but it will also disable it, set a code, and leave the subsystem disabled until the code is deleted by a technician. Various other problems, such as a slipping transaxle, can have the same effect.

Diagnostic & Service Tip

While the computers on these vehicles have become much more complex than earlier, simpler, automotive computers, they have also become much more reliable. Despite the widely held belief that almost any system problem, such as an engine performance problem, is caused by "the computer," the automotive computer itself is rarely the cause of problems associated with the system. Furthermore, many problems associated with the computer itself are caused by a technician performing tests improperly and, therefore, causing damage to the computer.

If you believe that the computer is responding incorrectly to its inputs, it is important to check for the 5 V reference signal at a sensor that uses it. This also tests how much voltage is getting to the computer's internal logic gates and memories. It is also important to test the voltage drop of the power and ground circuits that power up the computer. Positive-side voltage drop should not exceed 100 mV and ground side voltage drop should not exceed 50 mV. Checking voltage drop rather than resistance is a much more revealing test when trying to eliminate problems. Learn how to use voltage drop tests instead of resistance tests if you plan to do successful work on engine control computers.

Information Functions

The PCM also works with the BCM as well as other control modules to provide the driver with a variety of informational messages. These messages have to do with everything from engine and transmission oil life to remaining coolant level or windshield washer fluid level.

INPUTS

Heated Oxygen Sensor

Bias Voltage. Many GM systems use a PCM-generated 450 mV signal, known as a *bias voltage*. The PCM places the bias voltage on the oxygen sensor signal wire whenever the engine is started and until the oxygen sensor (O_2S) begins to generate its own voltage. The PCM's voltage-sensing circuits continue to watch this voltage signal. As the engine is warming up, eventually the voltage begins to exceed the 450 mV bias voltage placed on this circuit by the PCM, at which point the PCM knows that the O_2S is hot enough to begin generating its own voltage. The PCM will then remove the bias voltage signal and use the voltage that the O_2S is placing on this circuit. It is at this time that the PCM can slip into closed-loop fuel control. Therefore, on these systems, when watching the O_2S voltage values with a scan tool immediately after a cold engine start, expect to see a value of 450 mV, even before the oxygen sensor is hot enough to operate.

Oxygen Sensor Heater. Modern GM systems use heated O_2 sensors, as shown in Figure 12–3, so they can be brought up to operating temperature quickly. The PCM can control the temperature of the sensor's heating element by controlling the ground side of the heater with a high-frequency pulse-width-modulated signal.

Pre-Cat Oxygen Sensor. The primary O_2S is placed in front of the catalytic converter and is referred to as a *pre-cat* oxygen sensor. It is also sometimes referred to as an *upstream* oxygen sensor. The pre-cat O_2S has the greatest influence on air-fuel ratio control.

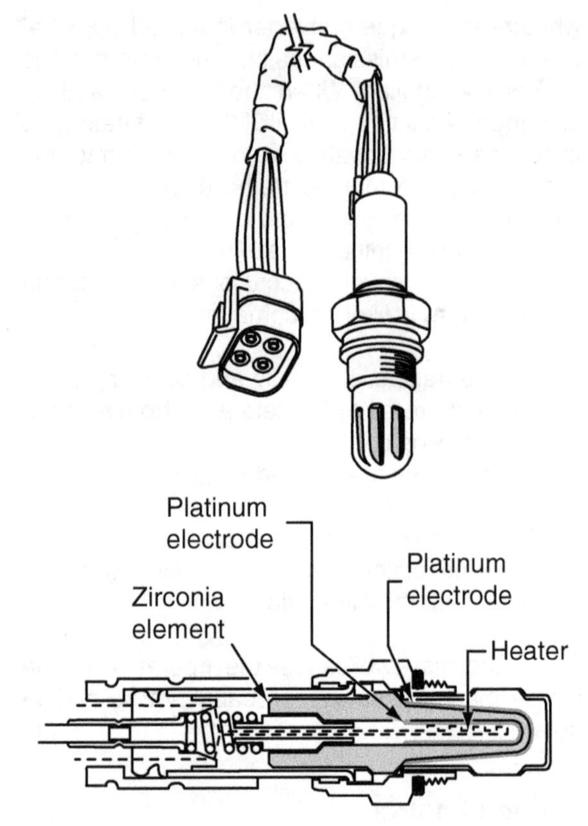

Figure 12–3 Heated oxygen sensor (HO_2S).

Post-Cat Oxygen Sensor. OBD II systems use a second oxygen sensor, functionally equivalent to the first, placed downstream of the catalytic converter. Its primary function is to check on the efficiency of the converter. If all is well with the catalytic reduction and oxidation processes, things will be fairly uneventful at the **post-cat oxygen sensor**. There should be little or no change of output voltage, since the converter will have used most of its oxygen to oxidize the carbon monoxide and hydrocarbons. When the post-cat oxygen sensor begins to mimic the signal of the pre-cat sensor, the catalytic converter is operating at a reduced efficiency. The PCM, at a predetermined level of inefficiency, will set a fault code and illuminate the MIL. An engine with dual exhaust pipes and dual catalytic converters will have dual post-cat sensors, one after each converter.

The post-cat O_2S also influences the PCM's control of the air-fuel ratio. This is known as *rear fuel control*. Therefore, once the engine is operating in closed-loop fuel control, any exhaust leak ahead of the post-cat sensor can cause the PCM to drive the air-fuel ratio overly rich, even to the point of an engine stall.

Multiple Upstream Oxygen Sensors. Many modern engines use two **primary oxygen sensors,** one on each cylinder bank. The individual oxygen sensors, obviously, report on the air-fuel mixture in the cylinders on their respective sides, but there is nothing in principle different from a single upstream sensor system.

O_2 Sensor Monitoring. The computer's trouble code recording capacity includes monitoring of each O_2 sensor, including both upstream sensors and the downstream sensor(s), and it also monitors each sensor's heater circuit separately. The PCM can also force the air-fuel ratio abnormally lean and rich in order to test the pre-cat sensor(s) and the post-cat sensor(s) for high voltage, low voltage, and response time in accordance with OBD II requirements. To test the O_2 sensor heaters, the PCM monitors how long following a cold engine start it takes for the sensors to begin generating a working voltage. The PCM also monitors the voltage drop across the heater element to determine the heater's electrical resistance and monitors the heater's current draw. Any failure will result in a specific DTC being set.

Wideband Air-Fuel Ratio Sensor. GM is using wideband air-fuel ratio sensors to report air-fuel ratios from 10:1 to ambient air. While these sensors can be found on their SFI engines, they are commonly used on their GDI engines that have the ability to operate with a stratified air-fuel charge. Initially, the wideband sensors replaced the pre-cat sensors only, but on newer engines they have now replaced the post-cat sensors as well. GM's reference voltage to these sensors is 2.6 V.

Intake Air Temperature Sensor

The intake air temperature (IAT) sensor is an NTC thermistor, a variable resistor that varies its resistance depending on its temperature. Receiving a 5 V reference voltage from the PCM, it modifies the signal corresponding to the temperature of the incoming air. It works similarly to other NTC temperature sensors. As the sensor's temperature increases, both the sensor's resistance and voltage signal decrease. The IAT sensor influences air-fuel mixture, spark timing, and idle speed control.

Engine Coolant Temperature Sensor

The engine coolant temperature (ECT) sensor, shown in Figure 12–4, informs the PCM of the temperature of the engine's coolant, which is important to the control of both the fuel management system and ignition timing. It can also affect the control of some emission systems. The PCM also monitors the ECT sensor during engine cranking to determine the required air-fuel ratio.

Because, on GM engines, the PCM can disable cylinders in the event of a total loss of coolant in order to protect the engine from high temperature (covered later in this chapter), many systems use a dual-range coolant temperature sensor. Within the PCM is a 3.65 kΩ resistor in series with a 348 Ω resistor (Figure 12–5). On a cold engine, 5 V (reference voltage) are passed through both of these resistors before being passed out to and through the ECT sensor itself. The circuit is then

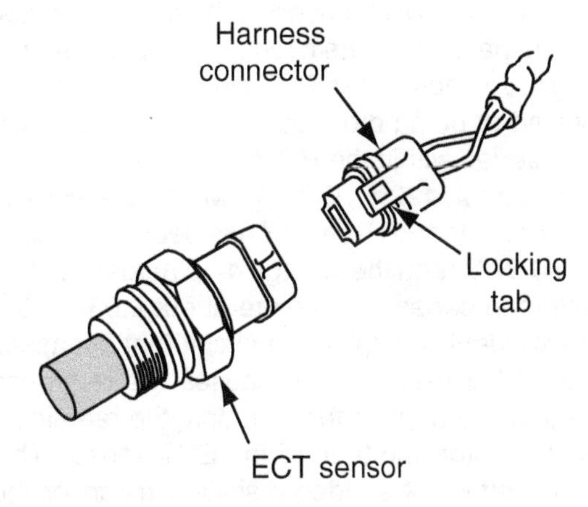

Figure 12–4 ECT sensor.

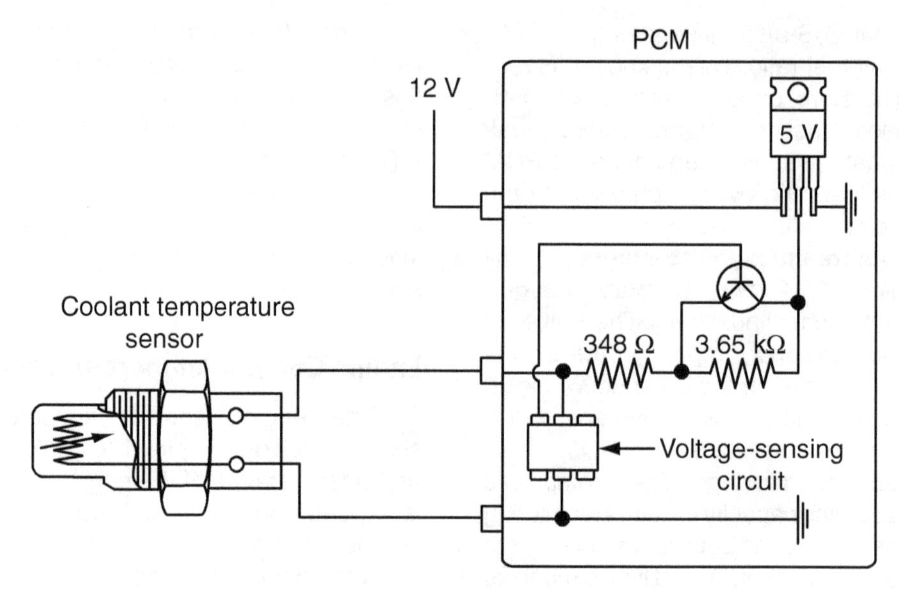

Figure 12–5 General Motors' dual-range temperature sensor circuit.

connected to ground through the PCM. As with a standard ECT circuit, the PCM's voltage-sensing circuit monitors the voltage drop across the ECT sensor. As the engine coolant warms, the resistance of the ECT sensor's thermistor decreases, thereby also decreasing the voltage drop across the ECT sensor as measured by the PCM. However, this also means that the voltage drop across the two resistors in series within the PCM is increasing. That is because all of the applied reference voltage will be used up by all of the resistance in the circuit if the circuit is complete. As the voltage drop across the ECT sensor decreases, it is shifted to being dropped across the two resistors in series within the PCM.

At about 122°F (50°C), when the voltage drop across the ECT sensor has been reduced to about 0.97 V (and the voltage drop across the two resistors in series is therefore about 4.03 V), the PCM suddenly completes a circuit that bypasses the 3.65 kΩ resistor. This causes the reference voltage to be applied through only the remaining 348 Ω resistor and then to the ECT sensor. This has the effect of suddenly shifting much of the voltage drop to across the ECT sensor, thereby suddenly increasing the PCM's measured voltage to a value closer to reference voltage (Figure 12–6). As the engine continues to warm, the PCM will see the voltage drop across the ECT sensor continue to decrease again.

Ultimately, this has the effect of extending the range of temperature that the PCM can measure into the *overtemp* range for the purpose of being able to control engine temperature in the

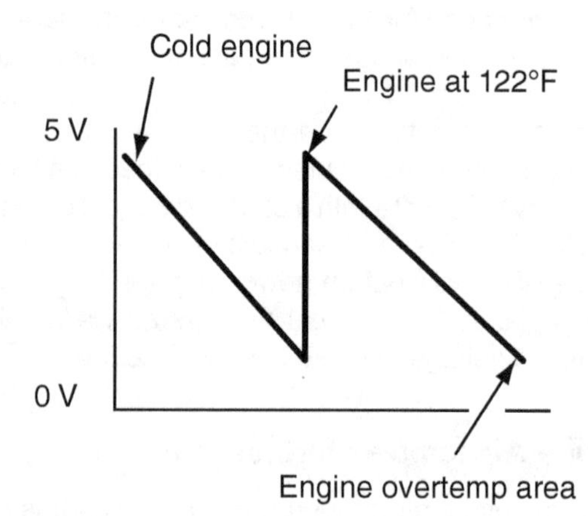

Figure 12–6 Dual-range ECT sensor waveform.

event of a total loss of coolant. The ECT sensor affects control of the air-fuel mixture, spark timing, spark knock control (on some engines), engine idle speed, and torque converter lockup clutch actuation. The information from the ECT sensor also determines the operation of the radiator cooling fan.

One of the internal audit tests the computer automatically performs is to check to see whether the engine is warming quickly enough. After the engine has run 4.25 minutes, it checks to see that the coolant temperature is 41°F (5°C) or greater. If it is not, or if the coolant temperature falls below that temperature for more than three seconds, a code is set. The timer ratchets backward during torque management or traction control activity. Ordinarily, the only reason that an engine does not reach this temperature within the specified time is that a thermostat is stuck open. This assumes, of course, that the engine has been run under moderate load during the elapsed time. If someone merely started an engine at –40° (F or C) and let it idle for 4.25 minutes, it might not gain the 80 degrees (F), and a code would be set.

Throttle Position Sensor

The PCM uses information from the throttle position sensor (TPS) for calculating idle speed, fuel mixture, spark advance, deceleration enleanment, and acceleration or WOT enrichment (Figure 12–7). The PCM also uses the TPS input during engine cranking to determine whether or not it needs to initiate *clear flood mode*.

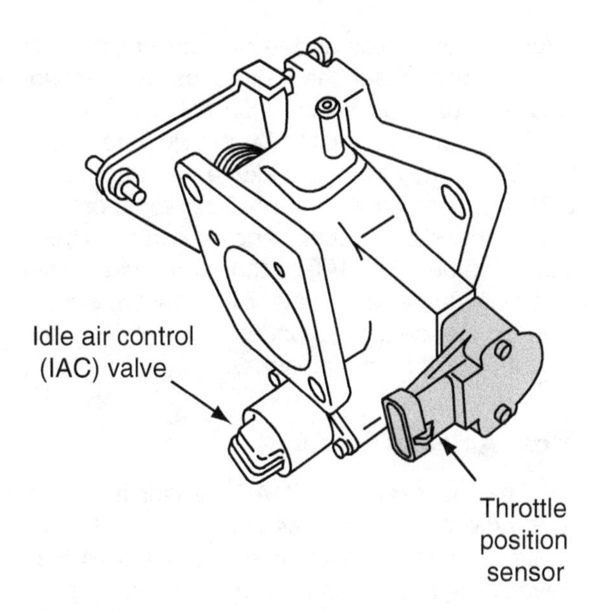

Figure 12–7 Throttle position sensor.

The TPS returns a proportionate signal: The lower the return voltage, the lower the throttle angle. The computer also compares (1) changes in throttle position with its internal clock to determine how quickly the throttle has been changed and (2) TPS with MAP/MAF sensor information: Any significant discrepancy indicates one or both circuits are faulty.

TPS Adjustment. The TPS on some older PFI systems is adjustable. Performing a correction of the minimum idle speed adjustment is cause to perform a TPS adjustment. Adjustment is checked by connecting a voltmeter between terminals A and B at the sensor connection with the ignition on. The technician can access the terminals for this test by disconnecting the three-wire weather-pack connector and inserting three short jumper leads to temporarily reconnect the circuit.

The TPS on modern GM systems is said to be self-adjusting. What that actually means is that it is not adjustable, but the computer learns what to expect from it and adjusts its calculation tables accordingly. For the self-adjustment, the computer assumes that the throttle is closed when the key is turned off with the throttle position signal at the same value twice in succession.

MAP and MAF Strategy

While modern vehicles of most makes other than GM use either a strict manifold absolute pressure (MAP) sensor and speed density calculation or a strict mass air flow (MAF) measurement with regard to engine load, GM vehicles are unique in this area. Many GM engines use a combination MAF/MAP strategy to monitor engine load. The

MAF sensor serves as the primary engine load sensor, but at times the PCM uses MAP sensor values to back up engine load calculations. The MAP sensor is also used by the PCM to help it test EGR flow rate and leakage as specified by OBD II standards. Also, an IAT sensor is present to help the PCM calculate engine load from MAP sensor values. The IAT sensor is not necessary when using the MAF sensor to measure engine load for fuel control purposes, but it is still used by the PCM for other reasons.

Mass Air Flow Sensor

The mass air flow (MAF) sensor is a **hot-wire-type sensor** that measures the total mass of air entering the intake manifold (Figure 12–8). It works by maintaining the hot wire at a specified temperature above ambient temperature. A second wire senses the ambient temperature. The hot wire is placed in the air-stream and cooled by the intake air. The current required to keep the wire at the specified temperature corresponds closely to the amount of air entering the intake manifold by mass. Internal circuits in the MAF sensor convert the current required to keep the hot wire at the specified temperature into a frequency signal sent to the PCM. The resulting output of this sensor can be measured with a Hertz (Hz) meter and will vary from around 1,500 to 1,600 Hz at idle to around 6,000 Hz under heavy engine load. This information is used in the calculation of fuel injector pulse width.

Manifold Absolute Pressure Sensor

The PCM gets engine load and barometric pressure information from the MAP sensor (Figure 12–9). It receives a 5 V reference signal from the PCM, reduces it in accordance with the sensed pressure, and returns the reduced voltage as an indication of manifold pressure. The MAP voltage signal is an analog voltage that is reduced

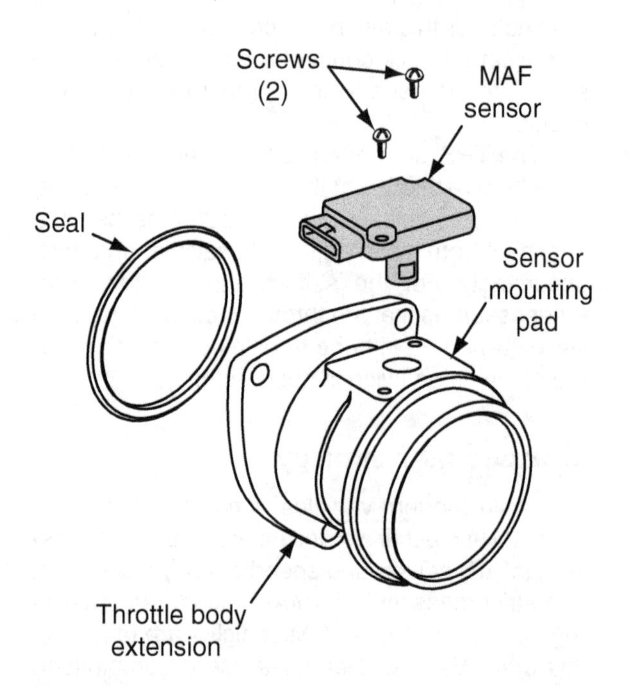

Figure 12–8 MAF sensor.

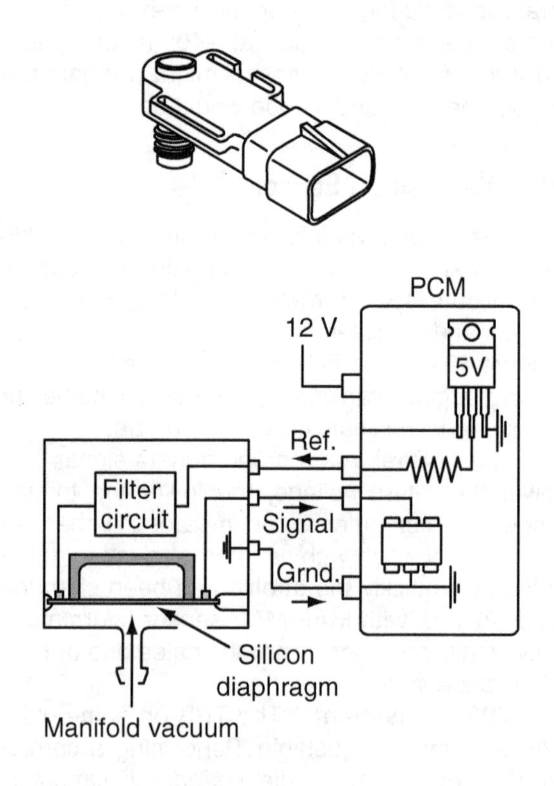

Figure 12–9 MAP sensor.

with reduced manifold pressure (high vacuum) and is increased with a rise in manifold pressure (loss of vacuum). The PCM compares the MAP signal against the throttle position sensor (TPS) signal, looking for discrepancies between them that would indicate one or the other is faulty. It will notice if throttle position changes as little as 3.2 degrees in 0.5 second. If the MAP sensor's signal does not indicate a change of at least 4 kilopascals (about 0.6 PSI) during the same period, the PCM will set a trouble code for the MAP sensor.

While this sensor is referred to as a MAP sensor, it is also used by the PCM to update BP values. When the ignition is first turned to run, but before it is turned to start, the PCM takes a reading from the MAP sensor. Since the engine is not running, the manifold pressure is identical to atmospheric pressure. The PCM records this reading as the base barometric pressure and uses it for calculations of the fuel mixture while operating in open loop and also for calculating spark timing. This BP reading is retained and used until the engine is restarted or until the throttle is opened to WOT. At that time manifold pressure increases to nearly atmospheric pressure again, and an updated BP reading is taken.

The computer uses MAP sensor information to calculate spark advance and fuel mixture, including acceleration enrichment, and sometimes to determine whether to employ torque limitation measures.

Tachometer Reference Signals

Tachometer reference (or *tach reference*) signals originate at a crankshaft position (CKP) sensor and/or a camshaft position (CMP) sensor. On earlier systems they originated at a distributor pickup. On an engine with an even number of cylinders (4-, 6-, 8-, or 10-cylinder engine) these signals are used as necessary to provide the following levels of information to the PCM:

- First level of tach reference information: This identifies for the PCM that *a pair of pistons is approaching TDC*. Only this level of information is required for the PCM to properly control a distributor ignition (DI) system, a throttle body fuel injection (TBI) system, or an electronic feedback carburetor. (On these systems the PCM does not need to identify which pair of pistons is approaching TDC.) This level of information originates at the CKP sensor.
- Second level of tach reference information: This identifies for the PCM *which pair of pistons is approaching TDC*. This level of information is needed for the PCM to properly control a waste spark electronic ignition (EI) system in order for the PCM to correctly choose which ignition coil to fire. (On this system the PCM does not need to identify which cylinder is on compression in order to fire the correct ignition coil.) This level of information most often originates at the CKP sensor.
- Third level of tach reference information: This identifies for the PCM *which cylinder is on the compression stroke*. This level of information is needed for the PCM to properly control a coil-on-plug (COP) or coil-near-plug (CNP) electronic ignition system. It is also required for the PCM to properly control a multipoint sequential fuel injection (SFI) system or a gasoline direct injection (GDI) system. This level of information originates at the CMP sensor because it cannot be obtained from a CKP sensor on a four-stroke engine. (The CKP signal will always look identical on both rotations.) During an engine start, an SFI system, when used with a waste spark EI system, may pulse the fuel injectors as a group initially from the CKP signal (known as a *primer pulse*) until the CMP sensor is seen to produce a signal and then switch over to sequential injection. However, due to timing requirements, a GDI engine, or an engine equipped with COP/CNP, must wait for the CMP signal before the PCM can start the engine.

The first level of information from the CKP sensor is the most critical signal. While the second and third levels of information are only used to identify which ignition coil to fire or which fuel injector to pulse, the actual firing/pulsing of these components is always timed and performed according to the first-level signal coming from the CKP sensor. If this signal is lost, the engine will immediately shut down and there will be no spark or fuel injection pulses present.

It should be noted that some GM engines equipped with a waste spark EI system and SFI (such as Saturn) artificially created a third-level signal without having a physical CMP sensor. On these systems, during the initial engine start, the PCM fired a waste spark ignition coil from the second-level CKP signal. As the coil fired, it monitored the voltage drop across each of the two spark plugs to determine which cylinder was on compression. The PCM then used this information to artificially create a third-level CMP signal and then switched the fuel injection system from group injection over to sequential injection. On this application, if the secondary wires were to become switched at the ignition coil, a DTC for a CMP signal would be set in the PCM's memory, even though the engine had no physical CMP sensor.

Early GM Tach Reference Sensors. Figure 12–10 shows a harmonic balancer with three equally spaced and equally sized vanes and a Hall effect sensor used on early GM Computer Controlled Coil Ignition (C3I) systems with an SFI fuel management system. This CKP sensor only provided the first level of tach reference information. The second and third levels of information originated at the CMP sensor. Therefore, on this application, the PCM could not fire an ignition coil until the CMP signal was seen. This could equate to as much as two rotations of the crankshaft.

Figure 12–11, Figure 12–12, Figure 12–13, and Figure 12–14 show a reluctor wheel and CKP sensor that identifies both the first and second levels of information for the PCM, due to an added synchronization notch (*sync* notch) designed into the reluctor wheel. The CKP sensor is a variable reluctance sensor (VRS) or a permanent magnet

Figure 12–10 Hall effect crankshaft sensor and harmonic balancer with three vanes and three windows.

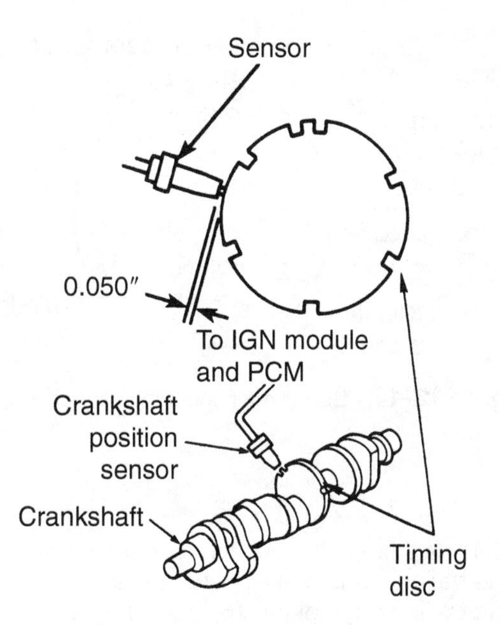

Figure 12–11 Crankshaft reluctor wheel.

design. However, the third level of information was still provided by a CMP sensor. GM used this design on the Direct Ignition System (DIS) and Integrated Direct Ignition (IDI) systems over many years on both GM and Isuzu vehicles.

Figure 12–15 shows a variation of the C3I CKP sensor that essentially moved the second level of information from the CMP sensor to originate at the CKP sensor. This allowed the engine to start faster since the information required to fire a waste spark ignition coil was now determined within 120 degrees of a crankshaft rotation and the fuel injectors were primer pulsed until the CMP signal was seen.

Figure 12–16 and Figure 12–17 show an optical pickup in an ignition distributor (used through 1997). This sensor allowed the PCM to determine the third level of information because a distributor is geared into the engine's camshaft and,

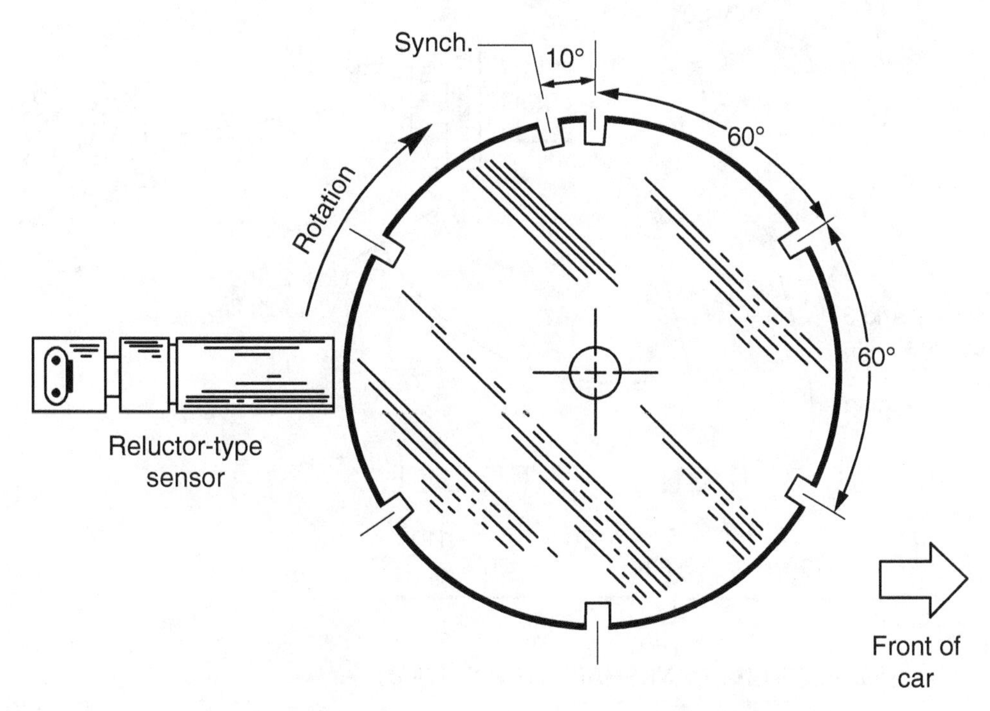

Figure 12–12 Crankshaft position sensor.

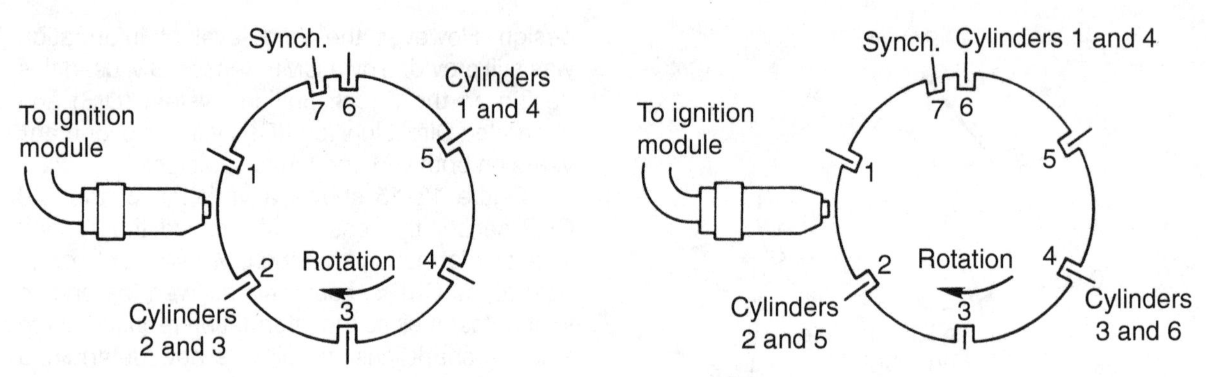

Figure 12–13 Four-cylinder coil-firing sequence.

Figure 12–14 Six-cylinder coil-firing sequence.

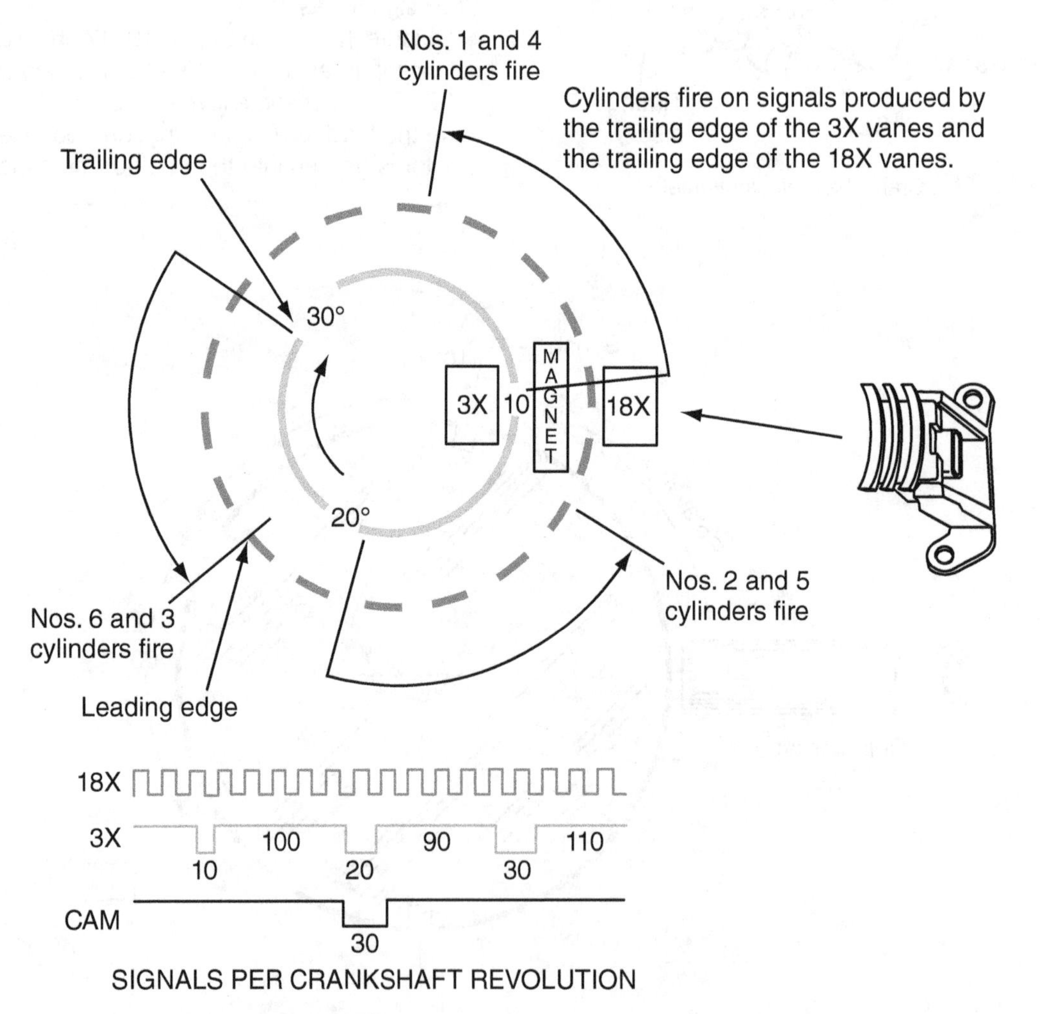

Figure 12–15 Combination Hall effect sensor for C3I (type III) fast start.

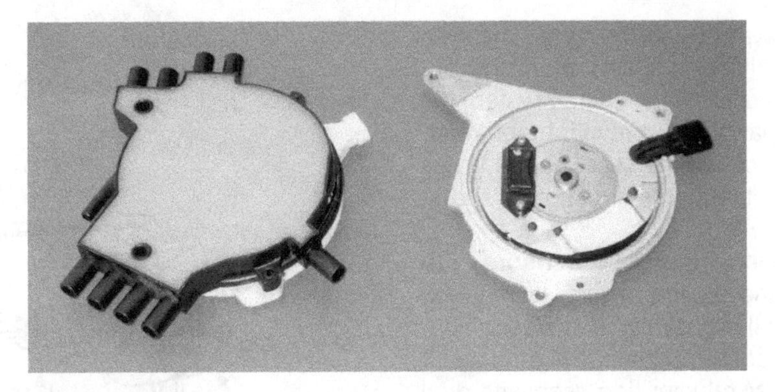

Figure 12–16 Opti-Spark distributor.

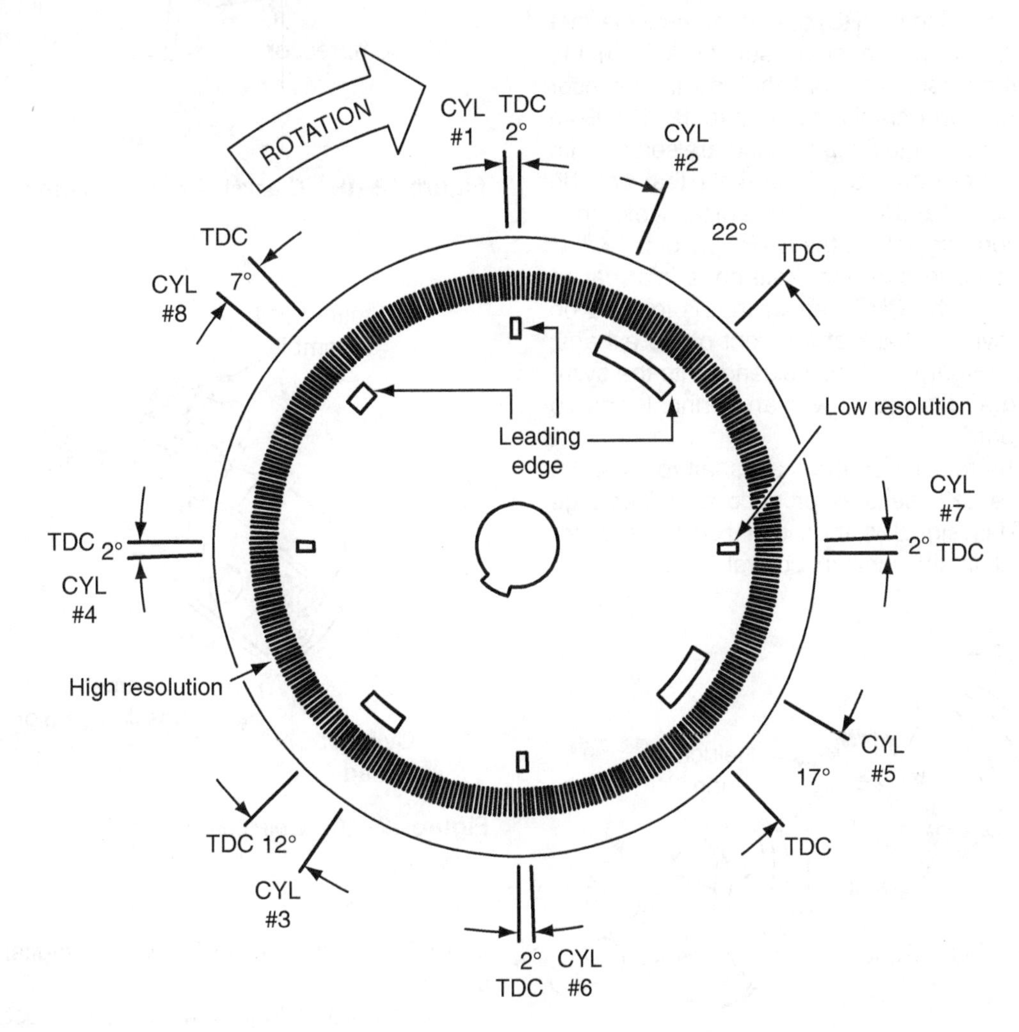

Figure 12–17 Opti-Spark sensor disk.

therefore, operates as a CMP sensor. The only purpose for the oddly sized slots was for the SFI fuel management system. The ignition system did not require more than the first level of tach reference information because the secondary components were responsible for distributing the spark, not the PCM.

Modern GM Tach Reference Sensors. To be able to determine engine misfire as required under current OBD II standards, GM uses a high-resolution CKP sensor arrangement. Figure 12–18 shows a modern GM reluctor ring used with dual CKP sensors. The CKP sensors are variable reluctance sensors (VRS) or a permanent magnet design. Crankshaft position sensor A is in the upper crankcase, and crankshaft position sensor B is in the lower crankcase (Figure 12–19). Both sensors extend into the block and are sealed with O-rings. There are no adjustments that can be made. The first and second levels of tach reference information originate at the CKP sensors.

The third level of tach reference information originates at the CMP sensor that is located on the rear cylinder bank at the front of the exhaust camshaft (Figure 12–20). It extends into the cylinder head and is sealed with an O-ring. It has no adjustments.

As the reluctor on the crankshaft rotates, the VRS-style CKP sensors produce an AC voltage signal. This signal is then digitized by a buffer circuit within the ignition control module (ICM) and is then sent to the PCM as a digital on/off signal.

As shown in Figure 12–21, the reluctor ring has 24 evenly spaced notches and an additional

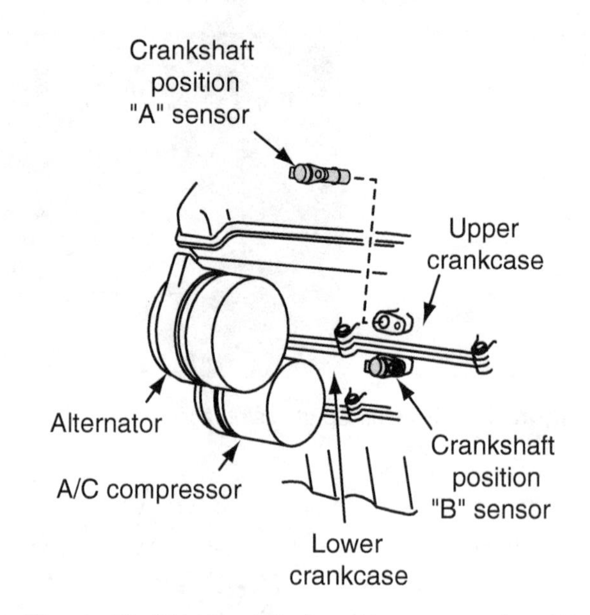

Figure 12–19 Crankshaft position sensors A and B.

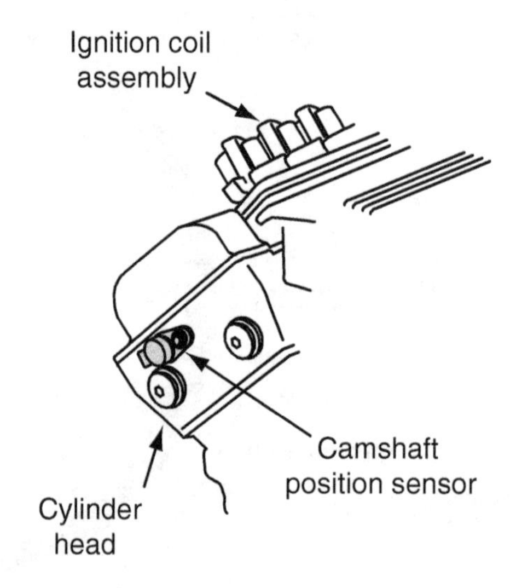

Figure 12–20 CMP sensor.

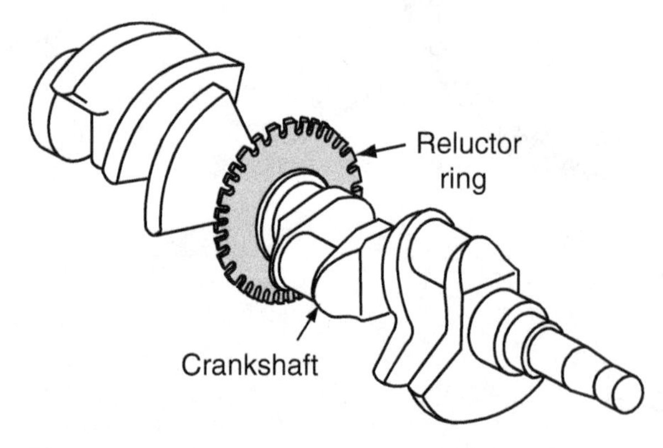

Figure 12–18 Crankshaft reluctor ring.

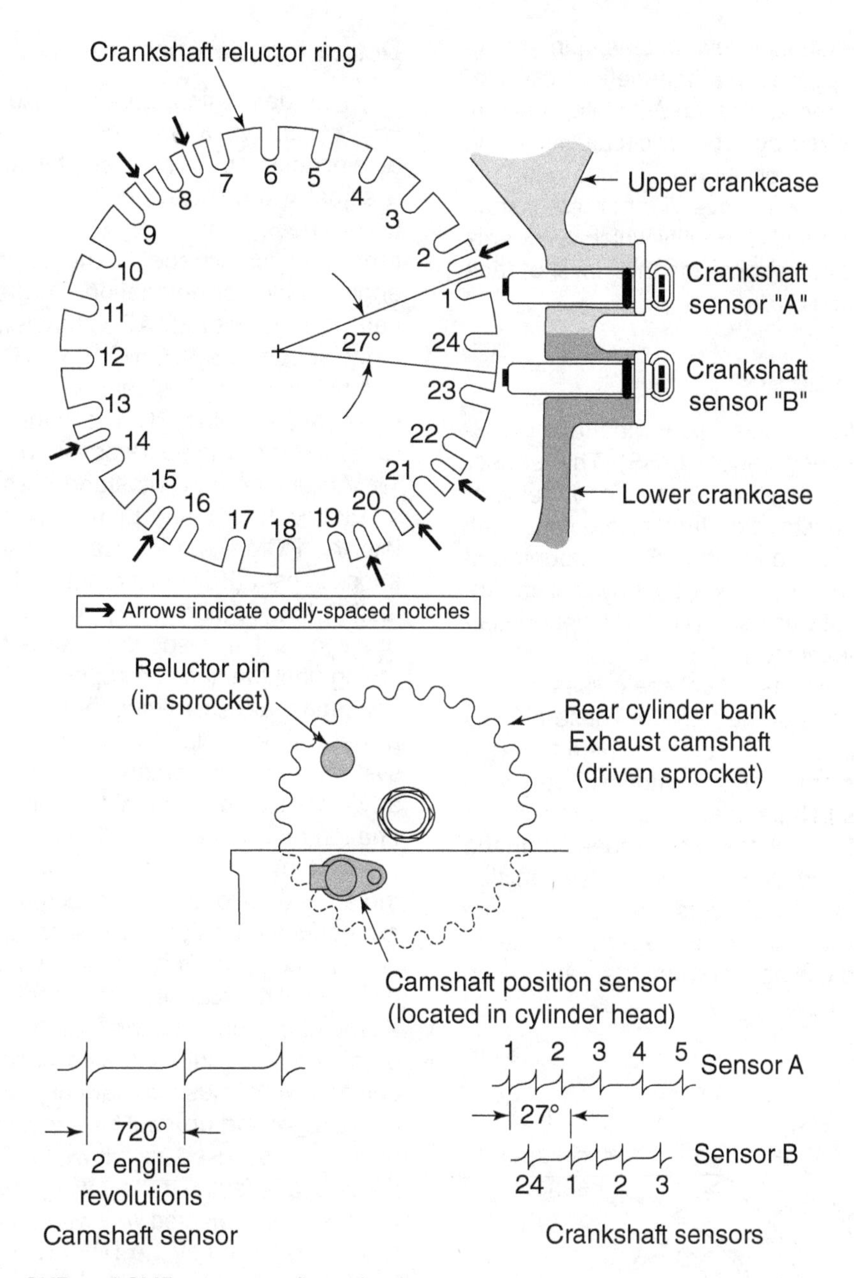

Figure 12–21 CKP and CMP sensor configurations.

eight unevenly spaced notches for a total of 32 notches. As the crankshaft rotates once, each sensor reports 32 on/off signals. The CKP sensors are positioned 27 degrees of crankshaft rotation apart and, therefore, create a pattern of on/off signals that enables the ignition control module (ICM) to identify the crankshaft's position in order to identify which pair of pistons are approaching TDC (the second level of tach reference information).

As the camshaft turns, a steel pin on its drive sprocket passes the magnetic pickup of the sensor. This generates an AC voltage signal that is also digitized by a buffer circuit within the ICM. This digital on/off signal, in time with the camshaft, occurs once every other crankshaft revolution and is used to identify which cylinder is on the compression stroke (the third level of tach reference information).

Vehicle Speed Sensor

Modern GM vehicles use a magnetic (VRS-style) vehicle speed sensor (VSS). This sensor is mounted in a position to monitor a reluctor located at the transmission/transaxle output shaft (Figure 12–22). (Because the VSS is mounted at the transmission or transaxle, the physical speedometer cable that was used with the older optical-style VSS is eliminated.)

The VSS produces AC voltage pulses proportional to the speed of the vehicle. These signals are fed to a buffer, which converts them to 4,004 digital pulses per mile. (This number will vary with specific models.) This signal is then transmitted to the PCM, which calculates vehicle speed from the frequency of the signal. Four-wheel-drive vehicles may have two similar sensors, one for transmission output and one for vehicle speed, to enable the computer to distinguish when the vehicle is in the low transfer case range.

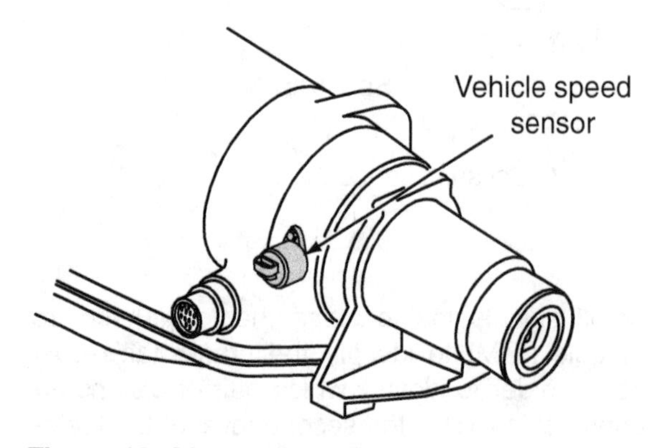

Figure 12–22 Vehicle speed sensor.

Detonation Sensor

The detonation sensor, also known as a *knock sensor (KS)*, used on GM systems is a piezoelectric silicon crystal that generates a signal when excited by vibrations characteristic of knock frequencies (Figure 12–23). It generates an AC voltage frequency in response to engine knock or detonation. The resulting signal can be read with an AC voltmeter, Hertz meter, or lab scope (DSO). On modern GM systems, the detonation sensor's signal is a direct input to the PCM, allowing the PCM to consider detonation when determining spark advance.

An engine can withstand a small amount of spark knock for a short time, and this is factored into the PCM's spark advance calculations. The knock sensor (KS) must report a detonation event lasting longer than 99.99 ms for three continuous cycles. The response then is to retard spark timing until the knock disappears. The basic ignition (spark advance) map for the engine assumes a fuel with an octane equivalence of 87, but the system can manipulate spark (more or less advanced) to accommodate a higher or lower detonation resistance with different fuels. In general, the system will drive the spark advance right to the edge of detonation to maximize fuel economy and reduce emissions problems.

The computer's KS is also used to detect KS *circuit* faults. Because the sensor is fundamentally a type of microphone, the PCM is programmed to be able to sort out what kind of noise is detonation and what is just ordinary engine noise, known as **background noise. The** expected background noise can be used to allow the PCM to detect some circuit faults. If the background noise does not rise in an expected way with engine speed and load, as reflected by the TPS and engine RPM for a given period of time, it will record a KS circuit fault even if there is no detonation occurring.

Power Steering Pressure Switch

A power steering pressure (PSP) switch is used on GM systems to alert the PCM when P/S

Figure 12–23 Detonation sensor.

pressure is high enough to significantly affect engine idle performance due to the momentary high load. Normally the PSP switch is closed, but when hydraulic boost pressure in the power steering system rises to 600 PSI, it opens. The purpose of the PSP switch is to instruct the PCM to increase idle speed slightly to compensate for the additional load during parking maneuvers. On some systems, the air-conditioning (A/C) compressor clutch is also disengaged to reduce the idle load on the engine.

The PCM monitors the switch information against its other inputs and sets a DTC if the switch is ever open at a vehicle speed above 45 MPH as sensed by the vehicle speed sensor. This is a speed at which little or no power steering boost should be required.

Park/Neutral Switch

The park/neutral (P/N) switch, which can be located near the shifter assembly or in the transmission itself, is normally open while driving. When the vehicle is shifted into park or neutral, it closes. The PCM monitors the P/N switch and thus knows whether the transmission is in gear. This information is used to help control engine idle and EGR control.

Air Conditioning Switch

On vehicles equipped with air conditioning, the A/C control switch on the control panel tells the PCM whether the driver has requested A/C system operation. The PCM uses this information to control the A/C relay and to adjust the engine's idle speed to compensate for the load of the A/C compressor.

Transmission Switches

On vehicles equipped with an electronic transmission, the transmission contains hydraulically operated electrical switches mounted on the valve body. These switches provide the PCM with signals indicating what gear the transmission is in. This information enables the PCM to control which gear the transmission is operating in as well to control the operation of the torque converter lockup clutch.

System Voltage

The PCM monitors system voltage through one of its battery voltage inputs. If voltage drops below a preprogrammed value, the system goes into a battery voltage correction mode.

Fuel Pump Feedback Circuit

GM fuel injection systems have a fuel pump feedback circuit that allows the PCM to monitor the output side of the fuel pump relay circuit that supplies battery voltage to the fuel pump. If the PCM determines that the fuel pump relay is defective, it will store a DTC in memory.

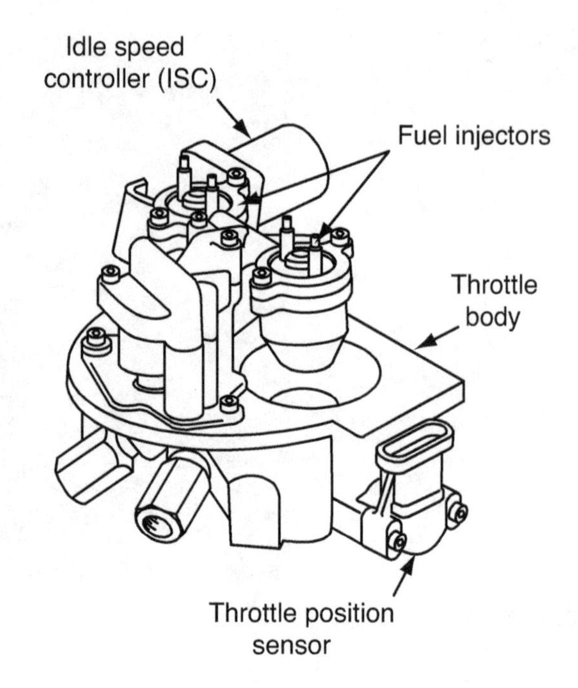

Figure 12–24 TBI assembly.

FUEL MANAGEMENT SYSTEMS

GM vehicles have used several different electronically controlled fuel injection systems: throttle body injection (TBI), port fuel injection (PFI), central multiport fuel injection (CMFI), central sequential fuel injection (CSFI), sequential fuel injection (SFI), and gasoline direct injection (GDI). GM also combines its variable displacement technology with its SFI systems to save fuel.

Throttle Body Injection

Throttle body injection (TBI) was initially used because it was a fairly simple bolt-on replacement for the carburetor and did not require reengineering of the intake manifold's design (Figure 12–24). This system was used through the mid-1990s. On a TBI system, a fuel injector sprays fuel *above* the throttle plate(s) and into atmospheric air pressure, which is fairly stable. Thus, the fuel rail pressure is held fairly stable as well.

Port Fuel Injection

Port fuel injection (PFI) systems use a fuel injector for every cylinder. The injector sprays fuel into the intake manifold runner and on the back side of the intake valve (Figure 12–25). On a

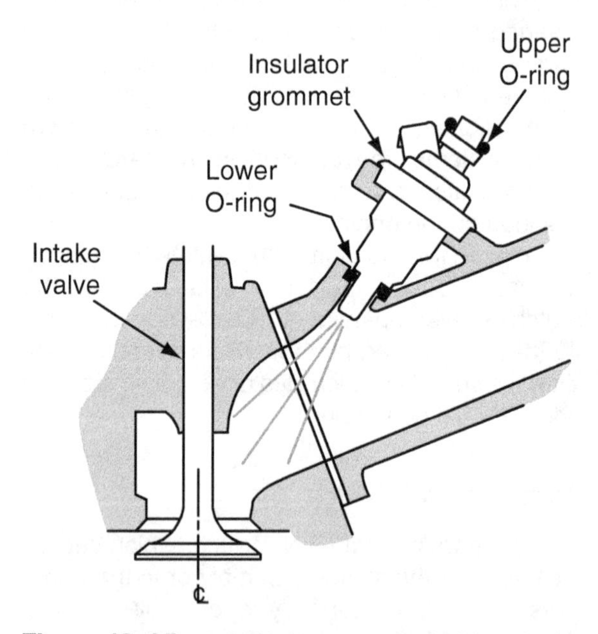

Figure 12–25 A PFI injector sprays fuel into the intake port and on the back side of the intake valve.

PFI system, a fuel injector sprays fuel below the throttle plate(s) and into intake manifold pressure, which varies according to the amount of manifold vacuum created by engine speed and the position of the throttle plate(s). Because of this, the pressure differential across the fuel injectors is subject to change which, in turn, changes the injectors' flow rates. Therefore, some method must be used either to maintain the pressure differential across the fuel injectors or to compensate for it.

On many GM PFI systems, the PCM pulsed all of the injectors simultaneously once per crankshaft rotation. As a result, a given injector was pulsed twice for each time the intake valve opened and for only 50 percent of the total fuel requirement for that cylinder. This was known as *group injection*. The design of the intake manifold is such that any fuel sprayed while the intake valve is closed simply sits and vaporizes until the intake valve opens.

On some GM PFI engines, the injectors were pulsed in pairs by the PCM. Each pulse was then for 100 percent of the cylinder's fuel requirement. This was known as *paired injection*.

Central Multiport Fuel Injection

In the mid-1990s, General Motors introduced a new port fuel injection system known as **Central Multiport Fuel Injection (CMFI)** on Vortec engines. This system uses an injector assembly that consists of a fuel metering body, a fuel pressure regulator, and a single TBI-style fuel injector, which is connected to individual poppet nozzles (one per cylinder) with nylon fuel tubes (Figure 12–26). This assembly is housed in the lower portion of a dual intake manifold assembly. The poppet nozzles are simple, nonelectric, spring-loaded valves (normally closed) that are opened by applied fuel pressure from the single fuel injector solenoid. When opened, each poppet nozzle sprays fuel to the back side of its intake valve until fuel pressure is reduced enough to allow it to close. The PCM controls fuel delivery

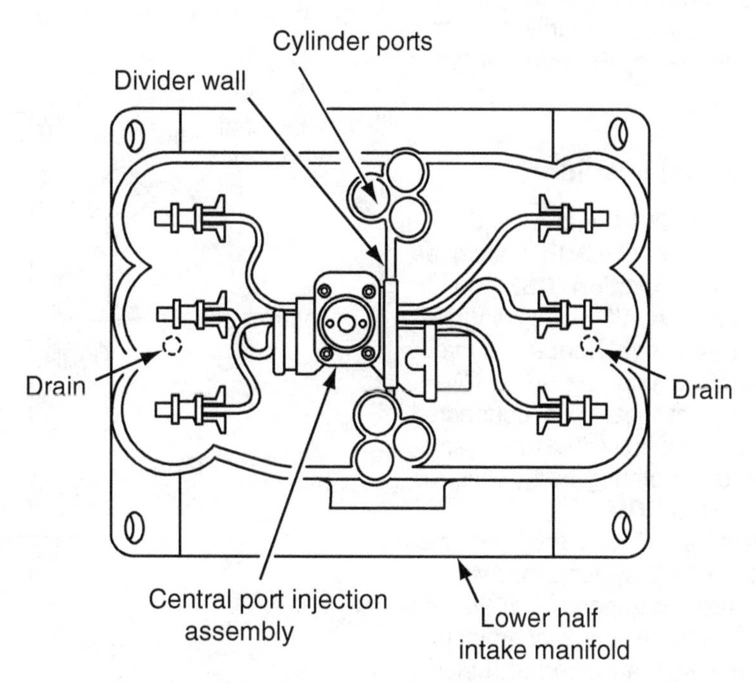

Figure 12–26 CMFI components in the lower half of the intake manifold.

through all poppet nozzles by controlling the on-time of the single injector solenoid.

Another interesting feature is that the fuel pressure regulator can maintain a steady pressure differential across the injector and poppet nozzles without the existence of the normal vacuum hose used to index most fuel pressure regulators to manifold pressure/vacuum. This is because it is located, as part of the fuel metering body, down in the lower intake manifold with its diaphragm already open to manifold pressure/vacuum.

The fuel injector solenoid in this assembly is operated as a peak and hold injector, as are most TBI-type fuel injectors. The injector's winding has low resistance, which allows more current to flow and thus provides faster response when it is energized by the PCM. Once opened, the PCM reduces the injector's current flow to prevent overheating of the solenoid winding, while maintaining enough current to keep the injector open for the remainder of its intended pulse.

Although the CMFI injector looks similar to a low-pressure TBI injector, CMFI is actually a high-pressure system. Fuel pressure is critical on this system, so be sure to check it early on in your diagnosis.

Central Sequential Fuel Injection

From 1996 through 2004 General Motors used a sequential version of CMFI known as **Central Sequential Fuel Injection (CSFI)** on the 4.3L V6 and 5.0L and 5.7L V8 Vortec engines. This was also known as *Central Sequential Injection (CSI)*. This design is very similar to CMFI, except that each poppet nozzle is connected through its nylon tube to an individual fuel injector solenoid within the fuel metering body, which is pulsed sequentially by the PCM.

CMFI/CSFI Demise. The Vortec engines that initially used the CMFI system, followed by the CSFI system, were redesigned in 1999 and were reverted back to a traditional sequential fuel injection (SFI) system with an external fuel rail and injectors, as is typical on most other vehicles.

Sequential Fuel Injection

Sequential fuel injection (SFI) systems were introduced by GM in the mid-1980s and became GM's mainstream fuel injection system by the early 1990s. They continued to be the most popular design for a fuel-injected engine until GM introduced the gasoline direct injection (GDI) system in the 2010 model year.

An SFI engine, like a PFI system, uses a fuel injector for every cylinder that also sprays fuel into the intake runner and on the back side of the intake valve (Figure 12–27). However, once the engine is running, the PCM in an SFI system will pulse each injector individually, in the engine's firing order, just before the intake valve opens, and for the cylinder's entire fuel requirement. Therefore, each injector is independently controlled by the PCM (Figure 12–28).

Because the timing of an SFI fuel injector is not critical due to the fact that each cylinder has its own dedicated intake manifold runner, during the initial engine start on an engine equipped with SFI and a waste spark ignition system, the PCM will fire an ignition coil from the CKP sensor

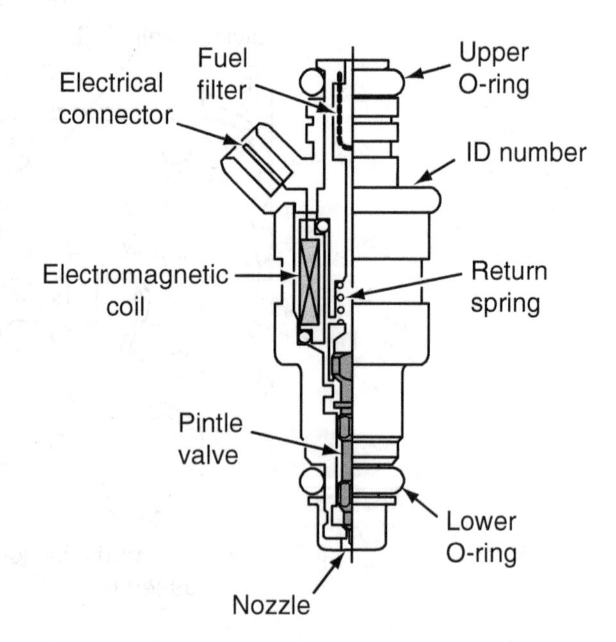

Figure 12–27 SFI injector (cutaway).

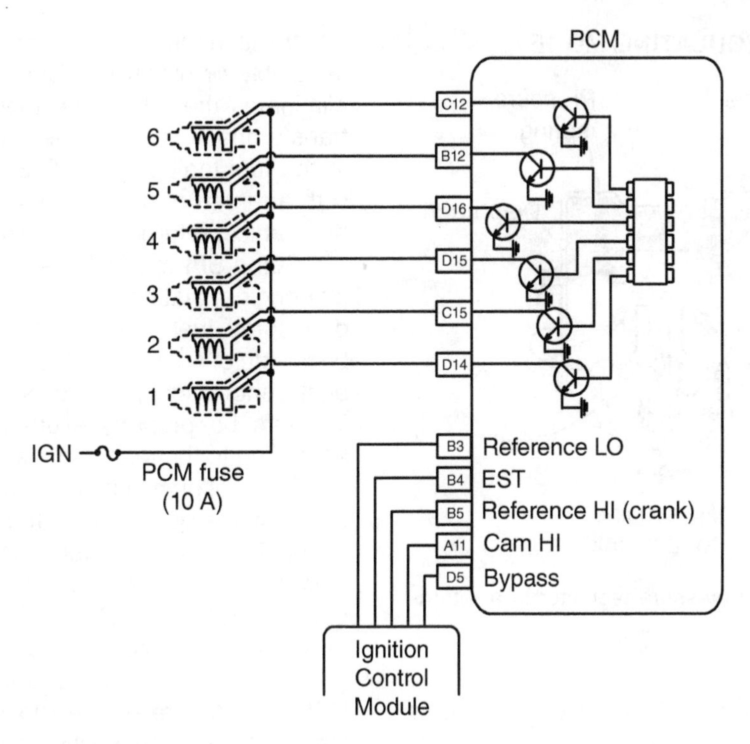

Figure 12–28 Electrical schematic of a typical SFI system.

(second level of tach reference information) and will pulse all the injectors as a group, known as a *primer pulse*, to get the engine started. Then, once the PCM sees a CMP signal occur, it will switch over to its sequential injection strategy. If the CMP sensor fails, the PCM will guess at injector sequencing and set a DTC in memory. If the injector sequence is incorrect, the engine will still run, but emissions might be slightly higher than normal if an injector is pulsed while the intake valve is open due to the increased pressure differential. On an SFI engine equipped with a COP ignition system, the engine will not start if the CMP signal is lost.

Early GM SFI engines had a return-type fuel injection system (Figure 12–29). The system used a mechanical fuel pressure regulator to maintain a constant pressure differential across the fuel injectors and, therefore, a constant flow rate

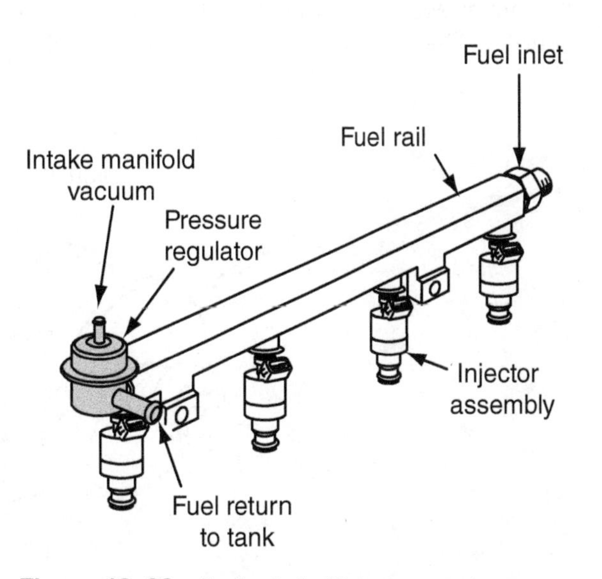

Figure 12–29 Fuel rail, fuel injectors, and pressure regulator in a return-type SFI system.

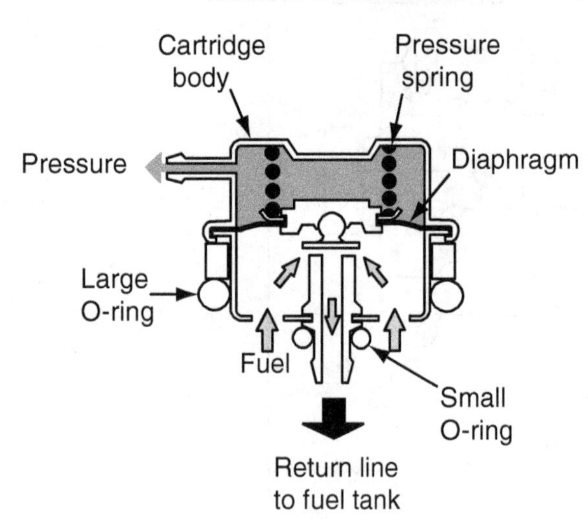

Figure 12–30 Fuel pressure regulator in a return-type SFI system.

(Figure 12–30). The signal to the pressure port on the regulator was an intake manifold pressure signal (manifold vacuum).

Newer SFI systems use a returnless system to reduce the fuel tank's evaporative emissions that are allowed to charge the evaporative system canister. The fuel pressure regulator is located either within the fuel tank as part of the fuel pump assembly or within the fuel filter assembly. This eliminates the return line that was continuously transferring heat from the engine compartment to the fuel tank. Figure 12–31 shows a fuel filter with an internal pressure regulator as used in a mechanical returnless SFI system.

Some SFI applications, such as GM's Northstar engine, that have four valves per cylinder use dual-spray fuel injectors that spray fuel on the back side of each cylinder's two intake valves. For best performance, it is recommended that these injectors be properly aligned with the fuel rail according to the manufacturer's specifications.

The fuel pump circuit used with GM fuel injection systems uses a relay that applies voltage to the positive side of the fuel pump when energized. However, unlike those of most other manufacturers, the PCM on a GM fuel injection system typically controls the positive side of the relay winding (Figure 12–32). If the control wire between the PCM and the relay were to become grounded, the short to ground would not keep the relay energized unintentionally. Also shown in Figure 12–32 is an oil pressure switch that is in parallel to the relay's contacts. When used on a GM vehicle, the oil pressure switch simply provides a redundant backup in case the fuel pump relay were to fail.

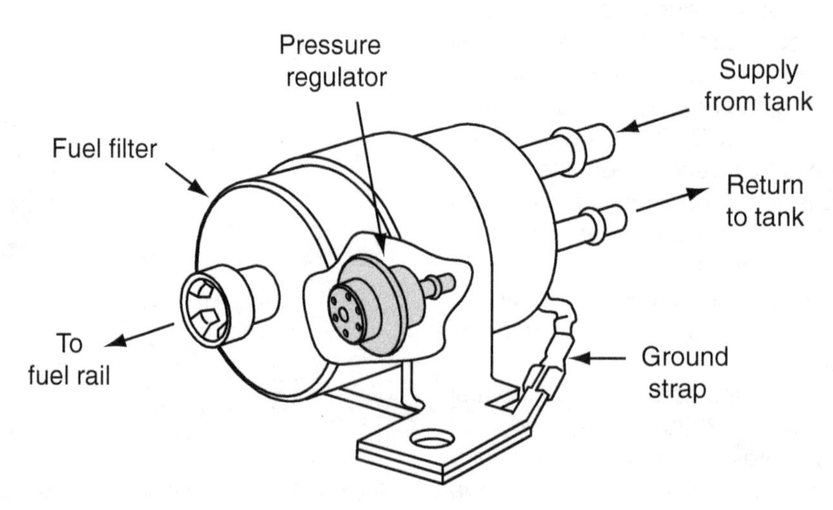

Figure 12–31 Fuel filter with an internal fuel pressure regulator.

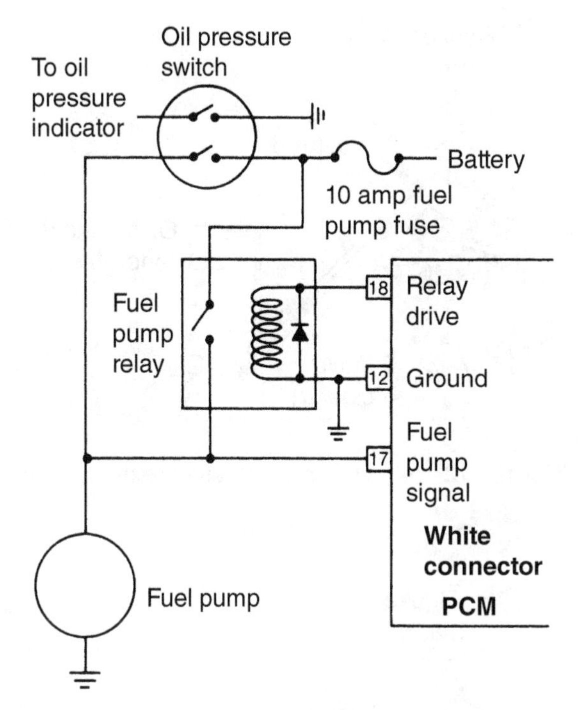

Figure 12–32 Typical GM fuel pump circuit.

Engine Coolant Temperature Fuel Disable Feature. The *engine coolant temperature fuel disable* feature (also known as **engine metal overtemp mode)** was advertised by GM as an "extended distance" feature that enabled the driver to "drive 50 additional miles after the engine has run out of coolant" to get the vehicle to a repair facility. This feature was first designed into the PCM's program on Northstar applications. Later, it also became a feature available on the Premium V6 and is now also available on other GM engines. If the PCM detects an ECT sensor value of 270°F, it disables the fuel injectors on every other cylinder in the engine's firing order, alternating with cylinders that are still contributing to engine performance. This allows the disabled cylinders to act as a heat transfer pump. The PCM then empowers the disabled cylinders and disables the empowered cylinders after a preprogrammed number of engine cycles, alternating them to control engine temperature until the driver can get the vehicle to a repair facility.

GM Active Fuel Management System

Used with an SFI system, this system is designed to increase fuel efficiency while cruising while giving the driver full access to the engine's performance potential. Engineered to accomplish what the Cadillac 8-6-4 engine of the early 1980s was supposed to do, the modern version of the system uses a more advanced technology.

GM introduced this system as the **Displacement on Demand (DOD) system** in 2005 on V6 applications, followed immediately by V8 applications. In 2007, GM changed the name to **Active Fuel Management (AFM)** in order to better promote the system's fuel-saving purpose. GM's initial goal was to have two million AFM engines on the road by 2008. Initially, AFM technology was to be used on V6 and V8 engines that were put into light trucks and SUVs. The fuel savings is rated at about 8 percent based on the EPA testing procedure, but it is estimated to be as high as 25 percent during sustained cruising conditions.

In actual operation, the AFM system allows the PCM to run a V8 on four cylinders when the need for power is small. Similarly, the V6 is cut to three cylinders when the AFM system kicks in. The engine always starts on all cylinders. When the vehicle is at cruise and the need for power is reduced, the PCM does two things:

1. It turns off the fuel injectors for every other cylinder in the engine's firing order, four cylinders on the V8 and three cylinders on the V6.
2. It simultaneously closes all intake and exhaust valves on the disabled cylinders, which simply allows the disabled cylinders to act as air springs. These cylinders use crankshaft power to compress the trapped air when the piston moves from BDC to TDC, but they also return power to the crankshaft when the piston moves from TDC to BDC. This avoids the energy losses that would occur if each disabled cylinder's piston were to continue pulling down against manifold vacuum and pushing out against ambient air pressure on the exhaust side; thus the *pumping losses* are reduced.

A specially designed "switching lifter" is used on the AFM system. It is designed with two portions, an inner body and an outer body, that can physically collapse on each other. A spring is designed as part of the lifter and tries to keep the lifter extended (Figure 12–33). This spring has less tension than the valve springs that are used to close the cylinders' valves. In early GM designs, the spring was external on the lifter. Beginning in 2006, the lifter design was changed so that the spring was internal, similar to Chrysler/Bosch designs. For normal operation, the lifter has a locking pin that keeps the two halves from collapsing on each other; the result is that the cam uses the lifter to open the valve. However, when the PCM wants to disable the valve, it energizes an oil solenoid that provides oil pressure to push the locking pin to "unlock the lifter" (Figure 12–34). At this point, the lifter collapses on its own spring tension. The spring is strong enough to keep the lifter following the cam lobe but not strong enough to overcome the valve spring tension, thus allowing the valve to stay closed. When the driver punches the throttle, the PCM instantaneously turns the fuel injector back on and de-energizes the oil control solenoid (Figure 12–35). This allows the lifter to expand, allowing the locking pin to lock the lifter in the extended position, and cylinder operation is thereby restored.

Only one-half of the cylinders are designed to be disabled and therefore use the specially

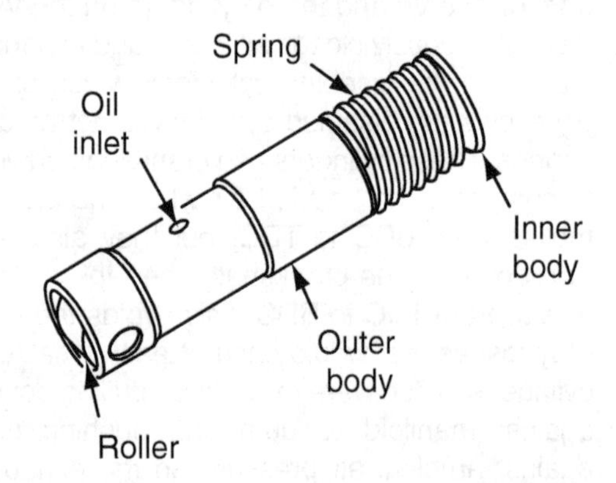

Figure 12–33 A switching lifter from an AFM system.

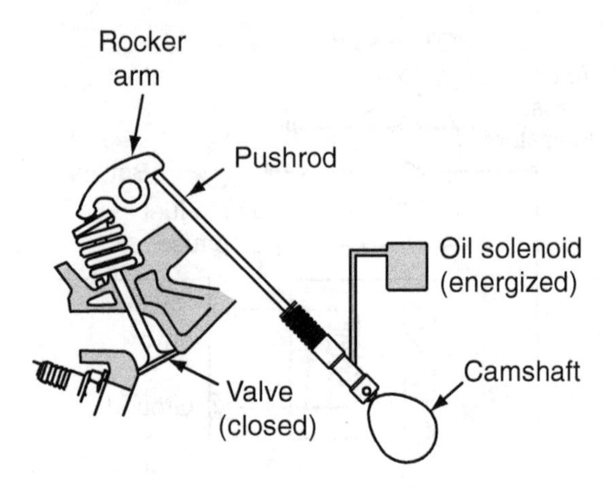

Figure 12–34 Oil solenoid energized and lifter collapsed.

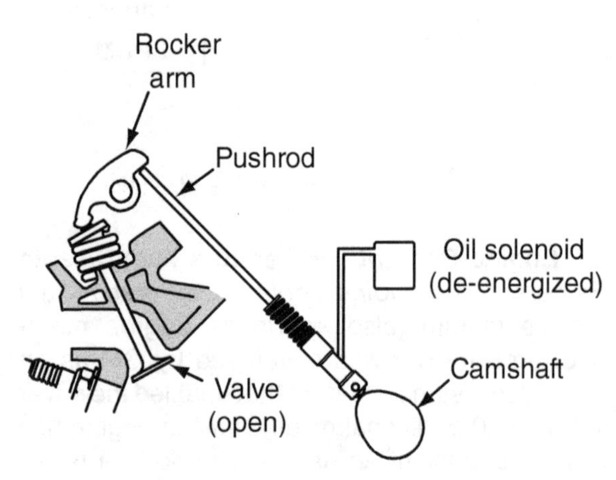

Figure 12–35 Oil solenoid de-energized and lifter extended.

designed lifters. The remaining cylinders are never disabled during AFM operation, and therefore they use standard hydraulic lifters. When operating in AFM mode, however, the PCM will re-enable the disabled cylinders once every 10 minutes for about one minute—this helps to minimize wear differences between cylinders.

Thus there are two major differences between an AFM system and the engine coolant *overtemp fuel disable* system described earlier. The engine coolant overtemp fuel disable system does not disable the valves but keeps operating them for pulling air through the cylinders to carry away heat, whereas the AFM system does disable the valves when a cylinder is disabled. Also, the engine coolant overtemp fuel disable system alternates which cylinders are disabled, whereas the AFM system always disables the same cylinders.

One of the advantages of current AFM technology, as compared to the Cadillac 8-6-4 system of the early 1980s, is that it is used with GM's throttle actuator control (TAC) technology. When such a system was used without TAC technology, the driver might find that he or she had to compensate slightly at the throttle pedal when the computer disabled or re-enabled cylinders. On a modern AFM system with a TAC system, the PCM can make this compensation automatically as it disables cylinders or re-enables them. Thus the driver can maintain a consistent throttle position as cylinders are disabled or re-enabled.

Gasoline Direct Injection

Gasoline direct injection (GDI) systems use a fuel injector for every cylinder, similar to an SFI system. However, the injector sprays fuel directly into the cylinder (Figure 12–36). Not only does this system use much higher fuel pressure than earlier systems, but the higher pressure, in turn, allows for better atomization of the fuel.

A GDI engine can also operate in three basic modes of air-fuel control as follows:

- Homogeneous charge mode with closed-loop fuel control: This mode is used for maintaining an overall stoichiometric air-fuel ratio.
- Homogeneous charge mode with open-loop fuel control: This mode is used for providing a rich air-fuel ratio when the vehicle is under moderate-to-heavy load.
- Stratified charge mode with closed-loop fuel control: This mode is used for maintaining

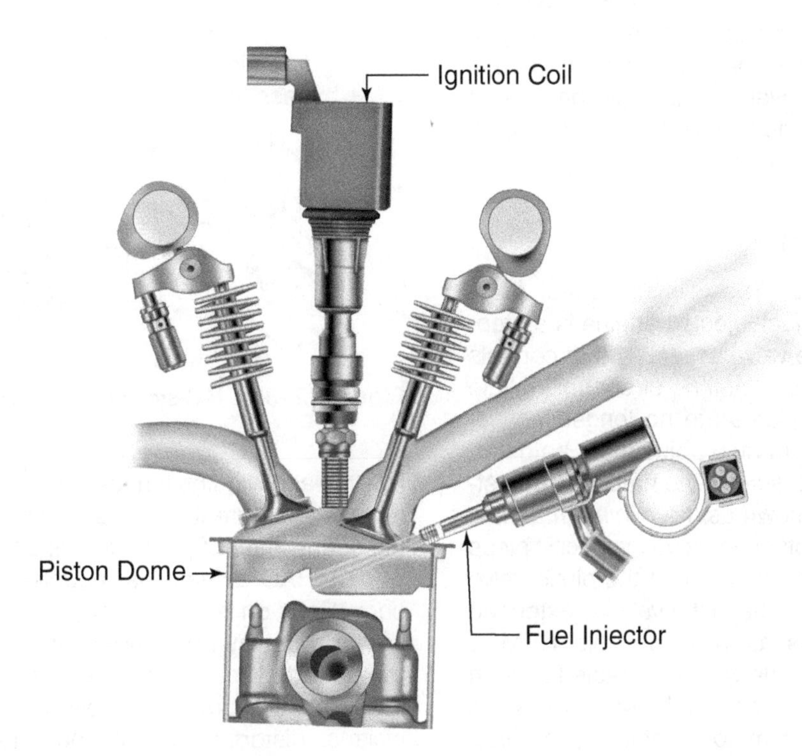

Figure 12–36 Gasoline direct injection.

a stoichiometric air-fuel ratio near the spark plug while providing an overall lean air-fuel ratio from 30:1 to 40:1 during cruise.

During cruise and while operating in the stratified charge mode the PCM can control engine speed and performance through control of fuel metering, similar to a diesel engine. This allows the PCM to further open the throttle plate(s) beyond what would normally be allowed without over-revving the engine. This, in turn, reduces the cylinder's pumping losses.

Additionally, while operating in the stratified charge mode, the engine's power can be reduced during cruise by simply reducing fuel delivery. Therefore, a GDI engine has the advantage of consuming less fuel when power is not needed, similar to a GM AFM system except that, with a GDI engine, the PCM does not have to disable cylinders to achieve this reduction in power and fuel. Therefore, the engine runs more smoothly and is less annoying to the driver while operating in the stratified charge mode than an AFM engine is when the PCM disables half of its cylinders. Another advantage over the AFM system is that GDI can be used in the stratified charge mode on smaller four-cylinder engines.

Idle Speed Control

Idle Air Control. The idle air control (IAC) assembly on a GM fuel-injected engine is a stepper motor and pintle valve assembly that controls throttle bypass air as a method of controlling idle speed. The throttle plates are no longer moved, but are left at the hard stop adjustment as determined by a hard stop screw. The hard stop adjustment sets what is known as the *minimum airflow rate.* The stepper motor can move in specific steps or increments to extend or retract the pintle valve (Figure 12–37). When the pintle valve is extended to block the bypass air passage, idle RPM is reduced. When the pintle valve is retracted to open up the bypass air passage, idle RPM is increased.

The IAC stepper motor contains two windings (Figure 12–38). The PCM can control current flow to each winding independently, but it can also reverse the current flow in each winding as part of the sequencing of power to this actuator.

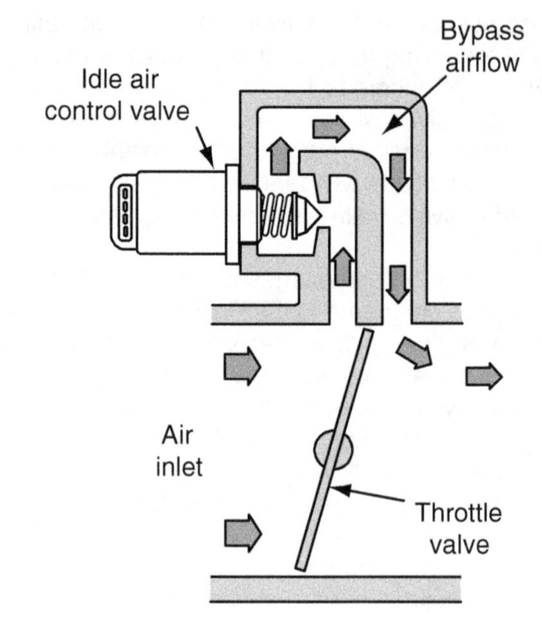

Figure 12–37 IAC stepper motor in a PFI/SFI throttle body.

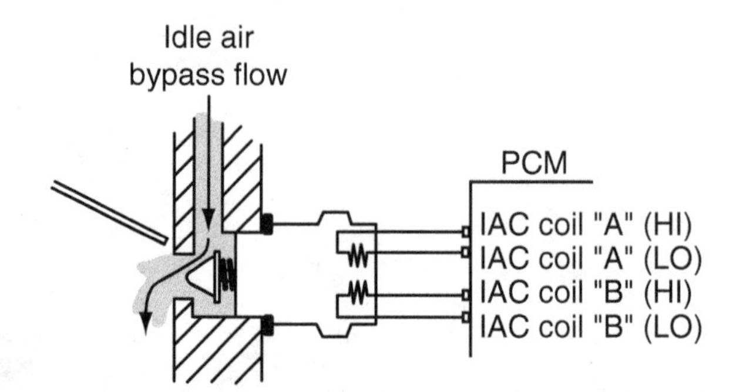

Figure 12–38 IAC stepper motor control circuit.

The IAC actuator is mounted to the throttle body. Early ones had a body that was threaded, allowing the stepper motor to be screwed into the threaded mounting hole in the throttle body. This style was sometimes overtightened by technicians, distorting the stepper motor's housing and causing it to stick. Newer ones are simply mounted to the throttle body with two small machine screws, thus eliminating the potential to overtighten the unit.

The IAC stepper motor can be controlled from step 0 (a PCM command to fully close the valve in order to reduce engine speed) to step 255 (a PCM command to increase engine speed to its full capacity). In order to know how to control the IAC stepper motor properly, the PCM must always keep track of the IAC counts. To do this, it must reference the IAC stepper motor and pintle valve to the fully closed position occasionally. On a GM PFI/SFI system, the PCM will reference it to the fully closed position each time the ignition is turned off. Following this, the PCM then issues a preprogrammed number of pulses to open the valve to a ready position for the next engine start. The PCM can compare the actual engine RPM with the position that it sets the IAC to. If the idle speed is higher or lower than the desired engine speed by 80 RPM or more, the test fails and a DTC is set in memory.

On any GM TBI, PFI, or SFI application, a scan tool can be used to read the *desired RPM*, the *actual RPM*, and the *IAC counts* from 0 through 255. This information is helpful to a technician in that it indicates what the PCM is trying to do with regard to engine speed. On a GM vehicle, a scan tool can also be used to command a specific engine speed from idle to about 2,000 RPM to test the PCM's ability to control the IAC.

Throttle Actuator Control

Known generically as *electronic throttle control (ETC)*, GM refers to this feature as their *throttle actuator control (TAC)* system. The TAC system allows the PCM to control throttle plate movement electronically and eliminates the need for the traditional throttle cable. The TAC system is often referred to generically as a "drive-by-wire" system. The TAC system offers several advantages, including lower emissions and quicker throttle response. This system was introduced on GM Generation III V8s in the 1999 model year and was added to the Cadillac Northstar LH2 in the 2004 model year.

The components of the TAC system include a DC motor that serves as the actuator used to control throttle plate position, a throttle position sensor (TPS) to monitor throttle plate position, and an accelerator pedal position sensor (APP sensor or APPS) (Figure 12–39). In basic computer theory for an engine with a cable-operated throttle plate, the TPS is said to give the PCM information about *driver demand*. In a TAC system, driver demand information now originates at the APPS.

Early TAC systems used potentiometers for both the TPS and the APPS. GM used a dual potentiometer for each sensor. With two potentiometers, a redundant backup is provided in the event of a potentiometer failure. (Potentiometers have a physical wear factor as the wiper passes across the resistive carbon material and, therefore, can wear out with repetitive use.)

In most applications, the dual potentiometers in the APP sensor operate electrically between different high and low values and/or in different directions. This allows the PCM to more precisely interpret exact throttle position, as opposed to all potentiometers delivering exactly the same voltage values at a given throttle position. The APPS is located in a throttle pedal assembly that is spring-loaded to give the driver a realistic pedal feel.

When dual potentiometers are used for the TPS and APPS functions, the following modes of operation will occur when a potentiometer fails:

- APPS Failure Mode 1: One sensor has failed. Operation will be fairly normal, but a DTC will set in memory and the PCM will illuminate the MIL.
- APPS Failure Mode 2: Both sensors have failed. Vehicle will have an *idle only* capability to provide a *limp home* function.
- TPS Failure Mode 1: One sensor has failed. The PCM will limit the vehicle's speed, a DTC will set in memory, and the PCM will illuminate the MIL.
- TPS Failure Mode 2: Both sensors have failed. Vehicle will have an *idle only* capability to provide a *limp home* function.

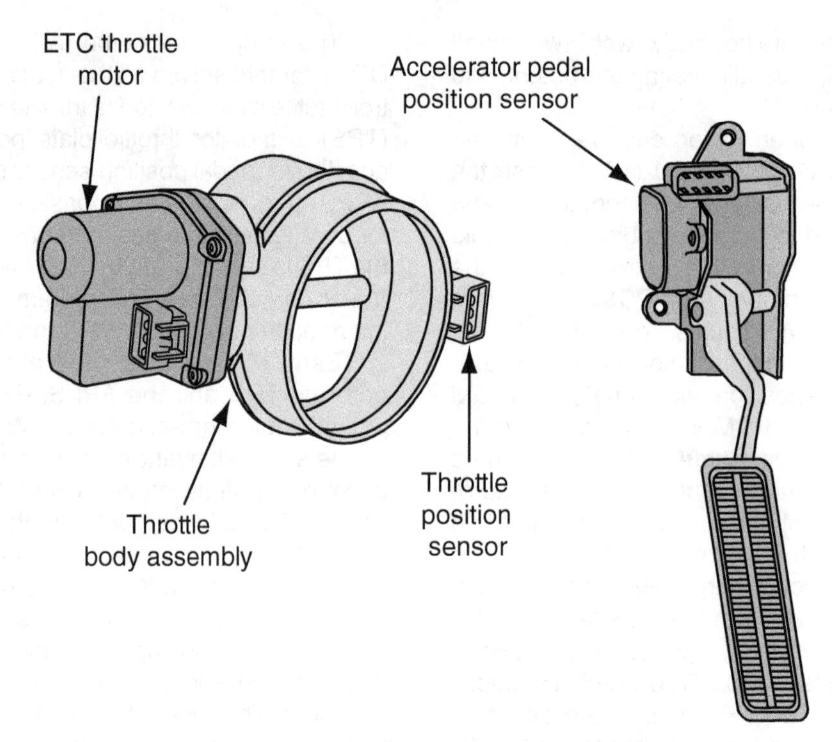

Figure 12–39 A throttle actuator control (TAC) system.

Modern GM engines use noncontact linear Hall effect sensors for both the TPS and the APPS functions. These sensors began replacing the traditional potentiometers in these applications due to their *no-wear* quality around 2010 or 2011. Today these sensors are used almost exclusively in TAC/ETC systems. Because of this no-wear quality, by 2012 some GM engines were using only one sensor to measure throttle position.

The PCM commands the DC motor to open or close the throttle plate(s) using bidirectional high-frequency pulse-width-modulated commands. That is, it can issue pulse-width commands that instruct the motor to move the throttle plate(s) slowly or rapidly. And bidirectional simply refers to the fact that polarity applied to the motor can be reversed to move the throttle plate(s) in either direction (opening versus closing).

In a TAC/ETC system, the PCM controls both air and fuel, unlike the cable-operated system in which the driver controls the air and the PCM controls only the fuel.

Figure 12–40 shows that the use of a TAC/ETC system also has some other benefits:

- The dedicated IAC stepper motor is eliminated. The PCM can now control idle RPM directly without the use of additional dedicated hardware.
- The dedicated cruise control (speed control) servo is eliminated. The speed control feature becomes a software function of the PCM and no longer requires additional hardware to control throttle position.
- The PCM can use the TAC/ETC system to reduce engine torque directly when the traction control feature mandates it.

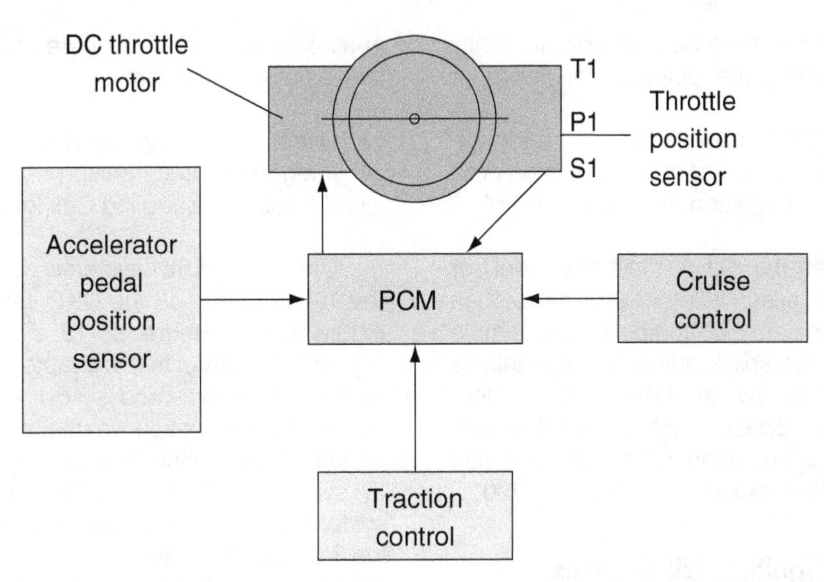

Figure 12–40 A TAC/ETC system also provides the cruise control and traction control features and eliminates the need for a dedicated idle air control actuator.

Diagnostic & Service Tip

When diagnosing a problem associated with a high idle speed on a GM vehicle with an IAC stepper motor, always use a scan tool to check *IAC counts* first. If the *IAC counts* show a value of 0, then you know the PCM is trying to reduce idle speed by completely closing the air bypass passage. In this case, look for a source of unmetered air such as a vacuum leak. If no vacuum leaks are found, then consider the possibility that the PCM cannot successfully control the IAC stepper motor. Next use a DMM to measure each winding's resistance. Each winding should have a resistance value that is typically between 55 and 60 Ω on GM vehicles. A point to note: The two green wires (green with a black tracer and green with a white tracer) represent one winding, and the two blue wires (blue with a black tracer and blue with a white tracer) represent the other winding.

SPARK MANAGEMENT SYSTEMS

Given the developments we have seen in computerized engine controls, there is no need to continue using a distributor with its tendencies toward wear and mechanical failure. Distributors are no longer needed due to the following advances in technology:

- An ignition module can be mounted anywhere in the engine compartment and does not necessarily have to be mounted on a distributor.
- Spark advance can be effectively controlled by a computer, eliminating the need for mounting mechanical advance mechanisms on a distributor.
- Tach reference signals can be taken directly from the crankshaft, thus eliminating the effects of timing chain stretch (on the ignition system).

- Spark timing can then be made permanently fixed, eliminating the potential for improper adjustment.
- Secondary spark distribution can be managed through the use of multiple ignition coils controlled by an ignition module and PCM.

After all, even the old points-and-condenser distributors were essentially nothing more than camshaft position sensors, combined with simple mechanical/vacuum spark advance mechanisms and a mechanical switch. GM's last distributors were the Opti-Spark distributor used as late as 1997 and a more traditional distributor that was used on light-duty trucks into the very early 2000s.

GM Electronic Ignition (EI) Systems

General Motors initially used several different electronic ignition (EI) systems on various engine applications, all of which were originally designed as *waste spark* systems:

- Direct Ignition System (DIS)
- Integrated Direct Ignition (IDI)
- Computer Controlled Coil Ignition (C3I)

GM DIS and IDI systems were introduced in the 1980s and used the CKP sensor depicted in Figure 12–11, Figure 12–12, Figure 12–13, and Figure 1214 earlier in this chapter. The DIS system was used on four- and six-cylinder engines and used separate ignition waste spark coils mounted on top of the ignition module (Figure 12–41). The IDI system used the same CKP sensor as the DIS system, but used a secondary conductor housing instead of secondary wires (Figure 12–42). When SFI was later added to these engines, both the DIS and IDI systems added a magnetic CMP sensor for injector sequencing.

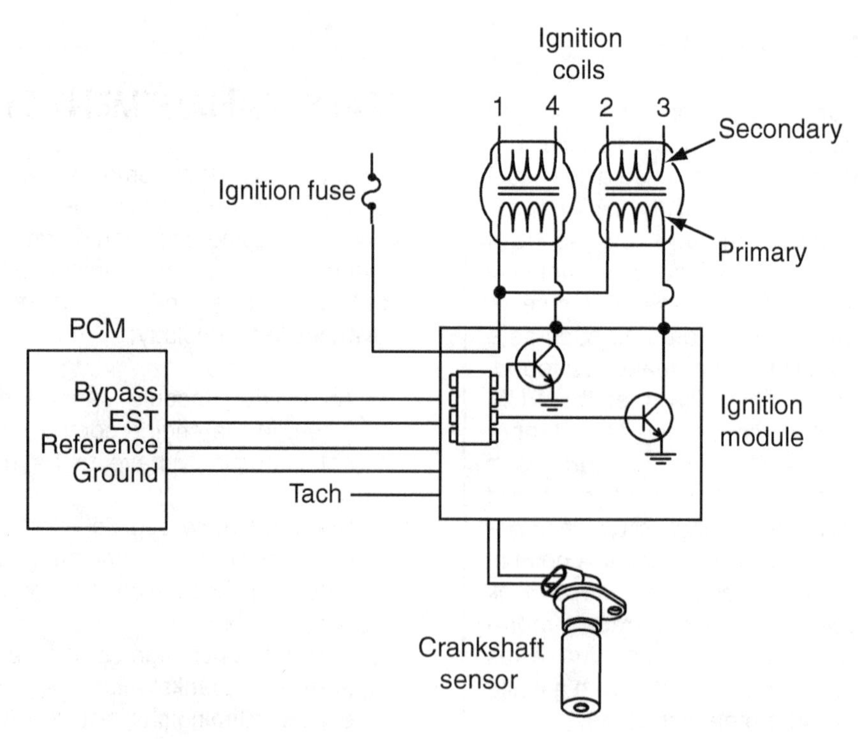

Figure 12–41 Four-cylinder DIS system schematic.

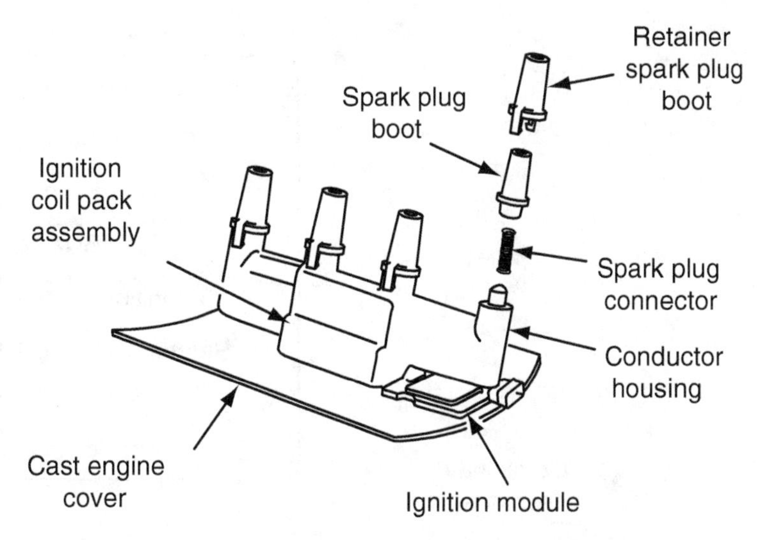

Figure 12–42 A Quad Four ignition system that does not use spark plug wires.

The C3I system was introduced in 1984 as either a type I or a type II system. The difference between the type I and type II C3I systems is that the type I system uses a triple-coil pack that mounts directly on top of the ignition module and connects with primary wires to the module, whereas the type II system uses individual coils that plug directly onto the ignition module's terminals, similar to those used with the DIS system. The original C3I system used a Hall effect CKP sensor depicted earlier in Figure 12–10. It also used a Hall effect CMP sensor for injector sequencing of the SFI injectors.

The type III C3I system is a "fast start" system that was introduced on 1988 model year vehicles with 3,800 engines. The Hall effect CKP sensor is depicted in Figure 12–15. With this system, the second level of tach reference information identifying which pair of pistons is at TDC is moved from the CMP sensor to the CKP sensor. The PCM now knew which waste spark coil to fire within 120 degrees of crankshaft rotation during an engine start. In a C3I ignition system, the PCM has a spark timing range capacity of 0 degrees to 70 degrees of spark advance.

Many GM engine use dual magnetic CKP sensors as depicted in Figure 12–18, Figure 12–19, and Figure 12–21. A magnetic CMP sensor is shown in Figure 12–20. Many newer GM engines now use noncontact linear Hall effect sensors as CMP and CKP sensors.

The GM ignition systems that use these sensors were initially designed with waste spark ignition coils (Figure 12–43). In model year 2000, GM began phasing in a coil-on-plug (COP) design to replace the waste spark ignition systems. COP ignition systems replaced waste spark ignition systems on most GM passenger cars by the 2012 model year and on GM light trucks by the 2014 model year. The COP-type ignition coils are, most often, individual coils, but a triple-coil cassette was used on the V6 version of the Northstar engine, the 3.5L LX5 (Figure 12–44).

Ignition Control Module

GM vehicles use a separate *ignition control module (ICM)*, which is controlled by the PCM. The ICM is usually mounted near the rear of the engine. It controls the waste spark coils or the COP-type coils using spark timing commands from the PCM.

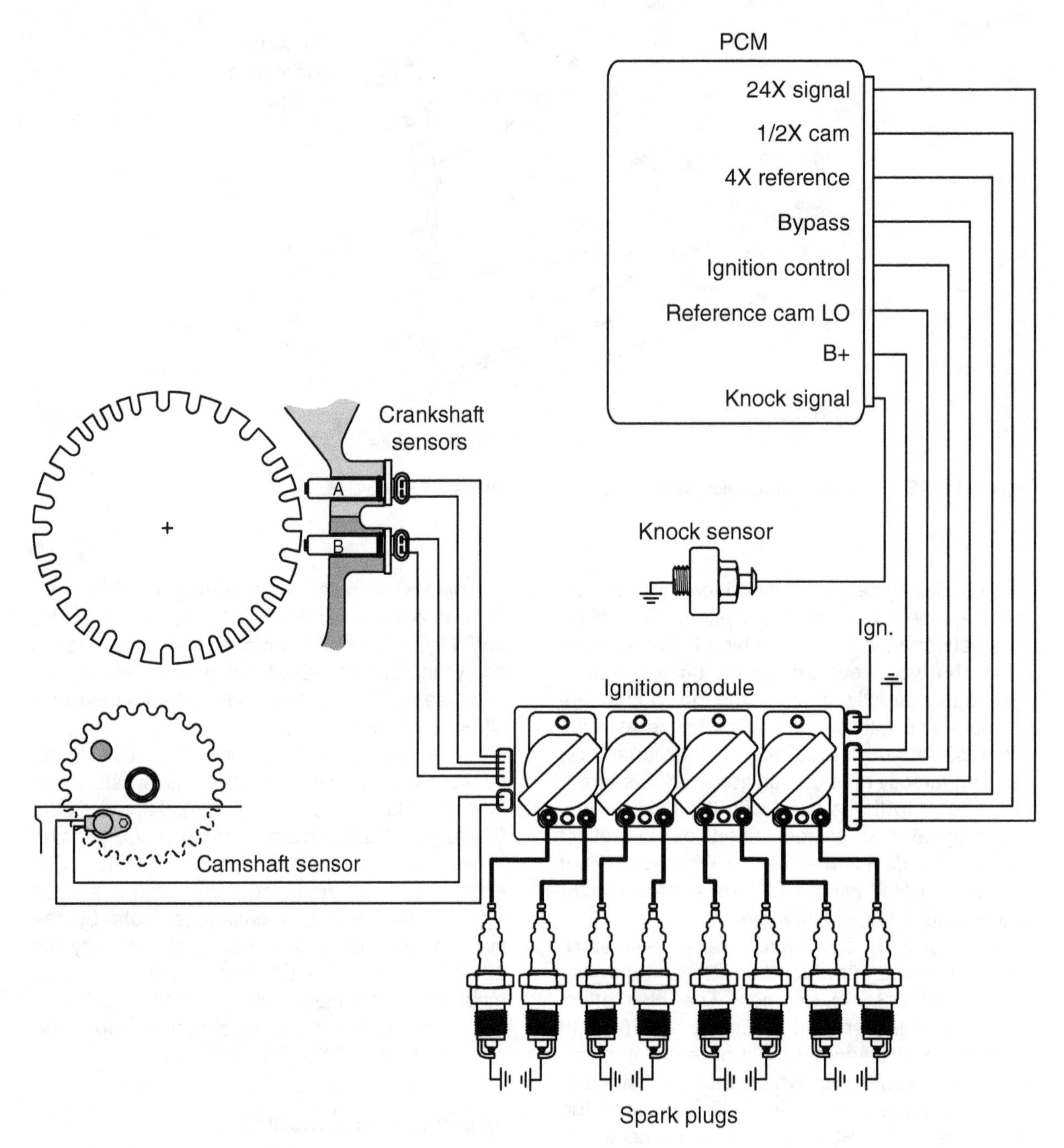

Figure 12–43 GM waste spark ignition system.

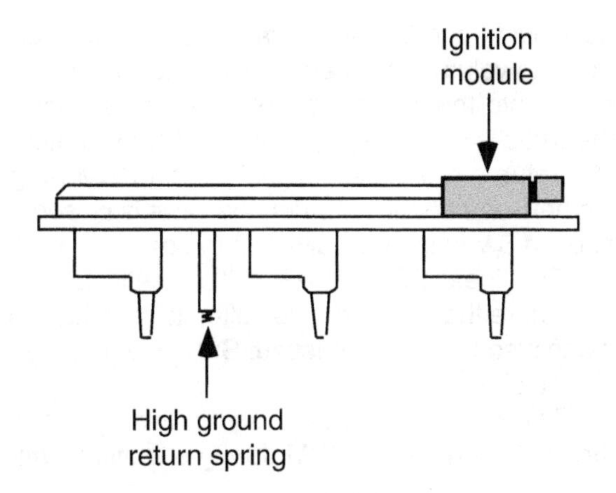

Figure 12–44 Ignition cassette containing three COP-type ignition coils.

EMISSION CONTROL SYSTEMS

Catalytic Converters

As is common with other manufacturers, GM vehicles use both a reduction catalytic converter and an oxidation catalytic converter. The reduction cat uses CO to reduce oxides of nitrogen (NO_x) back into N_2 and O_2. The oxidation cat oxidizes CO into CO_2 and HCs into both water and CO_2.

Close-Coupled Catalytic Converters. Beginning in the 2004 model year, some GM vehicles were equipped with a **close-coupled catalytic converter**. These converters are mounted directly adjacent to the exhaust manifold, thus allowing them to reach operating temperature more quickly following a cold start. This helps to reduce cold-start emissions.

Catalytic Converter Efficiency Test. The PCM compares the signal from the pre-cat oxygen sensor(s) with that of the post-cat oxygen sensor(s) downstream of the catalytic converter. If the converter is working properly, there should be little variation in the signal produced by the secondary oxygen sensor. If the post-cat oxygen sensor(s) produce a signal that is too similar to that generated by the pre-cat oxygen sensor(s), the PCM sets a DTC in memory.

Secondary Air Injection Reaction Systems

Modern GM secondary air injection reaction (AIR) systems use an electric air pump that is controlled by the PCM. In normal operation, the air pump is energized to pump ambient air upstream to the exhaust manifold during cold engine warmup only. Once the engine is warm enough to operate in closed-loop fuel control, the pump is shut down.

With an electrically operated pump, there is no longer any need for a provision to dump air to atmosphere as with the early belt-driven pumps. And there is no longer a provision to route air to the oxidation bed of the catalytic converter due to the influence of the post-cat oxygen sensor on fuel control. Therefore, modern electric air pumps are only turned on during cold engine warmup.

To test the AIR system, the PCM will wait until it is operating in closed-loop fuel control and then momentarily turn the pump on and off. The oxygen sensors should read the momentary ambient oxygen being pumped into the system.

Exhaust Gas Recirculation Systems

Like other manufacturers, GM has used exhaust gas recirculation (EGR) systems on many of their engines to help control NO_x. GM's first electric EGR valve was introduced in 1988 and was called the *digital EGR valve*. It consisted of three electrically operated solenoids that controlled three differently sized exhaust orifices (Figure 12–45). It was able to provide seven exhaust flow rates, depending upon the combination of three solenoids energized by the PCM (Figure 12–46).

Linear EGR Valve. In 1994, GM introduced the *linear EGR valve* (Figure 12–47). This valve, like the digital EGR valve, controls exhaust gas recirculation directly and without the use of vacuum as a control medium. The linear EGR valve consists of a single, normally closed, solenoid-operated pintle valve operated by the PCM. The PCM can precisely control the amount of physical valve opening by controlling the solenoid with a high-frequency

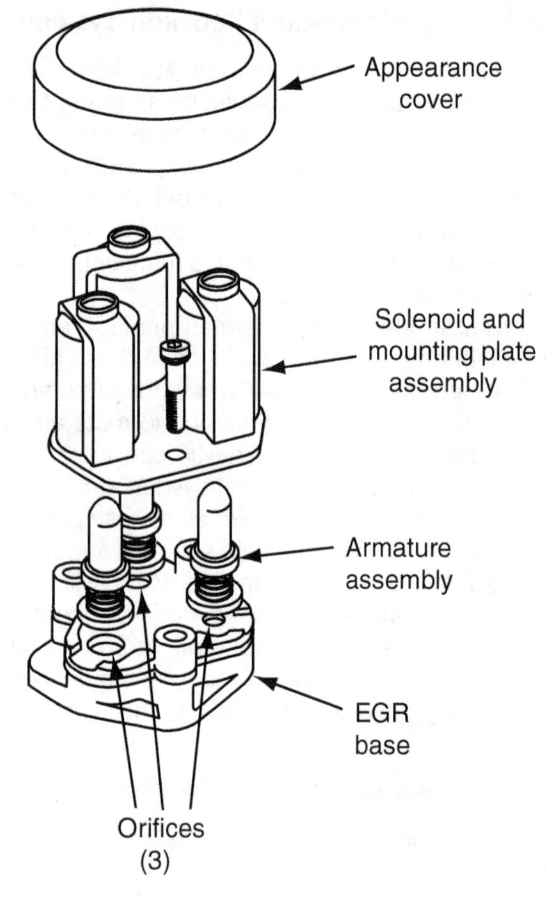

Figure 12–45 A digital EGR valve with three solenoids.

pulse-width-modulated signal. This, in turn, controls the volume of exhaust gases allowed to recirculate into the intake manifold. The movement of the pintle is monitored by an internal potentiometer that returns a feedback signal for the PCM ranging between about 1 V when the valve is closed to about 4.5 V when the valve is fully open.

The linear EGR valve is still currently used by GM, more than two decades after its introduction. It can also be found on Isuzu, Suzuki, and Chrysler vehicles.

The linear EGR valve used in this system is further tested by the PCM during specific driving

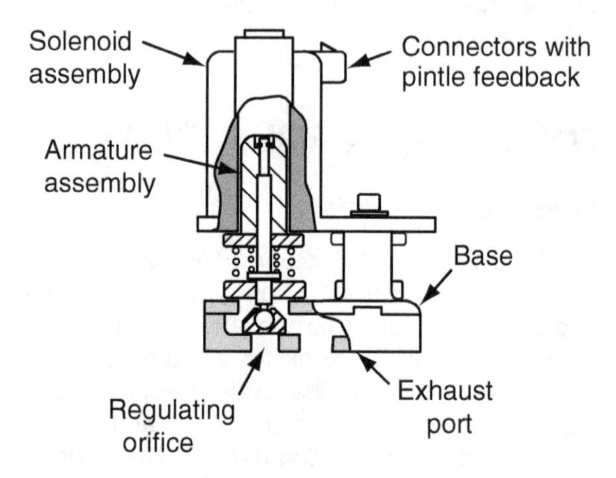

Figure 12–47 Linear EGR valve.

Increment	Orifice #1 (14%)	Orifice #2 (29%)	Orifice #3 (57%)	EGR flow (%)
0	closed	closed	closed	0
1	open	closed	closed	14
2	closed	open	closed	29
3	open	open	closed	43
4	closed	closed	open	57
5	open	closed	open	71
6	closed	open	open	86
7	open	open	open	100

Figure 12–46 Increments of EGR flow with GM's digital EGR valve.

operations, known as *enable criteria*. When the proper enable criteria are met, the PCM opens and closes the EGR valve while monitoring the MAP sensor signal. Obviously, opening another intake source should raise the intake manifold pressure (or decrease vacuum). The PCM repeats the test a number of times, comparing results to its program. If the results do not agree, it sets a DTC in memory.

Evaporative Emissions System

As with other manufacturers' vehicles, GM vehicles use an evaporative emissions system to contain the evaporative hydrocarbon emissions from the fuel tank while allowing the fuel tank to be vented to atmospheric pressure. A charcoal canister is used to capture and store the hydrocarbon emissions. It is then purged when the engine is operating in closed-loop fuel control. The purge solenoid is designed normally closed and is controlled by the PCM (Figure 12–48).

Evaporative Emissions System Tests

To meet OBD II requirements, the PCM performs various tests of the evaporative emissions system. By 1996 the PCM was required to perform a purge flow test and by 1999 the PCM was required to be able to determine a system leak. Some vehicles built from 1996 to 1998 were also able to perform the leak test. The evaporative system leak test was required to be able to identify any fuel tank and evaporative system leaks larger than 0.040 of an inch from 1996 through model year 2000. For model year 2001 and thereafter, it had to be able to identify any leaks larger than 0.020 of an inch

Evaporative Purge Flow Test. To perform a purge flow test, the PCM waits until it is operating in closed-loop fuel control and then suddenly turns the purge solenoid on while monitoring the oxygen sensors. If purging of the fuel vapors from the charcoal canister is successful, the oxygen sensors should respond by indicating a rich air-fuel ratio.

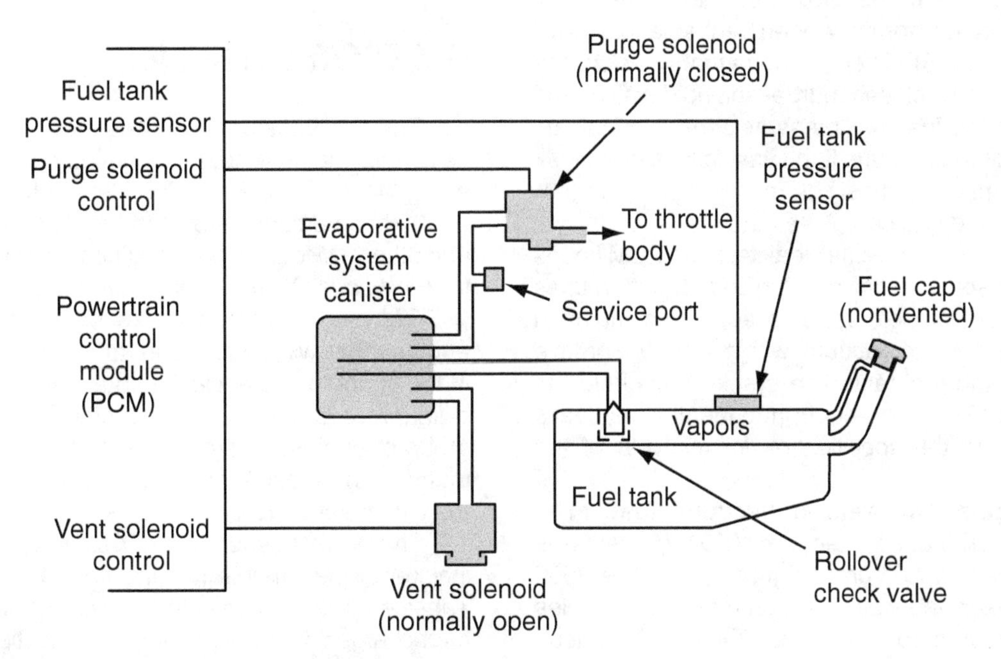

Figure 12–48 Evaporative system components.

Evaporative Vacuum Test. On early GM OBD II vehicles, the PCM causes a small amount of manifold vacuum to be applied to the evaporative system and fuel tank to check the system for leaks. The PCM will also verify the accuracy of the fuel tank pressure sensor by comparing it against the MAP sensor during the initial key-on cycle.

To run this test, the PCM checks ambient temperature as the engine is started. If it is 40°F or warmer, it is assumed that heat radiating up from the ground underneath the fuel tank will cause some fuel vaporization to take place. In turn, this will charge the charcoal canister with fuel vapors. The PCM will then wait until it is operating in closed-loop fuel control and then turn purging on while monitoring the oxygen sensors and STFT. If the oxygen sensors switch rich and the STFT values begin decreasing, the PCM knows that the purge function is working correctly and that the fuel tank's vapors are stored in the canister. However, if the oxygen sensors and STFT do not respond appropriately, the PCM performs a *gross leak test* and a *small leak test*.

To perform the gross leak test, the PCM energizes the normally open vent solenoid valve (Figure 12–48). Then it momentarily energizes the normally closed purge solenoid valve and watches the fuel tank pressure sensor to indicate a change in pressure. If this test fails, a large leak is presumed, the most common one being a problem with the gas cap.

To perform the small leak test, the PCM keeps the vent solenoid valve energized, and then uses the purge solenoid valve to apply and maintain a small level of vacuum within the evaporative system and fuel tank. The system must hold the vacuum signal long enough to identify any leaks larger than the specification for the year of the vehicle.

Engine Off Natural Vacuum Test. The *engine off natural vacuum (EONV)* test was introduced by GM on its light trucks in the 2003 model year and was found on most GM vehicles by the 2005 model year. The EONV test replaced the former vacuum test and required no hardware changes in the system. Minor electrical changes were implemented in order to keep the PCM powered up after the ignition switch is turned off. And the PCM's program was modified to allow it to perform this test.

When the vehicle is driven, heat from the engine and exhaust system will cause a slight increase in temperature of the fuel in the fuel tank which, in turn, causes it to expand. To perform the EONV test, the PCM is kept powered up for up to 40 minutes after the ignition switch is turned off. During this time the PCM keeps the purge solenoid valve de-energized and the vent solenoid valve energized, thus keeping both valves closed. As the fuel cools, it contracts and causes a *natural vacuum* to be formed within the evaporative system and fuel tank assembly, from about 3 inches to about 5 inches of water. During this time the PCM monitors the fuel tank pressure sensor to verify that the system is capable of holding this negative pressure.

With any sensed failure of the evaporative system, the PCM will illuminate the MIL and store the appropriate DTC in memory.

GM ECOTEC ENGINE

The GM Ecotec engine was initially designed as an inline four-cylinder all-aluminum engine with a displacement of 2.0 to 2.5 liters. Introduced in 2000, these engines replaced the Quad 4 engine and other GM four-cylinder engines. *Ecotec* stands for *Emissions Control Optimization Technology*.

GM has introduced a new series of Ecotec engines that will be found in its smaller vehicles in the years to come. Designed as either a three-cylinder or a four-cylinder engine, the newest group of engines in the Ecotec family will first be found in the Opel Adam and Chinese version of the Chevrolet Cruze.

The newest generation of Ecotec engines will include displacements ranging from 1.0L to 1.5L. These engines include both naturally aspirated and turbocharged versions. Power specifications vary from 75 to 165 HP and 70 to 184 ft.-lb of torque.

GM states that, by 2017, it will build as many as 2.5 million small-displacement Ecotec engines for 27 different vehicle models sold throughout the world and that, by 2017, one in every four of the engines GM produces will be part of the Ecotec family.

OTHER PCM-CONTROLLED SYSTEMS

Variable Valve Timing

Valve timing affects the cylinders' intake and exhaust flow. In turn, this affects intake manifold vacuum, volumetric efficiency, throttle response, and how much horsepower and torque the engine can develop. Most engines over many years have used a camshaft that has a fixed timing. This fixed timing is, at best, a compromise. Advancing valve timing improves idle quality and low-RPM torque. Retarding valve timing boosts high-end performance.

The variable valve timing (VVT) system on GM engines uses a mechanism called a "cam phaser" on the front end of each camshaft. Early GM cam phasers were designed with a helical spline that allows them to change their position in relation to the camshaft drive sprocket (Figure 12–49 and Figure 12–50). This in turn changes the camshaft's relationship to the crankshaft, thus effectively changing cam timing. The PCM controls a hydraulic solenoid, which can apply oil pressure to either side of the cam phaser *piston*, thus moving it in either direction (Figure 12–51). In one direction, cam phaser piston movement advances valve timing. In the other direction, cam phaser piston movement retards valve timing.

GM's newer design of cam phaser uses a vane-style rotor (Figure 12–52). A vane-style phaser responds more quickly to PCM commands. An oil pressure solenoid is used by the PCM to advance and retard cam timing.

With either design of cam phaser, the PCM controls the oil control solenoid on a pulse-width-

Figure 12–49 GM VVT cam phaser.

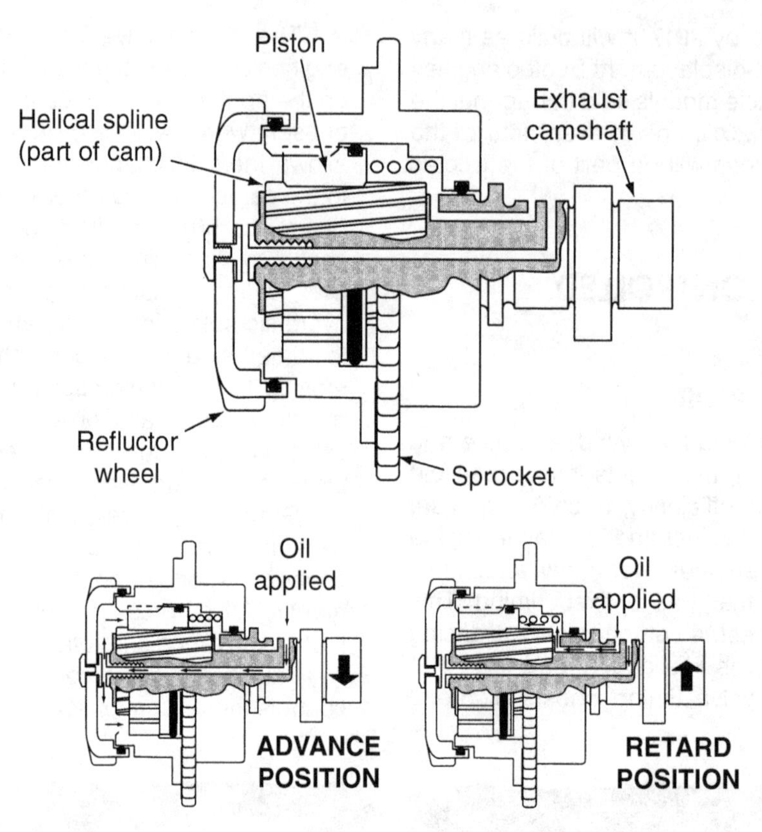

Figure 12–50 General Motors' VVT system.

modulated signal. This allows the PCM to precisely position the camshaft. The PCM, using input from the CMP and CKP sensors, can calculate the relationship of the camshaft to the crankshaft and therefore can determine the amount of advance or retard. Unlike most other manufacturers, GM's cam phaser is spring-loaded to the fully advanced position as the default setting.

On GM engines with independent control of the intake and exhaust cams, the PCM can control cam timing for the intake cams and the exhaust cams in a continuous fashion. The intake cam can be varied by as much as 40 degrees from one extreme to the other, and the exhaust cam can be varied by as much as 50 degrees from one extreme to the other. The PCM can either retard or advance the selected cam within the total potential range of adjustment. The PCM positions the cams for smooth engine operation when the engine is at low RPM. Under high RPM and load, the PCM positions the cams for best performance.

The PCM can increase the amount of inert exhaust gas retained within the cylinder using two methods. One method is to close the exhaust valve early, thus reducing the amount of time that the exhaust valve is open. Another method is to increase valve overlap by opening the intake valve earlier. This has the effect of allowing some of the exhaust gas to be drawn momentarily into the low-pressure area of the intake manifold and then be drawn back into the cylinder as the piston begins to move downward. This feature can reduce NO_x emissions without increasing hydrocarbon emissions as an EGR system sometimes does.

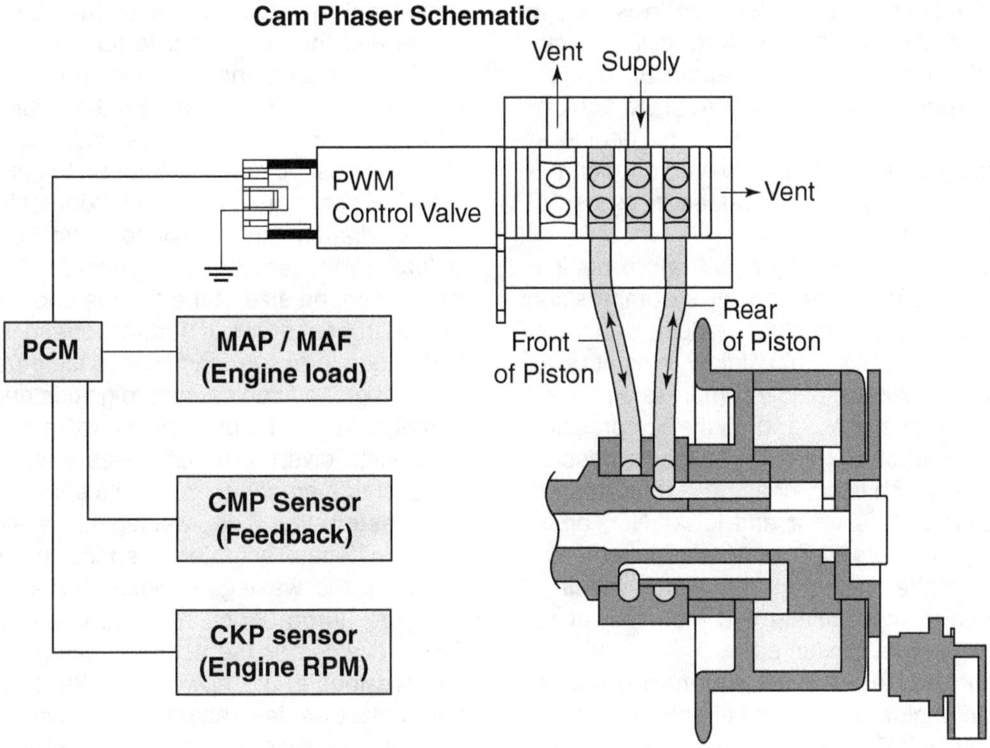

Cam Phaser Schematic

Vent
Supply
PWM
Control Valve
Vent
PCM
MAP / MAF
(Engine load)
Front
of Piston
Rear
of Piston
CMP Sensor
(Feedback)
CKP sensor
(Engine RPM)

Figure 12–51 Operation of a GM cam phaser.

Photo courtesy of William K. Bencini

Figure 12–52 Vane-type cam phaser.

Turbocharger

The major factor that determines engine power is the quantity of air and fuel that is delivered to the cylinders. Putting more fuel in is a fairly simple matter. Putting more air in is not as easy. The amount of air that can be put into the cylinder is known as *volumetric efficiency (VE)* and is limited by atmospheric pressure unless some means is used to force air in at greater than atmospheric pressure. VE can be increased with some type of supercharger or turbocharger. The turbocharger is the most popular method.

A turbocharger is a centrifugal, variable-displacement air pump driven by otherwise wasted heat energy in the exhaust stream. It pumps air into the intake manifold at pressures that are limited only by the pressure that can be developed without significantly increasing the air's temperature.

Most **normally aspirated engines** (those that rely on atmospheric pressure to fill the cylinder) only fill the cylinder to about 85 percent of atmospheric pressure. They actually achieve 85 percent VE at full throttle with the engine at its maximum torque speed. However, turbocharging potentially produces VE values in excess of 100 percent.

Because a turbocharger puts more air into the cylinder, it raises the engine's compression pressure. Because it does not produce significant boost pressure at low exhaust flow rates, it gives an engine the benefits of low compression during light throttle operation and high compression during heavy throttle operation. Low compression offers lower combustion chamber temperatures, which result in less wear and lower NO_x emissions. High compression provides higher combustion chamber temperatures, which result in more complete combustion and more power for the amount of fuel consumed.

Another way to look at it is in terms of displacement. If an engine is operating at an atmospheric pressure of 14.7 PSI, and the turbocharger is producing a boost of another 7.35 PSI, the cylinders are potentially charged at a pressure of approximately 22 PSI. The engine is consuming approximately 50 percent more air than it could without the turbocharger. We have effectively increased the engine's displacement by 50 percent. Admittedly, the engine does consume 50 percent more fuel as well. But the engine does not consume as much fuel as an engine with 50 percent more displacement because the higher compression pressure and temperature produce more complete combustion. The turbocharged engine also weighs less than an equivalently powerful naturally aspirated engine, thus saving fuel as well.

Turbocharger Operation. After leaving the manifold, the hot exhaust gases flow through the vanes of the exhaust turbine wheel and thus spin it (Figure 12–53). As the engine's production of exhaust increases, the turbine spins faster. It can achieve speeds in excess of 130,000 RPM. The exhaust turbine drives a short shaft that drives a compressor turbine. The high speed produces centrifugal force to move the air from between the vanes and thus causes it to flow out in a radial fashion. The air is then channeled into the intake manifold. As air is thrown from the turbine vanes, a low pressure develops in its place. Atmospheric pressure pushes more air through the air cleaner, the MAF sensor (if used), and the throttle body.

A criterion turbine speed must be reached before a pressure boost is realized; this speed varies with the size of the turbine and the design of the turbocharger. If turbine speed goes too high, boost pressure and charge temperature will go too high and can cause preignition and engine damage. To limit boost pressure, a **wastegate** is used to divert exhaust gases away from the exhaust turbine and thus limit its speed.

Wastegate. The wastegate is controlled by a wastegate actuator. A spring in the actuator holds the wastegate closed. When manifold pressure (turbo boost) reaches approximately 8 PSI, it overcomes the actuator spring and opens the wastegate. If, however, engine operating parameters are favorable (coolant temperature or incoming air temperature, for example), the PCM pulses a wastegate solenoid, which in turn bleeds off some of the pressure acting on the actuator. When this occurs, boost pressure is allowed to rise to about 10 PSI before the wastegate opens. The PCM will illuminate the MIL and set a DTC in memory if:

- An over-boost is sensed by the MAP sensor.
- The PCM, monitoring the wastegate solenoid circuit operation, sees a malfunction in the circuit.

GM GDI engines use a dual scroll turbo assembly with two exhaust feeds. The dual exhaust feed spins up the turbine faster and reduces the turbocharger's reaction time, known as *turbo lag*.

Torque Converter Clutch

The torque converter clutch (TCC) is a hydraulically actuated lockup clutch inside the

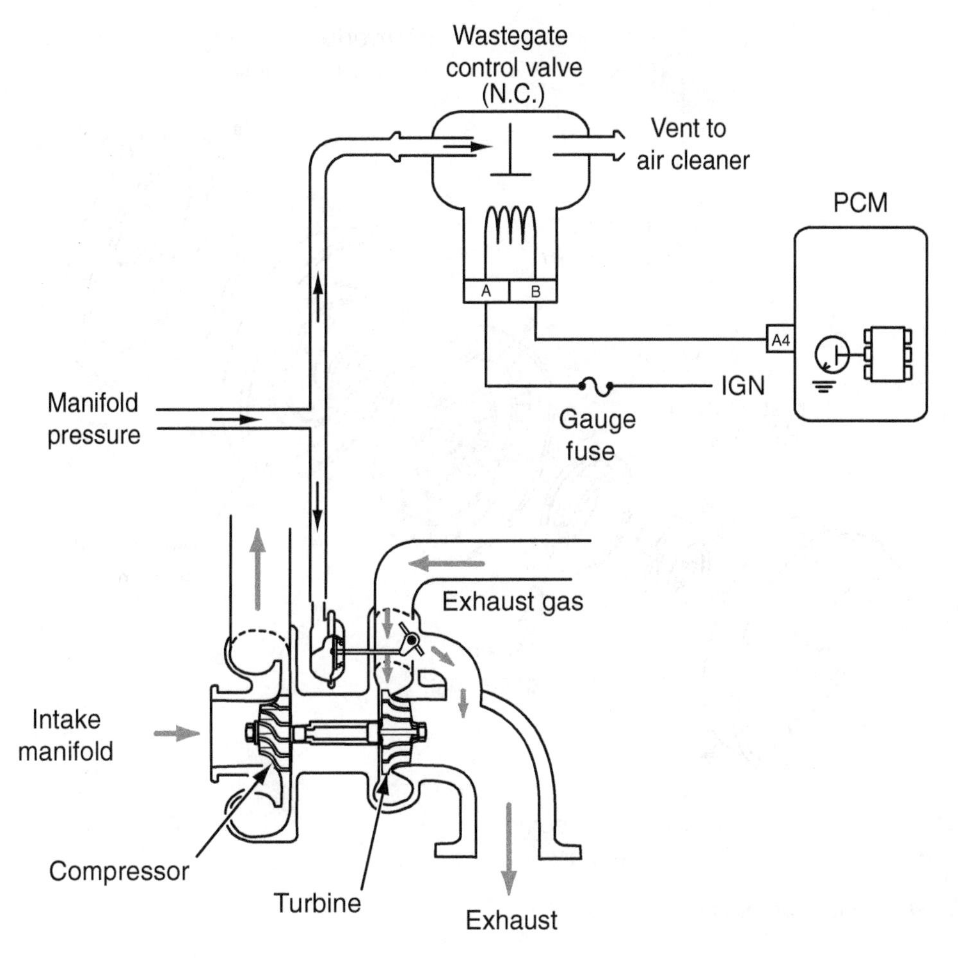

Figure 12–53 Turbocharger and wastegate control (typical).

transmission's torque converter. Once a cruise speed is achieved, the TCC is energized by the PCM. When energized, the TCC eliminates the slippage in the torque converter, which in turn improves the vehicle's fuel economy during cruise. This also reduces the heat produced within the torque converter, thus extending the life of the transmission fluid. When the PCM releases the lockup clutch the torque converter works normally. On some vehicles, engineers have used a torque converter with more converter slip than normal to maximize torque multiplication and smoothness, knowing they could eliminate all the slip once the vehicle was under way at cruise speed.

On early vehicles using a TCC lockup feature, the initial engagement could be felt by the driver and was sometimes misinterpreted as an extra gear shift. On modern applications the PCM controls the TCC on a high-frequency pulse-width-modulated signal in order to provide a smooth engagement of the torque converter clutch.

The major component of the TCC system is the lockup clutch itself. It becomes a fourth element added to a conventional torque converter (Figure 12–54). The clutch plate, splined to the turbine, has friction material bonded to its engine side near its outer circumference. The converter cover has a machined surface just inside, where

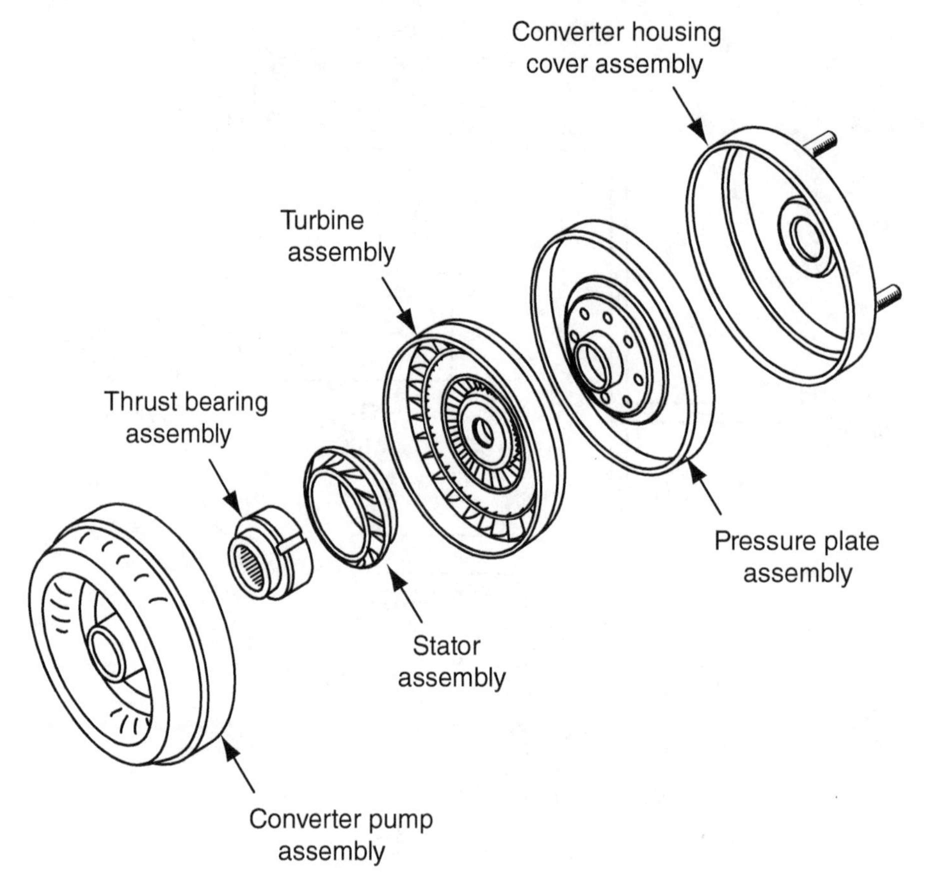

Figure 12–54 Torque converter with TCC.

the converter drive lugs attach. This machined surface mates with the friction material on the disc when the clutch is applied. When hydraulic pressure is applied to the turbine side of the disc, the disc is forced against the converter cover. The friction between the clutch's friction material and the cover locks up the unit and causes the torque converter to rotate as a solid unit. The pressure source is converter feed oil coming from the pressure regulator valve in the transmission valve body. In non-TCC applications, converter feed oil is used to charge and cool the converter by circulating through the converter and then the transmission cooler in the radiator. On TCC applications, however, converter feed oil must pass through a converter clutch apply valve on its way to and from the converter. The apply valve, a small valve in the transmission, controls the direction of oil flow through the torque converter (Figure 12–55).

With the apply valve in its de-energized position and the transmission in neutral or with the vehicle moving at low speed, converter feed oil is directed into the converter through the release passage by way of the hollow turbine shaft. This oil is fed into the converter between the converter cover and the clutch disc. The oil forces the disc away from the cover and thus releases the clutch. Converter feed oil flows over the circumference of the disc and circulates through the converter.

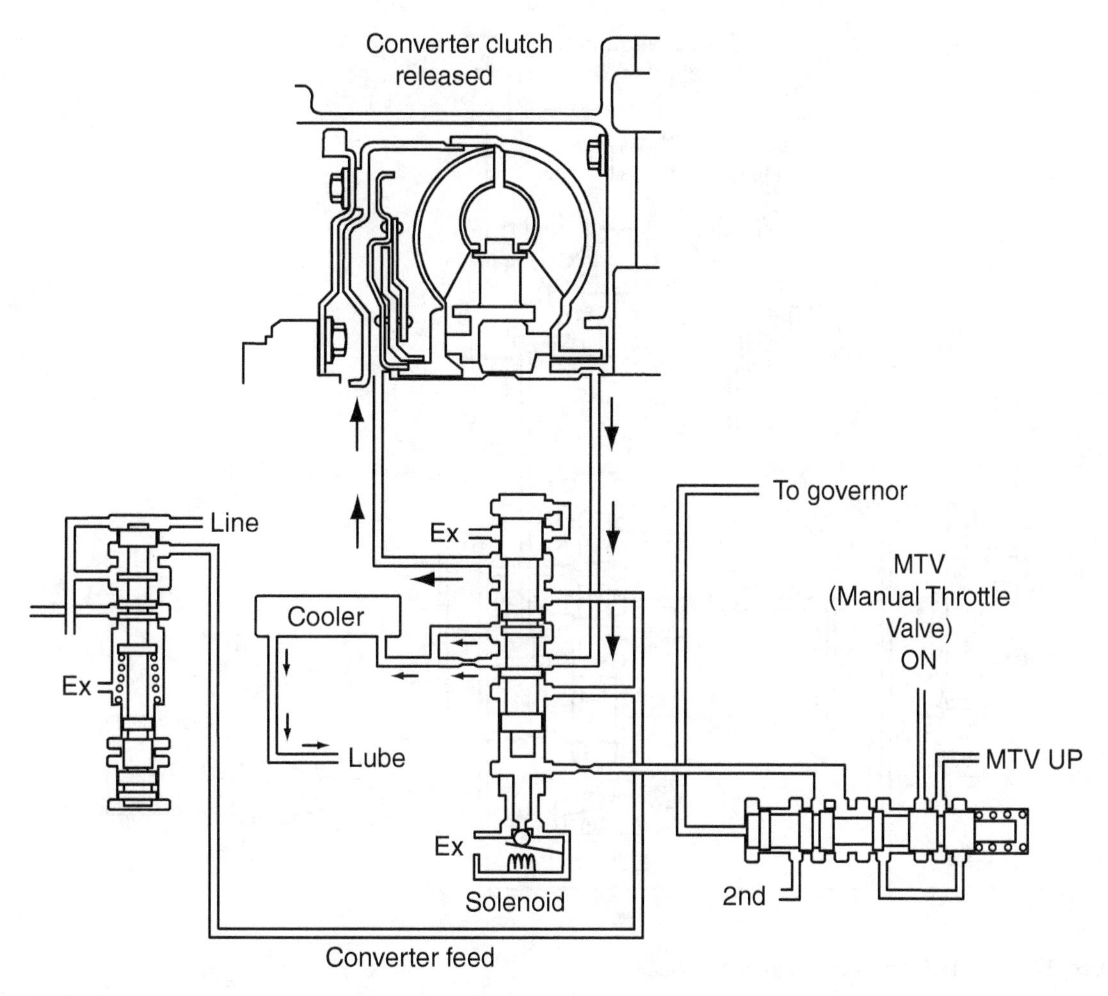

Figure 12–55 TCC apply circuit in release position.

It exits through the apply passage, a passage between the pump drive hub and the stator support shaft.

Several criteria must be met before the clutch applies: The transmission must be ready hydraulically, and the PCM must be satisfied with engine temperature, throttle position, engine load, and vehicle speed. When the transmission is in the right gear (this varies from transmission to transmission), hydraulic pressure is supplied at one end of the clutch apply valve (Figure 12–56). This applied hydraulic pressure can move the apply valve into the apply position. The TCC apply valve is moved, however, only if the PCM is ready for the lockup clutch to engage.

The PCM controls the position of the clutch apply valve with a solenoid that opens or closes an exhaust port at the converter apply signal end of the apply valve. If the solenoid is de-energized, the exhaust port is open and the converter apply signal oil exhausts from the signal end of the apply valve as fast as it arrives, and sufficient pressure does not develop to move the apply valve. Converter feed oil continues to flow

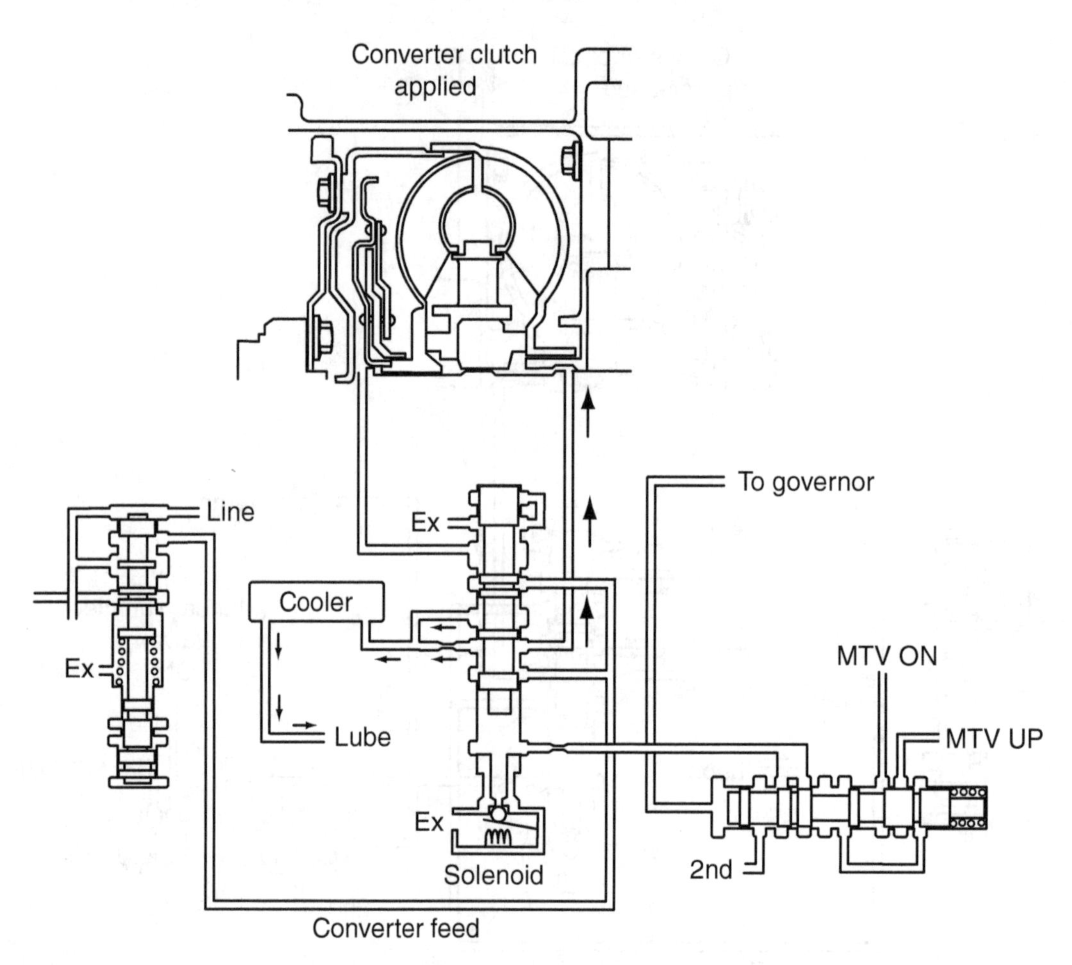

Figure 12–56 TCC apply circuit in apply position.

into the converter through the release passage (Figure 12–55). When the solenoid is energized, the exhaust port is blocked. The converter apply signal oil develops pressure, and the apply valve moves and is held in the apply position. Converter feed oil now flows into the converter through the apply passage (Figure 12–56). The clutch is applied and prevents oil from exhausting through the release pressure, holding static pressure.

Figure 12–57 shows a typical TCC solenoid control circuit. The power source is the ignition switch, through the gauge fuse. The brake switch is in series on the positive side of the circuit. Any time the brakes are applied, the solenoid is de-energized. The PCM controls the negative side of the circuit.

TCC Control Parameters. The PCM energizes the TCC apply solenoid circuit when:

- The criterion speed is reached according to the VSS. On some cars this can be as low as 24 to 36 MPH, depending on throttle position and transmission oil temperature.
- The engine has reached normal operating temperature. If the engine coolant temperature (ECT) is 64°F (18°C) or more when the engine starts, the PCM will apply the converter clutch when the coolant reaches 140°F

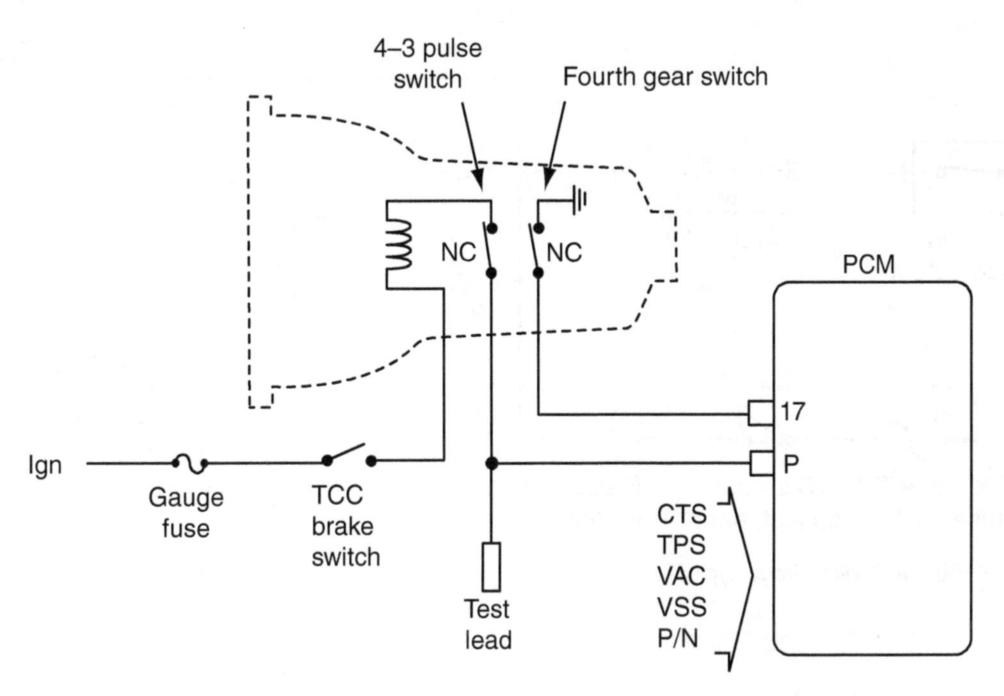

Figure 12–57 TCC electronic control circuit.

(60°C). If the coolant is colder than 64°F at startup, the PCM uses a fixed time delay before applying the clutch. The purpose of this delay is to improve low-temperature driveability until conditions allow lockup.

- The TPS signal is within a predetermined window that varies depending on what gear the transmission is in, among other factors. The TPS signal is compared to the vehicle speed signal and the PCM selects different apply speed thresholds for different combinations of TPS and VSS.

Early GM applications had a TCC test lead, which originated from the ground side of the solenoid circuit and could be used to indicate when the PCM had electrically applied the TCC or could be used by the technician to ground the solenoid circuit and energize the TCC manually for testing. GM has since deleted this circuit. Diagnosis of this circuit on a modern vehicle involves the use of a scan tool to retrieve DTCs, retrieve data stream information, and perform active commands.

Air Conditioning (A/C) Control

The PCM controls the relay that turns the A/C clutch on and off for two and in some cases three reasons:

1. When the A/C control switch (on the instrument panel) is turned on, the PCM delays A/C clutch engagement momentarily to allow time for it to adjust idle speed to compensate for the expected compressor clutch engagement.
2. The PCM disengages the A/C clutch during WOT operation.
3. On some applications, the PCM turns off the A/C clutch if power steering (P/S) pressure exceeds a specified value during idle. On others the P/S switch is in series with either the A/C clutch or the winding of the A/C relay and disengages the A/C clutch without relying on the PCM.

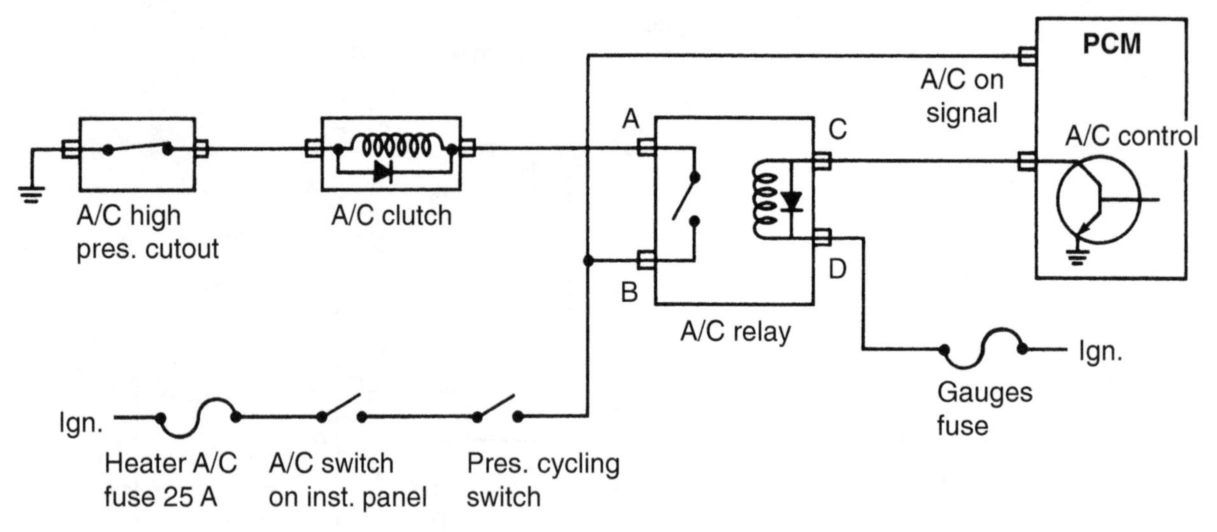

Figure 12–58 A/C relay circuit (typical).

The electrical schematic in Figure 12–58 illustrates a typical PCM-controlled A/C clutch.

Electric Cooling Fan

All GM vehicles with transverse-mounted engines and a few with longitudinal engines (parallel to the center line of the car) are equipped with an electric cooling fan to pull air through the radiator and A/C condenser. Many GM vehicles are equipped with dual cooling fans. With dual fans, operation may involve operating only one fan under some conditions and both fans under other conditions. Control of the fan(s) varies with engine application. In all cases, however, the fan(s) is/are turned on when:

- Coolant temperature exceeds a specified value. This is done by either the PCM or a coolant temperature override switch on some applications, and by the PCM only on others.
- A/C compressor output pressure (head pressure) exceeds a specified value. This is done by a switch in the high-pressure side of the A/C system on some engine applications. It is done by the PCM on other applications (on these applications, an A/C head pressure switch feeds head pressure information to the PCM).

On most applications, the PCM turns on the fan any time the A/C is on and vehicle speed is less than a specified value. Others turn it on under a specified speed whether the A/C is on or off. Once the vehicle reaches a criterion speed, enough air is pushed through the radiator without the aid of the fan, unless overheating or high A/C head pressure conditions exist. This is a fuel economy feature.

On many modern GM vehicles the cooling fan motor(s) is/are controlled by the PCM with the use of relays (Figure 12–59). On other modern GM vehicles the cooling fan motor is typically controlled by either the PCM or the body control module (BCM) on a high-frequency pulse-width-modulated command (Figure 12–60 and Figure 12–61). This PWM control provides the ultimate precision in control and provides enough fan operation to control coolant temperature while limiting current draw to a minimum to provide maximum fuel mileage.

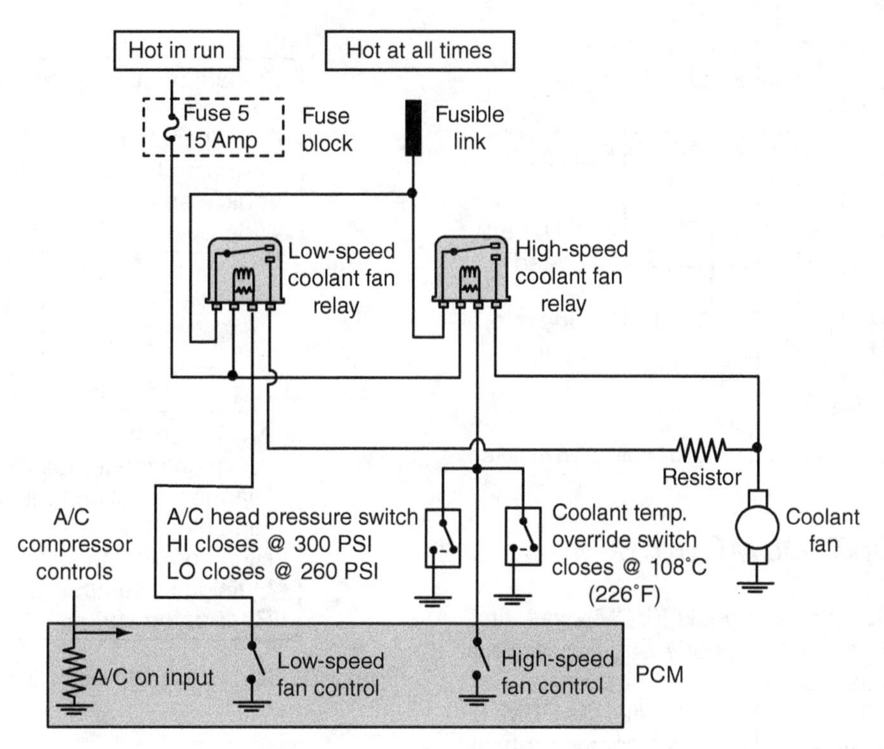

Figure 12–59 Cooling fan control circuit (typical).

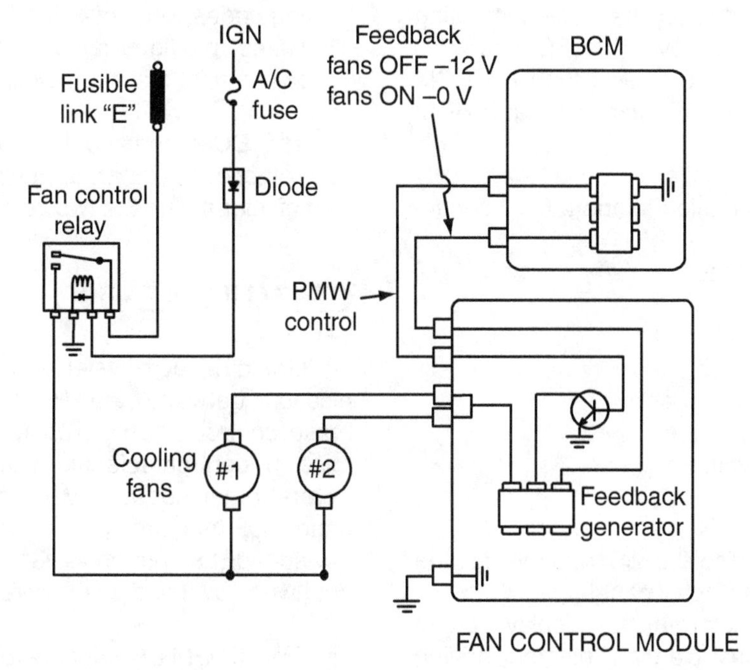

Figure 12–60 Cooling fan control.

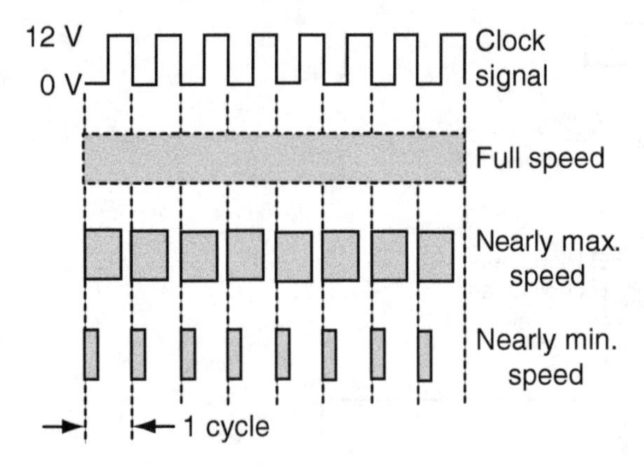

Figure 12–61 PWM command to cooling fan motors.

BODY CONTROL MODULE

The body control module (BCM) was first introduced on the 1985 C-body Cadillac. Its use has been expanded to include most other GM vehicles over the years. The BCM, like the PCM, is a microcomputer with all the basic components of the engine computer (ROM, RAM, PROM, input/output interfaces) and about the same computing power as the PCM. However, the BCM controls different functions on the vehicle than the PCM does. Following is a list of functions that a BCM may control:

- Electronic automatic temperature control (EATC)
- Electric cooling fans
- Power windows
- Power door locks
- Memory seats
- Heated and cooled seats
- Rear window defogger
- Heated outside mirrors
- Trunk release
- Electric sunroof
- Provides and displays information to the driver through a display panel
- Controls the information display panel dimming for visual clarity in different driving conditions

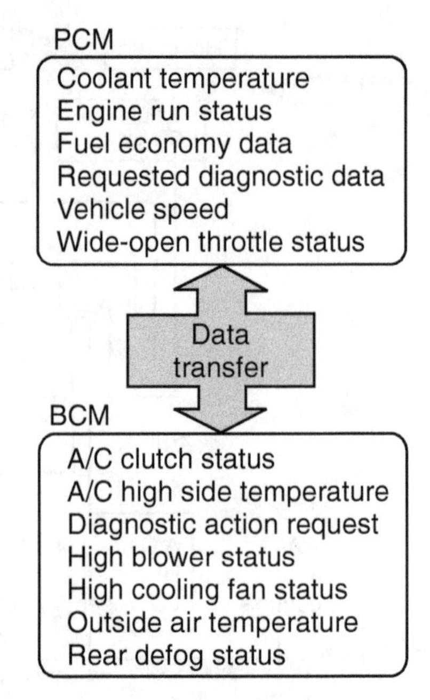

Figure 12–62 BCM/PCM interaction.

- Monitors the BCM system for faults, storing codes to identify the faults and, in some cases, provides fail-safe measures to compensate for a system failure

The BCM and the PCM communicate with each other over a data bus. Figure 12–62 shows the kind of information the two computers exchange.

GM MULTIPLEXING

GM data bus systems provide for communication between the PCM, BCM, electronic brake control module (EBCM), automatic transmission control solenoid valve assembly, and several other nodes. GM uses high-speed (HS) protocols, medium-speed (MS) protocols, and low-speed (LS) protocols. GM uses the following dual-wire (DW) and single-wire (SW) networks:

- DW HS GM LAN (local area network). This is GM's version of DW HS CAN. At a baud rate

of 500 Kb/s, it connects to terminals 6 and 14 of the DLC.
- DW MS GM LAN. This network connects to terminals 3 and 11 of the DLC.
- Object Detection Bus. This network connects to the rear view camera and front/rear exterior radar sensors. This network also connects to terminals 3 and 11 of the DLC. When present on the vehicle, the Object Detection Bus takes priority over and replaces the MS LAN bus.
- DW CAN Graphical Interface.
- DW MOST, 50 Mb/s (100 times faster than HS LAN).
- SW LS GM LAN. This is GM's version of SW LS CAN. This network connects to terminal 1 at the DLC.
- SW LS LIN.
- SW MS J1850, known by GM as its Class 2 protocol. This network, when used, communicates at a baud rate of 10.4 Kb/s and connects to terminal 2 at the DLC.
- KWP (Keyword Protocol 2000) was used by Cadillac for a while, but its use has been discontinued on all U.S. vehicles, although it is still found in Europe. When used, it connects to terminals 7 and 15 at the DLC.

GM's DW LAN networks use two 120 Ω termination resistors connected in parallel to each other. With the ignition turned off long enough for the nodes to stop communicating, the resistance at the DLC between terminals 6 and 14 (HS LAN), or 3 and 11 (MS LAN) should read about 60 Ω. If an open develops in one data bus wire, the resistance will read 120 Ω. Most opens in these circuits are wiring problems, not a problem with the termination resistor.

CAUTION: It is important to note that some GM vehicles use the radio as a gateway on the MOST data bus. On these vehicles, if the radio is removed from the vehicle, the engine will not start, and other symptoms will also be present.

ANTITHEFT SYSTEMS

As part of an antitheft system, the PCM gets information from the theft deterrent module to identify the key used for the ignition switch and steering lock. If the wrong key is used or if the system fails, the PCM disables the fuel injection system to prevent vehicle theft. On many GM vehicles, the theft deterrent module communicates over the LS LAN data bus with the BCM and then over the HS LAN data bus with the PCM.

ONSTAR

GM's OnStar system uses a module that constitutes a node on a data bus. It also adds satellite communication capabilities to the vehicle's control modules. The system provides many features, including:

- Automatic crash response
- Roadside assistance
- Stolen vehicle assistance
- Navigation and turn-by-turn directions
- RemoteLink mobile app
- Hands-free calling
- 4G LTE Wi-Fi hotspot

OnStar can also enable the manufacturer to retrieve vehicle diagnostic reports via satellite.

GM VOICE-RECOGNITION/ NAVIGATIONAL SYSTEM

General Motors introduced a combination voice recognition/navigational system on 2004 Cadillac SRX and XLR models. For model year 2005, this system was added to the Cadillac STS. For model year 2006, it was added to the Cadillac DeVille and the Buick Lucerne. Today, a voice recognition system is found on many GM vehicles.

The system allows the driver to press a button on the steering wheel labeled "VOICE" that activates a voice recognition system for issuing commands to the nodes on the data buses.

Voice entry of a desired command is achieved by depressing the Voice button on the steering wheel and then speaking the desired command aloud. The voice module then communicates the command to the appropriate module.

CADILLAC USER EXPERIENCE

Cadillac introduced an infotainment system on its 2013 model year vehicles known as the *Cadillac User Experience (CUE)* system. This system includes a navigational system and audio system with USB ports and a capacitive touch display screen. It also has a reconfigurable instrument cluster display through the use of control buttons on the steering wheel. The instrument cluster can be configured to provide an analog gauge display or a digital gauge display. And many driver information features can be selected to be displayed or can be deleted from the display. Whether an analog speedometer or a digital speedometer is selected for display in the instrument cluster, an additional heads-up display (HUD) provides a digital speedometer display on the windshield. This electronic display system can truly be personalized for each driver.

✔ SYSTEM DIAGNOSIS AND SERVICE

The first step in any diagnosis is to try to verify the customer complaint. A careful visual inspection should be performed next. One of the most important points of this inspection is to check the electrical engine grounds. Familiarize yourself with those locations (they vary with the vehicle) and learn to use a voltage drop test to check for resistance. While checking for grounds is often described as a visual inspection, it is, of course, nothing of the sort. Use voltage drop measurements to make these checks. Also use voltage drop tests to test the powers and grounds for the computer.

Many intermittent problems are the result of poor ground connections. Check for manifold vacuum leaks, loose connectors, and each of the ordinary problems that might disable any vehicle, computer controlled or not. Check for damaged or collapsed air ducts, as well as air leaks around the throttle body or sensor mating surfaces. Check any secondary ignition cables visually for cracks, hardness, and proper routing. Also check them for electrical leakage and continuity.

Check that the MIL is working properly. It should illuminate with the ignition switch turned on, but the engine not running, as a bulb check. If it does not illuminate, use a scan tool to verify the MIL command status.

The next step is to check for stored DTCs. Connect a scan tool to the DLC (which will be visible from a squatting position at the driver's door). Following the instructions on the tool, read out any stored DTCs and record them. Compare the stored codes to the code definitions found in the service manual. Also record the freeze frame data stored with a DTC.

Make sure the engine is not overheating. GM engines include a feature, called the *engine metal overtemp mode,* whereby at temperatures above 270°F the PCM will selectively disable half of the cylinders at a time on a V6 or V8 engine to keep the engine temperature from reaching destructive levels. This injector disabling will be felt by the driver as roughness and a loss of power.

Testing Fuel Pressure on an SFI Engine

Fuel pressure is critical to the performance of a fuel-injected engine. Most GM SFI systems have a Schrader valve that is found on the fuel line. For those systems that do not have a Schrader valve, the system must be depressurized and a pressure gauge must be connected to the fuel system using specialized hardware.

WARNING: While the fuel pressure gauge hose is screwed to the test fitting, a shop towel should be wrapped around the fitting to prevent gasoline from being sprayed on the engine. Remember, the fuel is under pressure, and the residual check valve in the fuel pump can hold pressure indefinitely after the engine is shut off. Spraying fuel can be not only a fire danger, but also a risk to the eyes, ears, and nose. Be sure to wear your safety glasses when performing any fuel system work.

Testing Fuel Pressure on an GDI Engine

Fuel pressure tests on a GM GDI engine should be made strictly with the use of a scan tool. The PCM uses pressure sensors on both the low-pressure and high-pressure sides of the system and is capable of showing both pressures over a scan tool in the data stream.

WARNING: If you see what appears to be leaking fuel on a GDI engine, DO NOT reach toward the engine and feel for the fuel spray with your fingers. The fuel pressure at idle is high enough to easily split skin, and if snap-throttled in the service bay the pressure will quickly rise to over 1,000 PSI.

SUMMARY

In this chapter, we presented an overview of the General Motors computerized engine control systems. We looked at GM's inputs and outputs. We discussed the systems that are controlled by the PCM, including electronic sequential fuel injection (SFI), the engine coolant temperature fuel disable feature, active fuel management, gasoline direct injection (GDI), idle speed control, and GM's version of an electronic throttle control (ETC) system known as *throttle actuator control (TAC).*

We also discussed GM's waste spark and coil-on-plug (COP) ignition systems and took a look at the major emission systems. A discussion of GM's Ecotec engines was included, along with a discussion of other PCM-controlled systems, including variable valve timing, turbochargers, the torque converter clutch lockup feature, A/C system operation, and cooling fan operation. Finally, a brief discussion was provided concerning some of the electronic systems that interact with the PCM, including GM's multiplexing networks, antitheft system, OnStar system, voice recognition system, and the Cadillac User Experience (CUE) system.

▲ DIAGNOSTIC EXERCISE

A sequential port fuel-injected vehicle has been towed into the shop with a customer complaint of "Cranks but won't start." As the technician is verifying the symptom, he notices that he does hear the fuel pump operate for 2 seconds when he first turns on the ignition switch. In other tests, he finds that although power is supplied to the fuel injectors with the ignition turned on, there is no PCM pulse to any of the fuel injectors. At this point, what faults could be responsible for the no-start condition? What test(s) would you recommend be performed next to narrow the list of possible faults?

Review Questions

1. *Technician A* says that modern PCMs have a more advanced learning capacity than previous PCMs did. *Technician B* says that disconnecting the battery will always erase all of the learned data from a modern PCM's memory. Who is correct?
   A. *Technician A* only
   B. *Technician B* only
   C. Both technicians
   D. Neither technician

2. On a GM vehicle, if the PCM senses that a front drive wheel has lost traction, it may individually apply the brakes at one wheel to regain traction. In extreme cases, it may also reduce engine torque by disabling up to seven of the fuel injectors *except* under which of the following conditions?
   A. When the engine coolant temperature is below –40°F
   B. When the engine coolant level is low
   C. When the engine speed is below 600 RPM
   D. All of the above
3. A GM engine in a vehicle is started cold and is then allowed to warm. The ECT sensor voltage is being monitored. While the engine is warming, the voltage is decreasing. Then the voltage suddenly increases. *Technician A* says that this indicates that the ECT sensor is faulty and should be replaced. *Technician B* says that this indicates that the PCM is faulty and should be replaced. Who is correct?
   A. *Technician A* only
   B. *Technician B* only
   C. Both technicians
   D. Neither technician
4. During cranking mode, input from the TPS is important to the PCM to determine which of the following?
   A. How much to advance the spark timing
   B. How far to open the EGR valve
   C. Whether to initiate clear flood mode
   D. Whether to energize the fuel pump relay
5. If the PCM detects cylinder misfire on a GM OBD II application, it will do which of the following?
   A. Disable the spark for the misfiring cylinder
   B. Disable the fuel injector for the misfiring cylinder
   C. Advance spark timing for the misfiring cylinder
   D. Retard spark timing for the misfiring cylinder
6. On a GM vehicle, the PCM will set a code in memory if the power steering pressure switch is open and vehicle speed is in excess of which of the following speeds?
   A. 10 MPH
   B. 25 MPH
   C. 30 MPH
   D. 45 MPH
7. On a GM vehicle, if the PCM sets an EEPROM defect code in memory, what must the technician do?
   A. Replace the EEPROM
   B. Replace the PCM
   C. Reset the EEPROM by disconnecting the vehicle battery for 10 minutes
   D. Reset the EEPROM by removing it from the PCM, then reinstalling it
8. What is meant by the term "close-coupled catalytic converters"?
   A. The NO_x reduction bed is located immediately downstream of the CO/HC oxidizing bed.
   B. The catalytic converters are mounted directly adjacent to the exhaust manifolds.
   C. The NO_x reduction bed is located immediately upstream of the muffler, and the CO/HC oxidizing bed is located immediately downstream of the muffler.
   D. The CO/HC oxidizing bed is located immediately upstream of the muffler, and the NO_x reduction bed is located immediately downstream of the muffler.
9. Regarding the addition of a TAC/ETC system to an engine, which of the following statements is *false*?
   A. If one APP sensor has failed, the engine will not start.
   B. The throttle cable is eliminated.
   C. The speed control servo is eliminated.
   D. The IAC stepper motor is eliminated.

10. On a GM AFM system, how does the PCM respond when the need for power is reduced?
    A. The PCM disables the fuel injectors for one-half of the cylinders and allows the valves to continue to operate normally.
    B. The PCM disables the fuel injectors for one-half of the cylinders and keeps the valves on those cylinders *open*.
    C. The PCM disables the fuel injectors for one-half of the cylinders and keeps the valves on those cylinders *closed*.
    D. The PCM reduces fuel delivery to all cylinders.
11. What other technology is used to help enable the success of a modern AFM system?
    A. TAC
    B. VVT
    C. 42 V
    D. Engine metal overtemp mode
12. Which of the following is the most reliable method of testing the PCM's power and ground circuits?
    A. Use a test light to test circuit continuity.
    B. Use a DMM to test voltage drop across each circuit.
    C. Use a logic probe to test circuit continuity.
    D. Use an ohmmeter to measure each circuit's resistance.
13. In a return-type PFI/SFI system, why is the fuel pressure increased as the throttle moves from idle to WOT?
    A. To provide more fuel flow through the injectors any time they are open under heavy load conditions
    B. To keep the pressure differential across the injectors constant under all engine operating conditions
    C. To keep the flow rate through the injectors constant any time they are open
    D. Both B and C
14. On a PFI/SFI system, when does the PCM reference the count of the IAC's pintle valve by momentarily moving the valve to the closed position?
    A. Each time the ignition is turned off
    B. Each time the engine is started
    C. When the PCM sees 30 MPH from the VSS
    D. When the PCM sees 600 RPM from the tach reference signal
15. When diagnosing a "Cranks but won't start" symptom on an SFI engine, a technician finds that there is no injector pulse from the PCM. Which of the following problems could cause this?
    A. The IAT sensor circuit is open, resulting in a 5 V signal to the PCM.
    B. The ECT sensor circuit is open, resulting in a 5 V signal to the PCM.
    C. The REF signal from the CKP sensor has been lost.
    D. The fuel pump relay is defective.
16. An EGR system is designed to allow some exhaust gas into the intake manifold to reduce which of the following?
    A. HC emissions
    B. CO emissions
    C. CO_2 emissions
    D. NO_x emissions
17. *Technician A* says that turbochargers increase engine performance by increasing the engine's volumetric efficiency. *Technician B* says that a wastegate is used to limit a turbocharger's boost pressure to avoid preignition and engine damage. Who is correct?
    A. *Technician A* only
    B. *Technician B* only
    C. Both technicians
    D. Neither technician

18. The TCC is designed to do all *except* which of the following?
    A. Increase fuel economy
    B. Increase performance potential during WOT operation
    C. Eliminate hydraulic slippage in the torque converter during cruise conditions
    D. Eliminate heat production in the torque converter
19. GM's DW HS LAN network connects to which two terminals at the DLC?
    A. Terminals 2 and 10
    B. Terminals 3 and 11
    C. Terminals 6 and 14
    D. Terminals 7 and 15
20. GM's DW HS LAN network uses two 120 Ω termination resistors connected in parallel to each other. With the ignition turned off long enough for the nodes to stop communicating, what should the resistance measure at the DLC between the two data bus terminals?
    A. 60 Ω
    B. 120 Ω
    C. 240 Ω
    D. 480 Ω

Chapter 13

Ford Motor Company Computerized Engine Controls

OBJECTIVES

Upon completion and review of this chapter, you should be able to:

- ❑ Describe the major aspects of the Ford PCM.
- ❑ Define the inputs used with the Ford EEC V system.
- ❑ Understand the differences between the return-type fuel injection system and the electronic returnless fuel injection system.
- ❑ Describe the various types of ignition systems used with the Ford EEC V system, including both types of distributorless systems.
- ❑ Define the major emissions systems used with the Ford EEC V system.
- ❑ Describe some of the advanced features that are used in conjunction with EEC V systems, including Variable Cam Timing, Electronic Throttle Control, Intake Manifold Runner Control, Fail-Safe Cooling, Adaptive Cruise Control, Ford's voice recognition and navigational system, and Ford's AdvanceTrac™ system.
- ❑ Define the differences between Ford's GEM, REM, and FEM electronic control modules.

KEY TERMS

Adaptive Cruise Control
Adaptive Fuel Control
Adaptive Strategy
AdvanceTrac™
Alternate Fuel Compatibility
Conversational Speech Interface Technology
Fail-Safe Cooling
Front Electronic Module (FEM)
Generic Electronic Module (GEM)
Inertia Switch
Intake Manifold Runner Control (IMRC)
Intelligent Architecture
Rear Electronic Module (REM)
Throttle Plate Position Controller (TPPC)
Variable Cam Timing (VCT)

Ford Motor Company took a unique approach to meeting OBD II standards. Rather than modifying an existing system to meet OBD II standards as most other manufacturers did, Ford designed a new engine control system known as *Electronic Engine Control V (EEC V)*, with OBD II specifically in mind. Ford Motor Company's EEC V system is the sixth generation of Ford's computerized engine control systems, following MCU (Microprocessor Control Unit), EEC I, EEC II, EEC III, and EEC IV. The EEC V system carries over from EEC IV many of the features with which you are already familiar and adds several other features under OBD II standards in order to improve a technician's diagnostic effectiveness.

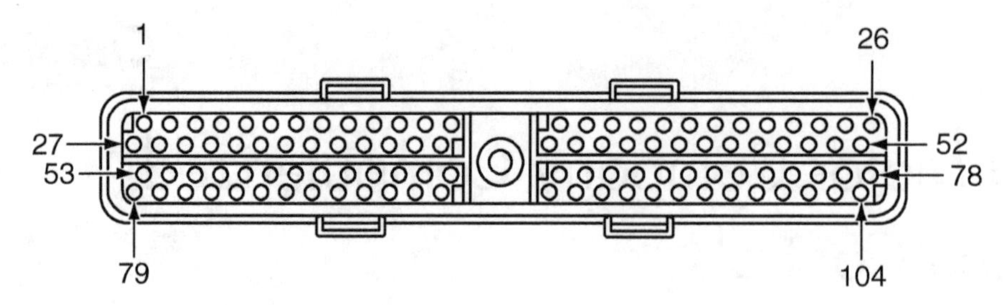

Figure 13–1 EEC V PCM with a 104-pin harness connector.

POWERTRAIN CONTROL MODULE

Initially, the EEC V powertrain control module (PCM) made its debut as a 104-pin processor (Figure 13–1). By making each of the terminals smaller and arranging them closer together in four rows, Ford engineers were able to design the 104-pin connector to be physically no larger than the 60-terminal EEC IV connector had been. The EEC V PCM quickly evolved to become a 150-terminal PCM (Figure 13–2).

The PCM controls fuel delivery and ignition timing throughout the range of the engine's operational capacity. It monitors the various sensors and switches whose information is relevant to calculating the proper values for fuel injector pulse widths and firing sequence, as well as to triggering actuation of the various components that execute its combustion commands.

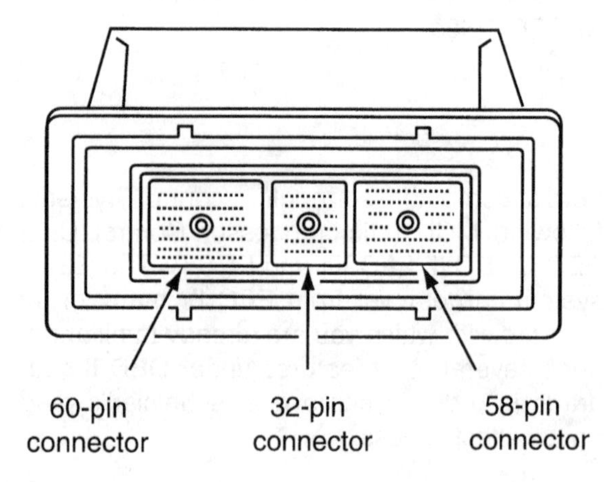

Figure 13–2 EEC V PCM with a 150-pin PCM harness connector.

The EEC V PCM assesses the gradual wearing and aging of the vehicle over time, as well as changes in altitude or barometric pressure; it can then make compensatory adjustments in its own programs to make whatever changes are needed due to the changing input values.

Adaptive Strategy

The **adaptive strategy** feature was first introduced by Ford in 1985. It enables the PCM to continuously adjust some of its original calibrations that engineers programmed into it based on ideal conditions. Whenever conditions are less than ideal, due to variations in manufacturing tolerances, wear, or deterioration of sensors or other components that affect the PCM's control of the most critical functions, the adaptive strategy function adds an adaptive factor into the calculation process. Adaptive strategy could also be called a *learning capacity* or a *self-correction ability*. The PCM learns from past experience so it can better control the present conditions. Learning starts when the engine is warm enough to go into closed loop and is in a stabilized mode.

Adaptive Fuel Control

Adaptive fuel control is an adaptive strategy that concerns the control of the air-fuel ratio. *Adaptive fuel control* is Ford's original name for what is known as *fuel trim* under the OBD II standardization of terminology. Fuel trim is divided into two functions: Short-term fuel trim (STFT) and long-term fuel trim (LTFT) and is discussed in greater depth in Chapter 4.

INPUTS

The principal inputs to the PCM are the heated exhaust gas oxygen (HEGO) sensors, the intake air temperature (IAT) sensor, the cylinder head temperature (CHT) sensor, the throttle position sensor (TPS), the mass airflow (MAF) sensor, the crankshaft position (CKP) sensor, the camshaft position (CMP) sensor, the vehicle speed sensor (VSS), the knock sensor (KS), and the power steering pressure (PSP) switch.

Heated Exhaust Gas Oxygen (HEGO) Sensors

Heated exhaust gas oxygen (HEGO) sensors are used to get today's engines into closed-loop fuel control earlier. This reduces cold start emissions. One HEGO sensor is placed before each catalytic converter and is used for closed-loop fuel control, and another is placed after each catalytic converter to monitor catalytic converter effectiveness (Figure 13–3). The post-cat HEGO sensor also influences closed-loop fuel control, known as *rear fuel control*.

If a post-cat HEGO sensor's signal is a reflection of the primary sensor's signal, the catalytic converter has lost its effectiveness at converting the exhaust gases. Because only a small inefficiency can result in a DTC for catalytic converter inefficiency, it is recommended that the technician not try to use the waveform of the downstream oxygen sensors as a diagnostic tool. If a catalyst efficiency DTC has been set for a catalytic converter, verify that there are no DTCs set for the oxygen sensors and that there are no exhaust leaks ahead of the either the pre-cat or post-cat sensors. Use a scan tool to check fuel trim values and to verify that the PCM is in proper control of the air-fuel ratio. Also, check for technical service

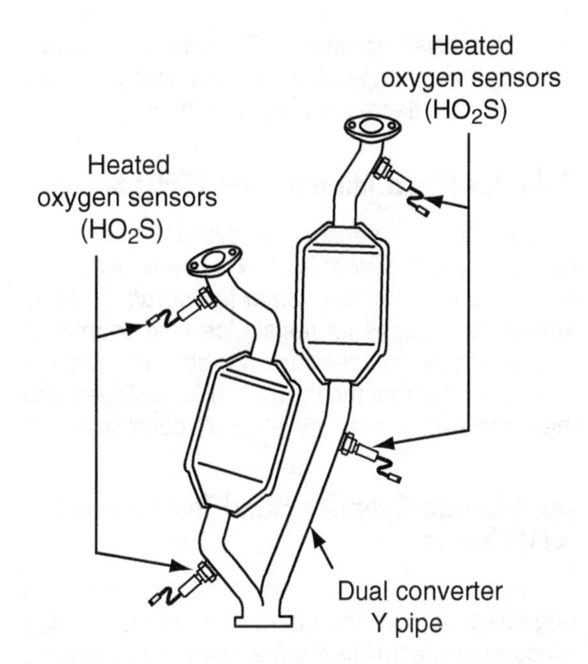

Figure 13–3 Heated exhaust gas oxygen sensors.

bulletins (TSBs). Otherwise, believe the DTC and replace the catalytic converter identified by the DTC.

Intake Air Temperature Sensor

The intake air temperature (IAT) sensor is an NTC thermistor that varies its resistance in response to the temperature of the intake air charge (Figure 13–4). As the sensor's temperature

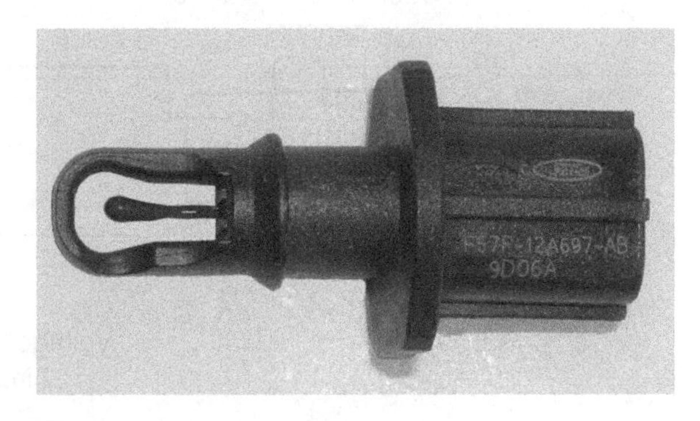

Figure 13–4 IAT sensor.

rises, the resistance of the IAT sensor decreases, causing the voltage drop measurement across the sensor to decrease proportionately.

Cylinder Head Temperature (CHT) Sensor

Like the IAT sensor, the cylinder head temperature (CHT) sensor is an NTC thermistor. Also known as the *engine coolant temperature (ECT)* sensor, it changes its resistance in response to engine coolant temperature. It is similar to the IAT sensor except that the thermistor is enclosed and therefore sealed from the engine's coolant.

Dual-Range Cylinder Head Temperature (CHT) Sensor

The CHT/ECT sensor that is used on engines that have the Fail-Safe Cooling strategy (described later in this chapter) also have a unique circuit. They are of a dual-range (dual temperature curve) type, similar to the dual-range ECT sensors being used by other manufacturers. The dual-range CHT/ECT sensor uses a circuit in the PCM that applies voltage to the sensor through a single fixed resistor within the PCM following a cold engine start. As the coolant begins to warm, the resulting reduction in the CHT/ECT sensor's resistance shifts some of the voltage drop from the sensor to the fixed resistor within the PCM. At about 170°F, the PCM completes a circuit through another fixed resistor in parallel to the circuit with the primary fixed resistor (Figure 13–5). This effectively lowers the resistance within the PCM, thereby shifting much of the voltage drop back out to the CHT/ECT sensor. As the engine continues to warm and the CHT/ECT sensor's resistance continues to decrease, the voltage drop is shifted back into the PCM's internal circuits for a second time. Ultimately, this has the effect of extending the measurable range of the sensor up into the over-temp area (Figure 13–6).

The PCM lowers its internal resistance in this circuit at about 170°F as the engine is warming, but it disconnects this circuit at about 130°F as the engine is cooling. This keeps the PCM from oscillating in and out of this mode if the engine is hovering around 170°F.

Throttle Position Sensor

On the side of the throttle body is a variable potentiometer that provides a return signal to the PCM directly proportional to the position of

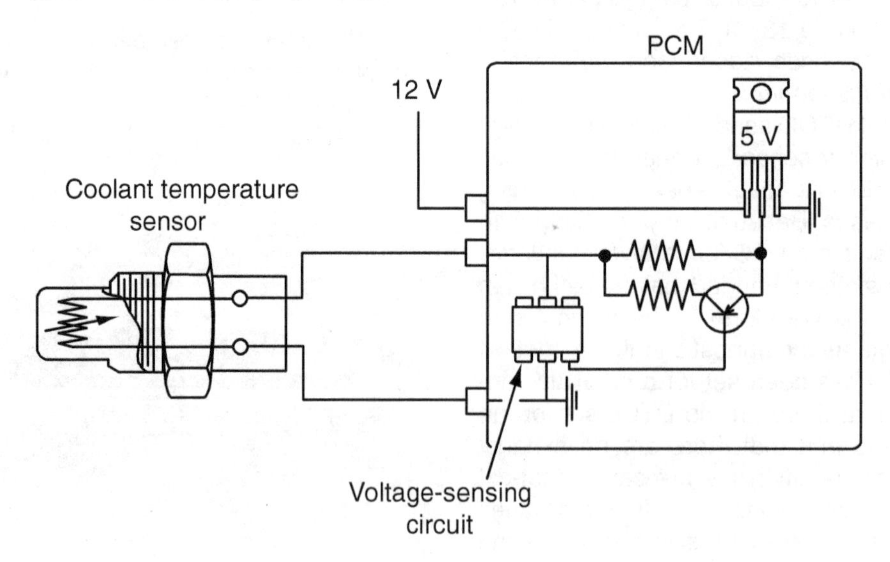

Figure 13–5 Ford dual-range CHT/ECT circuit.

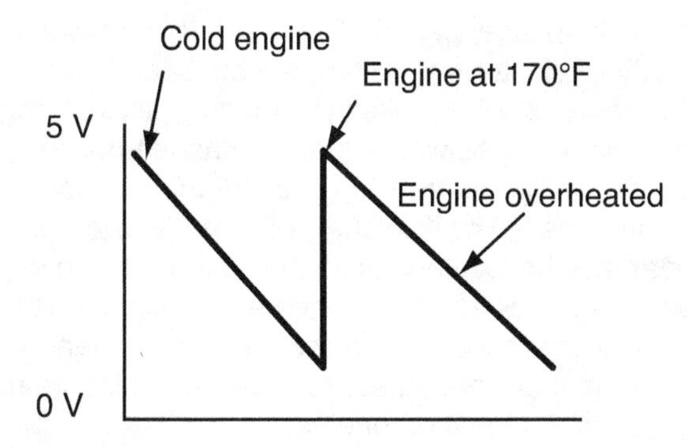

Figure 13–6 Dual-range CHT/ECT sensor waveform.

Figure 13–7 Throttle position sensor.

the throttle plate (Figure 13–7). This works the same way TP sensors have since 1981, and the potentiometer sweep tests are still quite similar—check for a smooth, continuous variation in the signal voltage throughout the throttle travel from idle to WOT.

Mass Air Flow Sensor

The Ford EEC V sequential fuel injection (SFI) system uses a hot-wire-type MAF sensor (Figure 13–8). The hot wire in the airstream is kept at a constant temperature above the ambient air, and a hot-wire-sensing element measures exactly how much current is required to maintain that heat. The Ford version of this sensor sends an analog voltage signal to the PCM corresponding to the mass of the intake air. Of course, as

Figure 13–8 MAF sensor.

with all such air mass sensors, it is very important to make sure there are no downstream leaks, as any intake air that does not go through the sensor will result in unmetered air. This, in turn, will cause the fuel trims to make a positive adjustment.

Manifold Absolute Pressure Sensor

Ford began phasing out the digital MAP sensor used with the EEC IV system when it introduced the MAF sensor in 1989. While the digital MAP sensor did survive on limited EEC IV applications such as F-Series trucks into the mid-1990s, it did not carry over into the EEC V system.

However, on many early Ford OBD II EEC V systems, the PCM communicates a digital frequency to the scan tool's data stream as a barometric pressure (BP) value. But, in the EEC V system, this frequency no longer originates at a digital MAP or BP sensor. Instead, the PCM uses idle and WOT values from the analog MAF sensor to artificially create the digital BP frequencies displayed in the data stream. As with the EEC IV system's digital MAP/BP sensor, the digital frequency that represents barometric pressure at sea level on an EEC V system is 159 Hz and is reduced approximately 3 Hz for each increase of 1,000 feet in altitude.

Therefore, Ford deleted the physical MAP sensor with the introduction of the EEC V system, and this system uses only a MAF sensor to measure engine load and to calculate barometric pressure. However, as explained later in this chapter, there are two exceptions to this:

- During the 2002 model year Ford modified the DPFE sensor to add an analog MAP sensor to it internally. The MAF sensor is retained on these systems as the engine load sensor, and the primary purpose of the analog MAP sensor is for emission system diagnostics.
- The Ford EcoBoost engine, introduced in the 2010 model year, uses a MAP sensor, instead of a MAF sensor, to determine engine load.

Crankshaft Position Sensor

The crankshaft position (CKP) sensor produces the tach reference signal for the PCM (Figure 13–9). It consists of a variable reluctance sensor (permanent magnet sensor) and produces an AC voltage signal. The sensor monitors a toothed reluctor ring with one tooth spaced every 10 degrees of crankshaft rotation, except for one blank space where a tooth was left out. This is commonly referred to as a "36-minus-1-tooth" arrangement. The tip of the sensor and the toothed gear may be located inside the timing gear cover, as on the 3.0L engine, or the sensor may monitor teeth that are located on the harmonic balancer, and it is then, of course, outside the timing gear cover, as on the 4.0L engine.

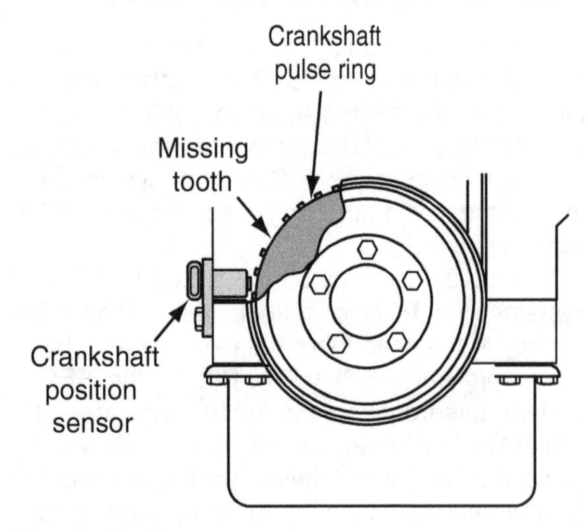

Figure 13–9 CKP sensor.

The missing tooth identifies a reference position of the crankshaft that identifies which pair of pistons is approaching TDC. Then the PCM can count off crankshaft rotation in 10-degree increments, identifying each piston pair as they approach TDC.

In the waste spark ignition system, this information allows the PCM to fire the correct ignition coil. In the coil-per-plug (CPP) ignition system, the signal from the CMP sensor is also needed to determine the appropriate ignition coil to fire, as well as to determine which fuel injector to pulse.

The CKP signal is considered a high-resolution signal, providing enough pulses per cylinder firing to allow the PCM to determine whether a cylinder is misfiring by watching the frequency of the pulses, as specified by OBD II standards. Even on an eight-cylinder engine, the PCM can count off nine teeth per cylinder firing (except, of course, for the missing tooth).

Camshaft Position Sensor

On the two-valve-per-cylinder engine, the CMP sensor is a single Hall effect sensor activated by a vane driven by the camshaft. On the four-valve-per-cylinder engine, the CMP sensor is a variable reluctance (permanent magnet) sensor triggered by the high point on the left exhaust camshaft sprocket. In both cases, the sensor provides the PCM with information used to fire the correct ignition coil in a COP ignition system and to synchronize the sequential fuel injection system with the cylinder firing order. On engines with **variable cam timing (VCT)**, each camshaft that is controlled by the PCM will be monitored, resulting in additional CMP sensors.

Vehicle Speed Sensor

The Ford VSS uses a magnetic pickup to send a rapid on/off pulse, the frequency of which corresponds to the vehicle's speed. The PCM uses this information, among other inputs, to calculate fuel metering, ignition timing, and control of the electronic transmission/transaxle.

Knock Sensor

The knock sensor (KS) is the usual piezoelectric crystal type, which produces an AC voltage signal in response to the occurrence of detonation (Figure 13–10). This information is used by the PCM to retard spark advance in order to eliminate the spark knock.

Power Steering Pressure Switch

The PSP switch is installed in the boost side of the power steering pump (Figure 13–11). Electrically, it is normally closed. When hydraulic boost builds to the point where it could be a load on the engine, the switch opens. This provides the computer with information relevant to setting the idle speed.

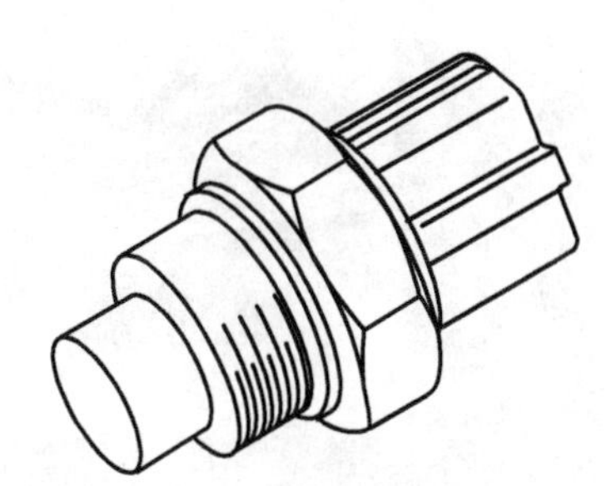

Figure 13–10 Knock sensor.

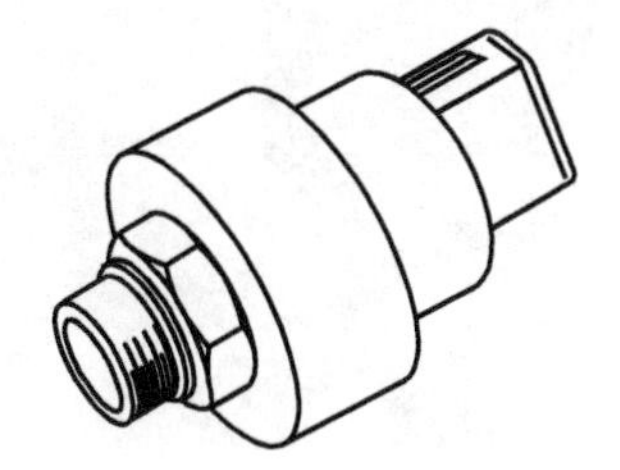

Figure 13–11 PSP switch.

FUEL MANAGEMENT SYSTEMS

As with other manufacturers, Ford Motor Company electronic fuel injection systems include the following systems:

- Throttle body injection (TBI), known by Ford as *Central Fuel Injection (CFI)*
- Port fuel injection (PFI), known by Ford as *Electronic Fuel Injection (EFI)*
- Sequential fuel injection (SFI), known originally by Ford as their *Sequential Electronic Fuel Injection (SEFI)* system, but currently referred to as *SFI*
- Gasoline direct injection (GDI), used currently on Ford EcoBoost engines

Electronic Fuel Injection System Safety Shutdown

Loss of Tach Reference Signal. As with other manufacturers' electronic fuel injection systems, if the tach reference signal to the PCM is lost, the PCM will de-energize the fuel pump relay within 2 seconds. As a result, if a fuel pressure line were to rupture during an accident, resulting in engine stall, the PCM would shut down the fuel pump to avoid the potential for emptying the fuel tank onto the ground through the ruptured line. Therefore, as a safety feature, the PCM will not keep the fuel pump relay energized when there is no tach reference signal coming from the CKP sensor.

There is, however, one exception to this: Even without a tach reference signal present, the PCM will always energize the fuel pump relay for 2 seconds whenever the ignition switch is first turned on and the PCM is initially powered up. This is used to run the fuel pump and top off the fuel system's residual pressure. As the starter motor begins cranking the engine, a tach reference signal is sent to the PCM from the CKP sensor, the PCM continues to keep the fuel pump relay energized, and the fuel pump continues to run.

Inertia Switch. Like earlier Ford EEC III and EEC IV fuel injection systems, the first 15 years of the EEC V fuel injection system employs an **inertia switch**, also called a "fuel pump shutoff switch." This component offers additional protection from gasoline fires in case of a collision. The inertia switch was implemented by Ford with the EEC III high-pressure CFI system in 1980 and continued to be used through the 2009 model year—a span of 30 years.

If a collision occurs, the jolt is likely to trip open the inertia switch, which in turn electrically removes power from the fuel pump. This is a safety feature used in addition to the "loss of the tach reference signal" concept just described. This helps to ensure that fuel from the fuel tank is not spilled onto the ground if a collision were to result in any of the following:

- Rupturing of the fuel return line. This would not result in an engine stall, but it could still spill fuel from the fuel tank onto the ground.
- Simultaneous rupturing of a fuel pressure line along with a shorting to ground of the fuel pump relay control circuit from the PCM, thereby negating the PCM's ability to disable the fuel pump. On Ford products, the PCM controls the ground side of the fuel pump. If, during an accident, this wire shorted to the metal of the vehicle, it could keep the fuel pump relay energized, even though the PCM had turned off the driver for this circuit.

Although Ford's inertia switches have been improved over the years, it is still possible to find one that has been tripped due to something less than an accident. For example, if the vehicle is bumped by another vehicle at slow speed, as in a parking lot, the inertia switch might trip even though no visible damage may have occurred. Conversely, an inertia switch may not trip in every accident. This is why the inertia switch is not used in place of the "loss of tach reference" concept, but is used in addition to it as a backup measure.

The inertia switch contains a steel ball and a cone (Figure 13–12). The ball is held in place in the center of the cone by a permanent magnet placed beneath it. An impact from any side dislodges the ball, allowing the cone to deflect the ball upward against a target plate, which opens a set of electrical contacts, pushing up a reset button. The inertia switch is mounted on a metal frame of the vehicle. It is located in the trunk or cargo area on early passenger cars. On newer passenger cars and most light trucks, minivans, and SUVs it is located either behind a kick panel or mounted on the interior side of the fire wall and just above the carpet, as shown in Figure 13–13.

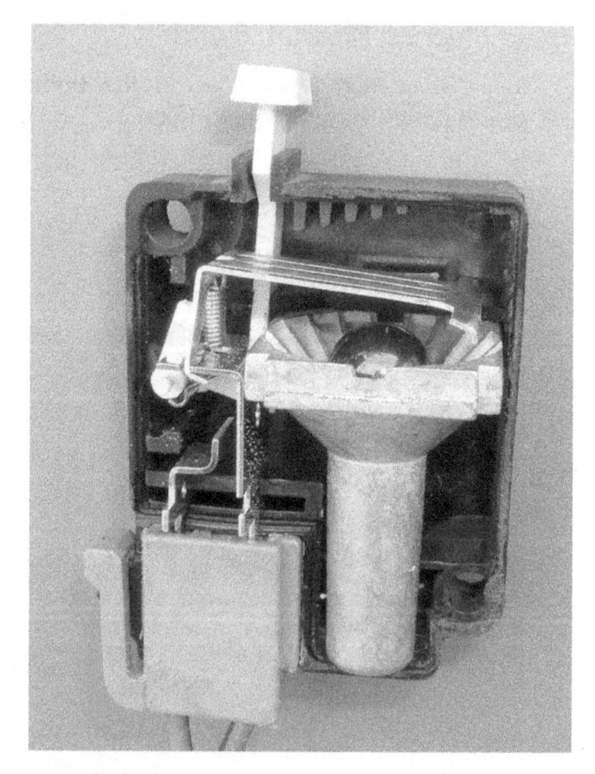

Figure 13–12 Inertia switch components.

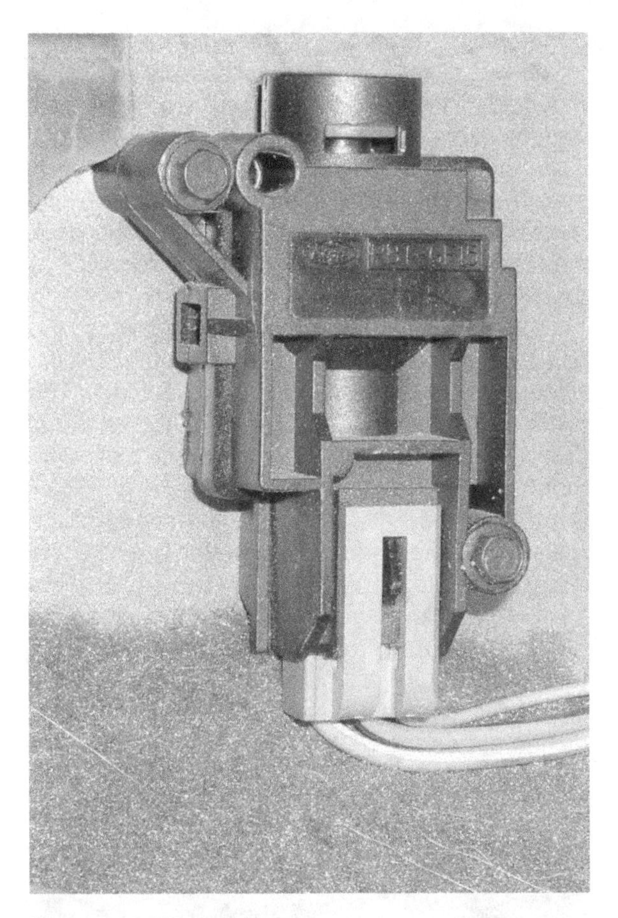

Figure 13–13 Inertia switch mounted to the interior side of the fire wall.

If the inertia switch has been tripped, the fuel pump can be re-enabled by depressing the reset switch on top. Experienced technicians will check the inertia switch's reset button as part of the initial visual inspection when diagnosing a no-start complaint.

CAUTION! When resetting a Ford inertia switch that has tripped, it is critically important for the technician to follow the resetting of the switch by pressurizing the fuel system and then performing a careful visual inspection of the entire fuel system to check for leaks. Failure to do this when resetting a Ford inertia switch can leave the technician liable for damage resulting from any existing fuel leaks.

The inertia switch was discontinued on most Ford vehicles in the 2010 model year due to the following two factors:

- The return line was eliminated when Ford began using the electronic returnless fuel injection system. Therefore, a rupture of the only remaining fuel line will result in engine stall and the loss of the tach reference signal.
- The fuel pump on the Ford electronic returnless system is no longer controlled through a relay. Both the positive and negative fuel pump control wires are controlled directly by an electronic control module. Therefore, no longer will a wire that is shorted to ground have the potential to keep the fuel pump circuit energized.

Sequential Fuel Injection

The sequential fuel injection (SFI) system used on Ford vehicles may perform a group injection pulse, called a *primer pulse*, during the initial engine start when this system is coupled with a waste spark ignition system. The PCM uses the CKP sensor signal to fire the first ignition coil, during which time the primer pulse delivers the fuel for the initial start. Once the PCM gets a CMP sensor signal, the PCM begins firing the injectors sequentially. This does not hold true when the system is coupled with a coil-on-plug ignition system because the PCM needs the CMP sensor signal before it can begin firing the ignition coils.

Fuel Injectors

The fuel injectors used with EEC V are designed with **alternate fuel compatibility,** meaning that they are compatible with both methanol and ethanol, as well as with gasoline. The O-rings that seal them in the manifold are likewise methanol and ethanol compatible, as should be any replacements. These injectors use a pintle designed to resist deposits from fuel additives as well.

Return-Type Fuel Injection System

Fuel Pressure Regulator. The fuel pressure regulator, which is also designed for alternate fuel compatibility, connects to the fuel rail (or *fuel injection supply manifold*, as Ford literature prefers to describe it) downstream of the injectors. It regulates the fuel pressure at the injectors through the interaction of its diaphragm-operated relief valve. One side of the diaphragm senses fuel pressure, and the other is subject to manifold pressure (vacuum). The initial fuel pressure is determined by the specially calibrated spring preload. The purpose of the arrangement is to maintain the fuel pressure drop across the fuel injectors at a constant pressure differential, regardless of changes in the intake manifold pressure, and to make it simpler to calculate the fuel injection pulse width. As long as the pressure differential is constant, a fixed amount of fuel will flow for a fixed pulse width. If the pressure differential were allowed to vary, the PCM would have to compensate for different fuel quantities.

Electronic Returnless Fuel Injection System

Like most other manufacturers' systems of the past, the Ford fuel pressure regulator has traditionally been located in the engine compartment on the fuel rail, and it releases excess fuel into the return line, which then carries it from the engine compartment back to the fuel tank. An unfortunate result of this design has always been that engine compartment heat was carried back into the fuel tank through the return line as well, thereby increasing the tendency of the fuel in the tank to vaporize and charge the charcoal canister excessively with fuel vapors.

Ford introduced an electronic returnless fuel injection system on 1998 model year vehicles. This system is designed to reduce the amount of engine compartment heat that gets carried back to the fuel tank. A differential pressure sensor is mounted on the fuel rail in the engine compartment and monitors fuel rail pressure (Figure 13–14). This sensor also monitors manifold absolute pressure (MAP) through a short vacuum hose. Thus, the resulting signal informs the PCM of the pressure differential across the fuel injectors and is therefore pertinent to the injectors' flow rate.

On early electronic returnless systems, the PCM controls a fuel pump driver module (FPDM) (Figure 13–15). The FPDM is mounted in the trunk on passenger cars. The FPDM, like a fuel pump relay, carries the current that the fuel pump motor draws. The difference is that the FPDM is used by the PCM to control the speed of the fuel pump motor through a high-frequency pulse-width-modulated (PWM) signal. The PCM sends the FPDM a low-current PWM command signal (Figure 13–16). In turn, the FPDM controls the fuel pump motor according to the same PWM signal but carries the higher current flow. This design allows the PCM to vary the speed of the

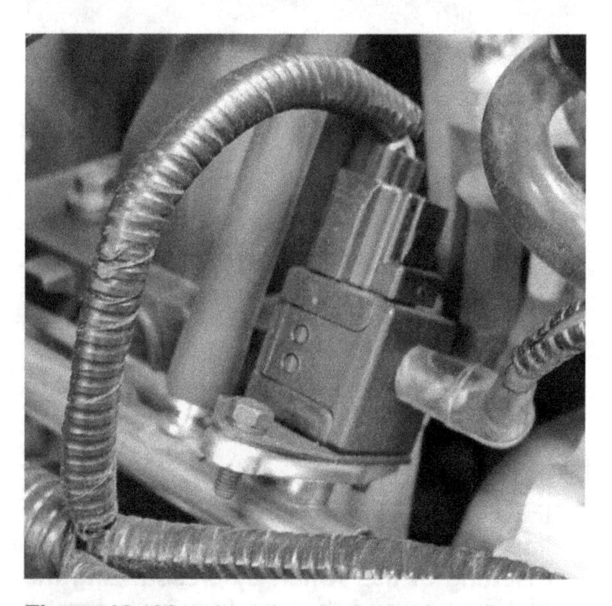

Figure 13–14 This fuel rail pressure sensor monitors the difference between fuel rail pressure and manifold absolute pressure.

Figure 13–15 Fuel pump driver module.

fuel pump to control the fuel pressure applied to the fuel injectors while watching the signal from the fuel pressure sensor on the fuel rail. The fuel pump assembly also contains a pressure relief valve (Figure 13–16 and Figure 13–17). An advantage of this design is that a scan tool may be used to read the system's fuel pressure.

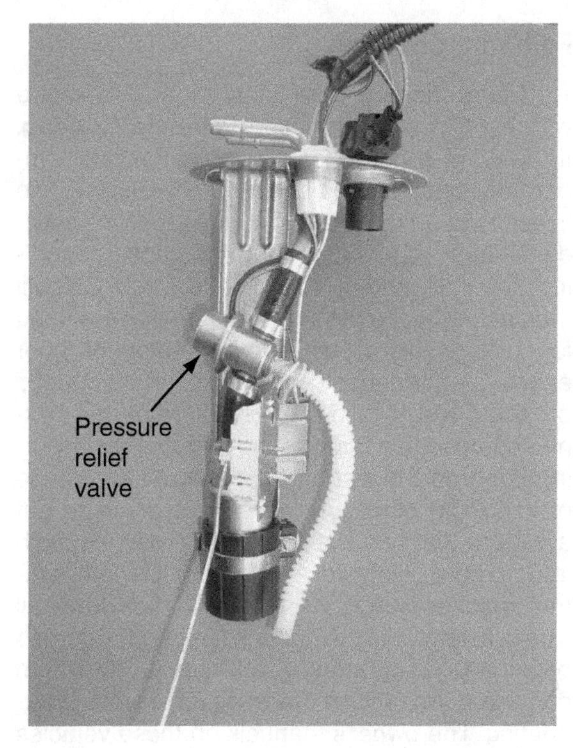

Figure 13–17 Fuel pump with fuel pressure relief valve.

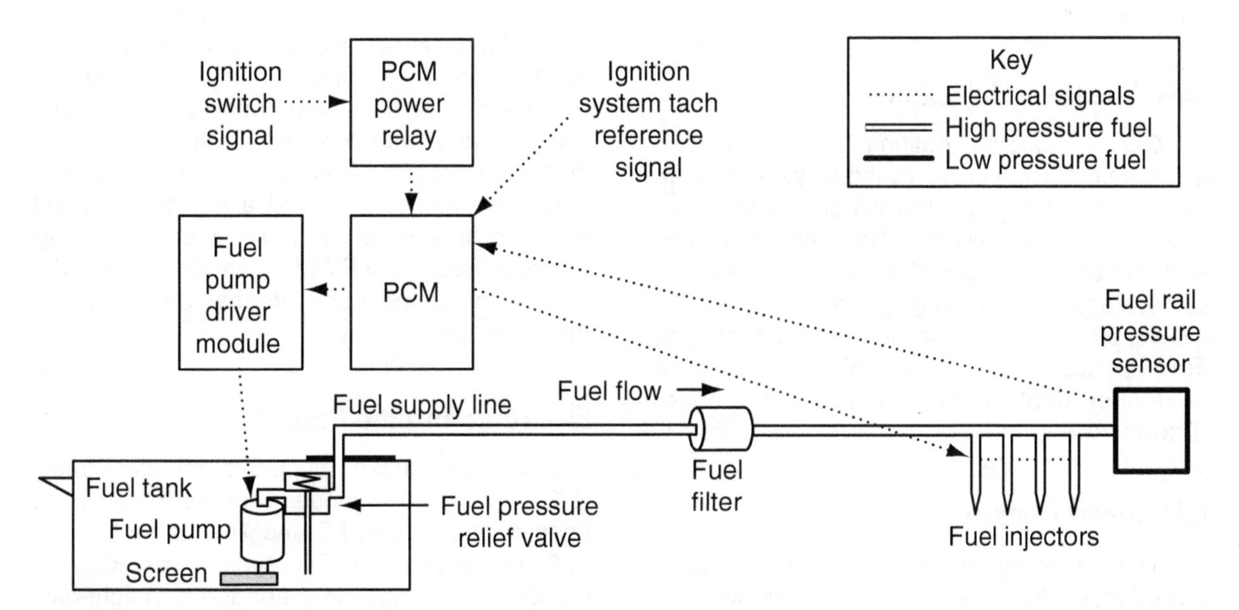

Figure 13–16 Electronic returnless fuel injection system.

Fail-Safe Cooling

Many modern Ford engines are equipped with a PCM strategy called **Fail-Safe Cooling.** If the PCM senses that the engine coolant temperature has reached 260°F, it will disable two injectors at a time and allow the respective cylinders simply to pump air for the purpose of cooling the block and cylinder heads. The PCM will then alternate the pairs of cylinders that are used for cooling. This will result in a symptom of "poor engine performance" and "engine runs rough," but the driver should also be aware of the overheated condition through the temperature-warning lamp on the instrument panel. The purpose of this PCM strategy is to allow a driver to get a vehicle that has lost coolant to a service facility for repair. However, if the engine's temperature reaches 330°F, the PCM will shut down all fuel injectors until the coolant temperature drops below 310°F. This strategy will keep a driver from damaging an engine because of severe overheating. The owner's manuals on these vehicles do instruct the driver to allow the stalled engine to cool for a short time, then—upon a successful restart of the engine—to drive the vehicle to a repair facility.

Gasoline Direct Injection

Gasoline direct injection (GDI) systems use a fuel injector for every cylinder, similar to an SFI system. However, the injector sprays fuel directly into the cylinder. Not only does this system use much higher fuel pressure than earlier systems, but the higher pressure, in turn, allows for better atomization of the fuel. The Ford application of a GDI engine is discussed in greater depth later in this chapter under "EcoBoost Engine."

Idle Speed Control

The idle air control (IAC) valve is a normally closed solenoid, similar to the one used with the EEC IV system, except that it is operated by the PCM on a high-frequency PWM signal in order to control throttle bypass air (Figure 13–18). The frequency of the signal is controlled by the PCM, varying from around 800 Hz to around 3,300 Hz. The solenoid is not expected to open and close at this extremely high cycling rate. Rather, the on-time of the signal controls the average voltage, and therefore the average current, in the circuit. This causes the solenoid's iron core/valve pintle to *float* at a designated point between fully closed and fully open. Ultimately, this high-frequency PWM signal determines the amount of atmospheric air that bypasses the throttle plates.

Figure 13–18 IAC valve.

Electronic Throttle Control

Ford introduced electronic throttle control (ETC) in the 2003 model year on the 3.9L V8 used on the Lincoln LS and Thunderbird. In the 2004 model year, the ETC system was added to the Explorer/Mountaineer and the new light-duty

F-series trucks. ETC was added to most other vehicles as the CAN protocol was implemented. The Ford ETC system, like the General Motors TAC system, physically decouples the throttle plate(s) from the throttle pedal. On the Ford ETC system, the PCM controls throttle plate movement electronically and directly while receiving input from the driver through an accelerator pedal position (APP) sensor, thus eliminating the traditional throttle cable. The ETC system is often referred to generically as a "drive-by-wire" system. The ETC system provides several advantages, including lowered emissions and quicker throttle response. Ford states that the use of an ETC system has the potential to increase fuel mileage due to a control strategy that allows the PCM to optimize fuel control with transmission shift schedules while continuing to deliver the driver-requested amount of torque at the wheels.

The components of Ford's ETC system include a **throttle plate position controller (TPPC),** a TPS to monitor throttle plate position, a DC motor that serves as the actuator used to control throttle plate position, and an accelerator pedal position (APP) sensor. With a cable-operated throttle plate, the TPS is said to give the PCM information concerning *driver demand.* In an ETC system, driver demand information now originates at the APP sensor.

The TPPC is a separate IC chip embedded within the PCM with the task of controlling the ETC system (Figure 13–19). Because safety is a major concern in such a system, a complex safety monitor strategy is used. This strategy involves additional hardware within the PCM, as well as additional programming strategies. The PCM's inputs are received by both the PCM's primary central processing unit (CPU) IC chip and by an additional IC chip called the Enhanced-Quizzer, or E-Quizzer, also shown in Figure 13–19. The primary CPU interprets the inputs and makes decisions regarding the output functions, but the E-Quizzer redundantly monitors the inputs and the primary CPU's handling of the information.

The primary CPU decides how the throttle plate(s) should be controlled and then instructs the TPPC where to position the throttle. The TPPC controls the throttle plate(s) with a DC motor attached to the throttle body (Figure 13–20). The control signal from the TPPC is a bidirectional pulse-width-modulated (PWM) signal, meaning that the polarity applied to the motor can be reversed for reversing the direction of motor movement and can also be operated on a wide pulse width or a narrow pulse width, depending on how quickly the throttle plate(s) must be opened or closed.

On early Ford ETC systems, the TPS was a two-channel potentiometer assembly. The TPS assembly contains two potentiometers and sends two signals to the PCM. The TPS signal is delivered not only to the primary CPU and the E-Quizzer but also to the TPPC. This allows the TPPC to watch the position of the throttle as it also controls it (a closed-loop feedback/control circuit).

On early Ford ETC systems, the APP sensor was a three-channel potentiometer assembly (Figure 13–21). The APP sensor assembly contains three potentiometers and sends three signals to the PCM (to both the primary CPU and the E-Quizzer). As the throttle pedal is moved from the idle position toward the WOT position, two of the APP sensor's channels increase the voltage signal from a low voltage up toward reference voltage, although the voltage signals are not an exact match at a given throttle position (they each move within a slightly different voltage range). The APP sensor's third channel has a voltage signal that moves downward as the throttle pedal is depressed, from near reference voltage at idle down to a low voltage at WOT (thus, electrically backward from the signals from the other two channels). This allows the PCM to more precisely interpret exact throttle position than would be the case if they all delivered exactly the same voltage values at a given throttle position. The APP sensor is located in a throttle pedal assembly that is spring-loaded to give the driver a traditional pedal feel.

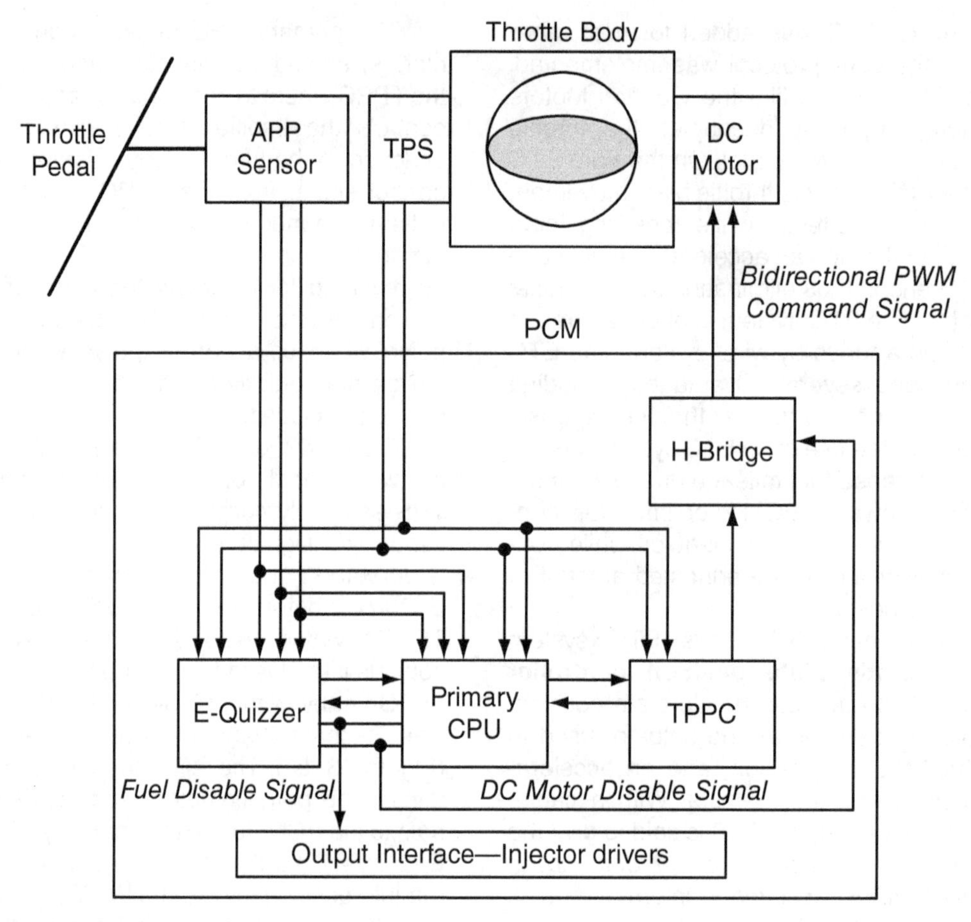

Figure 13–19 Ford's ETC system.

Figure 13–20 Throttle body with ETC unit and TPS.

Modern Ford ETC systems are using linear Hall effect sensors for the TPS and APP sensors. These sensors are known as *noncontact* sensors and therefore don't have the wear factor that a potentiometer does.

Beginning in the 2010 model year, Ford reduced the number of APP sensors from three to two. This change initially took effect on the 2010 3.5L EcoBoost GDI engine. By 2011 it was implemented on the non-GDI engines as well.

Compared to a cable-operated system, Ford advertises that the ETC system provides quicker throttle response. This is because a cable-operated system on a fuel-injected engine allows the driver

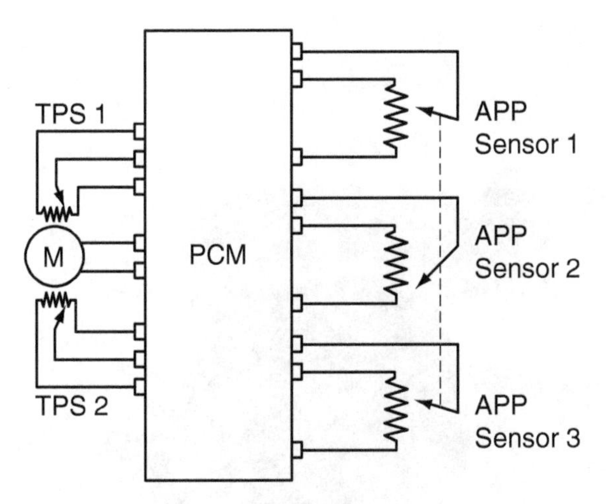

Figure 13–21 Ford ETC system showing the TPS and APP sensor circuits.

to control the volume of air allowed to enter the engine using the throttle cable, but the PCM controls the fuel. Bear in mind that to increase engine RPM and vehicle speed, both the volume of fuel and the volume of air entering the engine must be increased. In a cable-operated system, the following must happen for the PCM to know to add fuel:

- The driver must move the throttle pedal.
- The throttle pedal must move the throttle cable.
- The throttle cable must rotate the throttle shaft, thus opening the throttle plate(s) to allow additional air to enter the intake manifold. Simultaneously, this action physically rotates the TPS.
- The TPS must send the PCM a signal indicating that the throttle has opened the throttle plate(s).

In an ETC system, movement of the throttle pedal operates the APP sensor directly and immediately sends the signal to the PCM to increase engine speed. Ultimately, in an ETC system, the PCM controls both air and fuel, unlike the cable-operated system in which the driver controls the air and the PCM controls only the fuel.

The implementation of an ETC system also has some other benefits:

- The idle air control (IAC) solenoid is eliminated. The PCM can now control idle RPM directly without the use of additional dedicated hardware.
- The dedicated cruise control (speed control) servo is eliminated, and speed control becomes a software function of the PCM, no longer requiring additional hardware to control throttle position (other than the driver's control switches).
- The PCM can use the ETC system to control engine torque directly when the traction control feature mandates it.

IGNITION SYSTEMS

While a few TFI-IV distributor ignition (DI) systems carried over from EEC IV into some EEC V applications, by 1999 the only application that retained a distributor was the 3.3L Villager. With the exception of the 2.0L Probe (prior to 1999), the distributor applications continued to use a separate, dedicated ignition module to control the ignition coil. The 2.0L Probe allowed the PCM to directly control the ignition coil.

Ford's first distributorless ignition system, known under OBD II standards as an *electronic ignition (EI)* system, was introduced in the 1989 model year and was known as the Distributorless Ignition System (DIS). It used Hall effect CKP and CMP sensors and was strictly a waste spark ignition system.

Ford's second EI system was introduced in 1991 and was known as the Electronic Distributorless Ignition System (EDIS). The EDIS ignition system made its debut with the EEC IV computer system and carried over into EEC V.

The EDIS system originated as a waste spark system, using one ignition coil per two cylinders. All coils are formed into one component called a *coil pack* (*Figure* 13–22). Most 1996 and newer

Figure 13–22 EDIS coil pack.

EEC V EDIS applications (except for the 1996–1997 Windstar) allowed the PCM to directly control the ignition coil primary circuit, without the use of a separate external ignition module. By 1998, all EDIS applications allowed the PCM to directly control the ignition coil.

Modern EDIS ignition systems use one ignition coil for each cylinder (Figure 13–23). This is commonly known by Ford as *coil-per-plug (CPP)*. This system may also be referred to as *coil-on-plug (COP)*. Each coil sits directly on top of a spark plug so that the secondary ignition wires are eliminated. The advantage of this design is that the potential for induced voltages in nearby circuits from secondary wiring is reduced. This design was introduced on 1998 model-year vehicles and is used on most modern Ford gasoline engines. All of the CPP systems allow the PCM to control each ignition coil directly and without the use of an external dedicated ignition module. The advantage of this is that it enhances the efficiency of the antitheft disabling system.

EMISSION CONTROL SYSTEMS

Emission system components do not necessarily provide any advantage in terms of engine performance, but they are necessary to reduce toxic exhaust emissions. Some emissions components, like the PCV system, work independently of the engine control system, although the PCM may be able to monitor their performance. Others are controlled by the PCM as precisely as spark timing or the air-fuel ratio.

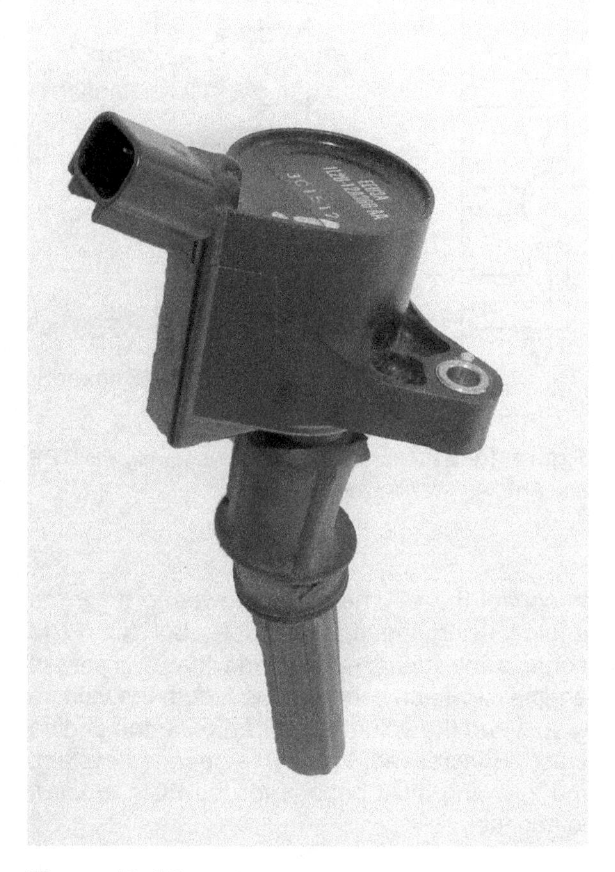

Figure 13–23 COP ignition coil.

Catalytic Converters

As is common with other manufacturers, Ford vehicles use both a reduction catalytic converter and an oxidation catalytic converter. The reduction cat uses CO to reduce NO_x back into N_2 and O_2. The oxidation cat oxidizes CO into CO_2 and HCs into both water and CO_2.

Catalytic Converter Efficiency Test. The PCM compares the signal from the pre-cat oxygen sensor(s) with that of the post-cat oxygen sensor(s) downstream of the catalytic converter. If the converter is working properly, there should be little variation in the signal produced by the secondary oxygen sensor. If the post-cat oxygen sensor(s) produce a signal that is too similar to that generated by the pre-cat oxygen sensor(s), the PCM sets a DTC in memory.

Exhaust Gas Recirculation System

To reduce the formation of oxides of nitrogen, or NO_x, the exhaust gas recirculation (EGR) system meters a measured amount of inert exhaust gas back through to the intake manifold. The exhaust gas contains no more than about one-half of a percent of oxygen when the system is in proper closed-loop control of the air-fuel ratio. This then displaces incoming ambient air, containing about 21 percent oxygen, which is drawn in through the throttle body. Thus, the amount of oxygen entering the cylinder is reduced. The fuel management system reduces the volume of fuel that is delivered to the engine proportionately. This ultimately reduces combustion chamber temperatures to below the temperature at which NO_x forms.

EVP Sensor. In 1980 Ford became the first domestic manufacturer to begin monitoring the EGR valve with a potentiometer placed on top of a vacuum-operated diaphragm-type valve. The potentiometer was called an EGR valve position (EVP) sensor. However, while the EVP sensor could inform the PCM of the EGR valve's physical position, it could not inform the PCM of EGR flow rates.

PFE Sensor. In 1987 Ford introduced a replacement for the EVP sensor known as the pressure feedback EGR (PFE) sensor. The PFE sensor was an analog absolute pressure sensor that monitored the exhaust pressure between the EGR valve and an orifice (restriction) placed in the EGR exhaust tube. This allowed the PCM to know EGR flow rates and was fairly accurate unless exhaust backpressure changed and was no longer at the original specification.

DPFE Sensor. The Ford EEC V system uses a differential pressure feedback EGR (DPFE) sensor (also known as a Delta Pressure Feedback EGR sensor) to continuously monitor the EGR flow rate (Figure 13–24). The DPFE sensor continues to monitor the pressure between the EGR valve and a metering orifice placed in the exhaust passage, just as the PFE sensor did. If the EGR valve were fully closed, this would likely be a positive pressure, above atmospheric pressure. But if the EGR valve were wide open, this would likely be a negative pressure, below atmospheric pressure. In addition to measuring this pressure, the DPFE sensor also indexes this pressure value to raw exhaust pressure just before the metering orifice (Figure 13–25).

The DPFE sensor actually allows the PCM to measure the pressure drop across the metering orifice in the exhaust passage. (This is similar in concept to how current flow in an electrical circuit affects the amount of voltage drop measured across a resistance.) This allows the PCM to accurately know the true EGR flow rate, even if exhaust backpressure were to change.

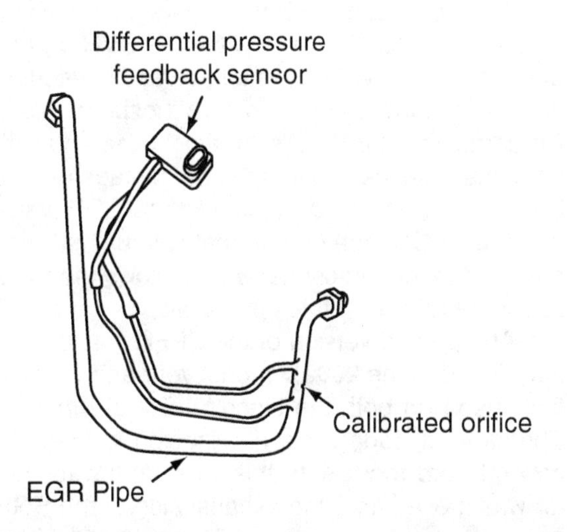

Figure 13–24 This diagram shows the DPFE sensor and the location of the calibrated orifice in the EGR pipe.

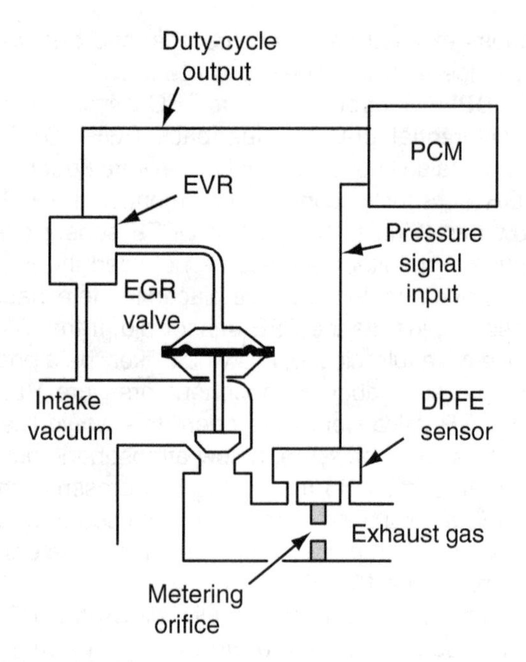

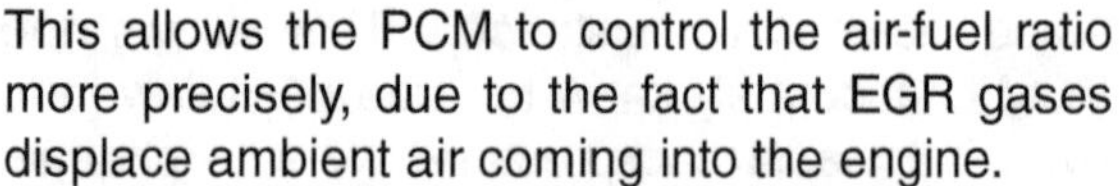

Figure 13–25 DPFE flow diagram.

Figure 13–26 EGR vacuum regulator (EVR).

This allows the PCM to control the air-fuel ratio more precisely, due to the fact that EGR gases displace ambient air coming into the engine.

To control the vacuum-operated EGR valve, the PCM controls an EGR vacuum regulator (EVR) on a duty cycle (Figure 13–26). The EVR carries over from EEC IV. It is normally closed to vacuum apply and normally open to an atmospheric vent. By controlling the EVR's on-time versus its off-time, the PCM can control the percentage of time that vacuum is being applied to the EGR valve and the percentage of time that vacuum is being released to the atmosphere. This allows the PCM to control EGR flow rates precisely.

An updated version of the DPFE sensor was introduced in the 2002½ model year and was initially used on both the Lincoln LS and the retro Thunderbird. Today, this new design is used on several Ford models. In this version, the restrictor was moved from the exhaust side of the EGR valve to the intake side (Figure 13–27). This DPFE sensor is both a *differential* (*delta*) pressure sensor and an *absolute* pressure sensor in one component (refer to Chapter 3 for an explanation of both types). In this design, the DPFE sensor is used by the PCM to measure the difference between MAP values in the intake manifold and the pressure on the EGR side of the metering orifice. The DPFE sensor actually measures the pressure drop across this metering orifice as it did when it was located on the exhaust side of the EGR. When the DPFE sensor sees that there is no pressure drop across the metering orifice (restriction), the PCM knows this indicates that there is no EGR flow. As EGR flow is increased, the PCM will see a greater pressure drop across the metering orifice. In addition to the DPFE signal, the DPFE sensor also operates as an analog MAP sensor and sends the PCM a MAP signal. To reduce costs, Ford manufactures the EGR valve, the DPFE sensor, and the EVR into one component called an EGR System Module (ESM) (Figure 13–28).

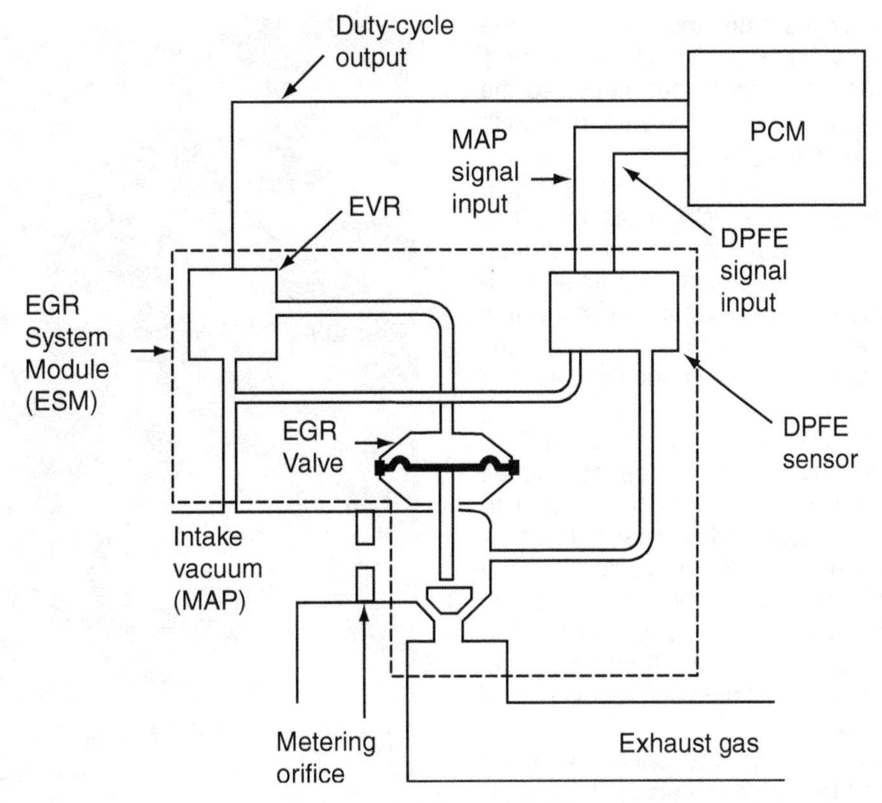

Figure 13–27 ESM assembly with an EGR valve, EVR, and DPFE sensor that monitors pressure drop across a metering orifice placed on the intake manifold side of the EGR valve.

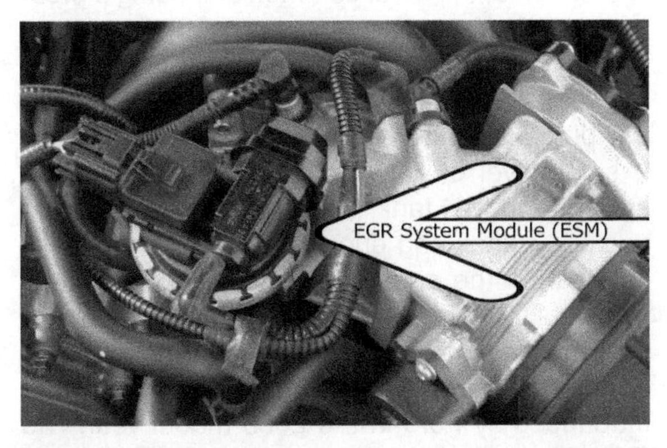

Figure 13–28 EGR System Module (ESM) with DPFE sensor (left) and EVR (right) attached to top of EGR valve.

Ford Motor Company also uses a vacuum-less EGR valve on 2.0L and 2.3L Ford and Mazda engines (that is, it does not require a vacuum signal for operation, similar to the General Motors digital and linear EGR valves). This is a stepper motor, located on the back end of the cylinder head, which can move a total of 52 discrete increments or "steps" from fully closed to fully open to precisely control EGR flow. Its operation is monitored through the use of a MAP sensor (similar to many General Motors vehicles).

Evaporative/Canister Purge System

Evaporative Purge Flow Test. Early Ford OBD II evaporative systems had a purge flow sensor located a few inches from the canister purge solenoid used by the PCM to control purging. This sensor was a thermistor that measures the cooling effect of fuel vapors as they are purged in order to know purge flow rate.

To perform a purge flow test on modern Ford evaporative systems, the PCM waits until it is operating in closed-loop fuel control and then suddenly turns the purge solenoid on while monitoring the oxygen sensors. If purging of the fuel vapors from the charcoal canister is successful, the oxygen sensors should respond by indicating a rich air-fuel ratio.

Evaporative Vacuum Test. On early Ford OBD II vehicles, the PCM would cause a small amount of manifold vacuum to be applied to the evaporative system and fuel tank to check the system for leaks. The PCM will also verify the accuracy of the fuel tank pressure sensor by comparing it against the MAP sensor during the initial key-on cycle. Figure 13–29 shows a fuel tank pressure sensor installed on the fuel pump/sender assembly.

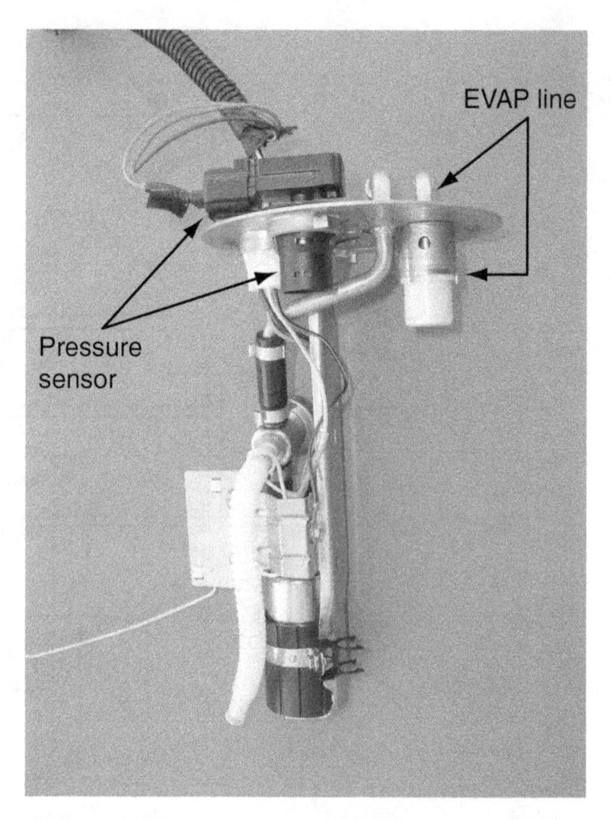

Figure 13–29 Fuel pump assembly with fuel tank pressure sensor and evaporative system line.

To run this test, the PCM checks ambient temperature as the engine is started. If it is 40°F or warmer, it is assumed that heat radiating up from the ground underneath the fuel tank will cause some fuel vaporization to take place. In turn, this will charge the charcoal canister with fuel vapors. The PCM will then wait until it is operating in closed-loop fuel control and then turn purging on while monitoring the oxygen sensors and STFT. If the oxygen sensors switch rich and the STFT values begin to decrease, the PCM knows that the purge function is working correctly and that the fuel tank's vapors are stored in the canister. However, if the oxygen sensors and STFT do not respond appropriately, the PCM performs a *gross leak test* and a *small leak test*.

To perform the gross leak test, the PCM energizes the normally open vent solenoid valve (Figure 13–30). Then it momentarily energizes the normally closed purge solenoid valve and watches the fuel tank pressure sensor to indicate a change in pressure. If this test fails, a large leak is presumed, the most common one being a problem with the gas cap.

To perform the small leak test, the PCM keeps the vent solenoid valve energized and then uses the purge solenoid valve to apply and maintain a small level of vacuum within the evaporative system and fuel tank. The system must hold the vacuum signal long enough to identify any leaks larger than the specification for the year of the vehicle.

Engine Off Natural Vacuum Test. The *engine off natural vacuum (EONV)* test was introduced by Ford in the 2005 model year. The EONV test replaced the former vacuum test and required no hardware changes in the system.

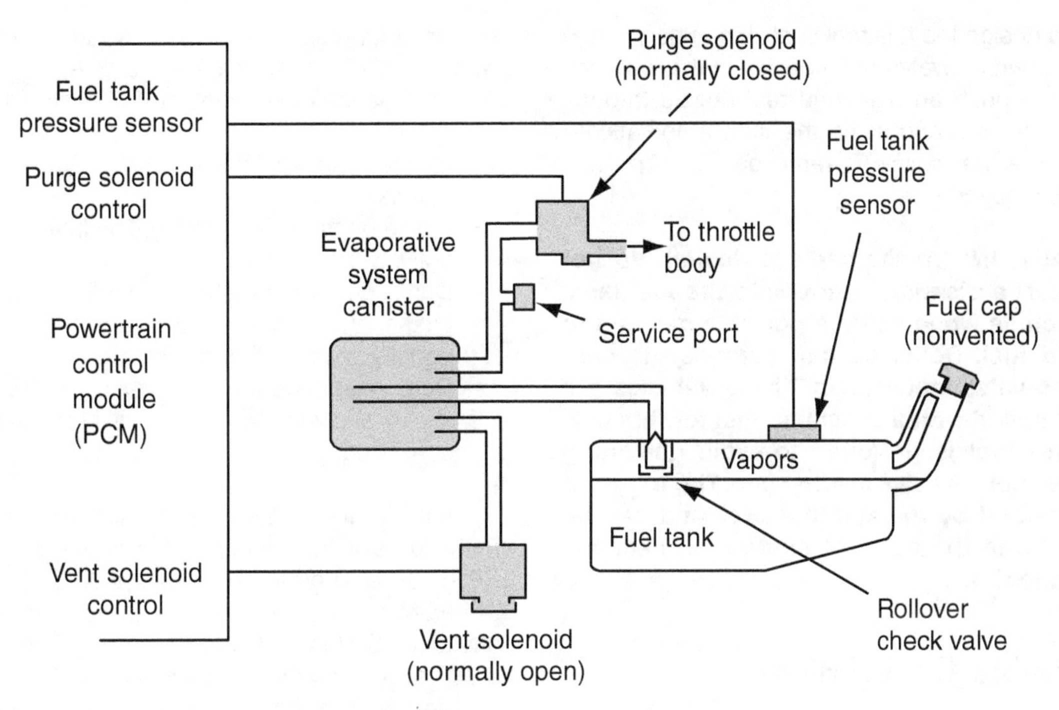

Figure 13–30 Evaporative system components.

Minor electrical changes were implemented in order to keep the PCM powered up after the ignition switch is turned off. And the PCM's program was modified to allow it to perform this test.

When the vehicle is driven, heat from the engine and exhaust system will cause a slight increase in temperature of the fuel in the fuel tank which, in turn, causes it to expand. To perform the EONV test, the PCM is kept powered up for up to 40 minutes after the ignition switch is turned off. During this time the PCM keeps the purge solenoid valve de-energized and the vent solenoid valve energized, thus keeping both valves closed. As the fuel cools, it contracts and causes a *natural vacuum* to be formed within the evaporative system and fuel tank assembly, from about 3 inches to about 5 inches of water. During this time the PCM monitors the fuel tank pressure sensor to verify that the system is capable of holding this negative pressure.

With any sensed failure of the evaporative system, the PCM will illuminate the MIL and store the appropriate DTC in memory.

Capless Fuel Tank. Many modern Ford vehicles use a capless fuel tank (Figure 13–31).

Figure 13–31 Capless fuel tank.

In this design the fuel tank's neck has been engineered with a valve mechanism that allows the driver to push an unleaded fuel nozzle through it for refueling and seals the tank automatically when the fuel nozzle is removed. Thus, no fuel cap is required.

CAUTION: On the capless design, do not insert a standard funnel into the fuel tank, such as when using a portable gas can to add fuel. Doing so can damage the capless valve mechanism. These vehicles are equipped with a special funnel for this purpose that is designed to avoid damaging the capless valve mechanism. The funnel is provided by the manufacturer and can be found in the trunk or cargo area near the spare tire.

VARIABLE CAM TIMING

Ford began using variable cam timing (VCT) in 1998 on a 2.0L engine that used a cam phaser on the front end of the camshaft to change the camshaft's position relative to the crankshaft. The cam phaser is a vane type and is controlled by the PCM using an oil control solenoid to apply engine oil pressure to one side or the other of a set of vanes connected to the camshaft. This action causes the camshaft to be rotated in one direction or the other to either advance or retard cam timing. Cam timing can be advanced to provide more low-end torque or it can be retarded to provide more high-end power. An adjustment of 60 degrees, from fully advanced to fully retarded, is provided. A pin locks the cam phaser in the fully advanced position for initial engine startup. The addition of the VCT system allowed Ford engineers to increase compression ratios while still using regular octane fuel. Additionally, the VCT design can be used to reduce the need for an EGR system and all of the associated hardware. However, certain engines, such as the 3.9L V8, retain the EGR system, in addition to the use of a VCT system, because of a combination of the higher compression ratio that is used, coupled with the PCM's performance strategies.

Ford defines four types of VCT systems:

- Exhaust phase shifting (EPS): The exhaust cam is actively retarded.
- Intake phase shifting (IPS): The intake cam is actively advanced.
- Dual equal phase shifting (DEPS): Both the intake and exhaust cams are actively and equally advanced or retarded.
- Dual independent phase shifting (DIPS): The intake and exhaust cams are independently controlled.

Initially, the 1998 2.0L engine allowed the PCM to vary the timing of the intake camshaft (IPS). Other engines, such as the 3.9L V8, allow the PCM to vary the timing of the exhaust camshafts (EPS). In the 2005 model year, Ford added a VCT system on the Triton 24-valve, 5.4L V8. This engine uses a three-valve-per-cylinder design with one overhead camshaft per bank. Each camshaft operates both the exhaust valves and the intake valves on the respective bank. The PCM varies the cam timing of these camshafts to vary the timing of both the intake and exhaust valves equally (DEPS). This is also referred to as "dual-equal variable cam timing"; it is the industry's first mass application of such a design.

Ford Motor Company suggests that, because the Triton 5.4L V8 uses electronically controlled VCT along with electronically controlled spark timing, the following advantages are realized:

- A special cold-start strategy is programmed into the PCM, which controls the spark timing system and the VCT system to bring the exhaust temperature and catalytic converters up to operating temperature more quickly, thereby reducing cold-start emissions.
- The PCM is programmed to control the VCT system to reduce pumping losses when engine operating conditions allow. (Pumping losses are defined as the cylinders' loss of power created by having to work against

engine vacuum to bring the air-fuel charge into the cylinder and having to work against atmospheric pressure to expel the burned air-fuel charge from the cylinder.)

- The PCM is programmed to use the VCT system to channel part of the exhaust gas back into the intake manifold during certain engine operating conditions, thus reducing the need for an EGR system.
- The PCM is programmed to use the VCT system to create a higher torque curve at lower RPM without sacrificing high-end power.

With the introduction of the EcoBoost GDI engine, Ford became the first manufacturer to introduce an electromagnetic cam phaser. This cam phaser does not use oil pressure to advance or retard valve timing. Instead, the PCM can precisely control valve timing by controlling the electromagnetic field strength of the windings. In turn, it does this by using a pulse-width-modulated command to control current flow through the windings of the cam phaser.

INTAKE MANIFOLD RUNNER CONTROL

The **Intake Manifold Runner Control (IMRC)** system has been used on Ford and Mazda engines in several forms. One early form (actually, a predecessor to the modern IMRC system) was first used on the Taurus 3.0L Super High Output (SHO) engine, beginning with its introduction in the 1989 model year, and then was used on the Taurus 3.4L SHO engine. This engine is a 24-valve V6 with two intake valves and two exhaust valves per cylinder. A long intake runner supplies air to one intake valve during normal engine RPM and load conditions. A short intake runner supplies air to the other intake valve, but it is allowed to do so only during high engine RPM and load conditions. During low RPM and load operation, a valve within the short runner is closed, inhibiting airflow through it. This design allows for good airflow velocity through the intake system and into the cylinders during low RPM and load conditions, resulting in good throttle response and greater fuel economy. However, by opening up the second runner during high RPM and load conditions, a greater quantity of air is allowed into the cylinders when it is needed for all-out performance without sacrificing the low-end throttle response.

Modern IMRC systems, as they are used on both Ford and Mazda engines, use a valve to lengthen or shorten the effective length of the intake manifold runner, depending upon engine RPM and load conditions. At low engine RPM and load conditions, the valve closes off a "shortcut," thereby effectively lengthening the intake manifold runner and making for good throttle response. At high engine RPM and load conditions, the valve opens up the shortcut, thereby effectively shortening the intake manifold runner and allowing additional air to be brought into the cylinders when more power is needed. The valve may be operated electrically by the PCM or may be operated with vacuum actuation.

Another variation of the IMRC system was a Split Port Induction (SPI) system that was used on 2.0L Escort engines beginning in the 1998 model year. This engine has intake manifold runners of different diameters. The larger-diameter runner was closed off at engine speeds below 3,000 RPM, using only the smaller-diameter runner to keep air velocity high. Above 3,000 RPM the larger-diameter runner is also opened up, allowing for additional air to fill the cylinders. Even though this engine only has two valves per cylinder, its performance approaches that of engines having multiple-valve heads.

The following are advantages of these designs:

- The ability to allow smaller engines to produce more power when it is needed without sacrificing low-end performance and throttle response
- The ability to sustain good low-end throttle response, greater fuel economy, and better emissions without sacrificing high-end all-out power

BI-FUEL SYSTEMS

Natural Gas/Propane

Beginning with the Triton 5.4 L V8 in the 1998 model year, Ford began offering bi-fuel versions of some engines. These engines have specially designed components that are designed to operate either with gasoline or with natural gas or propane. The redesigned components include the throttle body, the intake manifold, and the fuel injectors.

E85 Flex Fuel Vehicle

Ford offered its first E85 flexible fuel vehicle (FFV) in the 1998 model year with the introduction of an FFV option on the Taurus with the 3.0L engine. In the 1999 model year, the Ford Ranger was also offered with an FFV option on the 3.0L engine. These vehicles are designed to operate on E85, pure gasoline, or any combination of the two.

ECOBOOST ENGINE

Ford introduced a new 3.5L V6 engine in the 2010 model year known as *EcoBoost*. This engine was initially introduced on the Lincoln MKS and the Taurus SHO and is now offered in other models as well such as the Ford Flex. This engine is a gasoline-fueled engine with direct fuel injection and dual turbochargers.

The EcoBoost engine is rated at 355 hp at 5,500 RPM in the MKS and 365 hp at 5,500 RPM in the Taurus SHO. Both applications boast a flat torque curve that reaches and maintains a peak torque of 350 foot-pounds from 1,500 to 5,250 RPM.

Dual turbochargers are used on the 3.5L V6. This allows them to be made smaller which, in turn, reduces turbo lag. As a result, the twin turbochargers provide better response than a single turbocharger provides. The turbochargers are water-cooled. Because they are water-cooled, there is no need, after driving it hard, to let the engine idle before shutting it down. The turbochargers have a maximum 205,000 RPM and will typically produce 8 to 10 PSI of boost. Maximum boost pressure is 14 to 15 PSI and is controlled by a wastegate that is controlled by the PCM. An air-to-air intercooler is used.

The fuel injectors are installed in the cylinder heads and inject fuel directly into the cylinder just below the intake ports. Fuel pressure reaches a maximum of 2,200 PSI and is provided by a mechanical fuel pump that is driven by the left camshaft. The fuel is better atomized due to the higher fuel pressure—each drop is about 15 microns in size, compared to a droplet size of about 100 microns in a standard PFI system. (A micron equates to one one-millionth of a meter. Fifteen microns equates to 0.000590 inch. One hundred microns equates to 0.003937 inch.)

Warning: Due to the high fuel pressure, the technician should never attempt to use a fuel pressure gauge to test fuel pressure on this engine. Instead, because this engine uses fuel pressure sensors on both the low-pressure and high-pressure sides of the system, use only a scan tool to test fuel pressure. Likewise, if you think you might have spotted a liquid spray near the fuel injectors, do not reach in to feel for it. The idle pressure will easily split skin.

The EcoBoost engine is a speed density engine, using a MAP sensor and no MAF sensor. (It was decided not to use a MAF sensor because it would have been too far from the throttle plates due to the turbocharger plumbing.) There are two other similar sensors used as well: a full-time barometric pressure (BP) sensor and a throttle inlet pressure (TIP) sensor. All three sensors should correlate to each other during KOEO conditions. The system also uses three temperature sensors: an air cleaner temperature (ACT) sensor, a throttle charge temperature

(TCT) sensor, and manifold charge temperature (MCT) sensor.

The Taurus SHO application of the EcoBoost engine does boast one component that the Lincoln MKS lacks—a flapper valve in parallel with the intake air tube called a *noise generator.* It is simply designed to generate noise during a hard acceleration to increase the perception of high performance—not something that the typical Lincoln driver would appreciate.

The Taurus SHO boasts a 0-to-60 time that is three-tenths of a second faster than the Chrysler Hemi, eight-tenths of a second better than a Cadillac STX with the 4.6L engine, and six-tenths of a second better than a BMW 535i. Ford's claim is that the EcoBoost 3.5L V6 offers the fuel economy of a V6 with the performance of a V8.

Today, the EcoBoost system is offered in other smaller engines as well, including the 1.6L, 2.0L, 2.3L, and 2.5L four-cylinder engines. These engines are equipped with a single turbocharger. They use both a dedicated MAP sensor and a combination manifold absolute pressure/temperature (APT) sensor. The APT sensor has four wires: VREF, ground, MAP signal, and manifold temperature signal. The use of both a MAP sensor and an APT sensor provides a redundancy for monitoring engine load.

The EcoBoost engine can also operate in three basic modes of air-fuel control as follows:

- Homogeneous charge mode with closed-loop fuel control: This mode is used for maintaining an overall stoichiometric air-fuel ratio.
- Homogeneous charge mode with open-loop fuel control: This mode is used for providing a rich air-fuel ratio when the vehicle is under moderate-to-heavy load.
- Stratified charge mode with closed-loop fuel control: This mode is used for maintaining a stoichiometric air-fuel ratio near the spark plug while providing an overall lean air-fuel ratio from 30:1 to 40:1 during cruise.

During cruise and while operating in the stratified charge mode, the PCM can control engine speed and performance through its control of fuel metering, similar to a diesel engine. This allows the PCM to further open the throttle plate(s) beyond what would normally be allowed without over-revving the engine. Because of this, the EcoBoost engine can maintain as little as 1 inch of vacuum (Hg) in the intake manifold while cruising in the stratified charge mode. This, in turn, reduces the energy used by the crankshaft to pump air through the cylinders.

Additionally, while operating in the stratified charge mode, the engine's power can be reduced during cruise by simply reducing fuel delivery. Therefore, a GDI engine has the advantage of consuming less fuel when power is not needed, similar to a variable displacement system except that, with a GDI engine, the PCM does not have to disable cylinders to achieve this reduction in power and fuel. Therefore, the engine runs smoother and is less annoying to the driver while operating in the stratified charge mode than a variable displacement engine that is operating on half of its cylinders. Another advantage over a variable displacement system is that GDI can be used in the stratified charge mode on smaller four-cylinder engines.

ADAPTIVE CRUISE CONTROL

Ford has equipped several vehicles with a system called **Adaptive Cruise Control.** This system feeds information to the PCM and to a braking module from a radar "headway sensor" installed behind the vehicle's grille. When the cruise control is engaged, this system can reduce engine torque and even apply the brakes lightly to maintain a preset gap to a vehicle ahead. When a proper gap again appears, the vehicle automatically accelerates back to the set speed in the computer's memory. The initial system has been enhanced to provide drivers with both visual and audible warnings if a collision is imminent to let them know that harder braking or evasive steering is required to avoid an accident. Advantages of this system are:

- It maintains a safe following distance between vehicles.
- It maintains consistent performance, even in adverse weather conditions with poor visibility.
- It maintains consistent performance, even during cornering or when road elevation changes.
- It alerts the driver by way of automatic braking.

The Adaptive Cruise Control system was optional on the Ford-built 2005 Jaguar S-Type models. For the 2008 model year, it was also made available on the Ford-built Volvo S80, Volvo XC70, and Volvo V70.

BODY CONTROL MODULES AND FORD MULTIPLEXING SYSTEMS

In the mid-1990s, as Ford began introducing multiplexed systems on its vehicles, other control modules were added as well. A **Generic Electronic Module (GEM)** was added to control body functions such as interval wipers, the driver's door one-touch-down power window, the illuminated entry feature, and other features. In fact, Ford's GEM is most similar to the body control module on General Motors and Chrysler vehicles, except that it does not control the climate control system.

Beginning with the 2002 model year on the Lincoln LS and the Ford Thunderbird, Ford divided the GEM into two control modules: the **Front Electronic Module (FEM)** and the **Rear Electronic Module (REM).** The FEM controls the one-touch-down power driver's window, the horn relay, and the exterior lighting at the front of the vehicle. The REM controls the exterior lighting at the rear of the vehicle, the deck lid release solenoid, and, on the Thunderbird, the passenger power window. In the electronic returnless fuel injection system, the REM also controls the fuel pump on a pulse-width command from the PCM, thus replacing the FPDM that was used on the earlier systems.

On the Thunderbird, the FEM and REM control the driver's and passenger power windows, respectively, so as to lower them one-half inch when the respective door is opened. The module then raises the window fully when the door has been closed again to provide good sealing with either the hardtop when it is in place or with the soft convertible top when it is in place. The driver can calibrate this feature by lowering each front window fully and then continuing to depress the switch for at least 2 seconds after the window has hit bottom. Then the driver must raise the window fully and, again, continue to depress the switch for at least 2 seconds after the window is fully up. The FEM can determine the window's position by the fact that when the window reaches an extreme and the circuit continues to be completed, the current draw for that window's motor increases substantially due to the added mechanical resistance. This calibration allows the FEM to know the exact position of the window. Then, when a door is opened, the module counts the motor's current pulses in order to lower the window the prescribed amount.

GEM system modules previously used a similar concept. It is through the increase in current flow that the GEM knows when a one-touch-down window has hit bottom. If the window's track is misadjusted or lacks lubrication, causing current flow to rise prematurely, the GEM will deactivate the one-touch-down feature prematurely, resulting in a driver complaint.

On some newer Ford products, the multiple body modules have again been combined into one module, now called a *body control module (BCM)*, the name that most other manufacturers had already been using.

An additional control module, the instrument panel controller (IPC), not only controls the instrument cluster gauges and warning lamps but also controls the electronic tilt and telescopic steering column. On some models, such as the 2002–2005 Ford Thunderbird, the IPC moves the steering column fully up and in, as an easy-exit feature when the ignition key is removed from the lock cylinder.

Ford used a multiplexing protocol in the late 1990s, known as the *1850* protocol, which was designed to meet the OBD II requirements. The Ford J1850 bus was a dual-wire (DW) bus that connected to terminals 2 and 10 at the data link connector (DLC) at a baud rate of 41.6 Kb/s. Comparatively, GM and Chrysler used a single-wire (SW) J1850 data bus at a baud rate of 10.4 Kb/s. The Ford J1850 protocol was used until Ford replaced it with the DW high-speed (HS) CAN bus, the implementation of which began on the 2003 Lincoln LS and 2003 Thunderbird and was fully phased in by 2008.

Today, Ford vehicles commonly use two protocols. A DW HS CAN bus connects to terminals 6 and 14 at the DLC. A DW medium-speed (MS) CAN bus connects to terminals 3 and 11 at the DLC.

Diagnostic & Service Tip

If an exterior light bulb at the rear of the vehicle is replaced by one with an incorrect part number, resulting in more current draw than intended, the REM, seeing the initially high current draw, may shut down in a self-protective mode until the incorrect bulb is removed. This would cause a "cranks, no start" condition because the fuel pump would be shut down in addition to the symptoms associated with the inoperative bulbs at the rear of the vehicle. The deck lid release solenoid would also be inoperative. In diagnosing this type of problem, the use of an electrical schematic would show that the relationship between all inoperative components is that they are all controlled by one module—in this case, the REM.

VOICE RECOGNITION SYSTEMS

Ford Motor Company introduced a combination voice recognition/navigational system on 2004 model-year Lincoln Navigators and Aviators. This early system had a limited vocabulary and limited features. The 2005 model-year Lincoln Navigators and Aviators have a combination voice recognition/navigational system that is greatly enhanced in its capabilities. On this system, the driver operates a *voice command* button on the steering wheel and then follows with a verbalized command (Figure 13–32).

Ford Motor Company brought voice recognition technology to a new level with its advanced **Conversational Speech Interface Technology.** Equipped with a 50,000-word vocabulary, this technology allows the driver and/or passenger to carry on a normal dialogue with the vehicle. A text-to-speech system is used, but the resulting dialogue on the part of the vehicle sounds more like a real person than a robot. As a result, the driver/passenger does not have to memorize specific voice commands. The voice recognition system's control module is multiplexed to other computers on the vehicle. As a result, many systems can be controlled by voice, including the navigation system, the entertainment system, and the climate control

Figure 13–32 SyncMyRide™ voice command button.

system. For example, the driver can use voice commands to select climate control fan speed, desired temperature, and operating modes.

Personalization preferences for multiple drivers can also be configured through voice commands. These preferences include favorite music selections, style of the visually displayed graphics, and even the type of voice that is used by the vehicle to communicate with the driver/passenger.

The Conversational Speech Interface Technology allows everyday language to be used, but if the voice recognition control module does not fully comprehend a voice command, it is programmed to ask for more information. For example, when asked to look up a phone number, if it finds several results that seem to fit the original request, it will ask for additional information. However, the voice recognition system will only speak when spoken to.

If asked to play music, the control module will ask for a music type and then list the artists that fit the request. Volume can also be controlled through the voice recognition system. Music can be downloaded to an on-board computer in the MP3 format using Bluetooth™ wireless technology. The music may be downloaded from a PDA, cell phone, or laptop computer that is turned on and located within the vehicle. Similarly, personal files, including phone lists and Internet addresses, may also be downloaded to the on-board computer.

Intelligent Architecture

Because people "spend a great deal of time in their vehicles," Ford Motor Company engineers are designing software/hardware systems into the vehicle designed to make life easier. Known as **intelligent architecture,** early versions of this system first appeared on 2003 Lincoln Aviators. Today, these systems include the combination voice recognition and navigational systems, including the ability to personalize driver settings and to download information or music in the MP3 format to an on-board computer from a personal computer, PDA, or cell phone using Bluetooth wireless technology. Also included is real-time navigation, a feature that can alert drivers to an accident ahead or even remind them of a vehicle service that is due.

A diagnostic feature also allows the driver to connect the vehicle to an online diagnostic center, providing information about current problems found in the primary electronic control systems of the vehicle. It can also allow the driver to locate nearby dealerships and to schedule an appointment.

SyncMyRide™

Beginning in the 2008 model year, Ford Motor Company began offering a feature on vehicles that use Bluetooth wireless technology to enable a Bluetooth-enabled cell phone or media player to be controlled via the vehicle's voice recognition system. Ford advertises this as its *SyncMyRide*™ feature. The software that supports this feature is provided to Ford by Microsoft. The driver can use the voice recognition system to place a hands-free phone call or to request a particular genre of music stored in the media player.

SyncMyRide™ uses a capacitive touch display screen with a *four-corner*, *four-color* approach. The four corners of the display screen are identified with the colors red, blue, green, and yellow. If the driver touches a particular color, a menu is displayed for a particular system. The four colors represent various vehicle systems as follows:

- Red: Entertainment system
- Blue: Climate control system
- Green: Navigation system
- Yellow: Phone

Sync Connect™

Ford's recently-announced *Sync Connect*™ (to make its debut on the 2017 Ford Escape) will allow the driver to program engine remote start times on a smart phone using a Ford app. If you leave your workplace at 3:30 pm every afternoon, you can program the engine to start remotely at

3:20 pm, thus allowing the heating or air conditioning system to adjust the vehicle's interior temperature. The Sync Connect™ app will also allow you to access vehicle status information including tire pressures, fuel level, and the vehicle's location.

ADVANCETRAC™ SYSTEM

Added to model year 2003 Explorers and Expeditions (and their Mercury counterparts), the **AdvanceTrac™** system is a modified antilock brake and traction control system with collision avoidance in mind. Today, the AdvanceTrac™ system is used on many Ford vehicles.

This system receives input from seven sensors, monitoring such things as steering wheel angle, throttle position, wheel speed, and the vehicle's yaw rate (basically a determination of how far left or right a vehicle has moved from its intended course, and what might happen if the vehicle were to understeer or oversteer its intended path). If the vehicle's back end begins to fishtail (oversteer), the system detects this and applies a braking pulse at the outside front wheel to help the driver stabilize the car. This action allows the side of the vehicle on the inside of the turn to begin to pass the other side, thus reducing the level of oversteer. If the front end of the vehicle begins to drift to the outside of a turn (understeer), the system detects this and applies a braking pulse to the inside rear wheel. This action allows the side of the vehicle on the outside of the turn to begin to pass the other side, thus reducing the level of understeer. With either understeer or oversteer, the AdvanceTrac™ system can command the PCM to reduce engine torque as needed to help prevent the loss of vehicle control or rollover. If the AdvanceTrac™ system detects that the vehicle is beginning to roll over, it will work with the PCM to reduce the vehicle's speed by 10 MPH within 1 second to reduce the potential for rollover to occur.

Keep in mind that antilock brake systems, traction control systems, and AdvanceTrac™ systems cannot overcome the laws of physics. Just as the presence of an antilock braking system does not allow the driver to stop efficiently once the tires have already hydroplaned, the AdvanceTrac™ system may not be able to prevent loss of vehicle control if the driver's actions are severe. If the AdvanceTrac™ system has become activated, it is always an indication that one or more tires have lost traction.

✔ SYSTEM DIAGNOSIS AND SERVICE

Many of the diagnostic features that Ford used in the EEC IV system were carried over into the EEC V system and are still being used on Ford vehicles. EEC V also integrated the required OBD II standards. As a result, the EEC V system surpasses past systems in its ability to help the technician diagnose the system.

Manual methods of code pulling are eliminated with this OBD II system. That is, connector terminals at the data link connector (DLC) (located under the left end of the instrument panel as specified by OBD II standards) should never be shorted with a jumper wire. Instead, a scan tool should always be used.

Diagnosis should begin by connecting the manufacturer's scan tool or OBD II generic scan tool to the data link connector (DLC). Once the scan tool has been programmed with the VIN information, a menu displays all the diagnostic features available for the particular application. These features include a key-on, engine-off (KOEO) on-demand self-test, a key-on, engine-running (KOER) on-demand self-test, the ability to pull continuous memory DTCs from the PCM's memory, and the ability to use the scan tool to erase continuous memory DTCs. Continuous memory DTCs should always be erased with a scan tool and never by disconnecting the battery.

One difference between the EEC IV system and the current EEC V system is that the EEC V PCM will store a continuous memory DTC if an on-demand self-test is run and identifies that a fault is present. If the on-demand self-test is run more than once and identifies the same hard fault two or more times, the PCM will illuminate the MIL. With the older EEC IV system, on-demand self-test failures did not store in the PCM's memory.

Ford's Quick Test for the EEC V system includes the following items:

- Visual check
- Vehicle preparation and equipment hookup
- KOEO on-demand self-test
- KOER on-demand self-test
- Pulling continuous memory DTCs

One functional test that carried over from the EEC IV system is an active command mode (somewhat similar to the EEC IV output state check, though much more sophisticated) that allows a technician to command the PCM to energize/de-energize various actuators, although a scan tool must now be used. Several actuators have been added to what the EEC IV system allowed the technician to command.

The data stream allows technicians to access particular values, including inputs, output commands, calculated values, and system status information through the use of a scan tool. The accessed information is identified by a *parameter identification (PID)*. Other diagnostic features added include an On-Board System Readiness (OSR) test that allows the technician to view the status of all OBD II monitors and whether they are complete or incomplete. Incidentally, Ford has a unique DTC, P1000, that indicates if the OBD II monitors have not been completed since the last battery disconnect or erasing of codes. Also, per OBD II standards, freeze frame/snapshot data is also available with any stored DTCs by accessing the global/generic OBD II feature on the scan tool's menu. This data allows the technician to learn what many of the sensors were reporting when a fault occurred, as well as other pertinent information such as fuel trim values.

All 10 OBD II global modes (Mode $01 through Mode $0A) can be accessed through the generic OBD mode of a generic scan tool. For example, Mode $06 data for noncontinuous monitors may be accessed in this mode. Ford Mode $06 information is available on the International Automotive Technician's Network (IATN) website at http://www.iatn.net.

Also, most of the procedures used to retrieve continuous memory DTCs, run KOEO on-demand tests (or even KOER on-demand tests in some cases), and look at data stream PIDs and values apply to other control modules on the vehicle, including the GEM, FEM, REM, BCM, IPC, audio system, climate control system, and modules that control the antilock braking system and passive restraint system. In fact, some modules such as the FEM allow the technician to see four PIDs as they apply to each circuit they control. For example, if you look at the PIDs for the left front high beam headlight, the scan tool will indicate:

- Whether the circuit is activated
- Whether the circuit is open
- Whether the circuit is shorted to power
- Whether the circuit is shorted to ground

This is in addition to being able to run an on-demand test of the circuits that the FEM controls.

All this diagnostic programming is not limited to the PCM; it also applies to most other control modules on the vehicle.

SUMMARY

In this chapter, we have covered many aspects of the EEC V system, including its sensors and actuators, as well as the fuel injection, spark management, and emissions systems that are under control of the EEC V PCM. This includes updated sensors such as Ford's newest DPFE sensor and the ESM of which it is a part.

We took a look at some of the advanced features found on Ford vehicles, including variable cam timing and electronic throttle control. We also covered the Intake Manifold Runner Control system.

In this chapter, we also covered other advanced systems such as Ford's Fail-Safe Cooling, Adaptive Cruise Control, and the differences between a GEM, a REM, and a FEM. We also took a look at some of the newest technologies, including Ford's voice recognition and navigational system, the Intelligent Architecture technology, and the AdvanceTrac™ system.

▲ DIAGNOSTIC EXERCISE

A 2008 Ford vehicle with an EEC V system is towed into the shop. The complaint of "engine cranks, but won't start" is verified. One technician prepares to test the ignition system for spark using a spark tester. Another technician prepares to connect a fuel pressure gauge to the fuel rail to test fuel system pressure. What easier check are these technicians forgetting that should be performed first?

Review Questions

1. *Technician A* says that the EEC V PCM can determine the gradual wear and aging of the vehicle that occurs over time. *Technician B* says that the EEC V PCM can compensate for gradual wear and aging of the vehicle by making adjustments in its programs. Who is correct?
   A. *Technician A* only
   B. *Technician B* only
   C. Both technicians
   D. Neither technician
2. What type of sensor is used with an EEC V SFI fuel management system to measure engine load?
   A. VAF
   B. MAP
   C. MAF
   D. PFE
3. *Technician A* says that the purpose of the oxygen sensor located after the catalytic converter is to measure the amount of unburned hydrocarbons that were not oxidized by the converter. *Technician B* says that the purpose of the oxygen sensor located after the catalytic converter is to monitor the catalytic converter's effectiveness. Who is correct?
   A. *Technician A* only
   B. *Technician B* only
   C. Both technicians
   D. Neither technician
4. The power steering pressure (PSP) switch input primarily affects the PCM's control of which of the following?
   A. Idle speed
   B. Injector pulse width
   C. Spark timing
   D. The fuel pump relay
5. *Technician A* says that the crankshaft position (CKP) sensor allows the PCM to identify which pair of pistons is approaching TDC, enabling it to fire the correct ignition coil in the waste spark ignition system. *Technician B* says that the camshaft position (CMP) sensor provides the PCM with the information needed for fuel injection sequencing and CPP ignition system control. Who is correct?
   A. *Technician A* only
   B. *Technician B* only
   C. Both technicians
   D. Neither technician
6. What is the purpose of the inertia switch used in the EEC V system?
   A. To disable the starter circuit in case of impact
   B. To remove power from all fuel injectors in case of impact
   C. To shut down the ignition system in case of impact
   D. To shut off the fuel pump in case of impact

7. An EEC V vehicle is brought into the shop with a fuel injection system that vents fuel past a pressure regulator through a return line back to the fuel tank. *Technician A* says that the fuel pressure regulator is indexed to manifold pressure (vacuum) to increase the flow rate of the injectors while under heavy load. *Technician B* says that the fuel pressure regulator is indexed to manifold pressure (vacuum) to maintain a constant pressure differential across the injectors to maintain a constant flow rate any time the injectors are energized. Who is correct?
   A. *Technician A* only
   B. *Technician B* only
   C. Both technicians
   D. Neither technician
8. What is the primary purpose of designing a vehicle with an electronic returnless fuel injection system?
   A. To reduce vapor lock potential
   B. To increase fuel economy
   C. To increase engine performance
   D. To reduce how much engine compartment heat is transferred to the fuel tank
9. What is the primary advantage of a CPP (COP) distributorless ignition system?
   A. The potential for secondary voltage to induce a voltage in a nearby circuit has been reduced.
   B. The PCM has more accurate control of ignition timing than with a waste spark ignition system.
   C. Mounting ignition coils over the spark plugs makes the plugs easier to change at the recommended service intervals.
   D. CPP (COP) systems can continue to operate if the crankshaft position sensor were to fail.
10. *Technician A* says that the DPFE sensor measures the physical position of the EGR valve and cannot compensate for reduced EGR flow rate caused by carbon buildup. *Technician B* says that a DPFE sensor measures EGR flow rate and can compensate for changes in exhaust backpressure. Who is correct?
    A. *Technician A* only
    B. *Technician B* only
    C. Both technicians
    D. Neither technician
11. In what year did Ford introduce the engine off natural vacuum test for the evaporative system?
    A. 1996
    B. 2002
    C. 2005
    D. 2015
12. When performing an engine off natural vacuum test, the PCM will do which of the following?
    A. De-energize both the purge solenoid valve and the vent solenoid valve
    B. Energize both the purge solenoid valve and the vent solenoid valve
    C. De-energize the purge solenoid valve and energize the vent solenoid valve
    D. Energize the purge solenoid valve and de-energize the vent solenoid valve
13. If engine temperature rises above 260°F on a vehicle with the Fail-Safe Cooling strategy, the PCM will do which of the following?
    A. Disable one fuel injector at a time until engine temperature drops
    B. Disable two fuel injectors at a time until engine temperature drops
    C. Disable three fuel injectors at a time until engine temperature drops
    D. Disable all fuel injectors until engine temperature drops

14. Which of the following systems allows an engine to use a higher compression ratio while still being able to use regular octane fuel?
    A. VCT
    B. ETC
    C. IMRC
    D. Both B and C
15. In the Ford ETC system, what component within the PCM redundantly monitors the inputs and the primary CPU's handling of the information?
    A. E-Quizzer
    B. TPPC
    C. APP sensor
    D. H-Bridge
16. What type of sensor is used with an EEC V GDI EcoBoost four-cylinder engine to measure engine load?
    A. APT
    B. MAP
    C. MAF
    D. Both A and B
17. On a 2002–2005 Lincoln LS and Ford Thunderbird, what does Ford call the electronic control module that controls the one-touch-down power window, the horn relay, and the exterior lights at the front of the vehicle?
    A. BCM
    B. GEM
    C. REM
    D. FEM
18. If the *front end* of a vehicle with the AdvanceTrac system begins to drift to the *outside* of a turn, which wheel will the AdvanceTrac system apply a braking pulse to in order to correct the understeer?
    A. Inside rear wheel
    B. Inside front wheel
    C. Outside rear wheel
    D. Outside front wheel
19. *Technician A* says that when diagnosing an EEC V system, DTCs may be pulled manually by connecting a jumper wire between two terminals at the data link connector. *Technician B* says that when diagnosing an EEC V system, an OBD II scan tool must be used to initiate an on-demand selt-test or to pull continuous memory DTCs. Who is correct?
    A. *Technician A* only
    B. *Technician B* only
    C. Both technicians
    D. Neither technician
20. On the EEC V system, Ford's DTC P1000 indicates which of the following?
    A. The PCM has failed and should be replaced.
    B. The PCM has detected a fault on the input side that should be diagnosed and repaired.
    C. The PCM has detected a fault on the output side that should be diagnosed and repaired.
    D. The PCM has not seen all of the OBD II monitors completed since the last battery disconnect or erasing of codes.

Chapter 14

Chrysler Corporation Computerized Engine Controls

OBJECTIVES

Upon completion and review of this chapter, you should be able to:

- ❑ Describe the various Chrysler systems and PCMs used with fuel injection, as well as their operating modes.
- ❑ Define the inputs used with a Chrysler fuel-injected system.
- ❑ Define the outputs controlled by a Chrysler fuel-injected system.
- ❑ Identify the multiplexing systems used on Chrysler vehicles.
- ❑ Understand how a Chrysler fuel-injected system is designed to be put into self-diagnosis, as well as the various diagnostic features.

KEY TERMS

Adaptive Memory
Logic Module
Multipoint Fuel Injection
Power Module
Single-Point Fuel Injection

In 1972, Chrysler became the first auto manufacturer to switch from a mechanical points-and-condenser ignition system to an electronic transistorized ignition system. In 1976, Chrysler introduced an electronic spark timing control system. So the company was ready with a computer-controlled engine system in 1979 when it began systematic installation of an oxygen sensor feedback system on many vehicles. The first such system was on the California models with the slant-six engine, which combined the earlier Chrysler spark control system and oxygen sensor with a module that controlled an air-fuel mixture control solenoid. The next year the system was extended to four- and eight-cylinder engines. By 1981 it was applied to federal-specification vehicles, too. By that date, it also had idle speed control functions along with EGR and air-injection actuation. As the system developed, the shift indicator light, the radiator cooling fan, and the vacuum solenoid controlling the carburetor secondary barrel on cars built for high altitudes all followed.

Chrysler produced a few Imperials with an electronic fuel injection system in the early 1980s, but the first current versions of Chrysler fuel injection appeared on the 1984 2.2L engines. These were the **single-point fuel injection** system, which Chrysler calls Electronic Fuel Injection (EFI), and the **multipoint fuel injection** system, which Chrysler calls Multi-Point Injection (MPI). The 2.2L MPI system is often called the *turbo* system because it uses a turbocharger. Since the original engine offerings, however, other power plants have come with either the EFI or the MPI system. The two systems are quite similar, so this chapter covers them together, pointing out variations between the two.

Before 1986, the single-point system used high fuel pressure, about 36 PSI. In 1986, a low-pressure (14.5 PSI) injector with a ball-shaped valve replaced the previous unit with its pintle-style valve. This lower-pressure injector is found on many other fuel injection systems besides Chrysler's.

POWERTRAIN CONTROL MODULE

Logic Module/Power Module

From 1984 to 1988, two separate modules controlled the Chrysler fuel injection systems. The **logic module** with the central processor was inside the passenger compartment. The **power module** controls the actuators by switching the ground-side circuits for the ignition coil, fuel injector or injectors, and the automatic shutdown (ASD) relay.

Single-Module Engine Controller

In mid-1987 in some 3.0L engine applications, the logic module and the power module were both placed as separate printed circuit boards in a single unit known as the *Single-Module Engine Controller (SMEC).* Physically, the power module is placed in a cavity created within the logic module; the two modules are then contained within a single housing. Engine intake air flows through the modules' housing to remove heat generated by the modules. Jumper wires are run between the two modules' connectors. Together, of course, these two modules constitute the PCM. By 1988, most Chrysler vehicles used the SMEC system.

Single-Board Engine Controller

In 1989, Chrysler introduced a system that combined the two separate printed circuit boards of the SMEC system by forming all components onto a single circuit board. This system was given the name *Single-Board Engine Controller (SBEC).* The SBEC PCM is also located in the engine compartment to allow engine intake air to carry heat away from it. The SBEC PCM is recognizable by its single electrical connector. By 1990, most Chrysler vehicles used the SBEC system.

In 1991, Chrysler introduced its second generation of SBEC, known as SBEC II. By 1992, most Chrysler vehicles used the SBEC II system, including passenger cars, minivans, light trucks, and Jeeps. The SBEC II PCM still required intake air flowing through it for cooling.

In 1995, as OBD II standards were being set in place, Chrysler replaced the SBEC II system on its passenger cars and minivans with the SBEC III system. The SBEC III PCM is contained in a shielded case to reduce electromagnetic interference (EMI) and radio frequency interference (RFI). In addition, the SBEC III case dissipates heat through external aluminum fins so that it no longer requires intake air for cooling. In 1996, Chrysler also replaced the SBEC II system on Jeeps, Dodge trucks, and the Dodge Viper with a new system known as the Jeep/Truck Engine Controller (JTEC) system. In 1998, with enhanced OBD II standards requiring certain changes, Chrysler upgraded the SBEC III system on its passenger cars and minivans to an SBEC IIIA system. In each of these systems the PCM has different pin arrangements so that an incorrect replacement PCM cannot be used inadvertently. Chrysler continued to use both the SBEC IIIA system and the JTEC system on its vehicles through the 2001 model year.

On certain 2002 model-year vehicles, Chrysler introduced a new, state-of-the-art engine and transmission control system known as the Next Generation Controller, or NGC, system. This NGC PCM, shown in Figure 14–1, not only controls engine performance but also replaces the transmission control module (TCM) that had been used alongside the PCM on previous systems. As of the 2005 model year, the NGC system has replaced all SBEC IIIA and JTEC systems.

Figure 14–1 NGC PCM.

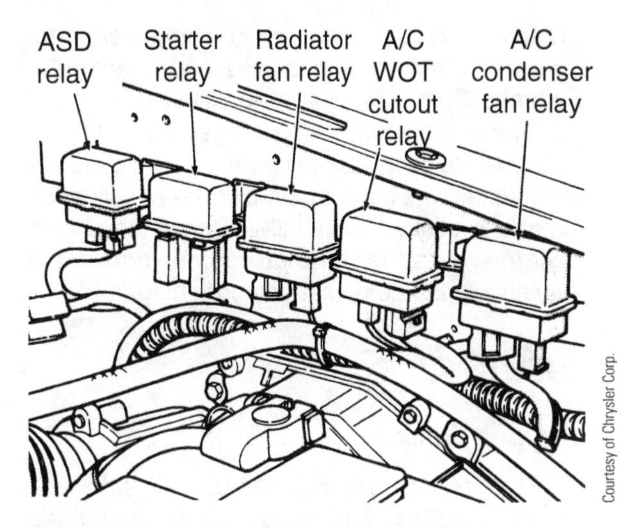

Figure 14–2 Typical location, ASD relay.

Features

Automatic Shutdown Relay. The ignition switch energizes the ASD relay when it is in the Run position. The power module energizes the ASD coil by grounding its return circuit, and the ASD relay powers the electric fuel pump, the ignition coil, and the injector(s).

The fuel pump has its own independent ground and turns on immediately. The ignition coil and injectors get power from battery positive, but no current flows through their circuits until the power module grounds their activation circuits. From 1985 through 1987 the ASD relay was internal to the power module. In 1984 vehicles and after 1987, it was a separate relay, outside the power module (Figure 14–2). Mounting it outside saves module space and improves module reliability by removing the heat it generates.

Beginning in 1990, the ASD relay was also used to provide power for the heating element of the heated oxygen sensor(s) on the vehicle. Beginning in 1996, SBEC systems began using a dedicated fuel pump relay to provide power for the fuel pump. But the system retained the ASD relay, which now provides power for the fuel injectors, the ignition coil primary winding, the heated oxygen sensor(s), and the alternator field windings. Both relays are deenergized if the tach reference signal is lost.

Adaptive Memory. Adaptive memory is the ability of an engine control computer to assess the success of its actuator and sensor signals and to modify its internal calculations to correct them if the desired result is not achieved. The PCM can modify some of its programmed calibrations

Diagnostic & Service Tip

Certain mechanical problems can throw a system out of closed loop repeatedly. Suppose a car develops a series of bad oxygen sensors in a relatively short time. An experienced technician will check for causes external to the system, too. This particular problem often comes when a head gasket is starting to fail and seeps coolant into the combustion chamber or exhaust. The silicone in the coolant forms a super-thin, impermeable layer on the exhaust surface of the sensor, "poisoning" it. Less often, coolant silicone from a burst hose or some other contaminant will get into the atmospheric side of the sensor. More rarely, silicone poisoning occurs when a technician uses older silicone-based gasket sealers.

("maps") for fuel metering to compensate for production tolerance variations and changes in barometric pressure or vehicle altitude.

There are limits to the range of adjustment available to the adaptive memory capacity. Sometimes mechanical problems—a collapsed exhaust system, a burnt valve—throw the engine performance so far off that no countermeasure available to the computer can correct for it.

INPUTS

Oxygen Sensor

Modern Chrysler vehicles use a heated zirconium dioxide oxygen sensor (O_2S) for both the pre-cat and post-cat sensors. Chrysler uses the post-cat oxygen sensor after each catalytic converter for catalytic converter efficiency monitoring. The post-cat sensor is also used for rear fuel control.

On many Chrysler vehicles the PCM biases the oxygen sensor voltage by 2½ V. This can be true for all four oxygen sensors, including both pre-cat sensors and both post-cat sensors. Depending upon the vehicle, the PCM may add 2½ V to the sensor's voltage on the physical oxygen sensor circuit or it may bias the oxygen sensor internally only. On either version, the data stream of a scan tool will show that the PCM is adding 2½ V to the voltage produced by the oxygen sensor. In closed-loop fuel control, the scan tool's data stream will show that the pre-cat oxygen sensor is cross-counting between 2½ and 3½ V.

On those vehicle in which the PCM adds 2½ V to the sensor's voltage on the physical oxygen sensor circuit, a DSO or DMM will also show that it is cross-counting between 2½ and 3½ V during closed-loop fuel control. However, on those vehicles in which the PCM is biasing the oxygen sensor internally only, if monitoring the pre-cat oxygen sensor's raw voltage during closed-loop fuel control with a DSO or DMM, the signal will cross-count between 0 V and 1 V, even though a scan tool will show that it is cross-counting between 2½ and 3½ V.

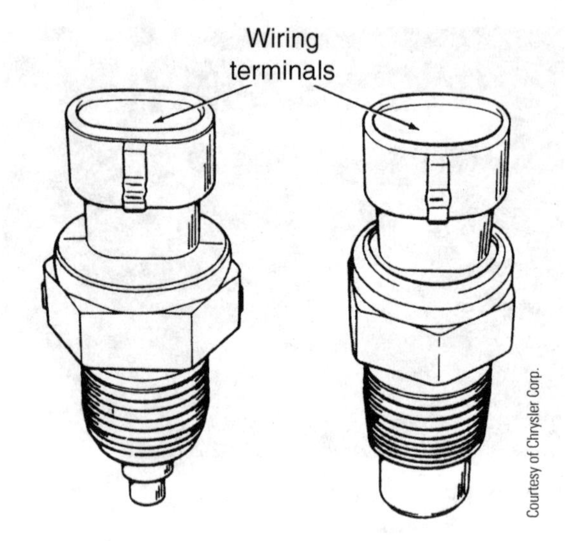

Figure 14–3 IAT and ECT sensors.

Engine Coolant Temperature Sensor

The ECT sensor is a negative temperature coefficient (NTC) thermistor screwed into the thermostat housing (Figure 14–3). As it warms up, its resistance goes down. Prior to OBD II standardization, the ECT sensor was known by Chrysler as a *coolant temperature sensor,* or *CTS.*

Intake Air Temperature Sensor

The IAT sensor is an NTC thermistor screwed into a runner of the intake manifold. It has about the same resistance range as the ECT sensor and looks similar to it (Figure 14–3). On early models its input contributes to cold engine enrichment calculations; otherwise it serves as a backup to the ECT sensor. On later multipoint systems, its input helps to control the air-fuel mixture when cold and to boost control at all times. Prior to OBD II standardization, the IAT sensor was known by Chrysler as a *charge temp sensor*.

Range-Switching Temperature Sensor (Dual-Range Temp Sensor)

To fine-tune the fuel-air mixture and spark timing even further, beginning in 1992 Chrysler

Temp.	Sensor resistance	Temp. change	Resistance change	Ohm change per degree	Volt drop across sensor	
– 20°F	156,667 Ω				4.7 V	With 10,000 Ω fixed resistance
		10°	50,388 Ω	5038.8		
– 10°F	106,279 Ω				4.57 V	
40°F	25,714 Ω				3.6 V	
		10°	6,302 Ω	630.2		
50°F	19,412 Ω				3.3 V	
110°F	4,577 Ω				1.57 V	
		10°	1,244 Ω	124.4		
120°F	3,333 Ω				1.25 V	
Sensor circuit shift (909-Ω fixed resistance rather than 10,000 Ω)						
140°F	2,338 Ω				3.6 V	With 909 Ω fixed resistance
		10°	406 Ω	40.6		
150°F	1,932 Ω				3.4 V	
200°F	839 Ω				2.4 V	
		10°	125 Ω	12.5		
210°F	714 Ω				2.2 V	
240°F	435 Ω				1.62 V	
		10°	64.5 Ω	6.45		
250°F	371 Ω				1.45 V	

Figure 14–4 Temperature, resistance, and voltage drop change with dual-range temperature sensor.

changed to range-switching temperature sensors for coolant and air temperatures (Figure 14–4).

The purpose of this change is to telescope the sensitivity of the sensor in the area of greatest importance, around operating temperature. It also accommodates the problem that the thermistor's reaction to changes in temperature is not linear—that is, there is more change in the cooler (lower) temperatures than at the upper end. Unfortunately, for engine management purposes, we are more interested in knowing about the higher temperatures in fine detail for air-fuel mixture control.

The sensor works like this: Below 125°F, there is a fixed 10,000 Ω resistor in series with the sensor's thermistor (Figure 14–5). At about 125°F, the PCM turns on a 1,000 Ω resistor in parallel with the 10,000 Ω fixed unit. This toggles the resistance to 909 Ω.

This spreads out the resistance more evenly over the range the computer is seeking. Of course, it is actually measuring the voltage drop across the thermistor. This range shift means, of course, that the computer must know that the change that occurs around 125°F is not a sudden cooling of the engine but the result of the resistance switch. This "expectation" is programmed into the PCM's memory.

When the second resistor is switched into parallel, the total fixed resistance is then considerably lower than that of the thermistor. There is a wider range of voltage drop available to be monitored, and greater accuracy is obtainable. This also extends the measurement range up into the over-temp area.

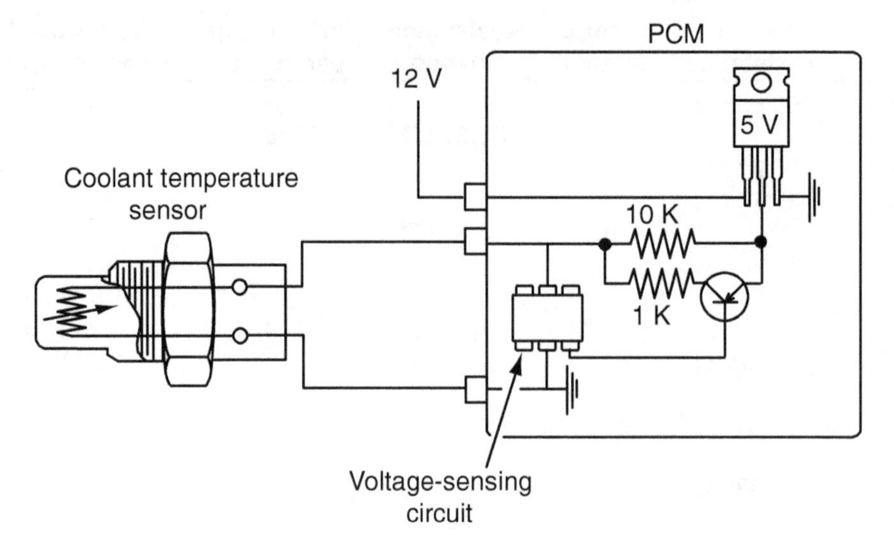

Figure 14–5 Dual-range temperature sensor electrical schematic.

Throttle Position Sensor

The TPS is a rotary potentiometer—a variable resistor—at the end of the throttle shaft on the side of the throttle body. Depending on the position of the throttle, the sensor modifies the 5 V reference signal to something less, corresponding with the throttle's current position, and it returns that signal to the computer. Engine management systems also keep track of how quickly the throttle moves, to more accurately provide acceleration enrichment when the pedal is suddenly floored or to shut off fuel if the throttle is suddenly closed at high engine speed.

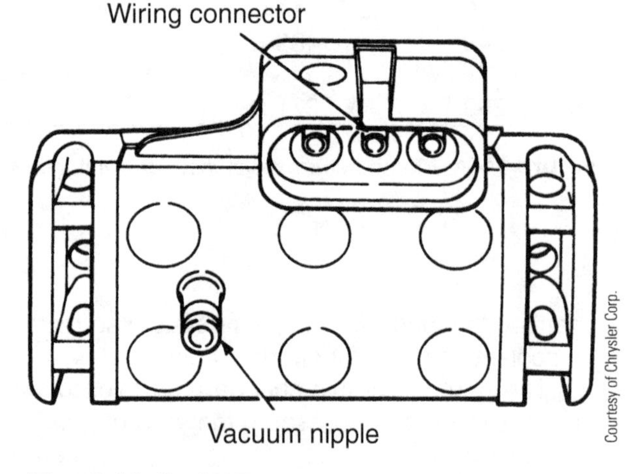

Figure 14–6 MAP sensor.

Manifold Absolute Pressure Sensor

The MAP sensor is a piezoresistive analog pressure sensor. A piezoresistive diaphragm changes its electrical resistance in response to a slight flexing caused by changes in pressure. The PCM uses its signal as a barometric pressure sensor during periods of key-on, engine-off or during certain other conditions such as wide-open throttle (WOT) operation (Figure 14–6).

The technician should understand clearly what manifold absolute pressure is. Formerly, the term *intake manifold vacuum* was used, but that can be misleading (particularly after the widespread introduction of turbocharged cars). The manifold absolute pressure is the actual pressure of the air in the intake manifold. Most often, of course, it is lower than ambient pressure, but it is still air pressure. If a turbocharger or supercharger increases the pressure above ambient pressure, it is still manifold absolute pressure. This value corresponds very closely to engine load.

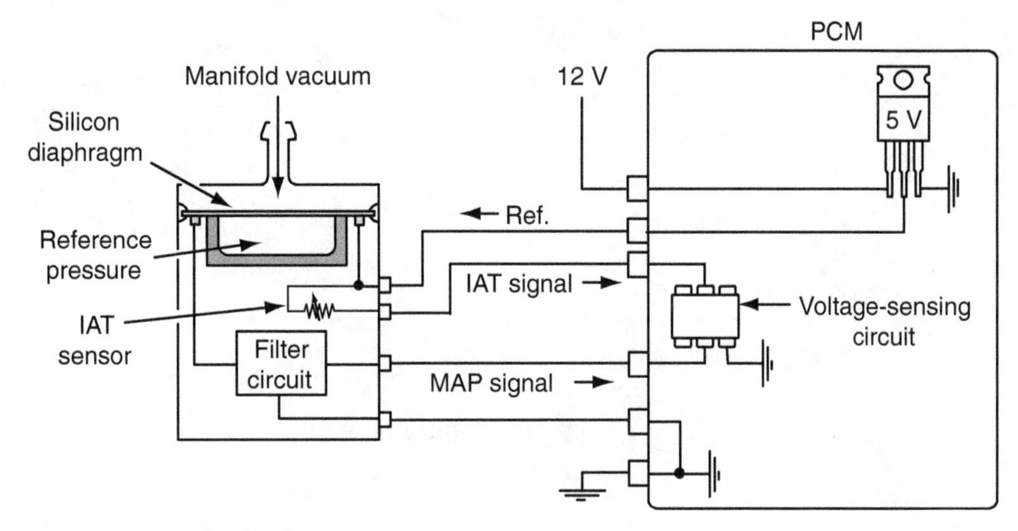

Figure 14–7 TMAP sensor electrical schematic.

TMAP Sensor

Chrysler 2.0L, 3.2L, and 3.5L engines use a sensor called the TMAP (Figure 14–7). This sensor combines the functions of the IAT sensor and the MAP sensor into one component. The TMAP sensor is located at the intake manifold.

TMAP Strategy

Due to slow MAP sensor response when the driver suddenly moves the throttle, some Chrysler PCMs follow a strategy that uses the TPS signal to initially calculate MAP values as the throttle is moved, resulting in a quicker PCM response to throttle movement. This strategy is referred to as the TMAP strategy; it should not be confused with the TMAP sensor.

Mass Air Flow Sensor

Chrysler introduced a mass air flow (MAF) sensor on the 2001 Chrysler Sebring and Dodge Stratus coupes with 2.4L and 3.0L engines. Chrysler's MAF sensor is made by Mitsubishi and is called the *Mitsubishi Ultimate Karman Air Flow Sensor,* or MUKAS. It replaces the speed density formula and measures the intake air's mass.

The Chrysler MAF sensor produces a square-wave variable digital frequency by grounding out a 5 V reference voltage from the PCM. The frequency at which it does this is measured by the PCM to determine measured air intake. A Hertz meter should be used to measure the resulting signal, similar to GM MAF sensors and Ford EEC IV digital MAP sensors. If the ignition is turned on but the engine is not running, the voltage is pulled to ground and does not oscillate. On the engines that Chrysler used a MAF sensor on, they also used a MAP sensor, similar to GM vehicles.

Chrysler's MAF sensor was short-lived. It was used on the Chrysler Sebring for just 4 years, from 2001 through mid-2004. It was used on the Dodge Stratus for just 5 years, from 2001 through 2005. By the 2006 model year, these engines had been converted back to a speed density system using a MAP sensor.

CKP and CMP Sensors

Modern Chrysler fuel injection and ignition systems use Hall effect crankshaft position (CKP) and camshaft position (CMP) sensors. These Hall effect sensors are of a unique design in that the body of the sensor contains a permanent magnet

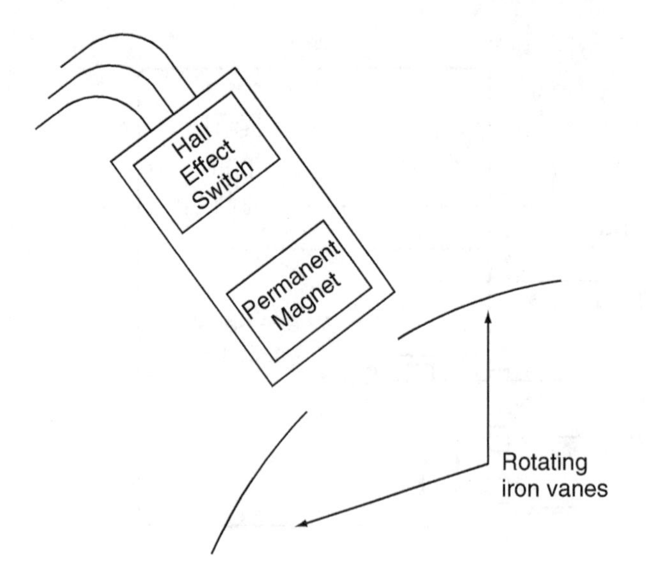

Figure 14–8 Chrysler Hall effect sensor.

near the front end and a Hall effect switch near the back end (Figure 14–8). As iron (metal) is passed near the front end, the magnet's magnetic field is enhanced enough to encompass the Hall effect crystal behind it, causing the Hall effect switch to turn on. When the iron moves away from the Hall effect switch, the magnet's magnetic field is weakened enough that it no longer encompasses the Hall effect switch behind it, causing the Hall effect switch to turn off.

The CKP sensor mounts on the bell housing and triggers off slots on the torque converter drive plate (Figure 14–9 and Figure 14–10). There are four slots per pair of companion cylinders.

The CMP sensor mounts on the timing cover and triggers off slots on the camshaft timing gear (Figure 14–10). The slots on the timing gear are coded so that signals from the sensor identify which pair of companion cylinders the ignition module should fire next in response to the computer's spark timing command. This signal can also determine which injector the computer pulses next. Because the camshaft sensor can identify individual cylinders (unlike the crankshaft sensor), the computer can use its signal to begin firing spark plugs and pulsing fuel injectors within the first crankshaft rotation.

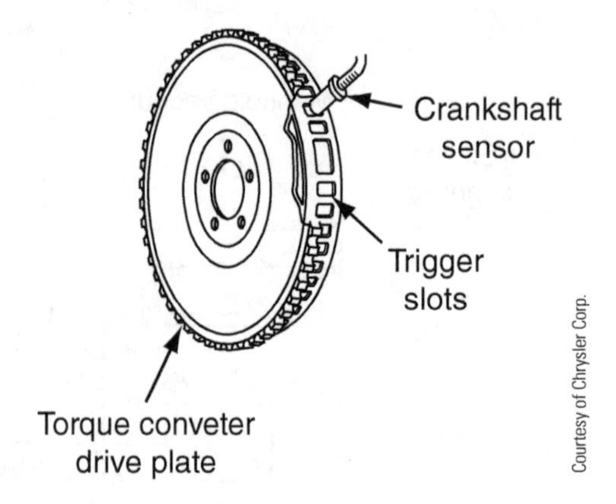

Courtesy of Chrysler Corp.

Figure 14–9 CKP sensor.

The computer uses these signals to determine CKP and engine speed. It then uses this information to determine ignition timing, injector timing, and injector pulse width.

Figure 14–11 shows the waveforms generated by Chrysler's typical Hall effect CMP and CKP sensors.

Vehicle Speed Sensor

In 1992 Chrysler began using a Hall effect VSS (Figure 14–12). To power up, this sensor receives power and ground, then delivers a digital square wave to the PCM at a rate of 8,000 pulses per mile. The power supply is delivered to the VSS from the PCM over the same circuit that carries power to the CKP and CMP Hall effect sensors. This voltage is 8 V on early models and 5 V on newer versions. Therefore, if the VSS were to short its power circuit to ground internally, the engine would not start due to the simultaneous loss of power at the CKP and CMP sensors.

Many Chrysler vehicles also use a permanent magnet (variable reluctance) VSS that monitors iron teeth and notches on the output shaft of the transaxle, producing AC voltage pulses that are sent to the PCM. As a result, depending on the application, Chrysler vehicles may use either a Hall effect VSS or a permanent magnet VSS.

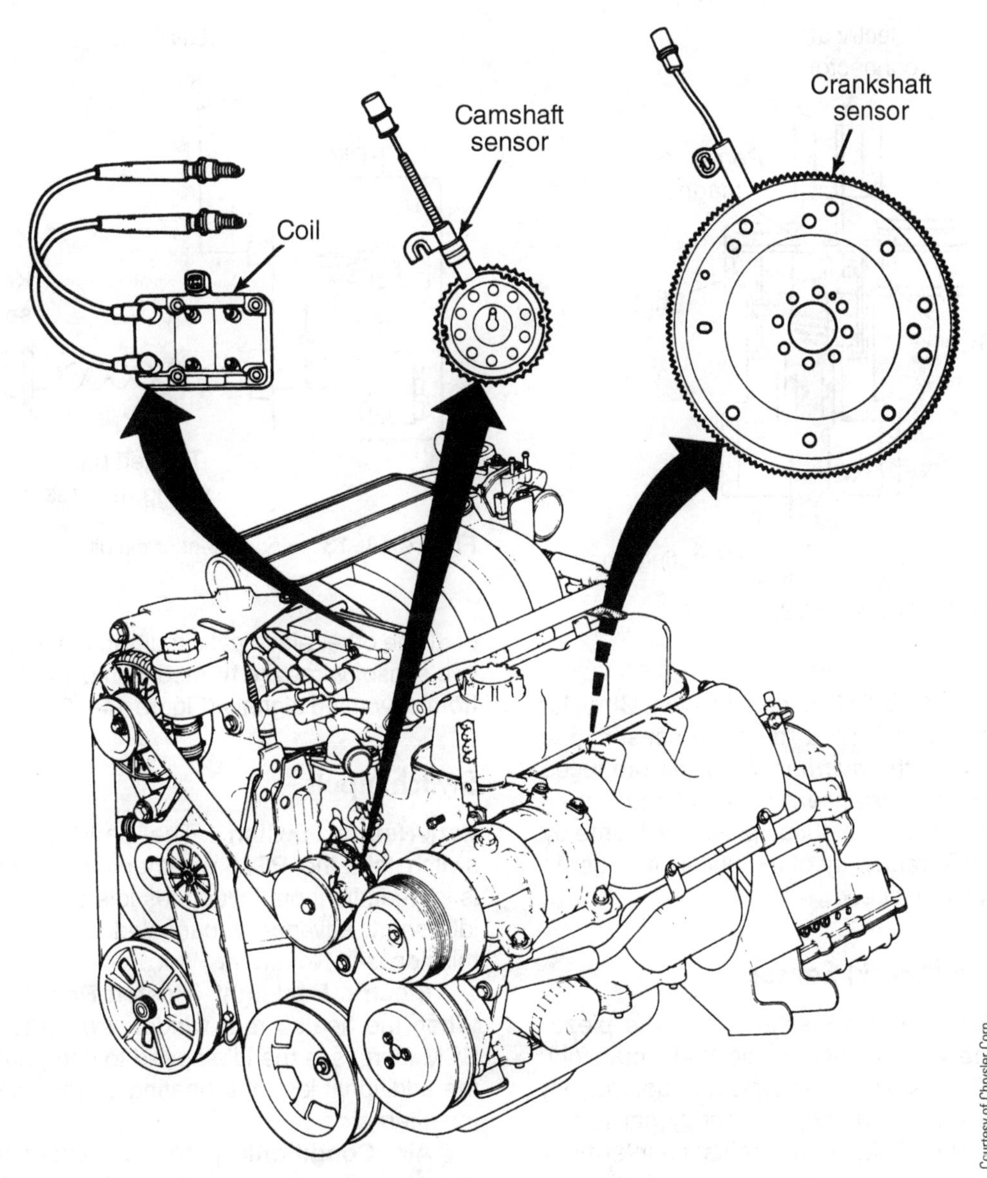

Figure 14–10 3.3L engine with CMP and CKP sensors.

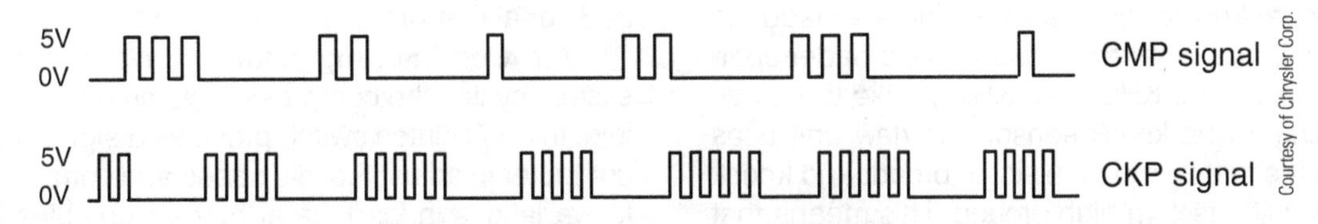

Figure 14–11 Typical waveforms generated by Chrysler's CMP and CKP Hall effect sensors.

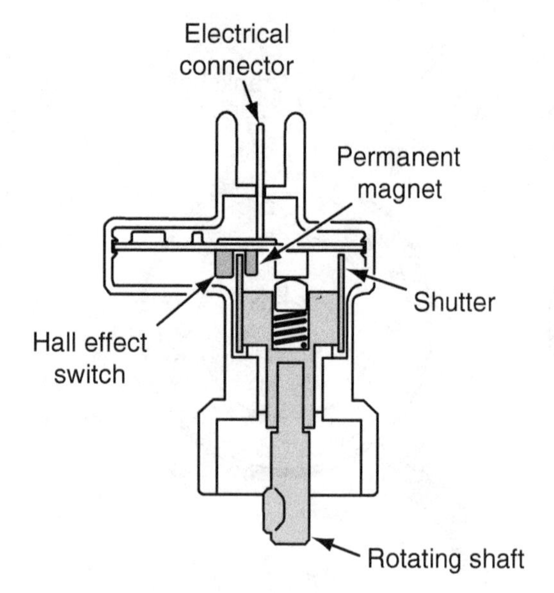

Figure 14–12 Hall effect VSS.

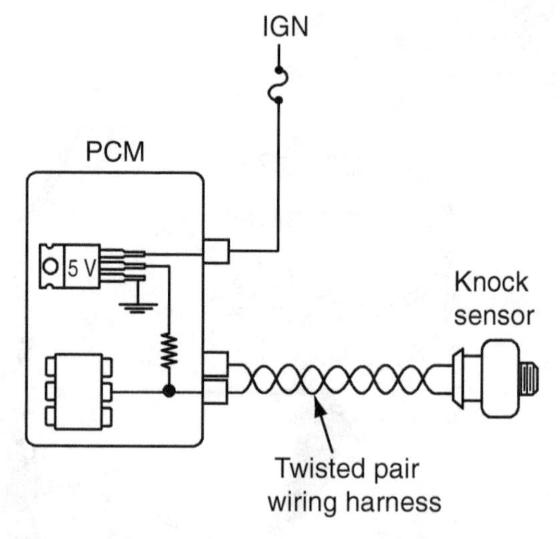

Figure 14–13 Knock sensor circuit.

Also, prior to OBD II standardization, Chrysler referred to the VSS as a "distance sensor" due to the fact that the *number* of pulses produced only represents distance. The PCM must then combine the number of these pulses with time by looking at the *frequency* of the pulses in order to calculate vehicle speed.

Detonation (Knock) Sensor

The Chrysler knock sensor is of the piezoelectric type; it is mounted on the intake manifold. When excited by vibrations typical of detonation, it sends a true alternating current signal to the computer. The PCM then begins countermeasures, which reduce spark advance by individual cylinders and/or reduce turbocharger boost.

Chrysler 3.2L and 3.5L engines have changed from dual-knock sensors to a single broadband knock sensor. This new sensor has a broader operating range of 5 KHz to 20 KHz. Unlike the previous single-wire knock sensor, the new unit uses two wires that provide the single broadband knock sensor with its own high ground. This means that the sensor no longer shares the common ground of all the other electrical circuits in the vehicle. The two sensor wires are twisted to improve the shield from unwanted signals (Figure 14–13).

Switch Inputs

Park/Neutral Switch. The park/neutral (P/N) switch tells the PCM whether the transmission is in gear. Its input influences idle speed. Normal idle spark advance is canceled when the transmission is in Neutral or Park.

Electric Backlight (Heated Rear Window). When the heated rear window switch is on, the PCM increases the idle speed to compensate for the additional load the heating current places on the alternator.

Air Conditioning (A/C). Whenever A/C operation is requested, the A/C control circuit sends the PCM computer a signal. The PCM then increases idle speed to compensate for the additional compressor load.

Air Conditioning Clutch. When the A/C system cycles the compressor clutch on and off at idle, the A/C clutch switch provides a signal for the computer to adjust the idle speed and compensate for variations in load. Later models combine both air conditioning inputs into one signal.

FUEL MANAGEMENT SYSTEMS

Injector/Injectors

Modern Chrysler fuel injection systems employ **sequential fuel injection (SFI),** a system in which each injector is pulsed individually in the engine's firing order. Figure 14–14 shows an SFI fuel injector. Not only does this design use an individual terminal of the PCM to operate each injector, but the PCM now needs to know TDC compression information for each cylinder. This information is obtained from the CMP sensor.

Electric Fuel Pump

A permanent magnet–type direct current electric motor drives the vane-type fuel pump, both of which are submerged in fuel at the bottom of the fuel tank. One check valve functions as a pressure relief valve; the other, in the outlet port, prevents fuel from running in either direction when the pump is off. This keeps the line full of pressurized fuel when the engine is off, reducing the potential for fuel volatility problems.

With the ignition switch on, the PCM grounds the fuel pump relay, powering the fuel pump. Unless there is a tach reference signal from the CKP sensor within 2 seconds, the PCM will de-energize the fuel pump relay and shut off the fuel pump. Once shut off, the relay stays off until the PCM sees a tach reference signal from the CKP sensor or the ignition switch is cycled off and on again. This design ensures that the PCM will de-energize the fuel pump relay if a fuel pressure line is ruptured, resulting in engine stall.

Return-Type Fuel Injection System Pressure Regulator

A return-type fuel injection system uses a fuel pressure regulator with a manifold vacuum hose attached to the spring side. This is designed to keep the difference between the fuel rail pressure and the intake manifold pressure constant (Figure 14–15). This is often described as a "pressure differential across the injector." This results in a constant volume whenever the injectors are energized. The PCM then has absolute control over fuel delivery through control of the injector's pulse width.

Chrysler SFI systems maintain a pressure difference of 55 PSI. Fuel rail pressure on a Chrysler multipoint system can vary from 42 to 62 PSI as the pressure in the intake manifold also varies with engine load and throttle plate position. Pressures above 55 PSI will be achieved when a turbocharged engine is under boost.

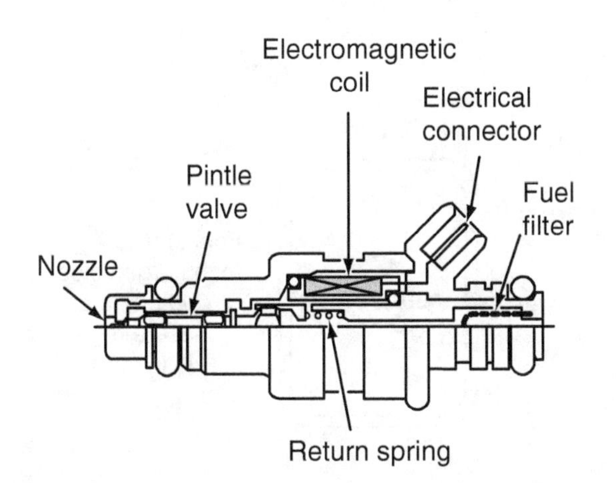

Figure 14–14 Typical SFI fuel injector.

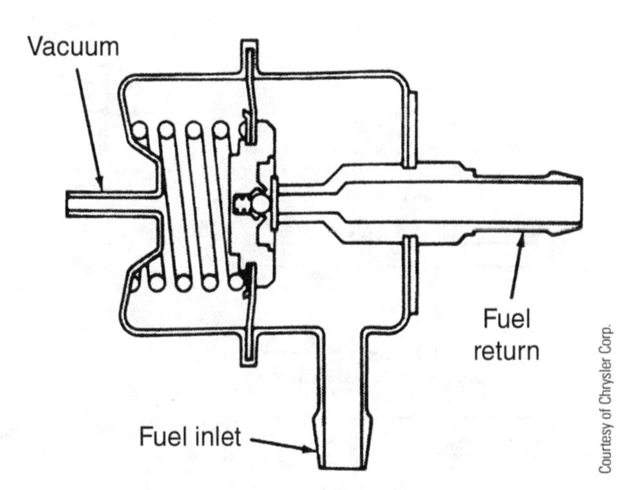

Figure 14–15 Fuel pressure regulator in a return-type fuel injection system.

Returnless Fuel System

By 1996 Chrysler fuel injection systems had changed further in two specific ways: First, fuel pressure is now held constant at approximately 49 PSI by a mechanical regulator independent of the PCM and of intake manifold vacuum. Second, the fuel rail plumbing consists of a fuel supply line only. There is no return line originating within the engine compartment.

Traditionally, the fuel pressure regulator has been located on the fuel rail or return line. Fuel in excess of engine requirements is returned to the fuel tank through a return line. Returned excess fuel may return the heat it absorbed while it was in the hot engine compartment. In an effort to reduce evaporative emissions from the fuel tank, modern Chrysler engines use a returnless fuel system. The fuel pressure regulator is located on top of the fuel pump module inside the fuel tank. Any fuel that exits the fuel tank assembly will not be returned to the fuel tank; therefore any excess fuel is simply returned into the tank at the point of the fuel pump assembly and never actually leaves the tank (Figure 14–16).

Semi-Returnless Fuel System

Based on the same concept as a returnless fuel system, a semi-returnless fuel system locates the fuel pressure regulator inside the fuel filter. The filter has three lines: a pressure line from the fuel tank, a pressure line to the fuel rail in the engine compartment, and a return fuel line that goes back to the fuel tank. The return line from the fuel filter is typically between 15 inches and 4 feet long but still serves to reduce the transport of engine compartment heat back to the fuel tank.

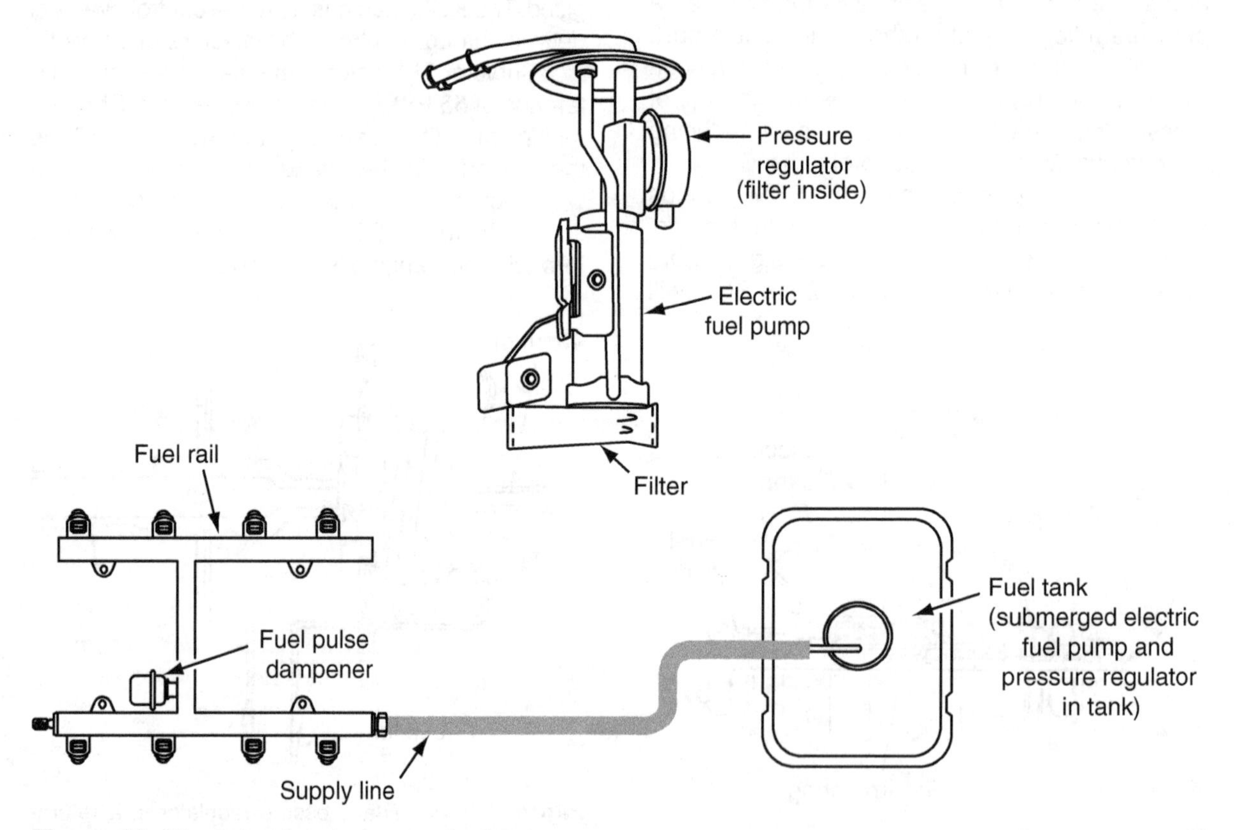

Figure 14–16 Returnless fuel system/tank unit.

PCM Compensation in a Returnless/Semi-Returnless Fuel System

In both forms of returnless systems, the fuel rail pressure is held constant. This means that under light engine load, when there is a larger pressure differential across the fuel injectors (high vacuum/low pressure in the intake manifold), the PCM must dramatically reduce the injectors' pulse width. Under high engine load, when there is a small pressure differential across the fuel injectors (low vacuum/high pressure in the intake manifold), the PCM must dramatically increase the injectors' pulse width. Ultimately, the PCM must calculate the pressure differential across the fuel injectors to reliably provide the proper injector pulse width.

IDLE SPEED CONTROL

Automatic Idle Speed Motor

The automatic idle speed (AIS) motor was originally a reversible electric motor (Figure 14–17). The PCM controls it based on information from the:

- TPS
- ECT
- VSS
- P/N switch
- Brake switch

The AIS motor moves its valve to control bypass air around the throttle plate in the throttle body. Even with the valve in its closed position, enough air gets past the throttle plate and through the throttle body to keep the engine idling at low speed with no load. The PCM directs the AIS motor to set the valve at different positions for different operating conditions. When the vehicle is decelerating, the valve opens to prevent engine stall and to prevent condensation of fuel on the intake port walls, as well as possible backfire.

The original DC motor was quickly replaced with a stepper motor version of the AIS actuator. This actuator moves a pintle valve to control throttle bypass air. The AIS stepper motor contains two windings that connect to the PCM with four wires. The PCM can apply current to the two windings in sequence; it can also reverse polarity to each winding as necessary. The PCM can move the AIS stepper motor through a total of 256 steps, from fully closed to fully open, to control idle speed. The stepper motor rotates a precise amount for each pulse that the computer sends it, so by tracking the number of pulses, the PCM knows the exact position of the air bypass valve. This provides the PCM with increased precision in controlling idle speed. The stepper motor version of the AIS actuator was used as late as the 2004 model year.

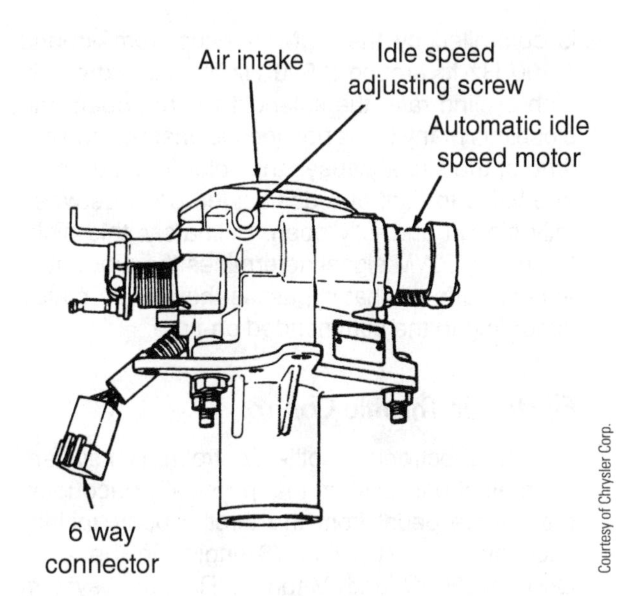

Figure 14–17 Multipoint throttle body showing the AIS motor.

Automatic Idle Speed Solenoid Valve

The AIS solenoid valve (identified as an actuator with only two wires) replaces the AIS stepper motor. It is a normally closed solenoid valve that is operated by the PCM on a high-frequency pulse-width-modulated (PWM) signal to control throttle bypass air. The frequency of the signal is controlled by the PCM, varying from around 1,200 Hz to around 2,500 Hz. At this extremely high cycling rate, the solenoid will not open and close this many times per second. Instead, the on-time of the signal causes the solenoid's iron core and valve to *float* at a designated point between fully closed and fully open. Ultimately, this high-frequency PWM signal determines the amount of atmospheric air that bypasses the throttle plates according to the commanded on-time.

Electronic Throttle Control

The Electronic Throttle Control (ETC) system is an electronic system that physically decouples the throttle pedal from the throttle body. Initially used on the 5.7L Hemi V8 engine in the 2005 Chrysler 300C and Magnum R/T, this system allows the PCM to directly control throttle plate movement electronically, thus eliminating the traditional throttle cable. The ETC system is often referred to generically as a "drive-by-wire" system. The ETC system provides several advantages, including lower emissions and quicker throttle response.

The components of the ETC system are a TPS to monitor throttle plate position, a DC motor that serves as the actuator used to control throttle plate position, and an accelerator pedal position (APP) sensor. In basic computer theory concerning an engine with a cable-operated throttle plate, the TPS is said to give the PCM information concerning *driver demand.* In an ETC system, driver demand information originates at the APP sensor.

The TPS and the APP sensor are both two-channel potentiometer assemblies, meaning that these sensors contain two potentiometers and send two signals to the PCM. In most applications, the potentiometers in the APP sensor operate electrically between different high and low values and/or in different directions. This allows the PCM to more precisely interpret exact throttle position than would be the case if they all delivered exactly the same voltage values at a given throttle position. The APP sensor is located in a throttle pedal assembly that is spring-loaded to give the driver a traditional pedal feel.

Newer Chrysler vehicles use noncontact linear Hall effect sensors instead of potentiometers to indicate both accelerator pedal position and throttle plate position. This increases the longevity of the sensor assemblies.

The PCM commands the DC motor to open or close the throttle plate(s) using bidirectional PWM commands. That is, it can issue pulse-width commands that instruct the motor to move the throttle plate(s) slowly or rapidly. "Bidirectional" refers simply to the fact that polarity applied to the motor can be reversed to move the throttle plate(s) in either direction (opening versus closing).

Compared to a cable-operated system, an ETC system potentially provides quicker throttle response. In a cable-operated system on a fuel-injected engine, the amount of both fuel and air entering the engine must be increased to increase engine RPM and vehicle speed. The driver controls the volume of air allowed to enter the engine using the throttle cable, but the PCM controls the fuel. In a cable-operated system, the following must happen for the PCM to know to add additional fuel:

- The driver must move the throttle pedal.
- The throttle pedal must move the throttle cable.
- The throttle cable must rotate the throttle shaft, thus opening the throttle plate(s) to allow additional air to enter the intake manifold. Simultaneously, this action physically moves the TPS.
- The TPS must send the PCM a signal indicating that the throttle has opened the throttle plate(s).

In an ETC system, movement of the throttle pedal operates the APP sensor directly and immediately sends the signal to the PCM to increase engine speed. In an ETC system, the PCM controls both air and fuel, unlike the cable-operated system, in which the driver controls the air and the PCM controls only the fuel.

The use of an ETC system also has some other benefits:

- The AIS actuator is eliminated. The PCM can now control idle RPM directly without the use of additional dedicated hardware.
- The dedicated cruise control (speed control) servo is eliminated, and speed control becomes a simple function of the PCM, no longer requiring additional hardware to control throttle position.
- The PCM can use the ETC system to control engine torque directly when the traction control feature mandates it.

SPARK MANAGEMENT SYSTEMS

Ignition Timing

The PCM calculates ignition timing from information reported by:

- The coolant temperature sensor
- The tach reference pulse (CKP/CMP sensors)
- The manifold absolute pressure (MAP) sensor
- The barometric pressure (start-up MAP reading)

At warm idle, the PCM first uses spark advance manipulation to control normal idle speed fluctuations. This action is referred to by Chrysler as *active timing.* If a specific amount of spark timing change does not put the idle speed at the design RPM, the PCM will use the AIS motor or ETC to change the amount of air entering the engine at idle.

Ignition Systems

Chrysler vehicles have used one of three ignition system designs: distributor ignition (DI), waste spark ignition, and coil-on-plug (COP) ignition. As technology advanced, DI systems were the first to disappear. Waste spark ignition systems were used extensively for many years, but now have given way to the COP ignition systems.

Distributor Ignition. A single coil was used with a distributor to distribute the spark to all cylinders. This is known as distributor ignition (DI). Chrysler DI systems used either permanent magnet, optical, or Hall effect pickups within the distributor.

Waste Spark Ignition System. In 1990, Chrysler introduced its own 3.3L, 60-degree V6 for some of its passenger cars and vans. This engine uses a waste spark ignition system, known by Chrysler as their *direct ignition system (DIS).* This system uses multiple coils. All ignition coils are formed into one component called a *coil pack* (Figure 14–10). Each coil within the coil pack is connected to the spark plugs on two companion cylinders. Both spark plugs on the same coil fire simultaneously: one at the end of its compression stroke just before the power stroke, the other at the end of its exhaust stroke. Because of the low resistance across the spark plug gap that is on the exhaust stroke, the *waste spark* requires less voltage than the *event spark* to the cylinder on compression.

Coil-on-Plug. First Introduced on the Chrysler 2.7L, 3.2L, and 3.5L engines, modern Chrysler vehicles use a coil-on-plug (COP) ignition system (Figure 14–18). The COP EI system uses the same Hall effect CMP and CKP sensors as the waste spark system does. Using one ignition coil for each cylinder and installing it on top of the spark plug eliminates secondary

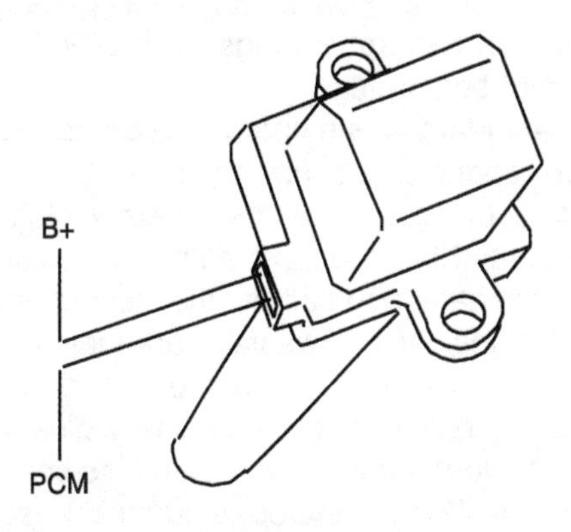

Figure 14–18 Coil-on-plug.

ignition wiring, thereby reducing the potential for the ignition system to induce voltages into nearby circuits.

In this system, battery voltage is fed directly to each ignition coil. The PCM controls the ground side of each coil. The primary winding of the coil has low impedance with a resistance of 0.5 Ω. This low resistance provides quicker magnetic saturation of the coil. However, it also provides current much higher than that normally used in a computer-controlled circuit. To prevent damage to the ignition coil primary winding or to the PCM's controlling circuitry, the PCM incorporates a current-limiting strategy. When engine temperature is below 176°F, the PCM limits primary current flow to 7.7 to 9.0 amps. When engine temperature is above 176°F, the PCM limits primary current flow to only 6.0 to 6.8 amps.

Dodge Hemi Ignition System

Another ignition coil design is used on the Dodge Hemi V8 engines—the 5.7L and 6.1L. (A side note is that the 5.7L, which equates to 345 cubic inches, puts out 375 ft.-lb. of torque at 4,400 RPM and 345 HP at 5,600 RPM—about 1 hp per cubic inch, as did the original Chrysler Hemi V8 engines. These figures do vary slightly depending upon the application.) These engines use a waste spark system in the secondary ignition system— one ignition coil for two spark plugs. However, the V8 Hemi engines use eight ignition coils for their eight cylinders. This is because the engine has two spark plugs each cylinder, for a total of 16 spark plugs.

A standard waste spark ignition coil is wired to two spark plugs on companion (opposite) cylinders by way of two secondary plug wires (Figure 14–19A). Some manufacturers' applications eliminate one-half of the secondary plug wires by designing a waste spark ignition coil to physically mount on top of one of the spark plugs, thus using only a single secondary plug wire to connect it to the companion cylinder's spark plug (Figure 14–19B). The Dodge Hemi V8 uses this concept as well, but with two spark plugs per cylinder. A waste spark ignition coil is mounted on top of one of the two spark plugs on one cylinder and uses a single secondary plug wire to connect to one of the two spark plugs on the companion cylinder (Figure 14–19C). Another waste spark ignition coil is mounted on top of the companion cylinder's remaining spark plug and is connected to the first cylinder's remaining spark plug by way of a single secondary plug wire. Therefore, each of the eight cylinders has an ignition coil mounted on top of one of its spark plugs, for a total of eight coils. Each cylinder's remaining spark plug receives spark through a secondary plug wire from the waste spark coil mounted at the companion cylinder. A newer design of this engine has the waste spark ignition coil mounted on top of the two spark plugs on the same cylinder.

This secondary ignition system can be a little confusing, with eight cylinders, 16 spark plugs, and 8 ignition coils mounted directly on top of the associated spark plugs. However, if the technician applies the principles of ignition system design covered in this textbook, it should not be too difficult to study a system as unique as this one and figure out how it is designed to operate.

Turbocharger Boost Control and Detonation Control

From the beginning of the turbocharged four-cylinder engine with multipoint fuel injection, the Chrysler multipoint system has been able to retard the timing on just the cylinder with spark knock occurring. If knock occurs, the PCM has already recorded which cylinder was just fired. It then slightly retards the spark for just that cylinder. If knock continues, the PCM will begin to reduce turbocharger boost.

Wastegate Control Solenoid

Before 1985, turbocharged engines used three types of boost control. The first was the wastegate. The wastegate actuator, shown in Figure 14–20, opens the wastegate at a boost pressure of 7.2 PSI. The second method uses the MAP sensor to

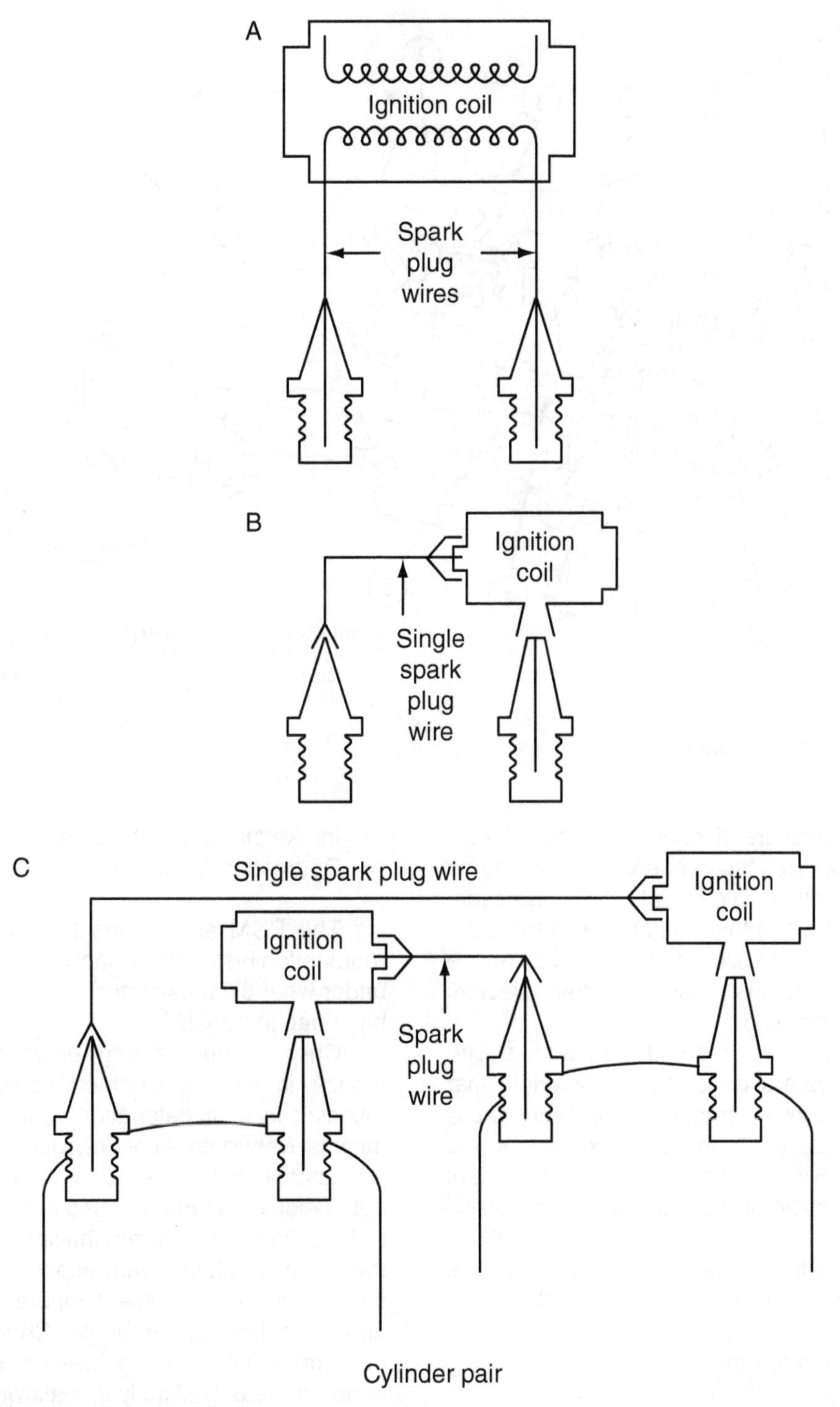

Figure 14–19 Waste spark configurations.

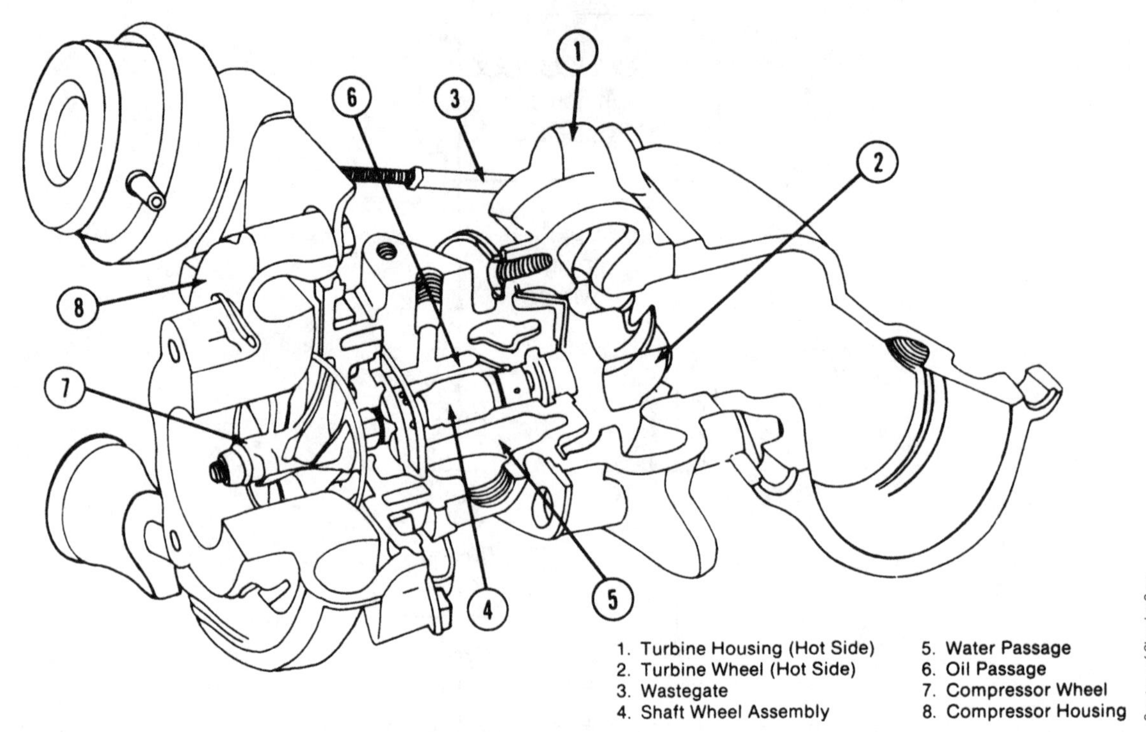

Figure 14–20 Turbocharger.

sense boost pressure. If boost exceeds 7.2 PSI, the PCM recognizes the over-boost pressure and skips fuel injection pulses until boost pressure drops to 7.2 PSI. The third method uses the electronic engine speed governor: If RPM goes above 6,650, the PCM stops pulsing the fuel injectors until the RPM drops to 6,100.

In 1985, a fourth method was added. The PCM can operate a boost control solenoid that vents the line conveying manifold pressure to the wastegate actuator. If operating conditions are favorable, the PCM will pulse the boost control and bleed off some of the pressure. Under these circumstances, the boost pressure can actually go to 10 PSI before the actuator opens the wastegate. The PCM reviews these inputs for this calculation:

- Barometric pressure
- Engine speed (RPM)
- Engine coolant temperature (ECT)
- Intake air temperature (IAT)
- Detonation (knock sensor)

The PCM also keeps track of the engine's "detonation history," how inclined it is to knock and under what circumstances, as well as how long it has been in boost.

Chrysler puts strong emphasis on performance in its turbocharged vehicles, which is reflected in their detonation control strategy. You must prevent detonation, of course, or the engine will destroy itself. There are only two ways to stop detonation on a turbocharged engine under boost: reduce the boost or retard the timing. If you reduce the boost, you lose performance; if you retard the timing, you raise exhaust temperature, which was already critical under boost. Chrysler's solution of retarding timing to one cylinder raises exhaust temperature only slightly and allows the other cylinders to keep producing at their full capacity.

Revised Boost Control

Beginning in 1988, the turbo boost actuator control on the engine worked directly from the turbocharger itself rather than from the intake manifold pressure. This keeps the wastegate open during part-throttle operation and eliminates boost completely except at wide-open throttle (WOT). The advantages are these:

- Exhaust backpressure is lower.
- The intake air-fuel charge is cooler.
- The tendency to knock is reduced.
- Part-throttle fuel economy improves.

EMISSION CONTROL SYSTEMS

Catalytic Converters

As is common with other manufacturers, Chrysler vehicles use both a reduction catalytic converter and an oxidation catalytic converter. The reduction cat uses CO to reduce NO_x back into N_2 and O_2. The oxidation cat oxidizes CO into CO_2 and HCs into both water and CO_2.

Catalytic Converter Efficiency Test. The PCM compares the signal from the pre-cat oxygen sensor(s) with that of the post-cat oxygen sensor(s) downstream of the catalytic converter. If the converter is working properly, there should be little variation in the signal produced by the secondary oxygen sensor. If the post-cat oxygen sensor(s) produce a signal that is too similar to that generated by the pre-cat oxygen sensor(s), the PCM sets a DTC in memory.

EGR System

Modern Chrysler vehicles use an EGR solenoid valve that is very similar in its design to the General Motors linear EGR valve (Figure 14–21). The electrical connector provides five terminals. Two terminals allow the PCM to control the EGR solenoid valve using a bidirectional, high-frequency, pulse-width-modulated command. The other three terminals are reference voltage, signal, and ground that provide a feedback signal to the PCM.

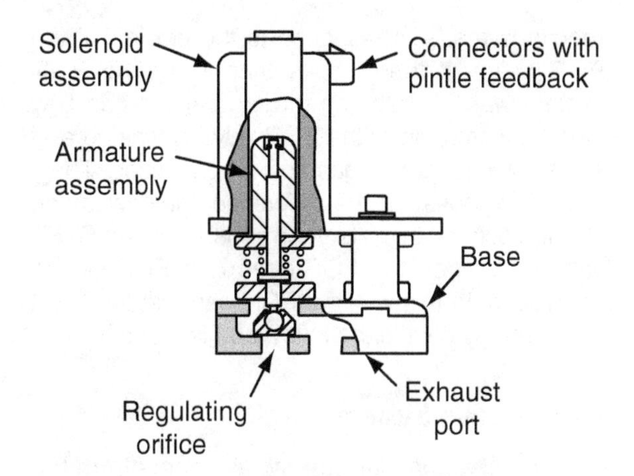

Figure 14–21 Linear EGR valve design.

OBD II EGR Monitoring Strategy. Chrysler OBD II PCMs monitor the ability of the EGR valve to both seal when closed and flow properly when opened using a strategy that is programmed into the PCM. When an EGR valve is opened, the exhaust gas flow that it allows into the intake manifold displaces ambient air that would otherwise be drawn into the cylinders, thereby reducing the amount of oxygen available for combustion. Fuel must also be reduced accordingly. With a carburetor, the appropriate fuel reduction is automatic, but with any form of electronically pulsed fuel injection, it is up to the PCM to make the appropriate fuel reduction. Conversely, if the EGR valve is then closed, because more ambient air is now drawn into the cylinders, the PCM must also appropriately increase the fuel delivery to compensate.

Chrysler uses a strategy that allows the PCM to close an already open EGR valve while cruising in closed loop, but without increasing fuel delivery. It then watches for the pre-cat oxygen sensor and short-term fuel trim (STFT) for a proper response. As the EGR valve is suddenly closed, this results in an increase in the volume of ambient air reaching the cylinders. The fact that a change in fuel delivery was not made during this test should cause the oxygen sensor to go lean. Simultaneously, the STFT values should go high, indicating the need to add fuel. If the PCM does not see the appropriate reactions made in response to the closing of the EGR valve, it then knows that either the EGR system was not allowing proper exhaust gas flow when it was open or the EGR valve is leaking and still allowing exhaust gas flow when closed. Thus, the PCM looks for the difference (or how much change occurs) in the oxygen sensor voltage/STFT values between the EGR-open and EGR-closed conditions. In this way the PCM can test the EGR valve and system for both flow and leakage.

Evaporative System

The charcoal canister stores vapors from the fuel tank. As long as the coolant temperature is below 180°F/82°C, the PCM blocks the vent line to the throttle body by energizing a solenoid on the line. If the canister is not purged, it will gradually fill and will not be able to store any more vapors. If it purges too early or at other inappropriate times, it can drive the air-fuel mixture richer than the mixture control strategies can correct for.

Modern Chrysler engines use a high-frequency pulse-width-modulated (PWM) purge solenoid valve to control the flow from the EVAP canister (Figure 14–22). The PCM's solenoid control circuit can operate at frequencies of up to 200 MHz.

Leak Detection Pump. Beginning in 1996 Chrysler vehicles used a leak detection pump (LDP) to apply a slight positive pressure to the fuel tank and evaporative system to check for leaks. For 1996 through the 2000 model year, the PCM had to be able to identify any leaks larger than 0.040 inch. In 2001, this requirement was tightened up to 0.020 inch. Chrysler used the LDP as late as the 2004 model year.

Engine Off Natural Vacuum Test. The *engine off natural vacuum (EONV)* test was introduced by Chrysler in the 2002 model year. The EONV test replaced the former LDP and required few hardware changes in the system. Minor electrical changes were implemented in order to keep the PCM powered up after the ignition switch is turned off. And the PCM's program was modified to allow it to perform this test.

When the vehicle is driven, heat from the engine and exhaust system will cause a slight increase in temperature of the fuel in the fuel

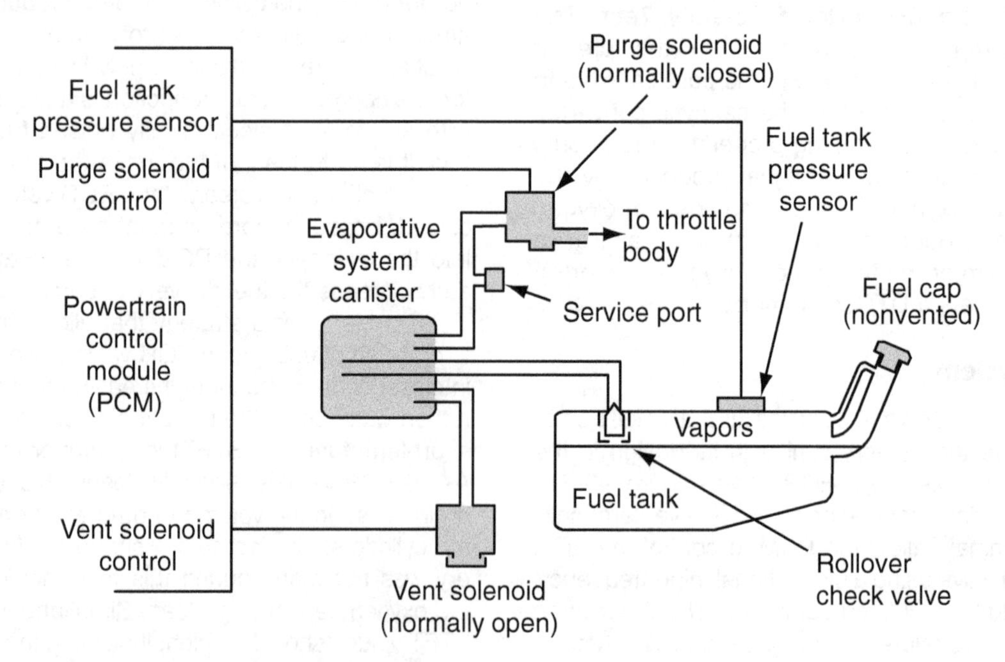

Figure 14–22 Evaporative system components.

tank which, in turn, causes it to expand. To perform the EONV test, the PCM is kept powered up for up to 40 minutes after the ignition switch is turned off. During this time the PCM keeps the purge solenoid valve de-energized and the vent solenoid valve energized, thus keeping both valves closed. As the fuel cools, it contracts and causes a *natural vacuum* to be formed within the evaporative system and fuel tank assembly, from about 3 inches to about 5 inches of water. During this time the PCM monitors the fuel tank pressure sensor to verify that the system is capable of holding this negative pressure.

With any sensed failure of the evaporative system, the PCM will illuminate the MIL and store the appropriate DTC in memory.

OTHER PCM-CONTROLLED SYSTEMS

Charging Circuit Control

Since 1985, Chrysler vehicles have had alternator output regulated by the PCM, replacing the voltage regulator (Figure 14–23). The PCM calculates the output based on the alternator's current output and on the battery temperature. The PCM controls the alternator output to between 12.9 and 15 V. When the key is turned on, an electronic voltage regulator (EVR) circuit inside the PCM uses pulse-width modulation to control the alternator field current. The PCM monitors the alternator output on a sensing wire from the ASD relay. If the charging rate cannot be monitored by the PCM, the field circuit duty cycle is limited to 25 percent so that there is no high-voltage surge that might damage computer and sensor components.

A/C Cutout Relay

The A/C cutout relay is on the ground side of the A/C compressor clutch circuit, between the clutch and the A/C switch. When open, it stops the A/C clutch circuit. The PCM closes the compressor engagement circuit *except:*

- When the engine is operated at WOT
- When engine speed is below 500 RPM
- When the engine is cranked with the A/C switch on

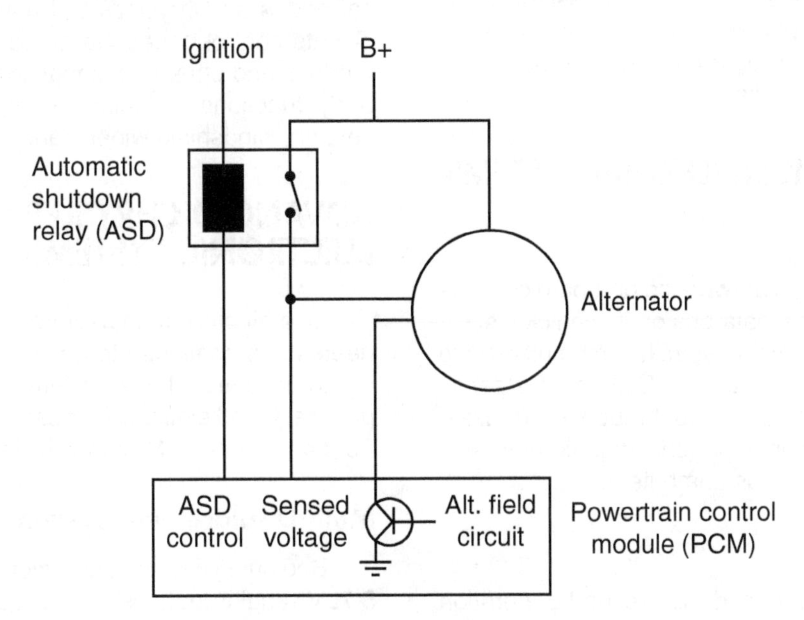

Figure 14–23 PCM alternator field control circuit.

The compressor cutout relay circuit remains open for 10 to 15 seconds after the engine starts.

Torque Converter Lockup Clutch

Chrysler first introduced a torque converter clutch lockup feature in 1988 on its Torqueflite transmissions. Modern Chrysler vehicles now implement this feature on their automatic-transmission vehicles. The PCM typically activates it under specific conditions when:

- the coolant temperature is above 150°F, *and*
- the park/neutral (P/N) switch indicates the transmission is in gear, *and*
- the brake switch indicates the brakes are not applied, *and*
- the throttle angle, as reported by the TPS or APP sensor, is above a certain minimum

The converter clutch works hydraulically, controlled by the computer. The actuation solenoid mounts on the valve body transfer plate and receives oil when the transmission goes into third gear. If the computer energizes the solenoid by grounding its circuit while oil pressure is applied, the torque converter locks the crankshaft to the transmission's input shaft. Otherwise, the solenoid bleeds off the oil pressure, and the torque converter clutch releases.

CHRYSLER MULTIPLEXING SYSTEMS

CCD Protocol

Chrysler began networking on-board computers together over a data bus on its vehicles starting with the 1988 model year. Its first multiplexing system, known as Chrysler Collision Detection (C2D), was used from 1988 through 2000, using a twisted pair of wires for serial data communication among the various computers.

PCI Protocol

Chrysler introduced its second-generation multiplexing system on 1998 models. This system is known as the Programmable Communication Interface (PCI) system. The PCI bus was installed on most Chrysler vehicles by 2001. The PCI system is an OBD II standardized protocol. It is a J1850 medium-speed (MS) variable-pulse-width (VPW) single-wire (SW) protocol and connects to terminal 2 at the DLC. The PCI data bus is similar to General Motors' Class 2 data bus.

CAN Protocols

Modern Chrysler vehicles now use both the high-speed (HS) CAN and medium-speed (MS) CAN protocols. The HS CAN protocol connects to terminals 6 and 14 at the DLC. The MS CAN protocol connects to terminals 3 and 11 at the DLC.

LIN Protocol

Modern Chrysler vehicles use a Local Interconnect Network (LIN) protocol to communicate between several nodes. The LIN protocol is a low-speed (LS) protocol and is used where high speed is not necessary. It operates on 12 V DC and is inexpensive to implement. The LIN protocol uses a master node with multiple slave nodes. The typical nodes on Chrysler's LIN network include the climate control panel, the air conditioning control module, and other nodes that control the vehicle's body functions including doors, seats, steering column, windshield wipers, and electric sunroof.

ADVANCED CHRYSLER ELECTRONIC SYSTEMS

Like all other manufacturers, Chrysler's engineers have continued to advance the technology on its vehicles. These systems are for the most part designed similarly to those systems found on Ford and General Motors vehicles.

Multi-Displacement System

Beginning with the 2005 model year, the Hemi 5.7L V8 engine features a cylinder deactivation strategy when mounted in the 300C and Magnum R/T.

Chrysler calls this feature the Multi-Displacement System (MDS), but it is also known in the industry as Displacement-on-Demand (DOD), Active Fuel Management (AFM), or Variable Cylinder Management (VCM). The MDS is designed to increase fuel efficiency while giving the driver full access to the engine's performance potential. The fuel savings is about three miles per gallon on the Chrysler/Dodge applications.

In actual operation, the MDS allows the PCM to run a V8 on four cylinders when the need for power is small. The engine always starts on all cylinders, but when cruising, when the need for power is reduced, the PCM does two things:

1. The PCM turns off the fuel injectors for four cylinders.
2. The PCM simultaneously closes all intake and exhaust valves on the disabled cylinders, which allows the disabled cylinders to act as an air spring does (using crankshaft power to compress the trapped air when the piston moves from BDC to TDC, but also returning power to the crankshaft when the piston moves from TDC to BDC). This avoids the energy losses that would occur if the disabled cylinder's piston were pulling down against manifold vacuum and pushing out against ambient air pressure on the exhaust side, known as "pumping losses."

A specially designed "switching lifter" is used on the MDS (Figure 14–24). It is designed with two portions—an inner body and an outer body—that can physically collapse in on each other. (This is similar to the manner in which a shock absorber operates.) A spring is designed as part of the lifter.

For normal operation, the lifter has a locking pin that keeps the two halves from collapsing in on each other; the result is that the cam uses the lifter to open the valve. When the PCM wants to disable the valve, it energizes an oil solenoid, which provides oil pressure to push the locking pin so as to "unlock the lifter." At this point, the lifter collapses on it's own spring tension. The spring is strong enough to keep the lifter following the cam lobe but not strong enough to overcome the valve-spring tension, thereby allowing the valve to stay closed. When the driver punches the throttle, the PCM instantaneously turns the fuel injector back on and de-energizes the oil control solenoid. The next stroke that allows the lifter to expand also allows the locking pin to lock the lifter in the extended position, and cylinder operation is restored.

Figure 14–24 Lifter used in an MDS.

Some of the noteworthy technologies enabling the Chrysler MDS to operate efficiently include the speed of electronic control systems, the sophistication of the algorithms programmed into the control modules that control the systems, and the use of an ETC system. The Hemi V8 can transition from eight cylinders to four cylinders in 40 ms. If the MDS were to be used without ETC technology, when the PCM disabled one-half of the cylinders the driver would suddenly find that the power had been reduced and would have to push the throttle farther down to compensate and maintain a consistent cruise speed—naturally, a cause of concern for the driver. (Opening the throttle plate[s] farther allows the remaining cylinders to work more efficiently.) On an MDS that is used with an ETC system, a "smarter" PCM can make this compensation automatically as it disables cylinders or re-enables them. Thus, the driver can maintain a consistent throttle position as cylinders are disabled or re-enabled. In fact, with an ETC system, it is not likely that the driver will be aware of how many cylinders are actually being used, resulting in fewer driver concerns while enabling a higher degree of fuel economy.

Ultimately, the MDS allows the 5.7L Hemi engine to combine the power and performance of a larger engine with the fuel efficiency of a smaller engine. By selectively deactivating half of the engine cylinders while cruising and operating all eight cylinders while idling and accelerating, the result is full power when you need it and increased fuel economy when power demands are reduced.

Electronic Stability Program

Chrysler is now using a feature on some of its vehicles that is designed to reduce the potential for a collision by aiding the driver's control of the vehicle if the vehicle's tires lose their grip with the road surface during cornering. Known as the Electronic Stability Program (ESP), this feature helps the driver to maintain vehicle directional stability, providing both oversteer and understeer control to maintain the vehicle's behavior on the road surface. In a cornering situation where the tires have lost traction with the road surface, the vehicle's yaw rate is changed abnormally, resulting either in oversteer or understeer. (Yaw rate is a determination of how far left or right a vehicle has moved from its intended course.) If the vehicle's back end begins to fishtail (oversteer) during a turn, the Electronic Stability Controller (ESC) detects this through a "yaw rate and lateral acceleration sensor" and applies a braking pulse at the outside front wheel to help the driver stabilize the car (Figure 14–25). (This action allows the side of the vehicle on the inside of the turn to begin to move ahead of the other side, thus reducing the oversteer.) If the front end of the vehicle begins to drift to the outside of a turn (understeer), the ESC detects this and applies a braking pulse to the inside rear wheel. (This action places a drag on the side of the vehicle at

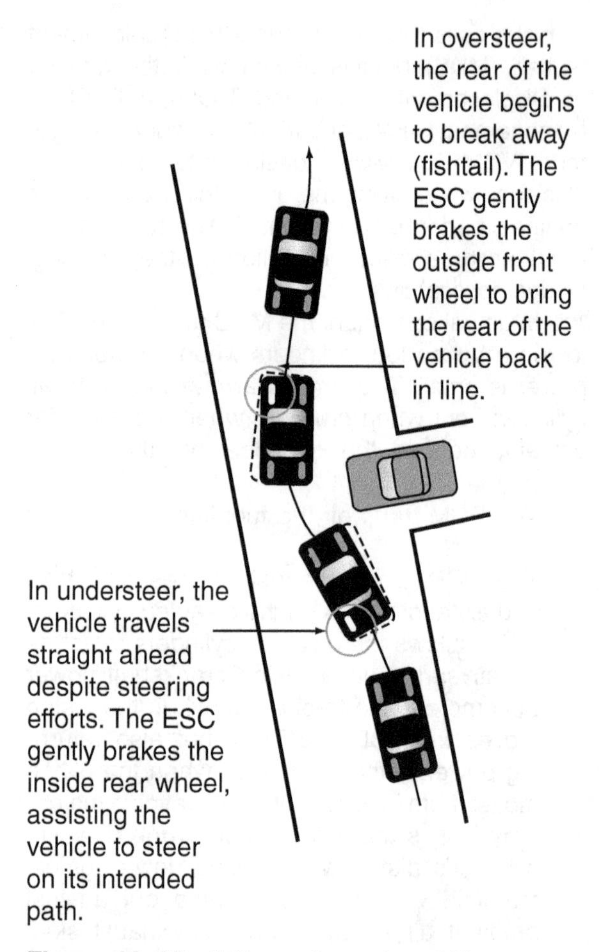

Figure 14–25 With an electronic stability program, the Electronic Stability Controller reacts to correct understeer or oversteer.

the inside of the turn and allows the side of the vehicle on the outside of the turn to move ahead, thus reducing the understeer.)

In both situations, it can also reduce engine torque as needed.

Fiat MultiAir Technology

Shortly after Fiat purchased a share of Chrysler Corporation, the U.S. government offered Fiat an additional 5 percent ownership if they could produce an American car by the 2013 model year that would get 40 MPG. This is the reason behind

Fiat putting the Fiat MultiAir system on the 2013 Dodge Dart. As of January 1, 2014, Fiat owned 41.5 percent of Chrysler Corporation.

Fiat bills their MultiAir system as the "ultimate air management strategy." This Fiat technology made its debut on the 2013 Dodge Dart with either the 2.0L or 2.4L engine.

The Dart is built in Belvidere, Illinois. The Dart became the first American car to have this technology.

Functional Concept. The PCM on the MultiAir system controls intake valve opening as the primary method of controlling engine speed. By controlling how far the intake valves open, the PCM can control the volume of air brought into the engine's cylinders. Oil pressure is used to control intake valve opening.

The MultiAir system four-cylinder engine has four valves per cylinder, two intake valves and two exhaust valves, for a total of 16 valves. Instead of the traditional intake cam lobes, each cylinder has one lobe that is used to pump oil.

Oil pressure is applied through an oil control solenoid to an actuator to open the intake valves (Figure 14–26). The system has a constant oil feed of 20 to 30 PSI. When the cam lobe "comes around," oil pressure potentially rises to 3,750 PSI. When the oil solenoid is de-energized, the oil pressure created by the cam lobe is vented. When the oil control solenoid is energized, oil pressure is applied to the actuator to open the intake valves.

When the high oil pressure is vented, it is vented to an upper metal oil cap located on top of the cylinder head and under the plastic valve

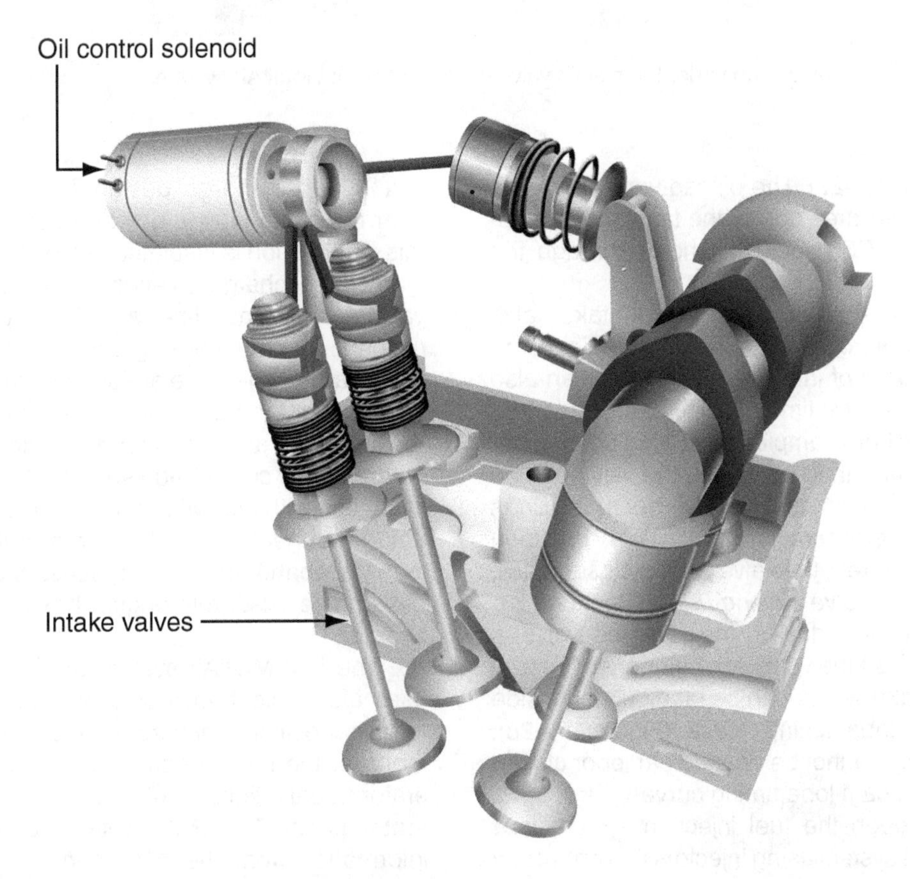

Figure 14–26 Fiat MultiAir system used on the 2013 Dodge Dart.

Figure 14–27 Metal oil cap under the plastic valve cover on the Fiat MultiAir system.

cover (Figure 14–27). The oil cap vents oil slowly into valve cover area due to the fact that oil pressure at 3,750 PSI would put holes through the plastic valve cover.

Oil pressure is used to control intake valve opening by controlling lift and duration. This controls the volume of intake air. The PCM can also change intake valve timing inside of the cam lobe timing curve. For example, by venting off oil pressure initially and then holding it, the valve can be opened late, known as *late valve opening (LVO)*. Or, if the oil pressure is initially held, but then vented prematurely, the valve van be closed early, known as *early valve closing (EVC)*. Because the valve is cam lobe driven, the later the valve is opened, the less the valve can be opened. Essentially, the intake valves can be manipulated inside of the cam lobe timing curve (Figure 14–28). However, they cannot be opened earlier or closed later than the cam lobe timing curve.

Even though the fuel injection system is a standard SFI system using injectors that spray fuel on the back side of the intake valves, the engine is a "stratified charge" design. The air-fuel charge near the spark plug is rich and burns quickly, but this combustion is insulated in a pocket of a super lean air-fuel charge. To accomplish this, the intake valves are opened late, after TDC. Because the piston is already "sucking down" when the intake valves are opened, the air-fuel mixture is released suddenly into the combustion chamber. As a result, the air and fuel spend less time in the cylinder and the combustion spends little, if any, time in contact with the cylinder walls. This reduces the amount of heat that is drawn out of the combustion. Wideband air-fuel ratio sensors are used on this engine. Also, this engine has a small turbocharger.

The Fiat MultiAir system on the Dodge Dart uses electronic throttle control (ETC). It has two throttle position sensors and a reversible DC motor at the throttle body. It also has two accelerator pedal position (APP) sensors at the accelerator pedal. The PCM obtains driver demand information from the APP sensors. But during cruise, it adjusts the ETC motor at the throttle

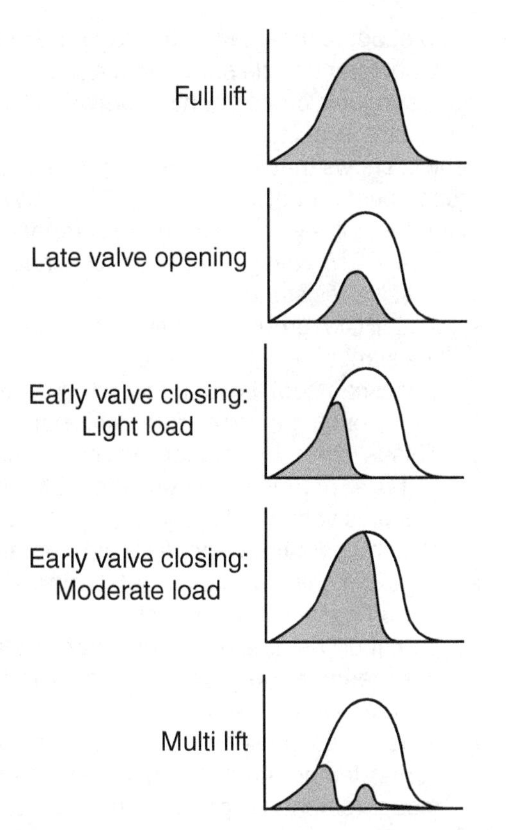

Figure 14–28 Examples of possible valve timing curves on the Fiat MultiAir system.

body in order to maintain a minimal amount of vacuum in the intake manifold for the PCV and evaporative purge systems to function.

✔ SYSTEM DIAGNOSIS AND SERVICE

A thorough diagnostic procedure is published by Chrysler Corporation for each model year and application. This driveability test procedure should be in the service manual you use on any particular car.

OBD II Diagnostics

Chrysler OBD II systems require an OBD II scan tool for all diagnostic features, although initially Chrysler did retain manual methods of code pulling in parallel to OBD II scan tool methods on early OBD II systems. That is, on Chrysler's SBEC III systems, the diagnostic trouble codes (DTCs) may be pulled manually by cycling the ignition switch and counting pulses of the MIL (as in pre-OBD II systems), or a scan tool may be used. However, cycling of the ignition switch produces only two-digit codes as did their pre-OBD II systems. To obtain the OBD II five-digit codes, an OBD II scan tool must be used. On SBEC IIIA systems, the manual method was eliminated, and all diagnostic functions must be done through an OBD II scan tool.

Once connected to the diagnostic link connector (DLC) under the driver's side of the instrument panel, Chrysler's *Wi-Tech wireless* scan tool, or an OBD II generic scan tool, may be used to access several on-board computers, including the powertrain control module (PCM), transmission control module (TCM), body control module (BCM), mechanical instrument cluster (MIC), and the antilock brake control module (ABCM). Each of these modules enables the technician to access memory DTCs, perform functional tests such as the Actuator Test Mode (ATM), and perform certain other tests, depending on the system that is selected. (The ATM is Chrysler's name for what is generically known as an *active command mode* or *bidirectional controls*.) For example, if the PCM is selected, the technician may adjust engine RPM in 100-RPM increments (typically from about 800 to about 2,000 RPM) through the scan tool's menu. And, in addition to DTCs, data stream information, and functional tests, PCM freeze frame data is also available. If the ABCM is selected, the technician may use the scan tool to bleed the ABS modulator assembly (part of the brake bleeding procedure). The TCM also allows the technician to select the vehicle's tire size from a menu to correct VSS calibration when tire size has been changed. To know what tests are offered with a particular scan tool, simply bring up the test menu for any of the modules that appear on the display screen.

SUMMARY

In this chapter, we covered sequential fuel injection systems used by Chrysler. We saw the role of the PCM in controlling combustion in the engine under all driving conditions. Likewise, we reviewed the way the computer controls the idle speed under various conditions, how it controls ignition timing, and how it prevents knock.

The text explained how the Chrysler system controls turbocharger boost to optimize power and engine life, first retarding spark to a specific cylinder, and only should that fail to end knock, by reducing turbocharger boost. The incorporation of control of the charging system output voltage was discussed. This accommodates battery condition when very cold and can also smooth the idle by reducing charge at low-torque engine speeds.

We also looked at some of the newer systems that Chrysler is using on its vehicles, including the Electronic Throttle Control (ETC), Multi-Displacement System (MDS), and Electronic Stability Program (ESP). We also discussed the Fiat MultiAir system that is now being used on Chrysler vehicles. And we took a look at diagnostic procedures as they apply OBD II systems.

▲ DIAGNOSTIC EXERCISE

Describe the PCM system problems that would be caused by a high-resistance battery negative connection to the engine block and the kinds of tests you would use to determine if this was the problem.

Review Questions

1. Which of the following engine control systems was introduced by Chrysler in 2002?
   A. The JTEC system
   B. The SBEC IIIA system
   C. The NGC system
   D. The SMEC system
2. A lab scope connected to the oxygen sensor on a Chrysler vehicle shows that a pre-cat oxygen sensor is cross-counting between 0 and 1 V. But a scan tool connected to the same vehicle shows that the same oxygen sensor is cross-counting between 2½ and 3½ V. What is the most likely cause for the discrepancy?
   A. The PCM voltage supply circuit has excessive resistance.
   B. The PCM ground circuit has excessive resistance.
   C. The scan tool is malfunctioning and is interpreting the serial data improperly.
   D. This is normal operation on many newer Chrysler products because the PCM adds in a bias voltage of 2½ V.
3. What is the primary purpose for incorporating a range-switching (dual-range) temperature sensor on some Chrysler vehicles?
   A. Its circuit uses less current than a standard temperature sensor, thereby reducing the load on the alternator.
   B. It increases the sensor's precision in measuring temperature and also extends the measurement range into the over-temp area.
   C. It allows the PCM to use the same temperature sensor to measure both engine coolant temperature and intake air temperature.
   D. Its circuitry is simpler and therefore less confusing to a technician.
4. *Technician A* says that, depending upon the year and model, the fuel pump on a Chrysler vehicle may be controlled by the ASD relay. *Technician B* says that, depending upon the year and model, the fuel pump on a Chrysler vehicle may be controlled by a dedicated fuel pump relay. Who is correct?
   A. *Technician A* only
   B. *Technician B* only
   C. Both technicians
   D. Neither technician

5. The ASD relay may provide power for all *except* which of the following?
   A. The fuel injectors
   B. The Hall effect CKP and CMP sensors
   C. The ignition coil primary winding
   D. The O_2 sensor's heating element
6. On some Chrysler vehicles, what two sensors are sometimes combined into a single component called a TMAP sensor?
   A. TPS and MAP
   B. IAT and MAP
   C. ECT and MAP
   D. MAF and MAP
7. In what model year did Chrysler begin using MAF sensors on the Chrysler Sebring and the Dodge Stratus?
   A. 1985
   B. 1996
   C. 2001
   D. 2012
8. What type of tool should be used to measure and interpret the resulting signal from Chrysler's MAF sensor?
   A. A Hertz meter
   B. A DC voltmeter
   C. An AC voltmeter
   D. A duty-cycle meter
9. What type of sensor is used on Chrysler waste spark and coil-on-plug ignition systems to monitor crankshaft position (CKP) and camshaft position (CMP)?
   A. Piezoelectric
   B. Permanent magnet
   C. Hall effect
   D. Optical
10. What type of detonation (knock) sensor is used on Chrysler applications?
    A. Piezoelectric
    B. Permanent magnet
    C. Hall effect
    D. Optical
11. What type of sensor is used on Chrysler products as a vehicle speed sensor?
    A. Variable reluctance (permanent magnet) sensor
    B. Hall effect sensor
    C. Optical sensor
    D. Either A or B
12. A return-type fuel injection system uses a fuel pressure regulator with a manifold vacuum hose attached to the spring side. What is the purpose of this vacuum hose?
    A. This increases fuel delivery when the engine is under load.
    B. This maintains a constant pressure differential across the injector.
    C. This results in a constant volume whenever the injectors are energized.
    D. Both B and C.
13. Why are returnless fuel injection systems being used on Chrysler vehicles?
    A. To reduce the potential for vapor lock
    B. To reduce evaporative emissions from the fuel tank
    C. To reduce the load on the fuel pump
    D. To improve cold engine performance
14. Regarding the addition of an ETC system to an engine, which of the following statements is *true*?
    A. The AIS actuator is eliminated.
    B. The speed control servo is eliminated.
    C. The PCM can use the ETC system to control engine torque directly when the traction control feature mandates it.
    D. All of the above.
15. The modern Dodge Hemi V8 engine uses a waste spark ignition system. This means that this engine has eight ignition coils and how many spark plugs?
    A. 32
    B. 16
    C. 8
    D. 4

16. Chrysler OBD II PCMs monitor the ability of the EGR valve to both seal when closed and flow properly when opened using which of the following sensors?
    A. Oxygen sensor
    B. MAP sensor
    C. DPFE sensor
    D. TMAP sensor
17. What test replaced the LDP beginning on 2002 model year Chrysler vehicles?
    A. EONV test
    B. KOEO test
    C. KOER test
    D. Cylinder imbalance test
18. How does a Chrysler MDS respond when the need for power is reduced?
    A. The PCM disables only the spark for one-half of the cylinders.
    B. The PCM disables the fuel injectors for one-half of the cylinders and allows the associated valves to continue to operate to allow the engine to pump air through these cylinders.
    C. The PCM turns off the fuel injectors for one-half of the cylinders and keeps the associated valves open.
    D. The PCM turns off the fuel injectors for one-half of the cylinders and closes the associated valves.
19. If the front end of a vehicle with the ESP feature begins to drift to the outside of a turn (understeer), to which wheel will the ESP system apply a braking pulse to correct the understeer?
    A. Inside rear wheel
    B. Inside front wheel
    C. Outside rear wheel
    D. Outside front wheel
20. On the Fiat MultiAir system used on the 2013 Dodge Dart, how much pressure is potentially developed when the cam lobe "comes around" to open the intake valves?
    A. 20 to 30 PSI
    B. 175 PSI
    C. 3,750 PSI
    D. 6,000 PSI

Chapter 15

European (Bosch) Computerized Engine Controls

OBJECTIVES

Upon completion and review of this chapter, you should be able to:

- ❑ Explain the distinction between the Motronic engine management system and the earlier Bosch dedicated fuel injection systems.
- ❑ Identify the various operating modes of the Motronic engine management system and the operating conditions to which they relate.
- ❑ Define the inputs used with a Motronic engine management system.
- ❑ Define the systems that are controlled by a Motronic engine management system.
- ❑ Describe what the Motronic ME7 engine management system adds to the Motronic engine management system.
- ❑ Describe what the Motronic MED engine management system adds to the Motronic ME7 engine management system.
- ❑ Explain the purpose and basic operation of a BMW Valvetronics system.

KEY TERMS

Jetronic
Lambda
Motronic

The Robert Bosch Corporation does not build cars or engines but rather components and control systems. Vehicle manufacturers buy these specially engineered components and control systems from the Bosch Corporation to use on their own vehicles. An early pioneer in fuel injection systems, the Bosch Corporation makes many of the products found on most German and Scandinavian cars, as well as those on many American and Japanese vehicles. Bosch also licenses several of its systems for manufacture by other firms, including Lucas, Hitachi, and the Denso Corporation (formerly Nippondenso), so an understanding of the Bosch systems in turn provides an understanding of many systems built by other sources. An English car with a Lucas control system, for example, is actually controlled by Bosch-designed components and controllers. Similarly, Bosch-designed fuel injection systems can be found on Asian vehicles.

Some of the more advanced technological systems engineered by Bosch include electronic throttle control (ETC) and the electronic stability program (ESP), already described in the General Motors, Ford, and Chrysler chapters of this textbook. These systems have been implemented on European vehicles as well. In fact, these systems tend to be implemented on the German vehicles several years ahead of their implementation on U.S. vehicles.

The purpose of this chapter is to discuss the Bosch engine performance control systems and components used over time. While this chapter generally refers to those systems found on European vehicles, many of the components discussed in this chapter also are found on both U.S. and Asian vehicles.

BOSCH FUEL MANAGEMANT SYSTEMS OVERVIEW

The Robert Bosch Corporation has been involved in fuel injection since the 1920s, when it introduced diesel fuel injection. Its experience with gasoline injection goes back to military vehicles in the late 1930s. One of its earliest systems for passenger cars was the direct, diesel-style injection system used on a few of the gasoline-fueled Mercedes-Benz SLs. That system was a mechanical forced-injection system very similar to diesel injection systems used on trucks.

All the Bosch engine management systems are computerized developments of earlier fuel injection systems, mostly mechanical. While our focus in this book is on computer-controlled systems, it would be difficult to understand what the Bosch engineers were doing without some explanation of the earlier mechanical systems on which the computerized systems are based.

While Bosch has built a version of throttle body injection for a limited number of European manufacturers, all of the systems used on vehicles imported into the United States have been multipoint injection systems.

Over many years, the Bosch fuel injection systems have always been defined in either of two basic categories: *electronically pulsed injection systems*, shown in Figure 15–1, and *mechanical continuous injection systems*, shown in Figure 15–2.

The term **Jetronic** applies to a fuel injection system in which the electronic control unit (ECU) is a dedicated module that controls only the fuel injection system. The Jetronic systems used an ECU that was an *analog* computer. This means that it made analog adjustments on the output side of the module in response to analog inputs.

The term **Motronic** applies to the more fully developed engine management systems that can control the fuel injection system, the spark management system, emissions systems, and other engine functions. In general, a Motronic system uses a single computer to manage all engine functions, while the earlier systems used independent controllers, typically one for ignition and another for fuel injection. Therefore, a Motronic ECU is a full function computer and, in today's terminology, a powertrain control module (PCM). The Motronic PCM is also a *digital* computer. *Digital electronics* is the basis for all of the modern technology around us today, from our cell phones and tablets to our personal computers, and is further defined in Chapters 1 and 2 of this textbook.

The term **lambda** applies to closed-loop fuel control. When the lambda value is 1, the air-fuel mixture in the cylinder is stoichiometric, and all of the fuel and oxygen is consumed in the cylinder, resulting in complete combustion.

Overview of Bosch Electronically Pulsed Fuel Injection Systems

Mono-Point Injection. A mono-point fuel injection system uses an electronically controlled single fuel injector solenoid valve that supplies fuel for several engine cylinders. This concept typically means that a four-cylinder engine will have one fuel injector and a V6 or V8 engine will have two. This is commonly known by many manufacturers, including GM and Chrysler, as

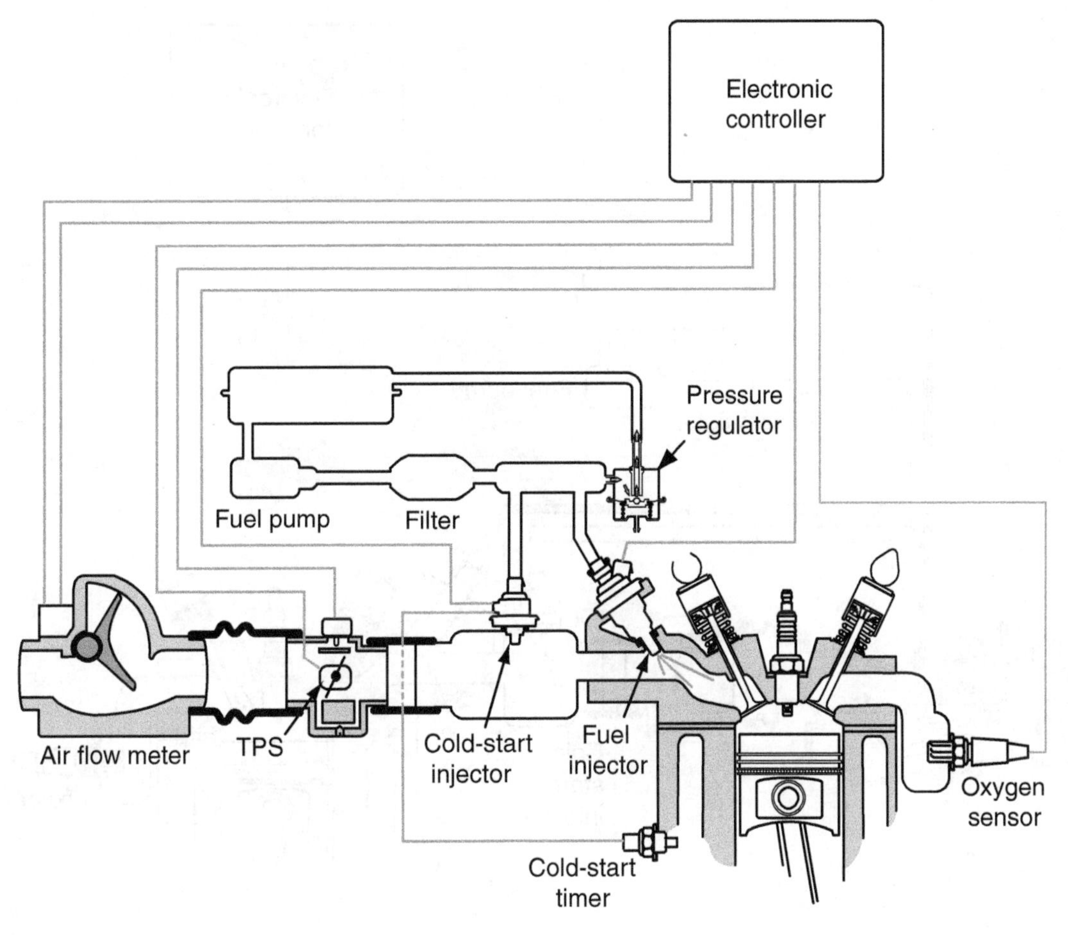

Figure 15–1 Pulsed injection system.

throttle body injection (TBI) and by Ford as *central fuel injection (CFI)*.

Multipoint Injection. A multipoint injection system uses an electronically controlled fuel injector solenoid valve for each cylinder of the engine. This includes port fuel injection (PFI) systems in which the injectors may be pulsed as a group, in pairs, or bank-to-bank (half the injectors at a time). This also includes sequential fuel injection (SFI) in which the injectors may be pulsed as a group during the engine's initial start (a primer pulse), but once the engine is running the injectors are pulsed individually in the engine's firing order. With this system, regardless of the firing strategy used, the fuel injectors spray fuel into the intake manifold runner and on the back side of the intake valve(s).

Gasoline Direct Injection. A gasoline direct injection (GDI) system is a modified multipoint injection system in that the fuel injectors spray fuel directly into the combustion chamber. These systems may operate in a homogeneous charge mode or in a stratified charge mode.

Overview of Bosch Fuel Injection and Engine Control Systems

K-Jetronic. This system is a mechanical port fuel injection system that was introduced in 1973 on European vehicles and sold in the

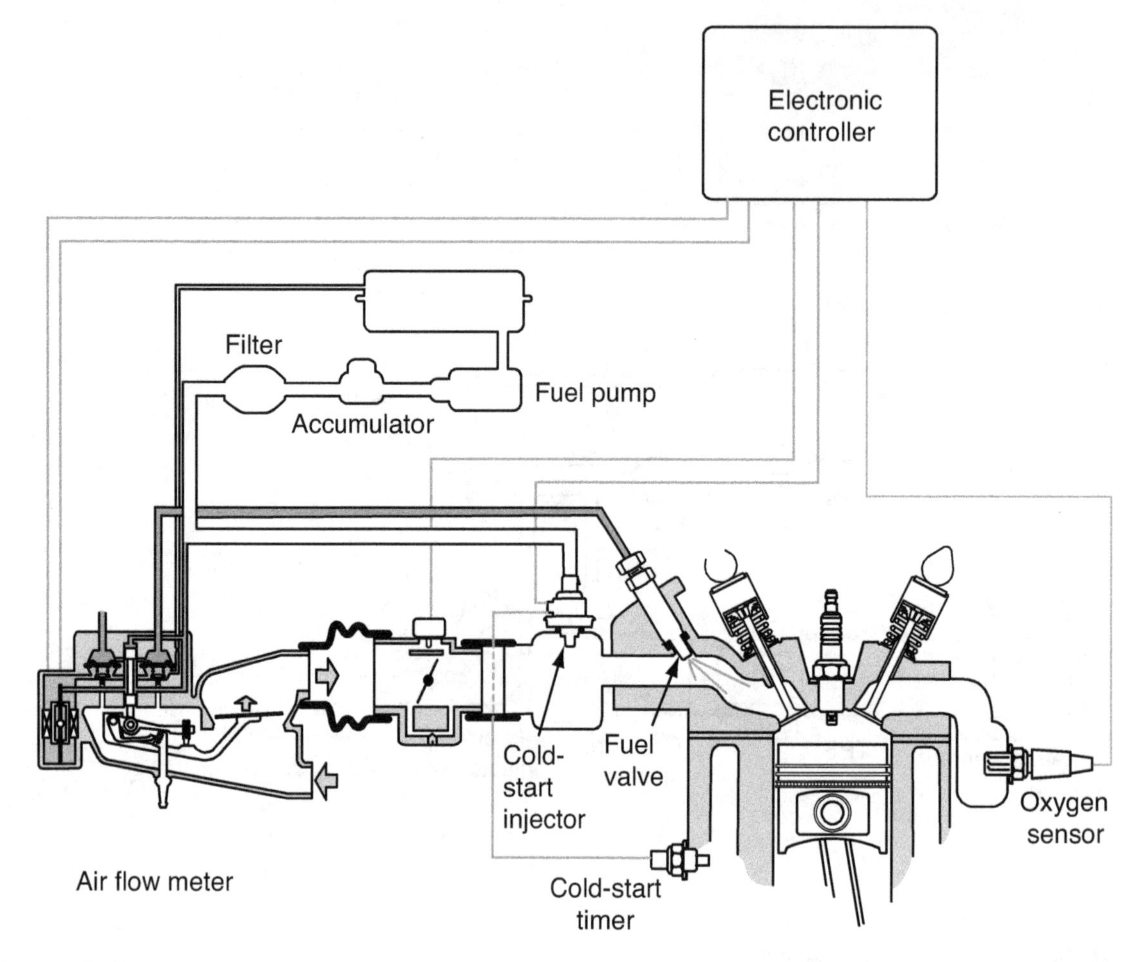

Figure 15–2 Continuous injection system.

United States. The *K* comes from the German word for *continuous*, *Kontinuerlich,* because the fuel flows continuously to the system's injectors as long as the engine is running. The fuel metering is controlled through an air valve that is controlled by the volume of intake air entering the engine. The opening of the air valve controls a fuel metering valve that applies fuel pressure to the fuel injectors. The injectors are not electrical solenoids, but rather spring-loaded valves that are opened by the fuel pressure applied from the fuel metering valve. The injectors spray fuel into the intake manifold runners and on the back side of the intake valves, similar to a modern electronically controlled PFI system. The system is commonly known as the *Continuous Injection System (CIS)*. It was produced from 1973 through 1994.

K-Jetronic and KU-Jetronic, Lambda Version. This system is a modified K-Jetronic system with closed-loop lambda control. It was also named KU-Jetronic in the United States (the letter "U" identifies a U.S. application.). The system later evolved into the KE-Jetronic systems. It was first introduced on a Volvo 265 in 1976.

KE-Jetronic. This system is a K-Jetronic/CIS system that has been adapted for electronic control of the air-fuel ratio. The letter E identifies the system as being an electronically controlled system. This system was produced from 1985 through 1993.

D-Jetronic. This is the first Bosch electronically pulsed fuel injection system. This system uses a MAP sensor to measure engine load. The *D* comes from the German word for (intake manifold) pressure, *Druck,* the major sensor in that system. The ECU was an analog design. It was produced from 1967 through 1976.

L-Jetronic. This system is similar to D-Jetronic, but uses a vane air flow (VAF) meter instead of a MAP sensor, to sense engine load. The *L* comes from the German word for air, *Luit, because* the VAF meter is that system's major sensor. The ECU was an analog design and controlled the air-fuel ratio only. This was not a full-function system. It was produced from 1974 through 1989.

LE-Jetronic. This system is an updated version of L-Jetronic. It used a port fuel injection (PFI) system design, but all injectors were always pulsed as a group; this was known as *group injection*. This was not a full function-system, and it had no self-diagnostic ability. It was produced from 1981 through 1991.

LU-Jetronic. This system is the U.S. version of L-Jetronic. It also used a VAF meter to sense engine load. The ECU was an analog design and controlled the air-fuel ratio only. This was not a full-function system, and it had no self-diagnostic ability. It was produced from 1983 through 1991.

LH-Jetronic. This system is a version of L-Jetronic that uses a hot-wire mass air flow (MAF) sensor, instead of a VAF meter, to sense engine load. The ECU controls the air-fuel ratio only; therefore this was not a full-function system. Initially, it had no self-diagnostic ability, but it was updated to provide some self-diagnostics in 1996. It was produced from 1982 through 1998.

Motronic. This system is Robert Bosch Corporation's first full-function system in that the PCM, formerly known as the ECU, controls the fuel injection system, the spark management system, and emission systems. The Motronic system was first introduced in 1979. The PCM was the first Bosch engine control system to use modern digital computer technology. (Bosch introduced its first modern *digital* computer as an ECU that was in control of an antilock brake system on the 1978 Mercedes Benz.) The initial Motronic system still used a VAF meter to sense engine load.

LH-Motronic. This system is a version of Motronic that uses a modern hot-wire MAF sensor to sense engine load. This system was also a full-function system and used a digital PCM.

Motronic ME7. This is a Motronic system that uses electronic throttle control (ETC) to control engine speed. *Driver demand* comes from the *electronic accelerator pedal (EAP)* sensor. This system also includes a torque-based engine management feature that uses the fuel injection and ignition timing systems to adjust engine torque based on the requirements of other systems including air-conditioning, traction control, and the antilock brake system (ABS).

Motronic MED. This is a Motronic ME7 system used with a GDI system. Robert Bosch Corporation first designed a direct fuel-injected engine for aircraft use in 1937 and then for automotive applications in 1952, so the concept has been around for many years. BMW introduced the Motronic MED system with a GDI system on its vehicles in the 2002 model year.

Overview of Bosch Alternative Fuel Systems

Flex Fuel. The Bosch Flex Fuel System uses a PCM that is programmed to be capable of recognizing and automatically adapting for any proportion of alcohol and gasoline in the fuel tank. This recognition is done through the use of the oxygen sensors and the fuel trim response. The PCM stores fuel level in memory when the ignition is turned off. When the ignition is again turned on, the PCM compares the current fuel level to the last fuel level recorded. If the driver refueled during the time the ignition was turned off, the PCM becomes aware of it. The PCM then waits until the engine is operating in closed-loop fuel control. If the fuel trim program has to make a sudden compensation at this time, either positive or negative, the PCM attributes this change to a change in the alcohol percentage in the fuel tank. It then adapts its fuel trim program accordingly.

It should be noted that the fuel trim adaptation may not occur on some vehicles unless the fuel level becomes less than a specified value. For example, if a fleet vehicle is topped every day at the end of the day after being driven a very short distance on a company's premises and not having used more than a couple gallons of fuel, the PCM may not make the fuel trim adaptation. This would not normally cause a driveability symptom as long as the gasoline being added always had the same alcohol content. But if the vehicle were driven off the company's premises on a special occasion for a couple of hundred miles and then the tank was refueled elsewhere and with a different alcohol content, upon returning to the company's base, a driveability symptom might develop in the coming days. This is because the PCM may not make the fuel trim adaptation back to the fuel that is used daily on the company's premises due to the frequent refueling.

Tri-Fuel. The Bosch Tri-fuel engine management system was launched in 2006. This system allows for the use of compressed natural gas (CNG), gasoline, and alcohol on the same vehicle. The PCM can switch transparently between CNG and any ratio of gasoline and alcohol.

THE BOSCH MOTRONIC AND LH-MOTRONIC SYSTEMS

The Motronic system is Bosch's first multifunction engine management system. In addition to the air-fuel mixture, it controls ignition timing and dwell. The system may also control idle speed, EGR operation, evaporative emissions, automatic transmission performance, and a turbocharger, depending on vehicle application.

Powertrain Control Module

The Motronic system's computer, or PCM, shown in Figure 15–3, is a digital computer with the ability to control many additional functions.

Bosch and Imports

Most of the imported European cars with a comprehensive computerized engine control system use a Bosch system. It first appeared on select BMWs. Most of the systems used on Japanese cars imported into the United States are built using Bosch patents. Practically all the gasoline engine multipoint injection systems for domestic vehicles use Bosch components or patents.

(It should be noted that the term *ECU* still continues to be used to generically refer to an *electronic control unit* that may be used to control any type of system, including an engine control system, while *powertrain control module [PCM]* is an OBD II standardized term that refers specifically to the module that controls the vehicle's powertrain, including engine and/or transmission performance.) An overview of the Motronic system is provided in Figure 15–4.

Main Relay

A main relay powers the PCM as soon as the ignition is turned on. This relay also includes a diode to protect the PCM against voltage surges and accidental polarity reversals.

On-Board Diagnostics

The first few model-year applications of Motronic do not feature any form of self-diagnosis, except that if a fault is detected in the closed-loop operating circuit, a calculated pulse width aimed at maintaining an air-fuel mixture near stoichiometric is used. Later versions, particularly those produced after the implementation of OBD II regulations, include extensive self-diagnostics.

Operating Modes

The PCM is programmed to use different operational strategies as driving conditions change.

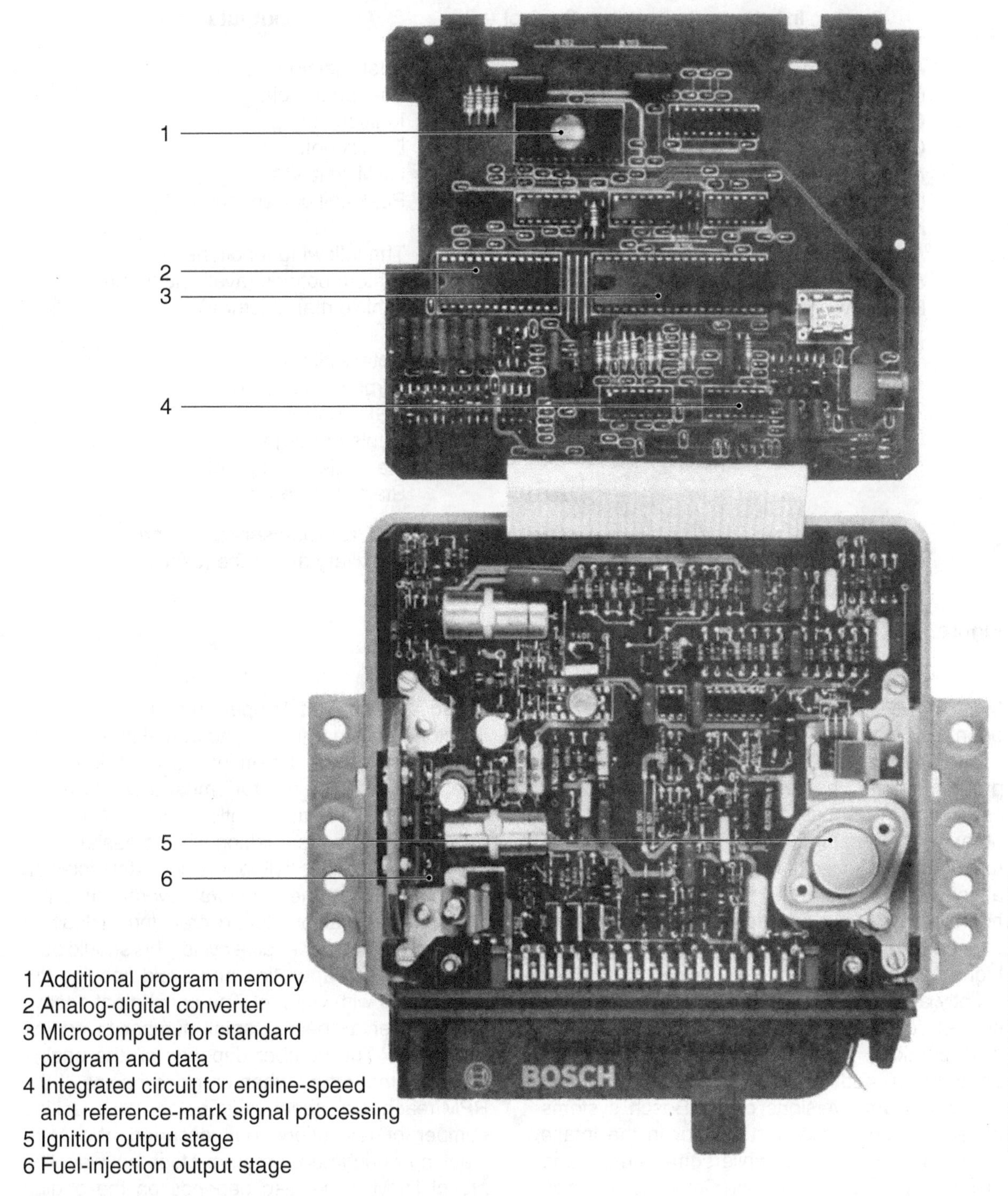

Figure 15–3 Bosch PCM.

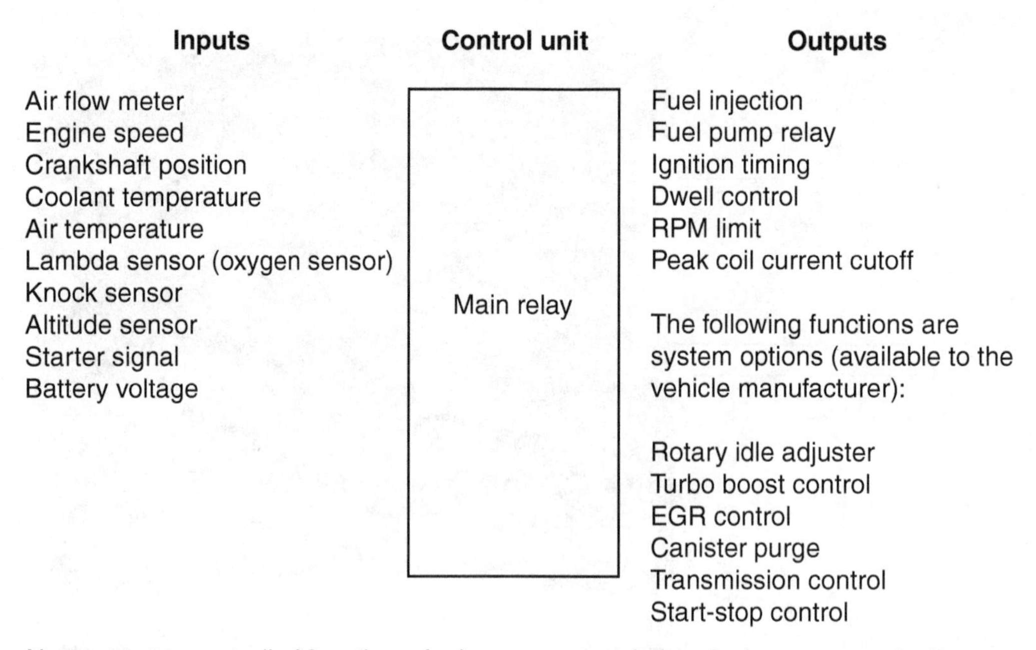

Figure 15–4 Overview of Motronic system.

Most of the differences in strategy occur during open-loop mode.

Cranking Mode. Two fuel-metering programs can be employed during engine cranking. One is based on cranking speed, the other on temperature. At lower crank speeds, the fuel quantity injected stays constant regardless of airflow fluctuations (input from the VAF meter is not reliable in these conditions because of pulsing caused by individual cylinder intake strokes). At higher cranking speeds, air intake diminishes slightly as a result of lower volumetric efficiency, so fuel quantity injected is also reduced. During cold cranking, the PCM adds an enrichment program in addition to the speed-dependent program. Some versions of the Bosch systems have included a cold-start injector in the intake manifold (Figure 15–5), while some merely add injection pulses at the port injectors. Ignition timing is also adjusted, depending on cranking speed and coolant temperature. The cold-start injector sprays a priming load of fuel into the cold intake manifold upstream of the main injectors. The extra distance the fuel must travel to reach the cylinders affords slightly more time for the gasoline to vaporize, making startup easier.

On applications without a cold-start injector, the PCM pulses the injectors several times per crankshaft revolution during cranking instead of the normal once per engine cycle. This should both enhance fuel evaporation and avoid flooding the spark plugs with wet fuel. The enrichment decays to zero over a specific, small number of engine revolutions. The number depends on the engine coolant temperature when cranking begins. If the RPM reaches a preset value before the specified number of revolutions has occurred, the cold-cranking enrichment program stops anyway. The preset RPM value also depends on the engine coolant temperature when cranking begins.

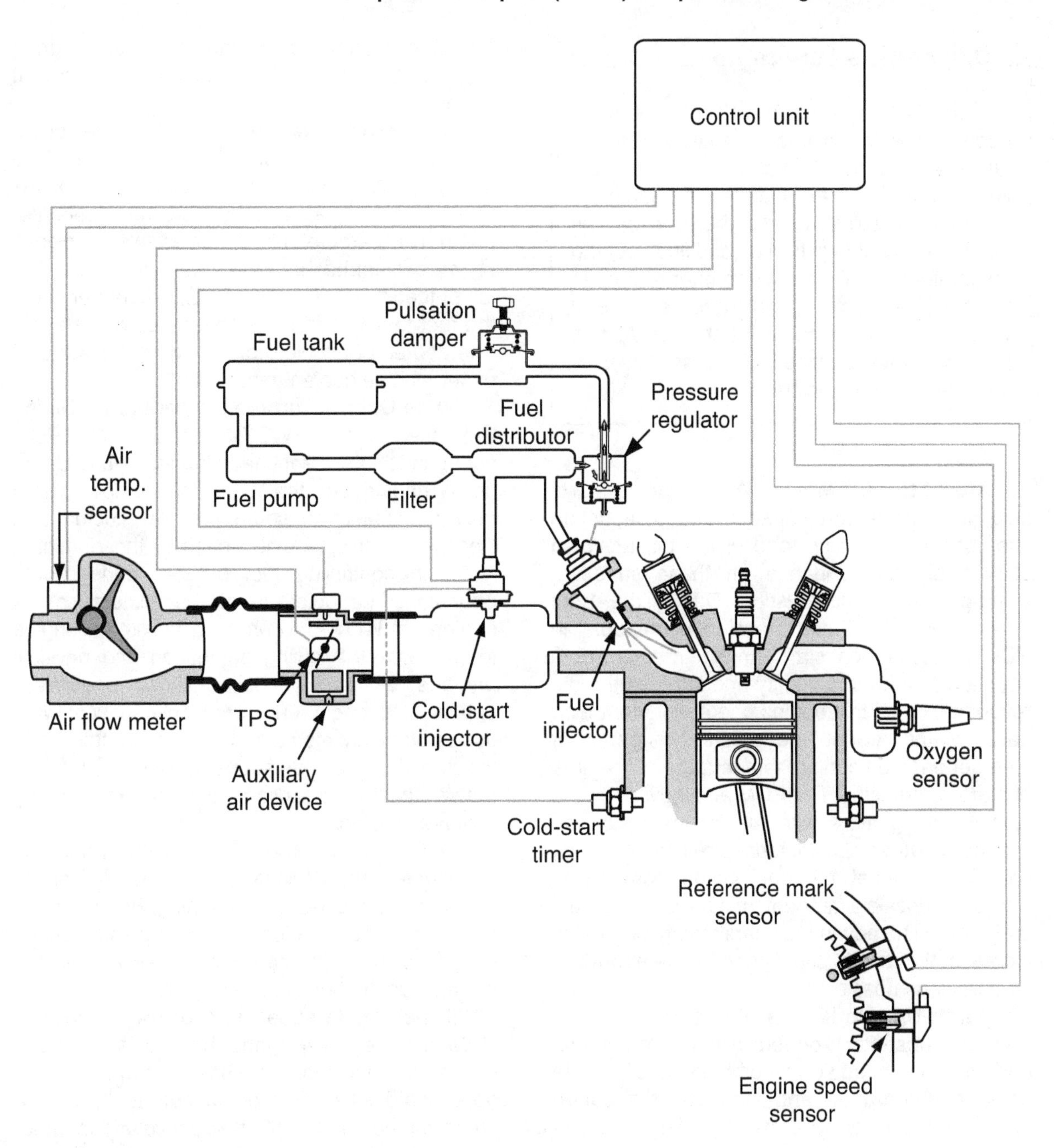

Figure 15–5 The Motronic system with cold-start injector.

Diagnostic & Service Tip

Sometimes when a vehicle has been used in a warm climate for a season, a certain amount of moisture can collect in the fuel line to the cold-start injector. This moisture can allow rust to form since the injector does not have occasion to turn on and flush out the contamination. When cold weather returns in the fall, the injector turns on but is plugged with rust, and the car will not start. At this point, you must service the cold-start injector by cleaning or replacing it.

After Start-Up Mode. As an engine starts cold, the intake port surfaces and combustion chamber walls, ceiling, and piston top are also cold. Fuel can condense on these surfaces, leading to poor combustion. During the brief time it takes for these areas to warm up, the PCM initiates a post-start enrichment program to maintain a good idle quality and improve throttle response. Once this begins, the enrichment also decays within a brief time to zero. This period of time depends on coolant temperature at the time of start-up cranking. The PCM also advances the ignition timing more with the engine cold than it does in other circumstances. All of this is to provide the most efficient engine power with the least possible amount of fuel and exhaust emissions. The advanced spark timing also helps increase the combustion chamber temperature as quickly as possible.

Warm-Up Mode. As the engine warms and the post-start period expires, additional enrichment is based on a combination of engine coolant temperature and load. An idle speed increase is applied to prevent stalling, improve driveability, and hasten warmup. The idle speed increase occurs during the post-start mode and continues into the warm-up mode. As an added means of bringing the oxygen sensor and catalytic converter to operating temperature as quickly as possible, some versions of the Bosch Motronic system retard ignition timing during warmup to make the exhaust gas somewhat hotter.

Acceleration Mode. Whenever the driver quickly opens the throttle from an earlier fixed position, the PCM briefly enriches the mixture. This serves the same purpose as the accelerator pump in a carburetor and solves two problems: (1) air accelerates into the combustion chamber faster than the fuel might and (2) it is harder to get the fuel to vaporize in higher-pressure intake air. The degree of accelerator enrichment is affected by the engine coolant temperature.

Wide-Open Throttle Operation. During wide-open throttle (WOT) operation, the PCM commands a fixed enriched air-fuel mixture. Under heavy engine load the VAF sensor's door (vane) may be subject to some irregular fluctuations. Because of this, engine speed is the dominant factor in controlling fuel delivery during WOT operation. The objective of the mixture control strategy at WOT is to produce maximum engine torque without allowing detonation and physical damage. Emissions quality is not specifically optimized during WOT operation because WOT use is ordinarily restricted to driving conditions in which safety considerations (passing, perhaps) outweigh the disadvantages of momentary adverse emissions.

Deceleration Mode. When the vehicle is decelerated, the PCM tapers the fuel delivery to zero and retards the ignition timing. Retarding the timing provides better engine braking and reduces the emission of hydrocarbons in the event there is still residual fuel on the intake port walls.

If the engine speed falls below a specified RPM or if the driver opens the throttle, the PCM returns fuel injection and spark advance to the appropriate level. This occurs gradually over a programmed number of engine revolutions rather than suddenly, for driveability and a smooth transition. On some Bosch systems, there is no fuel cutoff during deceleration, and in these systems a fuel-reduction program is used to avoid engine bucking and misfire, as well as high hydrocarbon emissions from misfires.

Closed-Loop Mode. The Bosch Motronic system, like other engine control systems, goes into closed loop once the engine coolant temperature and the oxygen sensor reach normal operating temperature. Bosch and other European manufacturers' technical literature frequently refers to the oxygen sensor as a "lambda sensor." While the oxygen sensors work the same way, a different description is used for the optimal air-fuel ratio. Until recently, European engineers identified the ideal ratio by the Greek letter *lambda*. In the United States, the ideal mixture has always been called the *stoichiometric ratio.* The two terms mean the same thing, and the oxygen sensors work similarly.

INPUTS

Oxygen Sensor (Lambda Sensor)

As previously explained, the earlier Bosch systems use oxygen sensors that work like familiar domestic units, but Bosch literature describes them as *lambda* sensors, using the term commonly used by German engineers to indicate air-fuel stoichiometry.

While European vehicles have traditionally used zirconium dioxide oxygen sensors, modern European vehicles now use wideband air-fuel ratio sensors (Figure 15–6). Robert Bosch Corporation patented the wideband air-fuel ratio sensor in 1994. They were first used as pre-cat sensors. Today, they are also being used as post-cat sensors. GDI applications tend to use them exclusively due to their ability to measure all the way lean to ambient air. The function of a wideband air-fuel ratio sensor is explained in greater detail in Chapter 3.

Coolant Temperature Sensor

The engine coolant temperature (ECT) sensor is an NTC thermistor bolted into the water jacket near the thermostat housing. Its operation is similar to those on domestic vehicles.

Intake Air Temperature Sensor

The intake air temperature (IAT) sensor is an NTC thermistor. When a VAF meter or MAF sensor is used, it is an integral part of the sensor. Otherwise, it is located in the intake airstream (Figure 15–7).

Throttle Position Sensor

Prior to the Motronic ME7 system, information concerning driver demand came from the throttle position sensor. With the Motronic ME7 and Motronic MED systems, driver demand information now comes from the electronic accelerator pedal (EAP) sensor. This sensor is known by most other manufacturers as the *accelerator pedal position (APP) sensor* or *APPS*.

Vane Air Flow Meter

The Bosch VAF meter used on the Motronic system is shown in Figure 15–8 and Figure 15–9. Sometimes also referred to as an air flow (AF) meter, it should never be confused with a hot-wire MAF sensor. The VAF meter has a swinging door that is spring-loaded to the closed position. Mounted in the intake airstream, as the engine draws in a greater volume of air, the door is pulled open against spring tension. The door's shaft rotates to move the wiper of a potentiometer, which returns a voltage signal to the PCM to indicate the door's physical position. This, in turn, is indicative of the volume of air entering the engine. The presence of a VAF meter, as with a MAP sensor, identifies the use of the *speed density formula* to *calculate* the mass of air entering the engine. The use of the VAF meter was discontinued in the mid-1990s.

Mass Air Flow Sensor

The MAF sensor, sometimes referred to as a hot-wire MAF sensor, measures the cooling effect of the intake air's molecules on a hot wire

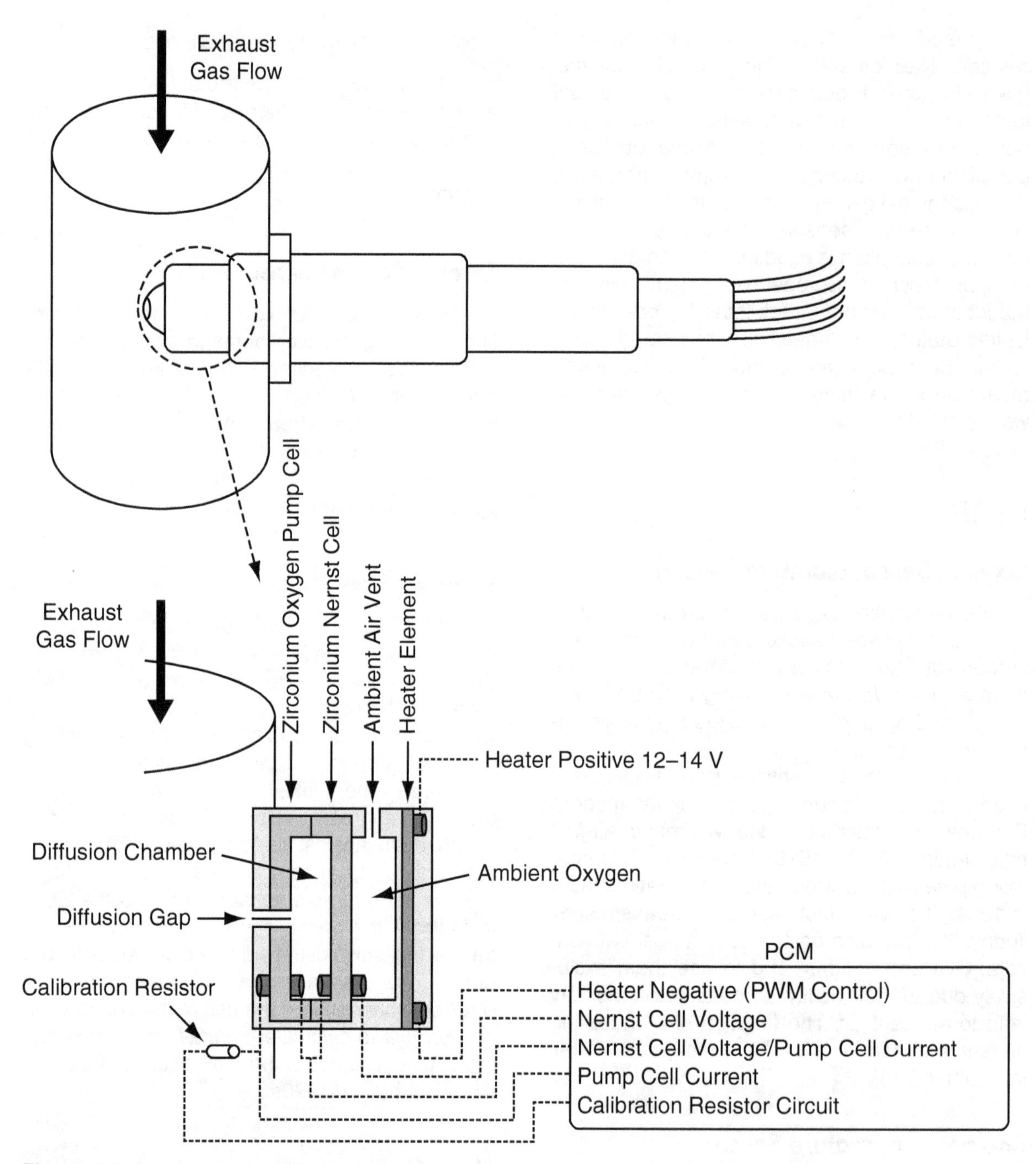

Figure 15–6 Wideband air-fuel ratio sensor.

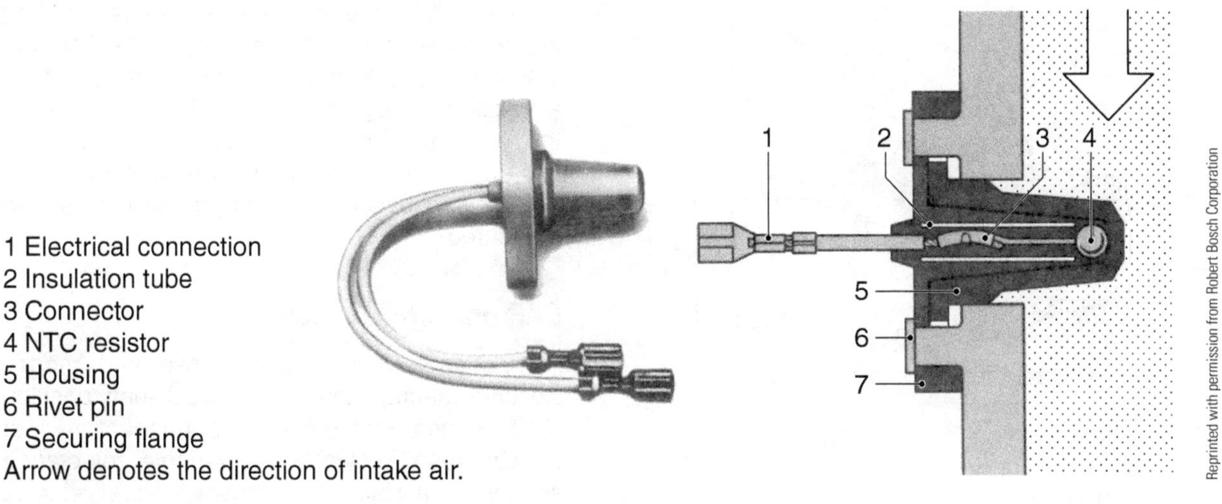

Figure 15–7 Air temperature sensor.

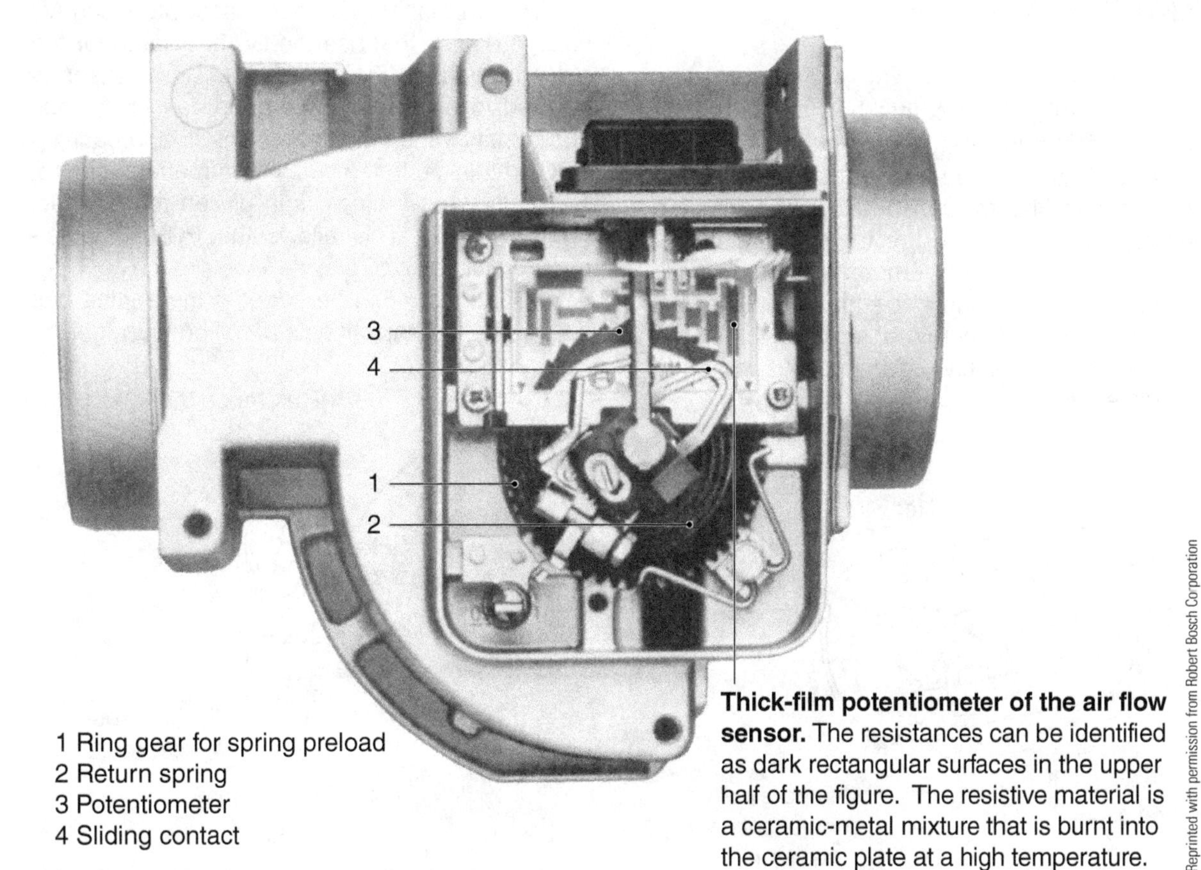

Figure 15–8 Air flow meter.

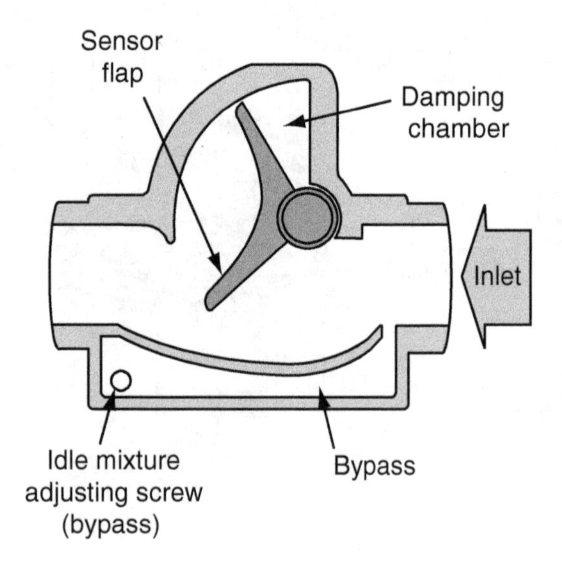

Figure 15–9 Air flow meter.

(Figure 15–10). It does this by measuring the amount of current that is required to keep the hot wire at its assigned temperature. A cold wire (thermistor) within the sensor allows the sensor's electronics to reference the temperature of the hot wire to ambient air temperature. Essentially a MAF sensor *measures* the mass of air entering the engine and, therefore, replaces the speed density formula that had been previously used to *calculate* this.

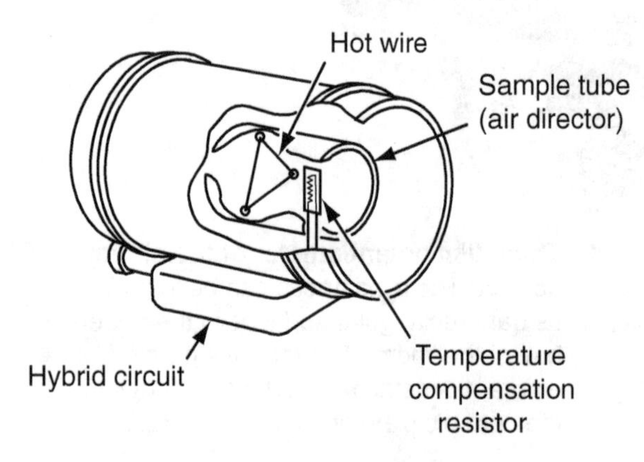

Figure 15–10 Mass air flow (MAF) sensor.

Like a VAF meter, a MAF sensor is mounted in the intake airstream, after the air filter box and before the throttle body so that all of the intake air passes through the sensor. The inlet tube for the positive crankcase ventilation system tees into the inlet air tube after the MAF sensor so that any air that passes through this system is also measured.

CKP and CMP Sensors

Depending upon year, make and model, Bosch systems may use a crankshaft position (CKP) sensor that is either a permanent magnet sensor, also known as a *variable reluctance sensor*, or a Hall effect sensor. Newer vehicles also use a camshaft position (CMP) sensor.

The permanent magnet sensor produces AC voltage pulses that use the flywheel teeth for the pulse-triggering (Figure 15–11). This sensor is referred to as the engine speed sensor. It does not identify which pair of pistons is approaching TDC. In this design, a second permanent magnet sensor senses a single iron pin on the flywheel (Figure 15–11). This allows the PCM to determine *which pair* of pistons is approaching TDC. A CMP sensor must be added if the engine has sequential fuel injection or a coil-on-plug ignition

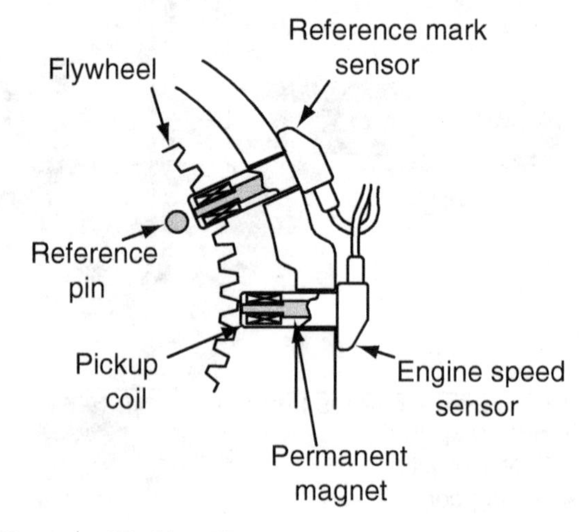

Figure 15–11 CKP sensors monitoring engine speed and reference marks.

system. The CMP sensor allows the PCM to determine which cylinder is on the compression stroke.

BMW uses Hall effect CKP and CMP sensors on many of their engines such as the N63 4.4L V8 engine with a GDI fuel management system. These Hall effect sensors are very similar to the ones used by Chrysler Corporation described in Chapter 14.

Knock Sensor

The Bosch Motronic system uses a piezoelectric knock sensor. The sensor generates a voltage signal in response to most engine vibration frequencies, but the signal is strongest at frequencies resonant with detonation knock frequencies. At a vibration between 5 and 10 KHz, the PCM recognizes knock and begins retarding the spark timing. The knock sensor is located on the block between two cylinders. On some engines, particularly V-type engines, two knock sensors are used. The BMW N63 4.4L V8 engine uses four knock sensors, one between each pair of cylinders. Combined with the information the computer has about crankshaft position, this can identify an individual cylinder with spark knock. On a modern Motronic system, this allows the system to retard the ignition timing to a cylinder selectively.

FUEL MANAGEMENT SYSTEM

Fuel Injectors

The fuel injectors used on Bosch systems are like those used in domestic-built multipoint fuel injection systems. In fact, domestic manufacturers have historically used Bosch injectors for their multipoint injection systems.

The Bosch injectors are solenoid-operated, pintle-type valves. In open-loop operation, the PCM calculates pulse width based on engine load and engine speed. Engine load is calculated from the MAF sensor and engine speed information. Figure 15–12 shows a three-dimensional graph or "map" correlating load, speed, and air-fuel ratio for a given engine application. This map is stored digitally in the PCM's read-only memory. It functions as the look-up table to determine pulse width. The arrows indicate increased load, speed, and enrichment.

The PCM can, however, modify the fuel injectors' pulse-width value represented by each line intersection in response to the following information:

- Engine temperature.
- Throttle position.

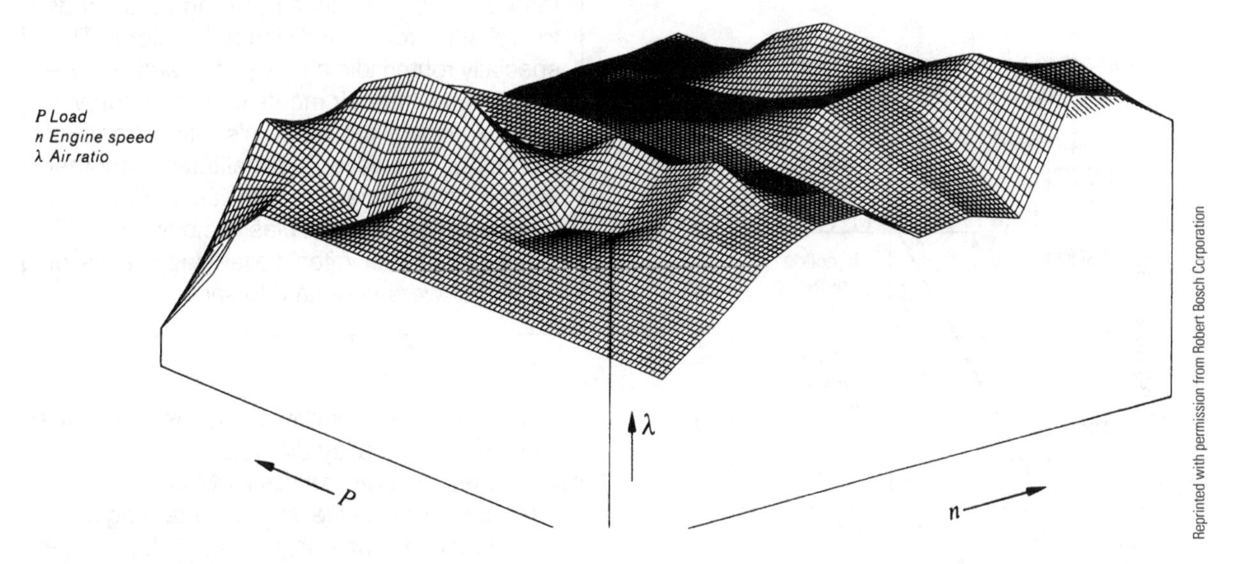

Figure 15–12 Pulse-width control map.

- Barometric pressure.
- Battery voltage.

Just as on domestic fuel injection systems, the amount of fuel that flows through the injector is affected by the injector's response time, which is directly reflective of charging system voltage. The PCM can detect this voltage and modify the pulse width within a certain range to correct the fuel delivery.

Engine Governor

Bosch Motronic systems, as with most modern vehicles produced by other manufacturers, have a rev limiter programmed into the PCM. The PCM disables fuel injectors to prevent the engine from over-revving and to limit the vehicle's top speed. The system uses fuel cutoff rather than spark override to keep from dumping excessive hydrocarbons into the atmosphere.

If the engine speed exceeds the programmed limit by more than 80 RPM, the PCM will shut off the fuel injectors until the engine falls to 80 RPM below the limit (Figure 15–13). This *redline* limit is to prevent mechanical damage to the engine from over-revving. This *rev limiter* also prevents damage to the belt-driven accessories that, depending upon pulley diameter, may spin up to seven times the crankshaft's RPM.

Please note, however, that the engine rev limiter function can only protect the engine from mechanical damage caused by powered

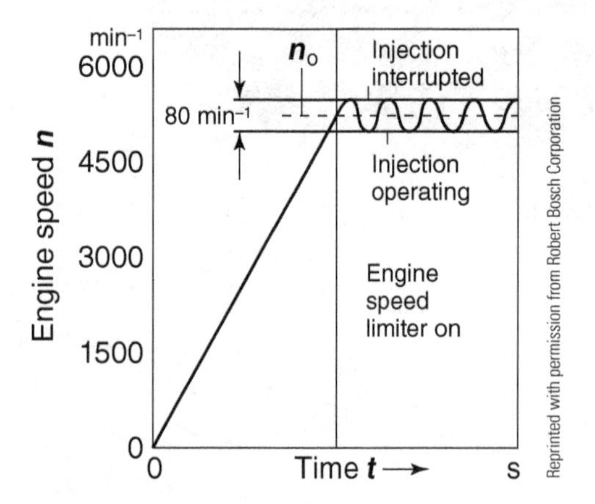

Figure 15–13 Engine speed governor.

Diagnostic & Service Tip

A common source of vacuum leaks, or "false air," as the Bosch literature sometimes calls it, because it does not go through the AF meter or MAF sensor, is around the injectors. The seal is usually a single O-ring. With heat and age, these O-rings can crack and cause vacuum leaks.

It is fairly common for an injector O-ring to leak after the injector is removed and reinstalled, even if it was not leaking air before, because the removal and replacement disturbed the seal. While it is not always necessary to replace O-ring seals when reinstalling an injector, they should be inspected carefully and replaced if necessary. Engine oil is generally regarded as the best O-ring installation lubricant, though at least one manufacturer favors 80-weight gear lubricant. Failure to inspect, and if necessary replace, the injector O-ring can allow a technician to build another new problem into a problem car.

Also, air-shrouded injectors are used by Volkswagen and several other car makers on some of their models. Idle air does not pass through the regular intake channels but through a special idle air passage that leads to collars around individual injectors. This specially routed idle air keeps the airflow's velocity high where air meets fuel and improves vaporization of the fuel. This idle air shroud plumbing, however, constitutes another source of potential vacuum leaks. On early Volkswagen models, a plastic elbow near the idle air controller often developed cracks or splits that were very hard to spot.

over-revving. If a driver downshifts into a lower gear at high speed, the governor can shut off the fuel, but, driven by vehicle inertia, the engine could still spin fast enough to damage engine accessories or even the engine itself.

The PCM also uses the rev limiter function to limit the vehicle's speed to the speed rating of the original manufacturer's tires that were installed on that vehicle.

Fuel Pump Relay

As on most fuel injection systems, the fuel pump is indirectly controlled on Motronic systems through a fuel pump relay. Power is available at the relay whenever the ignition switch is turned on. The PCM then energizes or de-energizes the relay's coil to activate or shut off the fuel pump. If the engine does not turn at a minimum specified speed, the PCM will shut off the relay to reduce the risk of fire. As on many other fuel injection systems, the high-pressure fuel pump is located inside the fuel tank and is submerged in fuel. Earlier systems employed a fuel pump just outside the tank, on the line to the engine compartment and just ahead of the fuel filter.

Fuel Pressure Regulator

The Bosch fuel pressure regulator, like most others, uses intake manifold absolute pressure to maintain a constant pressure differential between the fuel pressure in the fuel rail and the pressure in the intake manifold (Figure 15–14).

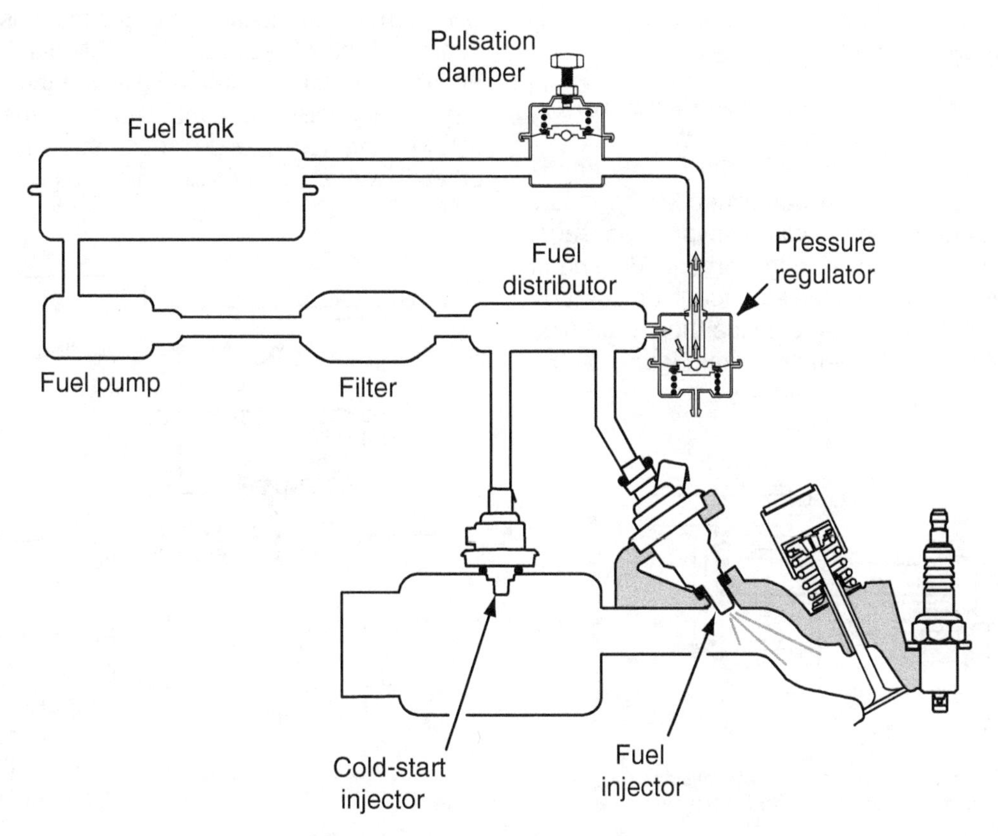

Figure 15–14 Fuel pressure regulator (right) and pulsation damper (left).

Fuel Pressure Pulsation Damper

The fuel pressure pulsation damper is also shown in Figure 15–14. It works to reduce the noise of fuel pulses from the fuel pump. While it looks much like the pressure regulator, it is not connected to manifold vacuum. It functions as a surge accumulator to absorb the pressure pulses from the opening and closing of the injectors. The pulsation damper is in the return line downstream from the pressure regulator. The higher-pressure Bosch systems, as well as those on four-cylinder engines, are more likely than the low-pressure systems or those on five-plus cylinder engines to use pressure dampers because the effects of the pulses are greater in them. Higher pressures are often preferred by car makers because the fuel is more finely atomized under higher pressure, other things being equal.

Idle Speed Control

Bosch Motronic systems have used two different methods to control idle speed.

Auxiliary Air Device. On some early applications, idle speed was controlled by an idle air bypass device in the idle air bypass passage. After a cold start and during warmup, idle speed is increased by routing the air through an auxiliary air valve. A bimetallic strip inside the valve holds the valve open at low temperature (Figure 15–15). With the passage open, additional air bypasses the throttle blade. This air has, however, passed through the AF meter, so the computer knows the correct amount of fuel to inject. In addition, once the ignition is turned on, voltage is applied to a heating element wrapped around the bimetallic strip. As the strip warms, it gradually closes the valve, blocking the flow of auxiliary air and slowing the idle as the engine warms. Once the engine is completely up to operating temperature, its own heat is enough to keep the valve closed.

Rotary Idle Control Valve. Other early applications used the rotary idle control valve to control throttle bypass air (Figure 15–16). This valve is controlled by the PCM and controls engine idle speed regardless of engine coolant temperature. At the end of an armature shaft, a rotary valve or slider valve can move in the idle air bypass passage (Figure 15–17). A reversible permanent magnetic DC electric motor is used to control throttle bypass air. The PCM can rotate the motor's armature incrementally and thus position the rotary valve to control engine speed. The PCM's memory has a programmed idle speed that it wants to maintain for any given engine

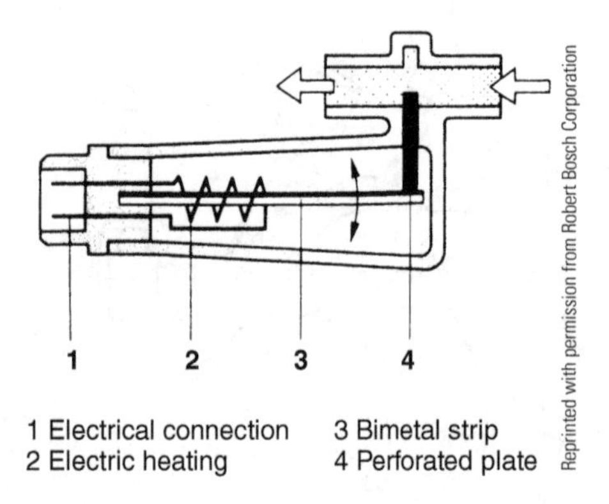

Figure 15–15 Auxiliary air bypass device schematic.

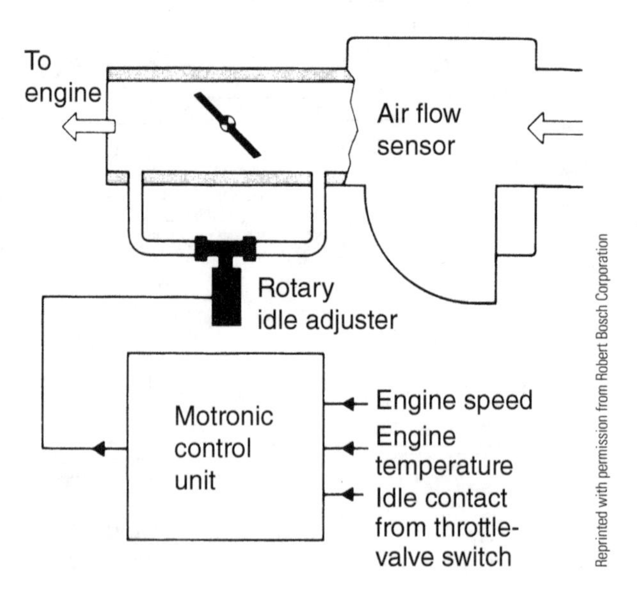

Figure 15–16 Rotary idle adjuster circuit.

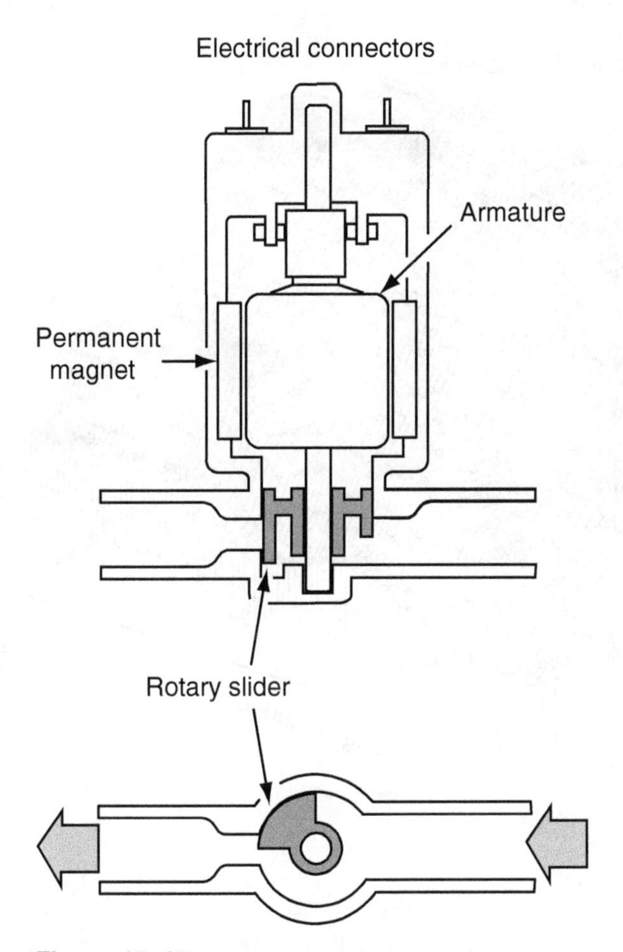

Figure 15–17 Rotary idle adjuster motor.

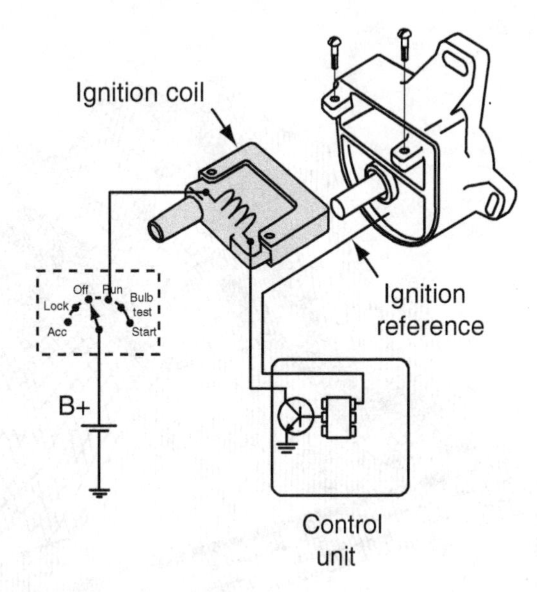

Figure 15–18 Ignition circuit.

coolant temperature. It first uses ignition timing to adjust idle speed; it then uses the rotary idle adjuster to bring idle speed to the desired RPM if the timing adjustments have been insufficient.

SPARK MANAGEMENT SYSTEM

Ignition Timing

The PCM controls the ignition coil(s) through ground switching of the primary winding(s) (Figure 15–18). Like fuel injection pulse width, ignition timing is based on engine load and speed. Figure 15–19 shows a three-dimensional map of spark advance for one engine. Arrows point in the directions of increased load, speed, and spark timing advance. The line intersections represent the spark advance for all combinations of engine speed and load. These values are stored digitally as a look-up table in the PCM's read-only memory.

The PCM reviews sensor inputs and recalculates spark timing after each cylinder firing. The PCM can modify the values in the read-only map in response to inputs from:

- ECT sensor
- IAT sensor
- TPS
- Barometric pressure (BP), calculated from the MAP or MAF sensor

The PCM modifies the spark advance according to the information from either the ECT sensor or the IAT sensor. This modification, of course, depends on driving conditions. At starting crank, spark timing depends entirely on engine speed and coolant temperature. At low cranking speeds and low temperatures, spark timing is close to TDC. At higher cranking speeds and with warmer

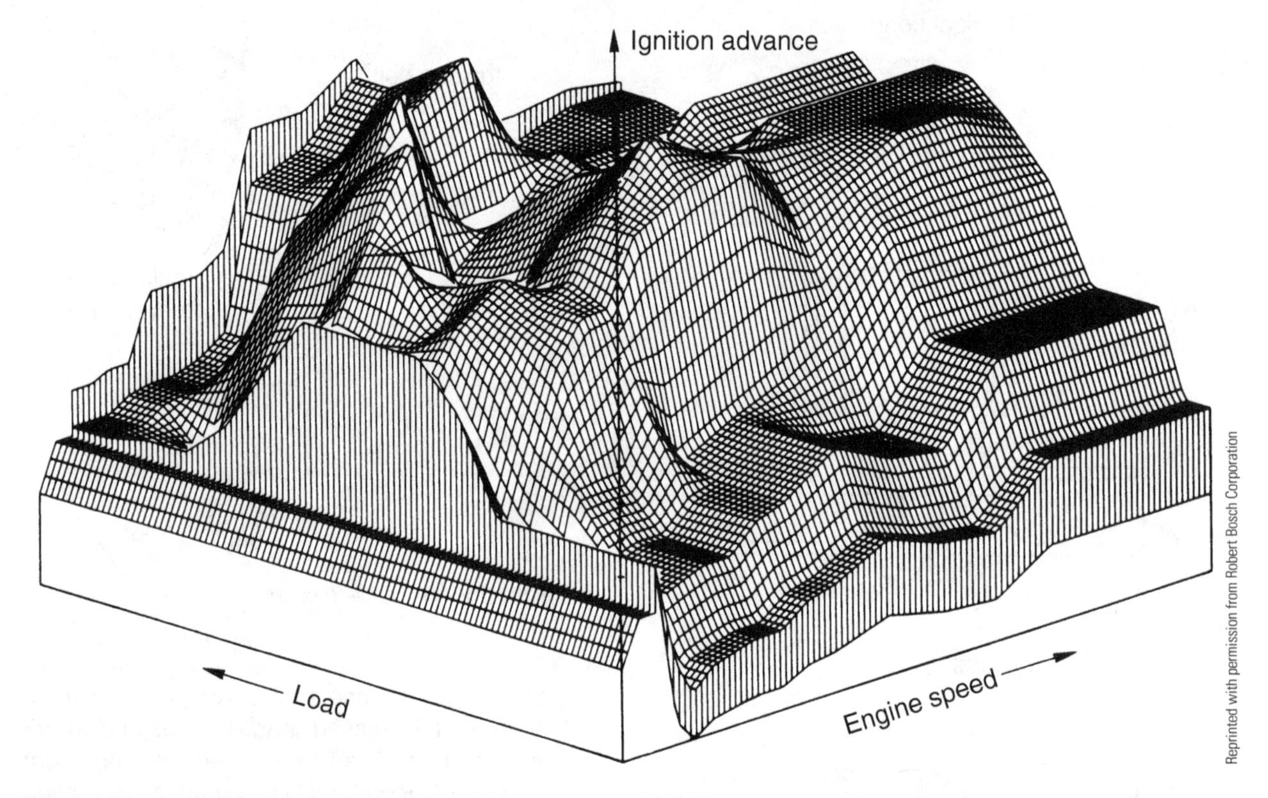

Figure 15–19 Ignition timing map.

engines, spark timing is slightly advanced. Each of these combinations improves starting and post-startup driveability.

Once the engine starts cold, spark timing is advanced significantly for a brief period to reduce the need for fuel enrichment. Once the engine is warm enough to reduce the likelihood of stalling, the spark timing is retarded on some vehicles to speed up engine warmup, even at the cost of lower delivered engine torque. Increased exhaust temperature helps warm the oxygen sensors and catalytic converters more quickly.

On earlier systems, during warm idle ignition timing is used to control idle speed. With no additional load, the transmission in neutral, and the air conditioning off, the timing can be retarded slightly and still maintain the desired idle speed. This keeps combustion chamber surfaces hotter to reduce the production of unburned hydrocarbons. If there is a load, such as the transmission going into gear, the idle speed begins to slow. The PCM responds by advancing ignition timing enough to achieve more torque and bring the idle speed back to the desired speed. This avoids having to increase throttle bypass air and reduces fuel consumption. The PCM will increase throttle bypass air if advancing the spark timing alone is not enough to keep the idle speed at the desired setting. (On ETC systems, the PCM will increase the throttle's opening.)

Spark Knock Control

At WOT, normal or higher engine coolant and/or intake air temperature causes the PCM to reduce spark timing at load combinations where there is a high probability of spark knock. If at a specific load there is acceleration short of WOT, the PCM retards the existing spark advance and then gradually begins to restore it. This reduces the potential for both spark knock and the production of NO_x.

On earlier systems the spark knock control was separate from the Motronic unit. Later, it was incorporated into the PCM. The newer version, controlled by the PCM, allows a more precise control of spark advance. When the knock sensor indicates a detonation signal, the PCM recognizes the knock and immediately reduces spark advance until the knock disappears. Then it gradually re-advances spark timing toward the original value.

Dwell Control

The Motronic PCM manages ignition coil dwell in the same way as the General Motors ignition system. As engine speed increases, dwell increases to ensure complete magnetic saturation of the coil. The Bosch system also considers charging system voltage when calculating dwell.

Coil Peak Current Cutoff

If the ignition coil remains energized too long, the ignition coil will overheat and short its windings. To avoid this problem, the PCM limits coil saturation to acceptable levels by controlling dwell.

Turbo Boost Control

Bosch Motronic systems use a combination of spark retard and boost pressure reduction to control spark knock on turbocharged engines. Particularly on smaller engines, Bosch engineers feel reducing boost alone puts unnecessary limits on engine performance. They also find that using spark retard alone to control knock is a poor choice because it raises exhaust temperature, which can be threatening to the already high-temperature turbocharger turbine. Using a combination of both measures is the strategy they prefer. Spark retard is the fastest measure, provided the knock is caused by spark timing and not by preignition. It can be applied to the very next power stroke after the detection of knock. It takes several intake strokes even after the wastegate opens before the manifold pressure begins to come down, because the turbocharger must slow and the intake air must start to cool. As boost pressure does come down, however, the Motronic system starts to re-advance the ignition timing to retain as much engine torque as possible. Later systems can retard ignition timing on individual cylinders.

EMISSION SYSTEMS

Catalytic Converters

As is common with other manufacturers, European vehicles use both a reduction catalytic converter and an oxidation catalytic converter. The reduction cat uses CO to reduce NO_x back into N_2 and O_2. The oxidation cat oxidizes CO into CO_2 and HCs into both water and CO_2.

Evaporative Emissions Control

Motronic system vehicles use an evaporative emissions system similar to those on domestic vehicles. Fuel vapors from the fuel tank are stored in a charcoal canister (Figure 15–20). Once the system goes into closed-loop fuel control, the PCM uses a purge solenoid to allow the charcoal canister vapors to purge to the intake manifold. The PCM operates this solenoid on a pulse-width-modulated command to precisely control purging while using the oxygen sensor signal for feedback.

EGR Valve Control

Early European Bosch-equipped vehicles did not use EGR systems to control the production of NO_x at high combustion temperatures. Instead, they used the reduction feature of the catalytic

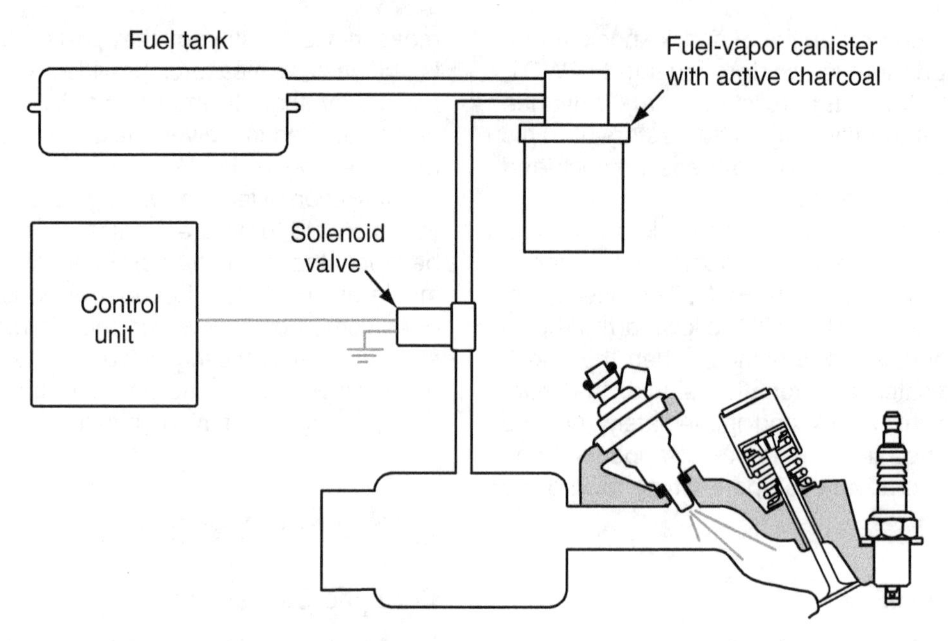

Figure 15–20 Evaporative emissions control system.

converter combined with the residual exhaust retained by the valve overlap. This method, however, does not allow the precise controls available with an electronically controlled EGR valve, so later versions employ EGR systems functionally similar to those used on most domestic-built systems.

ELECTRONIC TRANSMISSION CONTROL

Later versions of the Motronic system include sensors to indicate the transmission's status. A vehicle speed sensor, transmission gear selection sensor, and kick-down sensor provide the PCM with this information. On modern systems, the PCM controls the automatic transmission's hydraulic pressure and shift points using solenoid valves that replaced traditional spool-type shift valves (Figure 15–21). Shift point look-up tables, including one for economy, one for performance, and one for manual shifting, are stored in the PCM's read-only memory and provide optimum transmission shifting according to the programming look-up chart selected by the driver.

Smooth shifts with reduced clutch slip and wear are achieved, especially during WOT operation, by momentarily retarding ignition timing to reduce engine torque output during the shift. Motronic transmission control can thus improve fuel economy, shift quality, transmission torque capacity, and expected transmission life.

Start-Stop Control

Another optional feature that was added to some Motronic systems is the start-stop feature. This feature was initially incorporated on some European cars and is a predecessor of the modern systems such as the General Motors FAS and BAS systems. This feature helps conserve fuel and reduce emissions. The Bosch feature was initially applied only on vehicles with manual transmissions.

The Bosch start-stop control uses a vehicle speed sensor, a clutch pedal position sensing

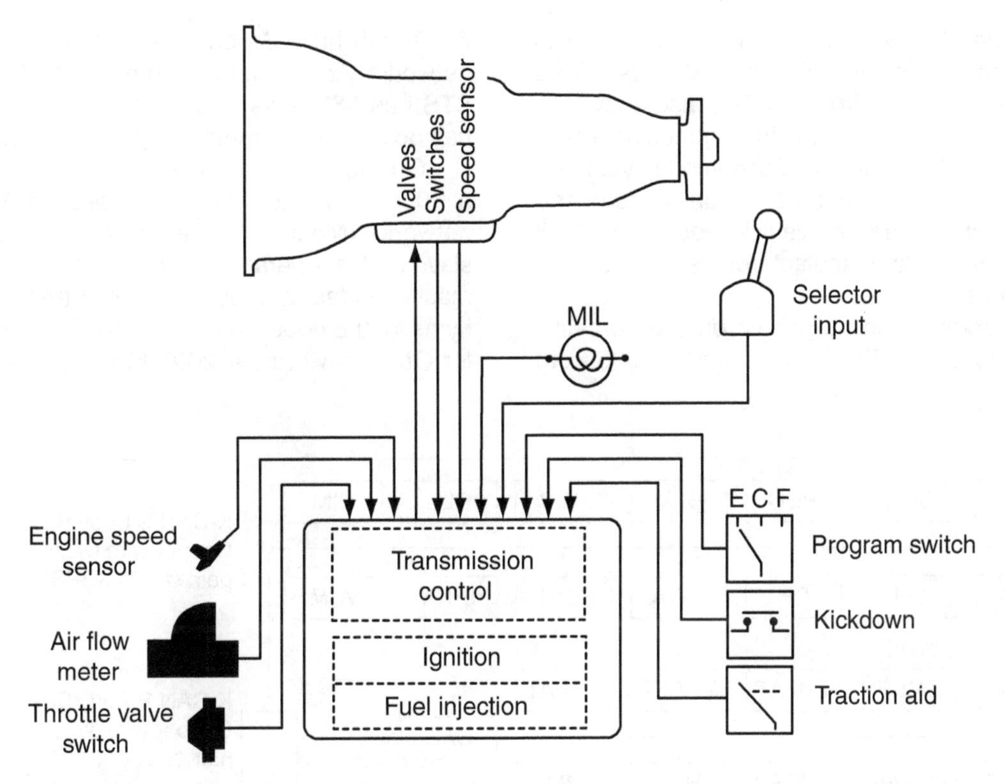

Figure 15–21 Electronic transmission control.

switch, and a separate control module. If vehicle speed falls below about 1 MPH with the clutch pedal depressed, the module signals the PCM, which shuts off the fuel. As long as the driver keeps the clutch pedal down, the module will activate the cranking circuit and fuel injection will resume as soon as the driver presses the accelerator. This system does not work until the engine is warm.

BOSCH MOTRONIC AND OBD II

Robert Bosch is the major European authority on the subject of computerized engine management systems, and its most highly evolved systems are in the Motronic family, used by Mercedes-Benz, Volkswagen, Volvo, and others. Motronic has always meant a combination of fuel and spark management since it was introduced in 1979. Now it has additional functions and numerical series designations like software.

The way the European car makers adapt Motronic to satisfy OBD II regulations is interesting. In Mercedes-Benz C-class cars, for instance, the Motronic control module (Mercedes-Benz calls the system HFM-SFI, for Hot Film Engine Management–Sequential Fuel Injection) is next to a proprietary diagnostic connector that is to be used with the company's own scan tool, which is used by dealership technicians. OBD II regulations are satisfied by a separate dedicated data link connector (DLC) that is electrically connected to the Motronic PCM, known under OBD II standards as the PCM. The connection to the DLC uses a controller area network (CAN) bus. This fulfills the EPA's legal requirements and sends emissions-related data to a standardized OBD II

DLC under the instrument panel for access with an aftermarket scan tool. The CAN bus allows many electronic control modules, also known as *nodes*, on a given vehicle to communicate with each other. While in the 2014 model year the average number of nodes communicating over a data network on a given vehicle was between 18 and 20, European manufacturers tended to have many more than this.

For comparison purposes, on the domestic side, a 2003 Ford Thunderbird and a 2005 Ford F150 both have 9 nodes communicating on the networks. At the other extreme, a 2006 Cadillac STS has 56 nodes and a 2011 Chevrolet Tahoe has 66 nodes communicating on the networks.

On the European side, Figure 15–22 shows a 2006 BMW E65 with 54 nodes on five data networks. Most of these nodes are dedicated slave nodes, operating under the control of a few master nodes. It should be noted that the acronyms for the nodes in Figure 15–22 are acronyms for German words. A 2007 Mercedes S600 has

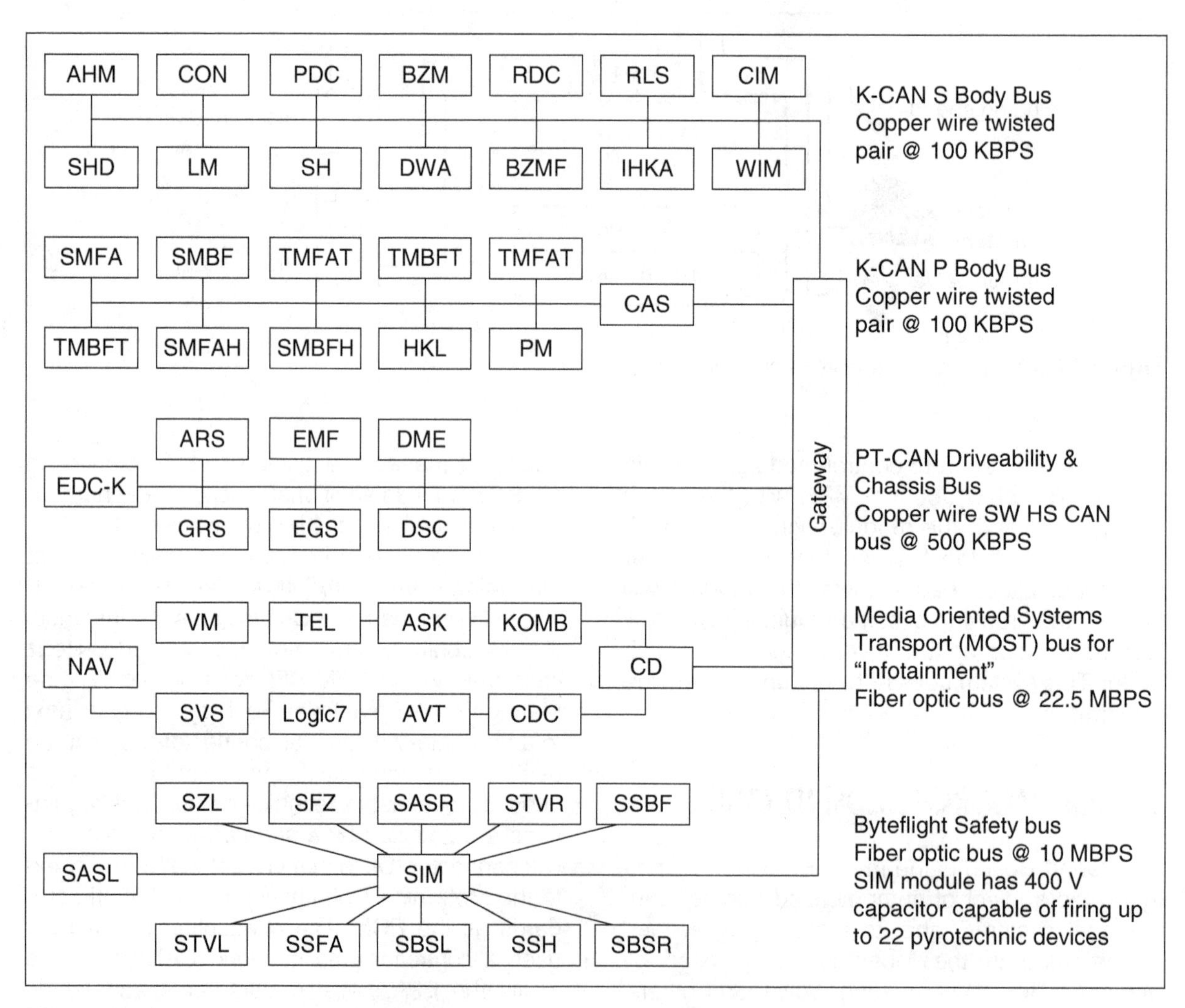

Figure 15–22 BMW E65 body electronics with 54 modules on-board.

64 nodes, a 2008 BMW 7 Series has 72 nodes, and a 2011 and newer Audi R8 has 124 nodes communicating on the networks.

On a Motronic system, just as with other computerized engine management systems regardless of the country of origin, there is a comprehensive list of sensors throughout the automobile. For example, the Motronic system uses a chassis accelerometer to differentiate between potholes and engine misfires. This is different from the approach of other makers, such as GM, which use signals from the ABS wheel speed sensors to indicate rough road surfaces.

The Motronic PCM also performs self-tests of many of its components and systems. The plausibility check of the MAF sensor is a good example of a system self-test. The computer makes continuous calculations from throttle angle and RPM to arrive at the probable injection duration (similar to the operation of a speed-density EFI system). Then it compares this value to that being obtained from the MAF sensor, and it stores a DTC if there is a large enough discrepancy.

MOTRONIC ME7

The Bosch Motronic ME7 engine management system incorporates an ETC system to control engine speed. The driver demand signal comes from the EAP sensor. While this may vary with the European brand of vehicle on which it is installed, on the early ETC systems the EAP sensor assembly will typically contain two potentiometers for redundancy, and the throttle body will typically contain two potentiometers that act as throttle position sensors for feedback. The throttle body also contains a DC motor that is controlled by the PCM on a bidirectional pulse-width-modulated command. Newer ETC systems use noncontact linear Hall effect sensors for both the EAP function and the TPS function.

The ETC system also includes an engine torque management feature that uses the fuel injection and ignition timing systems to adjust engine torque based on the requirements of other systems, including air-conditioning, traction control, and the antilock brake system (ABS). And the ETC system eliminates the need for a dedicated idle speed control actuator and a dedicated cruise control servo.

MOTRONIC MED

The Bosch Motronic MED engine management system incorporates ETC, as with the Motronic ME7 system, but it also incorporates a GDI system. The GDI system uses a fuel injector for every cylinder, similar to an SFI system. However, the injector sprays fuel directly into the cylinder. Not only does this system use much higher fuel pressure than earlier systems, but the higher pressure, in turn, allows for better atomization of the fuel.

A GDI engine can also operate in three basic modes of air-fuel control as follows:

- Homogeneous charge mode with closed-loop fuel control: This mode is used for maintaining an overall stoichiometric air-fuel ratio.
- Homogeneous charge mode with open-loop fuel control: This mode is used for providing a rich air-fuel ratio when the vehicle is under moderate-to-heavy load.
- Stratified charge mode with closed-loop fuel control: This mode is used for maintaining a stoichiometric air-fuel ratio near the spark plug while providing an overall super-lean air-fuel ratio during cruise.

In the stratified charge mode, the overall air-fuel ratio can be as lean as 60:1 on European vehicles, although a specialized NO_x reduction catalytic converter is required to control NO_x emissions when operating this lean. This catalytic converter has not yet been approved for use in the United States, so European vehicles built for the U.S. with this system operate with a 30:1 to 40:1 air-fuel ratio when operating in the stratified charge mode.

European vehicles, such as Audi, use swirl control valves in the intake manifold to help

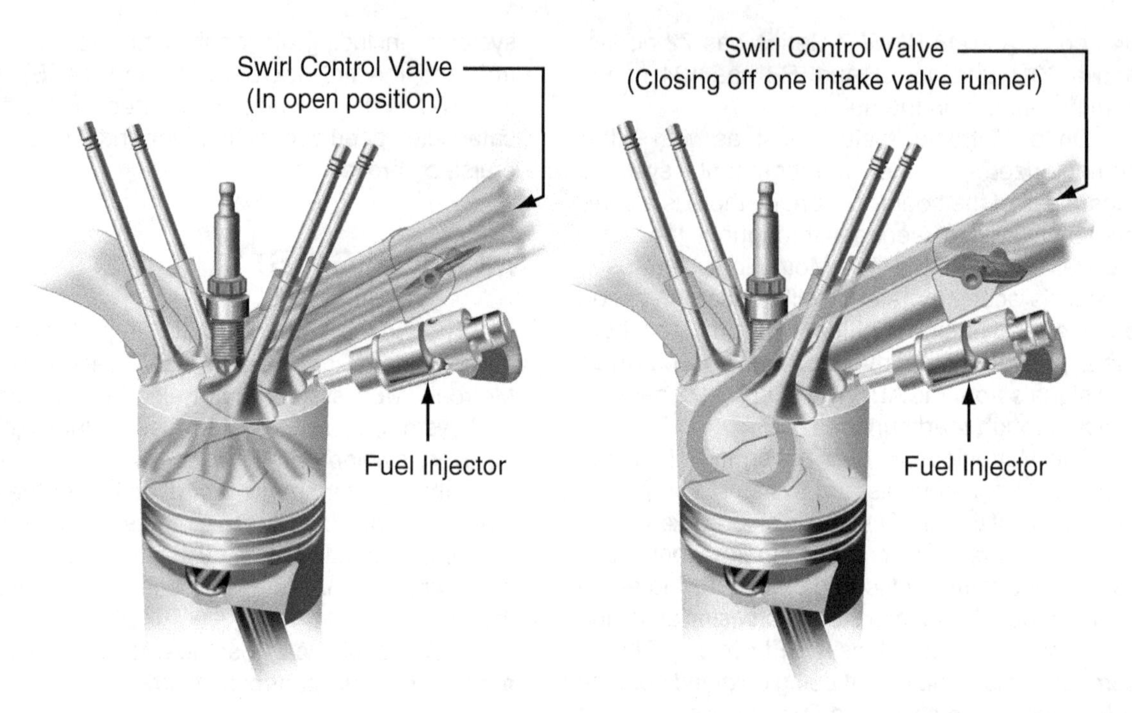

Figure 15–23 GDI cylinder with swirl control valves in the intake manifold runners.

control whether the engine is operating in homogeneous charge mode or stratified charge mode (Figure 15–23). The swirl control valves affect the patterned movement of air into the combustion chamber.

During cruise and while operating in the stratified charge mode, the PCM can control engine speed and performance through control of fuel metering, similar to a diesel engine. This allows the PCM to further open the throttle plate(s) beyond what would normally be allowed without over-revving the engine. This, in turn, reduces the cylinder's pumping losses.

Additionally, while operating in the stratified charge mode, the engine's power can be reduced during cruise by simply reducing fuel delivery. Therefore, a GDI engine has the advantage of consuming less fuel when power is not needed, similar to a variable displacement system except that, with a GDI engine, the PCM does not have to disable cylinders to achieve this reduction in power and fuel. Therefore, the engine runs smoother and is less annoying to the driver while operating in the stratified charge mode than an AFM engine is when the PCM disables half of its cylinders. Another advantage over the AFM system is that GDI can be used in the stratified charge mode on smaller four-cylinder engines.

BMW VALVETRONICS SYSTEM

The BMW Valvetronics system uses PCM control of intake valve opening, rather than a throttle plate, to control engine speed. This is done through a special mechanism that allows the PCM to control how much the camshaft's lobes open the intake valves (Figure 15–24). A PCM-controlled throttle plate is only retained to create a small amount of vacuum during cruise for both the PCV and evaporative purge systems to use. The reduced intake manifold vacuum reduces the engine's pumping losses. This system is used on the intake valves only; the exhaust valvetrain

Valvetrain

The valvetrain on the N52 uses the familiar 4 valve per cylinder arrangement used on previous engines. The N52 valvetrain also uses the proven Valvetronic technology which was introduced on the N62.

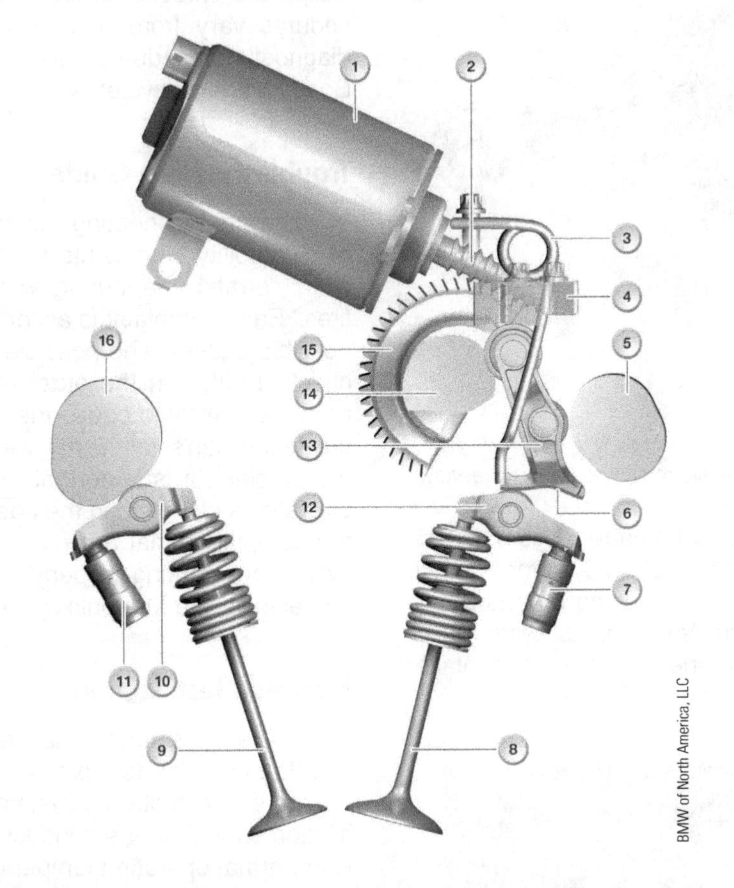

Index	Explanation	Index	Explanation
1	Actuator	9	Exhaust valve
2	Worm shaft	10	Roller cam follower
3	Return spring	11	HVA, exhaust
4	Gate block	12	Roller cam follower, intake
5	Intake camshaft	13	Intermediate lever
6	Ramp	14	Eccentric shaft
7	HVA intake	15	Worm gear
8	Intake valve	16	Exhaust camshaft

Figure 15–24 BMW Valvetronics mechanism controls the opening of the intake valves only.

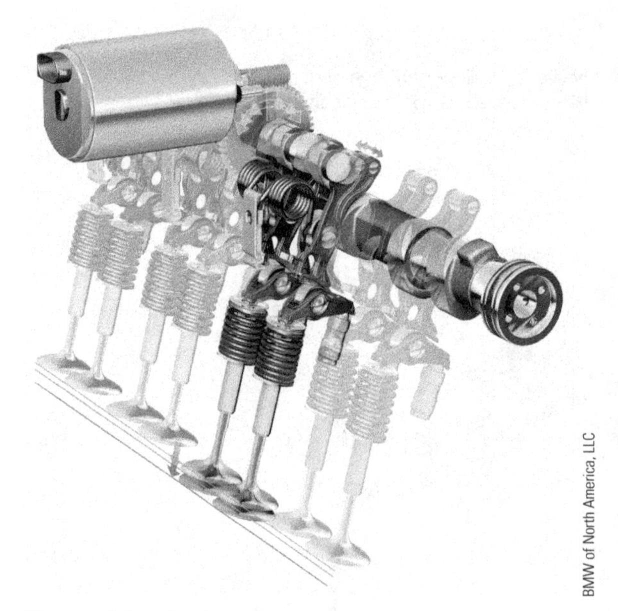

Figure 15–25 BMW Valvetronics operation overview.

operates in the traditional manner. Figure 15–25 shows that a single reversible DC motor is used to control the intake valve opening on one bank. A V8 engine with the Valvetronics system will have two DC motors, one controlling the intake valve opening on each bank (Figure 15–26).

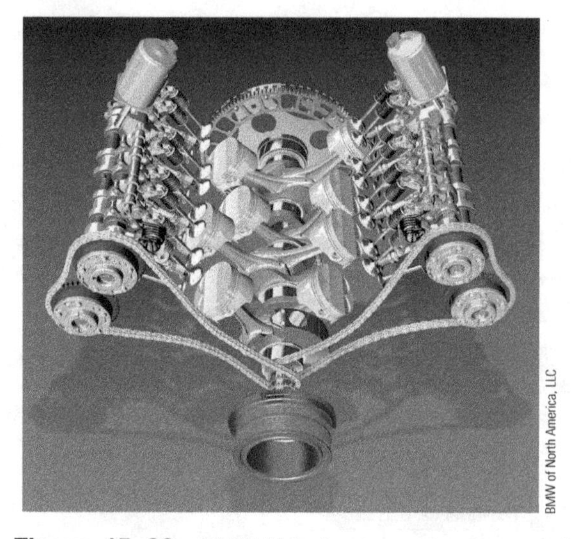

Figure 15–26 BMW Valvetronics system as used on a V8 engine.

✔ SYSTEM DIAGNOSIS AND SERVICE

The vehicle manufacturer establishes the diagnostic procedures for each vehicle, so procedures vary from one make to another. The diagnostic procedures here are representative of Bosch Motronic systems.

Troubleshooting Guide

The troubleshooting guide consists of a list of driveability complaints such as "will not start cold," "erratic idle during warmup," and "backfires." Each complaint is accompanied by a list of possible causes. The possible causes should be checked either in the order of easiest to check or of most likely to cause the problem, based on the technician's familiarity with the vehicle and the problem. It is important that all components or systems identified in the possible cause list for the complaints that are not part of the Motronic system are working properly before proceeding to the tests for the Motronic system.

Motronic Test Section

The service manual is likely to have a section that contains test procedures for each component of the Motronic system. The tests in this section should not be conducted until the engine is at normal operating temperature. When testing components and their circuits, you should remember that connections cause more problems than components.

SUMMARY

In this chapter, we reviewed the history of Bosch fuel injection systems, including the two distinct types of Bosch fuel management systems: mechanical continuous fuel injection and electronically pulsed injection.

We have covered the Motronic system as a full-function engine management system, including its primary inputs and the primary systems it controls. We also discussed the features that the Motronic ME7 and Motronic MED systems add to the basic Motronic system. Finally, we have looked at basic diagnostic approaches.

▲ DIAGNOSTIC EXERCISE

A vehicle with a Motronic system comes into the shop, having failed an emissions test for high CO readings. The oxygen sensor is verified to be working correctly and is reporting the rich exhaust condition to the engine computer. To check the computer's reaction to the oxygen sensor signal, the pulse width at the fuel injectors is measured. The pulse width on-time is low, about 1 ms at idle. What types of faults should be checked for?

Review Questions

1. The Robert Bosch Corporation has been involved with fuel injection since which of the following?
   A. 1920s
   B. 1950s
   C. 1970s
   D. 1980s
2. *Technician A* says that the fuel injectors in a Bosch D-Jetronic system are electronically controlled by a computer. *Technician B* says that the fuel injectors in a Bosch K-Jetronic system spray fuel continuously when the fuel distributor provides enough fuel pressure to the injectors. Who is correct?
   A. *Technician A* only
   B. *Technician B* only
   C. Both technicians
   D. Neither technician
3. *Technician A* says that a Motronic system computer may control the fuel injection system, the spark management system, emission control systems, and other vehicle functions such as the automatic transmission. *Technician B* says that the D-Jetronic, K-Jetronic, KE-Jetronic, L-Jetronic, and LH-Jetronic systems are designed to control the fuel injection system only. Who is correct?
   A. *Technician A* only
   B. *Technician B* only
   C. Both technicians
   D. Neither technician
4. Which of the following systems uses manifold pressure as the principal input for calculating engine load, and therefore for determining how much fuel to inject?
   A. D-Jetronic
   B. L-Jetronic
   C. LH-Jetronic
   D. KE-Jetronic
5. Which of the following systems uses an air flow meter with a movable door and potentiometer assembly as the principal input for calculating engine load, and therefore for determining how much fuel to inject?
   A. D-Jetronic
   B. L-Jetronic
   C. LH-Jetronic
   D. KE-Jetronic
6. Which of the following systems uses a hot-wire mass air flow sensor as the principal input for calculating engine load, and therefore for determining how much fuel to inject?
   A. L-Jetronic
   B. LH-Jetronic
   C. D-Jetronic
   D. KE-Jetronic

7. *Technician A* says that the primary objective of the Bosch Motronic PCM during wide-open throttle operation is to keep engine emissions low. *Technician B* says that the objective of the mixture control strategy at WOT is to produce maximum engine torque without allowing detonation and physical damage. Who is correct?
   A. *Technician A* only
   B. *Technician B* only
   C. Both technicians
   D. Neither technician
8. *Technician A* says that some Bosch Motronic PCMs are programmed to reduce fuel delivery to zero during vehicle deceleration until engine speed falls below a specified RPM. *Technician B* says that a Bosch Motronic PCM is programmed to advance spark timing during vehicle deceleration until engine speed falls below a specified RPM. Who is correct?
   A. *Technician A* only
   B. *Technician B* only
   C. Both technicians
   D. Neither technician
9. *Technician A* says that a Bosch Motronic system will not enter closed loop until the engine coolant and the oxygen sensor have both reached a specified operating temperature. *Technician B* says that a Bosch Motronic system is not a feedback system and operates in open loop under all driving conditions. Who is correct?
   A. *Technician A* only
   B. *Technician B* only
   C. Both technicians
   D. Neither technician
10. What is a lambda sensor?
    A. An air flow meter or an air mass sensor
    B. An oxygen sensor (O_2S)
    C. An engine coolant temperature (ECT) sensor
    D. An intake air temperature (IAT) sensor
11. On a Bosch Motronic system, the PCM responds to the knock sensor by retarding spark timing when the knock sensor recognizes vibration that is between which of the following amounts?
    A. 100 and 200 Hz
    B. 500 and 800 Hz
    C. 2 and 4 kHz
    D. 5 and 10 kHz
12. *Technician A* says that the fuel injectors used on Bosch Motronic systems are similar to those used on General Motors and Ford vehicles because Bosch generally purchases its fuel injectors from General Motors and Ford. *Technician B* says that to prevent engine damage caused by over-revving, the Bosch Motronic PCM will disable the spark from the ignition system if engine RPM exceeds its programmed limit by more than 80 RPM. Who is correct?
    A. *Technician A* only
    B. *Technician B* only
    C. Both technicians
    D. Neither technician
13. Why is the fuel pressure regulator on the Bosch Motronic system indexed to manifold absolute pressure?
    A. So that it can keep the fuel rail pressure constant under all driving conditions.
    B. So that it can maintain a constant pressure differential between the fuel pressure in the fuel rail and the pressure in the intake manifold under all driving conditions.
    C. So that it can vary the flow rate of the fuel injectors as a method of controlling the air-fuel ratio.
    D. So that it can compensate for poor fuel delivery at wide-open throttle in the event that the TPS were to fail.

14. To improve ease of starting, a Bosch Motronic PCM slightly advances the spark timing during engine start during which of the following conditions?
    A. At higher cranking speeds
    B. At lower cranking speeds
    C. With warmer engine temperature
    D. Both A and C
15. On a Bosch fuel-injected engine, the computer may control idle speed by all of the following methods *except* which of the following?
    A. Using a stepper motor to control throttle bypass air
    B. Using a DC motor-operated rotary control valve to control throttle bypass air
    C. Adjusting ignition timing to control idle RPM
    D. Using a DC motor to control the throttle plate
16. *Technician A* says that Bosch systems may retard spark timing to control spark knock on a turbocharged application. *Technician B* says that Bosch systems may reduce boost pressure to control spark knock on a turbocharged application. Who is correct?
    A. *Technician A* only
    B. *Technician B* only
    C. Both technicians
    D. Neither technician
17. *Technician A* says that on some Bosch Motronic systems, the PCM controls transmission shift points through the use of solenoid valves that replace the traditional spool-type shift valves. *Technician B* says that on some Bosch Motronic systems, the PCM can smooth automatic transmission shifts and reduce clutch slip by momentarily retarding ignition timing, particularly during wide-open throttle operation. Who is correct?
    A. *Technician A* only
    B. *Technician B* only
    C. Both technicians
    D. Neither technician
18. Which of the following is true of the Bosch Motronic ME7 engine management system?
    A. It provides a torque management function for the traction control system.
    B. It eliminates the need for a dedicated idle speed control actuator.
    C. It eliminates the need for a dedicated cruise control servo.
    D. All of the above.
19. When the Bosch Motronic MED engine management system is operating in the stratified charge mode, how lean can it run the air-fuel ratio on vehicles built to be sold in the United States?
    A. 14.7:1
    B. 17:1 to 20:1
    C. 30:1 to 40:1
    D. 60:1
20. How does the BMW Valvetronics system control engine speed?
    A. The driver controls the throttle plate mechanically to control engine speed.
    B. The PCM controls intake valve opening to control engine speed.
    C. The PCM controls exhaust valve opening to control engine speed.
    D. The PCM controls the throttle plate electronically to control engine speed.

Chapter 16

Asian Computerized Engine Controls

OBJECTIVES

Upon completion and review of this chapter, you should be able to:

- ❑ Describe the sensors and actuators associated with Nissan's ECCS system.
- ❑ Describe the sensors and actuators associated with Toyota's TCCS system.
- ❑ Describe the sensors and actuators associated with Honda's PGM-FI system.
- ❑ Explain the differences between the Toyota VVT-i system and the Honda VTEC and VTEC-E systems.
- ❑ Understand the basic diagnostic concepts associated with OBD II diagnosis of these systems.

KEY TERMS

Alpha
EVAP System Pressure Sensor
Fuel Temperature Sensor
Hot-Film Air Mass Sensor
Twin Injection
Two-Trip Malfunction Detection
Variable Cylinder Management (VCM)

In this chapter, we concentrate on Toyota, Nissan, and Honda computer control systems. Although these are not the only Asian cars imported to North America, they are representative and common, and their control systems are similar to those of other Asian manufacturers. Because of the widespread use of shared components by Japanese manufacturers, an understanding of one company's systems often provides an understanding of the systems used by others. Furthermore, while there are Asian vehicles manufactured in Korea, these control systems are basically Japanese, and the overwhelming number of such imports come unmodified (or simplified) from Japanese cars, so this is where our concentration will be. In each case, we consider only late-model vehicles, not the earlier and simpler systems. Many earlier vehicles used almost off-the-shelf Bosch controls or those built under Bosch license.

OBD II standardization of these systems has also minimized the differences among them as far as the technician is concerned: How the on-board computer reads its inputs, processes information, and controls outputs is essentially the same among different makes of vehicles. How a sensor or actuator works and how it should be pinpoint tested by a technician are also similar from one make to another. For example, the technician's ability to perform a sweep test on a throttle position sensor to check for glitches applies equally to all makes of vehicles. What is most often different from one make to another is the diagnostic program that has been programmed into a given PCM by the engineers. Prior to OBD II standards, the technician had to rely on the service manual for specific instructions on how to trigger the diagnostic procedures on a given vehicle. Under OBD II standards, the technician simply connects an OBD II–compliant scan tool to the diagnostic link connector (DLC) and then selects the appropriate function from the scan tool's menu.

NISSAN: ELECTRONIC CONCENTRATED CONTROL SYSTEM (ECCS)

Typical late-model Nissan ECCS applications use a **hot-film air mass sensor** as the principal input to the powertrain control module (PCM) to determine the fuel pulse width and spark advance. The PCM, of course, controls the fuel mixture and ignition based on information stored in its memory, as well as on inputs from all its other sensors. For a general overview of the entire system, see Figure 16–1.

Nissan uses both sequential injection and group injection, or simultaneous injection. The sequential mode injects fuel once every complete engine cycle (two crankshaft revolutions) during the corresponding cylinder's intake stroke. The group injection mode sprays all the injectors at once, twice per complete engine cycle. The group injection mode is used when the engine is starting, when the engine is in fail-safe mode, or when the camshaft position (CMP) sensor signal is lost and the engine is running on the crankshaft position (CKP) sensor signal alone.

The Nissan PCM does not keep ignition timing consistently at the edge of spark knock, using the knock sensor signal as a routine control input as with many other systems. Instead, the spark advance map is scaled to avoid knock under all but the extremes of load, engine temperature, and low fuel octane equivalency.

Besides normal conditions, the PCM also modifies ignition timing under special circumstances:

- During startup
- During warmup
- During idle
- When the engine is overheated
- During acceleration

The Nissan computer control system employs special fuel enrichment measures under many of the same circumstances:

- During warmup
- During cranking and starting
- During acceleration
- When the engine is overheated
- During high load and high vehicle speed conditions

The system employs special lean fuel mixtures (including complete fuel shutoff in some cases) under other circumstances:

- During deceleration
- During low-load, high-vehicle-speed conditions

Deceleration fuel cutoff will stop as engine speed falls below 2,200 RPM with no load. Fuel cutoff does not begin unless the unloaded engine is above approximately 2,700 RPM.

Mixture Ratio Self-Learning

The feedback control system employs a self-learning ability, similar to systems used on domestic cars. Like them, the Nissan system monitors the success of its pulse-width modification measures to keep the air-fuel mixture stoichiometric using input from the oxygen sensor. Known by most manufacturers as *fuel trim,* Nissan refers to this adaptive ability as **alpha**. The PCM can modify its working copy of the hardwired mixture-to-pulse-width map depending on the results it sees from the oxygen sensor. It uses both a short-term and a long-term *fuel trim compensation factor*, similar

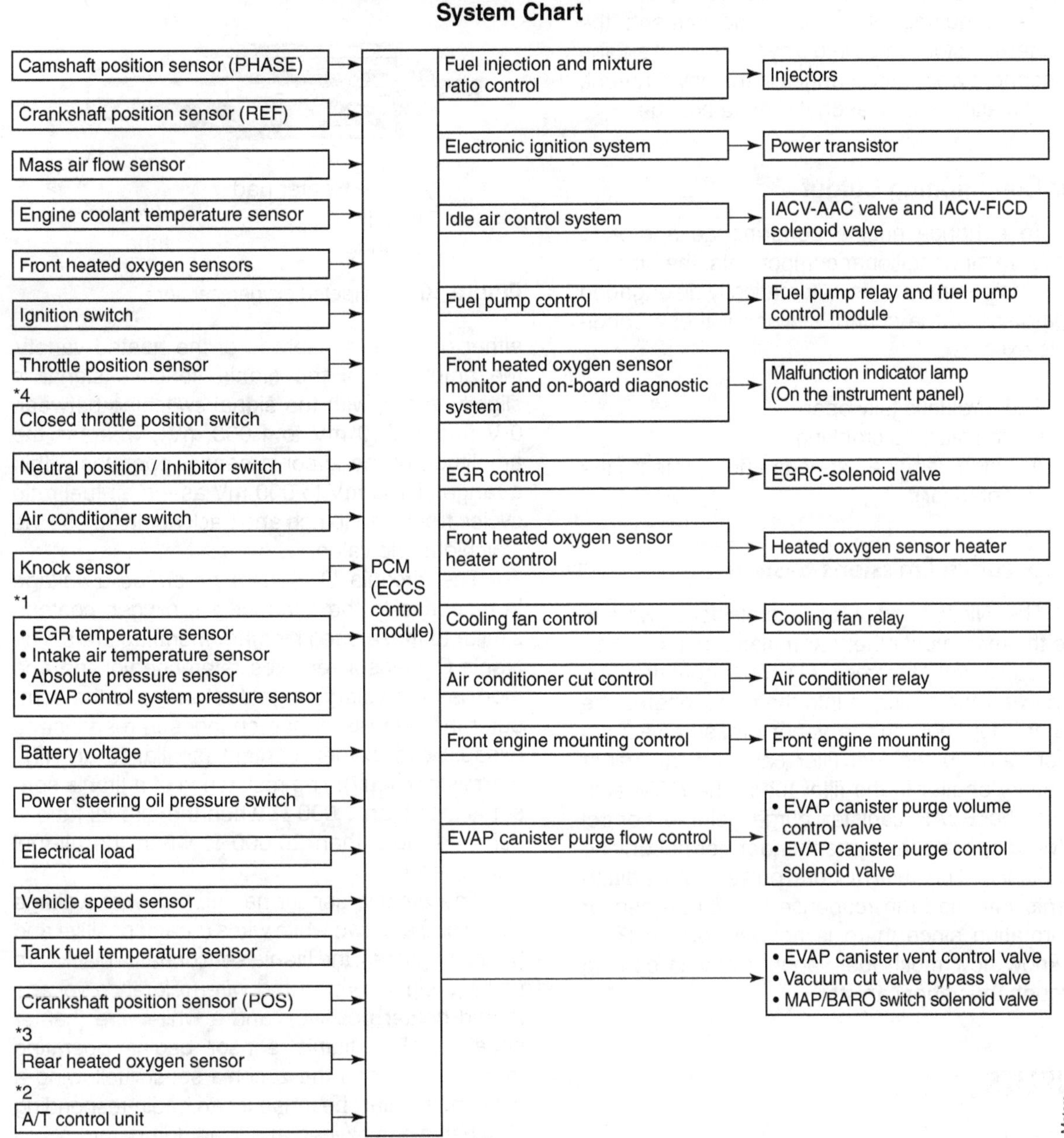

Figure 16–1 Nissan system overview.

to GM's fuel trim (integrator and block learn). The short-term factor corrects for temporary circumstances requiring fuel mixture adjustment; the long-term factor comes into play for gradual engine component wear and changes in the environment, such as altitude or seasonal climate change.

Air Conditioning Cutout

To enhance engine performance and/or to preserve air conditioner components, the air conditioning compressor is automatically disengaged regardless of the control settings if these conditions exist:

- The throttle is fully open.
- During start-up cranking.
- At high engine speeds (to protect the compressor).

Evaporative Emissions System

The Nissan evaporative emissions system, like those of most other car makers, uses a charcoal canister that stores fuel vapors from the tank to prevent their escape into the atmosphere. The system depends on the vapor pressure/vacuum relief valve in the fuel filler cap and the vapor recovery shutter in the filler tube. The PCM activates the EVAP canister purge volume control valve to operate the system under normal driving conditions. The fuel mixture pulse-width adjustments are made in response to oxygen sensor information since there is no way for the PCM to know how much fuel vapor, if any, is coming through the vapor canister.

INPUTS

Heated Oxygen Sensors

Nissans use either a ceramic zirconia oxygen sensor, as shown in Figure 16–2, or a ceramic titania oxygen sensor, with the latter being used on early to mid-1990s model year vehicles. While either of these sensors is of the heated variety, the amplitude of the titania sensor's signal is slightly larger, with the signal switching between 0 V and 1 V (0 mV to 1,000 mV), whereas the amplitude of the zirconia sensor operates within a range of 100 mV to 900 mV as the air-fuel ratio cycles from lean to rich and back again across the stoichiometric value.

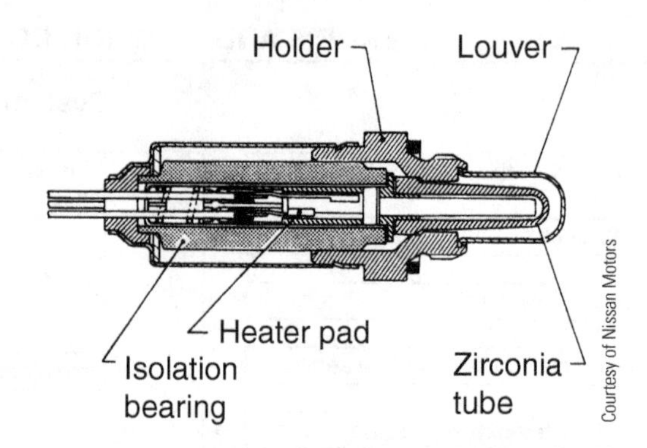

Figure 16–2 Heated oxygen sensor.

The zirconia O_2 sensor *produces* a voltage in response to the exhaust gas oxygen content, similar to those used by other manufacturers. The titania O_2 sensor receives reference voltage and *modifies* the voltage signal returning to the computer through resistance changes in response to exhaust gas oxygen content (similar to an NTC thermistor circuit). The resistance of a titania sensor is less than 1,000 Ω when the air-fuel ratio is rich and more than 20,000 Ω when the air-fuel ratio is lean.

The zirconia sensor has a black wire (voltage signal out) and two white wires (heater positive and heater negative); the titania sensor has a black wire (voltage signal out), a red wire (reference voltage in and heater positive), and a white wire (heater negative). The titania sensor begins operating more quickly than the zirconia sensor following a cold engine start because it can begin responding to exhaust gas oxygen at a lower temperature.

The PCM gets the sensor's output signal and constantly readjusts the injector pulse width to correct the air-fuel mixture and keep the signal cycling. On V-type engines, two front oxygen sensors are used, one on each bank; the PCM

Diagnostic & Service Tip

Often the problem with an oxygen sensor is not that it has failed completely but that it has begun to respond too slowly to changes in the residual exhaust oxygen. Once this happens, the PCM will not be able to properly cycle the injector pulse width to keep the mixture stoichiometric. There is no specific number of cycles per unit of time that an oxygen sensor should register, but the newer systems should work more quickly than the older ones. If in doubt, monitor the signal cycling of a "known good" sensor and compare it to one you suspect.

controls the mixture delivered to each bank independently of that delivered to the other. The oxygen sensors are heated to bring the system to closed loop as quickly as possible and to prevent the system from falling out of closed loop should the vehicle idle for a long time.

Nissan is also using air-fuel ratio sensors on some of its models. An air-fuel ratio sensor is a wideband oxygen sensor. These sensors will typically have five wires entering the sensor body. The sensor consists of two zirconium dioxide oxygen sensors in one assembly and is discussed in greater depth in Chapter 3.

Diagnostic & Service Tip

Oxygen sensors, particularly zirconia-based oxygen sensors, are relatively brittle and delicate components because of the ceramic element. If an oxygen sensor is dropped much more than a foot to a hard surface such as a concrete shop floor, it will probably sustain an invisible internal crack and will have to be replaced. Oxygen sensors are much more delicate than spark plugs and are also more expensive, so they should always be handled carefully.

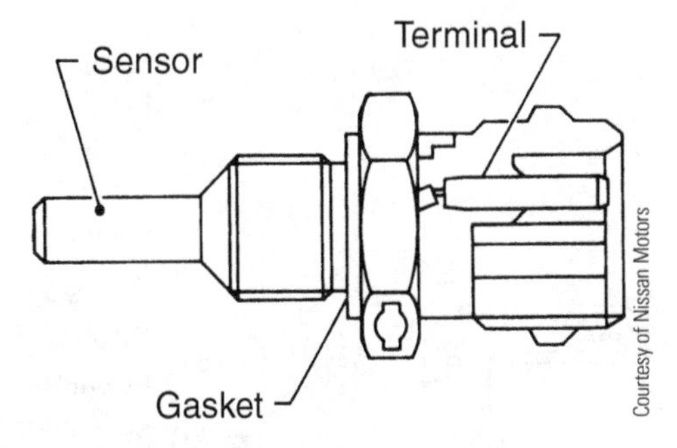

Figure 16–3 ECT sensor.

Engine Coolant Temperature Sensor

Like other sensors, the ECT sensor modifies a reference signal from the PCM to correspond to the monitored parameter (Figure 16–3). The resistance of the thermistor in the sensor decreases as the engine temperature increases. Information from the ECT sensor plays a significant role in engine management because a richer mixture is needed for cooler temperatures to ensure there is enough vaporized fuel to burn. At operating temperatures, the system switches to a closed-loop control system, using the signal from the oxygen sensor. The ECT sensor, however, is critical to the determination of when the system goes into closed loop.

Intake Air Temperature Sensor

The intake air temperature (IAT) sensor, shown in Figure 16–4, mounts in the air cleaner

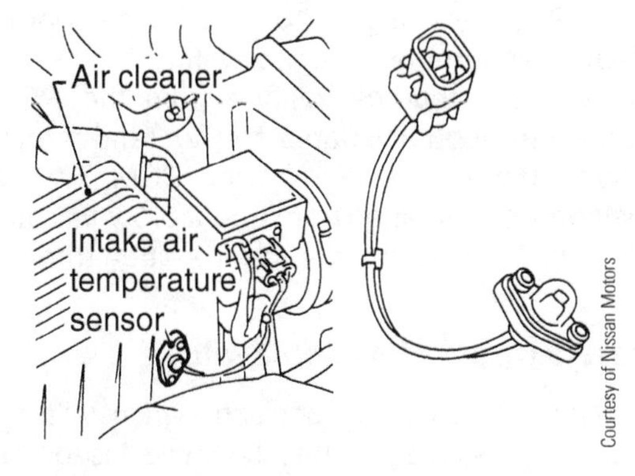

Figure 16–4 IAT sensor.

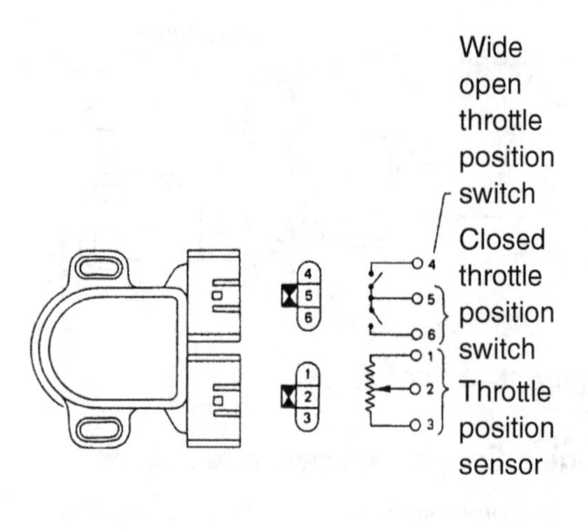

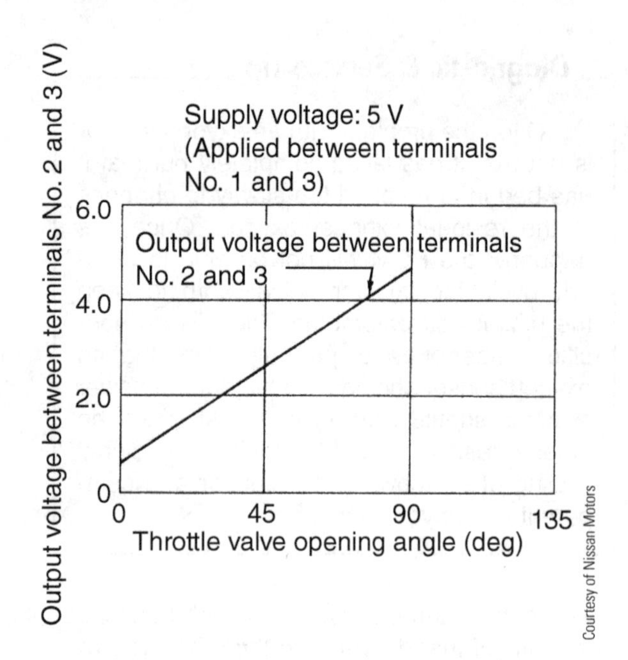

Figure 16–5 Throttle position sensor.

housing and transmits a voltage signal corresponding to the IAT. The sensor uses a thermistor that is responsive to changes in temperature. As temperature rises, the resistance of the thermistor goes down.

Throttle Position Sensor

The PCM learns about accelerator pedal movement and position from the throttle position (TP) sensor (Figure 16–5). Like many other linear measurement sensors, the TP sensor receives a reference voltage from the PCM, routes it through a variable resistor, and returns a corresponding voltage to the PCM. Throttle position signals range from about 0.5 V to about 4.6 V from closed throttle to wide-open throttle.

Closed-Throttle Position Switch

The closed-throttle position switch tells the PCM when the driver has removed his or her foot from the accelerator pedal (Figure 16–5). The PCM then knows to control the idle speed based on information from its other sensors. The closed-throttle switch is either on or off.

Manifold Absolute Pressure Sensor

The Nissan manifold absolute pressure (MAP) sensor, shown in Figure 16–6, connects to the map/baro switch solenoid through a hose. It detects both ambient pressure and intake manifold pressure. As the pressure rises, the voltage signal rises.

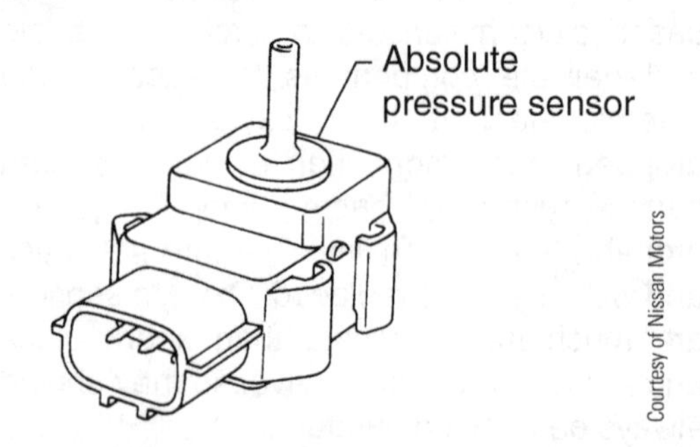

Figure 16–6 Manifold absolute pressure (MAP) sensor.

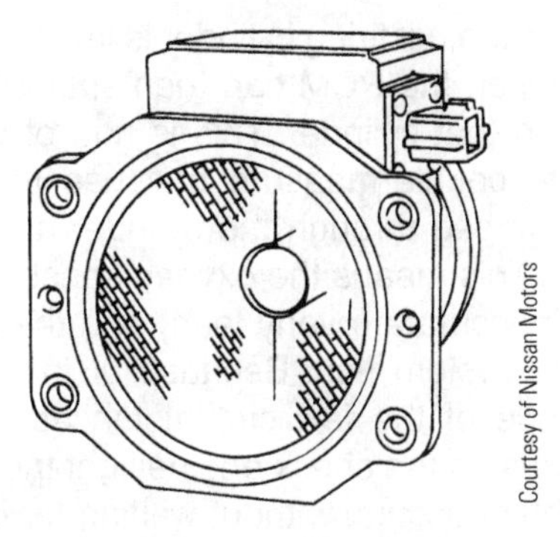

Figure 16–7 Mass airflow (MAF) sensor.

Mass Air Flow Sensor

The hot-film air mass sensor is located in the intake air duct between the air filter and the throttle (Figure 16–7). It should indicate a signal voltage of below 1 V with the ignition on and the engine off and should show between 1.0 and 1.7 V at idle. With the engine running at higher speeds and under load, the signal voltage can rise to as much as 4.0 V. Because of the sensitivity of the sensor to disturbances in the airflow, a frequent check should be done to make sure no dust or foreign material is on or near the sensor.

Camshaft Position Sensor

The CMP sensor is in the front engine cover facing the camshaft sprocket (Figure 16–8). Information from the CMP sensor distinguishes which cylinder is at the firing position. The CMP sensor is a coil-and-magnet pulse generator: It consists of a permanent magnet wound in an electrical coil. When the gap between the camshaft sprocket and the magnet changes, the magnetic field changes, generating an AC voltage pulse in the sensor output line. The PCM uses the CMP signal to sequence the fuel injectors. As the engine is initially started and prior to the PCM receiving a CMP signal, all fuel injectors are simultaneously pulsed. Once the PCM receives its first CMP signal it begins pulsing the injectors sequentially.

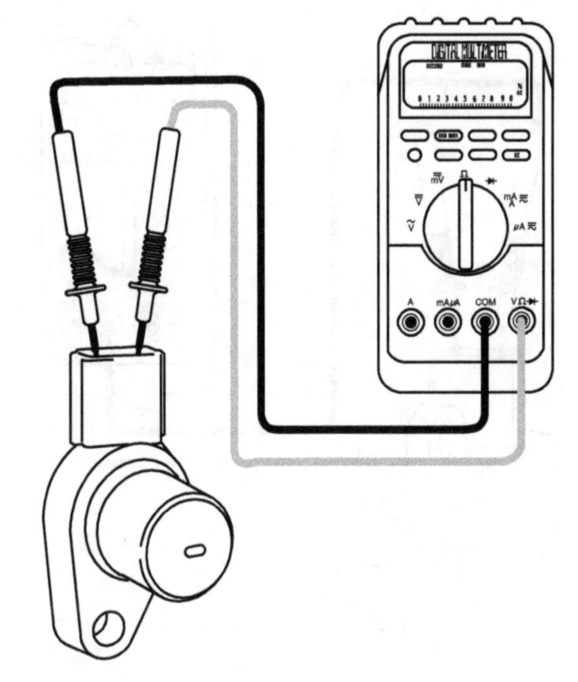

Figure 16–8 CMP sensor.

CMP sensors can be checked for shorts and opens with an ohmmeter, and for residual magnetism at the tip of the sensor with a strip of ferrous metal or a screwdriver tip.

Newer Nissan vehicles now use magnetic resistance element (MRE) sensors to monitor camshaft position. These are electronic sensors that monitor a rotor with alternating north and south magnetic poles embedded in it. (See the discussion later in this chapter on Toyota MRE vehicle speed sensors.)

Crankshaft Position Sensor

The CKP sensor is on the upper oil pan facing the flywheel (Figure 16–9). It informs the PCM when each pair of cylinders is at top dead center (TDC). This sensor is a coil-and-magnet inductive signal generator producing a signal that varies in voltage with engine speed. Resistance through its coil should range between 470 and 570 Ω at room temperature.

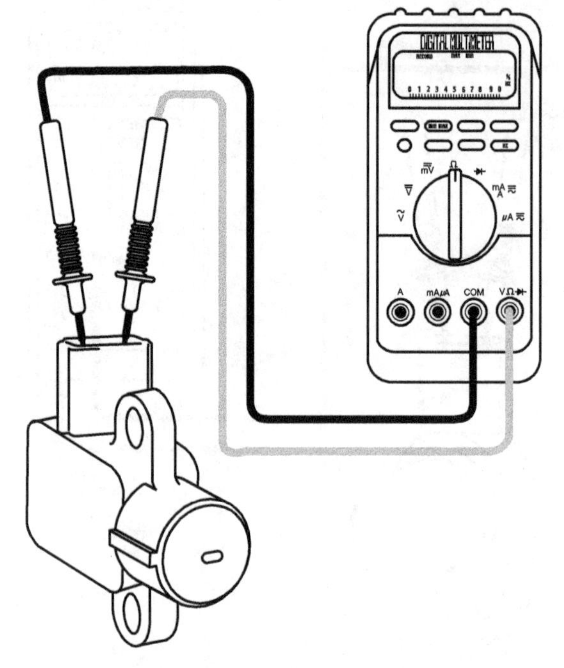

Figure 16–9 CKP sensor.

The CKP sensor plays a familiar role in spark advance timing, fuel-injection timing, and the newest vehicle emissions monitoring. It is located on the flywheel end of the oil pan and signals one-degree changes in the flywheel position to the PCM. In later cars, the CKP sensor is the principal source of information about cylinder misfire. When a cylinder misfires, whether from spark, fuel, or mechanical problems, the engine slows down slightly at the point when that cylinder's power stroke should occur. At idle, this slowing is perceptible as rough running, but the CKP sensor can report the miss to the PCM even at higher speeds.

The emissions-related reason for tracking cylinder miss is that if the cylinder does not fire, it does not consume the fuel that entered with the intake stroke. The first consequence would be a considerable overheating of the catalytic converter from all the increased fuel and oxygen. The second consequence could be a failed catalytic converter, and the third would be the dumping of raw hydrocarbons into the air after the catalytic converter could no longer oxidize them.

Once the misfiring cylinder is identified by the CKP sensor, the PCM can then shut off the fuel injector to that cylinder. This is not, of course, a complete countermeasure because the oxygen is still pumped through the cylinder and into the exhaust. This means the oxygen sensor's reports will be inaccurate (overly lean) and the PCM will drive the system rich. Because of this problem, this is one of the few circumstances when the PCM turns on the check engine light the first time the problem occurs, without waiting for it to recur on a subsequent trip.

Newer Nissan vehicles now use MRE sensors to monitor crankshaft position and RPM.

Ignition Switch

When the ignition switch is turned on, the PCM becomes active and sends reference voltage to the sensors that require it. It also begins monitoring the return signals and prepares to energize the appropriate actuators.

Park/Neutral Switch

The park/neutral switch allows the starter circuit to complete only with the transmission in one of these two gears. It also sends a signal to the PCM when the vehicle is in gear. This signal is used for idle control, spark timing, and sometimes fuel mixture.

Air Conditioner Switch

When the air conditioning is switched on, the PCM receives a signal that it uses to determine proper idle speed.

Knock Sensor

The Nissan knock sensor is a piezoelectric crystal type; it sends a voltage signal to the PCM in the event it sustains vibrations at a frequency characteristic of detonation. If the PCM receives such a signal, it retards ignition timing until the knock disappears. As already mentioned, the

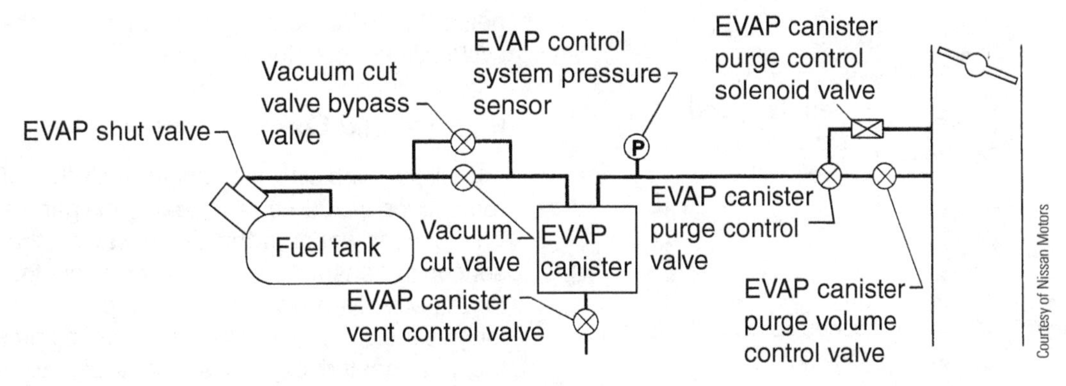

Figure 16–10 Evaporative emissions system diagram.

Nissan spark advance map is designed to prevent knock except under unusual fuel quality, load, or temperature conditions.

The knock sensor can be electrically inspected for shorts or opens. Nissan literature calls for an ohmmeter capable of measuring up to 10 MΩ. Like oxygen sensors, knock sensors are brittle and should not be reused if they have been dropped on a hard surface.

EVAP System Pressure Sensor

To monitor the effectiveness of the EVAP control system, the PCM receives a signal from the **EVAP system pressure sensor** (Figure 16–10). While this does not directly affect engine control functions, it does play a role in on-board emissions diagnosis. This sensor will indicate whether there is a leak in the vapor canister system.

This sensor cannot distinguish a leak in the system from a loose fuel filler cap, which will trigger the same trouble code and light the malfunction indicator lamp (MIL). Check for a properly functioning fuel filler cap first, including a pressure test of the cap, whenever this sensor indicates a problem.

Battery Voltage/Electrical Load

The PCM constantly monitors the voltage received from the battery and charging system and will illuminate the check engine light if it falls below proper voltages. This signal is used to modify the injector pulse width and the ignition coil dwell if system voltage falls below the standard charging system voltage.

Power Steering Oil Pressure Switch

The power steering oil pressure switch is on the power steering high-pressure tube (Figure 16–11). When a power steering load occurs, it triggers this switch, signaling the PCM to increase the idle speed setting. This switch conducts battery voltage and can be checked for continuity with an ohmmeter.

Vehicle Speed Sensor

The vehicle speed sensor is in the transaxle on the output shaft (Figure 16–12). Its pulse

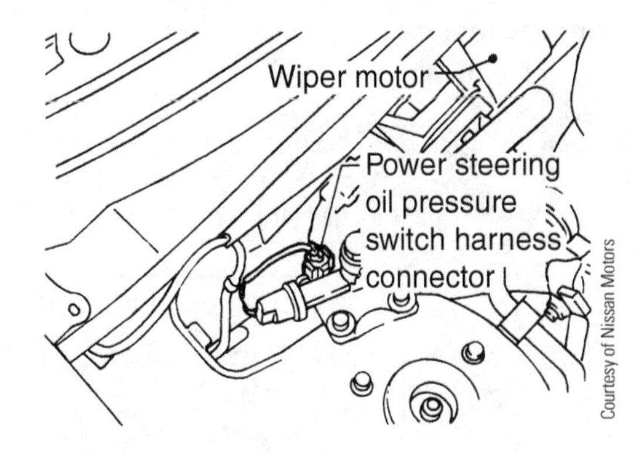

Figure 16–11 Power steering pressure switch.

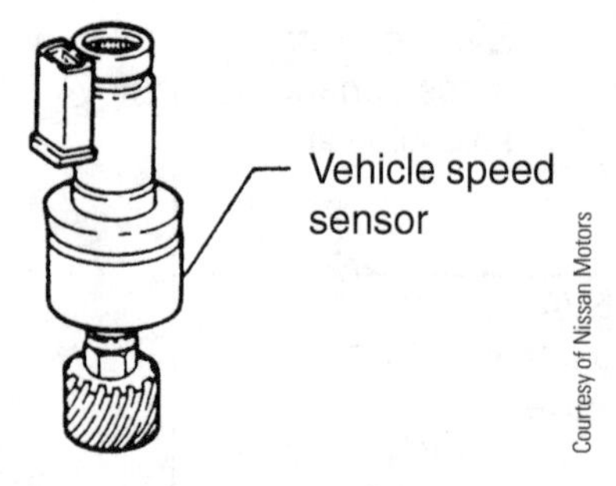

Figure 16–12 Vehicle speed sensor.

generator sends a signal corresponding to vehicle speed to the speedometer, which conveys the signal to the PCM. This information is used for both fuel pulse width and ignition timing calculations.

Fuel Temperature Sensor

The system uses a thermistor-type **fuel temperature sensor** to monitor the temperature of the fuel in the tank. The sensor is accessible under the rear seat (Figure 16–13). The fuel temperature sensor receives a reference voltage from the PCM and returns to it a voltage signal corresponding to the temperature of the fuel, from 3.5 V when the fuel is at 68°F to 2.2 V when the fuel reaches 122°F. Information from this sensor indicates changes in fuel viscosity and its inclination to vapor lock. Nissan uses this sensor in conjunction with the EVAP pressure sensor. In addition to fuel temperature information, this sensor also helps the PCM determine when to run the fuel system leak detection software.

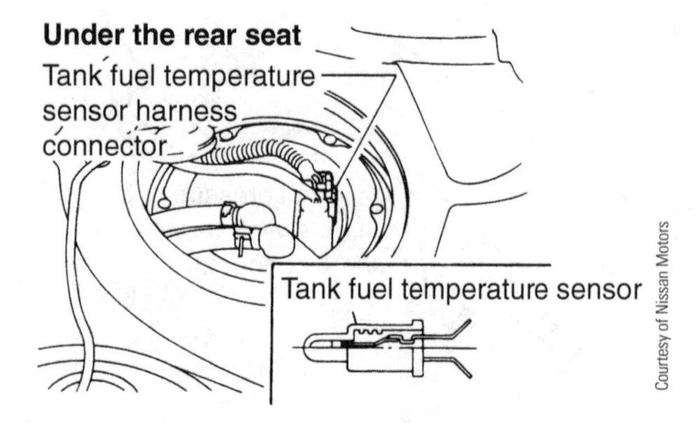

Figure 16–13 Tank fuel temperature sensor.

Rear Heated Oxygen Sensor

Nissan vehicles that conform to the OBD II regulations include a rear heated oxygen sensor. This sensor works in the same way as the front sensor or sensors but is used primarily to monitor the effectiveness of the catalytic converter. As long as the catalytic converter is working properly, the rear heated oxygen sensor's signal should remain almost flat or only slowly cycling over its output range. The PCM tracks the signal output and will set a code should the signal mirror that of the front sensor. If the front sensor fails, the PCM will use the rear sensor to control air-fuel mixture as closely as possible.

Automatic Transmission Diagnostic Communications Line

A communications link between the automatic transmission control unit and the engine control PCM conveys information related to transmission malfunctions. Any such information erased from the transmission unit must be separately erased from the engine control PCM.

OUTPUTS

Fuel Injectors

The multiport fuel injectors are the Bosch type used on many vehicles. Activated by a constant power source and grounded through the PCM, each injector opens by lifting its pintle off the seat, allowing fuel to spray through the nozzle for a fixed amount of time, which is calculated by the PCM on the basis of all the system inputs.

IACV-AAC Valve and IACV-FICD Solenoid Valve

The idle air control valve–auxiliary air control (IACV-AAC) valve controls throttle bypass

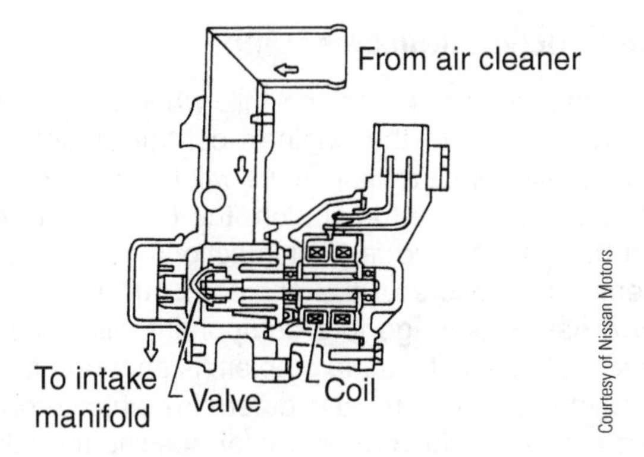

Figure 16–14 Idle air control valve.

air, which determines the engine's idle speed (Figure 16–14). The PCM activates this component according to inputs from the ECT sensor, electric load sensor, air-conditioning switch, and power steering pressure switch. Once removed from the throttle body, the operation of this valve can be checked by cycling the ignition switch on and off to observe whether the valve shaft moves smoothly forward and backward in accordance with the ignition switch position. Removal and reinstallation of the valve should only be done with the ignition switch off.

The idle air control valve/fast idle control device (IACV-FICD) solenoid combines with the IACV-AAC valve to constitute the idle air adjusting unit. The IACV-FICD solenoid is strictly an on/off solenoid moving a plunger in the air passage (Figure 16–15). It can be checked separately from the vehicle for smooth function.

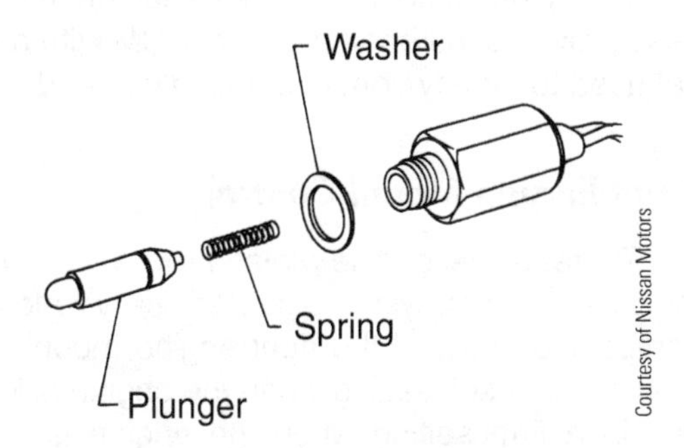

Figure 16–15 IACD-FICD solenoid valve.

Fuel Pump Relay and Fuel Pump Control Module

The fuel pump relay is activated for 1 second after the ignition switch is first turned on to pressurize the fuel system before starting the engine (Figure 16–16). The PCM will leave the relay energized as long as it receives input from the CKP sensor at the flywheel. The fuel pump control module adjusts voltage supplied to the fuel pump to control the amount of fuel flow (Figure 16–17). It reduces the voltage to approximately 9.5 V except in the following conditions:

- During engine start-up cranking
- When the engine coolant is below 45°F
- Within 30 seconds after startup with a warm engine
- Under high load and high speed conditions

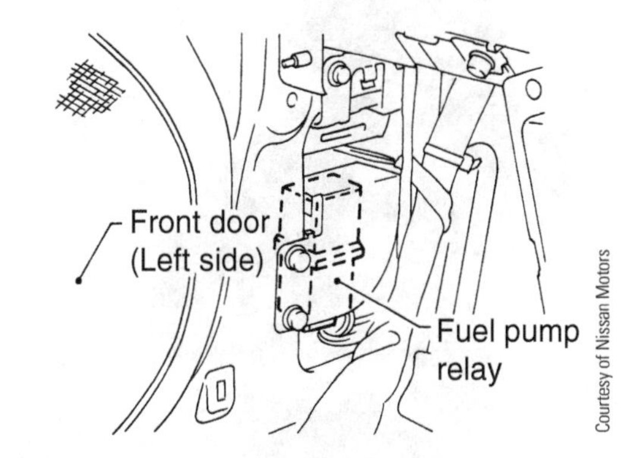

Figure 16–16 Fuel pump relay.

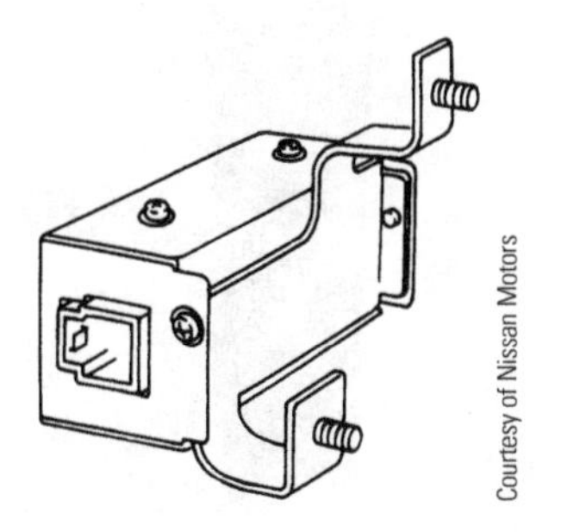

Figure 16–17 Fuel pump control module.

Nissan Electronic Throttle Control (ETC)

Nissan's electronic throttle control (ETC) system is similar to those made by other manufacturers. It uses potentiometers for both the throttle position sensors and the accelerator pedal position sensors. The ETC system eliminates the need for a dedicated idle air control actuator and a dedicated cruise control servo.

Ignition Coil Power Transistor

Late-model Nissans use one coil per cylinder in a direct ignition system (Figure 16–18). If a bad coil is suspected, moving the coil from one cylinder to another to see whether the cylinder miss moves with it is the accepted technique. No internal repairs are possible; replacement is the only service.

Figure 16–18 Nissan ignition coil.

Malfunction Indicator Lamp

The PCM turns on the MIL when it detects a fault in one of the engine's control or actuator circuits. Most malfunctions will have to occur in two successive trips before the PCM will turn on the MIL, but certain malfunctions that represent either failure of the engine to run properly or a risk of damage to a component will trigger the MIL the first time the problem is encountered. See the service manual to determine which problems fall into which category for specific models and years.

Fuel Pump Pressure

Fuel pump pressure with zero inches of vacuum at the MAP sensor should be approximately 43 PSI. At idle, the fuel pressure should read approximately 34 PSI.

Cooling Fan Relay

The engine control PCM operates the radiator cooling fans through a relay depending on the engine coolant temperature, vehicle speed, and whether the air conditioner is turned on. There are two fan speeds available to the PCM. There are also differences in the fan speed maps for cars built for the California market, though the rest of the cooling system is the same.

Air Conditioner Relay

When the air conditioner is turned on, the A/C relay provides the engine control PCM with a signal used to modify control of the idle speed.

Front Engine Mount Control

To keep the engine vibration to a minimum, the engine control system also electrically varies the dampening effect of the front engine mount. The mount is in a soft setting when the engine is idling and in a firm setting when the engine is doing work, moving the vehicle under load.

EVAP System Controls

The late-model Nissans use a more complex evaporative control system, storing fuel vapors in a charcoal canister as has been done for many years, but with a more complex set of controls and monitoring measures. Refer to Figure 16–10 for a complete system diagram. The basic actuator is the EVAP canister purge control valve (Figure 16–19). When the PCM sends an Off signal to the control solenoid valve, the vacuum between the intake manifold and the EVAP canister purge control valve is shut off. The EVAP canister purge volume control valve controls the passage of vapors from the EVAP canister purge control valve to the intake manifold (Figure 16–20). The system includes, as was described under the "Inputs" section in this chapter, an EVAP control system pressure sensor, shown in Figure 16–10, intended to detect leaks in the system or a faulty, loose, or missing fuel filler cap. While the EVAP control system differs somewhat from one year to the next and from one model to another, fundamentally they all work the same way, storing fuel vapors in the charcoal canister until the engine, running in closed loop, can vent the canister into the intake manifold and burn the stored vapors. Corrections to the fuel mixture, as required by the variable concentration of fuel vapors entering the engine, are made on the basis of input from the oxygen sensor feedback control system.

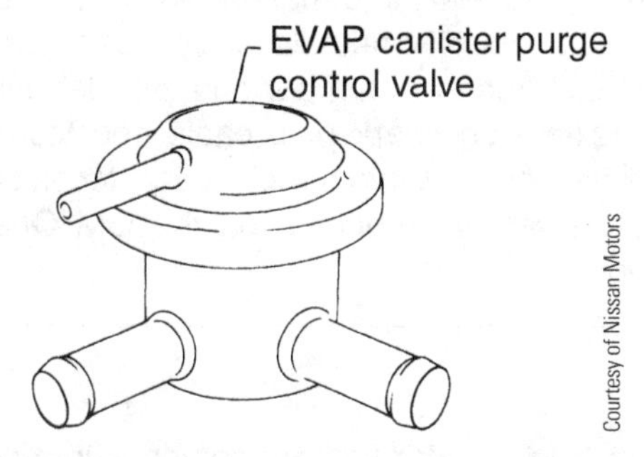

Figure 16–19 EVAP canister purge control valve.

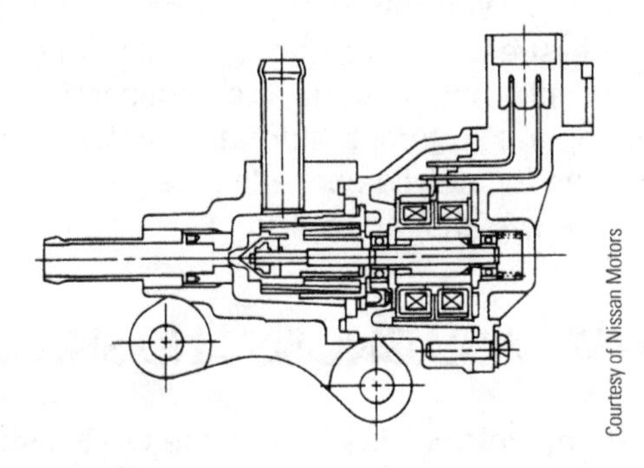

Figure 16–20 EVAP canister purge volume control valve.

Nissan Supercharger

The Nissan Frontier pickup truck 3.3L SOHC V6 engine introduced a Roots-type three-lobe supercharger, built by Eaton, for the 2001 model year. It mounts to the stock 3.3L intake manifold. It is limited to 6 pounds of boost (Figure 16–21).

The engine produces 210 horsepower with 245 ft.-lb of torque. A bypass valve diverts the air back into the supercharger during normal driving. The fuel pump flow rate has been increased by 80 percent to meet wide-open throttle demands. The injector flow rate also has been adjusted accordingly. This increased performance has been achieved with only a 10 percent reduction in fuel economy (Figure 16–22).

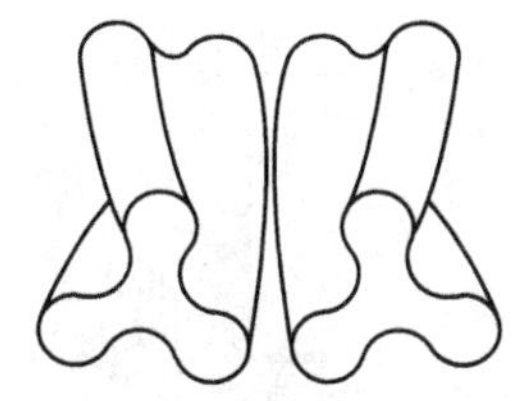

Figure 16–21 Nissan three-lobe helical cut rotors.

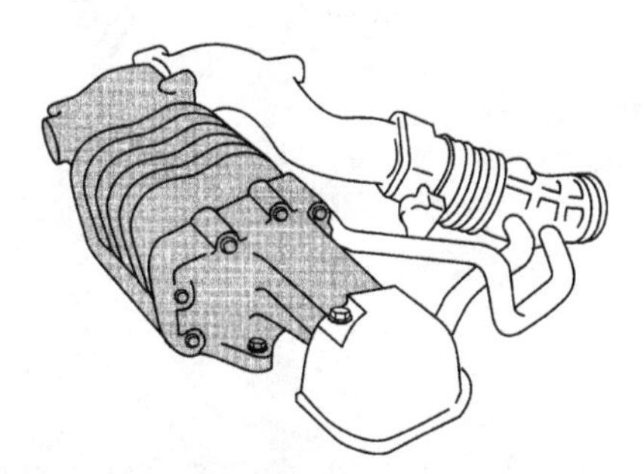

Figure 16–22 Supercharger system.

Nissan Intake Manifold Swirl Control System

The 2001 Nissan Sentra intake manifold has a door in each runner. All doors are mounted on a single shaft. The doors have two positions: rest and fully open (Figure 16–23).

When the door in the runner of the intake manifold is in the rest position, the intake manifold passageway is partially restricted, causing the moving airstream to swirl and increasing air velocity. The increased air velocity improves injected fuel vaporization. The doors are at rest at idle and during low engine speed. At higher engine speed, a single vacuum diaphragm opens the swirl control valves. The PCM turns a swirl control solenoid on or off to apply vacuum to the diaphragm. The restriction in the manifold runner is eliminated when the doors are wide open, improving intake efficiency (Figure 16–24).

Nissan Air-Cleaning Radiator

The California model of the Nissan Sentra is equipped with a new air-cleaning radiator to meet that state's super low emission vehicle (SULEV)

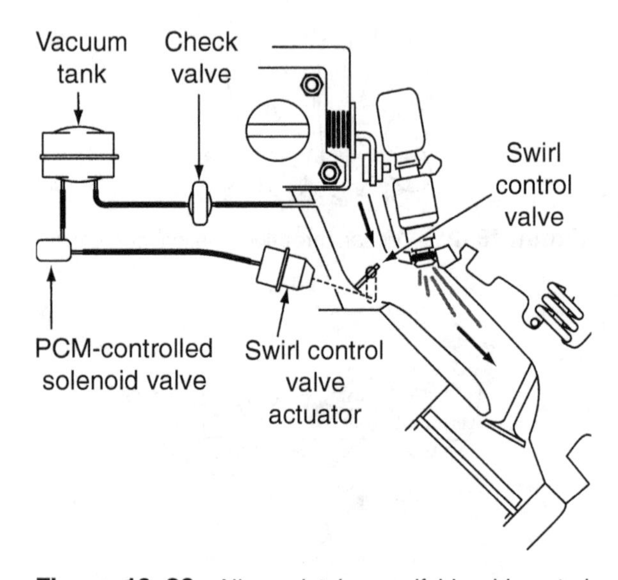

Figure 16–23 Nissan intake manifold swirl control system.

Swirl control operating conditions		
Throttle valve position	Engine RPM	Control solenoid
Closed	Below 3200 RPM	On
Open	Above 3200 RPM	Off
Note: Swirl control valves are kept open, at all engine speeds, below 50°F (10°C) and above 131°F (55°C)		

Figure 16–24 Nissan swirl control system operating conditions.

Diagnostic & Service Tip

With one coil per cylinder buried in the valve cover on late-model vehicles, it is somewhat more difficult to get spark advance and engine speed information without special Nissan tools. Nonetheless, you can remove the coil from the number one cylinder, attach an extension spark plug cable and test with conventional equipment. Many technicians use a similar technique on the GM Quad-4 engine.

rating. The radiator fins are coated with a special material that converts ozone molecules (O_3) into oxygen (O_2). When the driver drives the vehicle down the street, the cooling fan pulls dirty, polluted air into and through the radiator fins. At normal operating temperature, the radiator coating acts as a catalytic converter, cleaning the air before it leaves the radiator.

✔ SYSTEM DIAGNOSIS AND SERVICE

Late-model vehicles include the OBD II standardized diagnostic system. In addition, they have retained the earlier Nissan self-diagnostics. This older system uses a screwdriver-actuated potentiometer in the PCM itself (ordinarily under the front passenger seat, but behind the instrument panel near the glove box on newer vehicles), displaying trouble codes either by flashes of the MIL or, on the earliest systems, by flashing lights on the PCM itself. The older systems, of course, are simpler, rely on fewer inputs and outputs, and provide less information for diagnosis. Check the shop manual for the specific information corresponding to the model and year of the vehicle you are working on.

Nissans use a **two-trip malfunction detection** logic; that is, for most component circuit failures the PCM stores a fault on the first trip in which it is detected and turns on the MIL if it recurs on the next trip (a trip is defined as an engine start, followed by enough driving to bring the engine to operating feedback control system condition). There are several different ways to erase stored faults in Nissan systems, including using the PCM's potentiometer to switch from Test Mode II to Mode I or using the scan tool, so read the manual for the specific car before you perform these tests. Disconnecting the battery erases the codes (as well as the self-learning data), but it can take as long as 24 hours of disconnection before the energy stored in the capacitors discharges fully.

TOYOTA COMPUTER-CONTROLLED SYSTEM (TCCS)

As an example of a TCCS, we look at the 1MZ-FE V6 engine in a Toyota Camry. Other systems are similar, with appropriate modifications for engine type and vehicle style.

The Toyota system uses an MAF sensor to determine how much fuel to inject at a given engine condition. The six fuel injectors are sequentially operated in the firing order of the engine. Spark is delivered through individual coils at each spark plug, similar to the Nissan system described earlier in this chapter.

INPUTS

Heated Oxygen Sensors

Toyota uses front and rear oxygen sensors that may be standard zirconia sensors or air-fuel ratio sensors, depending upon year and model (Figure 16–25). On V6 engines, there are two front oxygen sensors, one for each bank. The front oxygen sensors generate the signals the computer uses for closed-loop fuel mixture control, and the rear sensor is used to determine the effectiveness of the catalytic converter. However, beginning on some models as early as 1988, the Toyota PCM was programmed to use the rear O_2 sensor for rear fuel control (as described in Chapter 4). Toyota, along with Saab, was one of the first manufacturers to begin using the rear O_2 sensor in this manner. The oxygen sensors all include a heater circuit to bring them up to operating temperature as quickly as possible in order to run the engine in closed loop.

Modern Toyotas use air-fuel ratio sensors on their models. An air-fuel ratio sensor is a wideband

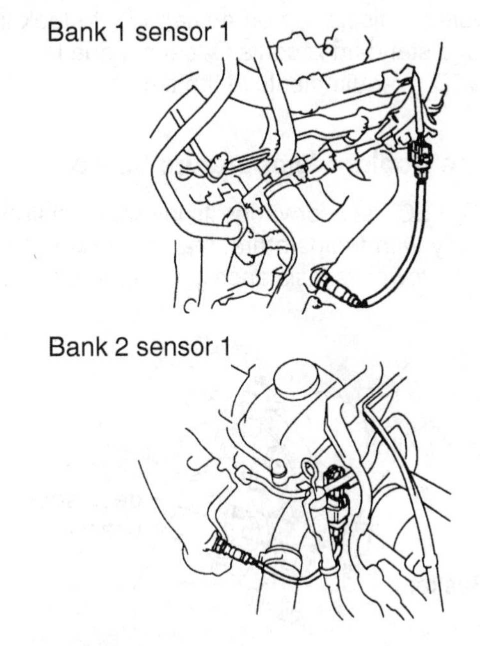

Figure 16–25 Heated oxygen sensors.

oxygen sensor. Toyota began using air-fuel ratio sensors on some models as early as 1999. While some early Toyota air-fuel ratio sensors may have only four wires, modern ones will typically have five wires entering the sensor body. The wideband air-fuel ratio sensor is discussed in greater detail in Chapter 3.

On Toyota applications, the control reference voltage will be about 3.3 V and will vary ever so slightly as the current varies during closed-loop operation. If the scan tool displays a voltage value for the sensor, anything less than the 3.3 V control reference voltage indicates a rich air-fuel ratio and anything greater than the 3.3 V control reference voltage indicates a lean air-fuel ratio. But the difference in voltage is actually created by a current flow that can be either positive (indicating a lean air-fuel ratio) or negative (indicating a rich air-fuel ratio). In reality, it is the small changes in current flow that allow the PCM to know the air-fuel ratio. Because the PCM reacts so quickly to changes in air-fuel ratios, the actual variance in current flow may be slightly less than 1 mA (1/1,000 of an amp) when the system is operating in closed loop.

On some of the early Toyota applications that used an air-fuel ratio sensor, the PCM converts the voltage displaying on a scan tool to look like that of a standard zirconia O_2 sensor due to existing OBD II requirements at the time.

Engine Coolant Temperature Sensor

The ECT sensor varies its internal resistance inversely with temperature; that is, the warmer it gets, the lower its resistance goes (Figure 16–26). It threads into the water jacket near the thermostat outlet and can be checked with an ohmmeter. At normal shop temperature, it should indicate approximately 2,000 Ω. The computer uses information from the ECT sensor to calculate fuel mixture, spark advance, and ignition timing. Its information also plays a role in determining when to go into closed loop, when to engage the EGR system, and when to open the passages to the evaporative emissions charcoal canister.

Throttle Position Sensor

The TP sensor, shown in Figure 16–27, provides the computer with information about where the driver's foot has positioned the accelerator pedal. On the Toyota system, the TP sensor is adjustable using a feeler gauge between the throttle stop screw and stop lever and an ohmmeter on the terminals, as shown in the shop manual. A defective TP sensor can lead to the wrong idle speed and poor driveability at other speeds.

Mass Air Flow Sensor

The MAF sensor provides the PCM with a fluctuating analog voltage signal corresponding to the amount of air passing through the air intake system (Figure 16–28).

Knock Sensors

V6 engines use two knock sensors, one on each bank (Figure 16–29). In-line engines use one. The knock sensor, as on other systems, is

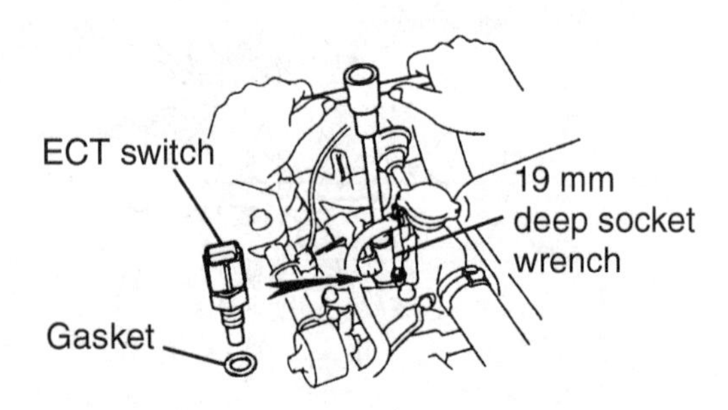

Figure 16–26 ECT sensor.

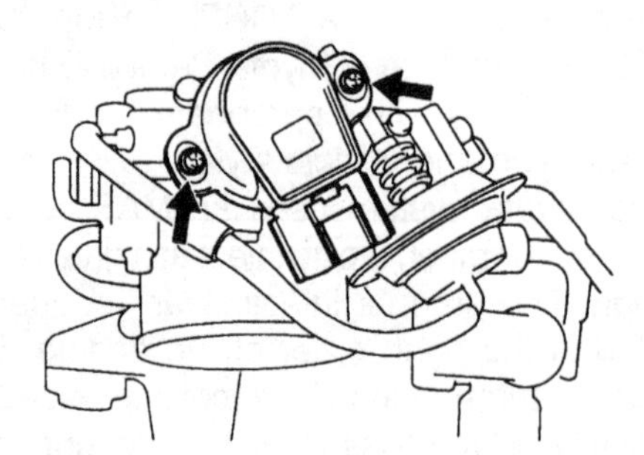

Figure 16–27 Throttle position sensor.

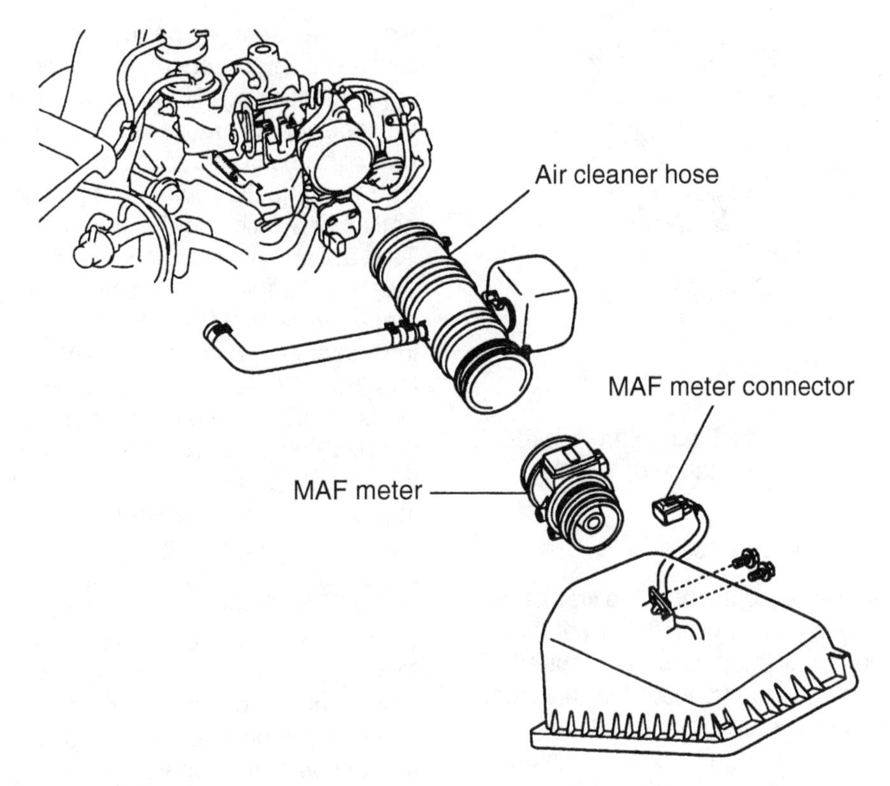

Figure 16–28 MAF sensor.

a piezoelectric unit that, when vibrated at a frequency characteristic of knock, sends a signal to the computer, which then retards spark advance until the knock disappears. Spark advance is then re-advanced until the knock just reappears. Like most knock sensors, its function can be checked by tapping on the engine block adjacent to the knock sensor with a tool while observing the spark advance (warm engine, closed-loop conditions). Knock sensors are relatively delicate and should never be dropped. If dropped to a hard surface, they will often sustain damage.

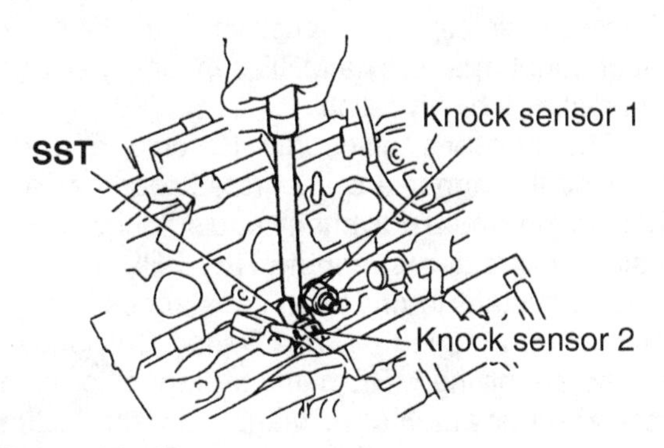

Figure 16–29 Knock sensors.

Camshaft Position Sensor

The latest Toyota V6 engines use a direct ignition system with one coil per cylinder rather than a distributor, and they therefore require a camshaft position (CMP) sensor (Figure 16–30). This sensor is on one of the camshaft sprockets on the accessory drive side of the engine. Its purpose is to enable the computer to determine which cylinder is approaching its power stroke firing position, information it uses both for spark firing and for fuel-injection sequencing. The CMP sensor is a coil-and-magnet inductive sensor, generating a current that alternates as the camshaft turns.

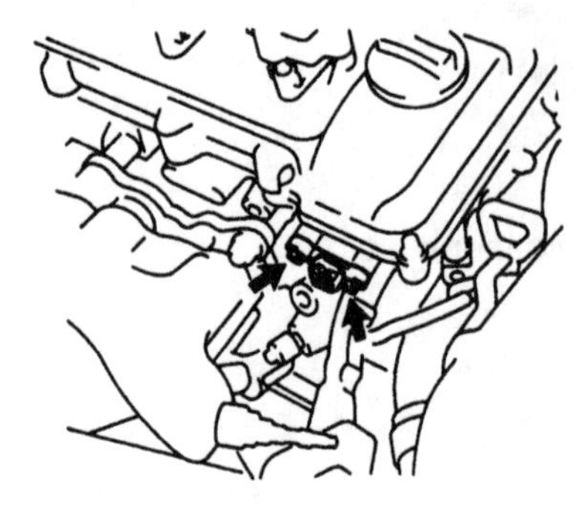

Figure 16–30 CMP sensor.

Electrical resistance should range from a cold value of 835 Ω to a hot resistance of 1,645 Ω.

Crankshaft Position Sensor

The CKP sensor is adjacent to the crankshaft pulley (Figure 16–31). It provides the computer with information about the position and speed of the crankshaft, information used for fuel injection, spark timing, and idle control, as well as functions restricted to on-board diagnostics. Resistance across the magnet-and-coil CKP sensor should range from a cold value of 1,630 Ω to a hot value of 3,225 Ω.

Vehicle Speed Sensor

In the 2007 model year, Toyota vehicles began using an electronic vehicle speed sensor (VSS) that, in essence, is a cross between a variable reluctance sensor (permanent magnet sensor) and a Hall effect sensor. This sensor uses a rotating magnetic ring placed on the output shaft of the transmission or transaxle. The rotating magnetic ring contains 10 magnetic north poles alternating with 10 magnetic south poles for a total of 20 magnetic poles built into the magnet (Figure 16–32). The alternating magnetic fields are monitored by an electronic sensor called a *magnetic resistance element (MRE)* (Figure 16–33). The rotating magnetic ring induces AC voltage pulses into the MRE sensor as the alternating magnetic fields pass by. The MRE sensor's electronics then convert these AC voltage pulses into crisp digital DC voltage pulses. The final output oscillates in a digital fashion between 0 V and battery voltage. Each time a north or south pole passes the MRE sensor, the output voltage signal pulls low to 0 V. In between the north and south poles, the voltage goes back to battery voltage momentarily. The MRE sensor has three wires: battery voltage, signal, and ground. As an electronic device, it must have both power and ground to power up and become functional.

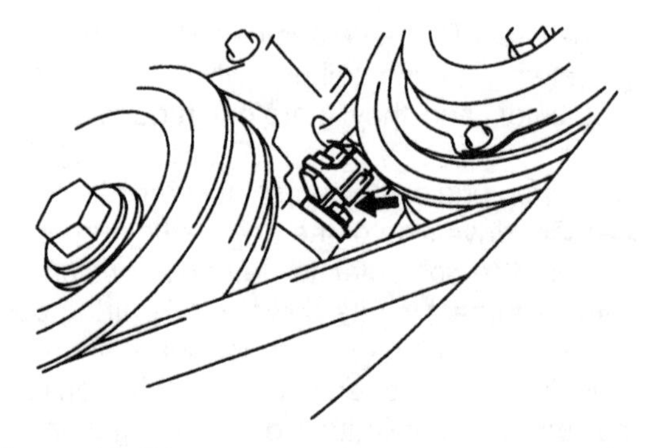

Figure 16–31 CKP sensor.

The Toyota MRE sensor is manufactured by the Denso Corporation. The Denso Corporation is a spin-off of the Toyota Motor Company. When it was part of Toyota, it was known as the Nippondenso Company. The Denso Corporation is Japan's leading producer of automobile parts and components. It is also the world's fourth-largest automotive parts supplier.

The Denso Corporation states that its MRE sensors have a high detection accuracy and are almost 10 times more sensitive than their Hall effect counterparts. The MRE sensor's construction makes it highly reliable.

The sensor's electronic IC chip and the cylindrical magnet are integrated together and then covered with polyphenylene sulphide (PPS) resin for greater strength. The PPS resin is highly resistant to chemical substances, including fuel and engine oil. The Denso Corporation is also supplying MRE cam/crank position sensors to other automotive manufacturers, including Nissan.

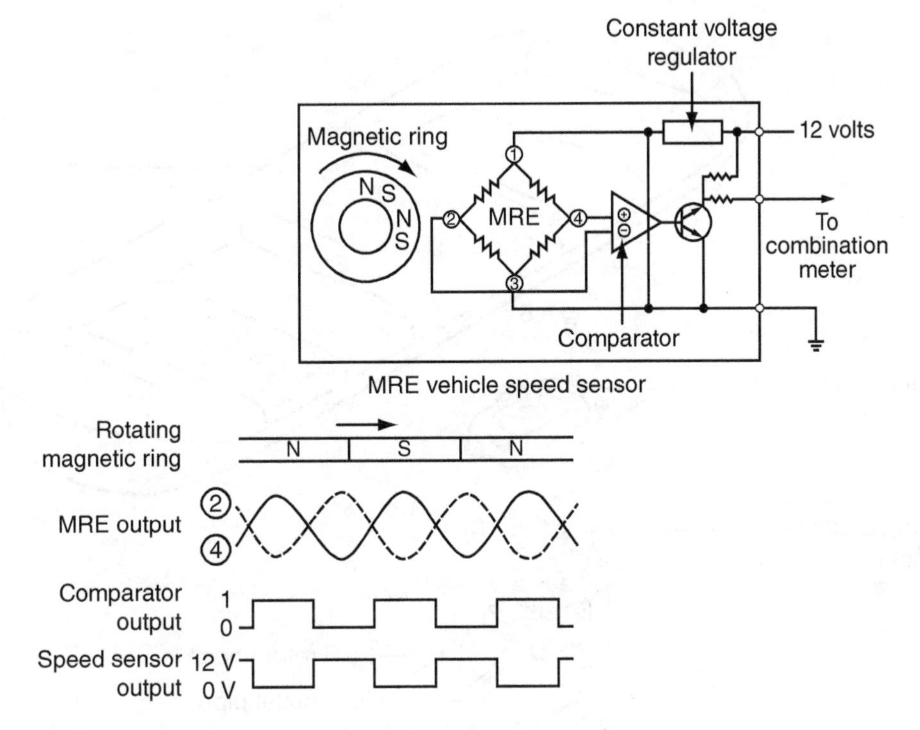

Figure 16–32 Toyota MRE vehicle speed sensor circuitry and waveforms.

Figure 16–33 Toyota MRE vehicle speed sensor.

OUTPUTS

Sequential Fuel Injection System

Fuel Injectors. The fuel injectors are Bosch-design units, one for each cylinder (Figure 16–34). The computer pulses them sequentially in the firing order of the engine during the corresponding

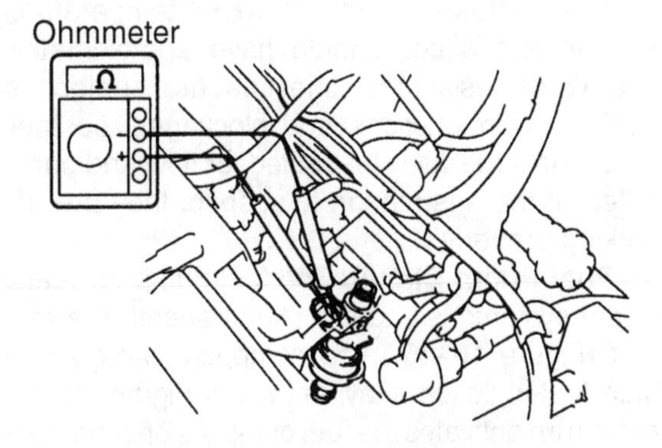

Figure 16–34 Fuel-injector resistance check.

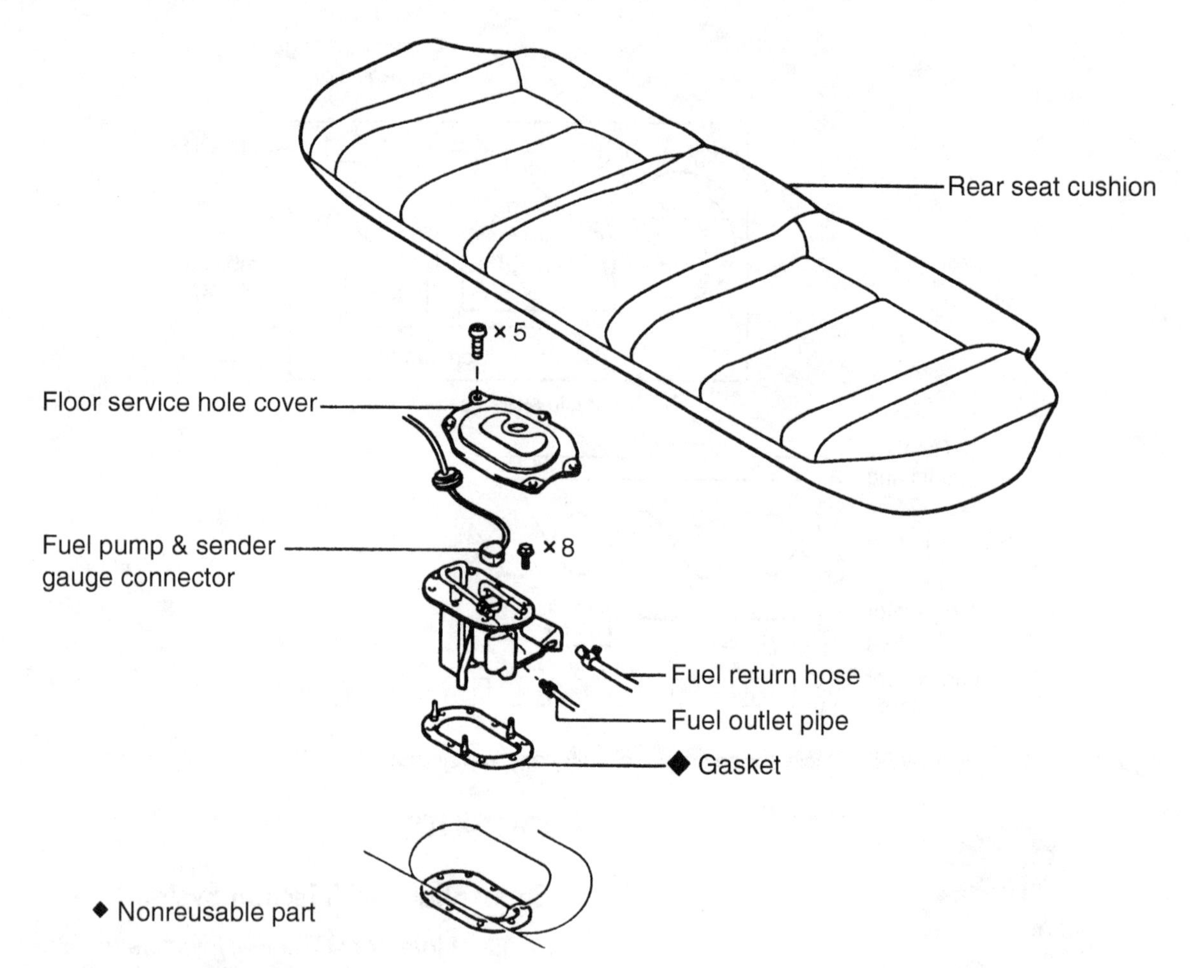

Figure 16–35 Fuel pump.

cylinder's intake stroke. At room temperature, each injector's coil should have approximately 13.8 Ω of resistance. Injectors are subject to mechanical restrictions and blockages, so they should be separately inspected for flow and spray pattern if there is reason to suspect they are not working properly.

Fuel Pump Circuit. The fuel pump on many Toyota systems is under a plate beneath the rear seat (Figure 16–35). The computer energizes a main fuel-injection relay, shown in Figure 16–36, that in turn activates the fuel pump. As on most systems, the fuel pump comes on for approximately

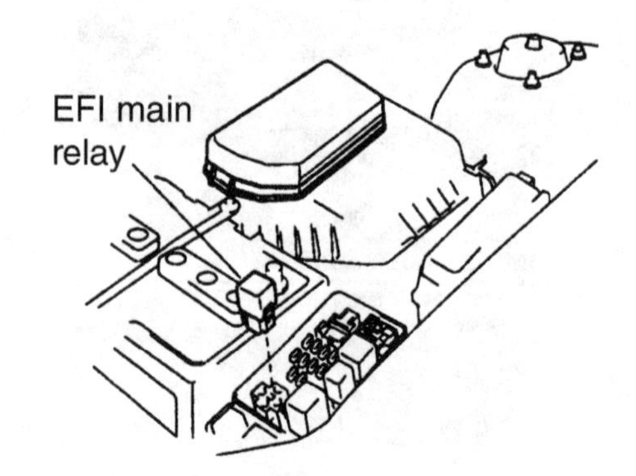

Figure 16–36 EFI (fuel pump) main relay.

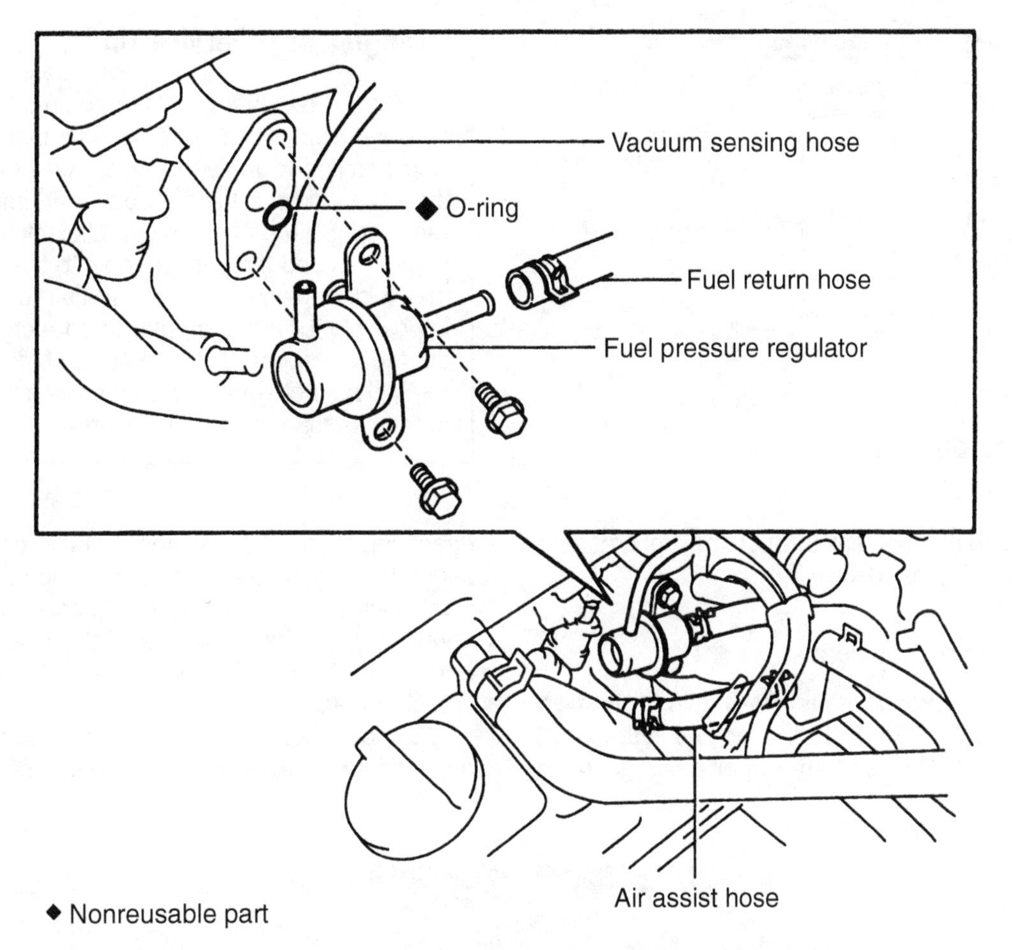

Figure 16–37 Fuel pressure regulator.

1 second after the ignition key is turned on to prime and pressurize the injectors and fuel rail, then shuts off if no CMP signal is received after the priming second.

The fuel from the pump fills the fuel rail; pressure is controlled by the fuel pressure regulator (Figure 16–37). The regulator bolts into the end of the fuel rail and reduces pressure corresponding to intake manifold vacuum, sending unused fuel back to the tank. The fuel pressure regulator, of course, is not directly controlled by the computer but responds to the pressure from the fuel pump and the partial vacuum in the intake manifold.

Returnless Sequential Fuel Injection System

Toyota has shifted to the returnless fuel system. This change reduces fuel tank emissions caused by the return of fuel from the engine compartment. As in other manufacturers' systems, this returned fuel contains heat absorbed when the fuel was in the engine-mounted fuel injector rail. Fuel in excess of that needed by the engine does not leave the fuel tank. A pressure regulator is located inside the tank above the submersible fuel pump. When the fuel moves outside the tank, it must go immediately to the fuel rail or it must be

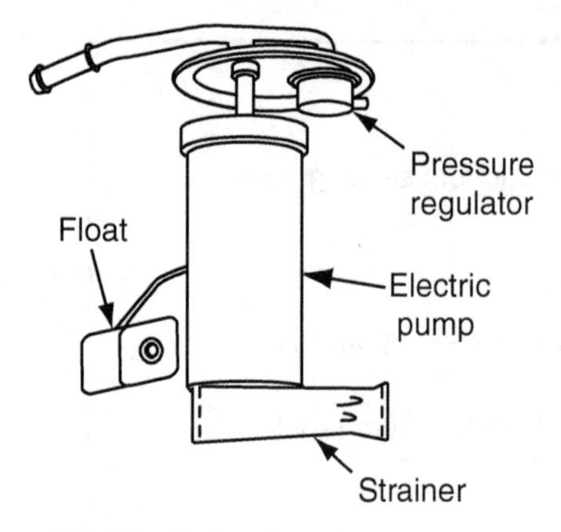

Figure 16–38 Fuel tank module.

Diagnostic & Service Tip

On the Toyota IAC valve and on others of similar design, it is almost always a better service practice to replace the gasket between the valve and the throttle body any time it is removed. Failure to do so can allow "false air" to pass around the gasket and into the intake manifold, making it more difficult or even impossible for the computer to correctly control the idle speed. This type of problem ordinarily does not set a code, so solving it takes more time than the repair is worth.

returned to the fuel tank through the fuel pressure regulator (Figure 16–38).

Twin Injection System

Toyota introduced its D4-S **twin injection** system on the 2GR-FSE 3.5L engine in the 2006 model year. This system combines a gasoline direct injection (GDI) system with a traditional sequential port fuel injection system (Figure 16–39). Each cylinder has two fuel injectors, one that sprays fuel on the back side of the intake valve and one that sprays fuel directly into the combustion chamber.

GDI decreases the potential for detonation to occur while increasing the engine's performance

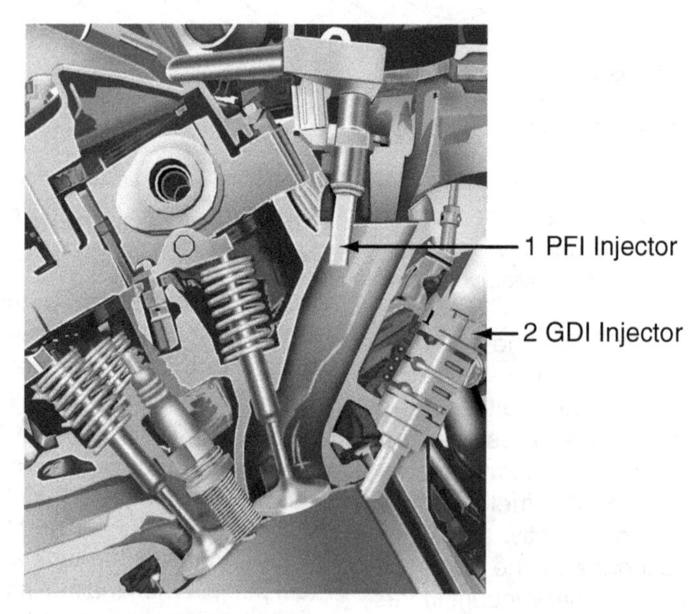

Figure 16–39 Toyota twin injection system.

potential. But GDI engines typically require a mechanism such as swirl control valves (tumble flaps) to control the air's turbulence in the cylinder. This helps to achieve a homogeneous air-fuel charge at low engine speeds and when under load. But the swirl control valves can also restrict airflow at full-throttle, high-end performance on a naturally aspirated engine. So instead of adding swirl control valves, Toyota added port SFI injectors to this GDI engine. This enhances the engine's ability to develop a fully homogeneous air-fuel charge at low engine speeds and when under load. The port SFI injectors also help to improve the full-throttle, high-end performance and are also used to improve cold-start emissions.

Toyota also uses a newly designed GDI fuel injector on this engine. This injector has a dual fan spray pattern that is perpendicular to the piston travel. The resulting spray pattern makes for a more complete fuel distribution throughout the combustion chamber and increases both the engine's performance and efficiency. This engine also uses Toyota's electronic throttle control (ETC-i) system.

Idle Air Control System

Figure 16–40 shows the IAC valve. It bolts to the throttle body and provides a bypass channel for air to pass the throttle blade for idle speed control, depending on the ECT, electric, and other accessory loads and whether the car is in a drive gear. The IAC valve is easily removed from the throttle body and can be inspected by jumping the connections to see whether the valve opens and closes properly.

Toyota Electronic Throttle Control (ETC-i)

Toyota's *electronic throttle control-intelligent (ETC-i)* system uses noncontact linear Hall effect sensors for both the throttle position sensors and the accelerator pedal position sensors. This system was first used on the Generation II Toyota Hybrid System in the 2004 model year and has been added to many more Toyota engines since. Toyota was the first manufacturer to begin using Hall effect–based sensors to monitor both throttle plate position and accelerator pedal position.

The ETC-i system eliminates the need for a dedicated idle air control actuator and a dedicated cruise control servo. It also provides the engine's torque management for both the Toyota Vehicle Stability Control (VSC) system and the Toyota Traction Control (TRC) system.

Ignition Coils

Many Toyota vehicles use a direct ignition system with one coil and driver per pair of cylinders (a waste spark system) (Figure 16–41). Ignition coil resistances should be 0.70 to 1.10 Ω

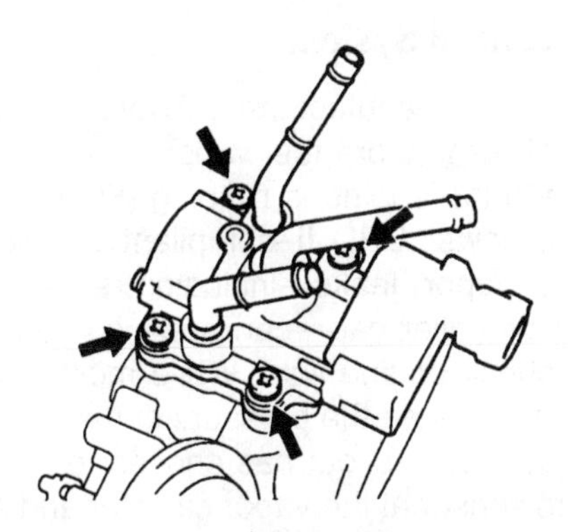

Figure 16–40 Idle air control valve.

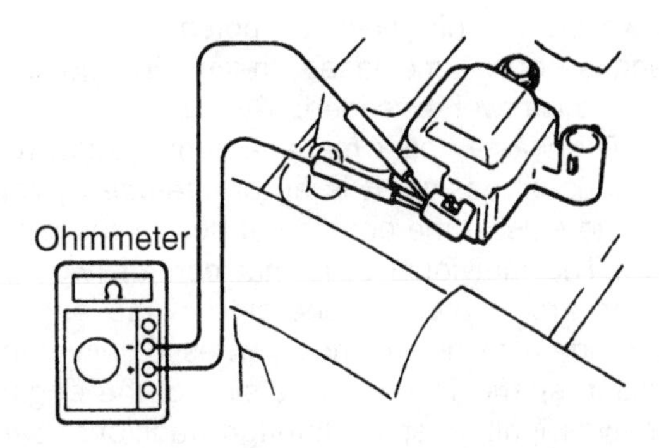

Figure 16–41 Ignition coil resistance check.

through the primary circuit and 10.8 to 17.5 KΩ through the secondary circuit, ranging from cold to hot temperatures. With individual coils of this design, the most time-efficient test for a bad coil is often to merely shift the coil from one cylinder to another to see whether the cylinder miss follows the coil. If it does not, of course, diagnosis should focus on fuel or mechanical aspects of the affected cylinder.

Toyota has a waste spark ignition system that looks different from most other waste spark ignition systems. Here is how it works:

This system gets its name from the exhaust cycle. As in the traditional waste spark system, one coil fires two cylinders at the same time. One cylinder is in the compression cycle while the other is in the exhaust cycle. This system uses the three principles of series circuitry. The first principle states that resistance adds together in a series circuit. This means the coil's secondary resistance, the spark plug's internal resistance, the spark plug's gap, and the secondary wire resistance all are added together. The second principle states that "current is consistent throughout all parts of a series circuit." Therefore, the current flow across both spark plug electrodes in the circuit is the same. The third principle states that voltage drops must equal the power supply. In other words, as electricity travels through each resistance it pays a price. That price is called *voltage drop.* The voltage required to power the spark plug during the compression cycle is higher than the voltage required to power the spark plug during the exhaust cycle. However, the coil's electrical potential is so high because of the increased on-time. The ignition reserve is never exceeded.

The waste spark management system has an unusual design. It is an *on-the-coil* system for one side of the engine but not for the other side. The individual coils that control two cylinders are located on one side of the engine. This side uses no secondary wires to deliver the ignition spark. The opposite side of the engine receives ignition spark through traditional secondary wires (Figure 16–42).

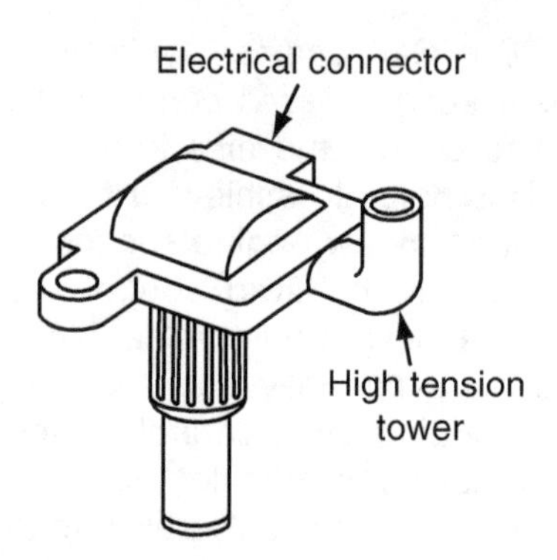

Figure 16–42 Waste spark coil.

Resistances for this coil are:

Primary	0.7 to 0.94 Ω (cold) 0.85 to 1.10 Ω (hot)
Secondary	
Asian made	10.8 to 14.9 KΩ (cold) 13.1 to 17.5 KΩ (hot)
Diamond made	6.8 to 11 KΩ (cold) 8.6 to 13.7 KΩ (hot)

As with other manufacturers, the newest Toyota vehicles use a coil-on-plug ignition system. Each ignition coil is mounted directly on top of the spark plug that it controls and with no secondary wires. Each ignition coil contains its own driver (power transistor) and receives a spark timing command from the PCM.

EVAP Control System

Like most manufacturers, Toyota uses a charcoal canister to store fuel vapors and burn them later when the engine is running (Figure 16–43). With the newer OBD II–compliant vehicles, any detected vapor leaks—including a leaking or missing fuel filler cap—can set a code and turn on the check engine light. While most manufacturers have placed the pressure sensor in the top of the fuel tank, Toyota has chosen to locate the pressure sensor at the vapor canister and associated hoses (Figure 16–44).

Figure 16–43 EVAP system charcoal canister.

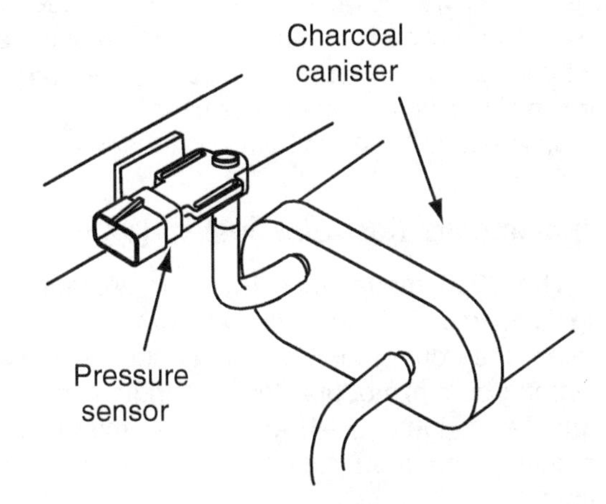

Figure 16–44 EVAP canister.

Variable Valve Timing with Intelligence

In 1998, Toyota introduced an engine equipped with variable valve timing (VVT). The recipient of this technology was the Supra and its normally aspirated 3.0L, dual overhead camshaft, 24-valve, inline six. (The Supra's twin turbocharger-fitted version of this 3.0L inline six retained fixed valve timing for the 1998 model year, but still managed to develop 320 horsepower and 315 ft.-lb of torque.) The Toyota system, named Variable Valve Timing with intelligence (VVT-i), was developed in response to ever more strict emissions requirements, but horsepower and torque are increased, too. The VVT-i-equipped engine develops 225 horsepower and 220 ft.-lb of torque at 4,000 RPM.

In the VVT-i system the engine management computer looks at a variety of inputs, including coolant temperature, engine speed, and engine load. The computer then controls valve overlap (the period of time during which both the intake valve and the exhaust valve are open) by advancing or retarding the intake camshaft. The VVT-i system control devices are an oil control valve (OCV) and a special intake camshaft timing pulley (VVT pulley) (Figure 16–45). Changing the intake camshaft "timing" while the engine is operating allows all of the typical factors that must be juggled during camshaft design—low-RPM torque, high-RPM power, engine smoothness, emissions, and fuel efficiency—to be optimized. A unique feature of the Toyota VVT-i system is that camshaft timing is continuously adjustable; competitive systems allow only two or three distinct camshaft "settings" that are strictly RPM dependent.

The VVT-i sprocket used in early systems has a piston with helical splines cut into both its inner and outer surfaces (Figure 16–46). The inner surface meshes with a splined gear attached to the end of the intake camshaft, while the outer surface meshes with a splined cup attached to the camshaft sprocket. Pressurized engine oil

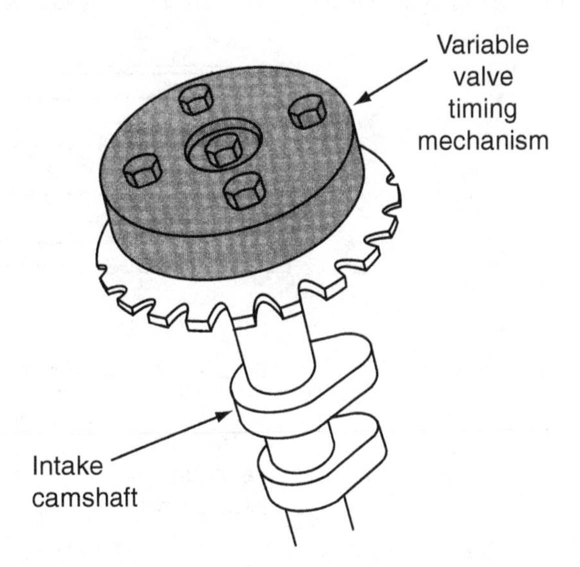

Figure 16–45 Intake camshaft and VVT unit.

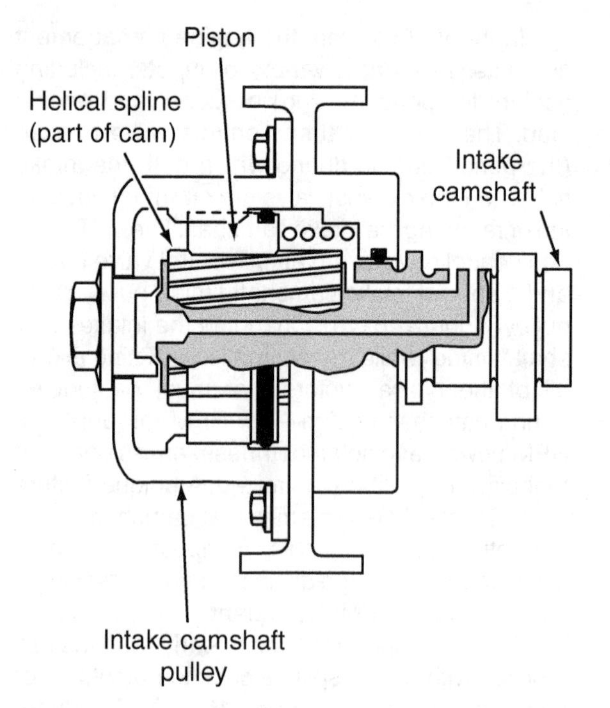

Figure 16–46 VVT-i camshaft timing pulley.

is applied to one side of the piston through the OCV, causing the piston to move axially. This axial movement changes the drive sprocket position relative to the camshaft. Modern VVT-i systems use a vane-style cam phaser, which is also controlled by the PCM through an OCV. In either style of system the PCM modulates the OCV to control camshaft timing, monitoring signals from the camshaft and crankshaft position sensors to determine actual camshaft position.

Two-Way Exhaust Control

Toyota recently introduced a method to control exhaust system backpressure. Exhaust gases are routed through the entire pathway of the muffler when the engine is at idle. This quiets exhaust noise. When the engine operates at higher speeds, the muffler provides a shorter route through a spring-loaded door, which reduces system pressure and improves engine performance. This route, shown in Figure 16–47, is both shorter and noisier.

Supplemental Restraint System (SRS)

The 2000 model-year Toyota integrates air bag deployment into the fuel pump drive program. If either the front or side air bag is deployed, the electronic control module stops grounding the fuel pump relay. Figure 16–48 shows that this opens the fuel pump feed circuit and stops the fuel pump.

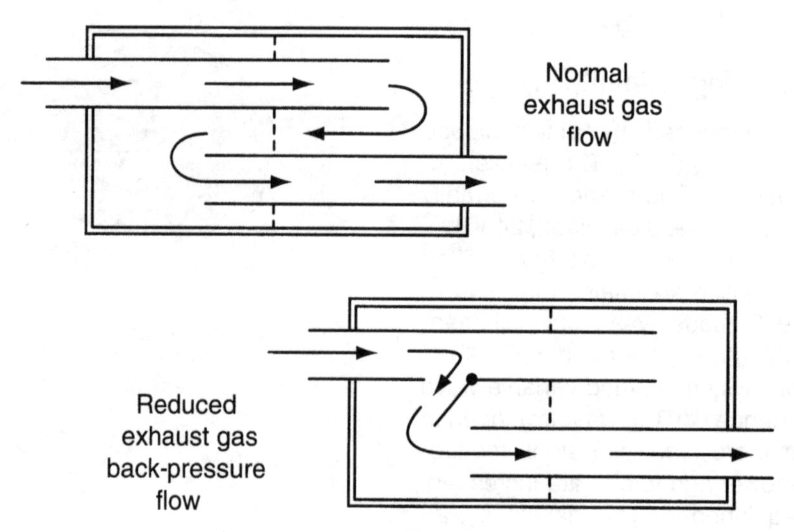

Figure 16–47 Typical two-way exhaust control muffler.

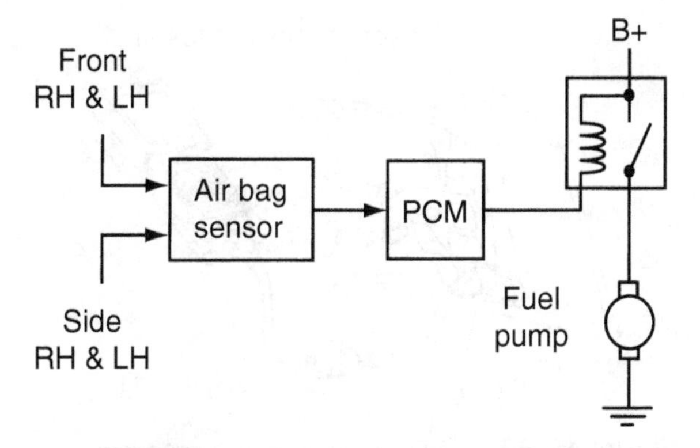

Figure 16–48 In case of an air bag deployment, the electric fuel pump is shut off.

Compressed Natural Gas (CNG)

The 2000 Toyota Camry introduced a natural gas version 5S-FNE-CNG engine. The CNG engine has a new fuel rail design, shown in Figure 16–49, that has an electric fuel shutoff valve at one end, a fuel temperature sender midway down the rail, and a fuel pressure sensor and discharge port at the other end.

The additional hoses, regulators, solenoids, and gas shutoff valve are not evident on the Camry until the vehicle is elevated. A large-diameter pressurized fuel tank is located at the rear of the vehicle. The fuel tank occupies most of the trunk space.

Active Control Engine Mount (ACEM)

The Avalon V6 engine has an active oil-filled engine mount. The mount contains a chamber divided by a rubber diaphragm. The chamber, shown in Figure 16–50, has computer-controlled engine vacuum on one side of the diaphragm; the other side of the diaphragm contains damping oil that is used to lessen engine vibration at idle.

At idle or at engine speeds of less than 900 RPM, the computer sends pulses to an electrically operated vacuum control valve. These signals are synchronized to engine RPM. The vacuum travels to the lower side of the rubber diaphragm inside the engine mount. The diaphragm vibrates, transferring that vibration through the damping

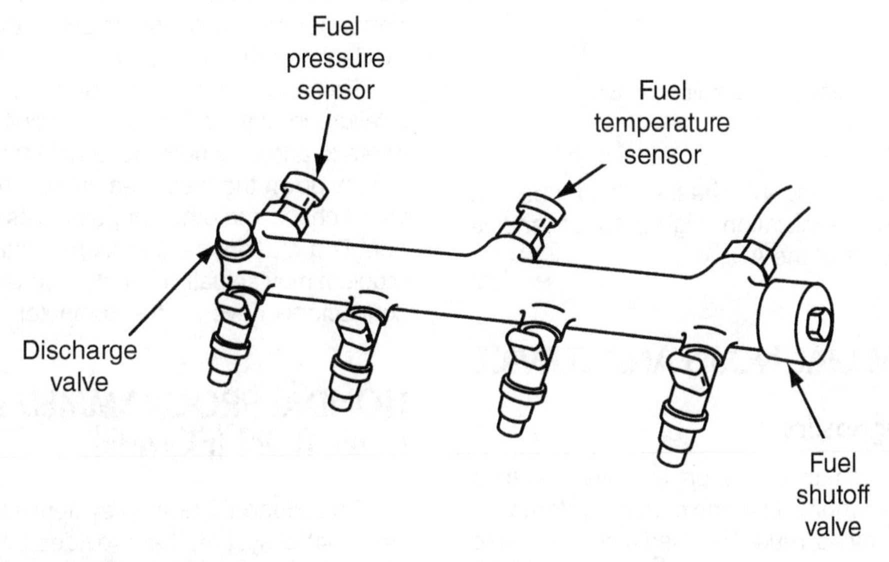

Figure 16–49 The fuel rail used on the CNG vehicle contains a pressure sensor, temperature sensor, shutoff valve, and a gauge test point.

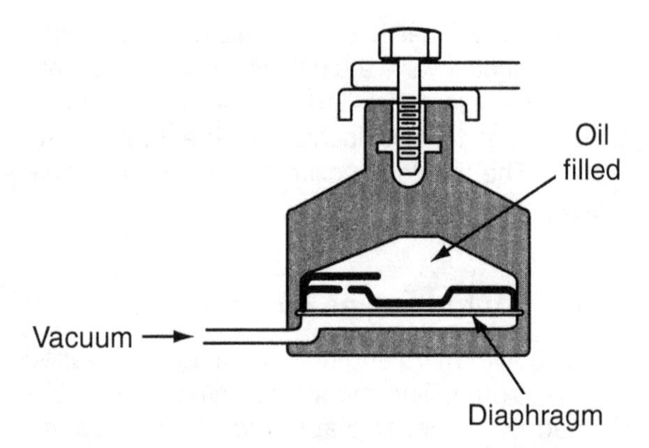

Figure 16–50 A cutaway active control engine mount (ACEM).

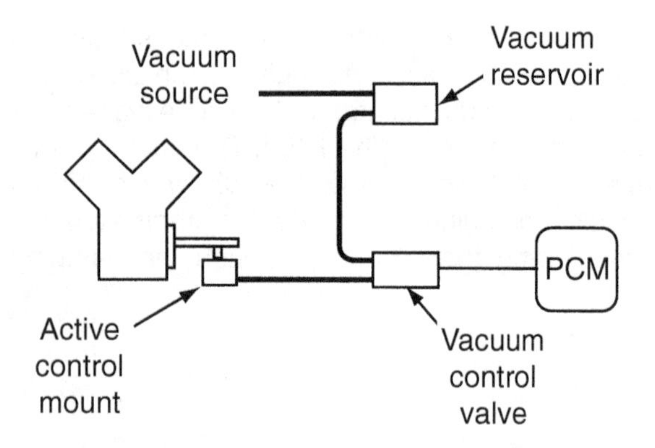

Figure 16–51 Electrical circuit of ACEM.

oil into the rubber mount. The mount cancels out unwanted engine vibration. Figure 16–51 shows the components of the ACEM.

✔ SYSTEM DIAGNOSIS AND SERVICE

OBD II Diagnostics

The first manufacturer on the street with a fully OBD II–compliant engine control system was Toyota, as early as 1994. This was a considerable engineering accomplishment, especially in light of the fact that some other import manufacturers

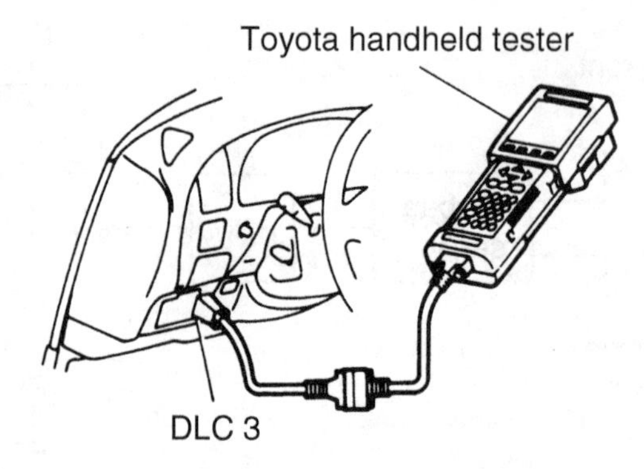

Figure 16–52 OBD II diagnosis with scan tool.

had stated early on in discussions about the OBD II requirements that they would have to use add-on modules to meet the standards.

Toyota's OBD II strategies closely resemble those of U.S. domestic vehicle manufacturers. Like other vehicle manufacturers, Toyota uses some proprietary DTCs as well as generic OBD II DTCs. With 1996 and newer Toyotas, generic OBD II scan tools may be used for diagnosis using the standardized data link connector in the lower left area of the instrument panel (Figure 16–52). Most component or signal failures that can affect the vehicle's emissions, fuel economy, or engine performance will be reflected in the information available on the scan tool.

Remember that the setting of a DTC represents a failure in any portion of the circuit that the DTC refers to and only helps to enable the technician to narrow down the fault area. Although performing a visual check and other pinpoint tests of the component is a reasonable first step (Figure 16–53), the problem may actually be in the circuitry's wiring and connectors or even in the computer.

HONDA: PROGRAMMED FUEL INJECTION (PGM-FI)

The Honda PGM-FI system is a computerized control system that provides optimum control of the engine's fuel, spark, and emissions control systems and also provides control of the automatic

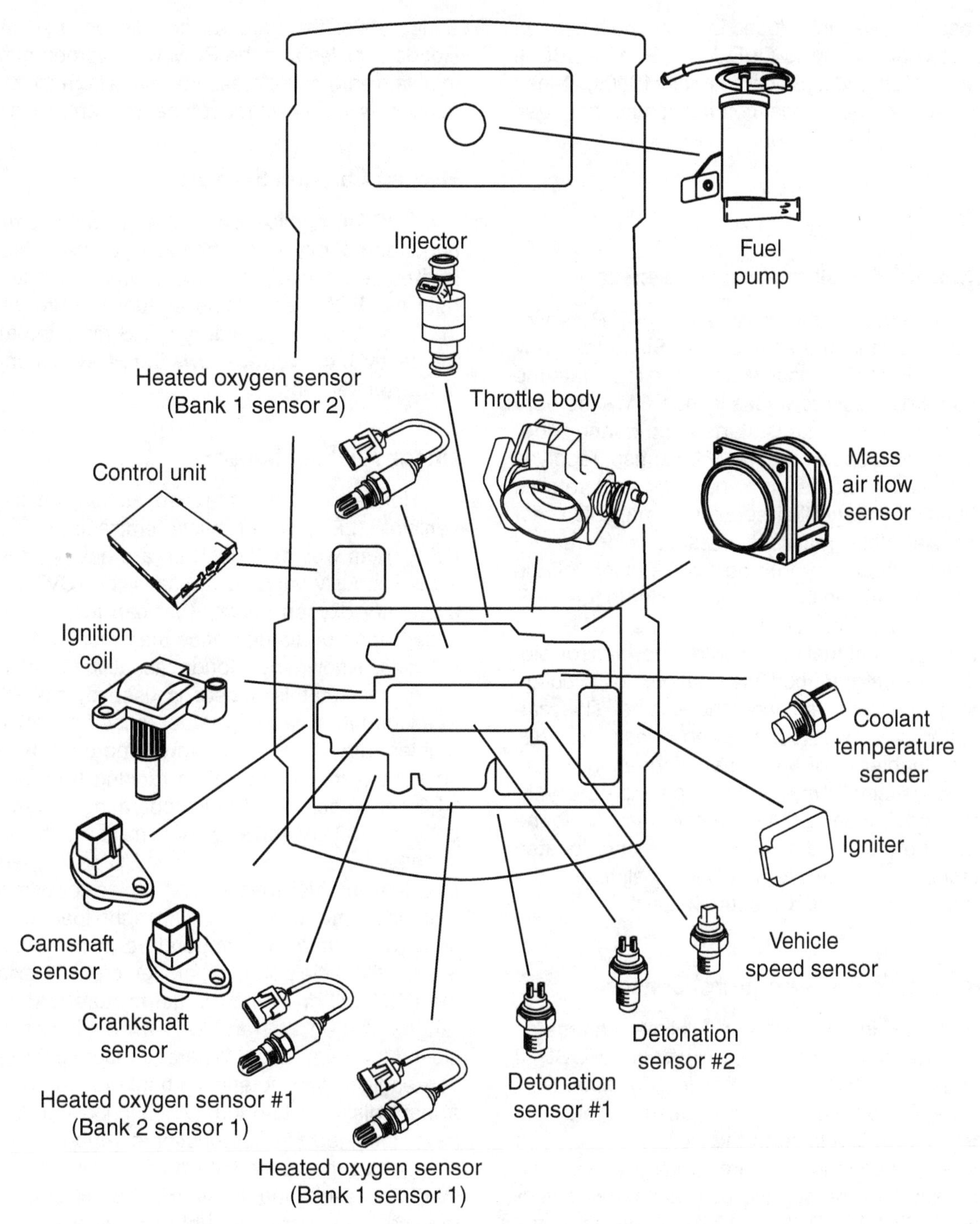

Figure 16–53 Component locations.

transaxle. It was introduced in 1985 and has been updated according to OBD I (1988) and OBD II (1996) standards. Since the mid-1990s, it has allowed for bidirectional communication between the PCM and a scan tool.

INPUTS

Manifold Absolute Pressure Sensor

On Honda vehicles, an analog MAP sensor is used to measure engine load. Since the early 1990s, the MAP sensor is also used to update the barometric pressure values in the PCM's memory whenever the engine is started and during wide-open throttle operation. The MAP sensor receives a reference voltage from the PCM and returns an analog voltage that equates to manifold pressure. As intake manifold pressure is decreased (as vacuum is applied), the MAP sensor voltage also is decreased in a manner similar to the MAP sensors found on GM vehicles. The PCM uses the resulting signal, along with other sensor signals and programmed information, to calculate how much air is entering the engine. This calculation is known as the *speed density formula.* (See Chapter 3 for additional information on the speed density formula.) Depending on the year and model, Honda MAP sensors may be found in a variety of locations. Newer ones are located attached to the intake manifold to eliminate the need for an attaching vacuum hose.

Barometric Pressure (Baro) Sensor

Honda initially used a full-time barometric pressure (baro) sensor, known prior to OBD II standardization as the *pressure absolute (PA) sensor.* This sensor is an analog sensor, returning an analog DC voltage to the PCM that indicates sensed barometric pressure. Originally designed as a separate sensor, the baro sensor was relocated within the PCM in 1988 (similar to some Chrysler vehicles of the mid-1980s). Then, in the early 1990s, the baro sensor disappeared from Honda vehicles and the PCM was reprogrammed to obtain barometric pressure values from the MAP sensor, as is done in most other modern vehicles.

Heated Oxygen Sensors

OBD II applications have a minimum of two heated oxygen sensors. A primary HO_2S is mounted ahead of the catalytic converter to help the PCM control the air-fuel mixture while in closed loop. A secondary HO_2S mounted after the catalytic converter allows the PCM to monitor converter efficiency.

Linear Air-Fuel Sensor

Hondas that are certified as low emission vehicles (LEV) or ultra-low emission vehicles (ULEV), as well as those that are equipped with a continuously variable transmission (CVT), use a primary oxygen sensor that can respond to a wider range of air-fuel ratios than those on other vehicles. Known by Honda as a *linear air-fuel (LAF) sensor*, this sensor is used by the PCM to accurately measure air-fuel ratios from 10:1 to ambient air. (Standard narrow-band oxygen sensors measure air-fuel ratios ranging from about 14.3:1 to about 15.1:1 while operating in a range between 100 mV and 900 mV.) The use of this sensor allows Honda to program its PCMs to operate at a lean air-fuel ratio during cruise to increase fuel economy. During certain light load conditions, and when the temperature of combustion is such that little potential for NO_x creation exists, the PCM will drive the air-fuel mixture lean to a range between 18:1 and 22:1 while still retaining closed-loop operation. Whereas most spark-ignition engines can potentially begin to experience a lean misfire at around 17:1, Honda engineers have designed the combustion chamber in such a way that the immediate area surrounding the tip of the spark plug tends to get a richer air-fuel mix than the overall air-fuel mix delivered to the cylinder. As a result, these engines can be run quite lean without experiencing any lean misfire symptoms and while continuing to control NO_x production successfully.

The LAF sensor is also known in the industry as a wideband air-fuel ratio sensor. Air-fuel ratio sensors are used on some Nissan and Toyota models as well. They are described in greater detail in Chapter 3. On Honda applications, the *control reference voltage* should be about 2.7 V and will vary ever so slightly with the current flow variances that occur during closed-loop operation.

The Honda version of this sensor has five wires entering the sensor body. But there is an eight-terminal connector in the LAF sensor harness. This connector contains a *calibration resistor* placed between terminals 3 and 4. Depending upon year and model, the measured resistance of this resistor could be 2.4 KΩ, 10 KΩ, or 15 KΩ. If this connector is damaged, be sure to replace it with a connector containing a calibration resistor of the same resistance as the one being replaced.

Engine Coolant Temperature Sensor

The ECT sensor is a negative temperature coefficient (NTC) thermistor, changing its resistance in response to changes in engine temperature. As the engine's coolant warms, the ECT sensor's resistance decreases, resulting in a decrease of the voltage drop signal that is measured across the sensor by the PCM. Prior to OBD II standardization, this sensor was referred to as the *temperature water (TW) sensor.*

Intake Air Temperature Sensor

The IAT sensor is an NTC thermistor that changes its resistance in response to changes in the temperature of the air entering the intake manifold. As the temperature of the intake air increases, the IAT sensor's resistance decreases, resulting in a decrease of the voltage drop signal measured across the sensor by the PCM, like the ECT sensor. Prior to OBD II standardization, this sensor was referred to as the *temperature air (TA) sensor.*

Throttle Position Sensor

The TP sensor is a rotary potentiometer that informs the PCM of driver demand. It is found at the throttle body on one end of the throttle shaft and is similar to those found on most other makes. Prior to OBD II standardization, this sensor was referred to as the *throttle angle (TA)* sensor.

Knock Sensor

The knock sensor is a piezoelectric crystal used to sense spark knock in the engine, similar to most other makes. The PCM reacts to input from the knock sensor by adjusting ignition timing until the knock disappears.

CKP and CMP Sensors

A CKP sensor is used to monitor a 12-tooth reluctor located behind the timing belt drive gear at the crankshaft (Figure 16–54). In addition, two camshaft position (CMP) sensors, known as TDC1 and TDC2, are used; they are located at the back cover of the cam gear housing (Figure 16–55).

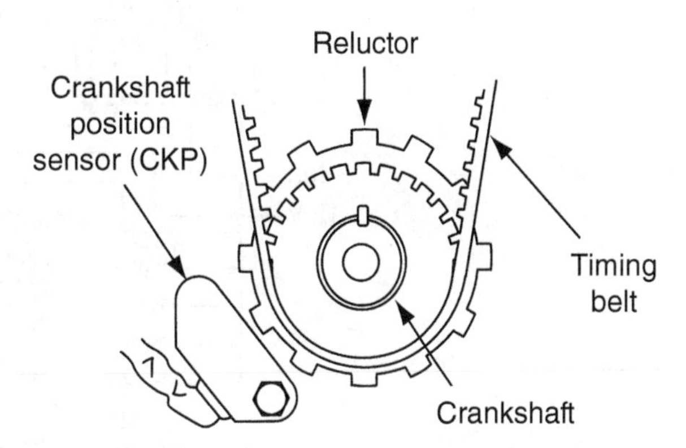

Figure 16–54 Reluctor gear on crankshaft and CKP sensor.

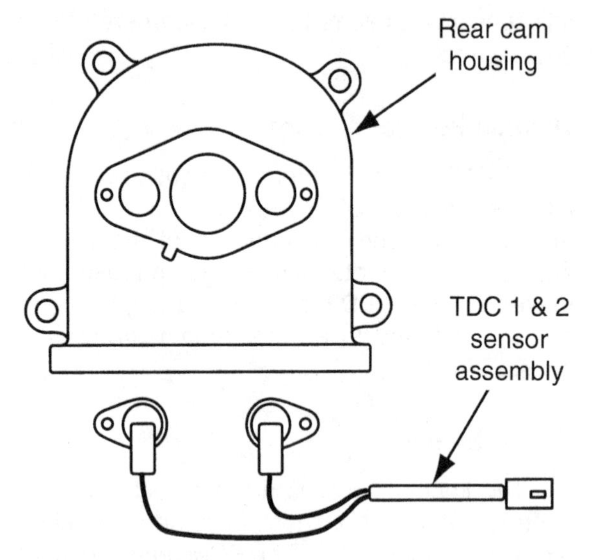

Figure 16–55 Cam housing back cover with TDC1 and TDC2 sensors.

OUTPUTS

Main Relay

The main relay assembly contains two relays. One relay, when energized by the ignition switch, provides power to the PCM and the fuel injectors (Figure 16–56). Beginning in 1988, it also was used to provide power to the IAC solenoid (not shown). The other is the fuel pump relay, which is under the control of the PCM.

Fuel Injectors

The Honda fuel injectors are low-resistance injectors, typically 1.5 to 3.0 Ω. When the ignition switch is turned to the Run position, the main relay provides power to the fuel injectors through external resistors, one per injector (Figure 16–56). The resistor assembly is typically mounted in front of the left front shock tower.

Fuel Pump Circuit

The PCM energizes the fuel pump relay portion of the main relay assembly (Figure 16–57) to run the fuel pump whenever the PCM is receiving a tach reference signal from the ignition system. The only exception to this, as with most other makes, is that the PCM energizes the fuel pump relay for 2 seconds when the PCM is initially powered up, whether or not a tach reference signal is present. This is used to top off

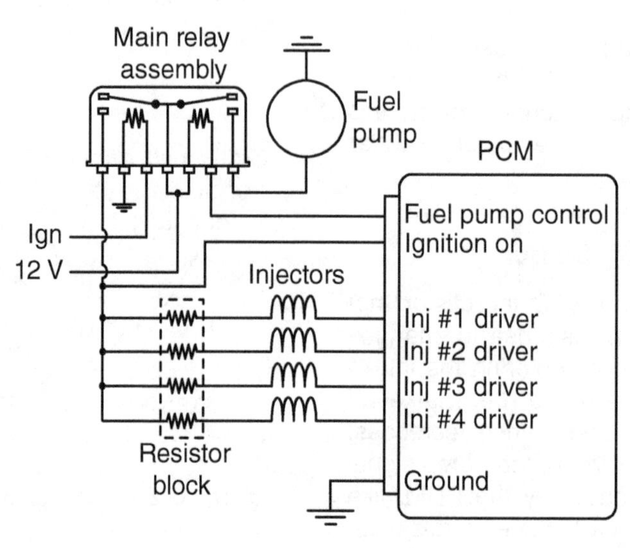

Figure 16–56 Honda main relay circuit.

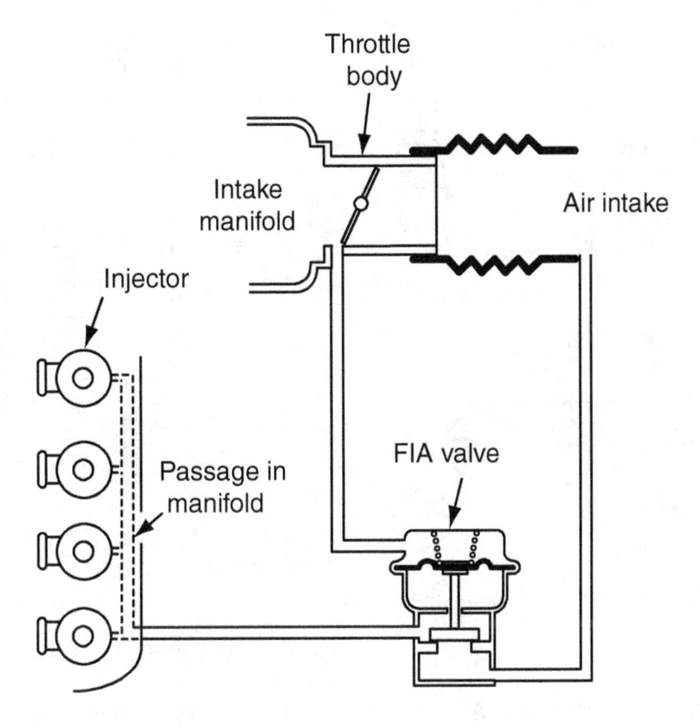

Figure 16–57 FIA control valve system.

the residual fuel rail pressure to provide an easy engine start.

Fuel-Injection Air Control Valve System

A fuel-injection air (FIA) control valve system is used on some Honda Accord 3.0L and Odyssey applications. The FIA control valve allows filtered air from the air cleaner to be directed into the path of the fuel injectors' spray patterns to increase fuel atomization (Figure 16–57). The FIA valve is opened by a ported vacuum signal to improve atomization just off idle.

Idle Air Control System

By 1988, Honda PGM-FI applications incorporated an IAC valve. Known prior to OBD II standardization as the Electronic Air Control Valve (EACV), the IAC valve is a normally closed, solenoid-operated valve that is controlled by the PCM on a duty cycle. The PCM increases the duty cycle on-time of the IAC valve to increase the flow of bypass air around the throttle plate(s).

Some early applications also used a mechanical fast idle valve to provide additional bypass air following a cold engine start. This valve was controlled through a wax pellet immersed in a coolant passage. When the engine was cold, the valve was opened to provide additional bypass air. As the engine warmed, the wax pellet expanded, closing the valve.

Ignition Coils

Honda applications used a single ignition coil and a distributor for many years. Many of these applications located the ignition coil within the distributor assembly. As Honda began to change over to EI ignition systems with multiple coils, a few early Honda Passports used a General Motors–style waste spark ignition system. Modern Honda

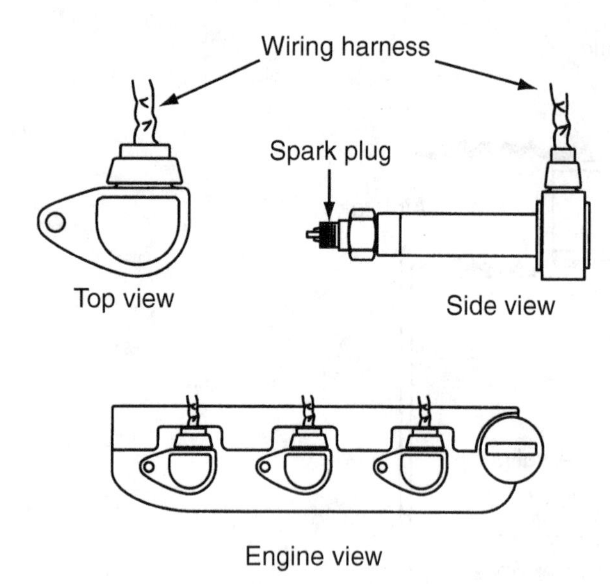

Figure 16–58 Top, side, and engine views of ignition coil used in at-plug system.

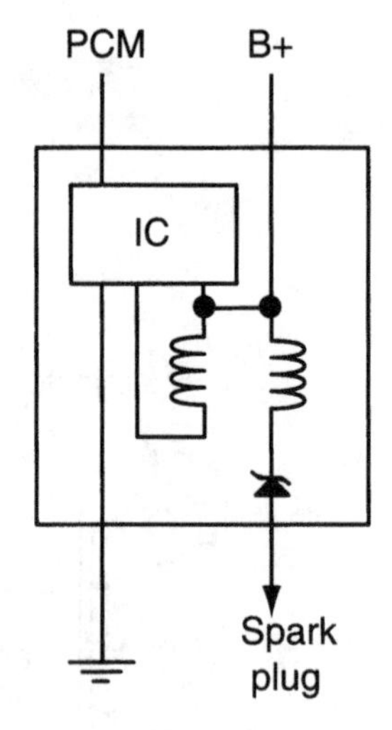

Figure 16–59 Internal schematic of ignition coil used in coil-at-plug system.

vehicles, including 3.0L engines in the Accord, 3.2L engines in the Acura, and 3.5L engines in the Acura, Odyssey, and Pilot, use a coil-at-plug design, similar to the coil-on-plug (COP) designs found on many other modern vehicles. In this design, each cylinder has its own ignition coil placed directly over the spark plug, eliminating the secondary plug wires (Figure 16–58). The ignition coils in Honda's coil-at-plug system cannot be tested in the traditional method using an ohmmeter because each ignition coil contains an electronic integrated circuit (IC) chip (Figure 16–59). The secondary winding also contains a diode. If the coil has proper power and ground and the PCM's trigger signal is present but the coil does not produce spark, the coil is at fault.

EVAP Control System

Honda's evaporative control system allows the PCM to monitor fuel tank pressure with a pressure sensor (Figure 16–60). When fuel tank pressure exceeds a predetermined value, a two-way valve opens to allow the fuel tank to vent to atmosphere by way of the charcoal canister. The PCM also controls canister purging through a purge solenoid.

In addition, Honda uses an ORVR system. During normal vehicle operation, when the fuel cap is installed, pressure in the filler neck area is equal to the pressure within the main body of the fuel tank. However, once the fuel cap is removed, atmospheric pressure is applied to the area in the neck of the gas tank, monitored by an ORVR signal tube that controls a normally closed ORVR vent shut valve (Figure 16–60). Meanwhile, pressure within the main body of the fuel tank rises due to the fuel that is entering the tank. The ORVR vent shut valve monitors this pressure as well. The pressure differential created opens the ORVR vent shut valve, allowing vapors to be routed to the charcoal canister during refueling.

Electronic Load Detection (ELD)

Alternator output on Hondas is controlled by the PCM. The PCM reduces field current, which in turn reduces alternator output. Output current is based on battery charge and electrical system loads. When the battery is low, or when high electrical demands are made by various systems in the car, the PCM signals the alternator to

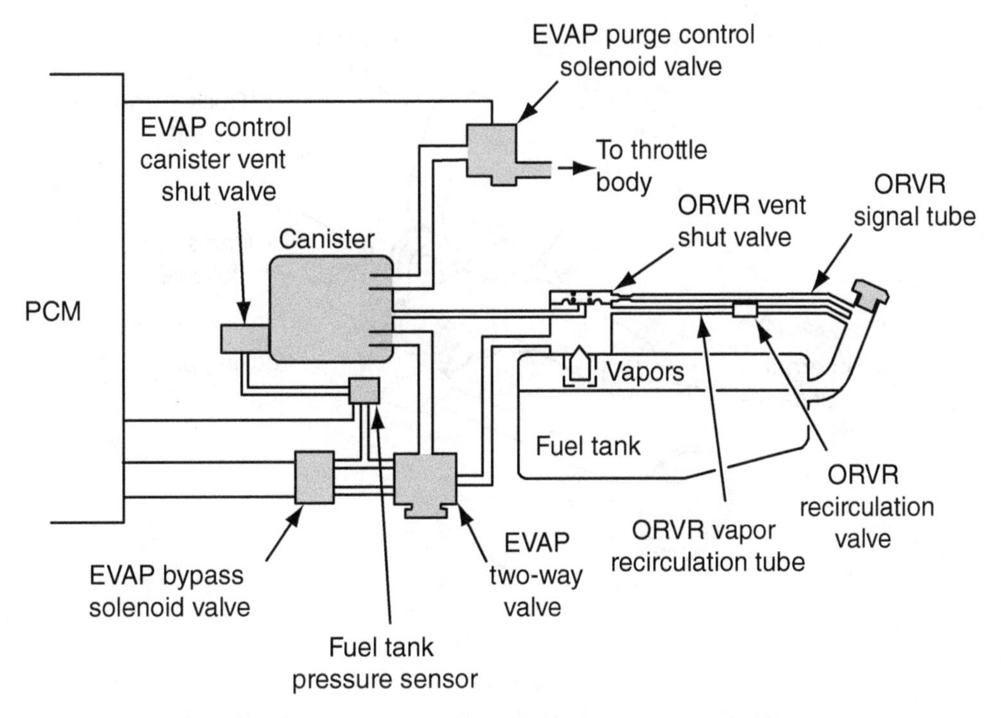

Figure 16–60 On-board refueling vapor recovery (ORVR) and evaporative system.

increase output; when the battery is charged up and there are few electrical system demands, the PCM reduces alternator output. Reduced alternator load results in better fuel economy. The PCM monitors alternator output through a field regulating (FR) signal that is returned to the PCM from the alternator.

HONDA'S VTEC SYSTEMS

Honda has used some form of its VTEC engine technology on all of its engines since 1992. By the original definitions, Honda applications used three different systems that allow the PCM to electronically control various combinations of VVT, lift, and duration. Known initially as the Variable Valve Timing and Lift, Electronic Control (VTEC) systems, the three versions of this system were initially referred to as VTEC, VTEC-E, and i-VTEC.

The Original VTEC Systems

VTEC. The original VTEC system uses two intake valves per cylinder, controlled by three rocker arms (Figure 16–61). At low RPM, the rocker arms operate independently. The primary rocker arm is used to open the primary intake valve; it is controlled by a low-lift, short-duration lobe of the intake camshaft. The secondary rocker arm is controlled by an ultra-low-lift, short-duration lobe that opens the secondary intake valve just enough to prevent fuel accumulation in the area of the intake port but provides little airflow. A mid (connecting) rocker arm is located between the primary and secondary rocker arms and is controlled by a high-lift, long-duration lobe. However, at low RPM, the mid rocker arm does not mechanically control any valve, but simply operates against the tension of a "lost motion" spring assembly. The primary and mid rocker arms each contain a synchronizing piston, but at low engine speeds this piston is

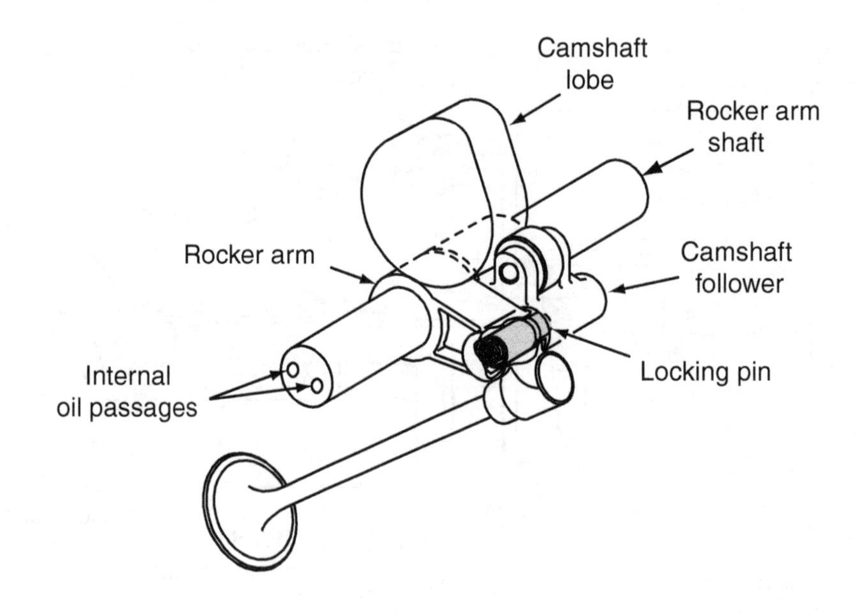

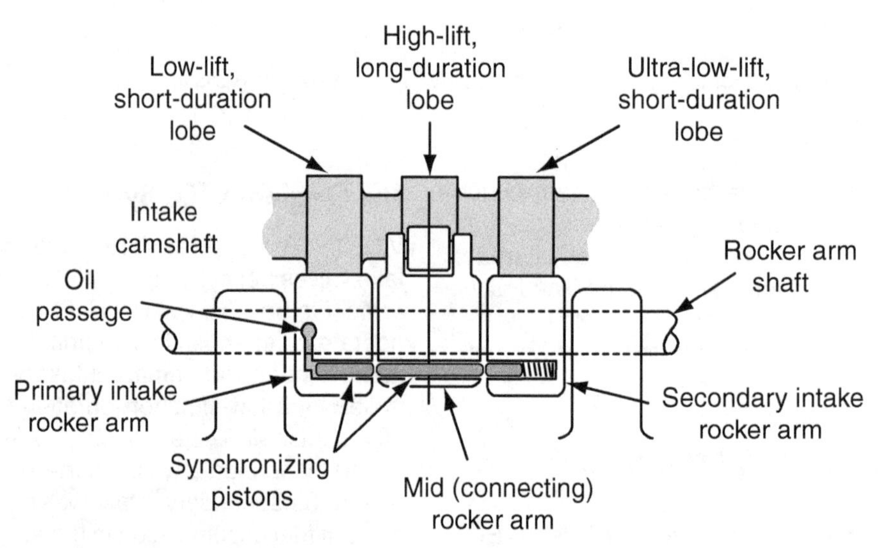

Figure 16–61 VTEC system rocker arms operating independently.

fully contained within the respective rocker arm due to a return spring and stopper pin assembly located in the secondary piston.

At high engine RPM, the PCM energizes a solenoid that moves a spool valve. The spool valve then directs oil pressure into an oil passage within the primary rocker arm. This oil pressure then moves both synchronizing pistons toward the spring-loaded stopper pin in the secondary rocker arm, thus locking all three rockers together mechanically (Figure 16–62). This causes both the primary and secondary intake valves to be

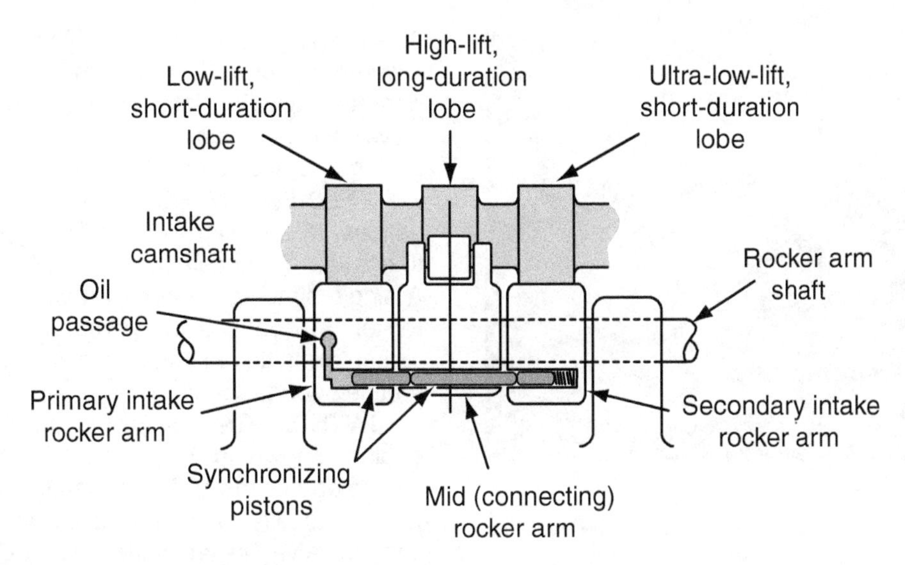

Figure 16–62 VTEC system rocker arms mechanically linked together.

controlled according to the mid rocker arm's high-lift, long-duration cam lobe. Once the engine speed slows down again, the PCM de-energizes the solenoid, allowing the applied oil pressure to be vented from the primary rocker arm. This allows the return spring to move both synchronizing pistons back into their respective rocker arms, again allowing independent movement of each rocker arm. Figure 16–63 shows the three rocker arms of a VTEC engine. Figure 16–64 shows the rocker arms with their metal pins and internal spring. Figure 16–65 shows the three cam lobes of a VTEC engine.

The PCM considers four inputs before energizing the VTEC oil solenoid (Figure 16–66). Engine RPM must be between 2,300 and 3,200 RPM on the four-cylinder application, must exceed 3,500 RPM on 3.0L applications, and must

Figure 16–63 The three rocker arms of a VTEC engine.

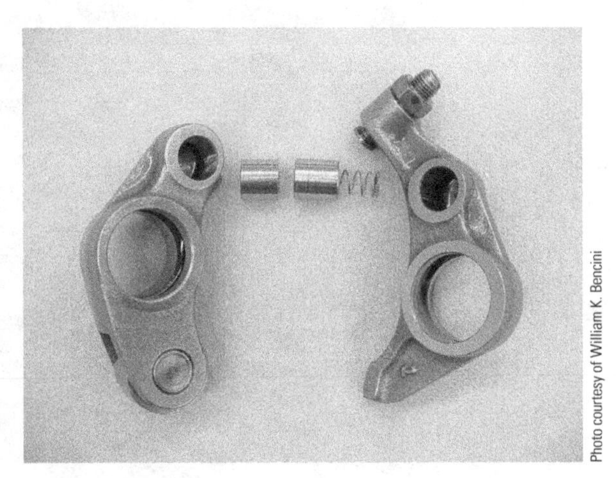

Figure 16–64 VTEC Rocker arms, pins, and internal spring.

Photo courtesy of William K. Bencini

Figure 16–65 The three cam lobes of a VTEC engine. Shown left to right: ultra-low lift/duration lobe, high-lift/long-duration lobe, and primary lobe.

exceed 4,900 RPM on certain Prelude applications. Vehicle speed must exceed 19 MPH, coolant temperature must exceed 140°F (60°C), and the engine must be under appropriate load. Ultimately, at lower RPM, the cylinder operates with a single intake valve and a single exhaust valve, thus keeping the airflow velocity high for improved low-end response. But, at high RPM, the cylinder operates using two intake valves and two exhaust valves, all opening and under the control of a high-lift, high-duration cam lobe for maximum airflow.

VTEC-E. A VTEC-E is used on some applications; it is similar to the VTEC system except that a mid (connecting) rocker arm is not used. A single synchronizing piston within the primary rocker arm causes the primary and secondary rocker arms to be mechanically linked at high engine speeds.

i-VTEC. The i-VTEC (or intelligent VTEC) engine, shown in Figure 16–67, combines the technology of the VTEC systems just discussed with the ability to rotate the intake camshaft to change valve overlap. Initially, i-VTEC technology was made available on two Honda vehicles: the 2002 Honda CR-V and the 2002 Acura RSX.

The variable valve-lift feature is accomplished by using two rocker arms and different camshaft profiles to change the opening of the valve. With a new "two rocker arm" design, only one valve is opened at lower engine speeds. This causes the intake air-fuel mixture to swirl, optimizing

Photo courtesy of William K. Bencini

Figure 16–66 VTEC oil solenoid.

Figure 16–67 i-VTEC 2.4L engine.

combustion for low-RPM operation. At a predetermined RPM, oil pressure is used to lock together the two rocker arms, resulting in both valves opening together. However, now both valves follow the higher camshaft profile that the second rocker arm normally follows.

The variable timing valve feature is accomplished by the use of a computer-controlled variable camshaft actuator, or cam phaser, which can vary the intake camshaft up to 50 degrees in relation to the crankshaft (Figure 16–68). The PCM can direct oil pressure to one side or the other of the extending arms of the central unit, which in turn is keyed to the intake camshaft. This changes the relationship of the center unit to the outside toothed gear, and thus changes the relationship of the intake camshaft to the crankshaft, ultimately advancing or retarding the camshaft's timing. Oil seals are installed on each of the two major components of this assembly where they contact each other.

Figure 16–68 VVT actuator, or cam phaser.

The Modern i-VTEC Systems

Honda's term i-VTEC was originally used to describe a particular function, as described in the previous paragraphs. Today Honda uses this term to describe all of its modern engines, regardless of the function or functions available on that engine. There are four possible i-VTEC functions, of which up to three may be used on any given engine. They are:

1. Variable lift and duration (VLD)
2. Valve timing control (VTC)
3. Progressive valve operation (PVO)
4. Disable valve operation (DVO)

Variable Lift and Duration. The VLD function is equivalent to the original VTEC system. During low-speed operation, the primary rocker arm is released from the secondary rocker arm, and the primary rocker arm operates its valve off a normal-lift-and-duration cam lobe. At high engine speeds, the primary rocker arm is locked to the secondary rocker arm so as to operate the primary rocker arm's valve off the secondary rocker arm's high-duration-and-lift cam lobe.

Valve Timing Control. The VTC function is equivalent to the VVT function of the i-VTEC system described previously. The VTC cam phaser, shown in Figure 16–68, is used to vary the timing of the intake camshaft in its relationship to the crankshaft. Honda engineers feel that their VTC system offers an advantage by varying the timing of the intake camshaft as compared to some manufacturers that vary the timing of the exhaust camshaft. Honda uses the VTC function to eliminate the need for an EGR system by retaining some of the exhaust gases in the cylinder when needed.

Progressive Valve Operation. The PVO function is the ability of the PCM to disengage a primary rocker arm that operates a valve from a secondary rocker arm that is operated according to a cam lobe in order to provide an *idle mode* whereby the valve is not opened. The PVO system will open and close only one intake valve and one exhaust valve at engine speeds below 2,500 RPM so as to keep the velocity of the airflow high, resulting in excellent low-end response. This also provides for *high swirl combustion* within the cylinder and allows for a leaner air-fuel ratio to be used during cruise. At engine speeds above 2,500 RPM, the PVO rocker arms are locked together to open and close a second intake valve and a second exhaust valve. This provides more airflow when the engine demands it.

Disable Valve Operation. The DVO function is similar to the PVO function in that it can disable a valve's ability to operate. However, unlike the PVO function, the DVO function is used by the PCM to disable all valves' ability to open for a specific cylinder, thus providing for Honda's **Variable Cylinder Management (VCM)** function. The DVO/VCM function uses rocker arms that are similar to those used in other Honda VTEC engines, except that the synchronizing piston allows a rocker arm to engage an adjacent rocker arm to provide "valve-lift mode" (normal operation), or it may be retracted into itself to provide a "rocker arm idle mode." The latter mode effectively fails to open the intake and exhaust valves, thus leaving them closed and allowing the deactivated cylinders' pistons to act as air springs. This reduces the normal pumping losses. The fuel injector for the deactivated cylinders is also de-energized. The DVO/VCM function is used with both Honda's hybrid and non-hybrid applications.

Direct Gasoline Injection

Honda has now added direct gasoline injection technology to the i-VTEC engine and refers to the new engine as "i-VTEC I." This engine, using special engineering features, can be run during a lean cruise at air-fuel ratios as lean as 65:1, far surpassing the 40:1 limit for direct-injection prototypes of the past. As with the original i-VTEC engine, only one intake valve is opened during low-load operation, causing a high-swirl effect as the fuel is injected into the cylinder. Ultimately, the i-VTEC I engine is capable of some markedly improved fuel economy while maintaining the responsiveness of a performance engine.

✔ SYSTEM DIAGNOSIS AND SERVICE

Prior to OBD II standardization, Honda PGM-FI systems could communicate memory DTCs to the technician through either one or four LEDs that were located on the PCM. With both LED designs, there was no need to trigger DTC output as with most other makes. Simply turning on the ignition switch to power up the PCM was all that was required to read DTC output on these early systems.

When one LED was used, turning on the ignition switch would result in a single flash that did not repeat if codes were not stored in the PCM's memory. If codes were stored in the PCM's memory, this was followed by repeated flashes of the LED to indicate the stored DTC(s). The total number of flashes represented the DTC. DTCs 1 through 15 were available, with 15 flashes of the LED representing DTC 15.

When four LEDs were used, they were numbered from left to right 8, 4, 2, and 1, similar to the base 10 values of binary code columns. The DTC number was interpreted by adding together the values of any illuminated LEDs. For example, if LEDs 8, 2, and 1 were illuminated, the DTC was code 11. The maximum DTC value was also a code 15.

In the early 1990s, Honda began making DTC output available through pulses of the MIL. This version required a jumper connection to be made at a test connector to trigger code output. This application allowed the PCM to illuminate the MIL if a fault within the system was sensed.

On OBD II applications, the PCM will illuminate the MIL by providing a ground and will set a DTC in an erasable memory if it senses a fault within the system. As with other OBD II vehicles, a scan tool may be connected to the DLC; it is then used to pull DTCs from the PCM's memory along with freeze-frame data. Additionally, a scan tool may be used to look at data stream values, including sensor values and output commands. Prior to OBD II applications, Honda vehicles had no data stream capability. Therefore, scan tools were unable to communicate with the Honda PCM until Honda's first OBD II vehicle was introduced: the 1995 Accord with a V6 engine.

SUMMARY

To focus on Asian engine control systems in this chapter, we have studied late-model Nissan, Toyota, and Honda vehicles.

We considered the different injection firing strategies (simultaneous and sequential) employed for startup, strategies to avoid the use of a cold-start injector, and strategies to optimize emissions during the engine's dirtiest phase: start-up cranking. The different fuel enrichment strategies, as well as spark advance map alterations, were considered. We saw how self-learning capabilities allowed the systems to adjust fuel mixture and spark timing from recorded experience. And we considered the more elaborate fuel canister vapor absorption and storage systems. We learned the use of an EGR temperature sensor to monitor the effectiveness of the bypass valve's circuit.

We also took a closer look at the various versions of the Honda VTEC series engines, including the newest i-VTEC functions: VLD, VTC, PVO, and DVO/VCM.

▲ DIAGNOSTIC EXERCISE

A customer brings in a late-model Nissan with the check engine light on. There are no driveability complaints, either from the customer or noticeable to a technician on a test drive. When checked with the scan tool at the OBD II DLC, the computer indicates a problem in the charcoal canister purge system. What is the most likely cause of this problem? How long should the repair take if the most likely cause is the actual cause?

Review Questions

1. *Technician A* says that a throttle position sensor on a Nissan ECCS application returns a voltage to the PCM from about 0.5 V at idle to as high as 4.6 V at WOT. *Technician B* says that some Nissan ECCS applications use an absolute pressure sensor to detect both ambient pressure and intake manifold pressure. Who is correct?
   A. *Technician A* only
   B. *Technician B* only
   C. Both technicians
   D. Neither technician
2. How much voltage should Nissan's analog MAF sensor return to the PCM at idle?
   A. 0.4 to 0.6 V
   B. 1.0 to 1.7 V
   C. 2.0 to 2.5 V
   D. About 4.0 V
3. Nissan's fuel temperature sensor will return a voltage of 3.5 V when the fuel in the fuel tank is at what temperature?
   A. –40°F
   B. 68°F
   C. 122°F
   D. 330°F
4. All except which of the following can result from a cylinder misfire?
   A. Production of too much CO_2 in the engine
   B. Overheating of the catalytic converter
   C. Catalytic converter failure
   D. Dumping of raw hydrocarbons into the atmosphere
5. *Technician A* says that some Nissan ECCS applications allow the PCM to electrically soften the dampening effect of the front engine mount to minimize engine vibration when the engine is idling. *Technician B* says on Nissan ECCS applications, the PCM may control the operation of the cooling fan. Who is correct?
   A. *Technician A* only
   B. *Technician B* only
   C. Both technicians
   D. Neither technician
6. Fuel pressure on port fuel–injected Nissan applications will typically vary between which of the following?
   A. 9 and 13 PSI
   B. 20 and 29 PSI
   C. 34 and 43 PSI
   D. 46 and 55 PSI
7. What component has been added only to California versions of the Nissan Sentra to help it meet the state's SULEV rating?
   A. Supercharger
   B. Electrically controlled front engine mount
   C. Rear heated oxygen sensor
   D. Air-cleaning radiator
8. Which of the following meters should be used to monitor a MAF sensor signal on a Toyota TCCS application?
   A. Voltmeter
   B. Hertz meter
   C. Duty-cycle meter
   D. Pulse-width meter
9. In the 2007 model year, Toyota vehicles began using an MRE sensor for which of the following?
   A. Vehicle speed sensor
   B. Manifold absolute pressure sensor
   C. Mass air flow sensor
   D. Intake air temperature sensor
10. An ohmmeter is used to test a camshaft sensor on a late-model Toyota TCCS application. With the meter's selector set to the 2K scale, a reading of 1.213 is seen on the display. *Technician A* says that this reading indicates that the camshaft sensor needs to be replaced. *Technician B* says that this reading indicates that the camshaft sensor's resistance is within specifications for a typical late-model Toyota. Who is correct?
    A. *Technician A* only
    B. *Technician B* only
    C. Both technicians
    D. Neither technician

11. *Technician A* says that the fuel pump on a Toyota TCCS application is controlled by the ignition switch only and should always run whenever the ignition switch is turned to the Run position. *Technician B* says that some late-model Toyotas use a returnless fuel injection system with the pressure regulator located within the fuel tank. Who is correct?
    A. *Technician A* only
    B. *Technician B* only
    C. Both technicians
    D. Neither technician
12. In the 2004 model year, Toyota vehicles began using noncontact linear Hall effect sensors for which of the following?
    A. Throttle position sensors
    B. Accelerator pedal position sensors
    C. Vehicle speed sensors
    D. Both A and B
13. *Technician A* says that Toyota's VVT-i system allows the PCM to advance or retard the timing of the intake camshaft. *Technician B* says that if a front or side air bag is deployed on a newer Toyota, the PCM will de-energize the fuel pump relay. Who is correct?
    A. *Technician A* only
    B. *Technician B* only
    C. Both technicians
    D. Neither technician
14. A 1998 Toyota comes into the shop with an engine performance complaint. *Technician A* says that the PCM on this vehicle has no ability to store DTCs in memory. *Technician B* says that the PCM on this vehicle will store DTCs in memory but that they have to be pulled using manual methods because this system is not designed to allow communication between a scan tool and the PCM. Who is correct?
    A. *Technician A* only
    B. *Technician B* only
    C. Both technicians
    D. Neither technician
15. Which of the following manufacturers introduced a twin injection system in the 2006 model year that combines a port fuel injection system with a gasoline direct injection system on the same engine?
    A. Toyota
    B. Nissan
    C. Honda
    D. Both B and C
16. The "main relay" on a Honda PGM-FI is used to provide power to all except which of the following?
    A. PCM
    B. Cooling fan
    C. Fuel injectors
    D. Fuel pump
17. What is the primary purpose of Honda's FIA control valve system?
    A. To allow the PCM to control idle speed
    B. To increase WOT engine performance
    C. To reduce NO_x emissions
    D. To increase fuel atomization
18. *Technician A* says that Honda's original VTEC system allows the primary and secondary intake valve rocker arms to be mechanically linked to a central high-lift rocker arm at high engine speeds. *Technician B* says that Honda's original VTEC system allows the PCM to control when the primary, secondary, and mid (connecting) rocker arms will be mechanically linked together. Who is correct?
    A. *Technician A* only
    B. *Technician B* only
    C. Both technicians
    D. Neither technician

19. *Technician A* says that the PCM on late-model Honda PGM-FI applications controls alternator field current to control charging system output. *Technician B* says that some Honda applications allow the PCM to control throttle position electronically based on input from an accelerator pedal position (APP) sensor. Who is correct?
    A. *Technician A* only
    B. *Technician B* only
    C. Both technicians
    D. Neither technician

20. Of the following Honda engines, which one is designed to operate during a lean cruise with an air-fuel ratio as lean as 65:1?
    A. VTEC
    B. VTEC-E
    C. i-VTEC
    D. i-VTEC I

Appendix A

Automotive-Related Internet Addresses

Publisher

Cengage Learning: www.cengagebrain.com

General Automotive-Related Web Addresses

California Air Resources Board: www.arb.ca.gov/homepage.htm

Environmental Protection Agency: www.epa.gov

Fuel Cell Store: www.fuelcellstore.com

How Stuff Works: www.howstuffworks.com

National Automotive Service Task Force: www.NASTF.org

Diagnostic Tools and Software

AES Wave: www.aeswave.com

AutoEnginuity: www.autoenginuity.com

Automotive Test Solutions: www.automotivetestsolutions.com

AutoTap: www.autotap.com

EASE Diagnostics: www.OBD2.com

Fluke Corporation: www.fluke.com

ScanTool.Net: www.scantool.net

Snap-on Tools: www.snapon.com

Diagnostic Information

AllData for the Do-It-Yourself person: www.alldatadiy.com

Diagnostic Trouble Code Definitions: www.dtcsearch.com

Professional Organizations

Automotive Service Excellence (ASE): www.ase.com

Identifix: www.identifix.com

International Automotive Technicians' Network: www.iatn.net

National Automotive Technicians Education Foundation (NATEF): www.natef.org

North American Council of Automotive Teachers (NACAT): www.nacat.org

Society of Automotive Engineers: www.sae.org/servlets/index

Training

ATech: www.atechtraining.com

Automotive Training Group: www.ATGtraining.com

CARQUEST Technical Institute: http://www.CTIonline.com

Lincoln College of Technology: www.lincolnedu.com

Lincoln Educational Services: www.lincolnedu.com

NAPA Training: www.NAPATraining.com

Appendix B

Automotive Apps for Smart Phones and Tablets

AUTO BUYING APPS

App	Features and/or Company
AutoTrader	Find new and used cars for sale
Cars.com	Find new and used cars for sale
Edmunds	Find new and used cars for sale
KBB.com	Kelley Blue Book prices and reviews
True Car	Never overpay for a new car
Bankrate	Auto app to calculate interest and payments

AUTOMOTIVE MAGAZINES

App	Features and/or Company
Automobile Magazine	For the automotive enthusiast
Car and Driver Magazine	For the automotive enthusiast
Car Craft	For the automotive enthusiast
Hot Rod Magazine	For the automotive enthusiast
Motor Age	For the automotive professional
Motor Magazine	For the automotive professional
Motor Trend Magazine	For the automotive enthusiast
Motor Trend Buyers Guide	Buyer’s guide for new cars, trucks, and SUVs
Road & Track Magazine	For the automotive enthusiast

AUTOMOTIVE PARTS

App	Features and/or Company
AutoZone	Automotive parts
NAPA	Automotive parts
O'Reilly Auto Parts	Automotive parts
RockAuto	Automotive parts

AUTOMOTIVE VIN DECODER

App	Features and/or Company
Advance Professional VIN Decoder	Advance Auto Parts
CARQUEST VIN Decoder	CARQUEST Technical Institute

OBD II DIAGNOSTIC TROUBLE CODE LOOK-UP

App	Features and/or Company
OBD2 Code Guide (Android only)	The Web's largest database of OBD2 codes
OBD Code Reference (Android only)	Contains over 14,000 DTCs

ON-BOARD COMPUTER SYSTEM DIAGNOSTIC TOOLS AND SOFTWARE

App	Features and/or Company	Required Hardware
BlueDriver	Premium diagnostic OBD II scan tool	BlueDriver wireless DLC receiver
DashCommand	Diagnostic OBD II scan tool and gauges	ELM327 wireless DLC receiver
Engine Link	Diagnostic OBD II scan tool	ELM327 or LELink wireless DLC receiver
Fusion	Diagnostic OBD II scan tool	LELink wireless DLC receiver
OBDWiz (www.scantool.net)	Premium diaggnostic OBD II scan tool	OBDLink LX, MX wireless DLC receiver
ScanXL Standard (www.scantool.net)	Premium diaggnostic OBD II scan tool	OBDLink LX, MX wireless DLC receiver
Scan XL Professional (www.scantool.net)	Premium diaggnostic OBD II scan tool	OBDLink LX, MX wireless DLC receiver
Torque Lite (Android only)	Diagnostic OBD II scan tool	ELM327 wireless DLC receiver
Torque Pro (Android only)	Diagnostic OBD II scan tool	ELM327 wireless DLC receiver

NAVIGATION

App	Features and/or Company
Scout GPS Navigation	Telenav, Inc.
Waze	Waze, Inc.

SAFETY APPS

App	Features and/or Company
Extraction Zones (& Pro)	Extraction Zones LLC
Hybrid Auto Extrication Guide	Field Applications, LLC
QRG (Quick Reference Guide)	National Alternative Fuels Training Consortium (NAFTC)

SPEEDOMETERS

App	Features and/or Company
Speedometer & HUD	Shows speed and heads-up display
Speedometer & Speed Limit Alert	Shows speed and speed limits
Speed Tracker	Shows and tracks speed
Speedometer View	Shows and tracks speed
Zilla Digital Dashboard & HUD	Shows speed and g-forces, heads-up display

MISCELLANEOUS AUTOMOTIVE APPS

App	Features and/or Company
Car Minder Plus	Car maintenance
Dayco App	Belts and accessories, serpentine belt routing
Gas Buddy	Find the lowest fuel prices nearby
Gates PIC Gauge 2	Determines serpentine belt wear
Goodyear KnowHow for iPhone	Goodyear Tire & Rubber Company
Goodyear SmartTech	Goodyear Tire & Rubber Company
Goodyear Truck Plus for iPhone	Goodyear Tire & Rubber Company
Hunter Touch Remote	Hunter Engineering/Alignment Equipment
My Firestone	Bridestone Retail Operations (BSRO)
RepairPal	Auto repair expert
SmoothRide Pro	Determines wheel and brake vibration

Glossary

AC synchronous motor—An AC motor, of which the rotation of the shaft and armature is synchronized to the frequency of the AC voltage that provides its power.

Active command—A command sent to a PCM or other control module that commands or instructs the module to energize and de-energize an actuator or other output circuit.

Active Fuel Management (AFM)—A variable-displacement system in which the PCM disables cylinders to improve fuel economy when engine load is light. The selected cylinders are disabled by turning off the fuel injector and spark and then disabling the ability of the valves to open through a specially designed lifter, thus reducing pumping losses by allowing the cylinder to act as an air spring.

Active voltage—A voltage that is actively pulled either high or low by a computer circuit.

Actuator—Any mechanism, such as a solenoid, relay, or motor, that when activated causes a change in the performance of a given system or circuit.

Adaptive cruise control—A cruise control system that can reduce engine torque and even apply the brakes lightly to maintain a preset gap to a vehicle ahead.

Adaptive fuel control—Ford's original name for what is known as *fuel trim* under the OBD II standardization of terminology.

Adaptive memory—The ability of an engine control computer to assess the success of its actuator and sensor signals and modify its internal calculations to correct them if the de-sired result is not achieved.

Adaptive strategy—The internal programs by which the Ford PCM learns from the results of its monitoring of sensors and activation of actuators and then modifies its programming to better achieve driveability, emissions, and fuel economy objectives.

AdvanceTrac™—A modified antilock brake and traction control system used by Ford that can pulse the brakes at any single corner of a vehicle to correct either understeer or oversteer. The control module can also reduce engine torque as needed.

Alpha—Nissan's name for what most manufacturers refer to as *fuel trim.*

Alternate fuel compatibility—The ability of an engine to run on various fuels. The first step is the use of methanol- and ethanol-compatible seals and other parts in the fuel delivery system. The use of entirely different fuels, such as natural gas, is predicted to be the next step toward fuel economy.

Amp or Ampere—A measurement of electrical current flow within a circuit; equal to 6.25 billion billion electrons passing a given point in one second.

Amperage—Current that is flowing in a circuit.

Amplitude—The difference between the high value and the low value in the voltage or current waveform of a signal.

Analog—A signal that continuously varies within a given range and with time taken for the change to occur.

Anode—A conductor with positive voltage potential applied.

Arbitration—The process of determining which node can continue to transmit data on a bus when two or more nodes begin transmitting data at the same time.

Armature—A winding of wire designed to create a strong electromagnetic field when current is flowing through it.

Asynchronous—Data sent on a data bus intermittently and as needed, rather than in a steady stream.

Background noise—Modern knock sensors are effective microphones informing the computer of all the engine's internal sounds, which the computer then sorts out for frequencies characteristic of knock. By tracking the increase in background noise with higher engine speed and load, the computer can test the effectiveness of the newer type of knock sensor.

Barometric pressure—The actual ambient pressure of the air at the engine. Because this pressure varies somewhat with the weather and altitude, the information is needed to correctly determine air-fuel mixture.

Baseline—Refers to the recording of diagnostic data prior to making any repairs. This allows the technician to compare the final diagnostic data to the baseline data to determine the effectiveness of the repairs that were made.

Baud rate—A variable unit of data transmission speed (as one bit per second). In a stream of binary signals, one baud is one bit per second. Baud refers to the speed at which information can be sent and received over a data bus. The limitation comes from the sending and receiving units, not from the wire itself. Many modern automotive computers can transmit and receive at a baud in excess of 10,000. The term *baud* is in honor of the Frenchman J. M. E. Baudot, who is responsible for much of the early development of unit data transmission.

Belt Alternator Starter (BAS)—A mild hybrid system on General Motors vehicles that uses an electric motor/generator (EMG) that operates as both a 42 V alternator and a 36 V starter. The EMG is belt-driven by the crankshaft and can also belt-drive the crankshaft.

Bidirectional controls—The ability of a scan tool to send commands to the PCM or other control module in order to initiate an active command.

Binary code—This is the form of all the signals transferred on a data bus. It consists of a series of on/off signals corresponding to the digits of the binary numbers, each with a designated meaning. Binary code transfer is very fast.

Bit—A single piece of information in a binary code system; somewhat comparable to a single letter within a word or a single digit within a number.

Breakout box (BOB)—A diagnostic tool using special connectors and a terminal box that enables the technician to check voltages on all circuits of a harness while the system is in operation.

Byte—Eight pieces of binary code information strung together to make a complete unit of information. Certain newer on-board computers recognize either 16 or 32 bits of information as representing a byte (known as 16-bit or 32-bit computers).

CAFE standards—See *Corporate Average Fuel Economy*.

Camshaft position (CMP) sensor—A sensor that monitors the rotation of the camshaft and informs the PCM and/or ignition module of camshaft position.

Capacitive-Touch Screen—A display screen that uses on the electrical properties of the human body to detect where the user touches the screen. Capacitive-touch displays can be controlled with a very light touch of a finger. Capacitive-touch displays require less pressure to activate than resistive-touch displays.

Capacitor—An electrical component that will store and give up electrons according to the voltages it is connected between. It does not have electrical continuity through it and, therefore, does not complete the circuit it is connected into. How many electrons can be stored in it is known as it's *capacitance*, usually rated in micro-farads.

Carbon dioxide (CO_2)—A molecule that contains one carbon atom and two oxygen atoms. A nontoxic emission produced by proper combustion taking place in the engine with enough oxygen present. CO_2 levels are an indication of combustion efficiency, with high CO_2 levels desired.

Carbon monoxide (CO)—An oxygen-starved molecule that contains one carbon atom and one oxygen atom. A toxic emission produced by combustion taking place in the engine with a shortage of oxygen present. A CO molecule will eventually seek out an oxygen atom and transform itself to CO_2. In the interim, exposure to CO can cause headaches, dizziness, nausea, and death. CO has no taste, odor, or color and cannot be detected by human senses.

Catalyst—An agent that causes a chemical reaction between other agents without being changed itself.

Cathode—A conductor with negative voltage potential applied.

Central Multiport Fuel Injection (CMFI)—A fuel injection system used by General Motors on Vortec engines that delivers fuel from a single, centrally located fuel injector through nylon tubes and spring-loaded poppet valves to spray fuel on the back side of each cylinder's intake valve. As a result, only air is moving through the intake manifold, similar to most port-injected engines.

Central Sequential Fuel Injection (CSFI)—A fuel injection system used by General Motors on Vortec engines that is similar to the CMFI system except that multiple injectors are used to deliver fuel to each cylinder by way of nylon tubes and poppet valves. The injector solenoids are centrally located between the upper and lower halves of the intake manifold.

Clamping diode—A diode placed across a coil or winding to trap the voltage spikes produced when the device is turned off.

Clock oscillator—The electronic device within every computer that produces a continuous stream of voltage pulses. The frequency of these pulses establishes the computer's clock speed.

Close-coupled catalytic converter—Catalytic converters that are mounted directly adjacent to the exhaust manifolds, thereby allowing them to reach operating temperature more quickly following a cold start.

Closed loop—The triangular relationship between a computer, an actuator, and a sensor that allows the computer to monitor the results of its own actuator control through a feedback signal from the sensor (or circuit). This concept may be used with certain computer-controlled actuators or systems such as air-fuel ratio control, EGR control, fuel pump and ignition system monitoring, Electronic Automatic Temperature Control (EATC) systems, antilock brake systems, and speed control systems.

Component ID (CID)—An identification code, usually a hexadecimal value, that identifies a component in a computerized engine control system. CIDs and their related information can be displayed using a scan tool. The related specifications often may be looked up at a manufacturer's website.

Compound—A molecule that contains atoms of two or more elements.

Concentration sampling—A method of gas analysis used today by most gas analyzers in automotive shops. It has a probe that is inserted into the exhaust pipe. The probe allows the analyzer to sample only a small percentage of the total exhaust gas volume. Each of the gasses is measured as a percentage, or concentration, of the total sample. The resulting measurements are expressed as either percent or parts per million (PPM), with PPM simply being a smaller proportion.

Confirmed Code—A Class B diagnostic trouble code (DTC) that has been identified by the PCM at least two times in trips and drive cycles under similar driving conditions. It will cause the PCM to illuminate the MIL. A confirmed code is also known as a *mature code*.

Constant volume sampling (CVS)—A method of gas analysis that requires that all of the exhaust be captured and measured along with enough ambient air to create a known volume of gasses. Then the density of this known volume is calculated. Finally, the concentration (percentage) of each gas is calculated to determine how many grams of each gas are being expelled from the exhaust system. The analytical computer is also able to use dynamometer readings to determine the distance that the vehicle has traveled in any instant in time or over a longer time period. At this point each gas is calculated as grams per mile (GPM).

Continuous monitor—A monitor that is monitored continuously by the PCM.

Continuously variable transmission (CVT)—A transmission that transfers optimum torque from either a gasoline engine or an electric motor (or from both) to the drive wheels regardless of vehicle speed or load. The driver never has to "shift gears" as with normal multispeed transmissions.

Conversational Speech Interface Technology—A voice recognition technology used by Ford that allows the driver and/or passenger to carry on a normal dialogue with the vehicle. The resulting dialogue on the part of the vehicle sounds more like a real person than a robot. Equipped with a 50,000-word vocabulary, the driver/passenger does not have to memorize specific voice commands, as do some voice recognition systems.

Corporate Average Fuel Economy (CAFE)—A set of regulations in the United States for automobile manufacturers that defines the minimum fuel economy standard of vehicles produced by each manufacturer. A monetary penalty is levied on the manufacturer in the event that the manufacturer fails to meet these standards. These standards are regulated by the National Highway Traffic Safety Administration (NHTSA) while the fuel efficiency is measured by the Environmental protection agency (EPA).

Cranking compression test—A compression test performed with the starter cranking the engine and the throttle held open. This test will verify *cylinder seal* for each cylinder.

Crankshaft position (CKP) sensor—A sensor that monitors the rotation of the crankshaft and informs the PCM and/or ignition module of crankshaft position and RPM.

Current ramping—The use of a lab scope (DSO) to obtain a current waveform of a load component or actuator. An inductive current probe is used with the scope to capture the waveform. The name comes from the fact that current builds slowly when turned on, causing an upslope, or upward ramp, to appear on the waveform.

Data bus network—The information transfer harness between the elements on a multiplexing system. In some vehicles, this consists of a special pair of twisted wires going to each element. The wires are twisted to prevent the introduction of spurious signals from electrical fields generated by speed sensors, horn and headlight circuits, radios (particularly high-power or CB), and spark plug cables. Many newer systems use a single-wire data bus design. The serial data on a single-wire bus has a waveform that is trapezoidal in shape. This allows the nodes that are communicating on the bus to distinguish between the sharp change of an induced voltage spike and a binary bit of information.

Data link connector (DLC)—The 16-terminal connector on an OBD II vehicle that allows a scan tool to be connected to communicate with the on-board computers. When connected to the DLC, the scan tool becomes a node on the data buses that connect to the DLC. It is sometimes referred to as a *diagnostic link connector*.

Data stream—The communication between a scan tool and an on-board computer when the scan tool is connected to the data link connector (DLC) and the ignition switch is turned on. The communication is in the form of binary code (serial data); it allows either one-way or two-way communication between the scan tool and the on-board computers.

Demultiplexor—An electronic chip that, when built into a node, gives it the ability to receive and decipher messages on a data bus.

Diagnostic trouble code (DTC)—A set of alphanumeric sequences the computer stores in its memory when it senses failures in certain sensor or actuator circuits. *DTC* is the term used in OBD II literature; *trouble codes* or *fault codes* were the terms used before OBD II.

Digital—A voltage signal that is either of two voltage values but does not vary in a continuous fashion between them. Typically these signals will be a voltage that is turned either on or off but may also vary between two voltage values that are both above ground. On a lab scope, the change of voltage (either high or low) shows a sharp, vertical edge.

Digital multimeter (DMM)—A diagnostic tool that allows the technician to view numerical readings relating to voltage, amperage, and resistance measurements. Many DMMs can also be used to measure frequency, duty cycle, pulse width, RPM, and temperature. High-end units also may include a bar graph to improve update resolution. Auto-scaling DMMs are designed to rescale automatically according to the values being measured. A DMM is designed to perform pinpoint tests of a circuit or component. It is sometimes referred to as a *digital volt/ohm meters (DVOM)*.

Digital storage oscilloscope (DSO)—A lab scope that displays a computer-generated waveform that may be frozen, measured, stored, downloaded to a PC, and printed out. The on-screen adjustments include coupling, ground level, volts per division, time per division, and trigger characteristics; they are similar to an analog lab scope. A DSO is most often used to capture voltage waveforms over the time shown on the screen but may also be configured with a low-current probe to capture current waveforms (known as *current ramping*). A DSO is designed to perform pinpoint tests of a circuit or component.

Digital volt/ohm meters (DVOM)—See *digital multimeter*.

Diode—The simplest semiconductor, consisting of only two semiconductor elements placed back to back. A diode operates as an electrical one-way check valve with no moving parts.

Displacement on Demand (DOD)—General Motors' original name for their Active Fuel Management (AFM) system, which is a variable displacement system in which the PCM disables cylinders to improve fuel economy when engine load is light. The selected cylinders are disabled by turning off the fuel injector and spark and then disabling the ability of the valves to open by use of a specially designed lifter, thus reducing pumping losses by allowing the cylinder to act as an air spring.

Distributor ignition (DI)—An ignition system using a single ignition coil. Spark distribution is *mechanically* controlled by a distributor. Spark distribution is a function of the secondary ignition circuit only.

Driveability—Those factors, including ease of starting, idle quality, and acceleration without hesitation, that affect the ease of driving and reliability.

Drive cycle—A series of operating conditions that allows the PCM to complete all of the OBD II emissions-related monitors.

Dual in-line package (DIP)—An integrated circuit chip with two rows of legs (terminals).

Duty cycle—The percentage of time that an actuator is turned on during an on/off cycle when the actuator is cycled at a fixed number of cycles per unit of time. (Example: Duty cycle might refer to the percentage of "on" time while turning on and off ten times per second.)

EEPROM—Electronically erasable programmable read-only memory, an improved and updated type of PROM. Like the PROM, it can retain tabular information for the PCM even if power is withdrawn from the computer. Unlike the PROM, the information in this memory can be changed by the computer. Also unlike the PROM, the EEPROM is a nonreplaceable part of the computer.

Efficiency monitoring—The diagnostic testing performed by a computer, specifically in its ability to compare input values to each other, or to operate an actuator while watching the change in a related input circuit. Also known as *rationality testing.*

Electrical—Refers to a circuit that performs physical work in the creation of heat, light, or magnetism. An electrical circuit operates at relatively high amperage. An electrical circuit may or may not be controlled electronically.

Electrical schematic—A map of an electrical/electronic system's circuitry that can be used by the technician as a diagnostic aid. It is most often used to increase the level of understanding of an unfamiliar system. An electrical schematic can be used to turn classroom theory into product-specific knowledge.

Electric motor/generator (EMG)—A component that can turn electrical energy into mechanical energy (as a motor does) and can also turn mechanical energy into electrical energy (as a generator does). Also sometimes referred to as a *motor/generator (MG)* or *electric machine*.

Electromotive force—Refers to the electrical pressure that pushes current flow. It is commonly called voltage and is measured in volts.

Electronic—Refers to solid-state control circuits that are placed in control of electrical circuits. Electronic circuits operate at relatively low amperage and consist of solid-state electronic parts such as diodes and transistors.

Electronic ignition (EI)—An ignition system using multiple ignition coils (either a waste spark or coil-on-plug system). Spark distribution is electronically controlled by the PCM and/or an ignition module. Spark distribution is a combined function of both the primary and secondary ignition circuits.

Electronic returnless fuel injection—A fuel injection system that eliminates the fuel return line from the fuel rail to the fuel tank by allowing the PCM to electronically control fuel pressure in the fuel supply line. To control fuel pressure, the PCM controls the fuel pump's speed by controlling the fuel pump on a pulse-width-modulated (PWM) signal. This is normally done through a secondary module that controls the fuel pump based on the command from the PCM. The PCM monitors the pressure drop across the fuel injectors using a pressure sensor mounted on the fuel rail that also has a vacuum hose attached to it.

Electronic Throttle Control (ETC)—A system, sometimes referred to as "Throttle Actuator Control (TAC)" or "drive-by-wire," that physically decouples the throttle pedal from the throttle plate(s), thus eliminating the traditional throttle cable. An accelerator pedal position (APP) sensor provides the driver demand information to the PCM; the throttle position sensor simply provides a feedback signal. The PCM then controls the position of the throttle plate(s) through pulse-width-modulated control of a DC motor.

Element—A molecule that has only one type of atom, as shown in the periodic table, defined by how many protons, neutrons, and electrons the atom contains.

Enable criteria—The specific operating conditions of the engine and the vehicle that are required to be present before the PCM will perform a test such as a drive cycle or a trip. When the preprogrammed enable criteria have been met, the PCM will perform a drive cycle to test one or more monitors. When the PCM needs to verify the presence or absence of a fault, it will perform a trip with the enable criteria aimed specifically at the particular fault.

Engine calibration—Adjustments or settings that produce desired results, such as spark timing, air-fuel mixture, and EGR control.

Engine metal overtemp mode—A unique feature of the Aurora/Northstar engine, whereby under circumstances of very high engine temperature (if, for example, the coolant has leaked out), the computer selectively disables four of the fuel injectors to prevent a temperature that would cause permanent mechanical damage to engine components.

Engine off natural vacuum (EONV) test—A test that uses a natural vacuum to test the evaporative system and fuel tank for leaks. The natural vacuum develops within the fuel tank and evaporative system as the fuel in the tank cools down after a drive cycle.

Environmental Protection Agency (EPA)—An agency established by the United States government to legislate federal programs designed to combat pollution and protect the environment. The EPA has no elected officials, makes its own laws, and sets and collects its own fines.

EVAP system pressure sensor—Used on many OBD II–compliant vehicles to test the fuel vapor canister system for leaks. On many vehicles, special plumbing allows the MAP sensor to provide this function during the period between ignition on and engine running.

Exhaust gas analyzer—A diagnostic machine that samples and displays measurements of automotive emissions readings, including HC, CO, CO_2, and O_2, for diagnostic purposes. Some analyzers also measure and display readings for NO_x.

Fail-Safe Cooling—Ford's name for the ability of the PCM to disable fuel injectors to allow the respective cylinders to simply pump air to cool the block and cylinder heads.

Feedback—Refers to a signal from a circuit or sensor that allows a computer to monitor the results of its own control in a closed-loop fashion.

Flexible fuel vehicle (FFV)—A Flexible Fuel Vehicle, or FFV, is one that allows for the use of alcohol concentrations in excess of 15 percent, such as E85. The vehicle is engineered, both from a hardware perspective and from a software perspective, to be able to burn E85 successfully.

Flowcharts—A series of diagnostic charts that are designed to guide the technician through the necessary pinpoint tests to identify the exact cause of a fault.

Flywheel Alternator Starter (FAS)—a mild hybrid system on General Motors vehicles that uses an electric motor generator (EMG) operating as both a 42 V alternator and a 36 V starter. The EMG is built into a redesigned torque converter in the area of the transmission's bell housing.

Footprint—A vehicle's footprint is defined as the vehicle's wheelbase times its average track width and is indicated in square feet (or square meters). The vehicle's footprint is being used in determining what set of CAFE standards the vehicle falls under.

Free electrons—Electrons that are loosely held in an atom's orbit. These electrons can be pushed from one atom to another if enough voltage is applied to overcome their resistance to moving.

Freeze frame data—Whenever an OBD II diagnostic trouble code is set, the computer records the information pertaining to the sensors, actuators, and operating conditions that are relevant to the fault. This information can be pulled from the computer's memory with a scan tool and is designed to aid the technician in diagnosis.

Front Electronic Module (FEM)—A Ford control module that controls the one-touch-down power window, the horn relay, and the exterior lighting at the front of the vehicle.

Fuel cell electric vehicle (FCEV)—A version of an electric vehicle that uses hydrogen fuel cells to provide the electrical power that the motor (or motor/generator) uses to accelerate the vehicle's drive wheels.

Fuel temperature sensor—On some cars, a sensor mounted in or near the fuel tank to inform the computer of the fuel temperature. This information is used to calculate fuel viscosity and susceptibility to vapor lock.

Fuel trim—The ability of the PCM to modify the base fuel calculation to maintain a stoichiometric air-fuel ratio while in closed loop. Fuel trim is divided into both a short-term fuel trim (STFT) and a long-term fuel trim (LTFT).

Functional test—A test (which has been programmed into a computer) designed to aid the technician in diagnosing faults in the controlled system. For example, a functional test may allow the technician to energize certain actuators as desired.

Gallium arsenate crystal—A semiconductor material that changes its conductivity when exposed to a magnetic field.

Gasoline Direct Injection (GDI)—A system whereby the fuel injectors spray fuel directly into the combustion chamber.

Generic Electronic Module (GEM)—A Ford control module that controls body functions such as interval wipers, the driver's door one-touch-down power window, the illuminated entry feature, and other features. Similar to the BCM (Body Control Module) on General Motors and Chrysler vehicles except that it does not control the climate control system. On newer Ford products, the GEM was separated into two control modules: the FEM and the REM.

Generic OBD II Mode—See *Global OBD II Mode*.

Global OBD II Mode—A scan tool diagnostic mode designed to allow an aftermarket generic scan tool access to the global diagnostic modes that were set aside by the EPA under OBD II standards. The EPA set aside 15 modes of which 10 are currently being used. This is also known as *Generic OBD II Mode* or *Generic Functions* by some scan tool manufacturers such as Snap-on.

Grade Logic Control—A control module strategy used by Honda, designed to reduce the driver's need to constantly compensate at the throttle pedal for transmission upshifts and downshifts when driving in mountainous terrain.

Grams-per-mile (GPM)—A measurement of exhaust gasses using a constant volume sampling (CVS) method in conjunction with a dynamometer.

Graphing multimeter (GMM)—A multimeter that is able to display any of its measurements as a graph over time. This provides the technician with a visual image of changing measurements.

Greenhouse Gas (GHG)—Any of the atmospheric gases that contribute to the Earth's greenhouse effect by absorbing solar infrared radiation. These gasses are thought to increase the warming of the Earth's surface. They include carbon dioxide (CO_2), nitrous oxide (NO_2), methane, and water vapor.

H-gate—Two pairs of transistors used to control current flow direction.

Hall effect sensor—A semiconductor sensor that is sensitive to magnetism. The "Hall effect" was discovered by Edwin Hall in 1879. A Hall effect sensor may be configured in two forms: A linear Hall effect sensor that outputs an analog voltage and a digital Hall effect switch that switches on and off in a digital fashion.

Hard fault—A fault that currently exists.

Hard wiring—Wiring designed to carry only one type of signal 100 percent of the time, unlike a data bus that can transmit multiple messages. For example, a throttle position sensor (TPS) signal wire is traditionally used to communicate information about throttle position to the PCM 100 percent of the time.

Hertz (Hz)—Cycles per second; a measure of frequency.

Hexadecimal—A system of assigning single digits to represent the total decimal value of four bits of binary code. The values 0, 1, 2, 3, 4, 5, 6, 7, 8, 9, A, B, C, D, E, and F are used, thus creating a "base-16" numbering system. Hexadecimal values are often represented using a dollar sign ($) ahead of the value. Also referred to as a "hex" value.

High-impedance—Refers to the effective high resistance that is built into computer-safe test equipment such as DVOMs, DSOs, and logic probes. This internal resistance ensures that any increase in current flow caused by the connection of the diagnostic tool to a circuit is so miniscule that there is little potential for the tool to damage a computer. The high resistance built into the diagnostic tool also increases the accuracy of voltage measurements when diagnosing high resistance computer circuits.

High-voltage (HV) battery—An on-board battery made up of many cells to create a voltage potential of several hundred volts. Used on gasoline/electric hybrid-drive vehicles.

History Code—The name given by General Motors to what is commonly known as a *confirmed code.*

Home energy station (HES)—A device that uses tap water with electricity from a wall outlet to create the hydrogen needed to refill a fuel cell vehicle's hydrogen tanks.

Homogeneous air-fuel charge—An air-fuel charge wherein the air and fuel are both evenly mixed and evenly distributed throughout the entirety of the combustion chamber.

Hot-film air mass sensor—A type of air mass sensor using changes in the temperature of the heated element to determine the mass of the air passing into the engine.

Hot-wire mass air flow sensor—An air flow sensor that keeps a special wire at a fixed number of degrees above ambient temperature in the intake air flow. The amount of current required to maintain the heat is converted into a voltage or frequency signal that conveys information about the mass of the intake air.

Hydrocarbons (HC)—A vehicle emission consisting of unburned gasoline or diesel fuel. The engine may produce excessive HCs due to a total misfire or rich air-fuel conditions. HCs may also be released into the atmosphere because of fuel evaporation.

Impedance—The total circuit resistance, including resistance and reactance.

Inertia switch—A Ford trademark safety feature and an occasional no-start puzzle. To protect against a gasoline fire following an accident, the inertia switch disables the fuel pump circuit upon impact.

Intake Manifold Runner Control (IMRC)—An intake air plenum design used on Ford and Mazda engines that can vary the effective length of the intake. When the engine is operating at low speed and under light load, it provides a longer effective length to keep the velocity of the air flow high so as to provide crisp throttle response. When engine speed and load are increased, it provides a shorter effective length to allow for additional air flow to meet the greater demand.

Integrated circuit (IC)—A circuit, typically placed on a semiconductor chip, that is minuscule in size and consists of semiconductor components such as diodes and transistors.

Intelligent Architecture—A system that includes a combination voice recognition and navigational system, as well as the ability to personalize driver settings or download information or music. The system also has several other features, including the ability to communicate with electronic devices using Bluetooth™ wireless technology. Also included is real-time navigation, a feature that can alert the driver to an accident ahead or serve as a reminder of a vehicle service that is due. Another feature, called MyHome, gives the driver the ability to communicate with Web-based equipment at home or elsewhere.

Intelligent Junction Box—A junction box that also functions as a node on a data bus.

Interface—Circuitry that converts both input and output information to analog or digital signals as needed and filters external circuit voltage to protect computer circuits.

Intrusive test—A self-test initiated by the PCM to test a non-continuous monitor. During this test the PCM energizes output actuators, not because engine/vehicle operating conditions warrant it, but only for the purpose of performing the test.

In-Use Performance Tracking—An ability added to OBD II global mode $09 that records how often the conditions are present for a monitor to run and how many times that monitor has actually run.

Jetronic—A Bosch fuel injection system that may be controlled mechanically or electronically. The electronic system uses a dedicated electronic control unit (ECU) to control only the fuel injection system.

KAM—Keep-alive memory, an electronic memory in the computer that must have a constant source of voltage to continue to exist.

Kilobyte—1024 bytes of information in binary code.

Lambda—A Greek letter (L) used to indicate how far the actual air-fuel mixture deviates from the ideal air-fuel mixture. Lambda equals actual inducted air quantity divided by the theoretical air requirement, which is 14.7 parts air to each one part of fuel as measured by weight. During closed loop fuel control, when the Lambda value is "1," the air-fuel mixture in the cylinder is stoichiometric and all of the fuel and oxygen is consumed in the cylinder, resulting in complete combustion.

Light-emitting diode (LED)—A semiconductor device that generates a small amount of light when forward biased (when voltage is applied with the polarity such that current flows across the PN junction).

Limp-in mode—A state of a computer engine management system in which one or more major component circuits have failed and a substitute value is used to retain driveability until the vehicle can be repaired.

Linear topology—A configuration of a multiplexing network that connects the data bus to all nodes on the bus in a parallel circuit.

Logic gates—Combinations of transistors within the computer that determine the computer's output reaction to a specific combination of input values.

Logic module and power module—Early- to mid-1980s Chrysler products separated the engine computer (PCM) into two components: the *logic module* and the *power module*. The logic module received data from the sensors and made the decisions concerning the output functions. The power module, under the control of the logic module, carried out the output commands and controlled the fuel injectors, the ignition coil, and the emission system solenoids.

Logic probe—A computer-safe test light using LEDs to indicate if a signal is present. Logic probes are typically used to provide quick checks of primary ignition and fuel injector circuits.

Long-term fuel trim (LTFT)—A long-term adjustment made within the PCM so that the performance of the short-term fuel trim (STFT) will cause the oxygen sensor signal to average between 400 and 500 mV. LTFT is designed to modify the base fuel calculation to compensate for wear and aging of the engine and other components that affect the air-fuel ratio. Originally known on General Motors' products as "block learn."

Loop topology—A configuration of a multiplexing network that connects the data bus to all nodes on the bus in a series circuit.

Magnetic resistance element (MRE) sensor—An electronic sensor that monitors a rotating magnetic ring contains several North poles alternating with several South poles. Essentially, a cross between a variable reluctance (permanent magnet) sensor and a Hall effect sensor, AC voltage pulses are initially induced in the *magnetic resistance element (MRE)*, which it then converts to crisp digital DC voltage pulses on the output side. The MRE sensor has three wires: battery voltage, signal, and ground. As an electronic device, it must have both power and ground to power up and become functional. The MRE sensor is manufactured by the Denso Corporation. They are used as vehicle speed sensors (VSS), camshaft position (CMP) sensors, and crankshaft position (CKP) sensors.

Malfunction indicator light (MIL)—A lamp on the instrument panel used to alert the driver when the PCM identifies a problem with the engine control system. Known commonly as a *check engine light* prior to OBD II standardization.

Master node—The dominant node that controls a data bus network.

Mature code—A Class B diagnostic trouble code (DTC) that has been identified by the PCM at least two times in trips and drive cycles under similar driving conditions. It will cause the PCM to illuminate the MIL. A mature code is also known as a *confirmed code*.

Megabyte—1,048,576 (1024 squared) bytes of information in binary code.

Memory—A computer's information storage capability.

Memory DTC—A diagnostic trouble code (DTC) that has been set in a computer's memory at some point in the past. This type of DTC may represent either a hard fault or a soft fault.

Micro-Electronic Mechanical System (MEMS)—Also known as *nano-technology*, these sensors are built on a silicone substrate and can be designed to detect movements such as yaw rate, rollover, G-force, and many other conditions. These sensors are very similar to the ones in modern smart phones that allow the phone to detect movement and tilt.

Microprocessor—A processor contained on an integrated circuit; a processor (central processing unit) that makes the arithmetic and logic decisions in a microcomputer.

Misfire detection—Part of the required OBD II program, misfire detection closely monitors engine RPM for the slight variations when combustion misfire occurs. In most cases, the computer disables the injector for a misfiring cylinder (to protect the catalytic converter from overheating) and sets an appropriate DTC.

Molecule—A small chemical building block that consists of two or more atoms. It may be an element or a compound.

Monitor—An on-board diagnostic test that an OBD II PCM performs to determine if a component or system is within design specifications.

Monitor ID (MID)—An identification code, usually a hexadecimal value, that identifies a monitor in a computerized engine control system. MIDs and their related information can be displayed using a scan tool. The related specifications often may be looked up at a manufacturer's website.

Motor/generator (MG)—A component that can turn electrical energy into mechanical energy (as a motor does) and can also turn mechanical energy into electrical energy (as a generator does). Also some-times referred to as an electric motor/generator (EMG).

Motor start-up in-rush current—The initial high current that an electrical motor draws when its circuit is first energized.

Motronic—A Bosch computerized engine control system that is a full function engine control system. The powertrain control module (PCM) controls the fuel injection system, the spark management system, emission systems and other engine functions.

Multimeter—A meter that can be used to measure voltage, amperage, resistance, and other electrical characteristics. It may be of any of the following designs: analog, digital, or graphing.

Multiplexing (MUX)—One of the most significant extensions of computer control technology in vehicles, multiplexing connects many or all of the control units in a vehicle over a single-wire network: the data bus. Information from all components is sent over the data bus and received by all.

Multiplexor—An electronic chip that, when built into a node, gives the node the ability to send messages on a data bus.

Multipoint fuel injection (MPFI or MFI)—A fuel injection system with one fuel injector for each cylinder. The injector sprays fuel on the back side of the intake valve. The PCM may energize the injectors as a group or individually in the engine's firing order and just before the intake valve opens.

National Highway Traffic Safety Administration (NHTSA)—An agency of the Executive Branch of the U.S. government and also part of the Department of Transportation (DOT). NHTSA (pronounced "NITS-uh") creates legislation with the intent of "saving lives, preventing injuries, and reducing vehicle-related crashes." NHTSA was officially established in 1970 by the Highway Safety Act of 1970.

NHTSA was the eventual result of Congressional hearings in 1966 regarding seat belts in vehicles.

Negative ion—An atom that has an extra electron, thus having more electrons than it has protons. It therefore has a negative electrical charge.

Nitrogen (N_2)—A gas that makes up approximately 78 percent of Earth's atmosphere. Ideally, the nitrogen that enters the engine's combustion chambers will go through unchanged.

Node—An electronic control module that is connected to a data bus and has the ability to send and/or receive messages on the bus.

Non-continuous monitor—An OBD II monitor that is not continually monitored by the PCM, but rather is tested by the PCM performing an intrusive test of the component or system. A non-continuous monitor is only run when specific preprogrammed enable criteria are met.

Non-powered test light—A tool that can be used to quickly check for a voltage potential (differential). Voltage potential should not be judged based on the brightness of the light. Unfortunately, due to its low internal resistance, a test light can act as a jumper wire and unexpectedly damage a computer or deploy an air bag.

Non-responsive unit (NRU)—A control module (automotive computer) that has stopped functioning.

Normally aspirated—An engine that uses ambient atmospheric pressure only to push air into the engine (as opposed to turbocharged and supercharged).

OBD II—A standardized computer control self-diagnostic program, mandated for most vehicles as of the 1996 model year. Nomenclature, trouble codes, and even the size, shape, and many circuits of the diagnostic connector are standardized across the industry.

Off-board reprogramming—The reprogramming of a control module (automotive computer) after it has been removed from the vehicle.

Ohm—A unit of measurement used to measure electrical resistance.

Ohm's law—An electrical law that states that it takes 1 V of electrical pressure to push one amp of current flow through 1 V of resistance.

On-board reprogramming—The reprogramming of a control module (automotive computer) while it is yet installed on the vehicle. This is usually done through the DLC.

On-demand DTC—A diagnostic trouble code (DTC) that results from a self-test (performance test), identifying any hard faults that were found within the computer system.

Open loop—Any operational mode in which a computer controls an output circuit but does not use a feedback signal from a sensor or circuit to monitor the results of its own control. However, the computer may monitor other input signals/sensors during open-loop control that do not pertain to a feedback signal. In the PCM's open-loop control of the air-fuel ratio, the air-fuel mixture is calculated based on coolant temperature, engine speed, and engine load without benefit of the oxygen sensor. *See also* Closed loop.

Oxides of nitrogen (NO_X)—A vehicle emission that is a combination of nitrogen (N_2) and oxygen (O_2), which are combined within the engine's combustion chamber or in the exhaust system due to temperatures that exceed 2500°F.

Oxygen (O_2)—A gas that makes up approximately 21 percent of Earth's atmosphere. Most of the oxygen that enters the engine's combustion chambers will be used up during combustion.

Palladium (Pd)—A silver-white, ductile, metallic element; number 46 on the periodic table.

Parallel hybrid—A vehicle that can be accelerated using either an electric motor/generator or a gasoline engine or both.

Parameter ID (PID)—A specific data stream parameter relating to a sensor value, an actuator command, or some other piece of information that a PCM or other control module may communicate to a scan tool through the data stream. The scan tool will show the PID as well as the related information.

Passive voltage—A voltage that is at rest, not being pulled either high or low by a computer circuit.

Pattern failures—Failures that tend to be common with a particular make, model, and year of vehicle.

Pending Code—A Class B diagnostic trouble code (DTC) that has been identified by the PCM only one time in a drive cycle. It will not cause the PCM to illuminate the MIL.

Permeability—A characteristic that demonstrates a material's responsiveness to magnetism. Iron is an example of a material that has high permeability.

Piezoelectric—A crystal that produces a voltage signal when subjected to physical stress such as pressure or vibration.

Piezoresistive—A semiconductor material whose electrical resistance changes in response to changes in physical stress.

Pinpoint test—A test that checks a computer circuit for the exact fault, whether it is a sensor/actuator, an open or short in the circuitry, a poor or open electrical connection, or the computer itself. Pinpoint tests may be performed either by following a flowchart or by following an electrical schematic. Following a pinpoint test series, the technician should know exactly what repair needs to be made to restore the system to its "new" condition.

Platinum (Pt)—A heavy, grayish white, ductile, metallic element; number 78 on the periodic table.

Polymer electrolyte membrane—A type of fuel cell that uses a solid polymer membrane (a thin plastic film) as an electrolyte. PEM fuel cells combine hydrogen (H_2) with oxygen (O_2) from atmospheric air to create electrical current flow. The only emissions resulting from fuel cell operation are water (H_2O) and heat. A PEM fuel cell is also commonly referred to as a Proton Exchange Membrane fuel cell.

Port fuel injection (PFI)—A fuel injection system in which the intake manifold has been designed with each cylinder having its own individual intake runner. There is one fuel injector for each cylinder. Each fuel injector sprays fuel on the back side of the cylinder's intake valve(s). Therefore, PFI injectors spray fuel below the throttle plate(s). Thus, the volume of fuel metered is affected both by the fuel rail pressure and by the pressure that is present in the intake manifold.

Positive ion—An atom that has a deficiency of one electron, thus having fewer electrons than it has protons. It therefore has a positive electrical charge.

Potentiometer—A three-terminal variable resistor that acts as a voltage divider. It is used to measure physical position and mechanical action. It is often used as a position sensor for such things as throttle angle. The potentiometer can be a linear or rotary style.

Potentiometer sweep test—The testing of a potentiometer throughout its range of operation to determine whether any electrical opens exist.

Power module—See *Logic module and power module.*

Powertrain control module (PCM)—The term and acronym now used (since OBD II) for the primary computer in charge of engine and transmission performance.

PROM—Programmable read-only memory.

Protocol—A standardized binary code that constitutes the language with which computers are able to communicate with each other over a data bus.

Proton Exchange Membrane (PEM)—*See* Polymer Electrolyte Membrane.

Pulse-width modulated (PWM)—A signal that is controlled according to a real-time measurement, usually in milliseconds or microseconds. This type of signal may be used to control fuel injectors, cooling fans, and blower motors. Also, when a serial data bus uses a fixed pulse width, typical with two-wire buses, the binary code is said to be a PWM signal.

Pumping losses—The load on the engine caused by the work needed to pump air through the engine. On gasoline engines, because of the vacuum developed behind the throttle, pumping losses are greatest at the most-closed throttle positions: idle and deceleration. The aerodynamic friction of air through the intake system also contributes to a lesser extent to the pumping losses. Diesel engines, because they have no throttle plate, have very low pumping losses when compared to gasoline engines.

Quad driver/output driver—A power transistor in the PCM, capable of working either four or seven different actuators (such as injectors or ignition coils). Each quad driver or output driver works when the PCM uses it to ground a component's circuit, completing it.

RAM—Random access memory.

Rationality testing—The diagnostic testing performed by a computer using its ability to compare input values to each other, or to operate an actuator while watching the change in a related input circuit. Also known as *efficiency monitoring.*

Rear Electronic Module (REM)—A Ford control module that controls the exterior lighting at the rear of the vehicle, the deck lid release solenoid, and the pulse-width-modulated fuel pump motor according to PCM command.

Rear Fuel Control—Refers to the use of the oxygen sensor to the rear of the catalytic converter as a primary input to the PCM regarding the control of the air-fuel ratio. The PCM adjusts the pulse width of the fuel injectors so as to affect the duty cycle of the front O_2 sensor in order to keep the lambda value as measured by the rear O_2 sensor within a range of 0.98 to 1.02.

Reference voltage regulator—The device within a computer that generates the 5 V reference voltage used by the computer's internal circuits, including the logic circuits. Its signal, sometimes referred to as VREF, is also sent to those sensors that require it.

Regenerative braking—The process of returning a moving vehicle's kinetic energy to an onboard battery when coasting or braking. This energy is then reused to get the vehicle moving again when reaccelerating, thus increasing fuel mileage. Brake life is also extended.

Reid Vapor Pressure (RVP)—The pressure produced by fuel when heated to 100°F in a confined space.

Reluctance—A material that does not demonstrate the characteristics associated with magnetism is said to have reluctance.

Resistance—Opposition to the flow of electrons; measured in ohms.

Resistive-Touch Screen—A display screen that is composed of multiple layers separated by thin spaces. As pressure is applied to the display's surface, the layers will touch. This completes specific electrical circuits which informs the display's electronics where the user is touching the screen. Resistive-touch displays require more pressure to activate than capacitive-touch displays.

Returnless fuel injection—A fuel injection system that eliminates the return line from the engine compartment to the fuel tank, therefore reducing the amount of engine compartment heat that is transferred to the fuel tank. Technically, there is still a short return line either within the fuel tank (returnless) or from the fuel filter (semi-returnless) to the fuel tank. Although the return line was originally added to help combat vapor lock, other system designs continue to keep vapor lock under control such as the location of the fuel pump at the rear of the vehicle and the higher fuel pressures that are now used in these systems, therefore allowing the elimination of the return line.

Return-type fuel injection—A fuel injection system that uses a return line to return excess fuel from the fuel rail at the engine to the fuel tank. This is done so that in periods of low fuel usage, such as when driving in heavy stop-and-go traffic, cooler fuel is continuously picked up from the fuel tank to help combat vapor lock.

Rhodium (Rh)—A white, hard, ductile, metallic element; a member of the noble metals family; number 45 on the periodic table.

ROM—Read only memory.

Running compression test—A compression test performed with the engine running. This test will verify each cylinder's *volumetric efficiency*.

Scan tool—A diagnostic tool used to retrieve from a vehicle's computer memory any malfunction codes it has retained. On earlier systems, each manufacturer used a scan tool connector and setup unique to that manufacturer (though there are aftermarket scan tools with adapters that can read many of them). With OBD II-compliant vehicles, the scan tool is standardized and can be used for the same purpose on any vehicle. Many scan tools can also be used to get direct measurements of various sensors and to exercise various actuators to check on their effectiveness; also known as *functional tests*.

Schmitt trigger—A device that trims an analog signal and converts it to a digital signal.

Self-test—The self-diagnostic portion of the Ford quick-test in which the computer tests itself and its circuits for faults.

Semiconductors—Solid-state devices, such as diodes and transistors, that are formed from semiconductor materials that have four electrons in the valence ring. Depending upon design and applied polarity, they may either conduct electrical current flow or insulate against electrical current flow.

Sensor—All the inputs and outputs to the computer are either sensors that provide information or actuators that do work. Most sensors are sent a reference voltage that is reduced in a predictable way by whatever parameter they monitor. Some generate voltages of their own.

Sequential fuel injection (SFI)—A form of port fuel injection (PFI) whereby the injectors are pulsed individually, in the firing order of the engine, and just before the intake valve opens. With an SFI/SMPI system, the PCM may pulse all injectors simultaneously during initial engine start, known as a *primer pulse*.

Sequential multi-point injection (SMPI)—See *Sequential fuel injection*.

Serial data—Digitized code known as *binary code,* which is a combination of high and low voltages known as ones and zeroes, used as a language to communicate between computers or computer components.

Series hybrid—A vehicle that uses a gasoline engine to power a generator (or motor/generator) to provide additional electrical energy to the primary motor/generator to accelerate the vehicle.

Short-term fuel trim (STFT)—The PCM's immediate response to the oxygen sensor signal. When the oxygen sensor indicates a lean mixture, STFT increases its value to increase injector on-time. This causes more fuel to be added to the base fuel calculation and enriches the air-fuel mixture. When the oxygen sensor indicates a rich mixture, STFT decreases its value to decrease injector on-time. This causes fuel to be subtracted from the base fuel calculation and leans out the air-fuel mixture. Originally known on General Motors vehicles as the "Integrator."

Signal—See *Voltage signal*.

Slave node—A dependent node on a data bus.

Snapshot—A computer recording of information from sensors and actuators.

Soft fault—An intermittent fault whose symptom does not appear to currently exist.

Solenoid—A device with a movable iron core inside an electromagnetic winding. The movable core is usually spring-loaded and will move against spring tension when the electromagnet is energized. It creates linear mechanical action and may be used to lock or unlock a door or engage a starter drive with a flywheel. It is often used to operate a valve to control the flow of air, vacuum, fuel vapors, or liquid, and may be spring-loaded normally closed or normally open.

Speed density formula—A mathematical formula used to *calculate* the mass of air entering the engine. This calculation involves input from several sensors. The use of this calculation is identified by the presence of a VAF meter or a MAP sensor.

Star topology—A configuration of a multiplexing network that connects the data bus to all nodes on the bus in a parallel circuit, although one node may be at the center hub of the data bus circuits where all nodes connect to it.

Stepper motor—A type of direct-current electric motor with the unique capacity to move an exact number of turns in response to a computer's signal. Typically, such a motor (used to adjust the idle speed bypass opening, for example) has a range of 256 different possible positions it can be set to. This allows the computer to open or close to an exact position. Such motors are also increasingly used in some heater door control systems.

Stoichiometric—In automotive terminology, this refers to an air-fuel ratio in which all combustible materials are used, with no deficiencies or excesses left over. In other words, there is neither any fuel nor any oxygen left over after combustion has completed within the cylinder. With straight gasoline (containing no alcohol) this equates to a 14.7 to 1 air-fuel ratio as measured by weight.

Stratified air-fuel charge—An air-fuel charge which is, overall, much leaner than a homogeneous air-fuel charge. This stratified charge is homogeneous and close to stoichiometric in the immediate area near the spark plug. But there is additional air in the cylinder outside of this area which allows the engine to run much leaner without producing excessive NO_X emissions and without the lean misfire generally associated with lean mixtures.

Tach reference signal—The basic timing signal sent to the computer, indicating crankshaft position and RPM. This signal may originate at a sensor within a distributor or at a crankshaft position (CKP) sensor. It is modified by the computer to create a computed timing signal with calculated spark advance added in.

Technical service bulletin (TSB)—A repair procedure issued by the manufacturer that addresses a particular problem and instructs the technician on how to perform a proper repair to correct the symptom.

Telematics —The ability of the vehicle, or something within the vehicle, to communicate with an outside entity. Telematics involves telecommunications and electronic technologies on the vehicle. The communication may take on several different forms including communication between people, between computers, or between any combination of people and computers.

Termination resistor—A resistor connected between Bus H and Bus L on a dual wire (DW) data bus. The CAN DW data bus uses two 120-ohm termination resistors, one at each end of the data bus.

Test ID (TID)—An identification code, usually given as a hexadecimal value, that identifies a test relating to a component in a computerized engine control system. TIDs and their related information can be displayed using a scan tool. The related specifications may be looked up, many times at a manufacturer's website.

Thermistor—A temperature-sensing device whose resistance changes dramatically as the temperature changes. Most thermistors have a negative temperature coefficient (the resistance goes up as the temperature goes down).

Throttle body injection (TBI)—Throttle body injection uses one or two fuel injectors in a throttle body mounted on an open air intake manifold (as a carburetor used to be) that delivers the fuel to the intake manifold from a single point. TBI injector(s) spray fuel above the throttle plate(s) into ambient air pressure.

Throttle plate position controller (TPPC)—An IC chip embedded within a Ford PCM, having the task of controlling the electronic throttle control (ETC) system.

Torque management—A set of capacities the engine control system has, allowing it to reduce engine torque output under specific circumstances. Its first strategy is to retard spark; in more extreme circumstances it can selectively shut off fuel injectors. The traction control system employs part of the torque management system to prevent front-wheel spin under power when the pavement is slippery.

Total fuel trim (TFT)—The sum of the average short-term fuel trim (STFT) added to the long-term fuel trim (LTFT) value.

Transistor—A semiconductor that consists of three semiconductor elements placed back to back. A transistor operates as an electronic switch with no moving parts and may be either PNP or NPN design.

Trigger—The characteristics that determine the instant in time that a DSO begins to display a waveform. These characteristics include trigger source, trigger level, trigger slope, and trigger delay (horizontal waveform position).

Trip—A key cycle consisting of Ignition On, Engine Run, Ignition Off, and PCM Power Down, during which the enable criteria for a particular diagnostic test are met and the diagnostic test is run by the PCM. A trip is used by the PCM to perform a diagnostic test to confirm a symptom or its repair.

Twin injection—A Toyota fuel injection system, known as the D4-S system, which was introduced on the 2GR-FSE 3.5 L engine in the 2006 model year. This system combines a gasoline direct injection (GDI) system with a traditional sequential port fuel injection system. Each cylinder has two fuel injectors, one that sprays fuel on the back side of the intake valve and one that sprays fuel directly into the combustion chamber.

Two-trip malfunction detection—On some vehicles, the computer waits to record certain faults until they appear on two separate driving trips.

Valance ring—The outermost electron shell of an atom.

Vapor lock—A condition that occurs when fuel vaporizes within the fuel management system prior to being metered into the engine. This results in engine stall due to the inability of the fuel management system to deliver enough volume of fuel to keep the engine running. After the system has cooled enough for the fuel in the system to re-condense, the engine can be restarted.

Variable cam timing (VCT)—A system that uses a cam phaser on the front end of the camshaft to change the camshaft's position relative to the crankshaft.

Variable Cylinder Management (VCM)—Honda's name for a variable displacement system to improve fuel economy, in which the PCM disables cylinders when engine load is light. The selected cylinders are disabled by turning off the fuel injector and then disabling the ability of the valves to open using a Honda VTEC-style rocker arm design, reducing pumping losses by allowing the cylinder to act as an air spring.

Variable pulse width (VPW)—The term assigned to serial data that consists of bits of information that are of more than one length. In this case, a "one" may be represented either as a short high-voltage pulse or as a long low-voltage pulse. Conversely, a number "zero" may be represented as a short low-voltage pulse or as a long high-voltage pulse. This allows the voltage to switch between high- and low-voltage values with every bit of information.

Variable reluctance sensor (VRS)—A magnetic sensor that produces an AC voltage signals as iron teeth and notches are rotated past it. A VRS contains a coil of wire wrapped around a permanent magnet. As iron teeth and notches are passed near the magnet, the coil's electrical reluctance changes, thus creating an alternating electron flow. This is also known as a *permanent magnet sensor* or *AC generator*.

Variable valve timing (VVT)—The ability of a computer to vary valve timing using an actuator designed for that purpose to get best performance under high load conditions while still getting good engine response at low RPM.

VIN entry mode—A diagnostic mode on a scan tool that allows the scan tool to recognize the particular make, model, and year of vehicle as well as other attributes such as engine identification and computer protocols using the vehicle identification number (VIN), which is entered into the scan tool. This mode generally allows the technician to access the most diagnostic information using a scan tool. Not all aftermarket scan tools support this mode.

Volatility—The ability of a liquid to vaporize, affected directly by the liquid's boiling point. A fuel's volatility is measured as its *Reid Vapor Pressure (RVP)*. The liquid's boiling point is inversely proportional to its volatility and RVP. In other words, as the boiling point is decreased, both the volatility and RVP are increased.

Volt—A unit of measurement of electrical pressure or electromotive force.

Voltage or voltage potential—A measurement of electrical pressure or electromotive force.

Voltage drop—The voltage that is used up as current flows through resistance in an electrical circuit.

Voltage drop test—A test that is a more precise substitute for resistance tests done with an ohmmeter. A voltage drop test checks the positive side or negative side of the circuit to measure precisely how much voltage is used up by the resistance in that portion of the circuit. It must be performed while current is flowing in a completed circuit. Excessive voltage drop indicates excessive unwanted resistance.

Voltage signal—The voltage values that a computer uses as communication signals.

Volumetric efficiency (VE)—A measure of cylinder-filling efficiency. VE is expressed as the percentage of atmospheric pressure to which the cylinder is filled at the completion of the compression stroke. This efficiency varies with engine speed; the point of highest volumetric efficiency (which can rise above 100 percent in some cases) is the point of highest engine output torque.

Wastegate—A valve or door that when opened allows the exhaust gas to bypass the exhaust turbine of a turbocharger. The wastegate is used to limit turbocharger boost pressure.

Waste spark—An ignition system using one coil between every complementary pair of cylinders, firing both plugs simultaneously, one at the end of the compression stroke and the other at the end of the exhaust stroke.

Water (H_2O)—One of the by-products of combustion of a hydrocarbon-based fuel within the engine.

Wattage—A measurement of total electrical power. Volts times amps equals watts (known as Watt's Law).

WOT—Wide-open throttle.

Yaw rate—A determination of how far left or right a vehicle has moved from its intended course.

Zirconium dioxide (ZrO_2)—A white crystalline compound that becomes oxygen-ion conductive at about 6008°F (3158°C).

Index

D

T